Important Physical Constants

Constant	Symbol	Value
Velocity of light (*vacuo*)	c	2.9979×10^{10} cm sec^{-1}
Planck constant	h	6.6256×10^{-27} erg sec
Avogadro number	N	6.0226×10^{23} particles mol^{-1}
Faraday constant	F	96491 C mol^{-1}
Gas constant	R	8.3143×10^{7} erg deg^{-1} mol^{-1}
		1.9872 cal deg^{-1} mol^{-1}
		8.2054×10^{-2} liter atm deg^{-1} mol^{-1}
Boltzmann constant	k	1.3805×10^{-16} erg deg^{-1}
Rest mass of the electron	m_c	9.1090×10^{-28} g
Electronic charge	e	-4.8033×10^{-10} esu
		-1.6021×10^{-19} C

Energy Conversion Factors

	Ergs	Joules	Calories	Liter atmospheres	Electron volts
1 erg =	1	10^{-7}	2.3901×10^{-8}	9.8687×10^{-10}	6.2418×10^{11}
1 joule =	10^7	1	2.3901×10^{-1}	9.8687×10^{-3}	6.2418×10^{18}
1 calorie =	4.1840×10^7	4.1840	1	4.1291×10^{-2}	2.6116×10^{19}
1 liter atmosphere =	1.0133×10^9	1.0133×10^2	24.218	1	6.3248×10^{20}
1 electron volt =	1.6021×10^{-12}	1.6021×10^{-19}	3.8291×10^{-20}	1.5811×10^{-21}	1

PRINCIPLES OF INSTRUMENTAL ANALYSIS

SECOND EDITION

Douglas A. Skoog *Stanford University*

Donald M. West *San Jose State University*

SAUNDERS GOLDEN SUNBURST SERIES

1980 **SAUNDERS COLLEGE**

PHILADELPHIA

Saunders College
West Washington Square
Philadelphia, PA 19105

Library of Congress Cataloging in Publication Data

Skoog, Douglas Arvid, 1918–
    Principles of instrumental analysis.

    Bibliography: p.
    Includes index.
    1. Instrumental analysis.   I. West, Donald M.,
joint author.   II. Title.
QD79.I5S58   1980   543′.08   79-26284
ISBN 0-03-021161-1

Printed in the United States of America

    2 3   144   9 8 7 6 5 4

PREFACE

Physical and biological scientists currently have available to them an amazing array of powerful tools for obtaining qualitative and quantitative information about the properties and composition of matter. For this reason it is our belief that students of chemistry, biochemistry, the health-related sciences, and mineral sciences should, very early in their careers develop an appreciation for these tools and a knowledge of how they can be employed to solve analytical problems.

We believe that the efficient use of analytical instruments requires that the scientist have an understanding of the fundamental principles upon which modern measuring devices are based. Only then can he make intelligent choices among the many possible ways of solving an analytical problem; only then can he appreciate the pitfalls that accompany any physical measurement; and only then can he develop a feel for the probable limitations of his measurements in terms of sensitivity and accuracy. It is the goal of this second edition of *Principles of Instrumental Analysis*, as it was the goal of the first, to provide the student with an introduction to these principles as well as to generate an appreciation of the kinds of instruments that are currently available from commercial sources and their strengths and limitations.

During the nine years that have elapsed since the appearance of the first edition of this book, major changes in instrumental design have occurred. Among these has been the nearly total displacement of the vacuum tube by the transistor. Second has been the appearance of tiny, inexpensive, operational amplifiers based on transistors. Such amplifiers have permitted scientists having a minimal understanding of electronics to design and build a vast array of useful circuits for measurement and computation. A third major development has been the introduction of the microprocessor—a self-contained and inexpensive computer etched on a tiny silicon chip. Microprocessors are found in ever-increasing numbers in modern instruments, where they serve to control instrument variables, to monitor signals, and to process data. Because of these electronic developments, two entirely new chapters have been added in this edition to present elementary treatments of electrical circuits, transistors, operational amplifiers, microprocessors, and computers. In addition, throughout the text, sections have been added that describe the applications of these devices, particularly to automation of various kinds of instruments.

Other new chapters include one dealing with thermal methods, another with components of instruments for optical spectroscopy, still another with general theoretical aspects of atomic and molecular absorption spectroscopy, and finally a chapter treating the general theory of chromatographic separations. Parts of the material for the latter three chapters are found scattered throughout several chapters in the first edition; other parts are entirely new. In the latter category is a section that treats the effects of instrumental and environmental noise on the precision and accuracy of spectroscopic absorption measurements. Also largely new is the development of general chromatographic theory from kinetic considerations. The chapter on optical instruments brings together information that was formerly found in chapters on infrared, ultraviolet and visible, and atomic absorption spectroscopy; flame and emission spectroscopy; and fluorescence and Raman spectroscopy.

Since the publication of the first edition, a host of modifications of well-established instrumental methods have emerged and are now included. Among these new developments are Fourier transform nuclear magnetic resonance and infrared methods, infrared photometers for pollutant measurements, laser modifications of spectroscopic measurements, nuclear magnetic resonance measurements on nuclei other than hydrogen, electron spin resonance spectroscopy, photon counting, inductively coupled plasma sources for emission spectroscopy, semiconductor detectors for X-rays and gamma rays, energy dispersive X-ray systems, the electron microprobe, X-ray photoelectric spectrometry, chemical, field ionization, and spark sources for mass spectrometry, gas electrodes, potentiostatic controls for polarography, differential pulsed and rapid scan polarography, cyclic and ac voltammetry, high performance liquid chromatography, new packings for gas-liquid chromatography, interfacing of chromatography with mass, infrared, and fluorescence detectors, instruments for simultaneous, multielement analysis, and automated instruments for spectroscopy, electroanalysis, titrimetry, and chromatography.

In addition to the foregoing, the new edition contains, as an appendix, a brief treatment of the propagation of random errors in analytical measurements.

It is worth noting that selected portions of the material found in this text have also appeared in the authors' two other titles.[1] The principal overlap of the material in this book with that in the earlier two occurs in the chapters on potentiometry, coulometry, and voltammetry. Duplication will be found to a lesser extent in the chapters on ultraviolet and visible absorption spectroscopy and atomic emission spectroscopy.

The authors wish to acknowledge with thanks the considerable contributions of Professor Alfred R. Armstrong, College of William and Mary, and Dr. James LuValle, Stanford University, who have read the entire manuscript in detail and offered many useful suggestions. We are also grateful to Professor R. deLevie of Georgetown University for his comprehensive review and numerous suggestions for Chapters 2 and 3, to Professor H. S. Mosher of Stanford University for his helpful comments on the chapter on optical activity, and to Professors S. R. Crouch of Michigan State University and J. D. Ingle, Jr., of Oregon State University for their detailed criticisms on and suggestions for the section on precision of spectral measurements. Finally, we offer our thanks to several others who have taken the time to comment, often in some detail, on portions

---

[1] D. A. Skoog and D. M. West, *Fundamentals of Analytical Chemistry*, 3d ed., 1976, and *Analytical Chemistry*, 3d ed., 1979, Holt, Rinehart and Winston, New York.

or all of the manuscript. Included are Professors R. R. Bessette, Southeastern Massachusetts University; M. F. Bryant, University of Georgia; J. F. Coetzee, University of Pittsburgh; P. Dumas, Trenton State College; E. T. Gray, Jr., University of Hartford; D. M. King, Western Washington University; P. F. Lott, University of Missouri, Kansas City; C. H. Lochmuller, Duke University; F. W. Smith, Youngstown State University; M. Thompson, University of Toronto; A. Timnick, Michigan State University; W. H. Smith, Texas Tech University; J. E. Byrd, California State College—Stanislaus; A. M. Olivares, Texas A & I University; and E. J. Billo, Boston College.

We also wish to thank Dr. Natalie McClure, now of Syntex Corporation, for preparing problem sets and their solutions for Chapters 8, 14, and 17, and Professor Daniel C. Harris, of University of California—Davis, for preparing problem sets and solutions for Chapters 5, 9–13, 23, and 25.

Stanford, California   Douglas A. Skoog
San Jose, California   Donald M. West
September, 1979

CONTENTS

PRINCIPLES
OF
INSTRUMENTAL
ANALYSIS

1

INTRODUCTION

A chemical analysis provides information about the composition of a sample of matter. The results of some analyses are qualitative and yield useful clues from which the molecular or atomic species, the structural features, or the functional groups in the sample can be deduced. Other analyses are quantitative; here, the results take the form of numerical data in units such as percent, parts per million, or milligrams per liter. In both types of analysis, the required information is obtained by measuring a physical property that is characteristically related to the component or components of interest.

It is convenient to describe properties which are useful for determining chemical composition as *analytical signals*; examples of such signals include emitted or absorbed light, conductance, weight, volume, and refractive index. None of these signals is unique to a given species. Thus, for example, all metallic elements in a sample will ordinarily emit ultraviolet and visible radiation when heated to a sufficiently high temperature in an electric arc; all charged species conduct electricity; and all of the components in a mixture contribute to its refractive index, weight, and volume. Therefore, all analyses require a separation. In some instances, the separation step involves physical isolation of the individual chemical components in the sample prior to signal generation; in others, a signal is generated or observed for the entire sample, following which the desired signal is isolated from the others. Some signals are susceptible to the latter treatment, while others are not. For example, when a sample is heated in an electric arc, the wavelength distribution for the radiation of each metallic species is unique to that species; separation of the wavelengths in a suitable device (a spectroscope) thus makes possible the identification of each component without physical separation. On the other hand, no general method exists for distinguishing the conductance of sodium ions from that of potassium ions. Here, a physical separation is required if conductance is to serve as the signal for the analysis of one of these species in a sample that also contains the other.

TYPES OF ANALYTICAL METHODS

Table 1-1 lists most of the common signals that are useful for analytical purposes. Note that the first six signals involve either the emission of radiation or the interaction of radiation with matter. The next three signals are electrical. Finally, five miscellaneous signals are grouped together. Also listed in the table are the names of analytical methods which are based on the various signals.

It is of interest to note that until perhaps 1920, most analyses were based on the last two signals listed in Table 1-1—namely, mass and volume. As a consequence, gravimetric and volumetric procedures have come to be known as *classical* methods of analysis, in contrast to the other procedures, which are known as *instrumental methods*.

Beyond the chronology of their development, few features clearly distinguish instrumental methods from classical ones. Some instrumental techniques are more sensitive than classical techniques, but many are not. With certain combinations of elements or compounds, an instrumental method may be more specific; with others, a gravimetric or a volumetric approach may be less subject to interference. Generalizations on the basis of accuracy, convenience, or expenditure of time are equally difficult to draw. Nor is it necessarily true that instrumental procedures employ more sophisticated or more costly apparatus; indeed, use of a modern automatic balance in a gravimetric analysis involves more complex and refined instrumentation than is required for several of the methods found in Table 1-1.

In addition to the methods listed in the second column of Table 1-1, there exists another group of analytical methods for the separation and resolution of closely related compounds. Common separation methods include chromatography, distillation, extraction, ion exchange, fractional crystallization, and selective precipitation. One of the signals listed in Table 1-1 is ordinarily used to complete the analysis following the separation step. Thus, for example, thermal conductivity, volume, refractive index, and electrical conductance have all been employed in conjunction with various chromatographic methods.

This text deals with most of the instrumental methods listed in Table 1-1, as well as with many of the most widely employed separation procedures. Little space is devoted to the classical methods, since their use is normally treated in elementary analytical courses.

Table 1-1 suggests that the chemist faced with an analytical problem may have a bewildering array of methods from which to choose. The amount of time spent on the analytical work and the quality of the results are critically dependent on this choice. In arriving at a decision as to which method to choose, the chemist must take into account the complexity of the materials to be analyzed, the concentration of the species of interest, the number of samples to be analyzed, and the accuracy required. His choice will then depend upon a knowledge of the basic principles underlying the various methods available to him, and thus their strengths and limitations. The development of this type of knowledge represents a major goal of this book.

INSTRUMENTS FOR ANALYSIS

In the broadest sense, an instrument for chemical analysis converts a signal that is usually not directly detectable and understandable by man to a form that is. Thus, an instrument can be viewed as a communication device between the system under study and the scientist.

TABLE 1-1 SOME ANALYTICAL SIGNALS

Signal	Analytical Methods Based on Measurement of Signal
Emission of radiation	Emission spectroscopy (X-ray, UV, visible, electron, Auger); flame photometry; fluorescence (X-ray, UV, visible); radiochemical methods
Absorption of radiation	Spectrophotometry (X-ray, UV, visible, IR); colorimetry; atomic absorption, nuclear magnetic resonance, and electron spin resonance spectroscopy
Scattering of radiation	Turbidimetry; nephelometry, Raman spectroscopy
Refraction of radiation	Refractometry; interferometry
Diffraction of radiation	X-Ray and electron diffraction methods
Rotation of radiation	Polarimetry; optical rotatory dispersion; circular dichroism
Electrical potential	Potentiometry; chronopotentiometry
Electrical current	Polarography; amperometry; coulometry
Electrical resistance	Conductimetry
Mass-to-charge ratio	Mass spectrometry
Rate of reaction	Kinetic methods
Thermal properties	Thermal conductivity and enthalpy methods
Mass	Gravimetric analysis
Volume	Volumetric analysis

Components of Instruments

Regardless of its complexity, an instrument generally contains no more than four fundamental components. As shown schematically in Figure 1-1, these components are a signal generator, an input transducer or detector, a signal processor, and an output transducer or readout. A general description of these components follows.

Signal Generators. Signal generators produce analytical signals from the components of the sample. The generator may simply be the sample itself. For example, the signal for an analytical balance is the mass of a component of the sample; for a pH meter, the signal is the activity of hydrogen ions in a solution. In many other instruments, however, the signal generator is more elaborate. For example, the signal generator of an infrared spectrophotometer includes, in addition to the sample, a source of infrared radiation, a monochromator, a beam chopper and splitter, a sample holder, and a radiation attenuator.

The second column of Table 1-2 lists some examples of signal generators.

Input Transducers or Detectors. A *transducer* is a device that converts one type of signal to another. An example is a thermocouple, which converts a radiant heat signal into an electrical voltage; the bellows in an aneroid barometer transduces a pressure signal into a mechanical motion signal. Most of the transducers that we will encounter convert analytical signals into an electrical voltage, current, or resistance, because of the ease with which electrical signals can be amplified and modified to drive readout devices. Note, however, that the transduced signals for the last two instruments in Table 1-2 are nonelectrical.

Signal Processors. The signal processor modifies the transduced signal in such a way as to make it more convenient for operation of the readout device. Perhaps the most common modification is that of amplification—that is, multiplication of the signal by a constant greater than unity. In a two-pan analytical balance, the motion of the beam is amplified by the pointer, whose displacement is significantly greater than that of the beam itself. Amplification by a photographic film is enormous; here, a single photon may produce as many as 10^{12} silver atoms. Electrical signals can, of course, be readily amplified by a factor of 10^6 or more.

A variety of other modifications are commonly carried out on electrical signals. In

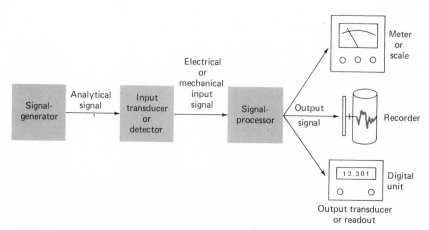

FIGURE 1-1 Components of a typical instrument.

addition to being amplified, signals are often multiplied by a constant smaller than unity (*attenuated*), integrated, differentiated, added or subtracted, or increased exponentially. Other operations may involve conversion into an alternating current, rectification to give a direct current, comparison of the transduced signal with one from a standard, and transformation from a current to a voltage or the converse.

Readout Devices. Readout devices employ an output transducer to convert the amplified signal from the processor to a signal that can be read by the human eye. Readout devices take the form of meters, strip chart recorders, oscilloscopes, pointer scales, and digital devices.

Electronics and Instrumentation

The growth of instrumental analysis has closely paralleled developments in the field of electronics, because the generation, transduction, amplification, and display of a signal can be rapidly and conveniently accomplished with electronic circuitry. Numerous transducers have been developed for the conversion of chemical signals to electrical form, and enormous amplification of the resulting electrical signals is possible. In turn, the electrical signals are readily presented on meters, on recorders, or in digital form.

Owing to the large amount of electronic circuitry in the laboratory, the modern chemist is faced with the question of how much knowledge of electronics is needed to make most efficient use of the equipment available for analysis. We believe that it is not only desirable but feasible for the chemist to possess a qualitative knowledge of how electronic circuits work. Thus, Chapters 2 and 3 are largely devoted to a presentation of this topic; subsequent chapters are concerned with the application of various instruments to the solution of chemical problems.

ABLE 1-2   SOME EXAMPLES OF INSTRUMENT COMPONENTS

strument	Signal Generator	Analytical Signal	Input Transducer	Transduced Signal	Signal Processor	Readout
notometer	Tungsten lamp, glass filter, sample	Attenuated light beam	Photocell	Electrical current	None	Current meter
tomic emission spectrometer	Flame, monochromator, chopper, sample	UV or visible radiation	Photomultiplier tube	Electrical potential	Amplifier, demodulator	Chart recorder
oulometer	dc source, sample	Cell current	Electrodes	Electrical current	Amplifier	Chart recorder
H meter	Sample	Hydrogen ion activity	Glass-calomel electrodes	Electrical potential	Amplifier, digitalizer	Digital unit
-Ray powder diffractometer	X-Ray tube, sample	Diffracted radiation	Photographic film	Latent image	Chemical developer	Black images on film
olor comparator	Sunlight, sample	Color	Eye	Optic nerve signal	Brain	Visual color response

2

ELECTRICITY AND ELECTRIC CIRCUITS

This chapter provides a brief and elementary review of the laws governing electrical currents, the properties of direct and alternating current circuits, and selected methods for measuring electrical quantities.[1]

INTRODUCTION

An electrical current is the motion of a charge through a medium. In metallic conductors, only electrons are mobile; here, the current involves motion of negative charges only. In media such as ionic solutions and semiconductors, both negative and positive species are mobile and participate in the passage of electricity.

Electrical Units

The unit of charge or quantity of electricity is the *coulomb*, C, which is the charge required to convert 0.00111800 g of silver ion to silver metal. Another unit, the *faraday*, F, corresponds to one equivalent of charge, that is, 6.02×10^{23} charged particles. The relationship between the two units is readily calculated from their definitions:

$$\frac{\text{coulombs}}{\text{faraday}} = \frac{107.87 \text{ g Ag}^+/\text{equivalent Ag}^+}{0.00111800 \text{ g Ag}^+/\text{coulomb}}$$

$$\times \frac{1 \text{ equivalent Ag}^+}{1 \text{ equivalent of charge}}$$

$$= 9.649 \times 10^4$$

Electric current I is the rate of flow of a charge. That is,

$$I = dQ/dt \qquad (2\text{-}1)$$

[1] Some general references, which cover these topics, include: A. J. Diefenderfer, *Principles on Electronic Instrumentation*, 2d ed. Philadelphia: Saunders, 1979; J. J. Brophy, *Basic Electronics for Scientists*, 3d ed. New York: McGraw-Hill, 1977; H. V. Malmstadt, C. G. Enke, S. R. Crouch, and G. Horlick, *Electronic Measurements for Scientists*. Menlo Park, CA: Benjamin, 1974; and R. J. Smith, *Circuits, Devices, and Systems*, 3d ed. New York: Wiley, 1976.

where Q is the charge. The *ampere*, A, the unit of current, corresponds to a rate of one coulomb per second.

The electrical potential V between two points in space is the work required to move an electrical charge from one of the points to the other. When the work is given in joules and the charge in coulombs, the unit for potential is the *volt*, V. That is, one volt is equal to one joule per coulomb.

The *ohm*, Ω, is the unit of resistance R to current flow and is the resistance through which a potential of one volt will produce a current of one ampere.

Electrical conductance G is the reciprocal of resistance and has the units of *reciprocal ohms*, Ω^{-1}, or siemens, S.

Electrical power P is the rate of electrical work in joules per second or *watts* W. The power dissipated during passage of electrons is given by

$$P = IV \qquad (2\text{-}2)$$

Substitution of Equation 2-4 yields the power loss in a resistance of R ohms; that is

$$P = I^2 R = V^2/R \qquad (2\text{-}3)$$

Electrical Currents

If a conducting path exists between two points with differing potentials, charge will flow until the difference in potentials becomes zero. For example, when the switch is closed in the direct current circuit shown in Figure 2-1, a conducting path is provided, through which charge can flow from one terminal of the battery to the other. Current will continue until the potential difference between A and B becomes zero—that is, until the battery is discharged.

By convention, the direction of a current is always from the positive terminal to the negative, regardless of the type of particle that carries the current. Thus, in the external circuit shown in Figure 2-1, current takes the form of

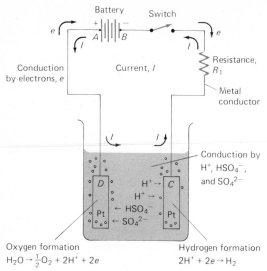

Oxygen formation
$H_2O \rightarrow \frac{1}{2}O_2 + 2H^+ + 2e$

Hydrogen formation
$2H^+ + 2e \rightarrow H_2$

Conduction by electrochemical reaction

FIGURE 2-1 Conduction through a dc circuit.

motion of electrons e through the metal conductors BC and AD. The direction of the current I, however, is said to be from A to D to C to B; that is, current is always treated as if it were a flow of positive charge.

The two platinum electrodes (C and D) and the dilute sulfuric acid solution in which they are immersed represent an electrochemical cell. The mechanism by which electricity passes through this cell is entirely different from that in the metallic conductor. Here, the flow involves migration of positive hydrogen ions toward electrode C, and of negative sulfate and hydrogen sulfate ions toward electrode D. Note that the excess negative charge that tends to accumulate around D as a result of the anionic migration is exactly offset by the postively charged hydrogen ions produced by the electrode process; similarly, the excess hydrogen ions that migrate to electrode C are removed by the electrochemical process. The consequence is that the solution remains homogeneous with respect to *charge*, but the region around electrode D is enriched

with sulfuric acid at the expense of the region around C.

Two electrochemical reactions make possible the transition from electronic conduction in the metal to conduction by ions in the solution. In the first of these two reactions, hydrogen ions consume electrons from C; in the second, water molecules give up electrons at D. Thus, three modes of passage of electricity exist in this simple circuit: electronic, ionic, and electrochemical.

Laws of Electricity

Ohm's Law. Ohm's law takes the form

$$V = IR \qquad (2\text{-}4)$$

where V is the potential in volts between two points[2] in a circuit, R is the resistance between the two points in ohms, and I is the resulting current in amperes.

Ohm's law applies to electronic and ionic conduction; it is not, however, applicable to conduction across interfaces such as the electrode surfaces in the cell shown in Figure 2-1.

Kirchhoff's Laws. *Kirchhoff's current law* states that the algebraic sum of currents around any point in a circuit is zero. *Kirchhoff's voltage law* states that the algebraic sum of the voltages around a closed electrical loop is zero.

The applications of Kirchhoff's and Ohm's laws to simple dc circuits are considered in the next section.

Series Circuits

Consider the simple electrical circuit in Figure 2-2, which consists of a battery, a switch, and three resistors in series. Applying

---

[2] In Chapters 2 and 3, the symbol V will be used to denote electrical potential in circuits. In Chapters 18 to 22, however, the electrochemical convention will be followed in which electromotive force is designated as E.

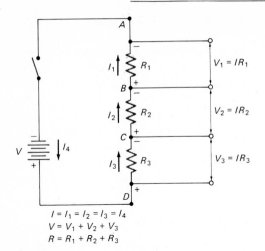

$$I = I_1 = I_2 = I_3 = I_4$$
$$V = V_1 + V_2 + V_3$$
$$R = R_1 + R_2 + R_3$$

FIGURE 2-2 Resistors in series; a voltage divider.

Kirchhoff's current law to point D in the circuit, we write

$$I_4 - I_3 = 0$$

or

$$I_4 = I_3$$

Note that the current out of D must be opposite in sign to the input current. Similarly, application of the law to point C gives

$$I_3 = I_2$$

Thus, it is apparent that the current is the same at all points in a series circuit or

$$I = I_1 = I_2 = I_3 = I_4$$

Application of the voltage law to the circuit yields the equation

$$V - V_3 - V_2 - V_1 = 0$$

or

$$V = V_1 + V_2 + V_3 \qquad (2-5)$$

Note that point D is positive with respect to point C, which in turn is positive with respect to point B; finally, B is positive with respect

to A. The three voltages thus oppose the voltage of the battery V and must be given opposite signs.

Substitution of Ohm's law into Equation 2-5 gives

$$V = I(R_1 + R_2 + R_3) = IR \qquad (2-6)$$

Note that the total resistance R of a series circuit is equal to the sum of the resistances of the individual components. That is,

$$R = R_1 + R_2 + R_3 \qquad (2-7)$$

Applying Ohm's law to the part of the circuit from points B to A gives

$$V_1 = I_1 R_1 = IR_1$$

Dividing by Equation 2-6 yields

$$\frac{V_1}{V} = \frac{\cancel{I}R_1}{\cancel{I}(R_1 + R_2 + R_3)}$$

or

$$V_1 = \frac{VR_1}{R_1 + R_2 + R_3} = V\frac{R_1}{R} \qquad (2-8)$$

It is clear that we may also write

$$V_2 = VR_2/R$$

and

$$V_3 = VR_3/R$$

Voltage Dividers. Series resistances are widely used in electronic circuits to provide potentials that are variable functions of an input voltage. Devices of this type are called *voltage dividers*. As shown in Figure 2-3a, one type of voltage divider provides voltages in discrete increments; the second type (Figure 2-3b), called a *potentiometer*,[3] provides a potential that is continuously variable.

---

[3] The word potentiometer is also used in a different context as the name for a complete instrument that employs a linear voltage divider for the accurate measurement of potentials.

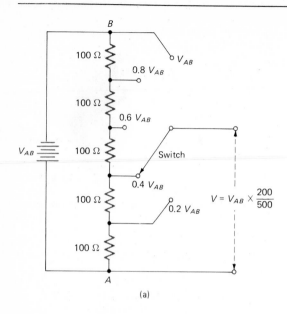

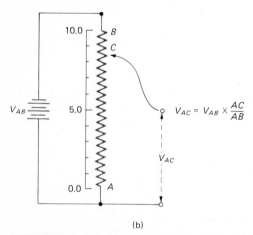

FIGURE 2-3   Voltage dividers: (a) selector type and (b) continuously variable type (potentiometer).

In most potentiometers, such as the one shown in Figure 2-3b, the resistance is linear—that is, the resistance between one end, A, and any point, C, is directly proportional to the length, AC, of that portion of the resistor. Then $R_{AC} = kAC$ where AC is expressed in convenient units of length and k is

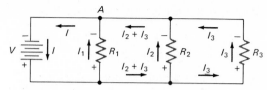

FIGURE 2-4   Resistors in parallel.

a proportionality constant. Similarly, $R_{AB} = kAB$. Combination of these relationships with Equation 2-8, yields

$$V_{AC} = V_{AB} \frac{R_{AC}}{R_{AB}} = V_{AB} \frac{AC}{AB} \qquad (2\text{-}9)$$

Parallel Circuits

Figure 2-4 depicts a *parallel* dc circuit. Applying Kirchhoff's current law to point A, we obtain

$$I_1 + I_2 + I_3 - I = 0$$

or

$$I = I_1 + I_2 + I_3 \qquad (2\text{-}10)$$

Kirchhoff's voltage law gives three independent equations. Thus, we may write, for the loop that contains the battery and R_1

$$V - I_1 R_1 = 0$$
$$V = I_1 R_1$$

For the loop containing V and R_2,

$$V = I_2 R_2$$

For the loop containing V and R_3,

$$V = I_3 R_3$$

We could write additional equations for the loop containing R_1 and R_2 as well as the loop containing R_2 and R_3. However, these equations are not independent of the foregoing three. Substitution of the three independent equations into Equation 2-10 yields

$$I = \frac{V}{R} = \frac{V}{R_1} + \frac{V}{R_2} + \frac{V}{R_3}$$

when R is the net circuit resistance, then

$$\frac{1}{R} = \frac{1}{R_1} + \frac{1}{R_2} + \frac{1}{R_3} \qquad (2\text{-}11)$$

In a parallel circuit, in contrast to a series circuit, it is the conductances that are additive rather than the resistances. That is, since $G = 1/R$,

$$G = G_1 + G_2 + G_3 \qquad (2\text{-}12)$$

Just as series resistances form a voltage divider, parallel resistances create a current divider. The fraction of the total current that is present in R_1 in Figure 2-4 is

$$\frac{I_1}{I} = \frac{V/R_1}{V/R} = \frac{1/R_1}{1/R} = \frac{G_1}{G}$$

or

$$I_1 = I\frac{R}{R_1} = I\frac{G_1}{G} \qquad (2\text{-}13)$$

EXAMPLE

For the accompanying circuit, calculate: (a) the total resistance; (b) the current drawn from the battery; (c) the current present in each of the resistors; and (d) the potential drop across each of the resistors.

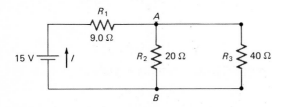

R_2 and R_3 are parallel resistances. Thus, the resistance $R_{2,3}$ between points A and B will be given by Equation 2-11. That is,

$$\frac{1}{R_{2,3}} = \frac{1}{20} + \frac{1}{40}$$

or

$$R_{2,3} = 13.3\ \Omega$$

We can now reduce the original circuit to the following *equivalent circuit*.

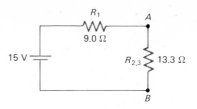

Here, we have the equivalent of two series resistances, and

$$R = R_1 + R_{2,3} = 9.0 + 13.3 = 22.3\ \Omega$$

From Ohm's law, the current I is given by

$$I = 15/22.3 = 0.67\ \text{A}$$

Employing Equation 2-8, the voltage V_1 across R_1 is

$$V = 15 \times 9.0/(9.0 + 13.3) = 6.0\ \text{V}$$

Similarly, the voltage across resistors R_2 and R_3 is

$$V_2 = V_3 = V_{2,3} = 15 \times 13.3/22.3$$

$$= 8.95 = 9.0\ \text{V}$$

Note that the sum of the two voltages is 15 V, as demanded by Kirchhoff's voltage law.

The current in R_1 is given by

$$I_1 = I = 0.67\ \text{A}$$

The currents through R_2 and R_3 are found from Ohm's law. Thus,

$$I_2 = 9.0/20 = 0.45\ \text{A}$$

$$I_3 = 9.0/40 = 0.22\ \text{A}$$

Note that the two currents add to give the net current, as required by Kirchhoff's law.

Alternating Current

The electrical output from transducers of analytical signals often fluctuate periodically. These fluctuations can be represented (as in Figure 2-5) by a plot of the instantaneous current or potential as a function of time. The *period*, τ, for the signal is the time required for the completion of one cycle.

The reciprocal of the period is the *frequency, f,* of the cycle. That is,

$$f = 1/\tau \qquad (2\text{-}14)$$

The unit of frequency is the hertz, Hz, which is defined as one cycle per second.

Sine-Wave Currents. The sinusoidal wave (Figure 2-5a) is the most frequently encountered type of periodic electrical signal. A common example is the alternating current produced by rotation of a coil in a magnetic field (as in an electrical generator). Thus, if the instantaneous current or voltage produced by a generator is plotted as a function of time, a sine wave results.

A pure sine wave is conveniently represented as a vector of length I_p (or V_p), which is rotating counterclockwise at a constant angular velocity ω. The relationship between the vector representation and the sine-wave plot is shown in Figure 2-6.

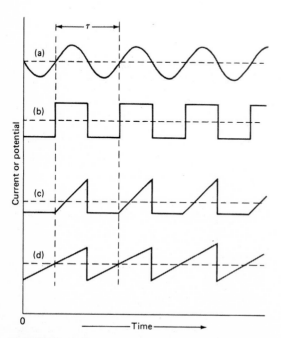

FIGURE 2-5 Examples of periodic signals: (a) sinusoidal, (b) square wave, (c) ramp, and (d) sawtooth.

The vector rotates at a rate of 2π radians in the period τ; thus, the angular velocity is given by

$$\omega = \frac{2\pi}{\tau} = 2\pi f \qquad (2\text{-}15)$$

If the vector quantity is current or voltage, the instantaneous current i or instantaneous voltage v at time t is given by[4] (see Figure 2-6)

$$i = I_p \sin \omega t = I_p \sin 2\pi ft \qquad (2\text{-}16)$$

or alternatively

$$v = V_p \sin \omega t = V_p \sin 2\pi ft \qquad (2\text{-}17)$$

where I_p and V_p, the maximum or peak current and voltage, are called the *amplitude A* of the sine wave.

Figure 2-7 shows two sine waves having different amplitudes. The two waves are also out of *phase* by 90 deg or $\pi/2$ radians. The phase difference is called the *phase angle,* which arises from one vector leading or lagging a second by this amount. A more generalized equation for a sine wave, then, is

$$i = I_p \sin(\omega t + \phi) = I_p \sin(2\pi ft + \phi) \quad (2\text{-}18)$$

where ϕ is the phase angle from some reference sine wave. An analogous equation could be written in terms of voltage.

The current or voltage associated with a sinusoidal current can be expressed in several ways. The simplest is the peak amplitude I_p (or V_p), which is the maximum instantaneous current or voltage during a cycle; the peak-to-peak value, which is $2I_p$ or $2V_p$ is also employed occasionally. The *root mean square* or *rms current* in an ac circuit will produce the same heating in a resistor as a direct cur-

---

[4] In treating currents that change with time, it is useful to symbolize the instantaneous current, voltage, or charge with the lower case letters i, v, and q. On the other hand, capital letters are used for steady current, voltage, or charge, or a specifically defined variable quantity such as a peak voltage current; that is, V_p and I_p.

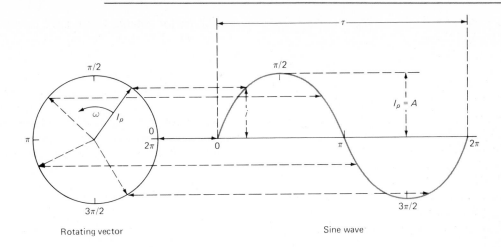

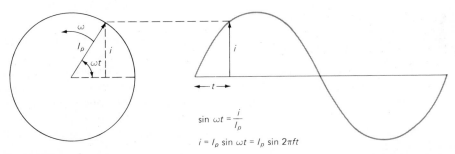

$$\sin \omega t = \frac{i}{I_p}$$

$$i = I_p \sin \omega t = I_p \sin 2\pi ft$$

FIGURE 2-6 Relationship between a sine wave of period τ and amplitude I_p and the corresponding vector of length I_p rotating at an angular velocity of ω or a frequency of f Hz.

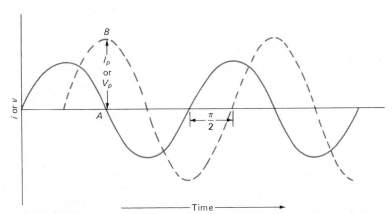

FIGURE 2-7 Sine waves with different amplitudes (I_p or V_p) and out of phase by 90 deg or $\pi/2$ radians.

rent of the same magnitude. Thus, the rms current is important in power calculations (Equations 2-2 and 2-3). The rms current is given by

$$I_{rms} = \sqrt{\frac{I_p^2}{2}} = 0.707 I_p \qquad (2\text{-}19)$$

Similarly,

$$V_{rms} = 0.707 V_p$$

REACTANCE IN ELECTRICAL CIRCUITS

Whenever the current in an electrical circuit is increased or decreased, energy is required to charge the electric and magnetic fields associated with the flow of charge. As a consequence, there develops a counterforce or *reactance* which tends to counteract the change. Two types of reactance can be recognized, namely, *capacitance* and *inductance*. When the rate of change in current is low, the reactance of most of the components in a circuit is sufficiently small to be neglected. With rapid changes, on the other hand, circuit elements such as switches, junctions, and resistors may exhibit a detectable reactance. Ordinarily, this type of reactance is undesirable, and every effort is made to diminish its magnitude.

Capacitance and inductance are often deliberately introduced into a circuit with *capacitors* and *inductors*. These devices provide a means for accomplishing such useful functions as converting alternating current to direct or the converse, discriminating among signals of different frequencies, or separating ac and dc signals.

In the section that follows, we shall consider the properties of inductors and capacitors. The behavior and uses of these two important circuit components frequently parallel one another. Thus, it will be instructive to note their similarities and differences as the discussion develops.

Capacitors and Inductors

Structurally, capacitors and inductors are quite different. A typical capacitor consists of a pair of conductors separated by a thin layer of a *dielectric* substance—that is, by an electrical insulator that contains essentially no mobile, current-carrying, charged species. The simplest capacitor consists of two sheets of metal foil separated by a thin film of a dielectric such as air, oil, plastic, mica, paper, ceramic, or metal oxide. Except for air and mica capacitors, the two layers of foil plus the insulator are usually folded or rolled into a compact package and sealed to prevent atmospheric deterioration.

In contrast, an inductor is ordinarily a coil of insulated wire of relatively large diameter to minimize resistance. Some inductors are wound around a soft iron core to enhance their reactance; others have an air core. The latter type is sometimes called a *choke*.

Capacitors are significantly less bulky than inductors and, in addition, can be fabricated as part of printed circuits. Thus, to the extent possible, capacitors are employed in preference to inductors in electronic circuit design. At high frequencies, however, the reactance of capacitors may be too small, and inductors are used preferentially.

Capacitance. In order to describe the properties of a capacitor, it is useful to consider the dc circuit shown at the top of Figure 2-8a, which contains a battery V_i, a resistor R, and a capacitor C in series. The capacitor is symbolized by a pair of parallel lines. A circuit of this kind is frequently called a *series RC circuit*.

When the switch S is closed to position 1, electrons flow from the negative terminal of the battery into the lower conductor or *plate* of the capacitor. Simultaneously, electrons are repelled from the upper plate and flow toward the positive terminal of the battery. This movement constitutes a momentary current, which decays to zero, however, owing to the potential difference that builds up

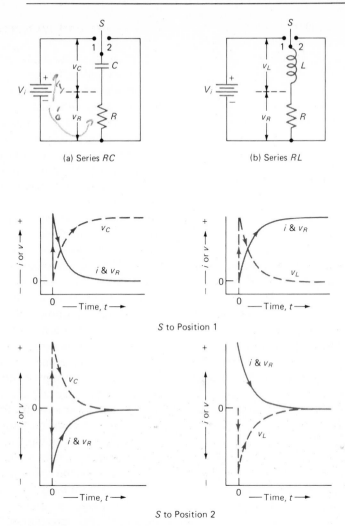

(a) Series RC

(b) Series RL

S to Position 1

S to Position 2

FIGURE 2-8 Behavior of (a) a series RC circuit and (b) a series RL circuit.

across the plates and prevents the continued flow of electrons. When the current ceases, the capacitor is said to be *charged*.

If the switch now is moved from 1 to 2, electrons will flow from the negatively charged lower plate of the capacitor through the resistor R to the positive upper plate. Again, this movement constitutes a current which decays to zero as the potential between

the two plates disappears; here, the capacitor is said to be *discharged*.

A useful property of a capacitor is its ability to store an electrical charge for a period of time and then to give up the stored charge as needed. Thus, if S in Figure 2-8a is first held at 1 until C is charged and is then moved to a position *between* 1 and 2, the capacitor will remain in a charged condition for an extended period. Upon moving S to 2, discharge occurs in the same way as it would if the change from 1 to 2 had been rapid.

The quantity of electricity, Q, required to charge a capacitor fully depends upon the area of the plates, their shape, the spacing between them, and the dielectric constant for the material that separates them. In addition, the charge, Q, is directly proportional to the applied voltage. That is,

$$Q = CV \qquad (2\text{-}20)$$

When V is the applied potential in volts and Q is the quantity of charge in coulombs, the proportionality constant C is the *capacitance* of a capacitor in *farads*, F. One farad, then, corresponds to one coulomb of charge per applied volt. Most of the capacitors used in electronic circuitry have capacitances in the microfarad (10^{-6} F) to picofarad (10^{-12} F) ranges.

Capacitance is important in ac circuits, particularly because a voltage that varies with time gives rise to a time-varying charge—that is, *a current*. This behavior is seen by differentiating Equation 2-20 to give

$$\frac{dq}{dt} = C\frac{dv}{dt} \qquad (2\text{-}21)$$

By definition (Equation 2-1), the current i is the rate of change of charge; that is, $dq/dt = i$. Thus,

$$i = C\frac{dv}{dt} \qquad (2\text{-}22)$$

It is important to note that the current in a capacitor is zero when the voltage is time independent—that is, for a direct current. Because a direct current refers to a steady state, the initial transient current that charges the condenser is of no significance in considering the overall effect of a capacitor on a current that has a dc component.

Inductance. It is found experimentally that a magnetic field surrounds any conductor as it carries an electric current. The work required to establish this field manifests itself

as a counter-potential v when the current first begins, and tends to oppose the flow of electrons. On the other hand, when the current in a conductor ceases, the magnetic field collapses; this process causes the development of a momentary potential which acts to continue the current.

The magnitude of the potential v that develops during the increase or decrease in current is found to be directly proportional to the rate of current change di/dt. That is,

$$v = -L\frac{di}{dt} \qquad (2\text{-}23)$$

Here, the negative sign indicates that the induced potential acts to oppose the change in current.

The proportionality constant L in Equation 2-23 is termed the *inductance* of the conductor; it has the units of *henrys*, H. One henry of inductance produces a counter-potential of one volt when the rate of change of current is one ampere per second. Inductors employed in electronic circuits have inductances that range from a few μH (microhenry) to several H or more.

Figure 2-8b shows a *series RL circuit*, which contains a battery V_i, a resistor R, and an inductor L connected in series. The magnitude of L depends upon the number of turns in the wire coil.

As was pointed out earlier, a capacitor stores energy as an electric field across a dielectric; an inductor, on the other hand, stores energy as a magnetic field surrounding a conductor. The electric field in the capacitor is proportional to the applied voltage; the magnetic field in the inductor is proportional to the current that exists in the inductor.

The difference between a capacitor and an inductor can also be seen by comparing Equations 2-22 and 2-23. The former indicates that a change in potential across a capacitor results in a current; the latter shows that a change in current through an inductor causes the development of a potential. Thus, the functions of the two components in an electric circuit tend to complement one another.

Rate of Current and Potential Changes in an *RC* Circuit. The rate at which a capacitor is charged or discharged is finite. Consider, for example, the circuit shown in Figure 2-8a. From Kirchhoff's voltage law, we know that at any instant after the switch is moved to position 1, the sum of the voltage across C and R (v_C and v_R) must equal the input voltage V_i. Thus,

$$V_i = v_C + v_R \qquad (2\text{-}24)$$

Because V_i is constant, the increase in v_C that accompanies the charging of the capacitor must be exactly offset by a decrease in v_R.

Substitution of Equations 2-4 and 2-20 into this equation gives, upon rearrangement,

$$V_i = \frac{q}{C} + iR \qquad (2\text{-}25)$$

Differentiating with respect to time t yields

$$0 = \frac{dq/dt}{C} + R\frac{di}{dt} \qquad (2\text{-}26)$$

Here again, we have used lower case letters to represent instantaneous charge and current.

As noted earlier, $dq/dt = i$. Substituting this expression into Equation 2-26 yields, upon rearrangement,

$$\frac{di}{i} = -\frac{dt}{RC}$$

Integration between the limits of the initial current I_{init} and i gives

$$i = I_{\text{init}} e^{-t/RC} \qquad (2\text{-}27)$$

In order to obtain a relationship between the instantaneous voltage across the resistor, Ohm's law is employed to replace i and I_{init} in Equation 2-27. Thus,

$$v_R = V_i e^{-t/RC} \qquad (2\text{-}28)$$

Substitution of this expression into Equation 2-24 yields, upon rearrangement,

$$v_C = V_i(1 - e^{-t/RC}) \qquad (2\text{-}29)$$

Note that the product RC that appears in the last three equations has the units of time; since $R = v_R/i$ and $C = q/v_C$,

$$RC = \frac{\text{volts}}{\text{coulombs/seconds}} \times \frac{\text{coulombs}}{\text{volts}}$$

$$= \text{seconds}$$

The term RC is called the *time constant* for the circuit.

The following example illustrates the use of the three equations that were just derived.

EXAMPLE

Values for the components in Figure 2-8a are $V_i = 10$ V, $R = 1000$ Ω, $C = 1.00$ μF or 1.00×10^{-6} F. Calculate: (a) the time constant for the circuit; and (b) i, v_C, and v_R after two-time constants ($t = 2RC$) have elapsed.

(a) Time constant $= RC = 1000 \times 1.00 \times 10^{-6} = 1.00 \times 10^{-3}$ s or 1.00 ms

(b) Substituting $t = 2.00$ ms in Equation 2-27 reveals

$$i = \frac{10.0}{1000} e^{-2.00/1.00}$$

$$= 1.35 \times 10^{-3} \quad \text{or} \quad 1.35 \text{ mA}$$

We find from Equation 2-28 that

$$v_R = 10.0 e^{-2.00/1.00} = 1.35 \text{ V}$$

and by substituting into Equation 2-29

$$v_C = 10.0 \, (1 - e^{-2.00/1.00}) = 8.65 \text{ V}$$

The center illustration in Figure 2-8a shows the changes in i, v_R, and v_C that occur during the charging cycle of an RC circuit. These plots were based upon the data given in the example just considered. Note that v_R and i assume their maximum values the instant the switch in Figure 2-8a is moved to 1. At the same instant, on the other hand, the voltage across the capacitor increases rapidly from zero and ultimately approaches a constant value. For practical purposes, a capacitor is considered to be fully charged after $5RC$s have elapsed. At this point the current will have decayed to less than 1% of its initial value ($e^{-5} = 0.0067 \cong 0.01$).

When the switch in Figure 2-8a is moved to position 2, the battery is removed from the circuit and the capacitor becomes a source of current. The flow of charge, however, will be in the opposite direction from what it was previously. Thus,

$$dq/dt = -i$$

The initial potential will be that of the battery. That is,

$$V_C = V_i$$

Employing these equations and proceeding as in the earlier derivation, we find that for the discharge cycle

$$i = -\frac{V_C}{R} e^{-t/RC} \tag{2-30}$$

$$v_R = -V_C e^{-t/RC} \tag{2-31}$$

and

$$v_C = V_C e^{-t/RC} \tag{2-32}$$

The bottom plot of Figure 2-8a shows how these variables change with time.

It is important to note that in each cycle, the change in voltage across the capacitor is *out of phase with and lags behind* that of the current and the potential across the resistor.

Rate of Current and Potential Change Across an *RL* Circuit. Equation 2-23 and techniques similar to those in the previous section can be used to derive a set of expressions for the *RL* circuit shown in Figure 2-8b that are analogous to Equations 2-27 through 2-32. For example, when the switch is closed to position 1,

$$v_R = V_i(1 - e^{-tR/L}) \tag{2-33}$$

and

$$v_L = V_i e^{-tR/L} \tag{2-34}$$

where L/R is the time constant for the circuit. These relationships for a typical *RL* circuit are shown in Figure 2-8b.

It is useful to compare the behavior of *RC* and *RL* circuits during a variation in signal. The plots in Figure 2-8 illustrate that the two reactances are alike in the sense that both show a potential change which is out of phase with the current. Note again the complementary behavior of the two; for an inductor, the voltage leads the current, while for a capacitor, the reverse is the case.

Response of Series *RC* and *RL* Circuits to Sinusoidal Inputs

In the sections that follow, the response of series *RC* and *RL* circuits to a sinusoidal ac voltage signal will be considered. The input signal is described by Equation 2-17; that is,

$$v_s = V_p \sin \omega t = V_p \sin 2\pi ft \tag{2-35}$$

Voltage, Current, and Phase Relationship for Series *RC* Circuit. Application of Kirchhoff's voltage law to the circuit shown in Figure 2-9a leads to the expression

$$v_s = v_C + v_R = V_p \sin \omega t$$

Substitution of Equations 2-20 and 2-4 yields

$$V_p \sin \omega t = \frac{q}{C} + Ri$$

Differentiating and replacing dq/dt with i gives

$$\omega V_p \cos \omega t = \frac{i}{C} + R\frac{di}{dt}$$

The solution of the differential equation can be shown to be[5]

$$i = \frac{V_p}{\sqrt{R^2 + (1/\omega C)^2}} \sin(\omega t + \phi) \tag{2-36}$$

where the phase angle ϕ is given by

$$\tan \phi = \frac{1}{\omega RC} \tag{2-37}$$

---

[5] J. J. Brophy, *Basic Electronics for Scientists.* New York: McGraw-Hill, 1966, pp. 57–59.

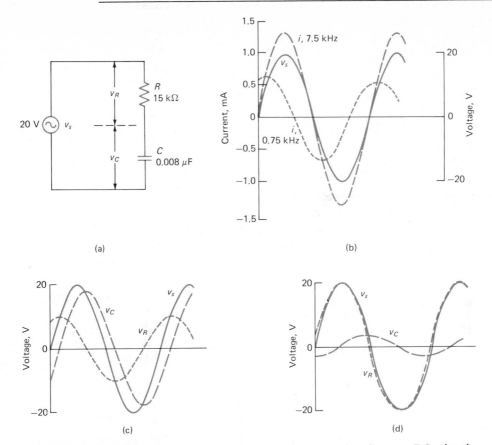

FIGURE 2-9 Current, voltage, and phase relationships for an *RC* circuit: (a) Circuit, (b) currents at 0.75 and 7.5 kHz, (c) voltages at 0.75 kHz, and (d) voltages at 7.5 kHz.

The peak current occurs whenever the $\sin(\omega t + \phi)$ is unity. Thus,

$$I_p = \frac{V_p}{\sqrt{R^2 + (1/\omega C)^2}} \qquad (2\text{-}38)$$

A comparison of Equations 2-36 and 2-35 reveals that the current in a series *RC* circuit is often not in phase with the input voltage signal. This condition is likely to prevail whenever the frequency (or the capacitance) is small; then, $\tan \phi$ and thus ϕ (Equation 2-37) becomes significant with respect to ωt in Equation 2-36. At sufficiently high frequen-

cies and capacitance, on the other hand, ϕ becomes negligible and, for all practical purposes, the current and input voltage are in phase. Under these circumstances, $(1/\omega C)$ becomes negligible with respect to R and Equation 2-38 reverts to Ohm's law.

EXAMPLE

Assume the components in the *RC* circuit shown in Figure 2-9a have values of $V_p = 20\,\text{V}, R = 1.5 \times 10^4\,\Omega,$ and $C = 0.0080\,\mu\text{F}.$ (a) Calculate the peak current (1) if $f = 0.75$ kHz, (2) $f = 75$ kHz, and (3) with the capacitor removed from the circuit. (b) Calculate the phase angles between the voltage and the current at the two frequencies.

(a) (1)

$$\frac{1}{\omega C} = \frac{1}{2\pi f C} = \frac{1}{2\pi \times 0.75 \times 10^3 \times 8.0 \times 10^{-9}}$$

$$= 2.65 \times 10^4 \ \Omega$$

Note that the quantity $1/2\pi f C$ has the dimensions of ohms. That is,

$$\frac{1}{fC} = \frac{1}{s^{-1} \cdot C \cdot V^{-1}}$$

But the number of coulombs C is equal to the product of the current in amperes and the time in seconds. Thus,

$$\frac{1}{fC} = \frac{1}{s^{-1} \cdot A \cdot s \cdot V^{-1}} = \frac{V}{A} = \Omega$$

Employing Equation 2-38, we write

$$I_p = \frac{20}{\sqrt{(1.5 \times 10^4)^2 + (2.65 \times 10^4)^2}}$$

$$= 6.6 \times 10^{-4} \ A$$

(2)

$$\frac{1}{2\pi f C} = \frac{1}{2\pi \times 75 \times 10^3 \times 8.0 \times 10^{-9}}$$

$$= 2.65 \times 10^2 \ \Omega$$

and

$$I_p = \frac{20}{\sqrt{(1.5 \times 10^4)^2 + (2.65 \times 10^2)^2}}$$

$$= 1.33 \times 10^{-3} \ A$$

(3) In the absence of C

$$I_p = \frac{20}{1.5 \times 10^4} = 1.33 \times 10^{-3} \ A$$

(b) Employing Equation 2-37, we find at 0.75 kHz

$$\tan \phi = \frac{2.65 \times 10^4}{1.5 \times 10^4} = 1.77$$

$$\phi = 60.5 \ \text{deg}$$

and at 75 kHz

$$\text{arc tan } \phi = \text{arc tan } \frac{2.65 \times 10^2}{1.5 \times 10^4}$$

$$= \text{arc tan } 0.0177 = 1.0 \ \text{deg}$$

Figure 2-9b, c, and d shows the instantaneous current, voltage, and phase relationships when the circuit characteristics are the same as those employed in the example. The time scales on the abscissas differ by a factor of 100 to accommodate the two frequencies. From Figure 2-9b and the example, it is evident that at the lower frequency, the current is significantly smaller than at the higher because of the greater reactance of the capacitor at the lower frequency. This larger reactance is also reflected in the large phase difference between the current and the input voltage shown in Figure 2-9c ($\phi = 60.5$ deg). At 75 kHz, this difference becomes considerably smaller (Figure 2-9d).

Figures 2-9c and 2-9d show the relationship between the input voltages and the voltage drops across the capacitor and the resistor. At the lower frequency, the drop across the capacitor is large with respect to that across the resistor. In contrast, at 75 kHz, v_C is small, indicating little reaction between the current and the capacitor. At frequencies greater than 1 MHz, essentially no interaction would occur, and the arrangement would behave as a purely resistive circuit.

Note also the phase relationships among v_s, v_C, and v_R shown in Figure 2-9c.

Voltage, Current, and Phase Relationships for Series *RL* Circuits. Applying Kirchhoff's voltage law to the circuit in Figure 2-10 gives

$$v_s = v_R + v_L = V_p \sin \omega t$$

Substituting Equations 2-4 and 2-23 yields

$$V_p \sin \omega t = Ri + L \, di/dt$$

The second term on the right-hand side of the equation is positive because the minus sign in Equation 2-23 means that the induced voltage

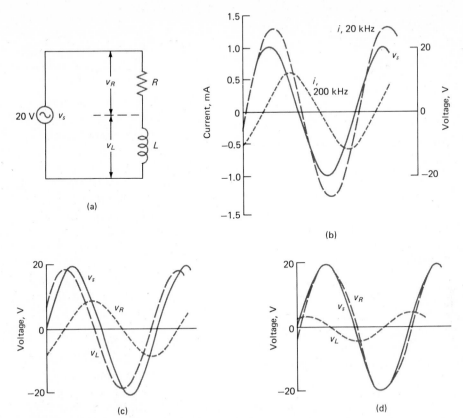

FIGURE 2-10 Current, voltage, and phase relationships in an *RL* circuit: (a) Circuit, (b) currents at 20 and 200 kHz, (c) voltages at 200 kHz, and (d) voltages at 20 kHz.

in an inductor *opposes* the current; thus the polarity of *Ri* and *L di/dt* must be the same. Integration of this equation yields[6]

$$i = \frac{V_p}{\sqrt{R^2 + (\omega L)^2}} \sin(\omega t + \phi) \quad (2\text{-}39)$$

where ϕ is given by

$$\tan \phi = \frac{\omega L}{R} \quad (2\text{-}40)$$

The peak current occurs where $R \sin(\omega t + \phi)$ is equal to unity, or

$$I_p = \frac{V_p}{\sqrt{R^2 + (\omega L)^2}} \quad (2\text{-}41)$$

Figure 2-10 shows the relationships among currents, voltages, and phases for an *RL* circuit. Comparison with Figure 2-9 again reveals the complementary nature of the two types of reactive components. For inductors, reaction to the current occurs at high frequencies but tends to disappear at low. Here, the current lags the input potential rather than leading it.

[6] J. J. Brophy, *Basic Electronics for Scientists*. New York: McGraw-Hill, 1966, pp. 67–69.

Capacitative and Inductive Reactance; Impedance. An examination of Equations 2-41 and 2-38 shows a similarity to Ohm's law, with the denominator terms being an expression of the impedance exerted by the circuit to the flow of electricity. Note that at high frequencies, where $1/\omega C \ll R$, Equation 2-38 reverts to Ohm's law; similarly, when $\omega L \ll R$, Equation 2-41 behaves in the same way.

The term $1/\omega C$ in the denominator of Equation 2-38 is termed the *capacitive reactance* X_C, where

$$X_C = 1/(\omega C) = 1/(2\pi f C) \qquad (2\text{-}42)$$

Similarly, the *inductive reactance* X_L is defined as (see Equation 2-41)

$$X_L = \omega L = 2\pi f L \qquad (2\text{-}43)$$

The *impedance Z* of the two circuits under discussion is a measure of the total effect of resistance and reactance and is given by the denominators of Equations 2-38 and 2-41. Thus, for the *RC* circuit

$$Z = \sqrt{R^2 + X_C^2} \qquad (2\text{-}44)$$

and for the *RL* circuit

$$Z = \sqrt{R^2 + X_L^2} \qquad (2\text{-}45)$$

Substitution of Equations 2-42 and 2-44 into Equation 2-38 yields

$$I_p = \frac{V_p}{Z}$$

A similar result is obtained when Equations 2-43 and 2-45 are substituted into Equation 2-41.

In one sense, the capacitive and inductive reactances behave in a manner similar to that of a resistor in a circuit—that is, they tend to impede the flow of electrons. They differ in two major ways from resistance, however. First, they are frequency dependent; second, they cause the current and voltage to differ in phase. As a consequence of the latter, the phase angle must always be taken into account in considering the behavior of cir-

cuits containing capacitive and inductive elements. A convenient way of visualizing these effects is by means of vector diagrams.

Vector Diagrams for Reactive Circuits. Because the voltage lags the current by 90 deg in a pure capacitance, it is convenient to let ϕ equal -90 deg for a capacitive reactance. The phase angle for a pure inductive circuit is $+90$ deg. For a pure resistive circuit, ϕ will be equal to 0 deg. The relationship among X_C, X_L, and R can then be represented vectorially as shown in Figure 2-11.

As indicated in Figure 2-11c, the impedance of a series circuit containing a resistor, an inductor, and a capacitor is determined by the relationship

$$Z = \sqrt{R^2 + (X_L - X_C)^2} \qquad (2\text{-}46)$$

Because the outputs of the capacitor and inductor are 180 deg out of phase, their combined effect is determined from the difference of their reactances.

EXAMPLE

For the following circuit, calculate the peak current and voltage drops across the three components.

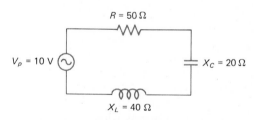

$R = 50\ \Omega$

$V_p = 10\ V$

$X_C = 20\ \Omega$

$X_L = 40\ \Omega$

$$Z = \sqrt{(50)^2 + (40 - 20)^2} = 53.8\ \Omega$$

$$I_p = 10\ V/53.8\ \Omega = 0.186\ A$$

$$V_R = 0.186 \times 50 = 9.3\ V$$

$$V_C = 0.186 \times 20 = 3.7\ V$$

$$V_L = 0.186 \times 40 = 7.4\ V$$

Note that the sum of the three voltages (20.4) in the example exceeds the source voltage. This situation is possible because the voltages are out of phase, with peaks occur-

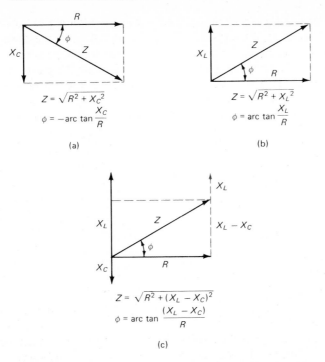

FIGURE 2-11   Vector diagrams for series circuits:
(a) *RC* circuit, (b) *RL* circuit, (c) series *RLC* circuit.

ring at different times. The sum of the *instantaneous* voltages, however, would add up to 10 V.

High-Pass and Low-Pass Filters. Series *RC* and *RL* circuits are often used as filters to attenuate high-frequency signals while passing low-frequency components (a low-pass filter) or, alternatively, to reduce low-frequency components while passing the high (a high-pass filter). Figure 2-12 shows how series *RC* and *RL* circuits can be arranged to give high- and low-pass filters. In each case, the input and output are indicated as the voltages $(V_p)_i$ and $(V_p)_o$.

In order to employ an *RC* circuit as a high-pass filter, the output voltage is taken across the resistor *R*. The peak current in this circuit is given by Equation 2-38. That is,

$$I_p = \frac{(V_p)_i}{\sqrt{R^2 + (1/\omega C)^2}}$$

Since the voltage drop across the resistor is in phase with the current,

$$I_p = \frac{(V_p)_o}{R}$$

The ratio of the peak output to the peak input voltage is obtained by dividing the first equation by the second and rearranging. Thus,

$$\frac{(V_p)_o}{(V_p)_i} = \frac{R}{\sqrt{R^2 + (1/\omega C)^2}} = \frac{1}{\sqrt{1 + (1/\omega RC)^2}} \tag{2-47}$$

A plot of this ratio as a function of frequency is shown in Figure 2-13a; here, a resistance of 1.0×10^4 Ω and a capacitance of 0.10 μF was employed. Note that frequencies below 50 Hz have been largely removed from the input signal. It should also be noted that because the input and output voltages are out of

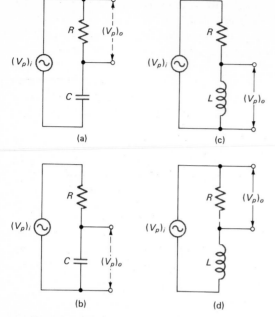

FIGURE 2-12   Filter circuits: (a) High-pass RC filter, (b) low-pass RC filter, (c) high-pass RL filter, and (d) low-pass RL filter.

phase, the peak voltages will occur at different times.

For the low-pass filter shown in Figure 2-12b, we may write

$$(V_p)_o = I_p X_C$$

Substituting Equation 2-42 gives

$$(V_p)_o = I_p/(\omega C)$$

Multiplying by Equation 2-38 and rearranging yields

$$\frac{(V_p)_o}{(V_p)_i} = \frac{1}{\omega C \sqrt{R^2 + (1/\omega C)^2}} = \frac{1}{\sqrt{(\omega CR)^2 + 1}}$$
$$(2\text{-}48)$$

Figure 2-13b was obtained with this equation.

As shown in Figure 2-12, RL circuits can also be employed as filters. Note, however, that the high-pass filter employs the potential across the reactive element while the low-pass filter employs the potential across the

resistor—behavior just opposite of that of the RC circuit. Curves similar to Figure 2-13 are obtained for RL filters.

Low- and high-pass filters are of great importance in the design of electronic circuits.

Resonant Circuits

A *resonant* or *RLC circuit* consists of a resistor, a capacitor, and an inductor arranged in series or parallel, as shown in Figure 2-14.

Series Resonant Filters. The impedance of the resonant circuit shown in Figure 2-14a is given by Equation 2-46. Note that the impedance of a series inductor and capacitor is the *difference* between their reactances and that the *net reactance of the combination is zero* when the two are identical. That is, the

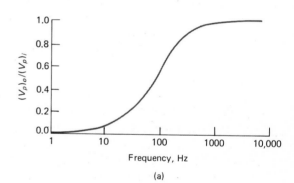

(a)

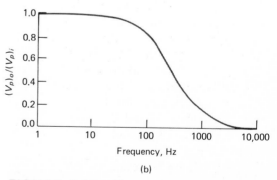

(b)

FIGURE 2-13   Frequency response of (a) the high-pass filter shown in Figure 2-12a and (b) the low-pass filter shown in Figure 2-12b; $R = 1.0 \times 10^4 \ \Omega$ and $C = 0.10 \ \mu F$.

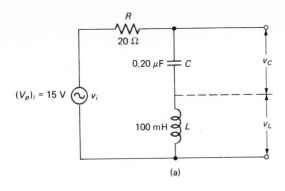

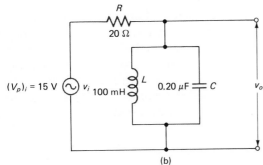

FIGURE 2-14 Resonant circuits employed as filters: (a) series circuit and (b) parallel circuit.

only impedance in the circuit is that of the resistor or, in its absence, the resistance of the inductor coil and the other wiring. This behavior becomes understandable when it is recalled that in a series RC circuit, the potentials across the capacitor and the inductor are 180 deg out of phase (see the vector diagrams in Figure 2-11). Thus, if X_C and X_L are alike, there is no *net* potential drop across the pair. Electricity continues to flow, however; therefore, the *net* reactance for the combination is zero. Also, when X_L and X_C are not identical, their *net* reactance is simply the difference between the two.

The condition for resonance is

$$X_L = X_C$$

Here, the energy stored in the capacitor during one half cycle is exactly equal to the energy stored in the magnetic field of the inductor during the other. Thus, during a one-half cycle, the energy of the capacitor forces the current through the inductor; in the other half cycle, the reverse is the case. In principle, a current induced in a closed resonant circuit would continue indefinitely except for the loss resulting from resistance in the leads and inductor wire.

The resonant frequency f_o can be readily calculated from Equations 2-42 and 2-43. That is, when $f = f_o$, $X_L = X_C$, and

$$\frac{1}{2\pi f_o C} = 2\pi f_o L$$

Upon rearrangement, it is found that

$$f_o = \frac{1}{2\pi\sqrt{LC}} \qquad (2\text{-}49)$$

The following example will show some of the properties of a series resonant circuit.

EXAMPLE

Assume the following values for the components of the circuit in Figure 2-14a: $(V_p)_i = 15.0$ V (peak voltage), $L = 100$ mH, $R = 20\ \Omega$, and $C = 0.200\ \mu$F. (a) Calculate the peak current at the resonance frequency and at a frequency 75 Hz greater than the resonance frequency. (b) Calculate the peak potentials across the resistor, inductor, and capacitor at the resonant frequency.

(a) We obtain the resonant frequency with Equation 2-49. Thus,

$$f_o = \frac{1}{2\pi\sqrt{100 \times 10^{-3} \times 0.2 \times 10^{-6}}}$$

$$= 1125 \text{ Hz}$$

By definition, $X_L = X_C$ at f_o; therefore, Equation 2-46 becomes

$$Z = \sqrt{20^2 + (0)^2} = 20\ \Omega$$

and

$$I_p = (V_p)_i/R = 15.0/20 = 0.75 \text{ A}$$

To calculate I_p at 1200 Hz, we substitute Equations 2-42 and 2-43 into Equation 2-46. Thus,

$$Z = \sqrt{R^2 + [2\pi f L - (1/2\pi f C)]^2}$$

$$= \sqrt{20^2 + [2\pi \times 1200 \times 100 \times 10^{-3}}$$

$$\overline{- 1/(2\pi \times 1200 \times 0.200 \times 10^{-6})]^2}$$

$$= \sqrt{400 + (754 - 663)^2} = 93 \ \Omega$$

and $I_p = 15.0/93 = 0.16$ A.

(b) At $f_o = 1125$ Hz, we have found that $I_p = 0.75$ A. The potentials across the resistor, capacitor, and inductor $(V_p)_R$, $(V_p)_L$, and $(V_p)_C$ are

$$(V_p)_R = 0.75 \times 20 = 15.0 \ \text{V}$$

$$(V_p)_L = I_p X_L$$

$$= 0.75 \times 2\pi \times 1125 \times 100 \times 10^{-3}$$

$$= 530 \ \text{V}$$

Since $X_L = X_C$ at resonance,

$$(V_p)_C = (V_p)_L = 530 \ \text{V}$$

Part (b) of this example shows that, at resonance, the peak voltage across the resistor is the potential of the source; on the other hand, the potentials across the inductor and capacitor are over 35 times greater than the input potential. It must be realized, however, that these peak potentials do not occur simultaneously. One lags the current by 90 deg and the other leads it by a similar amount. Thus, the *instantaneous* potentials across the capacitor and inductor cancel, and the potential drop across the resistor is then equal to the input potential. Clearly, in a circuit of this type, the capacitor and inductor may have to withstand considerably larger potentials than the amplitude of the input voltage would seem to indicate.

Figure 2-15, which was obtained for the circuit shown in the example, demonstrates that the output of a series resonant circuit is a narrow band of frequencies, the maximum of which depends upon the values chosen for L and C.

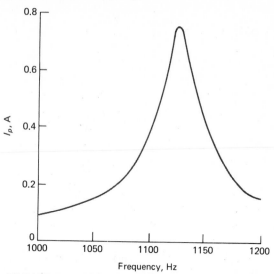

FIGURE 2-15 Frequency response of a series resonant circuit; $R = 20\,\Omega$, $L = 100\,\text{mH}$, $C = 0.2\ \mu\text{F}$, and $(V_p)_i = 15.0$ V.

If the output of the circuit shown in Figure 2-14a is taken across the inductor or capacitor, a potential is obtained which is several times larger than the input potential (see part b of the example). Figure 2-16 shows a plot of the ratio of the peak voltage across the inductor to the peak input voltage as a function of frequency. A similar plot is obtained if the inductor potential is replaced by the capacitor potential.

The upper curve in Figure 2-16 was obtained by substituting a 10-Ω resistor for the 20-Ω resistor employed in the example.

Parallel Resonant Filters. Figure 2-14b shows a typical parallel resonant circuit. Again, the condition of resonance is that $X_C = X_L$ and the resonant frequency is given by Equation 2-49.

The impedance of the parallel circuit is given by

$$Z = \sqrt{R^2 + \left(\frac{X_L X_C}{X_C - X_L}\right)^2} \qquad (2\text{-}50)$$

It is of interest to compare this equation with the equation for the impedance of the series

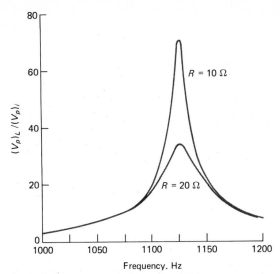

FIGURE 2-16   Ratio of output to input voltage for the series resonant filter; $L = 100$ mH, $C = 0.20$ μF, and $(V_p)_i = 15$ V.

filter (Equation 2-46). Note that the impedance in the latter circuit is a *minimum* at resonance, when $X_L = X_C$. In contrast, the impedance for the parallel circuit at resonance is a *maximum* and in principle is infinite. Consequently, a maximum voltage drop across (or a minimum current through) the parallel reactance is found at resonance. With both parallel and series circuits at resonance, a small initial signal causes resonance in which electricity is carried *back and forth between the capacitor and the inductor*. But current *from the source* through the reactance is minimal.

The parallel circuit, sometimes called a *tank circuit*, is widely used as for tuning radio or television circuits. Tuning is ordinarily accomplished by adjusting a variable capacitor until resonance is achieved.

Behavior of *RC* Circuits with Pulsed Inputs

When a pulsed input is applied to an *RC* circuit, the voltage outputs across the capacitor and resistor take various forms, depend-

ing upon the relationship between the width of the pulse and the time constant for the circuit. These effects are illustrated in Figure 2-17 where the input is a square wave having a pulse width of T_p seconds. The second column shows the variation in capacitor potential as a function of time, while the third column shows the change in resistor potential at the same times. In the top set of plots (Figure 2-17a), the time constant of the circuit is much greater than the input pulse width. Under these circumstances, the capacitor can become only partially charged during each pulse. It then discharges as the input potential returns to zero; a sawtooth output results. The output of the resistor under these circumstances rises instantaneously to a maximum value and then decreases essentially linearly during the pulse lifetime.

The bottom set of graphs (Figure 2-17c) illustrates the two outputs when the time constant of the circuit is much shorter than the pulse width. Here, the charge on the capacitor rises rapidly and approaches full charge near the end of the pulse. As a consequence, the potential across the resistor rapidly decreases to zero after its initial rise. When V_i goes to zero, the capacitor discharges immediately; the output across the resistor peaks in a negative direction and then quickly approaches zero.

These various output wave forms find applications in electronic circuitry. The sharply peaked voltage output shown in Figure 2-17c is particularly important in timing and trigger circuits.

SIMPLE ELECTRICAL MEASUREMENTS

This section describes selected methods for the measurement of current, voltage, and resistance. More sophisticated methods for measuring these and other electrical properties will be considered in Chapter 3 as well as in later chapters.

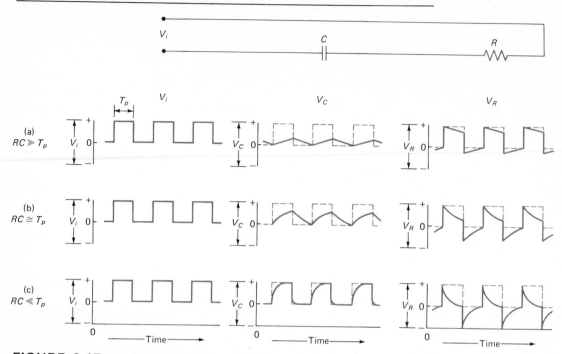

FIGURE 2-17   Output signals V_R and V_C for pulsed input signal V_i; (a) time constant $\gg$ pulse width T_p; (b) time constant $\cong$ pulse width; (c) time constant $\ll$ pulse width.

Galvanometers

The classical methods for measuring direct current, voltage, and resistance employ a galvanometer. This device, which was invented nearly a century ago, still finds widespread use in the laboratory. It is based upon the current-induced motion of a coil suspended in a fixed magnetic field. This arrangement is called a *D'Arsonval movement* or *coil*.

Figure 2-18 shows construction details of a typical D'Arsonval meter. The current to be measured passes through a rectangular coil wrapped about a cylindrical soft iron core that pivots between a pair of jeweled bearings; the coil is mounted in the air space between the pole pieces of a permanent magnet. The current-dependent magnetic field that develops in the coil interacts with the permanent magnetic field in such a way as to

cause rotation of the coil and deflection of the attached pointer; the deflection is directly proportional to the current present in the coil.

Meters with D'Arsonval movements are generally called *ammeters*, *milliammeters*, or *microammeters*, depending upon the current range they have been designed to measure. Their accuracies vary from 0.5 to 3% of full scale, depending upon the quality of the meter.

More sensitive galvanometers are obtained by suspending the coil between the poles of a magnet by a fine vertical filament that has very low resistance to rotation. A small mirror is attached to the filament; a light beam, reflected off the mirror to a scale, detects and measures the extent of rotation during current passage. Galvanometers can be constructed that will detect currents as small as 10^{-10} A.

The Ayrton Shunt. The *Ayrton shunt,* shown in Figure 2-19, is often employed to vary the range of a galvanometer. The example that follows demonstrates how the resistors for a shunt can be chosen.

EXAMPLE

Assume that the meter in Figure 2-19 registers full scale with a current of 50 μA; its internal resistance is 3000 Ω. Derive values for the four resistors such that the meter ranges will be 0.1, 1, 10, and 100 mA.

For each setting of the contact, the circuit is equivalent to two resistances in parallel. One resistance is made up of the meter and the resistors to the left of the contact in Figure 2-19; the other consists of the remaining resistor to the right of the contact. The potential across parallel resistances is, of course, the same. Thus,

$$I_M(R_M + R_L) = I_R R_R$$

where R_M is the resistance of the meter, R_L is the resistance of the resistors to the left of the contact, and R_R is the resistance to the right.

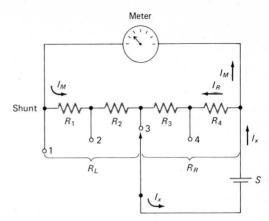

FIGURE 2-19   A current meter with an Ayrton shunt.

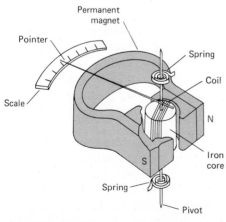

FIGURE 2-18   Current meter. (Taken from R. J. Smith, *Circuits, Devices, and Systems.* New York: Wiley, 1976. With permission.)

We may also write, from Kirchhoff's current law

$$I_x = I_M + I_R$$

which can be combined with the first equation to give, after rearrangement,

$$\frac{I_x}{I_M} = \frac{R_M + R_L + R_R}{R_R} = \frac{R_M + R_T}{R_R}$$

where R_T is the total resistance of the shunt ($R_L + R_R$).

With the contact in position 1, $R_L = 0$ and $R_R = R_T$; here, the greatest fraction of the current will pass through the meter. Therefore, setting 1 corresponds to the lowest current range, where the full meter reading ($I_M = 50$ μA) should correspond to 0.1 mA ($I_x = 0.1$ mA). Substituting these values into the derived equation yields

$$\frac{0.1 \times 10^{-3}\ \text{A}}{50 \times 10^{-6}\ \text{A}} = \frac{3000 + 0 + R_R}{R_R}$$

and

$$R_R = R_T = 3000\ \Omega$$

Substituting this value for R_T in the working equation gives

$$\frac{I_x}{I_M} = \frac{R_M + R_T}{R_R} = \frac{3000 + 3000}{R_R}$$

or

$$R_R = 6000 \times I_M/I_x$$

$$= 6000 \times 50 \times 10^{-6}/I_x = \frac{0.300}{I_x}$$

Employing this relationship, we obtain

Setting	I_x	R_R
1	0.1×10^{-3}	$3000 = (R_1 + R_2 + R_3 + R_4)$
2	1×10^{-3}	$300 = (R_2 + R_3 + R_4)$
3	10×10^{-3}	$30 = (R_3 + R_4)$
4	100×10^{-3}	$3 = R_4$

Thus, $R_4 = 3\ \Omega$, $R_3 = (30 - 3) = 27\ \Omega$, $R_2 = (300 - 30) = 270\ \Omega$, and $R_1 = (3000 - 300) = 2700\ \Omega$.

Measurement of Current and Voltage. Figure 2-20 presents circuit diagrams showing how a galvanometer can be employed to measure current and voltage in a circuit. Note that the meter is placed in series with the remainder of the circuit for a current measurement (Figure 2-20a). In contrast, a voltage measurement is obtained by placing the meter across or in parallel with the potential to be measured. Thus, in Figure 2-20b, the voltage

V_{AB} across the resistor R_2 is obtained by connecting the meter at points A and B. Under these circumstances,

$$V_{AB} = I_M R_M$$

where R_M is the resistance of the meter (and any resistors placed in series with it) and I_M is the current measured by the meter. Since R_M is a constant, the meter scale can be calibrated in volts rather than amperes.

Effect of Meter Resistance on Current Measurements. The coil of a meter or galvanometer offers resistance to the flow of electricity, and this resistance will reduce the magnitude of the current being measured; thus, an error is introduced by the measurement process itself. This situation is not peculiar to current measurements. In fact, it is a simple example of a general limitation to any physical measurement. That is, the process of measurement inevitably disturbs the system of interest so that the quantity actually measured differs from its value prior to the measurement. This type of error can never be completely eliminated; often, however, it can be reduced to insignificant proportions.

The circuit shown in Figure 2-20a can be used to illustrate how a simple measurement can perturb a system and thus alter the result

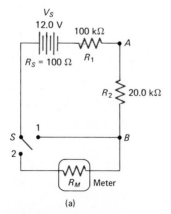

(a)

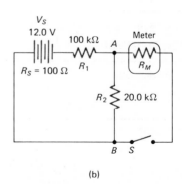

(b)

FIGURE 2-20   Circuits for measuring (a) current I and (b) voltage V_{AB}.

obtained. Here, it is desired to determine the current I in the resistor R_{AB} when the switch is thrown to position 1.

From Ohm's law, this current is given by

$$I = \frac{V_S}{R_1 + R_2 + R_S} = \frac{V_S}{R_T}$$

where V_S is the potential of the source, R_S is its internal resistance, and R_T is the total resistance of the circuit.

In order to measure the current, however, the meter must be brought into the circuit by moving the switch from position 1 to 2. Clearly, the measured current I_M will differ from I and will be given by

$$I_M = \frac{V_S}{R_1 + R_2 + R_S + R_M} = \frac{V_S}{R_T + R_M}$$

The relative error associated with the presence of the meter is given by

$$\text{rel error} = (I_M - I)/I$$

Substitution of the relationships for I and I_M yields, upon rearrangement,

$$\text{rel error} = -R_M/(R_T + R_M)$$

The data in Table 2-1 illustrate the magnitude of the disturbance to the system by the measurement process for various ratios of meter resistance to total resistance of the remaining components in the system.

Clearly, the error associated with this measurement can be minimized by choosing a meter having the lowest possible resistance. Alternatively, the meter can be replaced by a small precision resistor; the potential drop across this resistor can then be measured with a high resistance meter, and the current can be calculated from Ohm's law.

Often, a current meter or a precision resistor is made a permanent part of an instrument; under these circumstances, a somewhat smaller current is obtained, but its magnitude is given directly by the meter.

Effect of Meter Resistance on Voltage Measurements. As shown in the following example, the measurement of voltage by the circuit in Figure 2-20b may also alter the system in such a way as to create a significant error.

EXAMPLE

The voltage drop V_{AB} across the resistor R_2 in Figure 2-20b is to be measured by means of a meter having a resistance of 50.0 kΩ. (a) Calculate the relative error associated with the measurement. (b) Repeat the calculations assuming the meter resistance is still 50.0 kΩ, but that all other resistances

TABLE 2-1 EFFECT OF METER RESISTANCE ON CURRENT MEASUREMENTS[a]

Meter Resistance, R_M, kΩ	$\dfrac{R_M}{R_T}$	Measured Current, I_M, mA	Relative Error, %
meter absent	0.00	0.100	0
1	0.01	0.099	−0.8
10	0.08	0.092	−7.6
50	0.42	0.071	−29
100	0.83	0.055	−45

[a] For source of data, see Figure 2-20a. Here, $R_T = R_1 + R_2 + R_S$.

have values that are 1/10 that shown in the figure.

(a) With the switch S open so that the meter is out of the system, the potential drop V_{AB} can be obtained by use of Equation 2-8 (p. 9)

$$V_{AB} = \frac{12.0 \times 20.0}{100 + 20.0 + 0.100} = 2.00 \text{ V}$$

When the switch is closed, a system of two parallel resistors is created whose net resistance R_{AB} can be found by substitution into Equation 2-11 (p. 11). Thus,

$$\frac{1}{R_{AB}} = \frac{1}{R_2} + \frac{1}{R_M} = \frac{1}{20.0} + \frac{1}{50.0}$$

$$R_{AB} = 14.3 \text{ k}\Omega$$

The potential drop V_M, sensed by the meter, is now

$$V_M = \frac{12.0 \times 14.3}{100 + 14.3 + 0.100} = 1.50 \text{ V}$$

Therefore,

$$\text{rel error} = \frac{1.50 - 2.00}{2.00} \times 100 = -25\%$$

(b) Proceeding as in (a), we find

$$V_{AB} = \frac{12.0 \times 2.00}{10.0 + 2.00 + 0.0100} = 2.00 \text{ V}$$

$$\frac{1}{R} = \frac{1}{2.0} + \frac{1}{50.0}$$

and

$$R = 1.92 \text{ k}\Omega$$

$$V_M = \frac{12.0 \times 1.92}{10.0 + 1.92 + 0.0100} = 1.93 \text{ V}$$

$$\text{rel error} = \frac{1.93 - 2.00}{2.00} \times 100 = -3.5\%$$

Table 2-2 shows the effect of meter resistance on a potential measurement. Note that, in contrast to a current measurement, the accuracy of a potential measurement increases with meter resistance. As shown by

the second part of the example, however, the resistance of the potential source plays an important part in determining the meter resistance required for a given accuracy. That is, as the source resistance increases, so also must the meter resistance if accuracy is to be maintained. This relationship becomes particularly significant in the determination of pH by measurement of the potential of a cell containing a glass electrode. Here, the resistance of the system may be 100 megohms or more. Thus, a potential measuring device with an enormous internal resistance is required.

Resistance Measurements. Figure 2-21 gives a circuit for the measurement of resistance with a D'Arsonval meter. The source is usually a 1.5-V dry cell. To employ the circuit, the switch is placed in position 1 and the meter is adjusted to full scale by means of the variable resistor R_V. From Ohm's law, we may write

$$1.5 = I_1(R_M + R_V)$$

With the switch moved to position 2, the current decreases to I_2 because of the added resistance of the unknown R_x. Thus,

$$1.5 = I_2(R_M + R_V + R_x)$$

Division of the two equations and rearrangement yields

$$R_x = \left(\frac{I_1}{I_2} - 1\right)(R_M + R_V)$$

The value of $(R_M + R_V)$ is determined by the range of the meter. For example, if the meter range is 0 to 1 mA and the meter is set

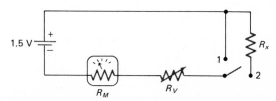

FIGURE 2-21 A circuit for measurement of resistance.

to 1 mA in the initial adjustment, we may write

$$1.5 = I_1(R_M + R_V) = 1.00 \times 10^{-3}(R_M + R_V)$$

$$(R_M + R_V) = 1500 \ \Omega$$

and

$$R_x = \left(\frac{I_1}{I_2} - 1\right)1500$$

$$= \left(\frac{1.00 \times 10^{-3}}{I_2} - 1\right)1500$$

Clearly, the relationship between the measured current and the resistance is nonlinear.

Comparison Measurements

Comparison or Null Measurements. Instruments based upon comparison of the system of interest with a standard are widely employed for chemical measurements; such devices are called *null* instruments. Null instruments are generally more accurate and precise than direct-reading instruments, in part because a null measurement generally causes less disturbance to the system being studied. Moreover, the readout devices employed can often be considerably more rugged and less subject to environmental effects than a direct-reading device.

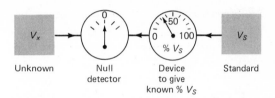

FIGURE 2-22 Block diagram for a comparison measurement system.

As shown in Figure 2-22, a typical comparison instrument consists of three components. One of these is a null detector, which indicates the inequality or equality of the two signals. The second component is a reference standard signal against which the unknown signal is compared. Finally, the instrument contains a device for continuously attenuating one of the two signals in such a way that the percent reduction is exactly known. In Figure 2-22, attenuation is performed on the reference signal; often, however, it is the unknown signal that is varied. Furthermore, attenuation may be performed mechanically rather than electrically. An example is a variable diaphragm that reduces the power of a beam of light used for photometric analysis.

Potentiometers. The potentiometer is a null instrument that permits accurate measurement of potentials while drawing a minimum current from the source under study. A typi-

TABLE 2-2 EFFECT OF METER RESISTANCE ON POTENTIAL MEASUREMENT[a]

Meter Resistance, R_M, kΩ	$\dfrac{R_M}{R_{AB}}$	Measured Voltage, V_M, V	Relative Error, %
10	0.5	0.75	−62
50	2.5	1.50	−25
500	25	1.935	−3.2
1×10^3	50	1.967	−1.6
1×10^4	500	1.997	−0.2

[a] For source of data, see Figure 2-20b.

cal laboratory potentiometer employs a linear voltage divider such as that shown in Figure 2-3b (p. 10) to attenuate a reference voltage until a null point is reached. In its simplest form, the divider consists of a uniform resistance wire mounted on a meter stick. A sliding contact permits variation of the output voltage. More conveniently, the divider is a precision wire wound resistor formed in a helical coil. A continuously variable tap that can be moved from one end of the helix to the other gives a variable voltage. Ordinarily, the divider is powered by dry cells or mercury batteries, which provide a potential somewhat larger than that which is to be measured.

Figure 2-23 is a sketch of a laboratory potentiometer employed for the measurement of an unknown potential V_x by comparing it with a standard potential V_s. The latter is ordinarily a *Weston standard cell* which can be represented as follows:

$$Cd(Hg)_x | CdSO_4 \cdot (8/3)H_2O(sat'd),$$

$$Hg_2SO_4(sat'd) | Hg$$

One of the electrodes is a cadmium amalgam represented as $Cd(Hg)_x$; the other is mercury. The electrolyte is a solution saturated with cadmium and mercury sulfates. The half reactions as they occur at the two electrodes are

$$Cd(Hg)_x \rightarrow Cd^{2+} + xHg + 2e$$

$$Hg_2SO_4 + 2e \rightarrow 2Hg + SO_4^{2-}$$

The potential of the Weston cell at 25°C is 1.0183 V.

A potential measurement with the instrument shown in Figure 2-23 requires two null-point measurements. Switch P is first closed to provide a current in the slide wire and a potential drop across AC. The Weston cell is next placed in the circuit by moving switch S to position 1. When the tapping key K is closed momentarily, a current will be indicated by the galvanometer G, unless the output from the standard cell is identical with the potential drop V_{AC} between A and C. If V_{AC} is greater than V_s, electrons will flow

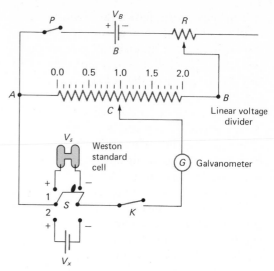

FIGURE 2-23   Circuit diagram of a laboratory potentiometer.

from the battery B through C and the circuit containing G and V_s. Electrons will flow in the opposite direction, of course, if V_{AC} is smaller than V_s. The position of contact C is then varied until a null condition is achieved, as indicated by the galvanometer when K is closed.

The foregoing process is repeated with S in position 2 so that the unknown cell is in the circuit. Equation 2-9 can then be employed to calculate V_x. Thus, with S at position 1

$$V_s = V_B \frac{AC_s}{AB}$$

and at position 2

$$V_x = V_B \frac{AC_x}{AB}$$

Dividing these equations gives, upon rearrangement,

$$V_x = V_s \frac{AC_x}{AC_s} \tag{2-51}$$

Thus, V_x is obtained from the potential of the standard cell and the two slide-wire settings.

The slide wire shown in Figure 2-22 is readily calibrated to read directly in volts, thus avoiding the necessity of calculating V_x each time. Calibration is accomplished by setting the slide wire to the voltage of the standard cell. Then, with that cell in the circuit, the potential across AB is adjusted by means of the variable resistor R until a null is achieved. The slide-wire setting will then give V_x directly when C is adjusted to the null position with the unknown voltage.

The need for a working battery B may be questioned. In principle, there is nothing to prevent a direct measurement of V_x by replacing B with the standard cell V_s. It must be remembered, however, that current is being continuously drawn from B; a standard cell would not maintain a constant potential for long under such usage.

The accuracy of a voltage measurement with a potentiometer equipped with a linear voltage divider depends upon several factors. Thus, it is necessary to assume that the voltage of the working battery B remains constant during the time required to balance the instrument against the standard cell and to measure the potential of the unknown cell. Ordinarily, this assumption does not lead to appreciable error if B consists of one or two heavy-duty dry cells (or mercury cells) in good condition or a lead storage battery. The instrument should be calibrated against the standard cell before each voltage measurement to compensate for possible changes in B, however.

The linearity of the resistance AB, as well as the precision with which distances along its length can be estimated, both contribute to the accuracy of the potentiometer. Ordinarily, however, the ultimate precision of a good-quality instrument is determined by the sensitivity of the galvanometer relative to the resistance of the circuit. Suppose, for example, that the electrical resistance of the galvanometer plus the unknown cell is 1000 Ω, a typical figure. Further, if we assume that the galvanometer is just capable of detecting a current of 1 μA (10^{-6} A), we can readily calculate from Ohm's law that the minimum distinguishable voltage difference will be $10^{-6} \times 1000 = 10^{-3}$ V, or 1 mV. By use of a galvanometer sensitive to 10^{-7} A, a difference of 0.1 mV will be detectable. A sensitivity of this order is found in an ordinary pointer-type galvanometer; more refined instruments with sensitivities up to 10^{-10} A are not uncommon.

From this discussion, it is clear that the sensitivity of a potentiometric measurement decreases as the electrical resistance of the cell increases. As a matter of fact, potentials of cells with resistances much greater than 1 MΩ cannot be measured accurately with a potentiometer employing a galvanometer as the current-sensing device.

Current Measurements with a Null Instrument. The null method is readily applied to the determination of current. A small precision resistor is placed in series in the circuit and the potential drop across the resistor is measured with a potentiometer.

Resistance Measurements; the Wheatstone Bridge. The Wheatstone bridge shown in Figure 2-24 provides another example of a null device, applied here to the measurement of resistance. The power source S provides an

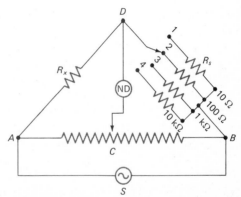

FIGURE 2-24 A Wheatstone bridge for resistance measurements.

ac current at a potential of 6 to 10 V. The resistances R_{AC} and R_{CB} are determined from the total resistances of the linear voltage divider, AB, and the position of C. The upper right arm of the bridge contains a series of precision resistors that provide a choice of several resistance ranges. The unknown resistance R_x is placed in the upper-left arm of the bridge.

A null detector ND is employed to indicate an absence of current between D and C. The detector often consists of a pair of ordinary earphones; an ac signal of about 1000 Hz is employed, the human ear being sensitive in this frequency range. Alternatively, the detector may be a cathode-ray tube or an ac microammeter.

To measure R_x, the position of C is adjusted to a minimum as indicated by the null detector. Application of Equation 2-9 to the voltage divider ACB yields

$$V_{AC} = V_{AB} \frac{AC}{AB}$$

For the circuit ADB, we may write

$$V_{AD} = V_{AB} \frac{R_x}{R_x + R_s}$$

But at null, $V_{AC} = V_{AD}$. Thus, the two equations can be combined to give, upon rearrangement,

$$R_x = \frac{R_s AC}{AB - AC} = R_s \frac{AC}{BC} \qquad (2\text{-}52)$$

PROBLEMS

1. It was desired to assemble the voltage divider shown below. Two of each of the following resistors were available: 50 Ω, 100 Ω, and 200 Ω.

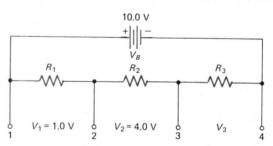

(a) Describe a suitable combination of the resistors that would give the indicated voltages.
(b) What would be the IR drop across R_3?
(c) What current would be drawn from the source?
(d) What power is dissipated by the circuit?

2. Assume that for a circuit similar to that shown in Problem 1, $R_1 = 200$ Ω, $R_2 = 500$ Ω, $R_3 = 1000$ Ω, and $V_B = 15$ V.
(a) Calculate the voltage V_2.
(b) What would be the power loss in resistor R_2?
(c) What fraction of the total power lost by the circuit would be dissipated in resistor R_2?

3. For a circuit similar to the one shown in Problem 1, $R_1 = 1.00$ kΩ, $R_2 = 2.50$ kΩ, $R_3 = 4.00$ kΩ, and $V_B = 12.0$ V. A voltmeter was placed across contacts 2 and 4. Calculate the relative error in the voltage reading if the internal resistance of the voltmeter was (a) 5000 Ω, (b) 50 kΩ, and (c) 500 kΩ.

4. A voltmeter was employed to measure the potential of a cell having an internal resistance of 750 Ω. What must the internal resistance of the meter be if the relative error in the measurement is to be less than (a) -1.0%, (b) -0.10%?

5. For the following circuit, calculate
   (a) the potential drop across each of the resistors.
   (b) the magnitude of each of the currents shown.
   (c) the power dissipated by resistor R_3.
   (d) the potential drop between points 3 and 4.

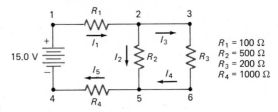

6. For the circuit shown below, calculate
   (a) the power dissipated between points 1 and 2.
   (b) the current drawn from the source.
   (c) the potential drop across resistor R_A.
   (d) the potential drop across resistor R_D.
   (e) the potential drop between points 5 and 4.

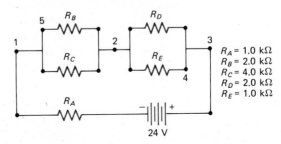

7. Assume that the resistor shown between A and B in Figure 2-23 (p. 34) is a linear slide wire exactly 1.00 m in length. With the standard Weston cell (1.018 V) in the circuit, a null point was observed when contact C was moved to a position 84.3 cm from point A. When the Weston cell was replaced with an unknown voltage, null was observed at 44.3 cm. Calculate the potential of the unknown.

8. A galvanometer with an internal resistance of 5000 Ω shows a full scale reading with a current of 0.100 mA. Design an Ayrton shunt such that the galvanometer will have a range of 0 to 10.0, 25.0, 50.0, and 100 mA.

9. Assume that the slide wire in the Wheatstone bridge shown in Figure 2-24 has a resistance of 500 Ω and that the switch is in position 3.
   (a) What is the resistance of the unknown R_x if a null is found when contact C is at 62.6% of the full length of the slide wire?

(b) If the source potential S is 15 V and the resistance of the null detector is 100 Ω, what is the current in the detector when C is set at 62.7%?

(c) If the switch is moved from 3 to 2, with R_x still in the circuit, where would C have to be moved in order to reach a new null?

10. The current in a circuit is to be determined by measuring the potential drop across a precision resistor in series with the circuit.

(a) What should be the resistance of the resistor in ohms if 1.00 V is to correspond to 50 μA?

(b) What must be the resistance of the voltage measuring device if the error in the current measurement is to be less than 1.0% relative?

11. An electrolysis at a nearly constant current can be performed with the following arrangement:

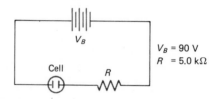

The 90-V source consists of dry cells whose potential can be assumed to remain constant for short periods. During the electrolysis, the resistance of the cell increases from 20 Ω to 40 Ω due to depletion of ionic species. Calculate the percent change in the current assuming that the internal resistance of the batteries is zero.

12. Repeat the calculations in Problem 11 assuming that $V_B = 9.0$ V and $R = 0.50$ kΩ.

13. A 24-V dc potential was applied across a resistor and capacitor in series. Calculate the current after 0.00, 0.010, 0.10, 1.0, and 10 sec if the resistance was 10 MΩ and the capacitance 0.20 μF.

14. How long would it take to discharge a 0.015 μF capacitor to 1% of its full charge through a resistance of (a) 10 MΩ, (b) 1 MΩ, (c) 1 kΩ?

15. Calculate time constants for each of the RC circuits described in Problem 14.

16. A series RC circuit consisted of a 25-V dc source, a 50 kΩ resistor, and a 0.035 μF capacitor.

(a) Calculate the time constant for the circuit.

(b) Calculate the current and potential drops across the capacitor and the resistor during a charging cycle; employ as times 0, 1, 2, 3, 4, 5, and 10 ms.

(c) Repeat the calculations in (b) for a discharge cycle.

17. Repeat the calculations in Problem 16 assuming that the potential was 15 V, the resistance was 20 MΩ, and the capacitance was 0.050 μF. Employ as times 0, 1, 2, 3, 4, 5, and 10 s.

18. A series *RL* circuit consisted of a 12 V source, a 1000 Ω resistor, and a 2 H inductor.
    (a) Calculate the time constant for the circuit.
    (b) Calculate the current and the voltage drops across the resistor and inductor for the following times after the current was initiated: 0, 1, 2, 4, 8, 10 ms.

19. Calculate the capacitive reactance, the impedance, and the phase angle ϕ for the following series circuits.

	Frequency, Hz	R, Ω	C, μF
(a)	1	20,000	0.033
(b)	10^3	20,000	0.033
(c)	10^6	20,000	0.033
(d)	1	200	0.033
(e)	10^3	200	0.033
(f)	10^6	200	0.033
(g)	1	2,000	0.33
(h)	10^3	2,000	0.33
(i)	10^6	2,000	0.33

20. Calculate the inductive reactance, the impedance, and the phase angle ϕ for the following series circuits.

	Frequency, Hz	R, Ω	L, mH
(a)	10^3	2000	50
(b)	10^4	2000	50
(c)	10^5	2000	50
(d)	10^3	200	50
(e)	10^4	200	50
(f)	10^5	200	50
(g)	10^3	200	5
(h)	10^4	200	5
(i)	10^5	200	5

21. Calculate the capacitive and inductive reactances and the impedance of the following series resonant circuits.

	Frequency, Hz	R, Ω	C, μF	L, mH
(a)	10^2	2000	0.20	1.2
(b)	10^4	2000	0.20	1.2
(c)	10^6	2000	0.20	1.2
(d)	10^2	20	0.20	1.2
(e)	10^4	20	0.20	1.2
(f)	10^6	20	0.20	1.2

22. For the circuits described in Problem 21,
    (a) calculate the frequency at which a current maximum would appear.

(b) calculate the current at this maximum if the input is a 15-V sinu-soidal signal.

(c) calculate the potential drop across each of the components if the input is a 15-V sinusoidal signal.

23. Derive a frequency response curve for a low-pass RC filter in which $R = 2.5 \times 10^5$ Ω and $C = 0.015$ μF. Cover a range of $(V_p)_o/(V_p)_i$ of 0.01 to 0.99.

24. Derive a frequency response curve for a high-pass RC filter in which $R = 5.0 \times 10^5$ Ω and $C = 200$ pF (1 pF $= 10^{-12}$ F). Cover a range of $(V_p)_o/(V_p)_i$ of 0.01 to 0.99.

25. Derive a frequency response curve for an RL low-pass filter in which $R = 120$ Ω and $L = 8.0$ H. Cover a range of $(V_p)_o/(V_p)_i$ of 0.01 to 0.99. Give a circuit design for an RC filter that would have the same characteristics.

26. Determine the current and the voltage across each component in a series resonant circuit in which $R = 100$ Ω, $C = 0.050$ μF, $L = 225$ mH, and $(V_p)_i = 12.0$ V
(a) at the current maximum.
(b) at a frequency 10% lower than the maximum.
(c) at a frequency 10% higher than the maximum.

27. Calculate the capacitive and inductive reactances and the impedance for parallel resonant circuits similar to that shown in Figure 2-14b assuming the values for f, R, C, and L given in Problem 21.

28. For the circuit shown in Figure 2-14b, calculate
(a) the frequency at which a current minimum would be expected.
(b) the current at a frequency that was 10% less than the minimum.
(c) the current at a frequency that was 10% more than the minimum.

3
ELEMENTARY ELECTRONICS

The circuit elements and circuits considered in Chapter 2 are linear in the sense that their input and output voltages and currents are proportional to one another. Electronic circuits contain one or more nonlinear devices such as transistors, vacuum tubes, gas-filled tubes, and semiconductor diodes.[1] Historically, electronic circuits were based upon vacuum tubes. Now, however, the vacuum tube has been almost totally replaced by semiconductor devices. The advantages of the latter include low cost, low power consumption, small heat generation, long life, and small size. Most modern analytical instruments use only semiconductors; we shall, therefore, limit our discussion to these devices.

SEMICONDUCTORS

A semiconductor is a crystalline material having a conductivity between that of a conductor and an insulator. Many types of semiconducting materials exist, including silicon and germanium, intermetallic compounds (such as silicon carbide), and a variety of organic compounds. Two semiconducting materials that find wide application for electronic devices are crystalline silicon and germanium; we shall limit our discussions to these.

Properties of Silicon and Germanium Semiconductors

Silicon and germanium are Group IV elements and thus have four valence electrons available for bond formation. In a silicon crystal, each of these electrons is immobilized by combination with an electron from another silicon atom to form a covalent bond. Thus, in principle, no free electrons should exist in crystalline silicon, and the material should be an insulator. In fact, however, sufficient thermal agitation occurs at room temperature to liberate an occasional electron, leaving it free to move through the crystal lattice and thus to conduct electricity. This thermal *excitation* of an electron leaves a positively charged region, termed a *hole*, associated with the silicon atom. The hole, however, is mobile as well, and thus also contributes to the electrical conductance of the crystal. The mechanism of hole movement is stepwise; a bound electron from a neighboring silicon atom jumps to the electron-deficient region and thereby leaves a positive hole in its wake. Thus, conduction by a semiconductor involves motion of thermal electrons in one direction and holes in the other.

The conduction of a silicon or germanium crystal can be greatly enhanced by *doping*, a process by which a tiny, controlled amount of an impurity is introduced into the crystal structure. Typically, a silicon semiconductor is doped with a Group V element such as arsenic or antimony or a Group III element such as indium or gallium. When an atom of a Group V element replaces a silicon atom in the lattice, one unbound electron is introduced into the structure; only a small thermal energy is needed to free this electron for conduction. Note that the resulting positive Group V ion does not provide a mobile hole inasmuch as there is little tendency for electrons to move from a covalent silicon bond to this nonbonding position. A semiconductor that has been doped so that it contains nonbonding electrons is termed an *n-type* (negative type) because electrons are the *majority carriers* of current. Positive holes still exist as in the undoped crystal (associated with silicon atoms), but their number is small with respect to the number of electrons; thus, holes represent *minority carriers* in an *n-type* semiconductor.

[1] For further information about modern electronics, see: A. J. Diefenderfer, *Principles of Electronic Instrumentation*, 2d ed., Philadelphia: Saunders, 1979; J. J. Brophy, *Basic Electronics for Scientists*, 3d ed. New York: McGraw-Hill, 1977; H. V. Malmstadt, C. G. Enke, S. R. Crouch, and G. Horlick, *Electronic Measurements for Scientists*. Menlo Park, CA: Benjamin, 1974; R. J. Smith, *Circuits, Devices, and Systems*, 3d ed. New York: Wiley, 1976; and B. H. Vassos and G. W. Ewing, *Analog and Digital Electronics for Scientists*. New York: Wiley-Interscience, 1972.

A *p-type* or positive type semiconductor is formed when silicon is doped with a Group III element which contains only three valence electrons. Here, positive holes are introduced when electrons from adjoining silicon atoms jump to the vacant orbital associated with the impurity atom. Note that this process imparts a negative charge to the Group III atoms. Movement of the holes from silicon atom to silicon atom, as described earlier, constitutes a current in which the majority carrier is positive. Positive holes are less mobile than free electrons; thus, the conductivity of a *p*-type semiconductor is inherently less than that of an *n*-type.

Semiconductor Diodes

A *diode* is a device that has greater conductance in one direction than in another. Useful diodes are manufactured by forming adjacent *n*- and *p*-type regions within a single germanium or silicon crystal; the interface between these regions is termed a *pn* junction.

Properties of a *pn* Junction. Figure 3-1a is a cross section of one type of *pn* junction,

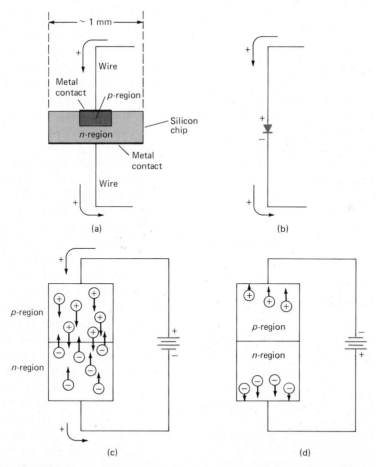

FIGURE 3-1 A *pn* junction diode. (a) Physical appearance of one type formed by diffusion of a *p*-type impurity into an *n*-type semiconductor; (b) symbol for; (c) current under forward bias; (d) resistance to current under reverse bias.

which is formed by diffusing an excess of a p-type impurity such as indium into a minute silicon chip that has been doped with an n-type impurity such as antimony. A junction of this kind permits ready flow of positive charge from the p region through the n region (or flow of negative charge in the reverse direction); it offers a high resistance to the flow of positive charge in the other direction and is thus a *current rectifier*.

Figure 3-1b illustrates the symbol employed in circuit diagrams to denote the presence of a diode. The arrow points in the direction of low resistance to positive currents.

Figure 3-1c shows the mechanism of conduction of electricity when the p region is made positive with respect to the n region by application of a potential; this process is called *forward biasing*. Here, the positive holes in the p region and the excess electrons in the n region (the majority carriers in both regions) move under the influence of the electric field toward the junction, where they can combine with and thus annihilate each other. The negative terminal of the battery injects new electrons into the n region, which can then continue the conduction process; the positive terminal, on the other hand, extracts electrons from the p region thus creating new holes which are free to migrate toward the pn junction.

When the diode is *reverse-biased*, as in Figure 3-1d, the majority carriers in each region drift away from the junction to leave a *depletion layer* that contains few charges. Only the small concentration of minority carriers present in each region drift toward the junction and thus carry a current. Consequently, conductance under reverse bias is typically 10^{-6} to 10^{-8} that of conductance under forward bias.

Current-Voltage Curves for Semiconductor Diodes. Figure 3-2 shows the behavior of a typical semiconductor diode under forward and reverse bias. With forward bias, the current increases nearly exponentially with voltage; often currents of several amperes are

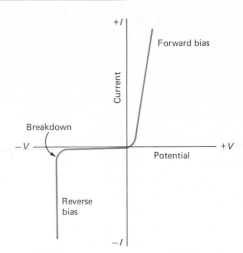

FIGURE 3-2 Current voltage characteristics of a silicon semiconductor diode. Note that, for the sake of clarity, the small current under reverse bias has been exaggerated.

observed. Under reverse bias, a current on the order of microamperes flows over a considerable voltage range; in this region, conduction is by the minority carriers. Ordinarily, this reverse current is of no consequence. As the reverse potential is increased, however, a *breakdown* voltage is ultimately reached where the reverse current increases abruptly to very high values. Here, holes and electrons, formed by the rupture of covalent bonds of the semiconductor, are accelerated by the field to produce additional electrons and holes by collision. In addition, quantum mechanical tunneling of electrons through the junction layer contributes to the enhanced conductance. This conduction, if sufficiently large, may result in heating and damaging of the diode. The voltage at which the sharp increase in current occurs under reverse bias is called the *Zener breakdown voltage*. By controlling the thickness and type of the junction layer, Zener voltages ranging from a few volts to several hundred volts can be realized. As we shall see, this phenomenon has practical application in electronics.

Transistors

The transistor is the basic semiconductor amplifying device and performs the same function as a vacuum amplifier tube—that is, it provides an output signal that is usually significantly greater than the input.

Bipolar Transistors. Bipolar transistors consist of two back-to-back semiconductor diodes. The *pnp* transistor consists of an *n*-type region sandwiched between two *p*-type regions; the *npn* type has the reverse structure. Bipolar transistors are constructed in a variety of ways, two of which are shown in Figure 3-3. The general features of all such

transistors are: (1) small size; (2) heavily doped outer layers, the smaller of which is called the *emitter* and the larger the *collector*; (3) a thin (~ 0.02 mm), lightly doped *base* layer that separates the emitter and the collector; and (4) metallic contacts and electrical leads to each of the three regions.

An alloy junction *pnp* transistor, such as that shown in Figure 3-3a, is often fabricated from a chip of *n*-type germanium. Small pellets of a Group III metal such as indium are fixed on both sides of the chip and heated until the metal melts and dissolves in the germanium. Upon cooling and crystallization, two heavily doped *p* regions are formed on the two sides of the thin layer of *n*-doped germanium.

A typical planar *npn* transistor, such as

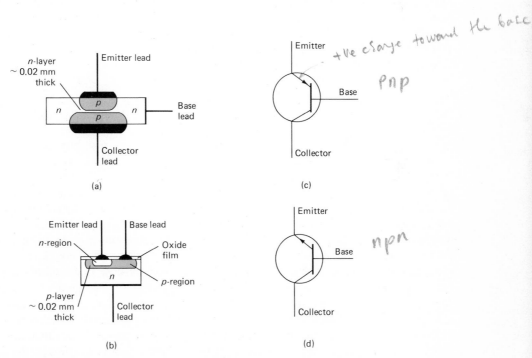

(a)

(b)

(c)

(d)

FIGURE 3-3 Two types of bipolar resistors. Construction details are shown in (a) for a *pnp* alloy junction transistor and in (b) for an *npn* planar transistor. Symbols for a *pnp* and an *npn* bipolar transistor are shown in (c) and (d), respectively. Note that alloy junction transistors may also be fabricated as *npn* types and planar transistors as *pnp*.

that shown in Figure 3-3b, is formed from a wafer of n-type silicon which has a surface coating of silicon dioxide; this layer is deposited by heating the wafer in an oxygen atmosphere. To fabricate the p-type base region, a circle is etched in the oxide surface; the crystal is then heated and exposed to vaporized boron to produce a lightly doped, p-type region by diffusion. The surface film is then reformed by oxidation, following which a smaller circle is etched within the original circle. This region is then heavily doped by diffusion of phosphorus into the heated crystal to give an n-type emitter.

The symbols for the pnp and the npn type of transistor are shown on the right in Figure 3-3. The arrow on the emitter lead indicates the direction of flow of positive charge. Thus, in the pnp type, positive charge flows from the emitter to the base; the reverse is true for the npn type.

Electrical Characteristics of a Bipolar Transistor. The discussion that follows will focus upon the behavior of a pnp-type bipolar transistor. It should be appreciated that the npn type acts analogously except for the direction of the flow of electricity, which is opposite.

When a transistor is to be used in an electronic device, one of its terminals is connected to the input and the second serves as the output; the third terminal is connected to both and is the *common* terminal. Three configurations are thus possible: a common-emitter, a common-collector, and a common-base. The common-emitter configuration has the widest application in amplification and is the one we shall consider in detail.

Figure 3-4 illustrates the current amplification that occurs when a pnp transistor is employed in the common-emitter mode. Here, a small dc input current I_B, which is to be amplified, is introduced in the emitter-base circuit; this current is labeled as the base current in the figure. As we shall show later, an ac current can also be amplified by introducing it in series with I_B.

After amplification, the dc component can then be removed by a high-pass filter (p. 23).

The emitter-collector circuit is powered by a dc source or power supply that may consist of a series of batteries or a rectifier. Typically, the power supply will provide a potential between 9 and 30 V.

Note that, as shown by the breadth of the arrows, the collector or output current I_C is significantly larger than the base input current I_B. Furthermore, the magnitude of the collector current is directly proportional to the input current. That is,

$$I_C = aI_B \qquad (3\text{-}1)$$

where the proportionality constant a is the *current gain*, which measures the current amplification that has occurred. The collector current is also proportional to the emitter current. Thus,

$$I_C = \alpha I_E \qquad (3\text{-}2)$$

where the proportionality constant α is called the *forward current transfer ratio*. Since $I_E = I_B + I_C$ (see Figure 3-4), it is readily shown that

$$a = \frac{\alpha}{1 - \alpha} \qquad (3\text{-}3)$$

Values for a for typical transistors range from 20 to 200.

Mechanism of Amplification with a Bipolar Transistor. It should be noted that the emitter-base interface of the transistor shown in Figure 3-4 constitutes a forward-biased pn junction similar in behavior to that shown in Figure 3-1c, while the base-collector region is a reverse-biased np junction similar to the circuit shown in Figure 3-1d. A current I_B readily flows through the forward-biased junction when an input signal of a few tenths of a volt is applied (see Figure 3-2). In contrast, passage of electricity across the reverse-biased collector-base junction is inhibited by the migration of majority carriers

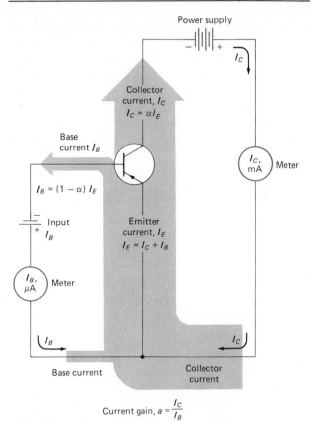

FIGURE 3-4   Currents in a common-emitter circuit with a *pnp* transistor. Ordinarily, $\alpha \cong 0.95$ to 0.995 or $\alpha \cong 20$ to 200.

away from the junction, as shown in Figure 3-1d.

In the forward-biased *pn* junction illustrated in Figure 3-1c, the number of holes in the *p* region is about equal to the number of mobile electrons in the *n* region. Thus, except for the small difference in their mobilities, conduction through the diode is shared more or less equally between the two types of charged bodies. Recall, however, that the *p* region of a *pnp* transistor is much more heavily doped than the *n* region. As a consequence, the concentration of holes in the *p* region is a hundredfold or more greater than the concentration of mobile electrons in

the *n* layer. Thus, the holes in this *pn* junction have a current-carrying capacity that is perhaps one hundred times greater than that for the electrons.

Turning again to Figure 3-4, it is apparent that holes are formed at the *p*-type emitter junction through abstraction of electrons by the two dc sources, namely, the input battery and the power supply. These holes can then move into the very thin *n*-type base region where some will combine with the electrons from the input source; the base current I_B is the result. The majority of the holes will, however, drift through the narrow base layer and be attracted to the negatively charged collector junction, where they can combine with electrons from the power supply; the collector current I_C is the result.

It is important to appreciate that the magnitude of the collector current is determined by the number of current-carrying holes available in the emitter. This number, however, is a fixed multiple of the number of electrons supplied by the input base current. Thus, when the base current doubles, so also does the collector current. This relationship leads to the current amplification exhibited by a bipolar transistor.

Field Effect Transistors

Two other types of transistors, field effect transistors (FET) and metal-oxide-semiconductor field effect transistors (MOSFET), are also encountered. An example of the former appears in Figure 3-5. Here, the *gate* is a cylindrical *p*-type semiconductor surrounding a center core of *n*-doped material called the *channel*. As with bipolar transistors, the *p*- and *n*-semiconductor regions can be reversed. The gate serves the same function as the base in a bipolar transistor; that is, the bias on the gate controls the flow of electrons between the *source* and *drain* leads of the transistor.

In the absence of a gate bias potential, electrons are free to flow from the power supply, through the channel and back to the power supply; conduction in the channel is by electrons. When a *reverse bias* is applied to the gate, however (as in Figure 3-5c), the supply of electrons in the channel is depleted (see Figure 3-1d, p. 43); thus, the resistance of the channel increases and the current decreases.

A major advantage of the field effect transistor arises from the fact that it is normally operated in the reverse-bias mode. As a consequence, the resistance of the gate circuit is very large, and the current drawn from the input signal small.

AMPLIFIERS EMPLOYING TRANSISTORS

The electrical signal from a transducer is ordinarily so small that it must be amplified, often millions of times, before it is sufficient to operate a meter or a recorder and thus be measured. In this section, we will consider the details of a few single-stage, practical amplifier circuits. It should be appreciated, however, that a single transistor seldom provides the amplification needed for most applications, and that a satisfactory output can only be achieved by cascading several amplifier stages.

A Simple Amplifier Circuit

Figure 3-6a is a circuit diagram for a simple amplifier capable of amplifying either an ac or a dc current by a factor of 100 or more.

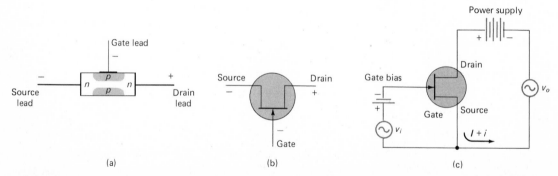

FIGURE 3-5 The field effect transistor. (a) Construction cross section, (b) symbol, (c) circuit.

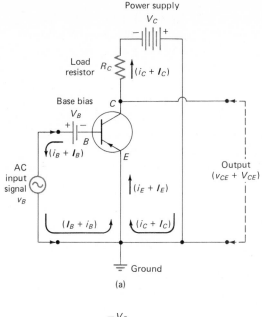

Power supply V_C

Load resistor R_C $(i_C + I_C)$

Base bias V_B

$(i_B + I_B)$

B

C

E

AC input signal v_B

$(I_B + i_B)$ $(i_C + I_C)$

$(i_E + I_E)$

Output $(v_{CE} + V_{CE})$

Ground

(a)

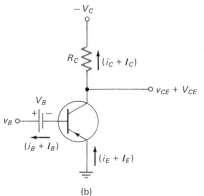

$-V_C$

R_C $(i_C + I_C)$

$v_{CE} + V_{CE}$

V_B

v_B

$(i_B + I_B)$

$(i_E + I_E)$

(b)

FIGURE 3-6   (a) A common-emitter circuit for amplification of a small ac signal. (b) Shorthand version of (a) wherein all indicated voltages are assumed to be with respect to the common ground.

With either, the input signal is placed in series with a constant *base bias* V_B, and is thus superimposed on the steady current I_B. The bias may be provided by a battery, as shown; more often, however, the bias is provided by the amplifier power supply.

In Figure 3-6a, the signal is assumed to be

ac. It is important to note that a dry cell (or other dc source generally) has, for all practical purposes, negligible resistance toward an ac signal; similarly, most ac sources can be assumed to offer virtually no resistance toward a dc current. Thus, the series arrangement shown will produce an ac signal that is superimposed upon the dc current. That is, the base current is given by $(I_B + i_B)$ where i_B is due to the ac signal.

Typically, an amplifier such as that shown in Figure 3-6a employs a power supply that has an output potential of 15 to 30 V dc; normally, the current is produced by rectification of 110-V ac line output. The power supply then serves as the source of the amplified collector current. Often this current is observed as the output voltage drop $(v_{CE} + V_{CE})$ across the emitter-collector terminals of the transistor.

Both the bias current and the signal current are, of course, amplified and appear as a dc component of the collector current I_C and an ac component i_C. Note that the collector current and the base current combine at the common junction to form an emitter current $(i_E + I_E)$, which also has both ac and dc character.

Ground Connections. Electronic circuits normally have a low-resistance wire or foil that interconnects one terminal of each signal source and the power supply; other components of the circuit may also be connected to this common line, called the *ground* of the system. The ground provides a common return for all currents to their sources. As a consequence, all voltages in the circuit are with reference to the common ground. Figure 3-6a shows that the plus terminal of the power supply is connected directly to the ground, which is indicated by the three parallel lines at the bottom of the figure. Note that a side of the input and output signals, as well as the emitter terminal of the transistor, are similarly connected and are thus at ground potential.

Figure 3-6b is a simplified diagram of the

amplifier shown in (a). Here, the lines connecting the power supply to the grounded sides of the output and input have been deleted, it being understood that one side of v_B, $(v_{CE} + V_{CE})$, and V_C is at ground potential. Note that positive electricity flows from ground to the power supply and the signal sources.

An Emitter-Follower Amplifier

Figure 3-7 is a diagram for an *emitter follower*. It differs from the common-emitter circuit in that the load resistor, R_L, is located between the emitter and the base. For the ac input, the collector terminal is at ground inasmuch as the dc power supply offers essentially no resistance toward the ac current; thus, the circuit is a common-collector circuit.

By applying Kirchhoff's voltage law to the loop that contains the input, the base, the emitter, and the load resistor R_L, we obtain

$$v_B = i_B X_{C_1} + v_{BE} + i_E R_L$$

The voltage v_{BE} across the base-emitter terminal is at most a few tenths of a volt and is generally insignificant when compared with the IR drop across R_L. In addition, the capacitance of C_1 is made large enough so that its

reactance to an alternating current is negligible in the frequency range for which the circuit is intended. Thus, typically, the approximation

$$v_B \cong i_E R_L \tag{3-4}$$

is valid to 1 to 2%. Clearly then, the emitter follower does not provide any voltage gain. As with the common-emitter circuit, however, good current gains are realized. It is also important to note that Equation 3-4 indicates that the output signal will be *in phase* with the input.

The emitter-follower circuit finds widespread use because it has a high input impedance and thus draws little current. High impedance is of importance in measuring electrical properties of devices whose output is disturbed by current passage. The input resistance is readily found by assuming that R_1 in Figure 3-7 is so large with respect to R_L that essentially all of the input signal passes through the latter. Then it is seen that

$$i_E = i_B + i_C$$

Substituting the ac form of Equations 3-1 and 3-4 yields

$$\frac{v_B}{R_L} = (1 + a)i_B$$

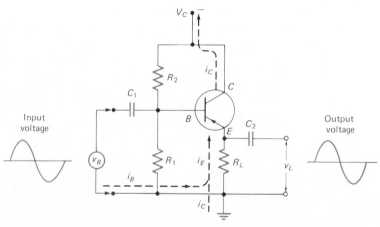

FIGURE 3-7 An emitter-follower circuit.

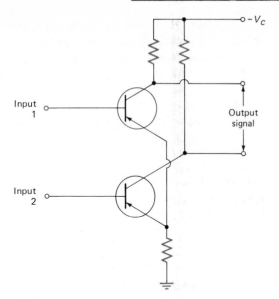

FIGURE 3-8 A differential amplifier.

transistors that are coupled through their emitters. As its name implies, this circuit amplifies the difference between input signals 1 and 2. If the two load resistors and transistors are perfectly matched, the output of this device will be zero for equal input signals. Otherwise, a linear response to signal *differences* is obtained.

The difference amplifier is useful for several reasons. It will, for example, reject an unwanted part of a signal that is common to both inputs. It is relatively unaffected by temperature changes and the passage of time. Finally, it is a useful device for comparing two signals by the null-point method. Because of its stability, this circuit is widely used to amplify dc signals.

Amplifiers with Negative Feedback. Often it is advantageous to return or feedback a fraction of the output signal from an amplifier to the input terminals. The fractional signal that is returned is called *feedback*. Figure 3-9 shows an amplifier with a feedback loop. Here, a fraction β of the output signal is sampled by means of a resistor-capacitor network (shown as the feedback box in the diagram); the resulting signal βv_o is then returned in such a way that it combines with the input signal v_i. The net input signal then is v_A, which is equal to $(v_i + \beta v_o)$. If the amplifier gain is a, we may write that

$$v_o = a v_A = a(v_i + \beta v_o)$$

which rearranges to

$$R_{\text{in}} = \frac{v_B}{i_B} = (1 + a)R_L \qquad (3\text{-}5)$$

where R_{in} is the resistance to an input current of i_B. Typically, then, the input resistance of an emitter follower is one hundredfold or greater than its output resistance, since a often ranges from 100 to 200.

A Differential Amplifier. Figure 3-8 is a circuit diagram for a *difference* or *differential amplifier*. It consists of a pair of matched

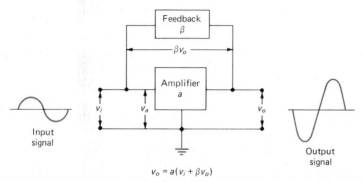

FIGURE 3-9 An amplifier with feedback.

which, upon rearrangement, gives

$$v_o = \frac{a}{1 - \beta a} v_i \qquad (3\text{-}6)$$

Thus, the net gain a' of the amplifier with feedback is

$$a' = \frac{a}{1 - \beta a} \qquad (3\text{-}7)$$

The feedback ratio β can have either a positive or negative sign. For example, the output from the amplifier shown in Figure 3-9 is out of phase with the input signal. The feedback from this signal, then, would interact destructively with the input, thus reducing the signal amplitude; here β is negative, and the circuit is said to have *negative feedback*. On the other hand, the output signal from the amplifier in Figure 3-7 is in phase with the input; here, the feedback would be positive. Negative feedback is of great importance in electronic circuits; we shall focus almost exclusively on this type of feedback.

The full advantage of negative feedback is realized when $\beta a \gg 1$ in Equation 3-7. Then

$$a' = -\frac{1}{\beta} \qquad (3\text{-}8)$$

Note that under these circumstances, the overall amplification *depends only upon the properties of the feedback network and not upon the properties of the amplifier system.* The feedback system, in contrast to the amplifier itself, can be made quite stable, inasmuch as the resistors and capacitors it contains are much less subject to temperature fluctuations and aging effects. For that matter, replacement of one or more transistors in an amplifier with good negative feedback often has no discernible effect on the overall amplification of the unit.

POWER SUPPLIES AND REGULATORS EMPLOYING TRANSISTORS

Generally, laboratory instruments require dc power to operate amplifiers and other reactive components. The most convenient source of electrical power, however, is 115-V ac furnished by public utility companies. As shown in Figure 3-10, laboratory power supply units increase or decrease the potential from the house supply, rectify the current so that it has a single polarity, and finally, smooth the output to give a signal that approximates dc. Most power supplies also contain a voltage regulator which maintains the output voltage at a constant desired level.

Transformers

Alternating current is readily increased or decreased in voltage by means of a power trans-

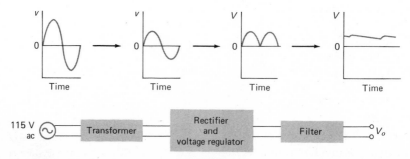

FIGURE 3-10 Diagram showing the components of a power supply and their effect on a signal.

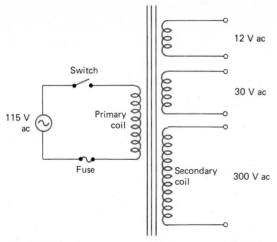

FIGURE 3-11   Schematic of a typical power transformer with multiple secondary windings.

former such as that shown schematically in Figure 3-11. The varying magnetic field arising in the *primary* coil in this device from the passage of the 115-V ac supply current induces currents in the *secondary* coils; the potential across each is given by

$$V_x = 115 \times N_2/N_1$$

where N_2 and N_1 are the number of turns in the secondary and primary coils, respectively. Power supplies with multiple taps, as in Figure 3-11, are available commercially; many voltage combinations can be had. Thus, a single transformer can serve as a power supply for several components of an instrument.

The output voltage of a simple transformer fluctuates only a few percent, even when the current drawn from the output varies widely.

Rectifiers

Figure 3-12 shows three types of rectifiers and their output-signal forms. Each uses transistor diodes (p. 43) to block current flow in one direction while permitting it in the other.

Filters

The output of each rectifier is also shown in the plots on the right of Figure 3-12. In order to minimize current fluctuations, the output is usually filtered by placing a large capacitance in parallel with the load as shown in Figure 3-13. The charge and discharge of the capacitor has the effect of decreasing the variations to a relatively small *ripple*. In some applications, an inductor in series and a capacitor in parallel with the load serve as a filter; this type of filter is known as an *L section*. By suitable choice of capacitance and inductance, the peak-to-peak ripple can be reduced to the millivolt range or lower.

Voltage Regulators

Often, instrument components require dc voltages that are constant and independent of the current being drawn. Voltage regulators serve this purpose.

Figure 3-14 illustrates two simple voltage regulators. Both employ a *Zener diode*, a *pn* junction which has been designed to operate under breakdown conditions; note the special

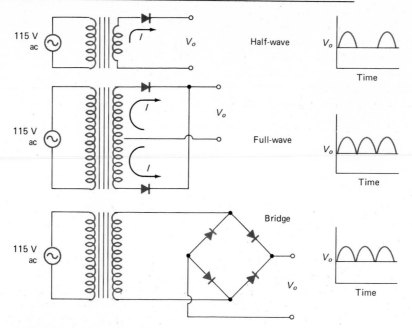

FIGURE 3-12   Three types of rectifiers.

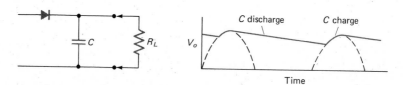

FIGURE 3-13   Filtering the current from a rectifier.

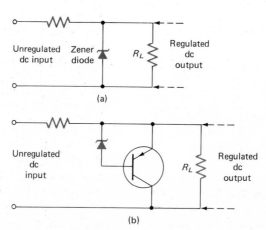

FIGURE 3-14   Voltage regulators.

symbol for this type of diode. In Figure 3-2 (p. 44), it is seen that at a certain reverse bias, a transistor diode undergoes an abrupt breakdown whereupon the current increases precipitously. For example, under breakdown conditions, a current change of 20 to 30 mA may result from a potential change of 0.1 V or less. Zener diodes with a variety of specified breakdown voltages are available commercially.

For voltage regulators, a Zener diode is chosen such that it always operates under breakdown conditions; that is, the input voltage to be regulated is greater than the breakdown voltage.

For the regulator shown in Figure 3-14a, an increase in voltage results in an increase in current through the diode. Because of the steepness of the current-voltage curve in the breakdown region (Figure 3-2), however, the voltage drop across the diode, and thus the load, is virtually constant.

The regulator circuit shown in Figure 3-14b provides the most effective and widely used type of voltage regulator. It serves to reduce the current, and thus the power that must be dissipated by the diode. Again, the voltage drop across the load is essentially that across the Zener diode.

OPERATIONAL AMPLIFIERS

Operational amplifiers derive their name from their original applications in analog computers, where they were employed to perform such mathematical operations as summing, multiplying, differentiating, and integrating. These operations also play an important part in modern instrumentation; as a consequence, operational amplifiers find widespread use in instrument design. In addition to their mathematic role, however, operational amplifiers find general application in the precise measurement of voltage, current, and resistance—the typical signals from the transducers employed for chemical measurements.[2]

General Characteristics of Operational Amplifiers

Operational amplifiers are a class of differential amplifiers that find wide application in instrumentation. The designs of these devices vary widely. Their performance characteristics, on the other hand, have a number of common features which include: (1) large gains (10^4 to 10^6); (2) high input impedances (often up to $10^{12}\,\Omega$ or greater); (3) low output impedances (typically 1 to 10 Ω); and (4) zero output for zero input.

Operational Amplifier Design. Operational amplifiers ordinarily contain several amplifying stages that contain resistors, capacitors, and transistors. Internal negative feedback is used extensively in their design. Early operational amplifiers employed vacuum tubes and were consequently bulky and expensive. In contrast, modern commercial operational amplifiers are tiny; typically, they are manufactured on a single thin silicon chip having surface dimensions that are perhaps 0.05 in. on a side. Through integrated-circuit technology, resistors, capacitors, and transistors are formed on the surface of the chip by photolithographic techniques. After leads have been soldered in place, the entire amplifier is encased in a plastic housing that has dimensions of a centimeter or less; the package, of course, does not include the power supply.

Current-day operational amplifiers, in addition to being compact, are remarkably reliable and inexpensive. Their cost ranges typically from less than a dollar to several dollars. Manufacturers offer a large number of operational amplifiers, each differing in

---

[2] For a more complete discussion of applications of operational amplifiers, see: R. Kalvoda, *Operational Amplifiers in Chemical Instrumentation*. New York: Halsted Press, 1975.

such properties as gain, input and output impedance, operating voltages, and maximum power.

Symbols for Operational Amplifiers. In circuit diagrams, the operational amplifier is ordinarily depicted as a triangle having, as a minimum, an input and an output terminal (Figure 3-15d). Often, additional terminals are indicated, as in (a), (b), and (c) of the figure. A symbol as complete as that in Figure 3-15a is seldom encountered. Here, the power supply leads are shown; ordinarily the power supply input is deleted, its presence being assumed.

Figures 3-15a and 3-15b depict the most common operational amplifier arrangement, in which the noninverting terminal (labeled +) is grounded. The input and output potentials are with respect to ground. Figure 3-15c shows an operational amplifier being used in the differential mode. Here, the difference between two input potentials is amplified and appears as an output with respect to ground.

The symbol shown in Figure 3-15d is often encountered. Unfortunately, it may lead to confusion because it is not evident whether the input is to the inverting or the noninverting terminal; thus, we shall avoid its use.

Inverting and Noninverting Terminals. It is important to realize that the *positive and negative signs show the inverting and noninverting terminals* of the amplifier and *do not* imply that they are necessarily to be connected to inputs of the corresponding signs. Thus, if the negative terminal of a battery is connected to the minus or inverting terminal, the output of the amplifier is positive with respect to it; if, on the other hand, the positive terminal of the battery is connected to the minus input of the amplifier, a negative output results. An ac

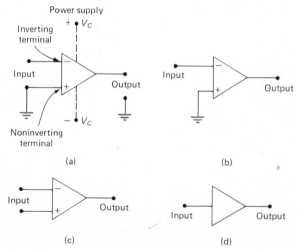

FIGURE 3-15 Symbols for operational amplifiers. In each case, the output is with respect to ground. Inverting circuits with grounded noninverting terminals are shown in (a) and (b). More detail than is usually provided is shown in (a). The circuit in (c) is a differential amplifier in which neither input is grounded. A shorthand representation of an operational amplifier circuit is shown in (d).

signal input into the inverting terminal yields an output *that is 180 deg out of phase*. The positive input of an amplifier, on the other hand, yields an in-phase output signal or a dc signal of the same polarity as the input.

Circuits Employing Operational Amplifiers

Operational amplifiers are employed in circuit networks that contain various combinations of capacitors, resistors, and other electrical components. Under ideal conditions, the output of the amplifier is determined entirely by the nature of the network and its components and is *independent of the operational amplifier itself*. Thus, it is important to examine some of the various networks that employ operational amplifiers.

Basic Inverting Circuit. Figure 3-16 is a circuit diagram of an inverting circuit employing an operational amplifier. The amplified output v_o is 180 deg out of phase with the input v_i. Note that part of the output is fed back through the resistor R_f; because of the phase shift, the feedback is negative. This arrangement is perhaps the most common of all networks associated with operational amplifier circuits.

The gain a of the operational amplifier shown in Figure 3-15a is given by the relationship

$$a = \frac{v_o}{v_+ - v_-} \tag{3-9}$$

Since the plus terminal is grounded, however, $v_+ = 0$; furthermore, $v_s = v_-$. Thus, the gain is given by

$$a = -v_o/v_s$$

The properties of the circuit in Figure 3-16 can be developed with the aid of Kirchhoff's current law; that is,

$$i_i = i_s + i_f$$

Recall, however, that the input impedance of an operational amplifier is always made ex-

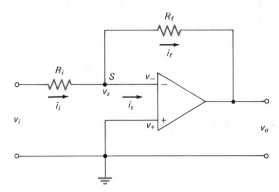

FIGURE 3-16 Inverting circuit with negative feedback.

tremely high (see criterion (2), p. 55); thus, $i_s \ll i_f$ and

$$i_i \cong i_f \tag{3-10}$$

Applying Ohm's law to this equality gives

$$\frac{v_i - v_s}{R_i} = \frac{v_s - v_o}{R_f} \tag{3-11}$$

Substituting the expression for the gain of the amplifier into Equation 3-11 and solving for the output voltage gives

$$v_o = \frac{-v_i(R_f/R_i)}{1 + (1 + R_f/R_i)/a}$$

Note, however, that a for an operational amplifier is usually larger than 10^4. Normally, the resistors R_f and R_i are chosen such that R_f/R_i is less than 100. Therefore, under usual conditions, $(1 + R_f/R_i)/a \ll 1$ and the foregoing equation becomes

$$v_o = -v_i \frac{R_f}{R_i} \tag{3-12}$$

Thus, when a high-gain amplifier is employed and when R_f/R_i is not too great, the circuit shown in Figure 3-16 provides a gain which depends only upon R_f/R_i and is *independent of fluctuations in* a, the gain of the amplifier itself.

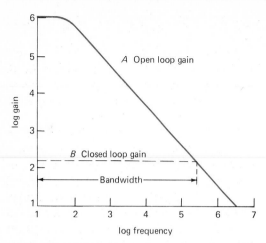

FIGURE 3-17 Frequency response of a typical operational amplifier. (a) Without negative feedback and (b) with negative feedback.

It is important to appreciate that when Equation 3-12 applies, the voltage at point S with respect to ground is necessarily negligible compared with either v_i or v_o. That this statement is correct can be seen by noting that the only condition under which Equation 3-12 follows from 3-11 is when $v_s \ll v_o$ and v_i. Thus, the potential at point S must approach ground potential; as a consequence, point S is said to be at *virtual ground* in the circuit.

It is of interest to determine the effective impedance Z_s between the point S and ground. This quantity is given by

$$Z_s = v_s/i_i$$

We have noted, however, that $i_i \cong i_f$, and from Figure 3-16, we see that

$$i_i = i_f = \frac{v_s - v_o}{R_f}$$

Combining the two equations gives

$$Z_s = \frac{v_s R_f}{v_s - v_o} = \frac{R_f}{1 - v_o/v_s}$$

As we have seen, however, the second term in the denominator is the amplifier gain. Thus,

$$Z_s = \frac{R_f}{1 + a} \qquad (3\text{-}13)$$

Typical values for R_f and a are $10^5\ \Omega$ and 10^4. Under these circumstances, the impedance between S and ground would be only $10\ \Omega$. This property of the circuit, shown in Figure 3-16, is of considerable importance in current measurement (see p. 30).

Frequency Response of the Basic Inverting Circuit. The gain of a typical operational amplifier decreases rapidly in response to high-frequency input signals. This frequency dependence arises from small capacitances that develop at *pn* junctions. The effect for a typical amplifier is shown in Figure 3-17, where the curve labeled open-loop gain represents the behavior of the amplifier *in the absence of the feedback resistor* R_f in Figure 3-16. Note that both the ordinate and abscissa are log scales and that the open-loop gain for this particular amplifier decreases rapidly at frequencies greater than about 100 Hz.

In contrast, an operational amplifier employing external negative feedback, as in Figure 3-16, has a constant gain or *bandwidth* that extends from dc to over 10^5 Hz. In this region, the gain depends only upon R_f/R_i, as shown by Equation 3-12. For many purposes, the frequency independence of the negative feedback inverting circuit is of great importance and more than offsets the loss in amplification.

The Voltage-Follower Circuit. Figure 3-18 depicts a *voltage follower*, a circuit in which

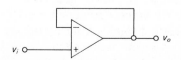

FIGURE 3-18 Voltage-follower circuit.

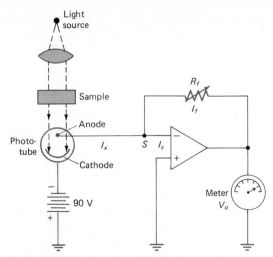

FIGURE 3-19 Application of an operational amplifier to the measurement of a small photocurrent, I_x.

the input is fed into the *noninverting* terminal, and a feedback loop is provided which involves the inverting terminal. The result is an amplifier with a gain of approximately unity. Because the input is not inverted, the output potential is the same as the input. The input impedance is the open-loop impedance for the amplifier, which can be very large (100 MΩ) when a field effect transistor is employed. The output impedance, on the other hand, is low ($< 1\ \Omega$). As will be shown later, this impedance transformation is valuable in the measurement of high-impedance sources with low-impedance measuring devices.

Amplification and Measurement of Transducer Signals

Operational amplifiers find general application to the amplification and measurement of the electrical signals from transducers. These signals, which are often concentration-dependent, include current, potential, and resistance (or conductance). This section includes simple applications of operational

amplifiers to the measurement of each type of signal.

Current Measurement. The accurate measurement of small currents is important to such analytical methods as voltammetry, coulometry, photometry, and gas chromatography.

As we pointed out in Chapter 2 (p. 30), an important concern that arises in all physical measurements, including that of current, is whether the measuring process will, in itself, alter significantly the signal being measured, thus leading to an error. It is inevitable that any measuring process will perturb a system under study in such a way that the quantity actually measured differs from its original value before the measurement. All that may be hoped is that the perturbation can be kept sufficiently small. For a current measurement, this consideration requires that the internal resistance of the measuring device be kept low.

A low-resistance current measuring device is readily obtained by deleting the resistor R_i in Figure 3-16 and using the current to be measured as the signal input. An arrangement of this kind is shown in Figure 3-19, where a small direct current I_x is generated by a phototube, a transducer that converts radiant energy such as light into an electrical current. When the cathode of the phototube is maintained at a potential of about -90 V, absorption of radiation by its surface results in the ejection of electrons, which are then accelerated to the wire anode; a current results that is directly proportional to the power of the radiant beam.

If the conclusions reached on page 57 are applied to this circuit, we may write

$$I_x = I_f + I_s \cong I_f$$

In addition, the point S is at virtual ground so that the potential V_o corresponds to the potential drop across the resistor R_f. Therefore, from Ohm's law,

$$V_o = -I_f R_f = -I_x R_f$$

and

$$I_x = -V_o/R_f = kV_o$$

Thus, the potential measurement V_o gives the current, provided R_f is known. By making R_f reasonably large, the accurate measurement of small currents is feasible. For example, if R_f is 100 kΩ, a 1-μA current results in a potential of 0.1 V, a quantity that is readily measured with a high degree of accuracy.

An important property of the circuit shown in Figure 3-19 is its low resistance toward the transducer, which minimizes the measuring error. Recall from the earlier discussion (p. 58) that S is at *virtual* ground, and the resistance of the measuring device is thus low.

EXAMPLE

Assume that R_f in Figure 3-19 is 200 kΩ, the internal resistance of the phototube is 5.0×10^4 Ω, and the amplifier gain is 1.0×10^5. Calculate the relative error in the current measurement that results from the presence of the measuring circuit.

Here, R_m, the resistance of the measuring device, is given by Equation 3-13

$$R_m = Z_s = R_f/(1 + a)$$
$$= 200 \times 10^3/(1 + 1.0 \times 10^5) = 2.0 \ \Omega$$

On page 31, it was demonstrated that the relative error in a current measurement is given by

$$\text{rel error} = -\frac{R_m}{R_T + R_m}$$

where R_T is the resistance of the circuit in the absence of the resistance of the measuring device R_m. Thus,

$$\text{rel error} = -\frac{2.0}{5.0 \times 10^4 + 2.0}$$
$$= -4.0 \times 10^{-5} \quad \text{or} \quad -0.004\%$$

The instrument shown in Figure 3-19 is a photometer, which measures the attenuation of radiation passing through a solution; this parameter is related to the concentration of the species responsible for the absorption. Photometers are described in detail in Chapter 7.

Potential Measurements. Potential measurements are used extensively for the measurement of temperature and for the determination of the concentration of ions in solution. In the former application, the transducer is a thermocouple; in the latter, it is a pair of electrodes.

In Chapter 2 (p. 32), it was demonstrated that accurate potential measurements require that the resistance of the measuring device be large with respect to the internal resistance of the voltage source to be measured. The need for a high-resistive measuring device becomes particularly acute in the determination of pH with a glass electrode. For this reason, the basic inverting circuit shown in Figure 3-16, which typically has an internal resistance of perhaps 10^5 Ω, is not satisfactory for voltage measurements. On the other hand, it can be combined with the voltage-follower circuit shown in Figure 3-18 to give a very high-impedance voltage-measuring device. An example of such a circuit is shown in Figure 3-20. The first stage involves a voltage follower, which provides an impedance of as much as 10^{12} Ω. An inverting amplifier circuit follows, which amplifies the input by R_f/R_i or 20. An amplifier with a resistance of

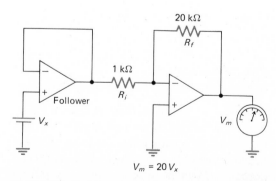

$$V_m = 20 V_x$$

FIGURE 3-20   A high-impedance circuit for voltage amplification.

100 MΩ or more is often called an *electrometer*.

Resistance or Conductance Measurements. Electrolytic cells and temperature-responsive devices, such as thermistors and bolometers, are common examples of transducers whose electrical resistance or conductance varies in response to a chemical signal. They are employed for conductometric and thermometric titrations, in infrared absorption and emission methods, and for temperature control in a variety of analytical measurements.

The circuit shown in Figure 3-16 provides a convenient means for measurement of the resistance or conductance of a transducer. Here, a constant potential source is employed for v_i and the transducer is substituted for either R_i or R_f in the circuit. The amplified output potential v_o is then measured with a suitable meter, potentiometer, or recorder. Thus, if the transducer is substituted for R_f in Figure 3-16, the output, as can be seen from rearrangement of Equation 3-12, is

$$R_x = -\frac{v_o R_i}{v_i} = kv_o \qquad (3\text{-}14)$$

where R_x is the resistance to be measured and k is a constant which can be calculated if R_i and v_i are known; alternatively, k can be determined by a calibration wherein R_x is replaced by a standard resistor.

If conductance rather than resistance is of interest, the transducer conveniently replaces R_i in the circuit. Here, from Equation 3-12, it is found that

$$\frac{1}{R_x} = G_x = -\frac{v_o}{v_i R_f} = k'v_o \qquad (3\text{-}15)$$

where G_x is the desired conductance. Note that in either type of measurement, the value of k, and thus the range of the measured values, can be readily varied by employing a variable resistor R_i or R_f.

Figure 3-21 illustrates two simple applications of operational amplifiers for the measure-

ment of conductance or resistance. In (a), the conductance of a cell for a conductometric titration is of interest. Here, an alternating-current input signal v_i of perhaps 5 to 10 V is provided from the secondary of a filament transformer. The output signal is then rectified and measured as a potential. The variable resistance R_f provides a means for varying the range of conductances that can be recorded or read. Calibration is provided by switching the standard resistor R_s into the circuit in place of the conductivity cell.

Figure 3-21b illustrates how the circuit in Figure 3-16 can be applied to the measurement of a ratio of resistances or conductances. Here, the absorption of radiant energy by a sample is being compared with that for a reference solution. The two photoconductance transducers replace R_f and R_i in Figure 3-16. A battery serves as the source of power and the output potential M, as seen from Equation 3-12, is

$$M = k\frac{R_o}{R} = k'\frac{P_o}{P} \qquad (3\text{-}16)$$

Typically, the resistance of a photoconductive cell is directly proportional to the radiant power, P, of the radiation striking it. Thus, the meter reading M is proportional to the ratio of the power of the two beams.

Comparison of Transducer Outputs. It is frequently desirable to compare a signal generated by an analyte to a reference signal, as in Figure 3-21b. A difference amplifier, such as that shown in Figure 3-22, can also be employed for this purpose. Here, the amplifier is being employed for a temperature measurement. Note that the two input resistors (R_i) have equal resistances; similarly, the feedback resistor and the resistor between the noninverting terminal and ground (both labeled R_k) are also alike.

Applying Ohm's law to the circuit shown in Figure 3-22 gives

$$I_1 = \frac{V_1 - V_-}{R_i}$$

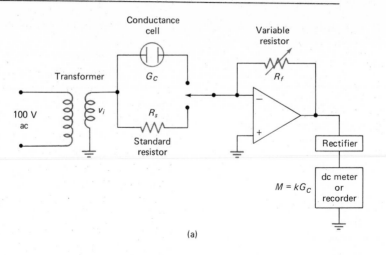

(a)

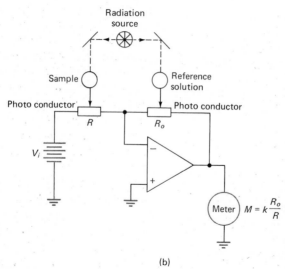

(b)

FIGURE 3-21 Two simple circuits for transducers with conductance or resistance outputs.

and

$$I_f = \frac{V_- - V_o}{R_k}$$

But Equation 3-10 reveals that I_1 and I_f are approximately equal. Thus,

$$\frac{V_1 - V_-}{R_i} = \frac{V_- - V_o}{R_k}$$

The argument demonstrating that point S in Figure 3-16 is at virtual ground (p. 58) can be applied to Figure 3-22, where it becomes apparent that

$$V_- = -V_+$$

Substitution of this relationship into the previous equation yields, upon rearrangement,

$$-V_+ = \frac{V_1 R_k + V_o R_i}{R_1 + R_k}$$

The potential V_+ can also be written in terms of V_2 by means of the voltage-divider equation. Thus,

$$-V_+ = \frac{V_2 R_k}{R_i + R_k}$$

Combining the last two equations gives, upon rearrangement,

$$V_o = \frac{R_k}{R_i}(V_2 - V_1) \qquad (3\text{-}17)$$

Thus, it is the difference between the two signals that is amplified.

Any extraneous potential *common to the two input terminals* shown in Figure 3-22 will be canceled and not appear in the output. Thus, any slow drift in the output of the transducers or any 60-cycle currents induced from the laboratory power lines will be eliminated from V_o. This useful property accounts for the widespread use of a differential circuit in the first amplifier stages of instruments.

The transducers shown in Figure 3-22 are a pair of *thermocouple junctions*, one of which is immersed in the sample and the second in a reference solution (often an ice bath) held at constant temperature. A temperature-dependent contact potential develops at each of the two junctions formed from wires made of copper and an alloy called constantan (other metal pairs are also employed). The potential developed is roughly 5 mV per 100°C temperature difference.

Application of Operational Amplifiers to Voltage and Current Control

Operational amplifiers are readily employed to generate constant-potential or constant-current signals.

Constant-Voltage Source. Several instrumental methods require a power source whose potential is precisely known and from which reasonable currents can be obtained without alteration of this potential. A circuit

that meets these qualifications is termed a *potentiostat*.

Several reference cells are available to provide an accurately known voltage; one of these, the Weston cell, was described on page 34. None of these sources will, however, maintain its potential when a significant current is required.

Figure 3-23 illustrates how a Weston cell or some other reference cell can be employed to provide a standard voltage source from which relatively large currents can be drawn. Note that in both circuits, the standard source appears in the feedback loop of an operational amplifier.

Recall (p. 58) that point S in Figure 3-23a is at virtual ground. For this to be so, it is necessary that $V_o = V_{std}$. That is, the current through the load must be such that $IR_L = V_{std}$. It is important to appreciate, however, that this current arises from the power source of the operational amplifier and *not the standard cell*. Thus, the standard cell controls V_o without the passage of a significant current through it.

Figure 3-23b illustrates a modification of the circuit in (a) which permits the output voltage of the potentiostat to be fixed at a level that is a known multiple of the output voltage of the standard cell.

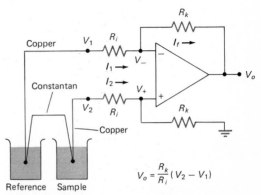

$$V_o = \frac{R_k}{R_i}(V_2 - V_1)$$

FIGURE 3-22 A circuit for the amplification of differences.

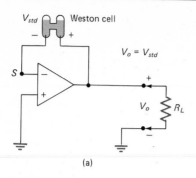

(a)

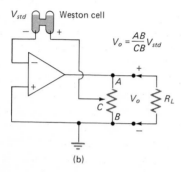

(b)

FIGURE 3-23   Constant potential sources.

Figure 3-24b is an amperostat which employs a standard cell (V_{std}) to maintain a constant current; no significant current is, however, drawn from the standard. The noninverting booster amplifier permits relatively large currents to be passed through the cell.

Application of Operational Amplifiers to Mathematical Operations

An important property of the basic inverting operational amplifier circuit shown in Figure 3-16 is that substitution of various combinations of resistors and capacitors for R_i and R_f permit mathematical operations to be performed on the input signal. Often, as a consequence, the output is more easily related to the concentration of the analyte. For example, we have shown in Figure 3-21b how the electrical output from a pair of photodiodes can be converted to a signal that is directly proportional to the ratio of the power of two beams of radiation. As will become

Constant Current Sources. Constant current sources, called *amperostats*, find application in several analytical instruments. These devices are usually employed to maintain a constant current through an electrochemical cell. An amperostat reacts to a change in input power or a change in internal resistance of the cell by altering its output potential in such a way as to maintain the current at a predetermined level.

Figure 3-24 shows two amperostats. The first requires a voltage input V_i whose potential is constant in the presence of a current. Recall from our earlier discussion that

$$I_L = I_i = \frac{V_i}{R_i}$$

Thus, the current will be constant and independent of the resistance of the cell, provided that V_i and R_i remain constant.

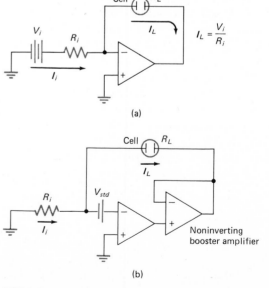

FIGURE 3-24   Constant current sources.

evident in Chapter 6, however, it is the logarithm of this ratio that is directly related to the analyte concentration. Suitable modification of the circuit shown in Figure 3-21b permits direct conversion of the output signal to a logarithmic function.

The outputs of several analytical methods, for example, chromatography, take the form of a Gaussian-shaped peak in which an electrical signal is plotted as a function of time. Integration to give the area under the peak is necessary in order to provide a parameter that is proportional to concentration. Operational amplifiers can be used to perform the integration process and thus give directly a concentration-dependent signal.

Multiplication and Division by a Constant. We have already seen that the output of the basic inverting amplifier circuit is given by

$$v_o = -v_i \frac{R_f}{R_i}$$

Here, the input signal is being multiplied by a constant whose magnitude can be varied by alteration of the resistance ratio. When this ratio is less than one, the operation is equivalent to division of the signal by a constant.

Addition and Subtraction. Figure 3-25 illustrates how an operational amplifier can be employed to yield an output that is the sum of several input signals. It is important to note again that the point S is at virtual ground. As a consequence, the various input signals do not interfere with one another; all flow independently to the virtual ground. Thus, the current at S is simply the sum of the individual currents. Because the impedance of the amplifier is large, however,

$$I_f \cong I_1 + I_2 + I_3 + I_4$$

But

$$I_f = -V_o/R_f$$

Therefore,

$$V_o = -R_f(I_1 + I_2 + I_3 + I_4) \quad (3\text{-}18)$$

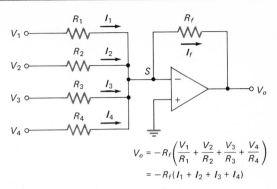

$$V_o = -R_f\left(\frac{V_1}{R_1} + \frac{V_2}{R_2} + \frac{V_3}{R_3} + \frac{V_4}{R_4}\right)$$

$$= -R_f(I_1 + I_2 + I_3 + I_4)$$

FIGURE 3-25 A summing circuit.

and

$$V_o = -R_f\left(\frac{V_1}{R_1} + \frac{V_2}{R_2} + \frac{V_3}{R_3} + \frac{V_4}{R_4}\right)$$
$$(3\text{-}19)$$

If the four input resistors and the feedback resistor are all alike ($R_f = R = R_1 = R_2 = R_3 = R_4$), then the output potential is the sum of the four input potentials. That is,

$$V_o = V_1 + V_2 + V_3 + V_4$$

Another interesting case arises when

$$R_1 = R_2 = R_3 = R_4 = R = 4R_f$$

Here, Equation 3-19 reduces to

$$V_o = -\frac{R}{4}\left(\frac{V_1}{R} + \frac{V_2}{R} + \frac{V_3}{R} + \frac{V_4}{R}\right)$$
$$= -\frac{(V_1 + V_2 + V_3 + V_4)}{4}$$

and the output gives the *average* value for the four signals. Clearly, a weighted average can be obtained by varying the values of the input resistors.

Subtraction becomes possible with the circuit by reversing the sign of one or more of the input signals.

Integration. Figure 3-26a shows a useful circuit for integration of a varying-input signal. Here, a capacitor C_f replaces the resistance R_f in the basic inverting circuit.

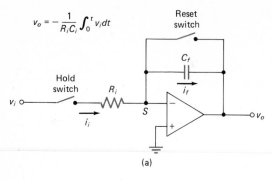

$$v_o = -\frac{1}{R_iC_i}\int_0^t v_i dt$$

(a)

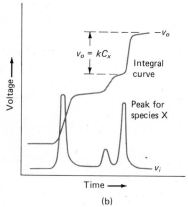

(b)

FIGURE 3-26 (a) Circuit for integration. (b) A chromatogram and its time integral, where v_o is proportional to the concentration of the species X.

To analyze this circuit, recall that for all practical purposes,

$$i_i = i_f$$

Because point S is at virtual ground, the output v_o, is the voltage across the capacitor. Thus, substituting Equation 2-22 (p. 16) and Ohm's law into the foregoing equality gives

$$\frac{v_i}{R_i} = -C\frac{dv_o}{dt}$$

or

$$dv_o = -\frac{v_i}{R_iC}dt$$

and

$$v_o = -\frac{1}{R_iC}\int_0^t v_i\,dt \qquad (3\text{-}20)$$

Thus, the output is the time integral of the input voltage. The definite integral is obtained by opening the reset switch and closing the hold switch (Figure 3-26a) at time zero. Then, when the hold switch is opened at time t, the integration is stopped, thus holding v_o at a constant level for measurement. Closing the reset switch then discharges the capacitor so that a new integration can be started.

Figure 3-26b shows a chromatogram (lower curve) and its time integral obtained with an operational amplifier circuit. Note that for each peak, there is a corresponding step in the integral curve. The height of this step corresponds to v_o in Equation 3-20 and is ordinarily proportional to the concentration of the species responsible for the peak.

Differentiation. Figure 3-27 is a simple circuit for differentiation. Note that it differs from the integration circuit only in the respect that the positions of C and R have been reversed. Here, we may write

$$C_i\frac{dv_i}{dt} = -\frac{v_o}{R_f}$$

or

$$v_o = -R_fC_i\frac{dv_i}{dt} \qquad (3\text{-}21)$$

The circuit shown in Figure 3-27 is not, in fact, practical for most chemical applications,

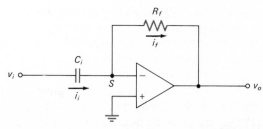

FIGURE 3-27 A simple circuit for differentiation.

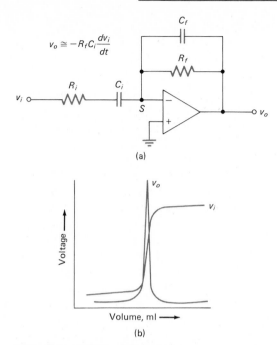

$$v_o \cong -R_f C_i \frac{dv_i}{dt}$$

(a)

(b)

FIGURE 3-28 (a) A practical circuit for differentiation. (b) Titration curve v_i and its derivative v_o.

where the rate of change in the transducer signal is low. For example, differentiation is a useful way to treat the data from a potentiometric titration; here, the potential change of interest occurs over a period of a second or more ($f \leq 1$ Hz). The input signal will, however, contain extraneous 60-, 120-, and 240-Hz potentials (see Figure 3-34), which are induced by the ac power supply. In addition, signal fluctuations resulting from incomplete mixing of the reagent and analyte solutions are often encountered. Unfortunately, the output of the circuit in Figure 3-27 has a strong frequency dependence; as a consequence, the output voltage from the extraneous signals often becomes as great or greater than that from the changing transducer signal, even though the magnitude of the former relative to the latter is small.

The modified circuit shown in Figure

3-28a overcomes this problem. Here, the introduction of a small capacitance C_f in the feedback circuit and a small resistor R_i in the input circuit filters the high-frequency voltages without significant attenuation of the signal of interest. Figure 3-28b shows the appearance of a typical titration curve v_i and its derivative v_o.

Generation of Logarithms and Antilogarithms. A number of transducers produce an electrical response which is exponentially related to concentration. Thus, it is the logarithm of the signal that is directly proportional to concentration. Figure 3-29a shows a circuit that will convert an input to its logarithm; Figure 3-29b shows a circuit that will provide the antilogarithm of an input. It should be noted that the circuits shown in Figure 3-29 require that the input voltage be of a fixed polarity.

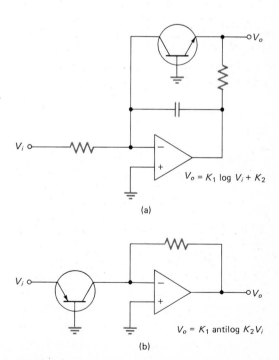

FIGURE 3-29 Circuits for generating (a) logarithms and (b) antilogarithms.

NOISE [3]

The ultimate limitation to the accuracy and sensitivity of an analytical method is the presence of extraneous and unwanted signals that are superimposed on the analyte signal being measured. These extraneous signals are called *noise*, the terminology being derived from radio engineering, where the presence of unwanted signals was recognizable to the ear as noise or static.

The effect of noise on an analyte signal is shown in the plots on the right-hand side of Figure 3-30. The graphs on the left are the theoretical signals in the absence of noise. Note that the analytical signal is a rectangular wave whose current or voltage amplitude is directly proportional to analyte concentration; the signals shown are for relative concentrations of $0 : 1 : 2 : 4$. The noise assumed to be associated with the measurement is shown in Figure 3-30b, where the analyte concentration is zero. For the purpose of illustration, this same noise has simply been added to the three analyte signals (c, e, and g). Thus, the representations in graphs d, f, and h are artificial because the fluctuations are random and would differ from measurement to measurement unless all four signals were observed simultaneously. Furthermore, it has been assumed that no significant noise is associated with the analyte signal itself; such a situation is unlikely in real measurements. Finally, noise is not necessarily random; it often has both random and nonrandom components. Nevertheless, the illustration is useful.

Examination of the signal plus noise in Figure 3-30d suggests that the presence of noise would make detection of the analyte signal difficult, or perhaps impossible, at this concentration level. At double the analyte concentration (Figure 3-30f), the existence of the rectangular wave becomes obvious; accurate determination of the ac amplitude would, however, be difficult. At the highest concentration of analyte, the presence of the repetitive analytical signal is clearly seen; the relative error associated with the measurement of its amplitude would be significantly smaller than at the lower concentrations.

Amplification of a noisy signal provides little improvement because typically both the noise and the information-bearing signal are amplified to the same extent; in addition, noise may be added by the amplification process. This effect is shown in Figure 3-31, where the noise shown in Figure 3-30b has been doubled and superimposed upon the pure signal in Figure 3-30e. Here, the output is effectively that of Figure 3-30d after doubling. Again, visual detection and measurement of the analyte signal would be unsatisfactory. It is worth noting here that lock-in amplifiers, which are discussed in a later section, do permit amplification and extraction of a signal from much of the accompanying noise.

It is apparent from these illustrations that noise becomes increasingly important as its magnitude approaches that of the analyte signal; thus, the *signal-to-noise ratio S/N* is a much more useful parameter than the absolute noise level in describing the quality of the output from a chemical measurement.

Signal-to-Noise Ratio

The signal-to-noise ratio associated with a measurement is often expressed in terms of a ratio of currents, voltages, or power associated with the signal and the noise. It may also be expressed in terms of the *decibel*; by definition,

$$\frac{S}{N} \text{ (decibels)}$$

$$= 20 \log \frac{\text{signal voltage or current}}{\text{noise voltage or current}}$$

[3] For a more detailed discussion of noise see: T. Coor, *J. Chem. Educ.*, **45**, A534 (1968) and A. J. Diefenderfer, *Principles of Electronic* Instrumentation. Philadelphia, Saunders, 1979, Chapter 13.

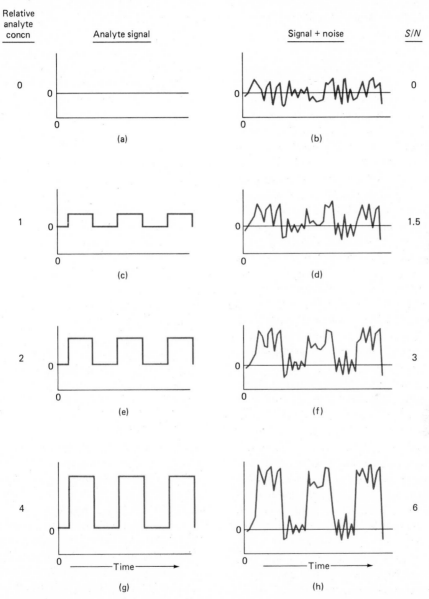

FIGURE 3-30   The effect of noise on an ac signal. In each instance, the noise shown in (a) has been superimposed on the signal from the first column.

The random noise for an instrument can often be obtained by performing 20 to 30 measurements on a blank. The noise then is taken as the standard deviation of the set (see Appendix 1). That is,

$$s = N = \sqrt{\frac{\sum_1^n (x_n - \bar{x})^2}{(n - 1)}} \qquad (3\text{-}22)$$

where x_n is the observed parameter and $\bar{x}$ is the mean for the n measurements. The S/N values shown in the last column in Figure 3-30 were based on this definition of N; that is, they are ratios of the amplitudes of the ac signals to the standard deviation derived from the blank. As a rule of thumb, visual observation of a signal becomes impossible when S/N is smaller than from two to three.

Sources of Noise

Noise is associated with each component of an instrument—that is with the source, the transducer, the signal processor, and the readout. Furthermore, the noise from each of these components may be of several types and arise from several sources. Thus, the noise that is actually observed is a complex composite, which usually cannot be fully characterized. Certain kinds of noise are recognizable, however, and a consideration of their properties is useful.

Instrumental noise can be divided into four general categories. Two, *thermal* or *Johnson noise* and *shot noise*, are well understood, and quantitative statements can be made about the magnitude of each. It is clear that neither Johnson nor shot noise can be totally eliminated from an instrument.

Two other types of noise are also recognizable, *environmental noise* and *flicker* or $1/f$ noise. Their sources are neither well defined nor understood. In principle, however, they can be eliminated by appropriate instrument design.

In addition to instrumental noise, the results of an analysis are often affected by a type of noise related to the analyte itself. For

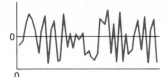

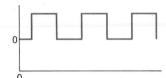

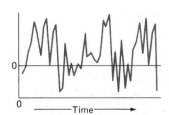

FIGURE 3-31 Effect of amplification of a noisy signal. Here, both the noise and the signal from Figure 3-30d have been doubled.

example, incomplete mixing of a reagent with an analyte solution during a titration often causes initial fluctuations (or noise) in the potential of an electrode system used to monitor the analyte concentration. Uncontrollable variations in the temperature of an arc source used to excite the atomic spectrum of an element often cause significant fluctuations in the intensity of the line that is employed as a measure of concentration. Noise of this type is often more significant than instrumental noise.

The sections that follow deal with instrumental noise only. Noise associated with the properties of analytes will be dealt with in the chapters devoted to specific instrumental methods.

Johnson Noise. Johnson noise owes its source to the thermal motion of electrons in resistors and other resistive elements of an electrical circuit. This agitation, which decreases as temperature is reduced, imparts a random motion to the electrons, which in turn causes inhomogeneities in electron densities throughout the resistor. Thus, momentary small voltage differences develop within regions of the resistor, and the net voltage drop across the resistor fluctuates slightly with time. Note that Johnson noise will occur even in the absence of a current.

The magnitude of Johnson noise is readily derived from thermodynamic considerations[4] and is given by

$$v_{rms} = \sqrt{4kTR\,\Delta f} \qquad (3\text{-}23)$$

where v_{rms} is the root-mean-square noise voltage lying in a frequency bandwidth of Δf Hz, k is the Boltzmann constant (1.38×10^{-23} J/deg), T is the absolute temperature, and R is resistance in ohms of the resistive element.

It is important to note that Johnson noise, while dependent upon the frequency bandwidth, is *independent* of frequency itself; thus, it is sometimes termed *white noise* by analogy to white light, which contains all visible frequencies. It is also noteworthy that Johnson noise is independent of the physical size of the resistor.

EXAMPLE
Calculate the Johnson noise at the input and the output of an amplifier having an input resistance of 1 MΩ and an amplification of 100. Assume the bandwidth of the input signal is 10^4 Hz. The temperature is 25°C.

Employing Equation 3-23, we find

$$v_{rms} = \sqrt{4 \times 1.38 \times 10^{-23} \times 298 \times 10^6 \times 10^4}$$

$$= 1.3 \times 10^{-5} \text{ V or } 0.013 \text{ mV}$$

---

[4] For example, see: T. Coor, *J. Chem. Educ.*, **45**, A533 and A583 (1968).

After amplification,

$$v_{rms} = 0.013 \times 100 = 1.3 \text{ mV}$$

Shot Noise. Shot noise arises wherever a current involves the transfer of electrons or other charged particles across junctions. In the typical electronic circuit, these junctions are found at p and n interfaces; in photocells and vacuum tubes, the junction consists of the evacuated space between the anode and the cathode. The currents in both result from a series of quantized events, namely, the transfer of individual electrons across the junction. Thus, the current results from a series of random events, the number of which is subject to fluctuation from instant to instant. Because the behavior is random, however, the magnitude of the current fluctuation is readily susceptible to statistical treatment using the equation

$$i_{rms} = \sqrt{2Ie\,\Delta f} \qquad (3\text{-}24)$$

where i_{rms} is the root-mean-square current fluctuation associated with the average direct current I, e is the charge on the electron (1.6×10^{-19} C), and Δf is again the bandwidth of frequencies being considered. Like Johnson noise, shot noise has a "white" spectrum.

Flicker Noise. Flicker noise is characterized as a noise whose magnitude is inversely proportional to the frequency of the signal being observed; it is sometimes termed $1/f$ (*one-over-f*) noise as a consequence. The causes of flicker noise are not well understood; its ubiquitous presence, however, is recognizable by its frequency dependence. Flicker noise becomes significant at frequencies lower than about 100 Hz. The long-term drift observed in dc amplifiers, meters, and galvanometers is considered to be a manifestation of flicker noise.

Environmental Noise. Environmental noise is a composite of noises arising from the surroundings. Figure 3-32 suggests typical sources of environmental noise in a university laboratory.

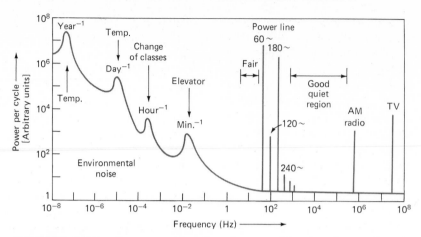

FIGURE 3-32   Some sources of environmental noises in a university laboratory. Note the frequency dependence. [From T. Coor, *J. Chem. Educ.*, **45**, 533 (1968). With permission.]

Much environmental noise occurs because each conductor in an instrument is potentially an antenna capable of picking up electromagnetic radiation and converting it to an electrical signal. Numerous sources of electromagnetic radiation exist, including ac power lines, ignition systems in gasoline engines, arcing switches, brushes in electrical motors, lightning, and ionospheric disturbances. Note that some of these sources, such as power lines, cause noises with limited-frequency bandwidths.

It is also noteworthy that the noise spectrum shown in Figure 3-32 contains a large, continuous noise region at low frequencies. This noise has the properties of flicker noise; its sources are unknown. Superimposed upon the flicker noise are noise peaks associated with yearly and daily temperature fluctuations and other periodic phenomena associated with the use of the laboratory building.

Finally, two quiet-frequency regions in which environmental noises are low are indicated in Figure 3-32. Often, signals are converted to these frequencies to reduce noise during signal processing.

Signal-to-Noise Improvement

Many laboratory measurements require only minimal effort to maintain the signal-to-noise ratio at an acceptable level, because the signals are relatively strong and the requirements for precision and accuracy are low. Examples include the weight determinations made in the course of a chemical synthesis or the color comparison made in determining the chlorine content of the water in a swimming pool. For both examples, the signal is large relative to the noise and the requirements for accuracy are minimal. When the need for sensitivity and accuracy increases, however, the S/N ratio often becomes the limiting factor of a measurement.

Two general methods exist for overcoming signal-to-noise limitations. The first entails reduction of the noise by appropriate instrument design. The second makes use of some type of signal averaging to isolate the signal from the noise.

The sections that follow deal briefly with some of the methods that are employed to overcome the effects of noise.

Grounding and Shielding. Noise arising from environmentally generated electromagnetic radiation can often be substantially reduced by shielding, grounding, and minimizing the lengths of conductors. Shielding consists of surrounding a circuit, or some of the wires in a circuit, with a conducting material that is attached to earth ground. Electromagnetic radiation is then absorbed by the shield rather than by the enclosed conductors; noise generation in the instrument circuit is thus avoided in principle.

Shielding becomes particularly important when the output of a high-impedance transducer, such as the glass electrode, is being amplified. Here, even minuscule induced currents give rise to relatively large voltage drops and thus to large voltage fluctuations.

Difference Amplifiers. Any noise generated in the transducer circuit is particularly critical because it appears in an amplified form in the instrument readout. To attenuate this type of noise, most instruments employ a difference amplifier, such as that shown on page 51, for the first stage of amplification. An ac signal induced in the transducer circuit generally appears in phase at both the inverting and noninverting terminals; cancellation, however, occurs at the output.

Filters

High, low, and resonant filters (pages 23 to 27) are often employed to remove noise that differs in frequency from the signal of interest. Figure 3-33 illustrates the use of a low-pass *RC* filter for reducing environmental and Johnson noise from a slowly varying dc signal. High-pass filters are also useful. Generally, this type of filter is employed to reduce drift and other flicker noise from an ac signal.

Resonant filters are also used to attenuate noise in electronic circuits. We have pointed out that the magnitude of fundamental noise is directly proportional to the frequency bandwidth of the signal. Thus significant noise reduction can be achieved by restricting

the input signal to a narrow band of frequencies and employing an amplifier that is tuned to this band. It should be appreciated, however, that if the signal generated from the analyte varies with time, the bandwidth must be sufficiently great to carry all of the information provided by the signal.

Modulation. Amplification of a low-frequency or dc signal is particularly troublesome because of amplifier drift and flicker noise. Often, this $1/f$ noise is several times larger than the types of noise that predominate at higher frequencies. For this reason, low-frequency or dc signals from transducers are often converted to a higher frequency, where $1/f$ noise is less troublesome. This process is called *modulation*. After amplification, the modulated signal can be freed from amplifier $1/f$ noise by filtering with a high-pass filter; demodulation and filtering with a low-pass filter then produces a dc signal suitable for driving a readout device.

Figure 3-34 is a schematic diagram showing the flow of a signal through such a system. Here, the original dc current is modulated to give a narrow-band 400-Hz signal

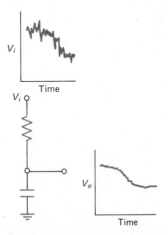

FIGURE 3-33 Use of a low-pass filter with a large time constant to remove noise from a slowly changing dc voltage.

that is amplified by a factor of 10^5. As shown, however, amplification introduces $1/f$ and power line noise; most of this noise can be removed with the aid of a suitable filter before demodulation.

Signal Chopping; the Chopper Amplifier. The *chopper amplifier* provides one means for accomplishing the signal flow shown in Figure 3-34. In this device, the input signal is converted to a rectangular wave form by an electronic or mechanical *chopper*. Chopping can be performed either on the source itself or on the electrical signal from the transducer. In general, it is desirable to chop the signal as close to its source as possible because it is only the noise arising *after chopping* that is removed by the process.

Infrared spectroscopy provides an example of the use of a mechanical chopper for signal modulation. Noise is a major concern in detecting and measuring infrared radiation because both source intensity and detector sensitivity are low in this region of the spectrum. As a consequence, the electrical signal from an infrared transducer is generally small and requires large amplification. Furthermore, infrared transducers, which are heat detectors, respond to thermal radiation from their surroundings; that is, they suffer from serious environmental noise effects.

In order to minimize these noise problems, the beams from infrared sources are generally chopped by imposition of a slotted rotating disk in the beam path. The rotation of this chopper produces a radiant signal that fluctuates periodically between zero and some maximum intensity. After interaction with the sample, the signal is converted by the transducer to an ac electrical signal whose frequency depends upon the size of the slots and the rate at which the disk rotates. Environmental noise associated with infrared measurement is generally dc or low-frequency ac; it can be largely eliminated by use of a high-pass filter prior to amplification of the electrical signal.

Another example of the use of a chopper is shown in Figure 3-35. This device is a *chopper amplifier*, which employs an ac-driven electromagnet to operate a switch that, in its closed position, shorts the input or the output signal to ground. The appearance of the signal at various stages is shown above the circuit diagram. The transducer input is assumed to be a 6-mV dc signal (*A*). The vibrating switch has the effect of forming an

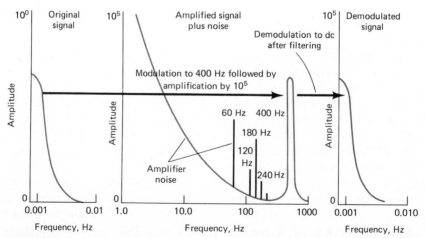

FIGURE 3-34 Amplification of a dc signal with a chopper amplifier.

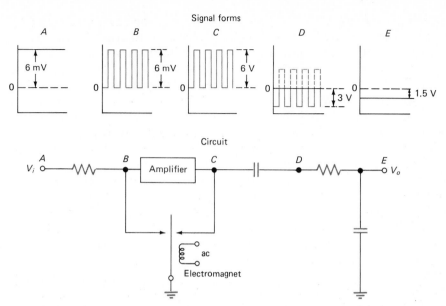

FIGURE 3-35 A chopper amplifier. The signal forms are idealized wave forms at the various indicated points in the circuit.

approximately rectangular wave signal with an amplitude of 6 mV (*B*). Amplification produces an ac signal with an amplitude of 6 V (*C*), which, however, is shorted to ground periodically; as shown in (*D*), shorting also reduces the amplitude of the signal to 3 V. Finally, the *RC* filter serves to smooth the signal and produce a 1.5-V dc output. The synchronous demodulation process has the effect of rejecting the noise generated within the amplifier.

Lock-in Amplifiers.[5] Lock-in amplifiers permit the recovery of small signals even when S/N is unity or less. Generally, a lock-in amplifier requires a reference signal that is coherent with the signal to be amplified. That is, the reference signal must be of the same frequency as the analytical signal and, equally important, must bear a fixed phase relationship to the latter. Figure 3-36a shows a system which employs an optical chopper to provide coherent analytical and reference signals. The reference signal is provided by a lamp and can be quite intense, thus freeing it from potential interferences. The reference and signal beams are chopped synchronously by the rotating slotted wheel, thus providing signals that are identical in frequency and have a fixed phase angle to one another.

The synchronous demodulator acts in a manner analogous to the double pole-double throw switch shown in Figure 3-36b. Here, the reference signal controls the switching so that the polarity of the analytical signal is reversed periodically to provide a rectified dc signal, as shown to the right in Figure 3-36c. The ac noise is then removed by a low-pass filtering system.

A lock-in amplifier is generally relatively free of noise because only those signals that are "locked-in" to the reference signal are amplified; that is, those that are coherent. All other frequencies are rejected by the system.

[5] See, T. C. O'Haver, *J. Chem. Educ.*, **49**, A131, A211 (1972).

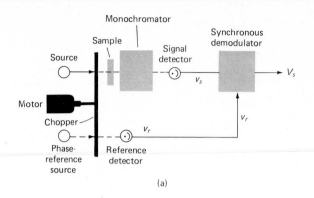

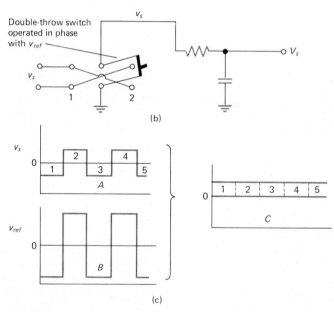

FIGURE 3-36   A lock-in amplifier system: (a) system for a spectrophotometer, (b) synchronous demodulation (schematic), and (c) signal form.

Signal Averaging[6]

Signal averaging provides a powerful tool for extracting a small signal from large amounts of noise. To apply this technique, it is necessary that the signal be repeatable, so that numerous measurements of its magnitude can be performed. If the measurements can all be made in exactly the same way, the signals will be additive, but the noise, insofar as it is random and tends to cancel, adds as the square root of the number of measurements n. Thus, the S/N ratio is improved by a factor of $\sqrt{n}$.

Many modern instruments are interfaced

---

[6] See: R. L. Rowell, *J. Chem. Educ.*, **51**, A71 (1974).

with computers in order to facilitate signal averaging. Here, the measuring process is triggered repetitively, and the signal is accumulated in one or more memory channels. After a suitable number of sweeps or measurements, an average for the signal is computed.

The effect of averaging is shown in Figure 3-37; the signal is part of the output from an NMR spectrometer. With a single scan, only a few of the peaks are discernible because they are roughly the same size as the random noise. After averaging the data from 50 scans, the peaks begin to emerge; averaging 200 scans provides an even better spectrum.

READOUT DEVICES

A variety of devices are employed for converting the output from a signal processor to a form that can be read and interpreted by man. We have already considered some of these in the section of Chapter 2 devoted to meters. We have also seen how the position of a sliding contact on a linear resistor can be employed to provide voltage data (p. 34) or resistance data (p. 35).

In this section, we consider two more sophisticated readout devices, namely, the cathode-ray tube and the laboratory recorder.

The Cathode-Ray Tube

The oscilloscope is a most useful and versatile laboratory instrument that employs a cathode-ray tube as a readout device. Figure 3-38 is a schematic drawing showing the main components of the tube. Here, the display is formed by the interaction of electrons in a focused beam with a phosphorescent coating on the interior of the large curved surface of the tube. Electrons are formed at a heated cathode, which is maintained at ground potential; an anode with a potential of the order of kilovolts accelerates the electrons through a control grid and a second anode that serves to focus the beam on the screen. In the absence of input signals, the beam

appears as a small bright dot in the center of the screen.

Input signals are applied to two sets of plates, one of which deflects the beam horizontally and the other vertically. Thus, x-y plotting of two related signals becomes possible. Because the screen is phosphorescent, the movement of the dot appears as a lighted continuous trace that fades after a brief period.

The most common way of operating the cathode tube is to cause the dot to be swept periodically across the horizontal axis of the tube. For this purpose, a sawtooth sweep signal (p. 12) is employed. The signal to be measured is then applied across the vertical plates. If the signal is dc, it will simply displace the horizontal line to a position above or below the central horizontal axis.

In order to have a repetitive signal, such as a sine wave, displayed on the screen, it is essential that each sweep begin at an identical place on the wave; that is, for example, at a maximum, a minimum, or a zero crossing. Synchronization is usually realized by mixing a portion of the test signal with the sweep signal in such a way as to produce a voltage spike for, say, each maximum or some multiple thereof. This spike then serves to trigger the sweep. Thus, the wave form can be observed as a continuous image on the screen.

Recorders[7]

The typical laboratory recorder is an example of a *servosystem*, a null device that compares two signals and then makes a mechanical adjustment that reduces their difference to zero; that is, a servosystem continuously seeks the null condition.

In the laboratory recorder, shown schematically in Figure 3-39, the signal to be recorded, V_x, is continuously compared with the output from a potentiometer powered by

---

[7] For a good discussion of laboratory recorders, see: G. W. Ewing, *J. Chem. Educ.*, **53**, A361, A407 (1976).

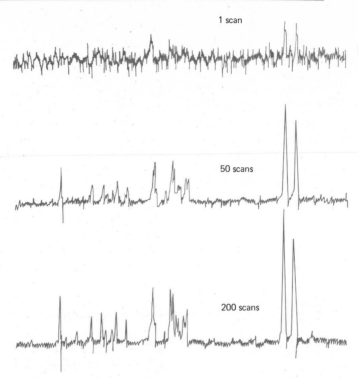

FIGURE 3-37   Effect of signal averaging. Note that the vertical scale is smaller as the number of scans increases. That is, the noise grows in absolute value with increased number of scans; its value relative to the analytical signal decreases, however.

a reference signal, V_{ref}. The latter may be one or more mercury cells, the potential of which remains constant throughout the lifetime of the battery; alternatively, a rectified ac signal, stabilized by a Zener diode (p. 53) serves as the reference. Any difference in potential between the potentiometer output and V_x is converted to a 60-cycle ac current by a mechanical chopper; the resulting signal is then amplified sufficiently to activate a small phase-sensitive electrical motor that is mechanically geared or linked (by a pulley arrangement in Figure 3-39) to both a recorder pen and the sliding contact of the potentiometer. The direction of rotation of the motor is such that the potential difference between the potentiometer and V_x is decreased to zero, whereupon the motor stops.

To understand the directional control of the motor, it is important to note that a reversible ac motor has two sets of coils, one of which is fixed (the stator) and the other of which rotates (the rotor). One of these, say, the rotor, is powered from the 110-V house line and thus has a continuously fluctuating magnetic field associated with it. The output from the ac amplifier, on the other hand, is fed to the coils of the stator. The magnetic field induced here interacts with the rotor field and causes the rotor to turn. The direction of motion depends upon the *phase* of the stator current with respect to that of the rotor; the phase of the stator current, however, differs by 180 deg, depending upon whether V_x is greater or smaller than the signal from V_{ref}. Thus, the amplified dif-

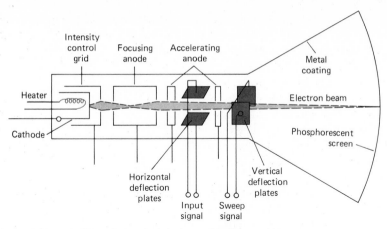

FIGURE 3-38   Schematic of a cathode-ray tube.

ference signal can be caused to drive the ser-vomechanism to the null state from either direction.

In most laboratory recorders, the paper is moved at a fixed speed. Thus, a plot of signal intensity as a function of time is obtained.

A good-quality laboratory recorder will provide a record of voltages that is accurate to a few microvolts.

INTERFACING OF INSTRUMENTS WITH COMPUTERS

Undoubtedly, one of the major developments in analytical chemistry during the last decade has been the interfacing of digital computers with the various instruments listed in Table 1-1. At least two reasons exist for connecting a computer to an analytical instrument. The

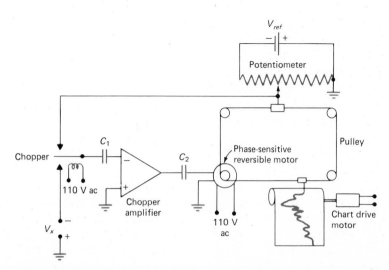

FIGURE 3-39   Schematic diagram of a self-balancing recording potentiometer.

first is that partial or complete automation of measurements becomes possible. Ordinarily, automation leads to more rapid data acquisition and shortens the time required for an analysis or increases its precision by providing time for additional replicate measurements to be made. Automation, moreover, frequently provides better control over experimental variables than a human operator can achieve; more precise and accurate data are the result.

A second reason for attaching computers to instruments is to take advantage of their tremendous computational and data-handling capabilities. These capabilities make possible the routine use of techniques that would ordinarily be impractical because of their excessive computational time requirements. Notable among such applications is the use of Fourier transform calculations in spectroscopy to extract small analytical signals from noisy environments.

The interfacing of computers with analytical instruments is too large and complex a field to be treated in detail here.[8] Thus, the discussion that follows is limited to a general summary of the properties possessed by computers and the advantages gained by interfacing them with analytical instruments.

For many purposes, it is convenient to categorize computers on the basis of size. Here, size refers not only to physical dimensions but more importantly to the quantity of information and instructions that can be stored. This discussion will focus on so-called *minicomputers* and *microcomputers* rather than on the large, centralized, general-purpose device.

---

[8] For monographs on computer interfacing, see: *Computers for Spectroscopists*, ed. R. A. G. Carrington. New York: Wiley, 1974; J. Finkel, *Computer-Aided Experimentation*. New York: Wiley-Interscience, 1974; J. S. Mattson, H. B. Mark, Jr., and H. C. MacDonald, Jr., *Computer Fundamentals for Chemists*. New York: Marcel Dekker, 1973; and A. J. Diefenderfer, *Principles of Electronic Instrumentation*, 2d ed. Philadelphia: Saunders, 1979, Chapters 10–12.

Binary Number System

The programs, instructions, mathematical operations, and memories of a digital computer are all based upon the binary system of numbers rather than on the more familiar decimal system. The decimal system is, of course, based on ten digits; the binary system uses but two, 0 and 1.

Table 3-1 illustrates the relationship between some decimal and binary numbers. The example that follows illustrates methods for binary-to-decimal conversion and the reverse.

EXAMPLE

(a) Convert 101011 to a decimal number. (b) Convert 710 to a binary number.

(a) To understand how this conversion can be made, it is useful to realize that a decimal number can be readily broken down into integer multiples of powers of 10. For example, the number 4753 can be written as

$4753 =$

$$4 \times 10^3 + 7 \times 10^2 + 5 \times 10^1 + 3 \times 10^0$$
$$4000 \quad + \quad 700 \quad + \quad 50 \quad + \quad 3$$

By analogy, binary numbers can be expressed in terms of base 2. Thus,

$$101011 = 1 \times 2^5 + 0 \times 2^4 + 1 \times 2^3$$
$$32 \quad + \quad 0 \quad + \quad 8$$
$$+ 0 \times 2^2 + 1 \times 2^1 + 1 \times 2^0$$
$$+ \quad 0 \quad + \quad 2 \quad + \quad 1 \quad = 43$$

(b) As a first step, we determine the largest power of 2 that is less than 710. Thus, since $2^{10} = 1024$,

$$2^9 = 512 \quad \text{and} \quad 710 - 512 = 198$$

The process is repeated for 198

$$2^7 = 128 \quad \text{and} \quad 198 - 128 = 70$$

Continuing, we find

$$2^6 = 64 \quad \text{and} \quad 70 - 64 = 6$$
$$2^2 = 4 \quad \text{and} \quad 6 - 4 = 2$$
$$2^1 = 2 \quad \text{and} \quad 2 - 2 = 0$$

The binary number is then derived as follows:

$$1 \quad 0 \quad 1 \quad 1 \quad 0 \quad 0 \quad 0 \quad 1 \quad 1 \quad 0$$
$$2^9 \quad - \quad 2^7 \; 2^6 \quad - \quad - \quad - \quad 2^2 \; 2^1 \quad -$$

Arithmetic with binary numbers is similar to, but simpler than, decimal arithmetic. For addition, only four combinations are possible.

0	0	1	1
+0	+1	+0	+1
0	1	1	10

Similarly, for multiplication,

0	0	1	1
×0	×1	×0	×1
0	0	0	1

The following example illustrates the use of these operations.

EXAMPLE
Perform the following calculations with binary arithmetic. (a) $7 + 3$, (b) $19 + 6$, (c) 7×3, (d) 22×2.

(a)
```
   7        111
 + 3      + 11
 ───      ────
  10      1010
```

(b)
```
  19       10011
 + 6         110
 ───       ─────
  25       11001
```

(c)
```
   7        111
 × 3         11
 ───       ────
  21        111
           111
          ─────
          10101
```

(d)
```
  22       10110
 × 5         101
 ───       ─────
 110       10110
          00000
          10110
          ───────
          1101110
```

Bits. The binary digits 0 and 1 are often referred to as *bits*. Thus, the decimal number 110 in part (d) of the example is called a three-digit number in the decimal system; the binary equivalent 1101110 is a seven-bit number.

Words. A *word* consists of one or more bits that are associated with one another. The number of bits per word depends upon the size of the computer; some common sizes include 8, 12, 16, 24, and 64. Minicomputers

---

TABLE 3-1   RELATIONSHIP BETWEEN SOME DECIMAL AND BINARY NUMBERS

Decimal Number	Binary Representation	Decimal Number	Binary Representation
0	0	8	1000
1	1	9	1001
2	10	10	1010
3	11	12	1100
4	100	15	1111
5	101	16	10000
6	110	32	100000
7	111	64	1000000

often employ 16-bit words and microcomputers 8. Computer memories are described in terms of the number of words they can retain. Thus, an 8K memory has space for approximately 8000 words.

A single word contains a pattern of bits that cause the computer to take a specific action. Generally, a word contains a datum or a command and an *address*, both written in binary notation. The latter gives the location of an instruction or a datum word in the memory. For example, for a computer employing 12-bit words, a word might take the form

$$101 \ 00 \ 1001010$$

Here, the first three bits could be the code for a command such as add, subtract, multiply, or store. The next two, in this example, are not used. The last seven bits give the location of the number upon which the command is to be exercised. For example, the word just shown might command the computer to add the number stored in the indicated location (storage number 74) to another number.

Major Components of a Computer

Figure 3-40 is a block diagram showing the arrangement of the major components of a digital computer.

The Central Processing Unit. The heart of a computer is the central processing unit, which contains a control unit and an arithmetic unit. The former controls the overall sequence of operations by means of a program, which is stored in the memory unit. The control unit receives instructions from the input device, fetches instructions and data from the memory unit, transmits instructions to the arithmetic unit for computation, and transmits the results of these computations to the output and often to memory as well.

The program, the instructions, the mathematical operations, and the memory of a digital computer are all based upon the binary system made up of just two states, on and off, or 0 and 1. The central processing unit of a computer contains a vast array of transistorized switches, each of which has only two settings, 0 or 1, corresponding to on or off. These switches are employed using binary

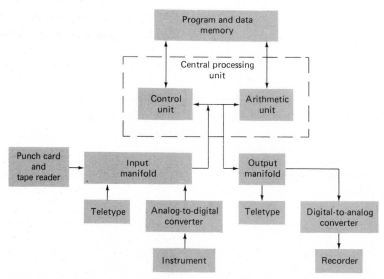

FIGURE 3-40   Components of a digital computer.

logic to perform the various functions of the computer; generally, the switches control voltages that range from 0 ± 0.5 V for the off or 0 position and 5 ± 2 V in the on or 1 position. The band between 0.5 and 3.0 V is sufficiently large to prevent interference by random noise.

Memory. Storage of data and instructions for a computer take two forms. One makes use of magnetic tapes or disks. Access to these is relatively slow, and they are used primarily for storage of programs and data for future use.

The main memory of computers is either a magnetic-core type or an integrated-circuit type. In both, instructions or data are stored in binary form with a separate core or circuit being required for each bit. Generally, the individual cores or circuits are arranged in square arrays of 32×32 or 1024 storage positions. For a computer employing words made up of 16 bits; one of these arrays would correspond to 64 words. The minimum memory capacity required for most computational applications is 4096 words (a 4K memory). Thus, the number of individual binary cores or circuits is enormous.

The individual storage unit or core in a magnetic-core memory is made of a ferrite ceramic, shaped in the form of a doughnut. The inside and outside diameters of the ring are 0.013 and 0.020 in., respectively. When an electrical signal passes through a wire that is threaded through the ring, magnetization of the ferrite material occurs; the direction of magnetization (clockwise or counter clockwise) depends upon the direction of the current and remains fixed after the current ceases. Thus, the core has two stable states and can store binary information. A sensing wire running through the ring is then employed to sense the direction of magnetization and thus retrieve the information stored in the element.

An integrated-circuit memory element generally consists of a circuit containing four transistors, two capacitors, and two switches.

Tiny single chips containing 1024 to 4096 of these elements are employed. Binary storage and retrieval depends upon which half of the circuit is conducting at any time. An integrated-circuit memory has the disadvantage of being volatile; that is, loss of electric power results in loss of the stored information.

Input-Output Systems. Input-output devices provide the means by which the user (or his instrument) communicates with the computer. Familiar input devices include the electric typewriter, punch cards or tapes, magnetic tapes or disks, and the electrical or mechanical signals from analytical instruments. Output devices include recorders, electric typewriters, cathode-ray screens, and meters. It is important to appreciate that most of these devices provide or use an *analog* signal while, as we have pointed out, the computer can respond only to digital signals. Thus, an important part of the input-output system is an *analog-to-digital converter* (ADC) for inputting data, which the computer can use, and a *digital-to-analog counter* (DAC) for converting the output from the computer to a usable signal.

By definition, an analog signal is one that varies continuously, whereas a digital signal is one that varies in discrete increments. An example of an analog signal is the voltage output of a glass electrode, which varies continuously from about 0 to about 1.0 V depending upon the pH of the solution in which it is immersed. Most, but not all, analytical instruments are analog; examples of exceptions include the output from the decay of a radioactive species or from a Geiger tube when it is employed to measure the intensity of an X-ray beam.

Most output devices such as recorders, plotters, and meters require analog rather than digital signals.

Where the analog signal is the rotary motion of a shaft, a digital signal is readily produced by means of a mechanical turn counter. One method of achieving a digital signal is to mount on the shaft a circular disk that is notched to give opaque and transparent slots. An array of light beams and photocells then produces a series of digital electrical pulses as the disk rotates. Such a device might be employed to monitor the wavelength setting of a monochromator or the null point of a potentiometer.

More commonly, analog-to-digital converters are electronic and are based upon counting square-wave pulses from a generator. Here, the number of pulses or their frequency is determined by the analytical signal. Alternatively, the time required for a ramp voltage to change from zero to a chosen level may be measured.

Digital-to-analog converters are electronic devices that convert a digital number into a voltage proportional to the number. The voltage can then be employed to drive a recorder, meter, or other output device.

Computer Programming

Communication with a computer entails setting an enormous aggregation of switches to appropriate off or on positions. A program consists of a set of instructions as to how these switches are to be set for each step in an instruction or a computation. Each of these instructions must be written in a form to which the computer can respond—that is, a binary *machine* code. Machine coding is tedious and time consuming and is often prone to errors. For this reason, *assembly languages* have been developed in which the switch-setting steps are assembled into groups which can be designated by code word. For example, the abbreviation for subtract might be SUB and might correspond to 101 in machine language. Clearly, SUB is a good deal easier for the programmer to remember than 101.

Assembly programming, while simpler than machine programming, is still difficult and tedious. As a consequence, a number of high-level languages, such as FORTRAN and Basic, have been developed. These languages, which are easily learned, have been designed to make communication with the computer relatively straightforward. Here, instructions in Basic or FORTRAN are translated by a computer program (called a *compiler*) into machine language which can then control the computer for computations. Unfortunately, loss of efficiency accompanies the use of higher-level languages.

Figure 3-41 illustrates the application of the three languages for obtaining a sum.

Minicomputers

A minicomputer is sufficiently small and inexpensive that its use can be limited to one laboratory or even to a single instrument. A minicomputer will have at least 4K of memory and often considerably more. Usually, it can be used on a *time-shared* basis. That is, its capabilities are such that it can be

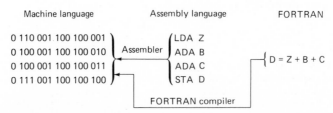

FIGURE 3-41 Relationships among machine, assembly, and a high-level language. (From S. Perone, *J. Chromatog, Sci.*, **7**, 715 (1969). With permission.)

used in more than one way and by more than one user at the same time without interferences. A single minicomputer is frequently used to control several instruments in one laboratory.

Minicomputers can usually be programmed in one or more high-level languages. Often, however, interfacing them with instrument signals may require the user to construct subroutines in assembly language.

The central processing unit in a minicomputer generally ranges in cost from $1000 to $10,000. The addition of storage, input and output devices, and other peripherals usually doubles these costs at a minimum and may multiply them by a factor as great as five to ten.

Microcomputers

The heart of a microcomputer is the *microprocessor*, a device that first appeared on the market in about 1973; the applications of microprocessors have grown exponentially since that time.[9] A microprocessor is a complete central processing unit, which has been miniaturized and formed on a single integrated-circuit chip having an area of a few square millimeters. Combination of a microprocessor with a memory and an input-output device leads to a remarkably inexpensive microcomputer, which can be conveniently employed in conjunction with a single instrument; such a computer is said to be *dedicated*. Most new analytical instruments are being sold with one or more programmed microprocessors as an integral part of the equipment. With these, no programming by the user is required to obtain automated instrument control.

Microcomputer memories tend to be smaller than those employed in minicomputers. Memory increments of 256 words are available; typically, eight of these are employed to provide a capacity of 2048 words. Only one or two of the memory increments is required for the control of many instruments. Word lengths for microcomputers are commonly eight bits.

Microcomputers, quite adequate for the control of a single instrument and with data-acquisition speeds of 1000 to 3000 words per second, are available for $1000 or less.

In more complex instruments, one or more microcomputers may be controlled by a minicomputer. The *slave* microcomputer then performs one or more control functions at the command of the minicomputer.

Applications of Computers

Computer interactions with analytical instruments are of two types, *passive* and *active*.[10] In passive applications, the computer does not participate in the control of the experiment but is used only for data handling, processing, storing, file searching, or display. In an active interaction, the computer is significantly involved in the control of the experiment. Instruments with computer control are said to be *automated*.

Passive Applications. Data processing by a computer may involve relatively simple mathematical operations such as calculation of concentrations, data averaging, least-square analysis, statistical analysis, and integration to obtain peak areas; as we have seen, several

[9] For a summary of the construction and applications of microprocessors and minicomputers, see: R. E. Dessy, P. Janse-Van Vuuren, and J. A. Titus, *Anal. Chem.*, **46** (11), 917A (1974); **46** (12), 1055A (1974); B. Soucek, *Microprocessors and Microcomputers.* New York: Wiley, 1976; and *An Introduction to Microcomputers*, Adam Osborne and Associates, P.O. Box 2036, Berkeley, CA 94710, 1977, vol. 0, 1, and 2.

[10] This classification has been suggested by Perone in his summary of computer applications. See: S. P. Perone and D. O. Jones, *Digital Computer in Scientific Instrumentation.* New York: McGraw-Hill, 1973, Chapter 12.

of these operations can also be performed with operational amplifiers. More complex calculations may involve solution of several simultaneous equations, curve fitting, and Fourier transformations.

Data storage is another important passive function of computers. For example, a powerful tool for the analysis of complex mixtures is obtained by linking gas-liquid chromatography (GLC) with mass spectrometry (MS). The former is capable of separating mixtures on the basis of the time required for the individual components to appear at the end of suitably packed columns. Mass spectrometry permits identification of each component according to the mass of the fragments formed when the compound is bombarded with a beam of electrons. GLC/MS equipment may produce as many as 100 spectra in less than an hour, with each spectrum being made up of tens to hundreds of peaks. Conversion of these data to an interpretable form is time consuming; thus, the data must be stored in digital form for subsequent printing.

Identification of a species from its mass spectrum involves a search of files of spectra for pure compounds until a match is found; this process is also time consuming, but can be accomplished quickly by using a computer. Here, a magnetic disc or magnetic tape file storage is required. For example, programs have been described which permit the search of several thousands of spectra in a minute or less. Such a search will frequently produce several possible compounds. Further comparison of spectra by the scientist usually makes identification possible.

Another important passive application utilizes the high-speed data fetching and correlating capabilities of the computer. Thus, for example, the computer can be called upon to display on a cathode-ray screen the spectrum of any one of the components after it has exited from a gas chromatographic column.

Active Applications. In active applications, the computer exercises at least some and sometimes all of the control of an instrument during an analysis. Most modern instruments contain one or more microcomputers that perform such functions. Examples would include control of the slit width and wavelength setting of a monochromator, the temperature of a chromatographic column, the potential applied to an electrode, the rate of addition of a reagent, and the time at which the integration of a peak is to begin. Referring again to the GLC/MS instrument considered in the last section, a computer is often used to initiate collection of mass spectral data each time a compound is sensed at the end of the chromatographic column.

Computer control can be relatively simple, as in the examples just cited, or more complex. For example, the determination of the concentration of elements by atomic emission involves the measurement of the heights of emission peaks that occur at wavelengths characteristic for each element. Here, the computer can cause a monochromator to sweep a range of wavelengths until a peak is located, and can then seek the exact wavelength at which the maximum output signal is obtained. Measurements are made at this point until an average is obtained that will give a suitable signal-to-noise ratio. The computer then causes the instrument to repeat this operation for each peak of interest in the spectrum. Finally, the computer processes the data and prints out the concentration of the elements present.

Because of its great speed, a computer can often control variables more efficiently than can a human operator. Furthermore, with some experiments, a computer can be programmed to alter the way in which the measurement is being made, according to the nature of the initial data. Here, a feedback loop is employed in which the signal output is fed back through the computer and serves to control and optimize the way in which additional measurements are made.

PROBLEMS

1. Design a circuit whose output V_o is given by

$$-V_o = 2V_1 + 4V_2 + 1V_3 + 3V_4$$

2. Design a circuit for calculating an average value for three voltage inputs multiplied by 100.

3. Design a circuit for performing the following calculation:

$$y = \tfrac{1}{5}(2x_1 + 3x_2)$$

4. For the following circuit:

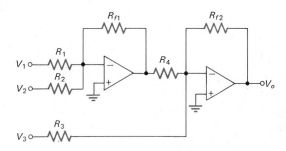

(a) write an expression giving the output voltage in terms of the three input voltages and the various resistances.

(b) indicate the mathematical operation performed by the circuit when $R_1 = R_{f1} = 100 \text{ k}\Omega$; $R_4 = R_{f2} = 50 \text{ k}\Omega$; $R_2 = 25 \text{ k}\Omega$; $R_3 = 10 \text{ k}\Omega$.

5. Show the algebraic relationship between the voltage input and output for the following circuit:

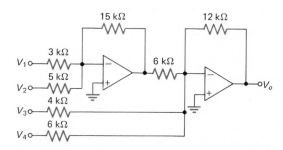

6. What would be the output voltage of the circuit shown in Figure 3-26a after 30 s, if the input voltage is 25 mV, the resistor is 200 kΩ, and the capacitor is 0.26 μF?

7. Derive an expression for the output potential of the following circuit:

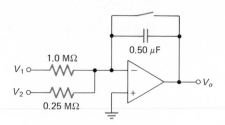

8. Show that when the four resistances are equal, the following circuit becomes a subtracting circuit.

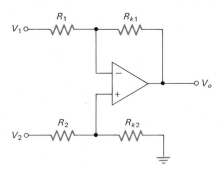

9. Derive a relationship between V_o, V_1, and V_2 for the circuit shown in Problem 8 when $R_{k1} = R_{k2}$ and $R_1 = R_2$.

10. Derive a relationship between V_o and V_i for the following circuit:

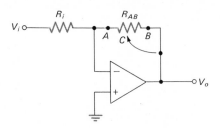

11. Derive a relationship between V_o and V_i for the following circuit:

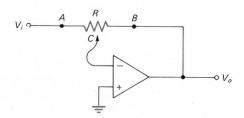

12. The linear slide wire AB has a length of 100 cm. Where along its length should contact C be placed in order to provide a potential of exactly 2.50 V?

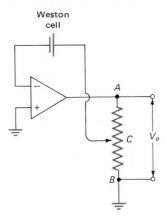

13. Design a circuit that will produce the following output:

$$v_o = 2 \int v_1 \, dt + 3 \int v_2 \, dt$$

14. Design a circuit that will produce the following output:

$$v_o = 3 \int v_1 \, dt - 4(v_2 + v_3)$$

15. The following data were obtained for repetitive weighings of a 1.254-g object with a top-loading balance. Calculate the signal-to-noise ratio for the measurements, assuming the noise to be random.

1.253 g	1.256 g
1.256	1.252
1.257	1.254
1.254	1.255

16. How many measurements would have to be averaged to increase S/N in Problem 15 to 1000?

17. The following data were obtained for a current measurement on a noisy system. What is the signal-to-noise ratio, assuming that the noise is random?

1.34 μA	1.10 μA
1.76	1.63
1.21	1.77
1.35	1.19

18. How many measurements would have to be averaged to obtain a signal-to-noise ratio of 10 for the mean in Problem 16; to obtain a ratio of 100?

19. The resistance of a dc circuit is 1500 Ω. What is the maximum resistance that a current-measuring device can have if the current in this circuit is to be measured with a relative error of less than 2%? The expected current range is 0 to 20 μA.

20. Devise an operational amplifier circuit employing a 0 to 10 mV meter that will meet the requirements specified in Problem 20.

21. The resistance of a dc circuit is 200 Ω. What is the maximum resistance that a current-measuring device can have if the current in this circuit is to be measured with a relative error of less than 5%? The expected current range is 0 to 50 μA.

22. Devise an operational amplifier circuit employing a 0 to 10 mV meter that will meet the requirements specified in Problem 21.

4

ELECTROMAGNETIC RADIATION AND ITS INTERACTIONS WITH MATTER

This chapter reviews several topics concerned with electromagnetic radiation, including the fundamental properties of this radiation and the mechanism of its interactions with matter. The material will serve as a general introduction to the various instrumental methods discussed in Chapters 5 through 14.[1]

PROPERTIES OF ELECTROMAGNETIC RADIATION

Electromagnetic radiation is a type of energy that is transmitted through space at enormous velocities. It takes numerous forms, the most easily recognizable being light and radiant heat. Less obvious manifestations include X-ray, ultraviolet, microwave, and radio radiations.

Many of the properties of electromagnetic radiation are conveniently described by means of a classical wave model which employs such parameters as wavelength, frequency, velocity, and amplitude. In contrast to other wave phenomena, such as sound, electromagnetic radiation requires no supporting medium for its transmission; thus, it readily passes through a vacuum.

The wave model fails to account for phenomena associated with the absorption or emission of radiant energy; for these processes, it is necessary to view electromagnetic radiation as a stream of discrete particles of energy called *photons*. The energy of a photon is proportional to the frequency of the radiation. These dual views of radiation as particles and waves are not mutually exclusive. Indeed, the duality is found to apply to the behavior of streams of electrons or other elementary particles as well and is readily rationalized by wave mechanics.

Wave Properties

For many purposes, electromagnetic radiation is conveniently treated as an oscillating electrical force field in space; associated with the electrical field, and at right angles to it, is a magnetic force field.

The electrical and magnetic fields associated with radiation are vector quantities; at any instant, they can be represented by an arrow whose length is proportional to the magnitude of the force and whose direction is parallel to that of the force. A graphical representation of a beam of radiation can be obtained by plotting one of these vector quantities as a function of time as the radiation passes a fixed point in space. Alternatively, the vector can be plotted as a function of distance, with time held constant.

In Figure 4-1, the electrical vector serves as the ordinate; a plot of the magnetic vector would be identical in every regard except that the ordinate would be rotated 90 deg around

[1] For a further review of these topics, see: E. J. Bair, *Introduction to Chemical Instrumentation*. New York: McGraw-Hill, 1962, Chapters 1–2; E. J. Bowen, *The Chemical Aspects of Light*, 2d ed. Oxford: The Clarendon Press, 1946; E. J. Meehan, *Treatise on Analytical Chemistry*, eds. I. M. Kolthoff and P. J. Elving, Part I, vol. 5, New York: Wiley, 1964, Chapter 53; and F. Grum in *Physical Methods of Chemistry*, eds. A. Weissberger and B. W. Rossiter. New York: Wiley-Interscience, 1972, Vol. I, Part III B, pp. 214–305.

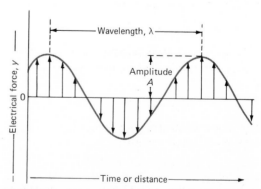

FIGURE 4-1 Representation of a beam of monochromatic, plane-polarized radiation. The arrows represent the electrical vector.

the zero axis. Normally, only the electrical vector is employed because it is the electrical force that is responsible for such phenomena as transmission, reflection, refraction, and absorption of radiation.

Figure 4-1 is a plot of the electrical vector of *monochromatic plane-polarized* radiation, the simplest type. The first term indicates that the radiation has a single frequency; the second implies that it oscillates in but a single plane in space.

Wave Parameters. The time required for the passage of successive maxima through a fixed point in space is called the *period p* of the radiation. The *frequency v* is the number of oscillations of the field that occurs per second[2] and is equal to $1/p$. It is important to realize that *the frequency is determined by the source and remains invariant* regardless of the media traversed by the radiation. In contrast, the *velocity* of propagation v_i, the rate at which the wave front moves through a medium, is *dependent* upon both the medium and the frequency; the subscript i is employed to indicate this frequency dependence. Another parameter of interest is the *wavelength* λ_i, which is the linear distance between successive maxima or minima of a wave.[3] Multiplication of the frequency in cycles per second by the wavelength in centimeters per cycle gives the velocity of propagation in centimeters per second; that is,

$$v_i = v\lambda_i \qquad (4-1)$$

In a vacuum, the velocity of radiation becomes independent of frequency and is at its maximum. This velocity, given the symbol c, has been accurately determined to be 2.99792×10^{10} cm/s. Thus, for a vacuum,

$$c = v\lambda \cong 3 \times 10^{10} \text{ cm/s} \qquad (4-2)$$

In any other medium, the rate of propagation is less because of interactions between the electromagnetic field of the radiation and the bound electrons in the atoms or molecules making up the medium. Since the radiant frequency is invariant and fixed by the source, the *wavelength must decrease* as radiation passes from a vacuum to a medium containing matter (Equation 4-1). This effect is illustrated in Figure 4-2.

It should be noted that the velocity of radiation in air differs only slightly from c (about 0.03% less); thus, Equation 4-2 is usually applicable to air as well as to a vacuum.

The *wavenumber* σ is defined as the number of waves per centimeter and is yet another way of describing electromagnetic radiation. When the wavelength *in vacuo* is expressed in centimeters, the wavenumber is equal to $1/\lambda$.

Radiant Power or Intensity. The *power P* of radiation is the energy of the beam that reaches a given area per second; the *intensity I* is the power per unit solid angle. These quantities are related to the square of the amplitude A (see Figure 4-1). Although it is not strictly correct to do so, power and intensity are often used synonymously.

Mathematical Description of a Wave. With time as a variable, the wave in Figure 4-1 can be described by the equation (see Equation 2-18, p. 12)

$$y = A \sin(\omega t + \phi) \qquad (4-3)$$

where y is the electric force, A is the *amplitude* or maximum value for y, t is time, and ϕ is the phase angle, a term which has been defined earlier (p. 12). The angular velocity of the vector, ω, is related to the frequency of the radiation v by the equation

$$\omega = 2\pi v$$

---

[2] The common unit of frequency is the *hertz* Hz, which is equal to one cycle per second.

[3] The units commonly used for describing wavelength differ considerably in the various spectral regions. For example, the ångström unit, Å (10^{-10} m), is convenient for X-ray and short ultraviolet radiation; the nanometer, nm (10^{-9} m), is employed with visible and ultraviolet radiation; the micrometer, μm (10^{-6} m), is useful for the infrared region.

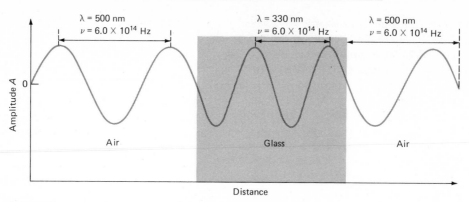

FIGURE 4-2   Effect of change of medium on a monochromatic beam of radiation.

Substitution of this relationship into Equation 4-3 yields

$$y = A \sin(2\pi vt + \phi) \qquad (4\text{-}4)$$

Superposition of Waves. *The principle of superposition* states that when two or more waves traverse the same space, a displacement occurs which is the sum of the displacements caused by the individual waves. This principle applies to electromagnetic waves, where the displacements involve an electrical force field, as well as to several other types of waves, where atoms or molecules are displaced. For example, when n electromagnetic waves of the same frequency but differing amplitudes and phase angles pass some point in space simultaneously, the principle of superposition and Equation 4-4 permits us to write

$$y = A_1 \sin(2\pi vt + \phi_1)$$
$$+ A_2 \sin(2\pi vt + \phi_2) + \cdots \quad (4\text{-}5)$$
$$+ A_n \sin(2\pi vt + \phi_n)$$

where y is the resultant force field.

The solid line in Figure 4-3a shows the application of Equation 4-5 to two waves of identical frequency but somewhat different amplitude and phase angle. Note that the resultant is a sine wave with the same frequency but different amplitude from the component

waves. Figure 4-3b differs from 4-3a in that the difference in phase angle is greater; here, the resultant amplitude is smaller than those of the component waves. Clearly, a maximum amplitude for the resultant will occur when the two waves are completely in phase; this situation prevails whenever the phase difference between waves $(\phi_1 - \phi_2)$ is 0 deg, 360 deg, or an integer multiple of 360 deg. Under these circumstances, maximum *constructive interference* is said to occur. A maximum *destructive interference* occurs when $(\phi_1 - \phi_2)$ is equal to 180 deg or 180 deg plus an integer multiple of 360 deg. The property of interference plays an important role in many instrumental methods based on electromagnetic radiation.

Figure 4-3c depicts the superposition of two waves with the same amplitude but different frequency. The resulting wave is no longer sinusoidal. It does, however, exhibit a periodicity. Thus, the wave from 0 to B in the figure is identical to that from B to C.

An important aspect of superposition is that the process can be reversed by a mathematical operation called a *Fourier transformation*. Jean Fourier, an early French mathematician (1768–1830), demonstrated that any wave motion, regardless of complexity, can be described by a sum of simple sine or cosine terms such as those shown in Equation 4-5. For example, the square wave form often employed in electronics can be represented by an equation having the form

$$y = A\left(\sin 2\pi vt + \tfrac{1}{3}\sin 6\pi vt \right.$$
$$\left. + \tfrac{1}{5}\sin 10\ \pi vt + \cdots + \frac{1}{n}\sin 2n\pi vt\right) \quad (4\text{-}6)$$

A graphical representation of the summation process is shown in Figure 4-4. The solid line in Figure 4-4a is the sum of three sine waves differing in amplitude in the ratio of 5 : 3 : 1 and in frequency in the ratio of 1 : 3 : 5. Note that the resultant is already beginning to approximate the shape of a square wave. As shown by the solid line in Figure 4-4b, the resultant more closely approaches a square wave when nine waves are incorporated.

Mathematically, resolution of a complex wave form into its sine or cosine components is tedious and time consuming; modern computers, however, have made it practical to exploit the power of the Fourier transformation on a routine basis. The application of this technique will be considered in the discussion of several types of spectroscopy.

Diffraction. Figure 4-5 is a schematic representation of *diffraction*, which occurs whenever waves of any type pass by a sharp barrier or through a narrow opening. Diffrac-

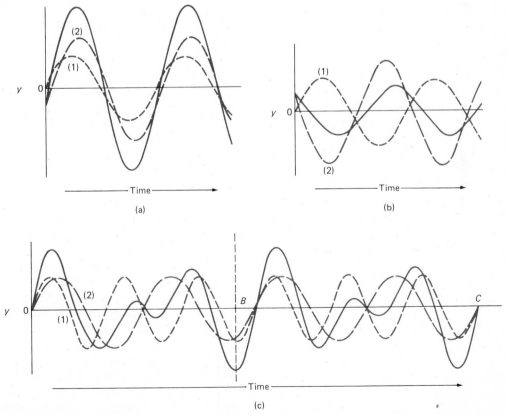

FIGURE 4-3 Superposition of sinusoidal waves: (a) $A_1 < A_2$, $(\phi_1 - \phi_2) = -20°$, $v_1 = v_2$; (b) $A_1 < A_2$, $(\phi_1 - \phi_2) = -200°$, $v_1 = v_2$; (c) $A_1 = A_2$, $\phi_1 = \phi_2$, $v_1 = 1.5\ v_2$. In each instance, the solid curve is the resultant from combination of the two dashed curves.

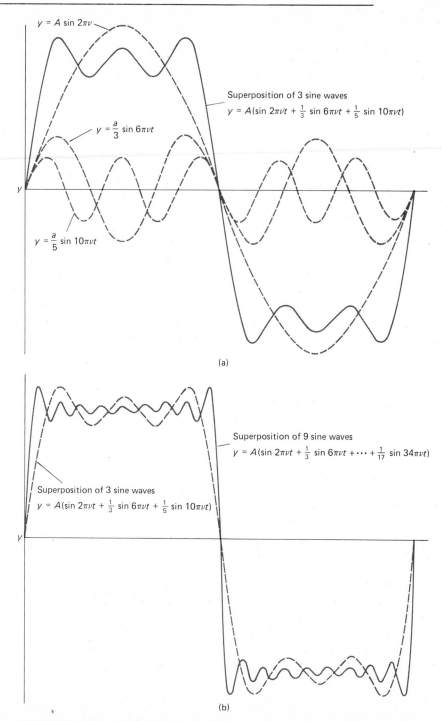

FIGURE 4-4 Superposition of sine waves to form a square wave: (a) combination of three sine waves, and (b) combination of three, as in (a), and nine sine waves.

tion is readily observed by generating waves of constant frequency in a tank of water and observing the wave crests before and after they pass through a rectangular opening or slit. When the slit is wide relative to the wavelength of the motion (Figure 4-5a), diffraction is slight and difficult to detect. On the other hand, when the wavelength and the slit opening are of the same order of magnitude, as in Figure 4-5b, diffraction becomes pronounced. Here, the slit or opening behaves as a new source from which waves radiate in a series of nearly 180-deg arcs. Thus, the direction of the wave front appears to bend as a consequence of passing the two edges of the slit.

Diffraction is a consequence of interference. This relationship is most easily understood by considering an experiment, performed first by Thomas Young in 1800, by which the wave nature of light was unambiguously demonstrated. As shown in Figure 4-6, a parallel beam of radiation is allowed to pass through a narrow slit A (or in Young's experiment, a pinhole) whereupon it is diffracted and illuminates more or less equally two closely spaced slits or pinholes B and C; the radiation emerging from these slits is then observed on the screen lying in a plane XY. If the radiation is monochromatic, a series of dark and light images perpendicular to the plane of the page is observed.

Figure 4-6b shows the intensities of the various bands reaching the screen. If, as in this diagram, the slit widths approach the wavelength of radiation, the band intensities decrease only gradually with increasing distances from the central band. With wider slits, the decrease is much more pronounced.

The existence of the central band E, which lies in the shadow of the opaque material separating the two slits, is readily explained by noting that the paths from B to E and C to E are identical. Thus, constructive interference of the diffracted rays from the two slits occurs, and an intense band is observed. With the aid of Figure 4-6c, the conditions for maximum constructive interference, which result

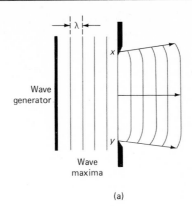

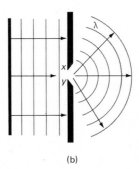

FIGURE 4-5   Propagation of waves through a slit. (a) $xy \gg \lambda$; (b) $xy \cong \lambda$.

in the other light bands, are readily derived. The angle of diffraction θ is the angle from the normal formed by the dotted line extending from a point, O, halfway between the slits, to the point of maximum intensity, D. The solid lines BD and CD represent the light paths from the slits B and C to this point. Ordinarily, the distance $\overline{OE}$ is enormous compared to the distance between the slits $\overline{BC}$; as a consequence, the lines BD, OD, and CD are, for all practical purposes, parallel. Line BF is perpendicular to CD and forms the triangle BCF, which is, to a close approximation, similar to DOE; consequently, the angle CBF is equal to the angle of diffraction θ. We may then write

$$\overline{CF} = \overline{BC} \sin \theta$$

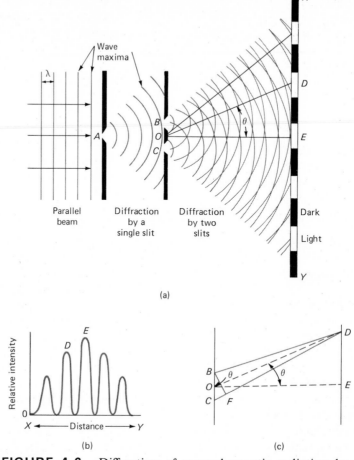

(a)

(b)   (c)

FIGURE 4-6   Diffraction of monochromatic radiation by slits.

Because $\overline{BC}$ is so very small compared to $\overline{OE}$, $\overline{FD}$ closely approximates $\overline{BD}$, and the distance $\overline{CF}$ is a good measure of the difference in path lengths of beams BD and CD. For the two beams to be in phase at D, it is necessary that $\overline{CF}$ correspond to the wavelength of the radiation; that is,

$$\lambda = \overline{CF} = \overline{BC} \sin \theta$$

Reinforcement would also occur when the additional path length corresponds to 2λ, 3λ, and so forth. Thus, a more general expression

for the light bands surrounding the central band is

$$\mathbf{n}\lambda = \overline{BC} \sin \theta \qquad (4\text{-}7)$$

where **n** is an integer called the *order* of interference.

The linear displacement $\overline{DE}$ of the diffracted beam along the plane of the screen is a function of the distance $\overline{OE}$ between the screen and the plane of the slits, as well as the spacing between the slits $\overline{BC}$; that is,

$$\overline{DE} = \overline{OD} \sin \theta$$

Substitution into Equation 4-7 gives

$$\mathbf{n}\lambda = \frac{\overline{BC}\,\overline{DE}}{\overline{OD}} \cong \frac{\overline{BC}\,\overline{DE}}{\overline{OE}} \qquad (4\text{-}8)$$

Equation 4-8 permits the calculation of the wavelength from the three measurable quantities.

EXAMPLE

Suppose that the screen in Figure 4-6 is 2.00 m from the plane of the slits and that the slit spacing is 0.300 mm. What is the wavelength of radiation if the fourth band is located 15.4 mm from the central band?

Substituting into Equation 4-8 gives

$$4\lambda = \frac{0.300 \times 15.4}{2 \times 1000}\ \text{mm}$$

$$\lambda = 5.78 \times 10^{-4}\ \text{mm} \quad \text{or} \quad 578\ \text{nm}$$

Coherent Radiation. In order to produce a diffraction pattern such as that shown in Figure 4-6a, it is necessary that the electromagnetic waves traveling from slits B and C to any given point on the screen (such as D or E) have sharply defined phase differences that remain entirely constant with time; that is, the radiation from slits B and C must be *coherent*. The conditions for coherence are: (1) the two sources of radiation must have identical frequency and wavelength (or sets of frequencies and wavelengths); and (2) the phase relationships between the two beams must remain constant with time. The necessity for these requirements can be demonstrated by illuminating the two slits in Figure 4-6a with individual filament lamps. Under these circumstances, the well-defined light and dark patterns disappear and are replaced by a more or less uniform illumination of the screen. This behavior is a consequence of the *incoherent* character of filament sources (many other sources of electromagnetic radiation are incoherent as well).

In incoherent sources, light is emitted by individual atoms or molecules, and the resulting beam is the summation of countless individual events, each of which lasts on the order of 10^{-8} second. Thus, a beam of radiation from this type of source is not continuous, as is the case in microwave or laser radiation; instead, it is composed of a series of *wave trains* that are a few meters in length at most. Because the processes that produce a train are random, the phase differences among the trains must also be variable. A wave train from slit B may arrive at a point on the screen in phase with a wave train from C, and constructive interference will occur; an instant later, the trains may be totally out of phase at the same point, and destructive interference will occur. Thus, the radiation at all points on the screen is governed by the random phase variations among the wave trains; uniform illumination, which represents an average for the trains, is the result.

Sources do exist which produce electromagnetic radiation in the form of trains with essentially infinite length and constant frequency. Examples include radio frequency oscillators, microwave sources, optical lasers, and various mechanical sources such as a two-pronged vibrating tapper in a riffle tank. When two of these are employed as sources for the experiment shown in Figure 4-6, a regular diffraction pattern is observed.

Diffraction patterns can be obtained from random sources, such as a tungsten filament, provided that an arrangement similar to that shown in Figure 4-6a is employed. Here, the very narrow slit A assures that the radiation reaching B and C emanates from the same small region of the source. Under these circumstances, the various wave trains exiting from slits B and C have a constant set of frequencies and phase relationships to one another and are thus coherent. If the slit at A is widened so that a larger part of the source is sampled, the diffraction pattern becomes less pronounced because the two beams are only partially coherent. If slit A is made sufficiently wide, the incoherence may become great enough to produce only a constant illumination across the screen.

Particle Properties of Radiation

Energy of Electromagnetic Radiation. An understanding of certain interactions between radiation and matter requires that the radiation be treated as packets of energy called *photons* or *quanta*. The energy of a photon depends upon the frequency of the radiation, and is given by

$$E = h\nu \qquad (4\text{-}9)$$

where h is Planck's constant (6.63×10^{-27} erg sec). In terms of wavelength,

$$E = \frac{hc}{\lambda} \qquad (4\text{-}10)$$

Thus, an X-ray photon ($\lambda \sim 10^{-8}$ cm) has approximately 10,000 times the energy of a photon emitted by a hot tungsten wire ($\lambda \sim 10^{-4}$ cm).

The Photoelectric Effect. The need for a particle model to describe the behavior of electromagnetic radiation can be seen by consideration of the *photoelectric effect.* When sufficiently energetic radiation impinges on a metallic surface, electrons are emitted. The energy of the emitted electrons is found to be related to the frequency of the incident radiation by the equation

$$E = h\nu - w \qquad (4\text{-}11)$$

where w, the *work function*, is the work required to remove the electron from the metal to a vacuum. While E is directly dependent upon the frequency, it is found to be totally independent of the intensity of the beam; an increase in intensity merely causes an increase in the *number of electrons emitted* with energy E.

Calculations indicate that no single electron could acquire sufficient energy for ejection if the radiation striking the metal were uniformly distributed over the surface; nor could any electron accumulate enough energy for its removal in a reasonable length of time. Thus, it is necessary to assume that the energy is not uniformly distributed over the beam front, but rather is concentrated at certain points or in particles of energy.

The work w required to cause emission of electrons is characteristic of the metal. The alkali metals possess low work functions and emit electrons when exposed to radiation in the visible region. Metals to the right of the alkali metals in the periodic chart have larger work functions and require the more energetic ultraviolet radiation to exhibit the photoelectric effect. As we shall note in later chapters, the photoelectric effect has great practical importance in the detection of radiation using phototubes.

Energy Units. The energy of a photon that is absorbed or emitted by a sample of matter can be related to an energy difference between two molecular or atomic states, or to the frequency of a molecular motion of a constituent of the matter. For this reason, it is often convenient to describe radiation in energy units, or alternatively in terms of frequency (Hz) or wavenumber (cm^{-1}), which are directly proportional to energy. On the other hand, the experimental measurement of radiation is most often expressed in terms of the reciprocally related wavelength units such as centimeters, micrometers, or nanometers. The chemist must become adept at interconversion of the various units employed in spectroscopy. Conversion factors commonly encountered for transformations are found inside of the front cover of this text.

The *electron volt* (eV) is the unit ordinarily employed to describe the more energetic X-ray or ultraviolet radiation. The electron volt is the energy acquired by an electron in falling through a potential of one volt. Radiant energy can also be expressed in terms of energy per mole of photons (that is, Avogadro's number of photons). For this purpose, units of kcal/mol or cal/mol are convenient.

The Electromagnetic Spectrum

The electromagnetic spectrum covers an immense range of wavelengths or energies. Figure 4-7 depicts qualitatively its major divi-

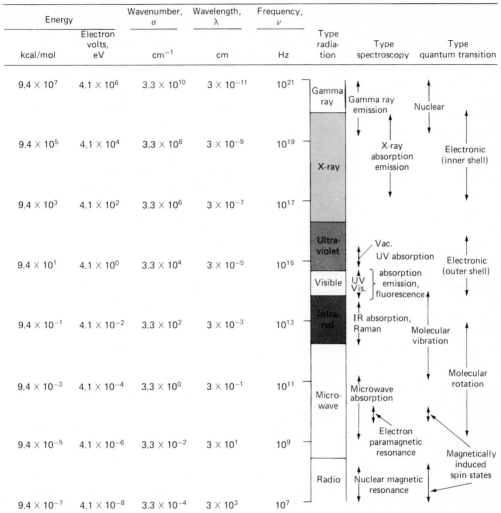

FIGURE 4-7 Spectral properties, applications, and interactions of electromagnetic radiation.

sions. A logarithmic scale has been employed in this representation; note that the portion to which the human eye is perceptive is very small. Such diverse radiations as gamma rays and radio waves are also electromagnetic radiations, differing from visible light only in the matter of frequency, and hence energy.

Figure 4-7 shows the regions of the spectrum that are useful for analytical purposes and the names of the spectroscopic methods associated with these applications. The molecular or atomic transitions responsible for absorption or emission of radiation in each region are also indicated.

THE INTERACTION OF RADIATION WITH MATTER

As radiation passes from a vacuum through the surface of a portion of matter, the electrical vector of the radiation interacts with the atoms and molecules of the medium. The nature of the interaction depends upon the properties of the matter and may lead to transmission, absorption, or scattering of the radiation.

Transmission of Radiation

It is observed experimentally that the rate at which radiation is propagated through a transparent substance is less than its velocity in a vacuum; furthermore, the rate depends upon the kinds and concentrations of atoms, ions, or molecules in the medium. It follows from these observations that the radiation must interact in some way with the matter. Because a frequency change is not observed, however, the interaction *cannot* involve a permanent energy transfer.

The *refractive index* of a medium is a measure of its interaction with radiation and is defined by

$$n_i = \frac{c}{v_i} \qquad (4\text{-}12)$$

where n_i is the refractive index at a specified frequency i, v_i is the velocity of the radiation in the medium, and c is its velocity *in vacuo*.

The interaction involved in the transmission process can be ascribed to the alternating electrical field of the radiation, which causes the bound electrons of the particles contained in the medium to oscillate with respect to their heavy (and essentially fixed) nuclei; periodic polarization of the particles thus results. Provided the radiation is not absorbed, the energy required for polarization is only momentarily retained (10^{-14} to 10^{-15} s) by the species and is reemitted without alteration as the substance returns to its original state. Since there is no net energy change in this process, the frequency of the emitted radiation is unchanged, but the rate of its propagation has been slowed by the time required for retention and reemission to occur. Thus, transmission through a medium can be viewed as a stepwise process involving oscil-

lating atoms, ions, or molecules as intermediates.

One would expect the radiation from each polarized particle in a medium to be emitted in all directions. If the particles are small, however, it can be shown that destructive interference prevents the propagation of significant amounts in any direction other than that of the original light path. On the other hand, if the medium contains large particles (such as polymer molecules or colloidal particles), this destructive effect is incomplete and a portion of the beam is *scattered* as a consequence of the interaction step. Scattering is considered in a later section of this chapter.

Dispersion. As we have noted, the velocity of radiation in matter is frequency-dependent; since c in Equation 4-12 is independent of this parameter, the refractive index of a substance must also change with frequency. The variation in refractive index of a substance with frequency or wavelength is called its *dispersion*. The dispersion of a typical substance is shown in Figure 4-8. Clearly, the relationship is complex; generally, however, dispersion plots exhibit two types of regions. In the

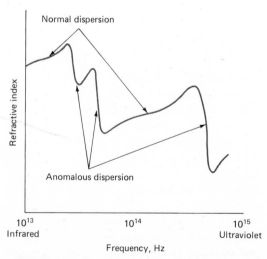

FIGURE 4-8 Typical dispersion curve.

normal dispersion region, there is a gradual increase in refractive index with increasing frequency (or decreasing wavelength). *Anomalous dispersion* regions are those frequency ranges in which a sharp change in refractive index is observed. Anomalous dispersion always occurs at frequencies that correspond to the natural harmonic frequency associated with some part of the molecule, atom, or ion of the substance. At such a frequency, permanent energy transfer from the radiation to the substance occurs and *absorption* of the beam is observed. Absorption is discussed in a later section.

Dispersion curves are important in the choice of materials for the optical components of instruments. A substance that exhibits normal dispersion over the wavelength region of interest is most suitable for the manufacture of lenses, for which a high and relatively constant refractive index is desirable. Chromatic aberration is minimized through the choice of such a material. In contrast, a substance with a refractive index that is not only large but also highly frequency-dependent is selected for the fabrication of prisms. The applicable wavelength region for the prism thus approaches the anomalous dispersion region for the material from which it was fabricated.

Refraction of Radiation. When radiation passes from one medium to another of differing physical density, an abrupt change in direction of the beam is observed as a consequence of differences in the velocity of the radiation in the two media. This *refraction* of a beam is illustrated in Figure 4-9. The extent of refraction is given by the relationship

$$\frac{\sin \theta_1}{\sin \theta_2} = \frac{n_2}{n_1} = \frac{v_1}{v_2} \qquad (4\text{-}13)$$

If the medium M_1 is vacuum, v_1 becomes c and n_1 is unity (Equation 4-12); thus, the refractive index n_2 of M_2 is simply the ratio of the sines of the two angles.

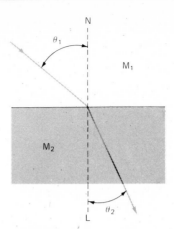

FIGURE 4-9 Refraction of light in passing from a less dense medium M_1 into a more dense medium M_2, where its velocity is lower.

Reflection and Scattering of Radiation

Reflection. When radiation crosses an interface between media with differing refractive indexes, reflection occurs. The fraction reflected becomes larger with increasing differences in refractive index; for a beam traveling normally to the interface, the fraction reflected is given by

$$\frac{I_r}{I_0} = \frac{(n_2 - n_1)^2}{(n_2 + n_1)^2} \qquad (4\text{-}14)$$

where I_0 is the intensity of the incident beam and I_r is the reflected intensity; n_1 and n_2 are the refractive indexes of the two media.

EXAMPLE

Calculate the percent loss of intensity due to reflection of a perpendicular beam of yellow light as it passes through a glass cell containing water. Assume that for yellow radiation the refractive index of glass is 1.5, of water is 1.33, and of air is 1.00.

In passing from air to glass, we find

$$\frac{I_r}{I_0} = \frac{(1.50 - 1.00)^2}{(1.50 + 1.00)^2} = 0.040 \quad \text{or} \quad 4.0\%$$

In passing from glass to water, we find the loss to be

$$\frac{I_r}{I_0} = \frac{(1.50 - 1.33)^2}{(1.50 + 1.33)^2} = 0.004 \quad \text{or} \quad 0.4\%$$

The same losses would occur as the beam passes from the water and through the glass. Thus,

$$\text{total percent loss} = 2(4.0 + 0.4) = 8.8$$

It will become evident in later chapters that losses such as those shown in this example are of considerable significance in various optical instruments.

Reflective losses at a polished glass or quartz surface increase only slightly as the angle of the incident beam increases up to about 60 deg. Beyond this figure, however, the percentage of radiation that is reflected increases rapidly and approaches 100% at 90 deg.

Scattering. As noted earlier, the transmission of radiation in matter can be pictured as a momentary retention of the radiant energy, which causes a brief polarization of the ions, atoms, or molecules present. Polarization is followed by reemission of radiation in all directions as the particles return to their original state. When the particles are small with respect to the wavelength of the radiation, destructive interference removes nearly all of the reemitted radiation except that which travels in the original direction of the beam; the path of the beam appears to be unaltered as a consequence of the interaction. Careful observation, however, reveals that a very small fraction of the radiation is transmitted at all angles from the original path and that the intensity of this *scattered radiation* increases with particle size. With particles of colloidal dimensions, scattering becomes sufficiently intense to be seen by the naked eye (the Tyndall effect).

Scattering by molecules or aggregates of molecules with dimensions significantly smaller than the wavelength of the radiation is called *Rayleigh scattering*; its intensity is readily related to wavelength (an inverse fourth-power effect), the dimensions of the scattering particles, and their polarizability. An everyday manifestation of scattering is the blueness of the sky, which results from the greater scattering of the shorter wavelengths of the visible spectrum.

Scattering by larger particles, as in a colloidal suspension, is much more difficult to treat theoretically; the intensity varies roughly as the inverse square of wavelength.

Measurements of scattered radiation can be used to determine the size and shape of polymer molecules and colloidal particles. The phenomenon is also utilized in *nephelometry*, an analytical method considered in Chapter 10.

Raman Scattering. The Raman effect differs from ordinary scattering in that part of the scattered radiation suffers quantized frequency changes. These changes are the result of vibrational energy level transitions occurring in the molecule as a consequence of the polarization process. Raman spectroscopy is discussed in Chapter 9.

Polarization of Radiation

Plane Polarization. Ordinary radiation can be visualized as a bundle of electromagnetic waves in which the amplitude of vibrations is equally distributed among a series of planes centered along the path of the beam. Viewed end-on, the electrical vectors would then appear as shown in Figure 4-10a. The vector in any one plane, say XY, can be resolved into two mutually perpendicular components AB and CD as shown in Figure 4-10b. If the two components for each plane are combined, the resultant has the appearance shown in Figure 4-10c. Note that Figure 4-10c has a

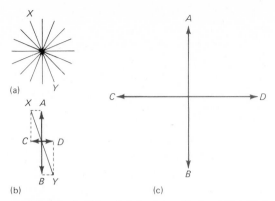

(a)

(b)

(c)

FIGURE 4-10 (a) A few of the electrical vectors of a beam traveling perpendicular to the page. (b) The resolution of a vector in plane XY into two mutually perpendicular components. (c) The resultant when all vectors are resolved (not to scale).

different scale from Figure 4-10a or 4-10b to keep its size within reason.

Removal of one of the two resultant planes of vibration in Figure 4-10c produces a beam that is *plane polarized*. The vibration of the electrical vector of a plane-polarized beam, then, occupies a single plane in space.

Plane-polarized electromagnetic radiation is produced by certain radiant energy sources. For example, the radio waves emanating from an antenna commonly have this characteristic. Presumably, the radiation from a single atom or molecule is also polarized; however, since common light sources contain large numbers of these particles in all orientations, the resultant is a beam that vibrates equally in all directions around the axis of travel.

The absorption of radiation by certain types of matter is dependent upon the plane of polarization of the radiation. For example, when properly oriented to a beam, anisotropic crystals selectively absorb radiations vibrating in one plane but not the other. Thus, a layer of anisotropic crystals absorbs all of the components of, say, CD in Figure 4-10c and transmits nearly completely the

components of AB. A polarizing sheet, regardless of orientation, removes approximately half of the radiation from an unpolarized beam, and transmits the other half as a plane-polarized beam. The plane of polarization of the transmitted beam is dependent upon the orientation of the sheet with respect to the incident beam. When two polarizing sheets, oriented at 90 deg to one another, are placed perpendicular to the beam path, essentially no radiation is transmitted. Rotation of one results in a continuous increase in transmission until a maximum is reached when molecules of the two sheets have the same alignment.

The way in which radiation is reflected, scattered, transmitted, and refracted by certain substances is also dependent upon direction of polarization. As a consequence, a group of analytical methods have been developed whose selectivity is based upon the interaction of these substances with light of a particular polarization. These methods are considered in detail in Chapter 13.

Absorption of Radiation

When radiation passes through a transparent layer of a solid, liquid, or gas, certain frequencies may be selectively removed by the process of *absorption*. Here, electromagnetic energy is transferred to the atoms or molecules constituting the sample; as a result, these particles are promoted from a lower-energy state to higher-energy states, or *excited states*. At room temperature, most substances are in their lowest energy or *ground state*. Absorption then ordinarily involves a transition from the ground state to higher-energy states.

Atoms, molecules, or ions have only a limited number of discrete, quantized energy levels; for absorption of radiation to occur, the energy of the exciting photon must exactly match the energy difference between the ground state and one of the excited states of the absorbing species. Since these energy differences are unique for each species, a study of the frequencies of absorbed radiation provides a means of characterizing the constituents of a sample of matter. For this purpose, a plot of absorbance as a function of wavelength or frequency is experimentally derived (absorbance, a measure of the decrease in radiant power, is defined on p. 149). Typical plots of this kind, called *absorption spectra*, are shown in Figure 4-11.

The general appearance of an absorption spectrum will depend upon the complexity, the physical state, and the environment of the absorbing species. It is convenient to recognize two types of spectra, namely, those associated with atomic absorption and those resulting from molecular absorption.

Atomic Absorption. The passage of polychromatic ultraviolet or visible radiation through a medium consisting of monatomic particles, such as gaseous mercury or sodium, results in the absorption of but a few well-defined frequencies (see Figure 4-11d). The relative simplicity of such spectra is due to the small number of possible energy states for the particles. Excitation can occur only by an electronic process in which one or more of the electrons of the atom is raised to a higher-energy level. Thus, with sodium, excitation of the $3s$ electron to the $3p$ state requires energy corresponding to a wavenumber of 1.697×10^4 cm^{-1}. As a result, sodium vapor exhibits a sharp absorption peak at 589.3 nm (yellow light). Several other narrow absorption lines, corresponding to other permitted electronic transitions, are also observed (see Figure 4-11d).

Ultraviolet and visible radiation have sufficient energy to cause transitions of the outermost or bonding electrons only. X-ray frequencies, on the other hand, are several orders of magnitude more energetic and are capable of interacting with electrons closest to the nuclei of atoms. Absorption peaks corresponding to electronic transitions of these innermost electrons are thus observed in the X-ray region.

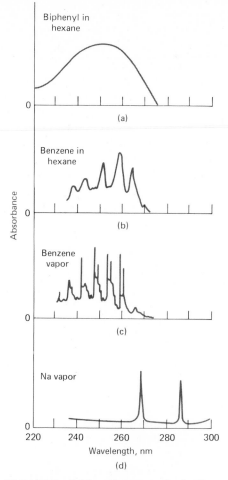

FIGURE 4-11 Some typical ultraviolet absorption spectra.

Regardless of the wavelength region involved, atomic absorption spectra typically consist of a limited number of narrow peaks. Spectra of this type are discussed in connection with X-ray absorption (Chapter 15) and atomic absorption spectroscopy (Chapter 11).

Molecular Absorption. Absorption by polyatomic molecules, particularly in the condensed state, is a considerably more complex process because the number of energy states is greatly enhanced. Here, the total energy of a molecule is given by

$$E = E_{electronic} + E_{vibrational} + E_{rotational} \quad (4\text{-}15)$$

where $E_{electronic}$ describes the electronic energy of the molecule, and $E_{vibrational}$ refers to the energy of the molecule resulting from various atomic vibrations. The third term in Equation 4-15 accounts for the energy associated with the rotation of the molecule around its center of gravity. For each electronic energy state of the molecule, there normally exist several possible vibrational states, and for each of these, in turn, numerous rotational states. As a consequence,

the number of possible energy levels for a molecule is much greater than for an atomic particle.

Figure 4-12 is a graphical representation of the energies associated with a few of the electronic and vibrational states of a molecule. The heavy line labeled E_0 represents the electronic energy of the molecule in its ground state (its state of lowest electronic energy); the lines labeled E_1 and E_2 represent the energies of two excited electronic states. Several vibrational energy levels (e_0, e_1, ..., e_n) are shown for each of these electronic states.

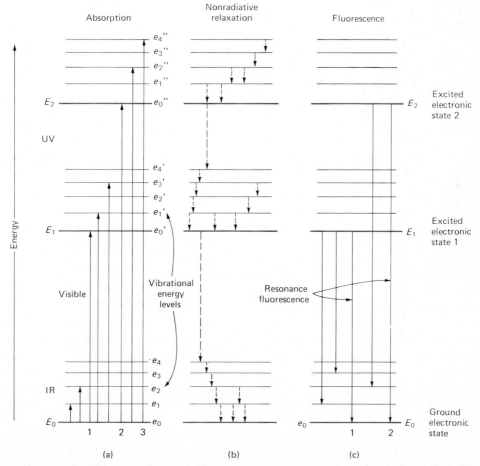

FIGURE 4-12 Partial energy level diagram for a fluorescent organic molecule.

As can be seen in Figure 4-12, the energy difference between the ground state and an electronically excited state is large relative to the energy differences between vibrational levels in a given electronic state (typically, the two differ by a factor of 10 to 100).

Figure 4-12a depicts with arrows some of the transitions which result from absorption of radiation. Visible radiation causes excitation of an electron from E_0 to any of the vibrational levels associated with E_1. The frequencies absorbed would be given by

$$v_n = \frac{1}{h}(E_1 + e'_n - E_0) \qquad (4\text{-}16)$$

Similarly, the frequencies of absorbed ultraviolet radiation would be given by

$$v_n = \frac{1}{h}(E_2 + e''_n - E_0) \qquad (4\text{-}17)$$

Finally, the less energetic near- and mid-infrared radiation can only bring about transition among the vibrational levels of the ground state.

$$v = \frac{1}{h}(e_n - e_0) \qquad (4\text{-}18)$$

Although they are not shown, several rotational energy levels are associated with each vibrational level. The energy difference between these is small relative to the energy difference between vibrational levels; transition to excited rotational states is brought about by radiation in the 500- to 100-cm^{-1} range.

In contrast to atomic absorption spectra, which consist of a series of sharp, well-defined lines, molecular spectra in the ultraviolet and visible regions are ordinarily characterized by absorption bands that often encompass a substantial wavelength range (see Figure 4-11a, b, c). Molecular absorption also involves electronic transitions. As shown by Equations 4-16 and 4-17, however, several closely spaced absorption lines will be associated with each electronic transition, owing to the existence of numerous vibrational states. Furthermore, as

we have mentioned, many rotational energy levels are associated with each vibrational state. As a consequence, the spectrum for a molecule ordinarily consists of a series of closely spaced absorption bands, such as that shown for benzene vapor in Figure 4-11c. Unless a high-resolution instrument is employed, the individual bands may not be detected, and the spectra will appear as smooth curves. Finally, in the condensed state, and in the presence of solvent molecules, the individual bands tend to broaden to give spectra such as those shown in Figure 4-11a and b. Solvent effects are considered in later chapters.

Pure vibrational absorption can be observed in the infrared region, where the energy of radiation is insufficient to cause electronic transitions. Here, spectra exhibit narrow, closely spaced absorption peaks resulting from transitions among the various vibrational quantum levels (see the bottom of Figure 4-12a). Variations in rotational levels may give rise to a series of peaks for each vibrational state. However, rotation is often hindered or prevented in liquid or solid samples; thus, the effects of these small energy differences are not ordinarily detected in such samples.

Pure rotational spectra for gases can be observed in the microwave region.

Absorption Induced by a Magnetic Field. When electrons or the nuclei of certain elements are subjected to a strong magnetic field, additional quantized energy levels are produced as a consequence of magnetic properties of these elementary particles. The difference in energy between the induced states is small, and transitions between the states are brought about only by absorption of long-wavelength (or low-frequency) radiation. With nuclei, radio waves ranging from 10 to 200 MHz are generally employed, while for electrons, microwaves with frequencies of 1000 to 25,000 MHz are absorbed.

Absorption by nuclei or by electrons in magnetic fields is studied by *nuclear magnetic resonance* (NMR) and *electron spin resonance* (ESR) techniques, respectively; these methods are considered in Chapter 14.

Relaxation Processes. Ordinarily, the lifetime of an atom or molecule excited by absorption of radiation is brief because several *relaxation processes* exist which permit its return to the ground state. As shown in Figure 4-12b, nonradiative relaxation involves the loss of energy in a series of small steps, the excitation energy being converted to kinetic energy by collision with other molecules. A minute increase in the temperature of the system results.

As shown in Figure 4-12c, relaxation can also occur by emission of fluorescent radiation. Still other relaxation processes are discussed in Chapters 10 and 14.

EMISSION OF RADIATION

Electromagnetic radiation is often produced when excited particles (ions, atoms, or molecules) return to lower-energy levels or to their ground states. Excitation can be brought about by a variety of means, including bombardment with electrons or other elementary particles, exposure to a high-potential alternating current spark, heat treatment in an arc or a flame, or absorption of electromagnetic radiation.

Radiating particles that are well separated from one another, as in the gaseous state, behave as independent bodies and often produce radiation containing relatively few specific wavelengths. The resulting spectrum is then *discontinuous* and is termed a *line spectrum*. A *continuous spectrum*, on the other hand, is one in which all wavelengths are represented over an appreciable range, or one in which the individual wavelengths are so closely spaced that resolution is not feasible by ordinary means. Continuous spectra result from excitation of: (1) solids or liquids, in which the atoms are so closely packed as to be incapable of independent behavior; or (2) complicated molecules possessing many closely related energy states. Continuous spectra also arise when the energy changes involve particles with unquantized kinetic energies.

Both continuous spectra and line spectra are of importance in analytical chemistry. The former are frequently employed in methods based on the interaction of radiation with matter, such as spectrophotometry. Line spectra, on the other hand, are important because they permit the identification and determination of the emitting species.

Thermal Radiation

When solids are heated to incandescence, the continuous radiation that is emitted is more characteristic of the temperature of the emitting surface than of the material of which that surface is composed. Radiation of this kind (called *black-body radiation*) is produced by the innumerable atomic and molecular oscillations excited in the condensed solid by the thermal energy. Theoretical treatment of black-body radiation leads to the following conclusions: (1) the radiation exhibits a maximum emission at a wavelength that varies inversely with the absolute temperature; (2) the total energy emitted by a black body (per unit of time and area) varies as the fourth power of temperature; and (3) the emissive power at a given temperature varies inversely as the fifth power of wavelength. These relationships are reflected in the behavior of several experimental radiation sources shown in Figure 4-13; the emission from these sources approaches that of the ideal black body. Note that the energy peaks in Figure 4-13 shift to shorter wavelengths with increasing temperature. It is clear that very high temperatures are needed to cause a thermally excited source to emit a substantial fraction of its energy as ultraviolet radiation.

Heated solids are used to produce infrared, visible, and longer-wavelength ultraviolet radiation for analytical instruments.

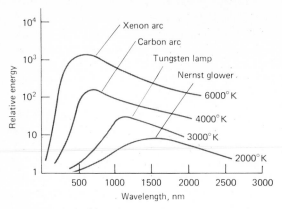

FIGURE 4-13   Black-body radiation curves.

Emission of Gases

Atoms, ions, or molecules in the gaseous state can often be excited, by electrical discharge or by heat, to produce radiation in the ultraviolet and visible regions. The process involves promotion of the outermost electrons of a species to an excited electronic state; radiation is then emitted as the excited species returns to the ground state. Atomic emission spectra consist of a series of discrete lines whose energies correspond to the energy differences between the various electronic states. For molecules, emission spectra are usually more complicated because there can exist several vibrational and rotational states for each possible electronic energy level; instead of a single line for each electronic transition, numerous closely spaced lines form an emission band.

A true continuous spectrum is sometimes produced from the excitation of gaseous molecules. For example, when hydrogen at low pressure is subjected to an electric discharge, an excited molecule is formed that dissociates to give two hydrogen atoms and an ultraviolet photon. The energetics of this process is described by the equation

$$E_{H_2} = \varepsilon_{H1} + \varepsilon_{H2} + h\nu$$

Here, E_{H_2} represents the *quantized* energy of the excited hydrogen atom, while ε_{H1} and ε_{H2} represent the kinetic energies of the atoms. The sum of the latter can vary continuously from zero to E_{H_2}. Thus, the frequency of radiation $h\nu$ can also vary continuously over this range. The consequence is a continuous spectrum from about 400 to 200 nm that is a useful source for absorption spectrophotometry.

Emission of X-ray Radiation

Radiation in the X-ray region is normally generated by the bombardment of a metal target with a stream of high-speed electrons. The electron beam causes the innermost electrons in the atoms of the target material to be raised to higher energy levels or to be ejected entirely. The excited atoms or ions then return to the ground state by various stepwise electronic transitions that are accompanied by the emission of photons, each having an energy $h\nu$. The consequence is the production of an X-ray spectrum consisting of a series of lines characteristic of the target material. This discrete spectrum is superimposed on a continuum of nonquantized radiation given off when some of the high-speed electrons are partially decelerated as they pass through the target material.

Fluorescence and Phosphorescence

Fluorescence and phosphorescence are analytically important emission processes in which atoms or molecules are excited by absorption of a beam of electromagnetic radiation; radiant emission then occurs as the excited species return to the ground state.

Fluorescence occurs much more rapidly than phosphorescence and is generally complete after about 10^{-5} s (or less) from the time of excitation; it is most easily observed at a 90-deg angle to the excitation beam. Phosphorescence emission takes place over periods longer than 10^{-5} s, and may indeed continue for minutes or even hours after irradiation has ceased.

Resonance fluorescence describes a process in which the emitted radiation is identical in frequency to the radiation employed for excitation. The lines labeled 1 and 2 in Figure 4-12a and 4-12c illustrate this type of fluorescence. Here, the species is excited to the energy states E_1 or E_2 by radiation having an energy of $(E_1 - E_0)$ or $(E_2 - E_0)$. After a brief period, emission of radiation of identical energy occurs, as depicted in Figure 4-12c. Resonance fluorescence is most commonly produced by atoms in the gaseous state which do not have vibrational energy states superimposed on electronic energy levels.

Nonresonance fluorescence is brought about by irradiation of molecules in solution or in the gaseous state. As shown in Figure 4-12a, absorption of radiation promotes the molecules into any of the several vibrational levels associated with the two excited electronic levels. The lifetimes of the excited vibrational states are momentary, however ($\sim 10^{-15}$ s), and, as shown in Figure 4-12b, relaxation to the lowest vibrational level (e_0'' and e_0') of a given electronic state takes place as a result of collision with other molecules. More often than not, the energy of the absorbed radiation is greater than that of the emitted radiation. Thus, for the absorption labeled 3 in Figure 4-12a, the absorbed energy would be equal to $(E_2 - E_0 + e_4'' - e_0'')$; the energy of the fluorescent radiation, on the other hand, would again be given by $(E_2 - E_0)$; the emitted radiation would necessarily have a lower frequency or greater wavelength than the radiation that excited the fluorescence. (This shift in wavelength to lower frequencies is sometimes called the *Stokes shift*.) Clearly, both resonance and nonresonance radiation can accompany fluorescence of molecules; the latter tends to predominate, however, because of the much larger number of vibrationally excited states.

Phosphorescence occurs when an excited molecule relaxes to a metastable excited electronic state which has an average lifetime of greater than about 10^{-5} s. The nature of this type of excited state is discussed in Chapter 10.

PROBLEMS

1. Calculate the frequency in hertz and the wavenumber for
   (a) an X-ray beam with a wavelength 4.2 Å.
   (b) an emission line for Cu at 211.0 nm.
   (c) the line at 694.3 nm produced by a ruby laser.
   (d) the output of a CO_2 gas laser at 10.6 μm.
   (e) an infrared absorption peak at 34.1 μm.
   (f) a microwave beam of wavelength 0.250 cm.

2. Calculate the wavelength in centimeters for
   (a) the radio frequency of an airport tower at 118.6 MHz.
   (b) a radio navigation beam having a frequency of 315.3 kHz.
   (c) an NMR signal at 42.6 MHz.
   (d) an EPR signal at 12,000 MHz.

3. Calculate the energy of each of the radiations in Problem 1 in
   (a) ergs/photon, (b) kcal/mol, and (c) eV.

4. Calculate the energy of each of the radiations in Problem 2 in
   (a) ergs/photon, (b) kcal/mol, and (c) eV.

5. Calculate the velocity, wavelength, and frequency of the sodium D line
   ($\lambda = 5890$ Å in a vacuum) in
   (a) acetone vapor ($n_D = 1.0011$).

    (b) carbon bisulfide ($n_D = 1.622$).

    (c) a flint glass ($n_D = 1.65$).

    (d) fused silica ($n_D = 1.44$).

6. Calculate the reflective loss when a 589-nm beam of radiation

    (a) passes through a piece of dense flint glass ($n_D = 1.63$).

    (b) passes through a cell having glass windows ($n_D = 1.46$) and containing an alcohol solution with a refractive index of 1.37.

5

COMPONENTS OF INSTRUMENTS FOR OPTICAL SPECTROSCOPY

The term *spectroscopy* was originally used to describe a branch of science based upon the resolution of visible radiation into its component wavelengths. With the passage of time, the meaning of the term has been expanded to include studies encompassing the entire electromagnetic spectrum (Figure 4-7).

The first spectroscopic instruments were developed for use in the visible region and were thus called *optical instruments*. This term has now been extended to include instruments designed for use in the ultraviolet and infrared regions as well; while not strictly correct, the terminology is nevertheless useful in that it emphasizes the many features that are common to the instruments used for studies in these three important spectral regions.

The purpose of this chapter is to describe the nature and properties of components of instruments employed for optical spectroscopy. The seven chapters that follow will provide details showing how the various components are assembled to produce optical instruments, and how these instruments in turn provide useful information. The instruments and techniques for spectroscopic studies in regions more energetic than the ultraviolet and less energetic than the infrared have characteristics that differ substantially from optical instruments and are considered separately in Chapters 14 and 15.

COMPONENTS AND CONFIGURATIONS OF INSTRUMENTS FOR OPTICAL SPECTROSCOPY

Spectroscopic methods are based upon the phenomena of *emission, absorption, fluorescence,* or *scattering.* While the instruments for each differ somewhat in configuration, their basic components are remarkably similar. Furthermore, the general properties of their components are the same regardless of

whether they are applied to the ultraviolet, visible, or infrared portion of the spectrum.[1]

Spectroscopic instruments contain five components, including: (1) a stable source of radiant energy; (2) a device that permits employment of a restricted wavelength region; (3) a transparent container for holding the sample; (4) a radiation detector or transducer that converts radiant energy to a usable signal (usually electrical); and (5) a signal processor and *readout.* Figure 5-1 shows the arrangement of these components for the four common types of spectroscopic measurements mentioned earlier. As can be seen in the figure, the configuration of components (4) and (5) is the same for each type of instrument.

Emission spectroscopy differs from the other three types in the respect that no external radiation source is required; the sample itself is the emitter. Here, the sample container is an arc, a spark, or a flame, which both holds the sample and causes it to emit characteristic radiation.

Absorption as well as fluorescence and scattering spectroscopy require an external source of radiant energy. In the former, the beam from the source passes through the sample after leaving the wavelength selector. In the latter two, the source induces the sample, held in a container, to emit characteristic fluorescent or scattered radiation, which is measured at an angle (usually 90 deg) with respect to the source.

Figure 5-2 summarizes the characteristics of the first four components shown in Figure 5-1. It is clear that instrument components

---

[1] For a more complete discussion of the components of optical instruments, see: R. P. Bauman, *Absorption Spectroscopy.* New York: Wiley, 1962, Chapters 2 and 3; E. J. Meehan, in *Treatise on Analytical Chemistry,* eds. I. M. Kolthoff and P. J. Elving. New York: Interscience, 1964, Part I, vol. 5, Chapter 55; and *Physical Methods of Chemistry,* eds. A. Weissberger and B. W. Rossiter. New York: Wiley-Interscience, 1972, Part IIIB, vol. 1, Chapters 1–5.

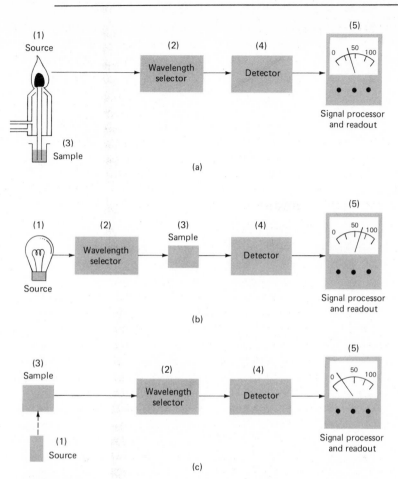

FIGURE 5-1 Components for various types of instruments for optical spectroscopy: (a) emission spectroscopy, (b) absorption spectroscopy, and (c) fluorescence and scattering spectroscopy.

differ in detail, depending upon the wavelength region within which they are to be used. Their design also depends on the primary use of the instrument; that is, whether it is to be employed for qualitative or quantitative analysis and whether it is to be applied to atomic or molecular spectroscopy. Nevertheless, the general function and performance requirements of each type of component are similar, regardless of wavelength region and application.

RADIATION SOURCES

In order to be suitable for spectroscopic studies, a source must generate a beam of radiation with sufficient power for ready detection and measurement. In addition, its output should be stable. Typically, the radiant power of a source varies exponentially with the electrical power supplied to it. Thus, a regulated power supply is often needed to provide the required stability. Alternatively, in some in-

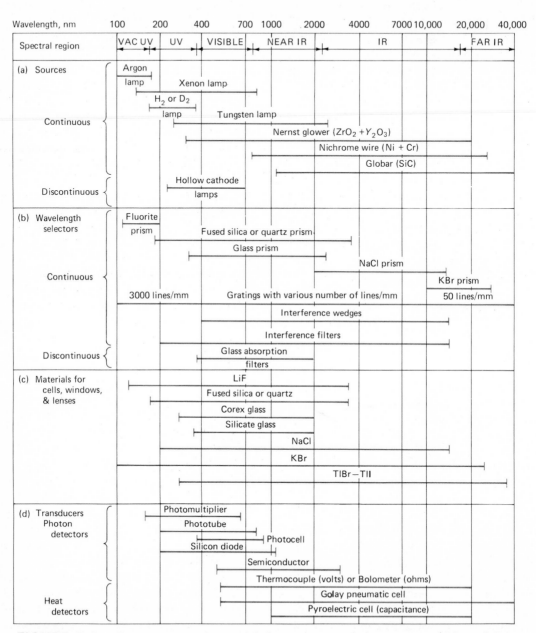

FIGURE 5-2 Components and materials for spectroscopic instruments. (Adapted from a figure by Professor A. R. Armstrong, College of William and Mary. With permission.)

struments, the output of the source is split into a reference beam and a sample beam. The first passes directly to a transducer while the second interacts with the sample and is then focused on a matched transducer. (Some instruments employ a single transducer which is irradiated alternately by the sample and reference beams.) The ratio of the outputs of the two transducers serves as the analytical parameter. The effect of fluctuations in the source output is largely canceled by this means.

Both continuous and line sources are used in optical spectroscopy. The former finds wide application in molecular absorption methods. The latter is employed in fluorescence and atomic absorption spectroscopy.

Figure 5-2a lists the most widely used spectroscopic sources.

Continuous Sources of Ultraviolet, Visible, and Near-Infrared Radiation

Continuous sources provide radiation whose power does not change sharply among adjacent wavelengths.

Hydrogen or Deuterium Lamps. As noted in Chapter 4 (p. 110), a continuous spectrum in the ultraviolet region is conveniently produced by the electrical excitation of hydrogen or deuterium at low pressure.

Two types of hydrogen lamps are encountered. The high-voltage variety employs potentials of 2000 to 6000 V to cause a discharge between aluminum electrodes; water cooling of the lamp is required if high radiation intensities are to be produced. In low-voltage lamps, an arc is formed between a heated, oxide-coated filament and a metal electrode. The heated filament provides electrons to maintain a dc current when a voltage of about 40 V is applied; a regulated power supply (p. 53) is required for constant intensities.

An important feature of hydrogen discharge lamps is the shape of the aperture between the two electrodes, which constricts the discharge to a narrow path. As a consequence, an intense ball of radiation about 1 to 1.5 mm in diameter is produced. Replacement of hydrogen by deuterium results in a somewhat larger light ball.

Both high- and low-voltage lamps produce a continuous spectrum in the region of 160 to 375 nm. Quartz windows must be employed in the tubes, since glass absorbs strongly in this wavelength region.

Tungsten Filament Lamps. The most common source of visible and near-infrared radiation is the tungsten filament lamp. The energy distribution of this source approximates that of a black body and is thus temperature dependent. Figure 4-13 illustrates the behavior of the tungsten filament lamp at 3000°K. In most absorption instruments, the operating filament temperature is about 2870°K; the bulk of the energy is thus emitted in the infrared region. A tungsten filament lamp is useful for the wavelength region between 320 and 2500 nm.

In the visible region, the energy output of a tungsten lamp varies approximately as the fourth power of the operating voltage. As a consequence, close voltage control is required for a stable radiation source. Constant voltage transformers or electronic voltage regulators are often employed for this purpose. As an alternative, the lamp can be operated from a 6-V storage battery, which provides a remarkably stable voltage source if it is maintained in good condition.

Xenon Arc Lamps. This lamp produces intense radiation by the passage of current through an atmosphere of xenon. The spectrum is continuous over the range between about 250 and 600 nm, with the peak intensity occurring at about 500 nm (see Figure 4-13, p. 110). In some instruments, the lamp is operated intermittently by regular discharges from a capacitor; high intensities are obtained.

Continuous Sources of Infrared Radiation

The common infrared source is an inert solid heated electrically to temperatures between 1500 and 2000°K. Continuous radiation approximating that of a black body results (see Figure 4-13). The maximum radiant intensity at these temperatures occurs between 1.7 and 2 μm (6000 to 5000 cm^{-1}). At longer wavelengths, the intensity falls off continuously until it is about 1% of the maximum at 15 μm (667 cm^{-1}). On the short wavelength side, the decrease is much more rapid, and a similar reduction in intensity is observed at about 1 μm (10,000 cm^{-1}).

The Nernst Glower. The Nernst glower is composed of rare earth oxides formed into a cylinder having a diameter of 1 to 2 mm and a length of perhaps 20 mm. Platinum leads are sealed to the ends of the cylinder to permit passage of current. This device has a large negative temperature coefficient of electrical resistance, and it must be heated externally to a dull red heat before a sufficient current passes to maintain the desired temperature. Because the resistance decreases with increasing temperature, the source circuit must be designed to limit the current; otherwise the glower rapidly becomes so hot that it is destroyed.

The Globar Source. A Globar is a silicon carbide rod, usually about 50 mm in length and 5 mm in diameter. It also is electrically heated and has the advantage of a positive coefficient of resistance. On the other hand, water cooling of the electrical contacts is required to prevent arcing. Spectral energies of the Globar and the Nernst glower are comparable except in the region below 5 μm (2000 cm^{-1}), where the Globar provides a significantly greater output.

Incandescent Wire Source. A source of somewhat lower intensity but longer life than the Globar or Nernst glower is a tightly wound spiral of nichrome wire heated by passage of current. A rhodium wire heater sealed in a ceramic cylinder has similar properties as a source.

Line Sources

Sources that emit a few discrete lines find use in atomic absorption spectroscopy, Raman spectroscopy, refractometry, and polarimetry.

Metal Vapor Lamps. Two of the most common line sources are the familiar mercury and sodium vapor lamps. A vapor lamp consists of a transparent envelope containing a gaseous element at low pressure. Excitation of the characteristic line spectrum of the element occurs when a potential is applied across a pair of electrodes fixed in the envelope. Conduction occurs as a consequence of electrons and ions formed by ionization of the metal. Ordinarily, initial heating is required to produce sufficient metal vapor; once started, however, the current is self-sustaining.

The mercury lamp produces a series of lines ranging in wavelength from 254 to 734 nm. With the sodium lamp, the pair of lines at 589.0 and 589.6 nm predominate.

Hollow Cathode Lamps. Hollow cathode lamps provide line spectra for a large number of elements. Their use has been confined to atomic absorption and atomic fluorescence spectroscopy. Discussion of this type of source is deferred to Chapter 11.

Lasers

The first laser was constructed in 1960.[2] Since that time, chemists have found numerous useful applications for these sources in high-resolution spectroscopy, kinetic studies of

---

[2] For a more complete discussion of lasers, see: B. A. Lengyel, *Lasers*. New York: Wiley, 1971; S. R. Leone and C. B. Moore, *Chemical and Biological Applications of Lasers*, ed. C. E. Moore. New York: Academic Press, 1974; and *Analytical Laser Spectroscopy*, ed. N. Omenetto. New York: Wiley, 1979.

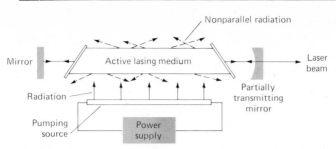

FIGURE 5-3   Schematic representation of a typical laser source.

processes with lifetimes in the range of 10^{-9} to 10^{-12} s, the detection and determination of extremely small concentrations of species in the atmosphere, and the induction of isotopically selective reactions.[3] In addition, laser sources have become important in several routine analytical methods, including Raman spectroscopy (Chapter 9), emission spectroscopy (Chapter 12), and Fourier transform infrared spectroscopy (Chapter 8).

The term laser is an acronym for *light amplification* by *stimulated emission* of *radiation.* As a consequence of their light-amplifying properties, lasers produce spatially narrow, extremely intense beams of radiation. The process of stimulated emission produces a beam of highly monochromatic (bandwidths of 0.01 nm or less) and remarkably coherent (p. 99) radiation. Because of these unique properties, lasers have become important sources for use in the ultraviolet, visible, and infrared regions of the spectrum. A limitation of early lasers was that the radiation from a given source was restricted to a relatively few discrete wavelengths or lines. Recently, however, dye lasers have become available; *tuning* of these sources provides a

narrow band of radiation at any chosen wavelength within the range of the source.

Figure 5-3 is a schematic representation showing the components of a typical laser source. The heart of the device is a *lasing medium.* It may be a solid crystal, such as ruby, a semiconductor such as gallium arsenide, a solution of an organic dye, or a gas. The lasing material can be activated or *pumped* by radiation from an external source so that a few photons of proper energy will trigger the formation of a cascade of photons of the same energy. Pumping can also be carried out by an electrical current or by an electrical discharge. Thus, *gas lasers* usually do not have the external radiation source shown in Figure 5-3; instead, the power supply is connected to a pair of electrodes contained in a cell filled with the gas.

A laser normally functions as an oscillator, in the sense that the radiation produced from the laser mechanism is caused to pass back and forth through the medium numerous times by means of a pair of mirrors as shown in Figure 5-3. Additional photons are generated with each passage, thus leading to enormous amplification. The repeated passage also produces a beam that is highly parallel because nonparallel radiation escapes from the sides of the medium after being reflected a few times (see Figure 5-3).

In order to obtain a usable laser beam, one of the mirrors is coated with a sufficiently

---

[3] For a review of some of these applications, see: S. R. Leone, *J. Chem. Educ.*, **53**, 13 (1976).

thin layer of reflecting material so that a fraction of the beam is transmitted rather than reflected (see Figure 5-3).

Laser action can be understood by considering the four processes depicted in Figure 5-4, namely, (a) pumping, (b) spontaneous emission (fluorescence), (c) stimulated emission, and (d) absorption. For purposes of illustration, we will focus on just two of several electronic energy levels that will exist in the atoms, ions, or molecules making up the laser material; as shown in the figure, the two electronic levels have energies E_y and E_x. Note that the higher electronic state is shown as having several slightly different vibrational energy levels, E_y, E_y', E_y'', and so forth. We have not shown additional levels for the lower electronic state, although such often exist.

Pumping. Pumping, which is necessary for laser action, is a process by which the active species of a laser is excited by means of an electrical discharge, passage of an electrical current, exposure to an intense radiant source, or interaction with a chemical species. During pumping, several of the higher electronic and vibrational energy levels of the active species will be populated. In diagram a-1 of Figure 5-4, one atom or molecule is shown as being promoted to an energy state E_y''; the second is excited to the slightly higher vibrational level E_y'''. The lifetime of excited *vibrational* states is brief, however; after 10^{-13} to 10^{-14} s, relaxation to the lowest excited vibrational level (E_y in diagram a-3) occurs with the production of an undetectable quantity of heat. Some excited electronic states of laser materials have lifetimes considerably longer (often 1 ms or more) than their excited vibrational counterparts; long-lived states are sometimes termed *metastable* as a consequence.

Spontaneous Emission. As was pointed out in the discussion of fluorescence, a species in an excited electronic state may lose all or part of its excess energy by spontaneous emission of radiation. This process is depicted in the diagrams shown in Figure 5-4b. Note that the wavelength of the fluorescent radiation is directly related to the energy difference between the two electronic states, $E_y - E_x$. It is also important to note that the instant at which the photon is emitted, and its direction, will vary from excited electron to excited electron. That is, spontaneous emission is a random process; thus, as shown in Figure 5-4, the fluorescent radiation produced by one of the particles in diagram b-1 differs in direction and phase from that produced by the second particle (diagram b-2). Spontaneous emission, therefore, yields *incoherent* radiation.

Stimulated Emission. Stimulated emission, which is the basis of laser behavior, is depicted in Figure 5-4c. Here, the excited laser particles are struck by externally produced photons having precisely the same energies $(E_y - E_x)$ as the photons produced by spontaneous emission. Collisions of this type cause the excited species to relax *immediately* to the lower energy state and to emit simultaneously a photon of exactly the same energy as the photon that stimulated the process. More important, and more remarkable, *the emitted photon is precisely in phase with the photon that triggered the event. That is, the stimulated emission is totally coherent with the incoming radiation.*

Absorption. The absorption process, which competes with stimulated emission, is depicted in Figure 5-4d. Here, two photons with energies exactly equal to $E_y - E_x$ are absorbed to produce the metastable excited state shown in diagram d-3; note that the state shown in diagram d-3 is identical to that attained by pumping diagram a-3.

Population Inversion and Light Amplification. In order to have light amplification in a laser, it is necessary that the number of photons produced by stimulated emission exceed the number lost by absorption. This condition will prevail only when the number of particles in the higher energy state exceeds the number in the lower; that is, a *population inversion* from the normal distribution of energy states must exist.

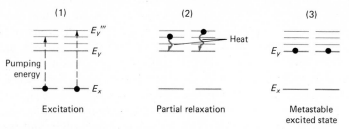

(a) Pumping (excitation by electrical, radiant, or chemical energy)

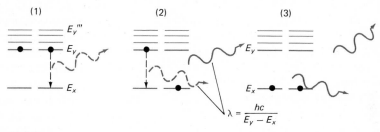

(b) Spontaneous Emission

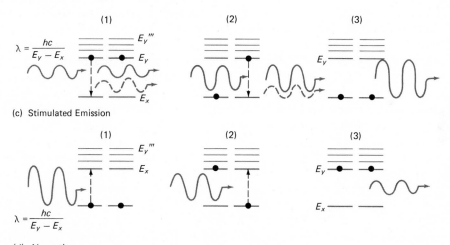

(c) Stimulated Emission

(d) Absorption

FIGURE 5-4   Four processes important in laser action: (a) pumping (excitation by electrical, radiant, or chemical energy), (b) spontaneous emission, (c) stimulated emission, and (d) absorption.

Population inversions are brought about by pumping. Figure 5-5 contrasts the effect of incoming radiation on a noninverted population with an inverted one.

Three- and Four-Level Laser Systems. Figure 5-6 shows simplified energy diagrams for the two common types of laser systems. In the three-level system, the transition responsible for laser radiation is between an excited state E_y and the ground state E_0; in a four-level system, on the other hand, radiation is generated by a transition from E_y to a state E_x that has a greater energy than the ground state. Furthermore, it is necessary that transitions between E_x and the ground state be rapid.

The advantage of the four-level system is that the population inversions necessary for laser action are more readily achieved. To understand this, note that at room temperature a large majority of the laser particles will be in the ground-state energy level E_0 in both systems. Sufficient energy must thus be provided to convert more than 50% of the lasing species to the E_y level of a three-level system. In contrast, it is only necessary to pump sufficiently to make the number of particles in the E_y energy level exceed the *number in E_x* of a four-level system. The lifetime of a particle in the E_x state is brief, however, because the transition to E_0 is fast; thus, the number in the E_x state will generally be negligible with respect to E_0 and also (with a modest input of pumping energy) with respect to E_y. That is, the four-level laser usually achieves a population inversion with a smaller expenditure of pumping energy.

Some Examples of Useful Lasers. The first successful laser, and one that still finds widespread use, was a three-level device in which a ruby crystal was the active medium. The ruby was machined into a rod about 4 cm in length and 0.5 cm in diameter. A flash tube was coiled around the cylinder to produce intense flashes of light. Because the pumping was discontinuous, a pulsed laser beam was produced. Continuous wave ruby sources are now available.

A variety of gas lasers are sold commercially. An important example is the argon ion

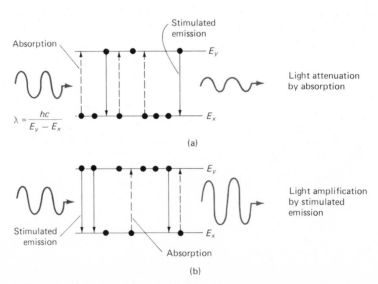

(a)

(b)

FIGURE 5-5 Passage of radiation through (a) a noninverted population and (b) an inverted population.

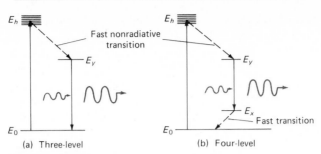

FIGURE 5-6   Energy level diagrams for two types of laser systems.

laser, which produces intense lines in the green (514.5 nm) and blue (488.0 nm) regions. This laser is a four-level device in which argon ions are formed by an electrical or radio frequency discharge. The input energy is sufficient to excite the ions from their ground state, with a principal quantum number of 3, to various $4p$ states. Laser activity then involves transitions to the $4s$ state.

Dye lasers[4] have become important radiation sources in chemistry because they are tunable over a range of 20 to 50 nm; that is, dye lasers provide bands of radiation of a chosen wavelength with widths of a few hundredths of a nanometer or less.

The active materials in dye lasers are solutions of organic compounds capable of fluorescing in the ultraviolet, visible, or infrared regions. In contrast to ruby or gas lasers, however, the lower energy level for laser action (E_x in Figure 5-6) is not a single energy but a band of energies arising from the superposition of a large number of small vibrational and rotational energy states upon the base electronic energy state. Electrons in E_y may then undergo transitions to any of these states, thus producing photons of slightly different energies.

Tuning of dye lasers can be readily accomplished by replacing the nontransmitting mirror shown in Figure 5-3 with a monochromator equipped with a reflection grating or a Littrow-type prism (p. 126) which will reflect only a narrow bandwidth of radiation into the laser medium; the peak wavelength can be varied by rotation of the grating or prism. Emission is then stimulated for only part of the fluorescent spectrum, namely, the wavelength reflected from the monochromator.

WAVELENGTH SELECTION; MONOCHROMATORS

With few minor exceptions, the analytical methods considered in the following seven chapters require dispersal of polychromatic radiation into bands that encompass a restricted wavelength region. The most common way of producing such bands is with a device called a *monochromator*.[5]

---

[4] For further information, see: R. B. Green, "Dye Laser Instrumentation," *J. Chem. Educ.*, **54** (9) A365, (10) A407, (1977).

[5] For a more complete discussion of monochromators, see: R. A. Bauman, *Absorption Spectroscopy*. New York: Wiley, 1962; F. A. Jenkins and H. E. White, *Fundamentals of Optics*, 3d ed. New York: McGraw-Hill, 1957; E. J. Meehan, in *Treatise on Analytical Chemistry*, eds. I. M. Kolthoff and P. J. Elving. New York: Wiley, 1964, Part I, vol. 5, Chapter 55; and J. F. James and R. S. Sternberg, *The Design of Optical Spectrometers*. London: Chapman and Hall, 1969.

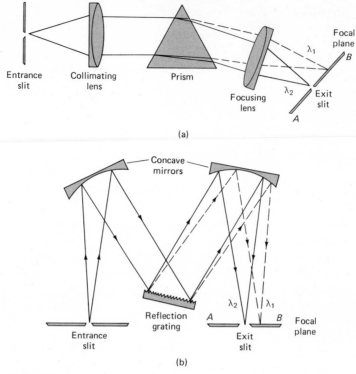

FIGURE 5-7 Two types of monochromators: (a) Bunsen prism monochromator, and (b) Czerney-Turner grating monochromator. (In both instances, $\lambda_1 > \lambda_2$.)

Monochromators for ultraviolet, visible, and infrared radiation are all similar in mechanical construction in the sense that they employ slits, lenses, mirrors, windows, and prisms or gratings. To be sure, the materials from which these components are fabricated will depend upon the wavelength region of intended use (see Figure 5-2b and c).

Components of a Monochromator

All monochromators contain an entrance slit, a collimating lens or mirror to produce a parallel beam of radiation, a prism or grating as a dispersing element,[6] and a focusing element

which projects a series of rectangular images of the entrance slit upon a plane surface (the *focal plane*). In addition, most monochromators have entrance and exit *windows*, which are designed to protect the components from dust and corrosive laboratory fumes.

Figure 5-7 shows the optical design of two typical monochromators, one employing a prism for dispersal of radiation and the other a grating. A source of radiation containing but two wavelengths, λ_1 and λ_2, is shown for purposes of illustration. This radiation enters the monochromators via a narrow rectangular opening or slit, is collimated, and then strikes the surface of the dispersing element at an angle. In the prism monochromator, refraction at the two faces results in angular dispersion of the radiation, as shown; for the

---

[6] Much less generally employed are *interference wedges*. These devices are described in the next section (p. 136).

grating, angular dispersion results from diffraction, which occurs at the reflective surface. In both designs, the dispersed radiation is focused on the focal plane *AB* where it appears as two images of the entrance slit (one for each wavelength).

Types of Instruments Employing Monochromators

The method of detecting the radiation dispersed by a monochromator varies. In a *spectroscope*, detection is accomplished visually with a movable eyepiece located along the focal plane. The wavelength is then determined by measurement of the angle between the incident and dispersed beams. In a *spectrograph*, a photographic film or plate is mounted along the focal plane; the series of darkened images of the slit along the length of the developed film or plate is called a *spectrogram*.

The monochromators shown in Figure 5-7 are also utilized for dispersing elements in *spectrometers*. Such instruments have a fixed exit slit located in the focal plane. With a continuous source, the wavelength emitted from this slit can be varied continuously by rotation of the dispersing element.

We need to define two other types of instruments to complete this discussion. A *spectrophotometer* is a spectrometer which contains a photoelectric device for determining the power of the radiation exiting from the slit. A *photometer* also employs a photoelectric detector but contains no monochromator; instead, filters are used to provide radiation bands encompassing limited wavelength regions. A photometer is incapable of providing a continuously variable band of radiation.

Prism Monochromators

Prisms can be used to disperse ultraviolet, visible, and infrared radiation. The material used for their construction will differ depending upon the wavelength region.

Construction Materials. Clearly, the materials employed for fabricating the windows, lenses, and prisms of a monochromator must transmit radiation in the frequency range of interest; ideally, the transmittance should approach 100%, although it is sometimes necessary to use substances with transmittances as low as 20%. The refractive index of materials used in the construction of windows and prisms should be low in order to reduce reflective losses (see Equation 4-14, p. 103). However, for lenses which function by bending rays of radiation, a material with a high refractive index is desirable in order to reduce focal lengths. Ideal construction materials for both lenses and windows should exhibit little change in refractive index with frequency, thus reducing chromatic aberrations. In contrast, just the opposite property is sought for prisms, where the extent of dispersion depends upon the rate of change of refractive index with frequency. In addition to the foregoing optical properties, it is desirable that monochromator components be resistant to mechanical abrasion and attack by atmospheric components and laboratory fumes.

Needless to say, no single substance meets all of these criteria and trade-offs are thus required in monochromator design. The choices here depend strongly upon the wavelength region to be used.

Table 5-1 lists the properties of the common substances employed for fabricating monochromator components. Clearly, no single material is suitable for the entire wavelength region. For the ultraviolet, visible, and near-infrared regions (up to about 3000 nm), a quartz prism is often employed; it should be noted, however, that glass provides better resolution for the same size prism for wavelengths between 350 and 2000 nm. As shown in column 5 of Table 5-1, several prisms are required to cover the entire infrared region.

Types of Prisms and Prism Monochromators. Figure 5-8 shows the two most common types of prism configurations. The first is a 60-deg design, which is ordinarily fabricated

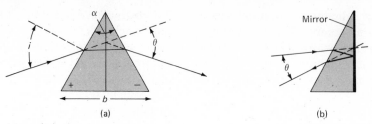

FIGURE 5-8 Dispersion by a prism: (a) quartz Cornu type, and (b) Littrow type.

from a single block of material. When crystalline (but not fused) quartz is employed, however, the prism is usually formed by cementing two 30-deg prisms together, as shown in Figure 5-8a; one is constructed from right-handed quartz and the second from left-handed quartz. In this way, the optically active quartz causes no net polarization of the emitted radiation; this type of prism is called a *Cornu prism*. Figure 5-7a shows a *Bunsen monochromator*, which employs a 60-deg prism, likewise often made of quartz.

Figure 5-8b depicts a *Littrow prism*, which is a 30-deg prism with a mirrored back. It is seen that refraction occurs twice at the same interface; the performance characteristics of the Littrow prism are thus similar to those of the 60-deg prism. The Littrow prism permits

TABLE 5-1 CONSTRUCTION MATERIAL FOR SPECTROPHOTOMETER OPTICS

Material	Transmittance Range, μm	Refractive Index	Angular Dispersion[a]	Useful Range for Prisms, μm	Hardness and Chemical Resistance
Fused silica	0.18–3.3	1.46	0.52×10^{-4}	0.18–2.7	Excellent
Quartz	0.20–3.3	1.54	0.63×10^{-4}	0.20–2.7	Excellent
Flint glass	0.35–2.2	1.66	1.70×10^{-4}	0.35–2	Excellent
Calcium fluoride (fluorite)	0.12–12	1.43	0.33×10^{-4}	5–9.4	Good
Lithium fluoride	0.12–6	1.39	0.29×10^{-4}	2.7–5.5	Poor
Sodium chloride	0.3–17	1.54	0.94×10^{-4}	8–16	Poor
Potassium bromide	0.3–29	1.56	1.45×10^{-4}	15–28	Poor
Cesium iodide	0.3–70	1.79	—	15–55	Poor
KRS-5 (TlBr-TlI)	1–40	2.63	—	24–40	Good

[a] Radians/μm at 0.589 μm

somewhat more compact monochromator designs. In addition, when quartz is employed, polarization is canceled by the reversal of the radiation path. A typical Littrow-type monochromator is shown in Figure 7-10.

Angular Dispersion of Prisms. The *angular dispersion* of a prism is the rate of change of the angle θ in Figure 5-8 as a function of wavelength; that is, $d\theta/d\lambda$. The spectral purity of the radiation exiting from a prism monochromator is dependent upon this quantity.

The angular dispersion of a prism can be resolved into two parts

$$\frac{d\theta}{d\lambda} = \frac{d\theta}{dn} \cdot \frac{dn}{d\lambda} \qquad (5\text{-}1)$$

where $d\theta/dn$ represents the change in θ as a function of the refractive index n of the prism material and $dn/d\lambda$ expresses the variation of the refractive index with wavelength (that is, the dispersion of the substance from which the prism is fabricated).

The magnitude of $d\theta/dn$ is determined by the geometry of the prism and the angle of incidence i (Figure 5-8a). In order to avoid problems of astigmatism (double images), this angle should be such that the path of the beam through the prism is within a few degrees of being parallel to the base of the prism. Under these circumstances, $d\theta/dn$ depends only upon the prism angle α (Figure 5-8a) and increases rapidly with this quantity. Reflection losses, however, impose an upper limit of about 60 deg for α. For a prism in which $\alpha = 60$ deg, it can be shown that

$$d\theta/dn = (1 - n^2/4)^{-\frac{1}{2}} \qquad (5\text{-}2)$$

The term $dn/d\lambda$ in Equation 5-1 is related to the dispersion of the substance from which the prism is constructed. We have noted (p. 102) that the greatest dispersion for a given material is near its anomalous dispersion region, which in turn is close to a region of absorption. The dispersion of some substances employed for construction of prisms is shown in Figure 5-9. Note that the rapid rise in the refractive index for glass below 400 nm corresponds to the sharp increase in absorption that prevents the use of this material below 350 nm. In the region of 350 to 2000 nm, however, glass is greatly superior to quartz for prism fabrication because of its larger change in refractive index with wavelength ($dn/d\lambda$).

Focal-Plane Dispersion of Prism Monochromators. The focal-plane dispersion of a monochromator refers to the variation in wavelength as a function of y, the linear distance along the line AB of the focal plane of the instrument (see Figure 5-7). That is, the focal-plane dispersion is given by $dy/d\lambda$.

Figure 5-10 shows the focal-plane dispersion of two prism monochromators and one grating instrument. Note that the dispersion for the two prism monochromators is highly nonlinear, the longer wavelengths being bunched together over a relatively small distance. The two prism monochromators were Littrow types, each having a height of 57 mm. Note the much larger dispersion exhibited by the monochromator with a glass prism, in the region of 350 to 800 nm.

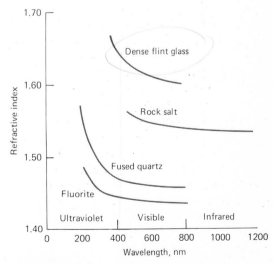

FIGURE 5-9   Dispersion of several optical materials.

Figure 5-10 illustrates one of the important advantages of grating monochromators—linear dispersion along the focal plane.

Resolving Power of Prism Monochromators. The *resolving power R* of a prism gives the limit of its ability to separate adjacent images having slightly different wavelengths. Mathematically, this quantity is defined as

$$R = \lambda/d\lambda \qquad (5\text{-}3)$$

where $d\lambda$ represents the wavelength difference that can just be resolved and λ is the average wavelength of the two images. It can be shown that the resolving power of a prism is directly proportional to the length of the prism base b (Figure 5-8) and the dispersion of its construction material. That is,

$$R = b(dn/d\lambda) \qquad (5\text{-}4)$$

Thus, high resolution demands large prisms.

Grating Monochromators for Wavelength Selection

Dispersion of ultraviolet, visible, and infrared radiation can be brought about by passage of a beam through a *transmission grating* or by reflection from a *reflection grating*. A transmission grating consists of a series of parallel and closely spaced grooves ruled on a piece of glass or other transparent material. A grating suitable for use in the ultraviolet and visible region has between 2000 and 6000 lines per millimeter. An infrared grating requires considerably fewer lines; thus, for the far infrared region, gratings with 20 to 30 lines per millimeter may suffice. It is vital that these lines be equally spaced throughout the several centimeters in length of the typical grating. Such gratings require elaborate apparatus for their production and are consequently expensive. Replica gratings are less costly. They are manufactured by employing a master grating as a mold for the production of numerous plastic replicas; the products of this process, while inferior in performance to an original grating, are adequate for many applications.

When a transmission grating is illuminated from a slit, each groove scatters radiation and thus effectively becomes opaque. The nonruled portions then behave as a series of closely spaced slits, each of which acts as a new radiation source; interference among the multitude of beams results in diffraction of

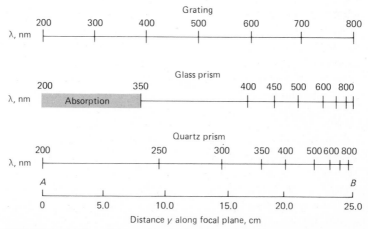

FIGURE 5-10   Dispersion for three types of monochromators. The points A and B on the scale correspond to the points shown in Figure 5-7.

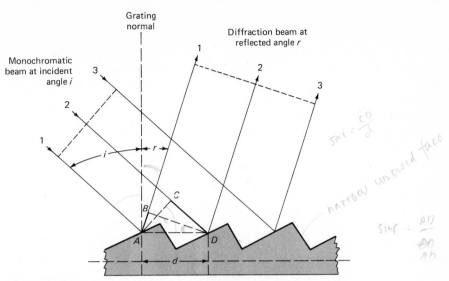

FIGURE 5-11 Schematic diagram illustrating the mechanism of diffraction from an echellette-type grating. (From R. P. Bauman, *Absorption Spectroscopy*. New York, Wiley, 1962, p. 65. With permission.)

the radiation as shown in Figure 4-6a (p. 98). The angle of diffraction, of course, depends upon the wavelength.

Reflection gratings are used more extensively in instrument construction than their transmitting counterparts. Such gratings are made by ruling a polished metal surface or by evaporating a thin film of aluminum onto the surface of a replica grating. As shown in Figure 5-11, the incident radiation is reflected from one of the faces of the groove, which then acts as a new radiation source. Interference results in radiation of differing wavelengths being reflected at different angles *r*.

Diffraction by a Grating. Figure 5-11 is a schematic representation of an *echellette-type* grating, which is grooved or *blazed* such that it has relatively broad faces from which reflection occurs and narrow unused faces. This geometry provides highly efficient diffraction of radiation. Each of the broad faces can be considered to be a point source of radiation; thus interference among the reflected beams 1, 2, and 3 can occur. In order for constructive interference to occur between adjacent beams, it is necessary that their path lengths differ by an integral multiple **n** of the wavelength λ of the incident beam.

In Figure 5-11, parallel beams of monochromatic radiation 1 and 2 are shown striking the grating at an incident angle *i* to the *grating normal*. Maximum constructive interference is shown as occurring at the reflected angle *r*. It is evident that beam 2 travels a greater distance than beam 1 and that this difference is equal to $(\overline{CD} - \overline{AB})$. For constructive interference to occur, this difference must equal **n**λ. That is,

$$\mathbf{n}\lambda = (\overline{CD} - \overline{AB})$$

Note, however, that angle *CAD* is equal to angle *i* and that angle *BDA* is identical with angle *r*. Therefore, from simple trigonometry, we may write

$$\overline{CD} = d \sin i$$

where d is the spacing between the reflecting surfaces. It is also seen that

$$\overline{AB} = -d \sin r$$

The minus sign, by convention, indicates that reflection has occurred. The angle r, then, is negative when it lies on the opposite side of the grating normal from the angle i (as in Figure 5-11), and is positive when it is on the same side. Substitution of the last two expressions into the first gives the condition for constructive interference. Thus,

$$n\lambda = d(\sin i + \sin r) \qquad (5\text{-}5)$$

Equation 5-5 suggests that several values of λ exist for a given diffraction angle r. Thus, if a first-order line ($n = 1$) of 800 nm is found at r, second-order (400 nm) and third-order (267 nm) lines also appear at this angle. Ordinarily, the first-order line is the most intense; indeed, it is possible to design gratings that concentrate as much as 90% of the incident intensity in this order. The higher-order lines can generally be removed by filters. For example, glass, which absorbs radiation below 350 nm, eliminates the higher-order spectra associated with first-order radiation in most of the visible region. The example which follows illustrates these points.

EXAMPLE

An echellette grating containing 2000 blazes per millimeter was irradiated with a polychromatic beam at an incident angle 48 deg to the grating normal. Calculate the wavelengths of radiation that would appear at an angle of reflection of $+20$, $+10$, 0, and -10 deg (angle r, Figure 5-11).

To obtain d in Equation 5-5, we write

$$d = \frac{1 \text{ mm}}{2000 \text{ blazes}} \times 10^6 \frac{\text{nm}}{\text{mm}} = 500 \frac{\text{nm}}{\text{blaze}}$$

When r in Figure 5-11 equals $+20$ deg

$$\lambda = \frac{500}{n} (\sin 48 + \sin 20) = \frac{542.6}{n}$$

and the wavelengths for the first-, second-, and third-order reflections are 543, 271, and 181 nm, respectively.

Similarly, when $r = -10$ deg (here r lies to the right of the grating normal)

$$\lambda = \frac{500}{n} [\sin 48 + \sin(-10)] = \frac{284.7}{n}$$

Similar calculations yield the following data:

	Wavelength (nm) for		
r	$n = 1$	$n = 2$	$n = 3$
20	543	271	181
10	458	229	153
0	372	186	124
-10	285	142	95

A concave grating can be produced by ruling a spherical reflecting surface. Such a diffracting element serves also to focus the radiation on the exit slit and eliminates the need for a lens.

Dispersion of Gratings. The angular dispersion of a grating can be obtained by differentiating Equation 5-5 while holding i constant; thus, at any given angle of incidence

$$\frac{dr}{d\lambda} = \frac{n}{d \cos r} \qquad (5\text{-}6)$$

Note that the dispersion increases as the distance d between rulings decreases or as the number of lines per millimeter increases. Over short wavelength ranges, $\cos r$ does not change greatly with λ, so that the *dispersion of a grating is nearly linear*. By proper design of the optics of a grating monochromator, it is possible to produce an instrument that for all practical purposes has a linear dispersion of radiation along the focal plane of the exit slit. Figure 5-10 shows the contrast between a grating and a prism monochromator in this regard.

Resolving Power of a Grating. It can be shown[7] that the resolving power R of a grating is given by the very simple expression

$$R = \frac{\lambda}{\Delta\lambda} = nN \qquad (5\text{-}7)$$

where **n** is the diffraction order and N is the number of lines illuminated by the radiation from the entrance slit. Thus, as with a prism (Equation 5-4), the resolving power depends upon the physical size of the dispersing element.

EXAMPLE

Compare the size of (1) a 60-deg fused silica prism; (2) a 60-deg glass prism; and (3) a grating with 2000 lines/mm, that would be required to resolve two lithium emission lines at 460.20 and 460.30 nm. Average values for the dispersion $(dn/d\lambda)$ of fused silica and glass in the region of interest are 1.3×10^{-4} and 3.6×10^{-4} nm^{-1}, respectively.

For the two prisms, we employ Equation 5-4

$$R = \frac{\lambda}{\Delta\lambda} = b\frac{dn}{d\lambda}$$

or

$$b = \frac{\lambda}{\Delta\lambda} \times \frac{1}{dn/d\lambda} = \frac{460.25}{460.30 - 460.20} \times \frac{1}{dn/d\lambda}$$

For fused silica,

$$b = \frac{460.25\ nm}{0.10\ nm} \times \frac{1}{1.3 \times 10^{-4}\ nm^{-1}} \times \frac{10^{-7}\ cm}{nm}$$

$$= 3.5\ cm$$

and for glass,

$$b = \frac{460.25}{0.10} \times \frac{1}{3.6 \times 10^{-4}} \times 10^{-7} = 1.3\ cm$$

For the grating, we employ Equation 5-7

$$\frac{\lambda}{\Delta\lambda} = nN$$

For the first-order spectrum $(n = 1)$

$$N = \frac{460.25}{0.10 \times 1} = 4.60 \times 10^3\ lines$$

$$grating\ length = \frac{4.6 \times 10^3\ lines}{2000\ lines/mm} \times \frac{0.1\ cm}{mm}$$

$$= 0.23\ cm$$

Thus, in theory, the separation can be accomplished with a silica prism with a base of 3.5 cm, a glass prism with a base of 1.3 cm, or 0.23 cm of grating.

Gratings versus Prisms. Gratings offer several advantages over prisms as dispersing elements. One of these is that the dispersion is nearly constant with wavelength; this property makes the design of monochromators considerably more simple. A second advantage can be seen from the foregoing example, namely, that better dispersion can be expected for the same size of dispersing element. In addition, reflection gratings provide a means of dispersing radiation in the far ultraviolet and far infrared regions where absorption prevents the use of prisms.

Gratings have the disadvantage of producing somewhat greater amounts of stray radiation as well as higher-order spectra. These disadvantages are ordinarily not serious because the extraneous radiation can be minimized through the use of filters and by proper instrument design.

For a number of years, high-quality gratings were more expensive than good prisms. This disadvantage has now largely disappeared.

Monochromator Slits

The slits of a monochromator play an important part in determining its quality. Slit jaws are formed by carefully machining two pieces

---

[7] R. A. Sawyer, *Experimental Spectroscopy*, 2d ed. New Jersey: Prentice-Hall, 1951, p. 130.

of metal to give sharp edges, as shown in Figure 5-12. Care is taken to assure that the edges of the slit are exactly parallel to one another and that they lie on the same plane.

In some monochromators, the openings of the two slits are fixed; more commonly, the spacing can be adjusted with a micrometer mechanism.

The entrance slit (see Figure 5-7) serves as a radiation source; its image is focused on the surface containing the exit slit. If the radiation source consists of a few discrete wavelengths, a series of rectangular images appears on this surface as bright lines, each corresponding to a given wavelength. A particular line can be brought to focus on the exit slit by rotating the dispersing element. If the entrance and exit slits are of the same size (as is usually the case), the image of the entrance slit will in theory just fill the exit-slit opening when the setting of the monochromator corresponds to the wavelength of the radiation. Movement of the monochromator mount in one direction or the other results in a continuous decrease in emitted intensity, zero being reached when the entrance-slit image has been displaced by its full width.

Figure 5-13 illustrates the situation in which monochromatic radiation of wavelength λ_2 strikes the exit slit. Here, the monochromator is set for λ_2 and the two slits are identical in width. The image of the entrance slit just fills the exit slit. Movement of the monochromator to a setting of λ_1 or λ_3 re-

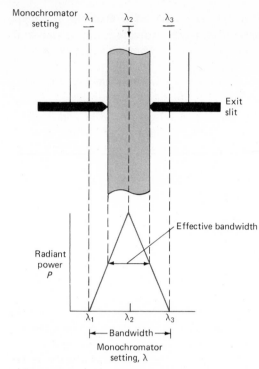

FIGURE 5-13 Illumination of an exit slit by monochromatic radiation λ_2 at various monochromator settings. Exit and entrance slits are identical.

sults in the image being moved completely out of the slit. The lower half of the figure shows a plot of the radiant power emitted as a function of monochromator setting. Note that the *bandwidth* is defined as the span of monochromator settings (in units of wavelength) needed to move the image of the entrance slit across the exit slit. If polychromatic radiation were employed, it would also represent the span of wavelengths emitted from the exit slit for a given monochromator setting.

The *effective bandwidth* for a monochromator is also defined in Figure 5-13 as one-half the wavelength range transmitted by the instrument at a given wavelength setting, or the range of wavelengths encompassed at half

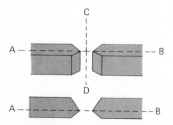

FIGURE 5-12 Construction of slits.

peak power. Figure 5-14 illustrates the relationship between the effective bandwidth of an instrument and its effectiveness in resolving spectral peaks. Here, the exit slit of a monochromator is illuminated with a beam composed of just three wavelengths, λ_1, λ_2, and λ_3; each beam has the same intensity. In the top figure, the effective bandwidth of the instrument is exactly equal to the difference in wavelength between λ_1 and λ_2 or λ_2 and λ_3. When the monochromator is set at λ_2, radiation of this wavelength just fills the slit.

Movement of the monochromator in either direction diminishes the transmitted intensity of λ_2, but increases the intensity of one of the other lines by an equivalent amount. As shown by the solid line in the plot to the right, no spectral resolution of the three wavelengths is achieved.

In the middle drawing of Figure 5-14, the effective bandwidth of the instrument has been reduced by narrowing the openings of the exit and entrance slits to three-quarters that of their original dimensions. The solid

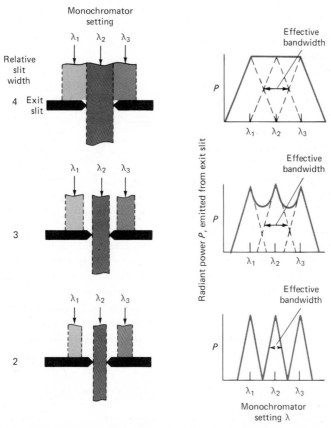

FIGURE 5-14 The effect of the slit width on spectra. The entrance slit is illuminated with λ_1, λ_2, and λ_3 only. Entrance and exit slits are identical. Plots on the right show changes in emitted power as the setting of monochromator is varied.

line in the plot on the right shows that partial resolution of the three lines results. When the effective bandwidth is decreased to one-half the difference in wavelengths of the three beams, complete resolution is obtained, as shown in the bottom drawing.

The effective bandwidth of a monochromator depends upon both the dispersion of the prism or grating as well as the width of the entrance and exit slits. Most monochromators are equipped with variable slits so that the effective bandwidth can be changed. The use of minimal slit widths is desirable where the resolution of narrow absorption or emission bands is needed. On the other hand, there is a marked decrease in the available radiant power as the slits are narrowed, and accurate measurement of this power becomes more difficult. Thus, wider slit widths may be used for quantitative analysis than for qualitative work, where spectral detail is important.

As noted previously, the dispersion of a prism is not linear (see Figure 5-10). Much narrower slits must, therefore, be employed at long wavelengths than at short in order to obtain radiation of a given effective bandwidth. The slit width in millimeters required to maintain an effective bandwidth of one nanometer for a monochromator equipped with a Littrow prism is shown in Figure 5-15. One of the advantages of a grating monochromator is that a fixed slit width produces radiation of nearly constant bandwidth, regardless of wavelength.

Stray Radiation in Monochromators

The exit beam of a monochromator is usually contaminated with small amounts of radiation having wavelengths far different from that of the instrument setting. There are several sources of this unwanted radiation. Reflections of the beam from various optical parts and the monochromator housing contribute to it. Scattering by dust particles in the atmosphere or on the surfaces of optical

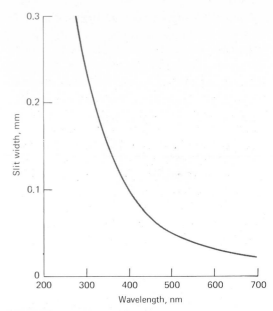

FIGURE 5-15   The slit width needed to maintain a constant effective bandwidth of 1 nm from a quartz Littrow prism monochromator.

parts also causes stray radiation to reach the exit slit. Generally, the effects of spurious radiation are minimized by introducing baffles in appropriate spots in the monochromator and by coating interior surfaces with flat black paint. In addition, the monochromator is sealed with windows over the slits to prevent entrance of dust and fumes. Despite these precautions, however, some spurious radiation is still emitted; we shall find that its presence can have serious effects on absorption measurements under certain conditions.

Double Monochromators

Many modern monochromators contain two dispersing elements; that is, two prisms, two gratings, or a prism and a grating. This arrangement markedly reduces the amount of stray radiation and also provides greater dispersion and spectral resolution. Furthermore, if one of the elements is a grating, higher-order wavelengths are removed by the second element.

WAVELENGTH SELECTION; FILTERS

Absorption and interference filters are employed for wavelength selection. The former are restricted to the visible region of the spectrum. Interference filters, on the other hand, are available for ultraviolet, visible, and infrared radiation.

Interference Filters

As the name implies, interference filters rely on optical interference to provide relatively narrow bands of radiation (see Figure 5-17). An interference filter consists of a transparent dielectric (frequently calcium fluoride or magnesium fluoride) that occupies the space between two semitransparent metallic films coated on the inside surfaces of two glass plates. The thickness of the dielectric layer is carefully controlled and determines the wavelength of the transmitted radiation. When a perpendicular beam of collimated radiation strikes this array, a fraction passes through the first metallic layer while the remainder is reflected. The portion that is passed undergoes a similar partition upon striking the second metallic film. If the reflected portion from this second interaction is of the proper wavelength, it is partially reflected from the inner side of the first layer in phase with incoming light of the same wavelength. The result is that this particular wavelength is reinforced, while most other wavelengths, being out of phase, suffer destructive interference.

Figure 5-16 illustrates an interference filter. For purposes of clarity, the incident beam is shown as arriving at an angle θ from the perpendicular. In ordinary use, θ approaches

zero so that the equation accompanying Figure 5-16 simplifies to

$$\mathbf{n}\lambda' = 2t \qquad (5\text{-}8)$$

where λ' is the wavelength of radiation *in the dielectric* and t is the thickness of the dielectric. The corresponding wavelength in air is given by

$$\lambda = \lambda'n$$

where n is the refractive index of the dielectric medium. Thus, the wavelengths of radiation transmitted by the filter are

$$\lambda = \frac{2tn}{\mathbf{n}} \qquad (5\text{-}9)$$

The integer $\mathbf{n}$ is the *order* of interference. The glass layers of the filter are often selected to absorb all but one of the reinforced bands; transmission is thus restricted to a single order.

Figure 5-17 illustrates the performance characteristics of typical interference filters.

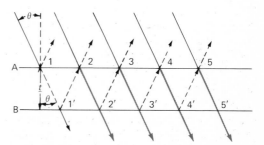

FIGURE 5-16   The interference filter. At point 1, light strikes the semitransparent film at an angle θ from the perpendicular, is partially reflected, partially passed. The same process occurs at 1', 2, 2', and so forth. For reinforcement to occur at point 2, the distance traveled by the beam reflected at 1' must be some multiple of its wavelength in the medium λ'. Since the path length between surfaces can be expressed as $t/\cos\theta$, the condition for reinforcement is that $\mathbf{n}\lambda' = 2t/\cos\theta$ where $\mathbf{n}$ is a small whole number.

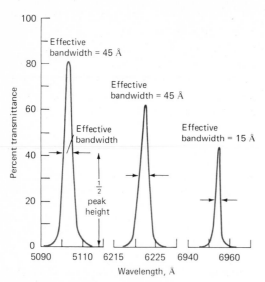

FIGURE 5-17 Transmission character-istics of typical interference filters.

Ordinarily, filters are characterized, as shown, by the wavelength of their transmittance peaks, the percentage of incident radiation transmitted at the peak (their *percent trans-mittance*, p. 149), and their bandwidths at a transmittance equal to one-half the maximum (the *effective bandwidth*).

Interference filters are available through-out the ultraviolet and visible regions and up to about 14 μm in the infrared. Typically, half bandwidths are about 1.5% of the wavelength at peak transmittance, although this figure is reduced to 0.15% in some narrow-band filters; these have maximum transmittances of about 10%.

Interference Wedges

An interference wedge consists of a pair of mirrored, partially transparent plates sepa-rated by a wedge-shaped layer of a dielectric material. The length of the plates ranges from about 50 to 200 mm. The radiation trans-mitted varies continuously from one end to the other as the thickness of the wedge varies. By choosing the proper linear position along the wedge, one can isolate a bandwidth of about 20 nm.

Interference wedges are available for the visible region (400 to 700 nm), the near-infra-red region (1000 to 2000 nm), and for several parts of the infrared region (2.5 to 14.5 μm).

Absorption Filters

Absorption filters, which are generally less ex-pensive than interference filters, have been widely used for band selection in the visible region. These filters limit radiation by absorb-

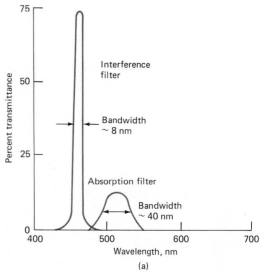

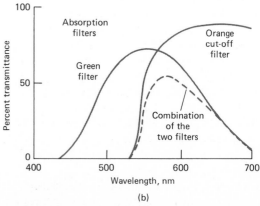

FIGURE 5-18 Comparison of various types of filters for visible radiation.

ing certain portions of the spectra. The most common type consists of colored glass or of a dye suspended in gelatin and sandwiched between glass plates. The former have the advantage of greater thermal stability.

Absorption filters have effective bandwidths that range from perhaps 30 to 250 nm (see Figure 5-18). Filters that provide the narrowest bandwidths also absorb a significant fraction of the desired radiation, and may have a transmittance of 0.1 or less at their band peaks. Glass filters with transmittance maxima throughout the entire visible region are available commercially.

Cut-off filters have transmittances of nearly 100% over a portion of the visible spectrum, but then rapidly decrease to zero transmittance over the remainder. A narrow spectral band can be isolated by coupling a cut-off filter with a second filter (see Figure 5-18b).

It is apparent from Figure 5-18a that the performance characteristics of absorption filters are significantly inferior to those of interference-type filters. Not only are their bandwidths greater, but the fraction of light transmitted is also less. Nevertheless, absorption filters are totally adequate for many applications.

SAMPLE CONTAINERS

Sample containers are required for all spectroscopic studies except emission spectroscopy. In common with the optical elements of monochromators, the *cells* or *cuvettes* that hold the samples must be made of material that passes radiation in the spectral region of interest. Thus, as shown in Figure 5-2, quartz or fused silica is required for work in the ultraviolet region (below 350 nm); both of these substances are transparent in the visible region and to about 3 μm in the infrared region as well. Silicate glasses can be employed in the region between 350 and 2000 nm. Plastic containers have also found application in the visible region. Crystalline sodium chloride is the most common substance employed for cell windows in the infrared region; the other infrared transparent materials listed in Table 5-1 (p. 126) may also be used for this purpose.

The best cells have windows that are perfectly normal to the direction of the beam in order to minimize reflection losses. The most common cell length for studies in the ultraviolet and visible regions is 1 cm; matched, calibrated cells of this size are available from several commercial sources. Other path lengths, from 0.1 cm (and shorter) to 10 cm, can also be purchased. Transparent spacers for shortening the path length of 1-cm cells to 0.1 cm are also available.

Cells for infrared studies of liquids and solutions generally have path lengths of less than 1 mm.

For reasons of economy, cylindrical cells are sometimes employed in the ultraviolet and visible regions. Care must be taken to duplicate the position of the cell with respect to the beam; otherwise, variations in path length and reflection losses at the curved surfaces can cause significant errors.

The quality of absorbance data is critically dependent upon the way the matched cells are used and maintained. Fingerprints, grease, or other deposits on the walls alter the transmission characteristics of a cell markedly. Thus, thorough cleaning before and after use is imperative; the surface of the windows must not be touched during the handling. Matched cells should never be dried by heating in an oven or over a flame—such treatment may cause physical damage or a change in path length. The cells should be calibrated against each other regularly with an absorbing solution.

RADIATION DETECTION

Radiation Detectors

Early instruments for the measurement of emission and absorption of radiation required visual or photographic methods for detection.

These detectors have been largely supplanted by transducers which convert radiant energy into an electrical signal; our discussion will be confined to detectors of this kind. Brief consideration of photographic detection will be found in Chapter 12.

Properties of Photoelectric Detectors. To be useful, the detector must respond to radiant energy over a broad wavelength range. It should, in addition, be sensitive to low levels of radiant power, respond rapidly to the radiation, produce an electrical signal that can be readily amplified, and have a relatively low noise level (for stability). Finally, it is essential that the signal produced be directly proportional to the beam power P; that is,

$$G = KP + K' \qquad (5\text{-}10)$$

where G is the electrical response of the detector in units of current, resistance, or emf. The constant K measures the sensitivity of the detector in terms of electrical response per unit of radiant power. Many detectors exhibit a small constant response, known as a *dark current K'*, even when no radiation impinges on their surfaces. Instruments with detectors that have a dark-current response are ordinarily equipped with a compensating circuit that permits application of a countersignal to reduce K' to zero. Thus, under ordinary circumstances, we may write

$$G = K''P \qquad (5\text{-}11)$$

Types of Photoelectric Detectors. As indicated in Figure 5-2, two general types of radiation transducers are employed; one responds to photons, the other to heat. All photon detectors are based upon interaction of radiation with a reactive surface to produce electrons (photoemission) or to promote electrons to energy states in which they can conduct electricity (photoconduction). Only ultraviolet, visible, and near-infrared radiation have sufficient energy to cause these processes to occur. Photoelectric detectors also differ from heat detectors in that their electrical signal results from a series of individual events (absorption of one photon), the probability of which can be described by the use of statistics. In contrast, heat transducers, which are required for the detection of infrared radiation, are nonquantized sensors.

As will be shown in Chapter 6, the distinction between photon and heat detectors is important because shot noise (p. 71) often limits the behavior of the former while Johnson noise may limit the latter. As a consequence, indeterminate errors associated with the two types of detectors are different.

Photon Detectors

Several types of photon detectors are available, including: (1) *photovoltaic cells*, in which the radiant energy generates a current at the interface of a semiconductor layer and a metal; (2) *phototubes*, in which radiation causes emission of electrons from a photosensitive solid surface; (3) *photomultiplier tubes*, which contain a photoemissive surface as well as several additional surfaces that emit a cascade of electrons when struck by electrons from the photosensitive area; (4) *semiconductor detectors*, in which absorption of radiation produces electrons and holes, thus leading to enhanced conductivity; and (5) *silicon diodes*, in which photons increase the conductance across a reverse biased *pn* junction.

Photovoltaic or Barrier-Layer Cells. The photovoltaic cell is used primarily to detect and measure radiation in the visible region. The typical cell has a maximum sensitivity at about 550 nm; the response falls off to perhaps 10% of the maximum at 350 and 750 nm. Its range approximates that of the human eye.

The photovoltaic cell consists of a flat copper or iron electrode upon which is deposited a layer of semiconducting material, such as selenium or copper(I) oxide (see Figure 5-19). The outer surface of the semiconductor is coated with a thin transparent metallic film of gold, silver, or lead, which serves as the second or collector electrode;

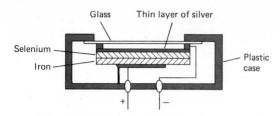

Glass — Thin layer of silver

Selenium —
Iron —
Plastic case

+ —

FIGURE 5-19 Schematic of a typical barrier-layer cell.

the entire array is protected by a transparent envelope. When radiation of sufficient energy reaches the semiconductor, covalent bonds are broken, with the result that conduction electrons and holes are formed. The electrons then migrate toward the metallic film and the holes toward the base upon which the semiconductor is deposited. The liberated electrons are free to migrate through the external circuit to interact with these holes. The result is an electrical current whose magnitude is proportional to the number of photons striking the semiconductor surface. Ordinarily, the currents produced by a photovoltaic cell are large enough to be measured with a galvanometer or microammeter; if the resistance of the external circuit is kept small, the magnitude of the photocurrent is directly proportional to the power of the radiation striking the cell. Currents on the order of 10 to 100 μA are typical.

The barrier-layer cell constitutes a rugged, low-cost means for measuring radiant power. No external source of electrical energy is required. On the other hand, the low internal resistance of the cell makes the amplification of its output less convenient. Thus, although the barrier-layer cell provides a readily measured response at high levels of illumination, it suffers from lack of sensitivity at low levels.

Another disadvantage of the barrier-type cell is that it exhibits *fatigue*; that is, its current output decreases gradually during continual illumination; proper circuit design and

choice of experimental conditions minimize this effect.

Barrier-type cells are widely used in simple, portable, instruments where ruggedness and low cost are important. For routine analyses, these instruments often provide perfectly reliable analytical data.

Phototubes. A second type of photoelectric device is the phototube, which consists of a semicylindrical cathode and a wire anode sealed inside an evacuated transparent envelope. The concave surface of the electrode supports a layer of photoemissive material (p. 100) that tends to emit electrons upon being irradiated. When a potential is applied across the electrodes, the emitted electrons flow to the wire anode, generating a photocurrent. The currents produced are generally about one-tenth as great as those from a photovoltaic cell for a given radiant intensity. In contrast, however, amplification is easily accomplished since the phototube has a high electrical resistance. Figure 3-19 (p. 59) is a schematic diagram of a typical phototube arrangement.

The number of electrons ejected from a photoemissive surface is directly proportional to the radiant power of the beam striking that surface. As the potential applied across the two electrodes of the tube is increased, the fraction of the emitted electrons reaching the anode rapidly increases; when the saturation potential is achieved, essentially all of the electrons are collected at the anode. The current then becomes independent of potential and directly proportional to the radiant power (see Figure 5-20). Phototubes are usually operated at a potential of about 90 V, which is well within the saturation region.

A variety of photoemissive surfaces are used in commercial phototubes. Some consist of one or more of the alkali metals intermixed with alkali metal oxides and silver oxide. This mixture is formed on the surface of the cathode. Other surfaces contain such semiconductors as Cs_3Sb or Na_2KSb mixed with small amounts of the alkali metals. Some

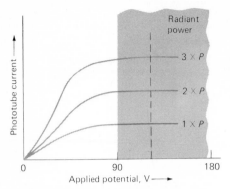

FIGURE 5-20 Variation of phototube response with the applied potential at three levels of illumination. The shaded portion indicates the saturation region. Note the linearity of response at the higher applied potentials.

cathodes are semitransparent, with the photoemissive material being present on the surface opposite that which faces the radiation.

As shown in Figure 5-21, the photoemissive coating on the cathode determines the spectral response of a phototube. When sensitivity to radiation below 350 nm is required, the tube must be supplied with an envelope or window of quartz. Phototubes are available for various ranges within the region between 180 and 1000 nm.

Phototubes frequently produce a small current in the absence of radiation; this *dark current* (see Equation 5-10) results from thermally induced electron emission.

Photomultiplier Tubes. For the measurement of low radiant power, the *photomultiplier* tube offers advantages over the ordinary phototube. Figure 5-22 is a schematic diagram of such a device. The cathode surface is similar in composition to that of a phototube, electrons being emitted upon exposure to radiation. The tube also contains additional electrodes (nine in Figure 5-22) called *dynodes*. Dynode 1 is maintained at a potential 90 V more positive than the cathode, and

electrons are accelerated toward it as a consequence. Upon striking the dynode, each photoelectron causes emission of several additional electrons; these, in turn, are accelerated toward dynode 2, which is 90 V more positive than dynode 1. Again, several electrons are emitted for each electron striking the surface. By the time this process has been repeated nine times, 10^6 to 10^7 electrons have been formed for each photon; this cascade is finally collected at the anode. The resulting current is then electronically amplified and measured.

Semiconductor Detectors. The most sensitive detectors for monitoring radiation in the near-infrared region of about 0.75 to 3 μm are photoconductors whose resistances decrease when radiation within this range is absorbed.

Crystalline semiconductors are formed from the sulfides, selenides, and stibnides of such metals as lead, cadmium, gallium, and indium. Absorption of radiation by these

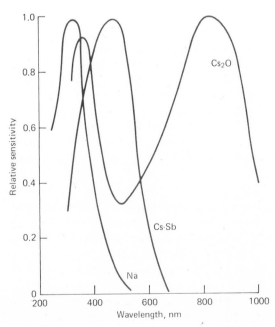

FIGURE 5-21 Spectral responses of some photoemissive surfaces.

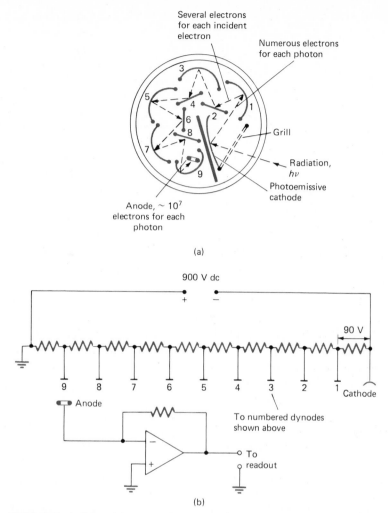

FIGURE 5-22 Photomultiplier tube: (a) cross section of the tube; (b) electrical circuit.

materials promotes some of their bound electrons into an energy state in which they are free to conduct electricity. The resulting change in conductivity can then be measured with a Wheatstone bridge arrangement (Figure 2-24, p. 35) or, alternatively, with a circuit such as that shown in Figure 3-21b (p. 62).

The most widely used photoconductive material is lead sulfide, which is sensitive in the region between 0.8 and 2 μm (12,500 to 5000 cm^{-1}). A thin layer of this compound is deposited on glass or quartz plates to form the cell. The entire assembly is then sealed in an evacuated container to protect the semiconductor from reaction with the atmosphere.

Silicon Diode Detectors. A silicon diode detector consists of a reverse-biased *pn* junction formed on a silicon chip. As shown in Figure 3-1d (p. 43), the reverse bias creates a depletion layer which reduces the conductance of the junction to nearly zero. If radiation is allowed to impinge on the *n* region, however, holes and electrons are formed. The former can diffuse through the depletion layer to the *p* region where they are annihilated; an increase in conductance, which is proportional to radiant power, is the consequence.

A silicon diode detector is not so sensitive as a photomultiplier tube. It has, however, become important because arrays of these detectors can be formed on the surface of a single silicon chip. These chips are an important component of *vidicon tubes*.[8]

Figure 5-23a is a schematic diagram of a vidicon tube. It is similar in construction to a television tube in which a stream of electrons is employed to scan a target area with a series of successive horizontal sweeps. The target in Figure 5-23 is a silicon diode array with a diameter of 16 mm and containing over 15,000 photodiodes per square millimeter.

Figure 5-23b is a magnified cross section and end view of a part of the target array of photodiodes. Each photodiode consists of a cylindrical section of *p*-type silicon surrounded by an insulating layer of silicon dioxide. Each diode is thus electrically independent of neighboring diodes; all are connected to the common *n*-type layer. It should be emphasized that the schematic has been greatly magnified; indeed, the distances between centers of the *p* regions is only about 8 μm.

When the surface of the target is swept by the electron beam, each of the *p*-type cylinders becomes successively charged to the potential of the beam and acts as a small charged capacitor, which remains charged (except for leakage) until a photon strikes the *n*-type surface opposite to it. Then, holes are created which diffuse toward the other surface and discharge several of the capacitors that lie along the extension of the path of the incident beam. The electron beam then recharges these capacitors when it again reaches them in the course of its sweeps (see the end view, Figure 5-23b). This charging current is then amplified and serves as the signal.

The width of the electron beams is such (~ 20 μm) that the surface of the target is effectively divided into 500 lines or channels. The signal from each of these channels can be stored separately in a computer memory. Thus, if the vidicon tube is placed at the focal plane of a monochromator, the signal from each channel corresponds to radiation of a different wavelength. It therefore becomes possible to obtain a spectrum without movement of the monochromator. That is, scanning is accomplished electronically rather than optically, and channel numbers rather than monochromator settings serve as a measure of wavelength for spectra.

An instrument consisting of a monochromator, vidicon tube, and computer, such as that just described, is called an *optical multichannel analyzer*. It is of considerable importance because it allows a portion or all of a spectrum to be recorded essentially simultaneously. A disadvantage is that the physical size of the detector places a limit on either the wavelength range or the resolution of such an analyzer. For example, when a vidicon tube was mounted at the focal plane of a typical grating monochromator, the face of the tube corresponded to only about 27.0 nm for any given wavelength setting.[9] The resolution, however, was better than 1 nm and allowed the simultaneous determination of lead, tin, nickel, and iron in a brass, based upon the emission lines of these elements between 280

[8] For a discussion of vidicon tubes and other analogous devices, see: Y. Talmi, *Anal. Chem.*, **47**, 697A, 658A (1975); and G. Horlick and E. G. Codding, in *Contemporary Topics in Analytical and Clinical Chemistry*, eds. D. M. Hercules, G. M. Hieftje, L. R. Snyder, and M. A. Evenson. New York: Plenum Press, 1977, vol. 1, Chapter 4.

[9] E. G. Codding and G. Horlick, *Spectrosc. Lett.*, **7** (1), 33 (1974).

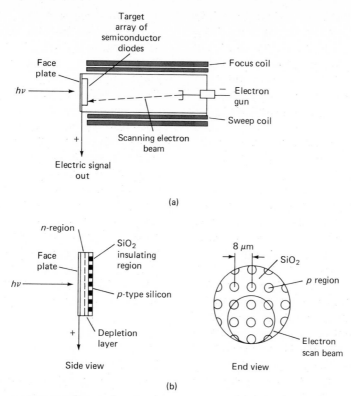

FIGURE 5-23 A vidicon tube detector: (a) vidicon tube and (b) target array of silicon diodes.

and 307 nm. By sacrificing resolution, a much larger wavelength range can be covered. Thus, a compact multichannel spectrometer for field use has been designed that covers a 650 nm wavelength range[10]; its resolution, however, is only 10 nm. Instruments have also been designed to provide a two dimensional display of spectra on the surface of a vidicon tube. In one of these, an optical system is employed which divides the ultraviolet-visible spectrum into six 100 nm segments which are displayed along the horizontal axis of the tube at different vertical positions.[11] A spectral resolution of about 1 nm is achieved over the range from 200 to 800 nm.

Heat Detectors

The measurement of infrared radiation is difficult as a result of the low intensity of available sources and the low energy of the infrared photon. As a consequence of these properties, the electrical signal from an infrared detector is small, and its measurement requires large amplification. It is usually the

[10] G. A. H. Walker, *et al., Rev. Sci. Instrum.*, **45**, 1349 (1974).

[11] R. M. Hoffman and H. L. Pardue, *Anal. Chem.*, **51**, 1267 (1979).

detector system that limits the sensitivity and the precision of an infrared instrument.[12]

The convenient phototubes discussed earlier are generally not applicable in the infrared because the photons in this region lack the energy to cause photoemission of electrons. Thus, thermal detectors and detection based upon photoconduction are required. Neither of these is as satisfactory as the photocell.

Thermal detectors, whose responses depend upon the heating effect of the radiation, can be employed for detection of all but the shortest infrared wavelengths. With these devices, the radiation is absorbed by a small black body and the resultant temperature increase is measured. The radiant power level from a spectrophotometer beam is minute (10^{-7} to 10^{-9} W), so that the heat capacity of the absorbing element must be as small as possible if a detectable temperature rise is to be produced. Every effort is made to minimize the size and thickness of the absorbing element and to concentrate the entire infrared beam on its surface. Under the best of circumstances, temperature changes are confined to a few thousandths of a degree Celsius.

The problem of measuring infrared radiation by thermal means is compounded by thermal effects from the surroundings. The absorbing element is housed in a vacuum and is carefully shielded from thermal radiation emitted by other nearby objects. In order to minimize the effects of extraneous heat sources, a chopped beam is always employed in infrared measurements. Thus, the analyte signal, after transduction, has the frequency of the chopper; with appropriate circuitry, this signal is readily separated from extraneous radiation signals, which ordinarily vary only slowly with time.

Thermocouples. In its simplest form, a thermocouple consists of the junctions formed when two pieces of a given metal are fused to either end of a dissimilar metal or a semiconducting metal alloy. A potential develops between the two thermocouple junctions which varies with the temperature *difference* between the two junctions. For most applications, one of the junctions (the *reference junction*) is held at a constant temperature (often in an ice bath) and the second junction serves as the temperature-sensitive detector.

The junction for infrared detection is formed from very fine wires of such metals as platinum and silver or antimony and bismuth. Alternatively, the detector junction may be constructed by evaporating the metals onto a nonconducting support. In either case, the junction is usually blackened (to improve its heat absorbing capacity) and sealed in an evacuated chamber with a window that is transparent to infrared radiation.

The reference junction is designed to have a relatively large heat capacity and is carefully shielded from the incident radiation. Because the analyte signal is chopped, only the difference in temperature between the two junctions is important; therefore, the reference junction does not need to be maintained at constant temperature.

A well-designed thermocouple detector is capable of responding to temperature differences of 10^{-6} °C. This figure corresponds to a potential difference of about 6 to 8 $\mu V/\mu W$. The thermocouple of an infrared detector is a low-impedance device; it is frequently connected to a high-impedance preamplifier such as that shown in Figure 5-24.

Bolometers. A bolometer is a type of resistance thermometer constructed of strips of metals such as platinum or nickel, or from a semiconductor; the latter devices are sometimes called *thermistors*. These materials exhibit a relatively large change in resistance as a function of temperature. The responsive element is kept small and blackened to

[12] For further information on infrared detectors, see: G. W. Ewing, *J. Chem. Educ.*, **48** (9), A521 (1971); and H. Levinstein, *Anal. Chem.*, **41** (14), 81A (1969).

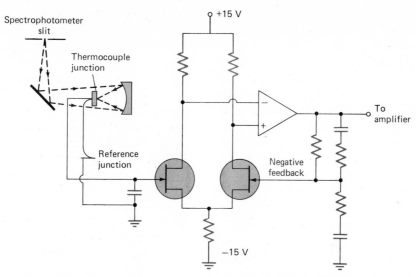

FIGURE 5-24   Thermocouple and preamplifier. (Adapted from G. W. Ewing, *J. Chem. Educ.*, **48**(9), A521, (1971). With permission.)

absorb the radiant heat. Bolometers are not so extensively used as other infrared detectors.

Pyroelectric Detectors. Certain crystals, such as lithium tantalate, barium titanate, and triglycine sulfate, possess temperature-sensitive dipole moments. When placed between metal plates, a temperature-sensitive capacitor is formed which can be employed for measuring the power of infrared radiation. Here, the transduced signal is capacitance.

The Golay Detector. The Golay detector is essentially a sensitive gas thermometer in which xenon is contained in a small cylindrical chamber, which contains a blackened membrane. One end of the cylinder is sealed with an infrared window; the other consists of a flexible diaphragm which is silvered on the outside. A light beam is reflected from this silvered surface to the cathode of a vacuum phototube. When infrared radiation enters the cell, the blackened membrane is warmed, which in turn heats the xenon by conduction. The resulting increase in pressure causes distortion of the silvered diaphragm. As a consequence, the fraction of the reflected light beam that strikes the active surface of the phototube is changed; a change in photocurrent results, which can be related to the power of the infrared beam.

The Golay cell is more expensive than other heat detectors and no more sensitive to near- and mid-infrared radiation; thus, it is seldom employed for these spectral regions. On the other hand, the Golay detector is significantly superior for radiation with wavelengths greater than 50 μm (200 cm^{-1}); it finds general use in instruments designed for the far-infrared region.

SIGNAL PROCESSORS AND READOUTS

The signal processor is ordinarily an electronic device that amplifies the electrical signal from the detector; in addition, it may alter the signal from dc to ac (or the reverse), change the phase of the signal, and filter it to remove unwanted components. Furthermore, the signal processor may be called upon to perform such mathematical operations on the signal as differentiation, integration, or conversion to a logarithm.

Several types of the readout devices described in Chapters 2 and 3 are found in modern instruments. Some of these include the d'Arsonval meter, digital meters, the scales of potentiometers, recorders, and cathode-ray tubes.

Photon Counting

Ordinarily, the output from the photoelectric detectors described in the previous section is processed and displayed by analog techniques. That is, the average current, potential, or conductance associated with the detector is amplified and recorded or fed into a suitable meter. As was indicated in Chapter 3 (p. 83), analog signals such as these vary continuously; in spectroscopy, they are generally proportional to the average radiant power of the incident beam.

In some instances, it is possible and advantageous to employ direct digital techniques wherein electrical pulses produced by individual photons are counted. Here, radiant power is proportional to the number of pulses rather than to an average current or potential.

Counting techniques have been used for many years for measuring the power of X-ray beams and of radiation produced by the decay of radioactive species (these techniques are considered in detail in Chapter 15). Only recently, however, has photon counting been applied to ultraviolet and visible radiation.[13] Here, the output of a photomultiplier tube is employed. In the previous section, it was indicated that a single photon striking the cathode of a photomultiplier ultimately leads to a cascade of 10^6 to 10^7 electrons or a pulse of current which can be amplified and counted.

Generally, the equipment for photon counting includes a pulse-height discriminator (see Figure 15-13), which rejects pulses unless they exceed some predetermined minimum voltage. Such a device is useful because dark current and instrument noise are often significantly smaller than the signal pulse and are thus not counted; an improved signal-to-noise ratio results.

Photon counting has a number of advantages over analog signal processing, including improved signal-to-noise ratio, sensitivity to low radiation levels, improved precision for a given measurement time, and lowered sensitivity to voltage and temperature changes. The required equipment is, however, more complex and expensive; the technique has not yet been widely applied for routine measurements in the ultraviolet and visible regions.

---

[13] For a review of photon counting, see: H. V. Malmstadt, M. L. Franklin, and G. Horlick, *Anal. Chem.*, **44** (8), 63A (1972).

PROBLEMS

1. State the differences between and the relative advantages of Littrow and Bunsen prism monochromators.

2. What are the relative advantages and disadvantages of gratings and prisms for dispersion of light?

3. Why must the slit width of a prism monochromator be varied to provide constant resolution whereas a nearly constant slit width may be used with a grating monochromator?

4. Why do quantitative and qualitative analyses often require different monochromator slit widths?

5. For a grating, how many lines per centimeter would be required in

order for the first-order diffraction line for $\lambda = 500$ nm to be observed at a reflection angle of -40 deg when the angle of incidence is 60 deg?

6. Consider an infrared grating with 72.0 lines per millimeter and 5.00 mm of illuminated area. Calculate the first-order resolution ($\lambda/\Delta\lambda$) of this grating. How far apart (in cm^{-1}) must two lines centered at 1000 cm^{-1} be if they are to be resolved?

7. Calculate the angular dispersion ($d\theta/d\lambda$) at 500 nm for a Cornu type prism having a prism angle α of 60 deg cm^{-1} assuming the material of construction is
   (a) a dense flint glass.
   (b) fused quartz.

8. Use the curves in Figure 5-9 to estimate $dn/d\lambda$ and n at 500 nm for prisms constructed from the materials in Problem 7.

9. A common infrared source consists of an inert solid that behaves as a black-body radiator when electrically heated. The energy density of black-body radiation is given by the equation

$$\text{energy density at wavelength } \lambda = \frac{8\pi ch}{\lambda^5}\left(e^{ch/\lambda kT} - 1\right)^{-1}$$

where c is the speed of light, h is Planck's constant, k is Boltzmann's constant, and T is temperature in °K. The total amount of radiant energy per second per square centimeter of surface of a black body is given by $E = (5.70 \times 10^{-5} \text{ erg cm}^{-2}\text{ s}^{-1}\text{ °K}^{-4})T^4$, where T is temperature. The wavelength of maximum emission energy is given by $\lambda_{max} = 0.289$ cm °K$/T$ °K. A graph of black-body radiation intensity is shown in Figure 4-13.
   (a) Calculate λ_{max} for black body radiators at 1500°K and at 1000°K.
   (b) Calculate the ratio of energy density at 1500 cm^{-1} compared to that at λ_{max} for black bodies at 1500°K and at 1000°K.
   (c) The center of a typical infrared spectrum is at a wavenumber near 1500 cm^{-1}. Why is it advantageous in infrared spectroscopy for the source to be heated to 1500°K instead of 1000°K?

6

AN INTRODUCTION TO ABSORPTION SPECTROSCOPY

This chapter provides introductory material which is applicable to ultraviolet-visible, infrared, and atomic absorption spectroscopy.[1] Detailed discussion of these methods appears in later chapters.

TERMS EMPLOYED IN ABSORPTION SPECTROSCOPY

Table 6-1 lists the common terms and symbols employed in absorption spectroscopy. In recent years, a considerable effort has been made by the American Society for Testing Materials to develop a standard nomenclature; the terms and symbols listed in the first two columns of Table 6-1 are based on ASTM recommendations. Column 3 contains alternative symbols that will be found in the older literature. A standard nomenclature seems most worthwhile in order to avoid ambiguities; the reader is, therefore, urged to learn and use the recommended terms and symbols.

Transmittance

Figure 6-1 depicts a beam of parallel radiation before and after it has passed through a layer of solution with a thickness of b cm and a concentration of c of an absorbing species. As a consequence of interactions between the photons and absorbing particles, the power of the beam is attenuated from P_0 to P. The *transmittance T* of the solution is the fraction

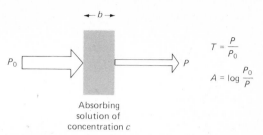

FIGURE 6-1 Attenuation of a beam of radiation by an absorbing solution.

of incident radiation transmitted by the solution. That is,

$$T = P/P_0 \qquad (6\text{-}1)$$

Transmittance is often expressed as a percentage.

Absorbance

The absorbance of a solution is defined by the equation

$$A = -\log_{10} T = \log \frac{P_0}{P} \qquad (6\text{-}2)$$

Note that, in contrast to transmittance, the absorbance of a solution increases as the attenuation of the beam becomes larger.

Absorptivity and Molar Absorptivity

As will be shown presently, absorbance is directly proportional to the path length through the solution and the concentration of the absorbing species. That is,

$$A = abc \qquad (6\text{-}3)$$

where a is a proportionality constant called the *absorptivity*. The magnitude of a will clearly depend upon the units used for b and c. When the concentration is expressed in moles per liter and the cell length is in centimeters, the absorptivity is called the *molar absorptivity* and given the special symbol ε.

[1] For more detailed treatment of absorption spectroscopy, see: R. P. Bauman, *Absorption Spectroscopy*. New York: Wiley, 1962; J. R. Edisbury, *Practical Hints on Absorption Spectrometry*. New York: Plenum Press, 1968; F. Grum, in *Physical Methods of Chemistry*, eds. A. Weissberger and B. W. Roositer. New York: Wiley-Interscience, 1972, Vol. I, Part III B, Chapter 3; G. F. Lothian, *Absorption Spectrophotometry*, 3d ed. London: Adam Hilger Ltd., 1969; and J. E. Crooks, *The Spectrum in Chemistry*. London: Academic Press, 1978.

Thus, when b is in centimeters and c is in moles per liter,

$$A = \varepsilon bc \qquad (6\text{-}4)$$

Experimental Measurement of P and P_0

The relationship given by Equation 6-1 or 6-2 is not directly applicable to chemical analysis. Neither P nor P_0, as defined, can be conveniently measured in the laboratory because the solution to be studied must be held in some sort of container. Interaction between the radiation and the walls is inevitable leading to a loss by reflection at each interface; moreover, significant absorption may occur within the walls themselves. Finally, the beam may suffer a diminution in power during its passage through the solution as a result of scattering

by large molecules or inhomogeneities. Reflection losses can be appreciable; for example, about 4% of a beam is reflected upon vertical passage of visible radiation across an air-to-glass or glass-to-air interface (see example, p. 103).

In order to compensate for these effects, the power of the beam transmitted through the absorbing solution is generally compared with that which passes through an identical cell containing the solvent for the sample. An experimental absorbance that closely approximates the true absorbance of the solution can then be obtained; that is,

$$A = \log \frac{P_{\text{solvent}}}{P_{\text{solution}}} = \log \frac{P_0}{P}$$

The terms P_0 and P, when used henceforth, refer to the power of a beam of

TABLE 6-1   IMPORTANT TERMS AND SYMBOLS EMPLOYED IN ABSORPTION MEASUREMENT

Term and Symbol[a]	Definition	Alternative Name and Symbol
Radiant power, P, P_0	Energy of radiation (in ergs) impinging on a 1-cm^2 area of a detector per second	Radiation intensity, I, I_0
Absorbance, A	$\log \dfrac{P_0}{P}$	Optical density, D; extinction, E
Transmittance, T	$\dfrac{P}{P_0}$	Transmission, T
Path length of radiation, in cm b	—	l, d
Absorptivity,[b] a	$\dfrac{A}{bc}$	Extinction coefficient, k
Molar absorptivity,[c] ε	$\dfrac{A}{bc}$	Molar extinction coefficient

[a] Terminology recommended by the American Chemical Society. Reprinted with permission from *Anal. Chem.*, **24**, 1349 (1952); **48**, 2298 (1976). Copyright by the American Chemical Society.
[b] c may be expressed in g/liter or other specified concentration units; b may be expressed in cm or in other units of length.
[c] c is expressed in units of mol/liter.

radiation after it has passed through a cell containing the solvent or the solution of the analyte.

Measurement of Transmittance and Absorbance

Transmittance (or absorbance) measurements are made with instruments having the components arranged as shown in Figure 5-1b (p. 115). The electrical output G of the detector of such an instrument is given by (see p. 138).

$$G = KP + K' \qquad (6\text{-}5)$$

where K' is the dark current that often exists when no radiation strikes the transducer and K is a proportionality constant.

In many instruments, the readout device will consist of a linear scale calibrated in units of 0 to $100\% \ T$. A transmittance measurement with such an instrument requires three steps. First, with the light beam blocked from the transducer by means of a shutter, an electrical adjustment is made until the pointer of the readout device is at exactly zero; this step is called the *dark current* or *$0\% \ T$ adjustment*. A *$100\% \ T$ adjustment* is then carried out with the shutter open and the solvent in the light path. This adjustment may involve increasing or decreasing the power output of the source electrically; alternatively, the power of the beam may be varied with an adjustable diaphragm or by appropriate positioning of a comb or optical wedge. After this step, we may write Equation 6-5 in the form

$$G_0 = 100 = KP_0 + 0.00$$

In the final step of the measurement, the solvent cell is replaced by the sample. The meter reading is then given by

$$G = KP + 0.00$$

Dividing this equation by the previous one yields

$$G = \frac{P}{P_0} \times 100 = T \times 100$$

The meter is thus made direct reading in percent transmittance. Obviously, an absorbance scale can also be scribed on the meter face.

QUANTITATIVE ASPECTS OF ABSORPTION MEASUREMENTS

The remainder of this chapter is devoted to an examination of Equation 6-4 ($A = \varepsilon bc$), with particular attention to causes of deviations from this relationship. In addition, consideration is given to the effects that uncertainties in the measurement of P and P_0 have on absorbance (and thus concentration).

Beer's Law

Equations 6-3 and 6-4 are statements of *Beer's law*. These relationships can be rationalized as follows.[2] Consider the block of absorbing matter (solid, liquid, or gas) shown in Figure 6-2. A beam of parallel monochromatic radiation with power P_0 strikes the block perpendicular to a surface; after passing through a length b of the material, which

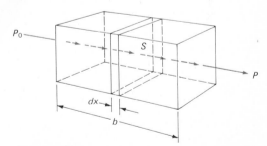

FIGURE 6-2 Attenuation of radiation with initial power P_0 by a solution containing c moles per liter of absorbing solute with a path length of b cm. $P < P_0$.

---

[2] The discussion that follows is based on a paper by F. C. Strong, *Anal. Chem.*, **24**, 338 (1952). For a rigorous derivation of the law, see: D. J. Swinehart, *J. Chem. Educ.*, **39**, 333 (1972).

contains n absorbing particles (atoms, ions, or molecules), its power is decreased to P as a result of absorption. Consider now a cross section of the block having an area S and an infinitesimal thickness dx. Within this section there are dn absorbing particles; associated with each particle, we can imagine a surface at which photon capture will occur. That is, if a photon reaches one of these areas by chance, absorption will follow immediately. The total projected area of these capture surfaces within the section is designated as dS; the ratio of the capture area to the total area, then, is dS/S. On a statistical average, this ratio represents the probability for the capture of photons within the section.

The power of the beam entering the section, P_x, is proportional to the number of photons per square centimeter per second, and dP_x represents the quantity removed per second within the section; the fraction absorbed is then $-dP_x/P_x$, and this ratio also equals the average probability for capture. The term is given a minus sign to indicate that P undergoes a decrease. Thus,

$$-\frac{dP_x}{P_x} = \frac{dS}{S} \tag{6-6}$$

Recall, now, that dS is the sum of the capture areas for particles within the section; it must therefore be proportional to the number of particles, or

$$dS = a\,dn \tag{6-7}$$

where dn is the number of particles and a is a proportionality constant, which can be called the *capture cross section*. Combining Equations 6-6 and 6-7 and summing over the interval between zero and n, we obtain

$$-\int_{P_0}^{P} \frac{dP_x}{P_x} = \int_{0}^{n} \frac{a\,dn}{S}$$

which, upon integration, gives

$$-\ln \frac{P}{P_0} = \frac{an}{S}$$

Upon converting to base 10 logarithms and inverting the fraction to change the sign, we obtain

$$\log \frac{P_0}{P} = \frac{an}{2.303S} \tag{6-8}$$

where n is the total number of particles within the block shown in Figure 6-2. The cross-sectional area S can be expressed in terms of the volume of the block V and its length b. Thus,

$$S = \frac{V}{b} \text{ cm}^2$$

Substitution of this quantity into Equation 6-8 yields

$$\log \frac{P_0}{P} = \frac{anb}{2.303V} \tag{6-9}$$

Note that n/V has the units of concentration (that is, number of particles per cubic centimeter); we can readily convert n/V to moles per liter. Thus,

$$c = \frac{n \text{ particles}}{6.02 \times 10^{23} \text{ particles/mol}}$$

$$\times \frac{1000 \text{ cm}^3/\text{liter}}{V \text{ cm}^3} = \frac{1000n}{6.02 \times 10^{23}V} \text{ mol/liter}$$

Combining this relationship with Equation 6-9 yields

$$\log \frac{P_0}{P} = \frac{6.02 \times 10^{23}abc}{2.303 \times 1000}$$

Finally, the constants in this equation can be collected into a single term ε to give

$$\log \frac{P_0}{P} = \varepsilon bc = A \tag{6-10}$$

which is, of course, a statement of Beer's law.

Application of Beer's Law to Mixtures

Beer's law also applies to a solution containing more than one kind of absorbing substance. Provided there is no interaction among the various species, the total absorbance for a multicomponent system is given by

$$A_{\text{total}} = A_1 + A_2 + \cdots + A_n \qquad (6\text{-}11)$$
$$= \varepsilon_1 b c_1 + \varepsilon_2 b c_2 + \cdots + \varepsilon_n b c_n$$

where the subscripts refer to absorbing components 1, 2, ..., n.

Limitations to the Applicability of Beer's Law

The linear relationship between absorbance and path length at a fixed concentration of absorbing substances is a generalization for which no exceptions are known. On the other hand, deviations from the direct proportionality between the measured absorbance and concentration when b is constant are frequently encountered. Some of these deviations are fundamental and represent real limitations of the law. Others occur as a consequence of the manner in which the absorbance measurements are made or as a result of chemical changes associated with concentration changes; the latter two are sometimes known, respectively, as *instrumental deviations* and *chemical deviations*.

Real Limitations to Beer's Law. Beer's law is successful in describing the absorption behavior of dilute solution only; in this sense, it is a limiting law. At high concentrations (usually $> 0.01M$), the average distance between the species responsible for absorption is diminished to the point where each affects the charge distribution of its neighbors. This interaction, in turn, can alter their ability to absorb a given wavelength of radiation. Because the extent of interaction depends upon concentration, the occurrence of this phenomenon causes deviations from the linear relationship between absorbance and concentration. A similar effect is sometimes encountered in solutions containing low absorber concentrations and high concentrations of other species, particularly electrolytes. The close proximity of ions to the absorber alters the molar absorptivity of the latter by electrostatic interactions; the effect is lessened by dilution.

While the effect of molecular interactions is ordinarily not significant at concentrations below $0.01M$, some exceptions are encountered among certain large organic ions or molecules. For example, the molar absorptivity at 436 nm for the cation of methylene blue is reported to increase by 88% as the dye concentration is increased from 10^{-5} to $10^{-2}M$; even below $10^{-6}M$, strict adherence to Beer's law is not observed.

Deviations from Beer's law also arise because ε is dependent upon the refractive index of the solution.[3] Thus, if concentration changes cause significant alterations in the refractive index n of a solution, departures from Beer's law are observed. A correction for this effect can be made by substitution of the quantity $\varepsilon n/(n^2 + 2)^2$ for ε in Equation 6-10. In general, this correction is never very large and is rarely significant at concentrations less than $0.01M$.

Chemical Deviations. Apparent deviations from Beer's law are frequently encountered as a consequence of association, dissociation, or reaction of the absorbing species with the solvent. A classic example of a chemical deviation occurs in unbuffered potassium dichromate solutions, in which the following equilibria exist:

$$Cr_2O_7^{2-} + H_2O$$
$$\rightleftharpoons 2HCrO_4^- \rightleftharpoons 2H^+ + 2CrO_4^{2-}$$

At most wavelengths, the molar absorptivities of the dichromate ion and the two chromate species are quite different. Thus, the total absorbance of any solution depends upon the concentration ratio between the dimeric and the monomeric forms. This ratio, however, changes markedly with dilution, and causes a

---

[3] G. Kortum and M. Seiler, *Angew. Chem.*, **52**, 687 (1939).

pronounced deviation from linearity between the absorbance and the total concentration of chromium(VI). Nevertheless, the absorbance due to the dichromate ion remains directly proportional to its molar concentration; the same is true for the chromate ions. This fact is easily demonstrated by measuring the absorbance of chromium(VI) solutions in strongly acidic solution, where dichromate is the principal species, and in strongly alkaline solution, where chromate predominates. Thus, deviations in the absorbance of this system from Beer's law are more apparent than real because they result from shifts in chemical equilibria. These deviations can, in fact, be readily predicted from the equilibrium constants for the reactions and the molar absorptivities of the dichromate and chromate ions.

EXAMPLE

Two solutions of the acid-base indicator HIn were prepared by diluting $1.02 \times 10^{-4}F$ solutions of HIn ($K_a = 1.42 \times 10^{-5}$) with equal volumes of: (a) $0.2F$ NaOH; and (b) $0.2F$ HCl. When measured in a 1.00-cm cell, the following absorbance data were obtained at 430 and 570 nm:

Solution Diluted With	A_{430}	A_{570}
$0.2F$ NaOH	1.051	0.049
$0.2F$ HCl	0.032	0.363

Derive absorbance data at the two wavelengths for unbuffered solutions having indicator concentrations in the range of 2×10^{-5} to 16×10^{-5} F. Plot the data.

The absorbance of the various solutions depends upon the equilibrium

$$HIN \rightleftharpoons H^+ + In^-$$

for which

$$K_a = 1.42 \times 10^{-5} = \frac{[H^+][In^-]}{[HIn]}$$

Substitution of the hydrogen ion concentration of the strong acid or base solutions

into the expression for K_a makes it apparent that, in the presence of strong base, essentially all of the indicator is dissociated to In^-; therefore,

$$[In^-] \cong (1.02 \times 10^{-4})/2 = 5.10 \times 10^{-5}$$

Similarly, in the strong acid $[HIn] \cong 5.10 \times 10^{-5}$. The molar absorptivities for the two species can thus be determined. At 430 nm,

$$\varepsilon'_{In} = \frac{A}{bc} = \frac{1.051}{1.00 \times 5.10 \times 10^{-5}}$$

$$= 2.06 \times 10^4$$

$$\varepsilon'_{HIn} = \frac{0.032}{5.10 \times 10^{-5}} = 6.3 \times 10^2$$

and at 570 nm,

$$\varepsilon''_{In} = \frac{0.049}{5.10 \times 10^{-5}} = 9.6 \times 10^2$$

$$\varepsilon''_{HIn} = \frac{0.363}{5.10 \times 10^{-5}} = 7.12 \times 10^3$$

Let us calculate the concentration of HIn and In^- in an unbuffered $2.00 \times 10^{-5}F$ solution of HIn. From the equation for the dissociation reaction, we see that

$$[H^+] = [In^-]$$

Furthermore,

$$[In^-] + [HIn] = 2.00 \times 10^{-5}$$

Substitution of these relationships into the expression for K_a gives

$$\frac{[In^-]^2}{2.00 \times 10^{-5} - [In^-]} = 1.42 \times 10^{-5}$$

Rearrangement yields the quadratic expression

$$[In^-]^2 + 1.42 \times 10^{-5}[In^-]$$
$$- 2.84 \times 10^{-10} = 0$$

which gives

$$[In^-] = 1.12 \times 10^{-5}$$

$$[HIn] = 2.00 \times 10^{-5} - 1.12 \times 10^{-5}$$

$$= 0.88 \times 10^{-5}$$

We are now able to calculate the absorbance at the two wavelengths. Thus,

$$A_{430} = \varepsilon'_{In} b[In^-] + \varepsilon'_{HIn} b[HIn]$$

$$= 2.06 \times 10^4 \times 1.00 \times 1.12 \times 10^{-5}$$

$$+ 6.3 \times 10^2 \times 1.00 \times 0.88 \times 10^{-5}$$

$$= 0.236$$

$$A_{570} = 9.6 \times 10^2 \times 1.12 \times 10^{-5}$$

$$+ 7.12 \times 10^3 \times 0.88 \times 10^{-5}$$

$$= 0.073$$

The accompanying data were obtained similarly.

ually, we may write for radiation λ'

$$A' = \log \frac{P'_0}{P'} = \varepsilon' bc$$

or

$$\frac{P'_0}{P'} = 10^{\varepsilon' bc}$$

Similarly, for λ''

$$\frac{P''_0}{P''} = 10^{\varepsilon'' bc}$$

F_{HIn}	[HIn]	[In$^-$]	A_{430}	A_{570}
2.00×10^{-5}	0.88×10^{-5}	1.12×10^{-5}	0.236	0.073
4.00×10^{-5}	2.22×10^{-5}	1.78×10^{-5}	0.381	0.175
8.00×10^{-5}	5.27×10^{-5}	2.73×10^{-5}	0.596	0.401
12.00×10^{-5}	8.52×10^{-5}	3.48×10^{-5}	0.771	0.640
16.00×10^{-5}	11.9×10^{-5}	4.11×10^{-5}	0.922	0.887

Figure 6-3 is a plot of the data derived in the foregoing example and illustrates the kind of departures from Beer's law that arises when the absorbing system is capable of undergoing dissociation or association. Note that the direction of curvature is opposite for the two wavelengths.

Instrumental Deviations with Polychromatic Radiation. Strict adherence to Beer's law is observed only when the radiation employed is truly monochromatic; this observation is yet another manifestation of the limiting character of the law. Unfortunately, the use of radiation that is restricted to a single wavelength is seldom practical; devices that isolate portions of the output from a continuous source produce a more or less symmetric band of wavelengths around the desired one (see Figures 5-14 and 5-17, for example).

The following derivation shows the effect of polychromatic radiation on Beer's law.

Consider a beam comprising of just two wavelengths λ' and λ''. Assuming that Beer's law applies strictly for each of these individ-

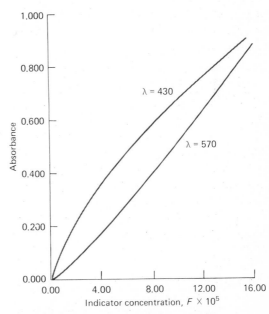

FIGURE 6-3 Chemical deviations from Beer's law for unbuffered solutions of the indicator HIn. For data, see example on this page.

When an absorbance measurement is made with radiation composed of both wavelengths, the power of the beam emerging from the solution is given by $(P' + P'')$ and that of the beam from the solvent by $(P_0' + P_0'')$. Therefore, the measured absorbance is

$$A_M = \log \frac{(P_0' + P_0'')}{(P' + P'')}$$

which can be rewritten as

$$A_M = \log \frac{(P_0' + P_0'')}{(P_0' 10^{-\varepsilon' bc} + P_0'' 10^{-\varepsilon'' bc})}$$

or

$$A_M = \log(P_0' + P_0'')$$
$$- \log(P_0' 10^{-\varepsilon' bc} + P_0'' 10^{-\varepsilon'' bc})$$

Now, when $\varepsilon' = \varepsilon''$, this equation simplifies to

$$A_M = \varepsilon' bc$$

and Beer's law is followed. If the two molar absorptivities differ, however, the relationship between A_M and concentration will no longer be linear; moreover, greater departures from linearity can be expected with increasing differences between ε' and ε''. This derivation can be expanded to include additional wavelengths; the relationships remain the same.

It is an experimental fact that deviations from Beer's law resulting from the use of a polychromatic beam are not appreciable, provided the radiation used does not encompass a spectral region in which the absorber exhibits large changes in absorbance as a function of wavelength. This observation is illustrated in Figure 6-4.

Instrumental Deviations in the Presence of Stray Radiation. We have noted earlier (Chapter 5) that the radiation exiting from a monochromator is ordinarily contaminated with small amounts of scattered or stray radiation which reaches the exit slit owing to reflections from various internal surfaces. Stray radiation often differs greatly in wavelength from that of the principal radiation and, in addition, may not have passed through the sample.

When measurements are made in the presence of stray radiation, the observed absorbance is given by

$$A' = \log \frac{P_0 + P_s}{P + P_s}$$

where P_s is the power of the stray radiation. Figure 6-5 shows a plot of A' versus concen-

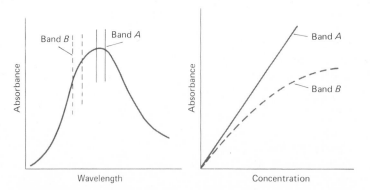

FIGURE 6-4   The effect of polychromatic radiation upon the Beer's law relationship. Band A shows little deviation since ε does not change greatly throughout the band. Band B shows marked deviations since ε undergoes significant changes in this region.

tration for various levels of P_s as compared to P_0.

Note that the instrumental deviations illustrated in Figures 6-4 and 6-5 result in absorbances that are smaller than theoretical. It can be shown that instrumental deviations always lead to negative absorbance errors.[4]

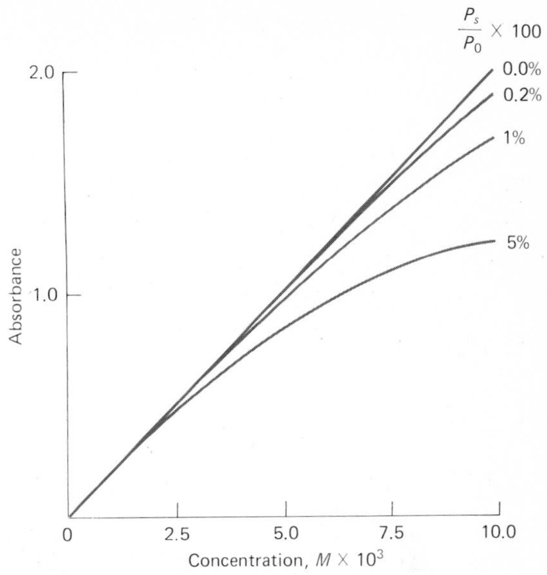

The Effect of Instrumental Noise on the Precision of Spectrophotometric Analyses [5]

The accuracy and precision of spectrophotometric analyses are often limited by the uncertainties or noise associated with the instrument. A general discussion of instrumental noise and signal-to-noise optimization is found in Chapter 3; the reader may find it helpful to review this material before undertaking a detailed study of the effect of instrumental noise on the precision of spectrophotometric measurements. In addition, a review of the contents of Appendix 1 on the use of standard deviation as a measure of precision may also prove worthwhile.

As was pointed out earlier, a spectrophotometric measurement entails three steps: a 0% T adjustment, a 100% T adjustment, and a measurement of % T with the sample in the radiation path. The noise associated with each of these steps combines to give a net uncertainty for the final value obtained for T. The relationship between the noise encountered in the measurement of T and the

FIGURE 6-5   Apparent deviation from Beer's law brought about by various amounts of stray radiation.

uncertainty in concentration can be derived by writing Beer's law in the form

$$c = -\frac{1}{\varepsilon b} \log T = \frac{-0.434}{\varepsilon b} \ln T$$

The partial derivative of this equation, holding b and c constant, is given by

$$\partial c = \frac{-0.434}{\varepsilon b T} \partial T$$

where ∂c is the variation in c that results from the noise or uncertainty ∂T in T. Dividing by the first equation yields

$$\frac{\partial c}{c} = \frac{0.434}{\log T} \times \frac{\partial T}{T} \qquad (6\text{-}12)$$

Here, $\partial T/T$ is the net *relative* uncertainty or noise in T that arises from the three measurement steps and $\partial c/c$ is the resulting relative uncertainty in concentration.

The best and most useful measure of the noise ∂T is the standard deviation s_T, which

[4] I. M. Kolthoff and P. J. Elving, eds., *Treatise on Analytical Chemistry*. New York: Interscience Publishers, 1964, Part I, vol. 5, pp. 2767–2773.

[5] See: L. D. Rothman, S. R. Crouch, and J. D. Ingle, Jr., *Anal. Chem.*, **47**, 1226 (1975); J. D. Ingle, Jr. and S. R. Crouch, *Anal. Chem.*, **44**, 1375 (1972); J. O. Erickson and T. Surles, *American Laboratory*, **8** (6), 41 (1976); *Optimum Parameters for Spectrophotometry*. Varian Instrument Division, Palo Alto, CA.

for a series of replicate measurements is given by (see Appendix 1)[6]

$$s_T = \partial T = \sqrt{\frac{\sum_{i=1}^{N} (T_i - \bar{T})^2}{N - 1}} \quad (6\text{-}13)$$

In this expression, T_i represents the individual values of T obtained for a set of N replicate measurements, and $\bar{T}$ is the average or mean for the set. From Equation 6-12, the *relative* standard deviation in terms of concentration (s_c/c) is then

$$\frac{s_c}{c} = \frac{0.434}{\log T} \times \frac{s_T}{T} \quad (6\text{-}14)$$

Experimentally, s_T can be evaluated by making, say, 20 replicate measurements $(N = 20)$ of the transmittance of a solution in exactly the same way and substituting the data into Equation 6-13.

It is clear from an examination of Equation 6-14 that the uncertainty in a photometric concentration measurement varies in a complex way with the magnitude of the transmittance. The situation is even more complicated than is suggested by Equation 6-14, however, because the uncertainty s_T is, under many circumstances, also *dependent upon T.*

In a detailed theoretical and experimental study, Rothman, Crouch and Ingle have described several sources of instrumental uncertainties and shown their net effect on the precision of absorbance or transmittance measurements. These uncertainties fall into three categories: those for which the magnitude of s_T is (1) proportional to T, (2) proportional to $\sqrt{T^2 + T}$, and (3) independent of T. Table 6-2 summarizes information about these sources of uncertainty. Clearly, each type of source has a different effect upon the magnitude of the concentration error. An error analysis for a given spectrophotometric

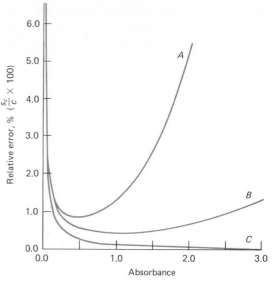

FIGURE 6-6   Error curves for various instrumental uncertainties. Data from Table 6-3. Curve A: 0% T and Johnson noise. Curve B: Signal shot noise. Curve C: Flicker noise.

method thus requires an understanding of the likely type (or types) that predominate in a given set of circumstances.

0% T Uncertainties. As shown in Table 6-2, 0% T noise includes three types of noise associated with transducers and amplifiers; as the name implies all are associated with the uncertainties in making the 0% T adjustment. Sources would include light leaks around the shutter that screens the radiation from the detector, vibrational effects associated with movement of the shutter, and temperature fluctuations in the transducer and amplifier systems. In modern spectrophotometers, 0% T noise is seldom important except at the wavelength extremes of the instrument, where K' in Equation 6-5 may approach KP and KP_0. Note that 0% T noise is independent of the magnitude of T.

Johnson Noise. As noted in Chapter 3 (p. 71), Johnson noise results from thermally induced motion of electrons in resistive circuit elements. This motion momentarily in-

---

[6] For a discussion of the standard deviation as a measure of experimental uncertainties, see: D. A. Skoog and D. M. West, *Fundamentals of Analytical Chemistry,* 3d ed. New York: Holt, Rinehart and Winston, 1976, Chapter 4.

creases or decreases the average number of electrons moving in a given direction in the resistor element. Johnson noise does not depend on the magnitude of the current; indeed, it occurs even in the absence of a current.

Johnson noise exists in all transducers and electronic components of photometers and spectrophotometers. With one important exception, however, its magnitude is so small relative to other sources of noise that for all practical purposes it can be neglected. The exception is for the thermal detectors for infrared radiation. With such detectors, Johnson noise may predominate and limit the precision of spectral measurements. This situation is in distinct contrast to the ultraviolet and visible regions, where Johnson noise in phototubes, photomultiplier tubes, and amplifiers is inconsequential.

Effects of 0% T and Johnson Noise. As we have noted, the magnitude of Johnson noise and 0% T noise is independent of the transmittance. Thus, s_T in Equation 6-14 is *independent of* T and can be assumed to be constant in determining the effect of s_T on s_c. For a given instrument, a good estimate of s_T can be derived by a standard deviation calcu-lation for 20 or 30 independent transmittance measurements of an absorbing solution.

Column 3 of Table 6-3 shows the relative error in concentration that would be encountered with a spectrophotometer for which the performance was limited by Johnson or 0% T noise. Here, an absolute uncertainty s_T in transmittance of ± 0.003, or $\pm 0.3\%$ T, was assumed. These data are plotted in Figure 6-6, curve A, where it is seen that the relative standard deviation of c reaches a minimum at an absorbance of about 0.4. (By setting the derivative of Equation 6-14 equal to zero, it can be shown that this minimum occurs at a transmittance of 0.368 or an absorbance of 0.434.)

The standard deviation in transmittance measurements for most commercial spectrophotometers will range from ± 0.0001 to ± 0.01 ($T = 0.01$ to 1%); the lower figure is found in high-quality ultraviolet-visible instruments, whereas the upper figure might be expected in a low-cost infrared instrument. The figure of ± 0.003 in Table 6-3 is perhaps typical for medium quality 0% T or Johnson noise-limited instruments; concentration errors of about 1 to 2% relative are thus inherent when measured absorbances lie in the

TABLE 6-2   SOURCES OF UNCERTAINTY IN A SPECTROPHOTOMETRIC MEASUREMENT

s_T Independent of T	$s_T = k'\sqrt{T^2 + T}$	$s_T = k''T$
0% T Noise	Detector shot noise (photon detectors only)	Source flicker noise
Dark current shot noise Dark current excess noise Amplifier excess noise		Cell positioning uncertainty
Detector Johnson noise (thermal detectors only)		
Limited readout resolution		
Amplifier shot noise		

range of 0.15 to 1.0. The higher error associated with more strongly absorbing samples can usually be avoided by suitable dilution prior to measurement. Relative errors greater than 2% are inevitable, however, when absorbances are less than 0.1.

It is important to emphasize that the conclusions just described are based on the assumption that the limiting uncertainty in an analysis is a constant indeterminate error s_T that is independent of transmittance; this type of uncertainty is most commonly encountered in instruments employing thermal detectors.

Limited Readout Resolution. The precision of some commercial spectrophotometers is limited by the resolution of their readout devices. For example, an instrument with a meter readable to $\pm 0.5\%$ full scale will have an uncertainty (s_T) of $\pm 0.005T$ due to this source. This uncertainty is also independent of T;

thus, its effect on the relative error in concentration will be similar to $0\% \ T$ and Johnson noise and will be described by Equation 6-14. Substitution of 0.005 into this equation reveals that such an instrument will yield concentration data that are uncertain to the extent of 1.5 to 3% relative, provided percent transmittances are limited to a range of 10 to 80%.

Signal Shot Noise. As was pointed out earlier (p. 71), shot noise must be expected whenever the passage of electricity involves transfer of charge across a junction, such as the movement of electrons from the cathode to the anode of a photomultiplier tube or across the junction of a barrier-type photocell. Here, the current results from a series of discrete events (emission of electrons from a cathode); the number of these events per unit time is distributed in a random way

TABLE 6-3 RELATIVE CONCENTRATION ERROR AS A FUNCTION OF TRANSMITTANCE AND ABSORBANCE FOR VARIOUS SOURCES OF UNCERTAINTY

		Relative Error, $\dfrac{s_c}{c} \times 100$		
		For Measurements Limited by:		
Transmittance, T	**Absorbance, A**	**$0\%\ T$ or Johnson Noise[a]**	**Signal Shot Noise[b]**	**Flicker Noise[c]**
0.95	0.022	$\pm\ 6.2$	± 8.4	± 5.8
0.90	0.046	$\pm\ 3.2$	± 4.1	± 2.8
0.80	0.097	$\pm\ 1.7$	± 2.0	± 1.3
0.60	0.222	$\pm\ 0.98$	± 0.96	± 0.59
0.40	0.398	$\pm\ 0.82$	± 0.61	± 0.33
0.20	0.699	$\pm\ 0.93$	± 0.46	± 0.18
0.10	1.00	$\pm\ 1.3$	± 0.43	± 0.13
0.032	1.50	$\pm\ 2.7$	± 0.50	± 0.09
0.010	2.00	$\pm\ 6.5$	± 0.65	± 0.06
0.0032	2.50	± 16.3	± 0.92	± 0.05
0.0010	3.00	± 43.4	± 1.4	± 0.04

[a] From Equation 6-14 employing $s_T = \pm 0.003$.
[b] From Equation 6-16 employing $k' = \pm 0.003$.
[c] From Equation 6-18 employing $k'' = \pm 0.003$.

around a mean value. Here, the magnitude of the current fluctuations is proportional to the square root of current (see Equation 3-24, p. 71).

In contrast to the types of noise we have thus far considered, shot noise increases in magnitude with increases in the size of the quantity being measured. As a consequence, s_T in Equation 6-14 varies as a function of T. The effect of shot noise on s_T and s_c is readily derived, as shown in the following example.

EXAMPLE

Derive an expression relating s_T and s_c to shot noise.

Shot noise is associated with each of the three steps required for measuring transmittance or absorbance. Generally, however, the noise associated with the 0% T adjustment is small relative to the measurement of P and P_0 because the magnitude of the noise increases as the square root of the signal (p. 71). Thus, the uncertainty in T is generally determined largely by noise associated with the measurement of P and P_0. Since T is the quotient of these two numbers, its *relative* standard deviation can be obtained from the relative standard deviation of P and P_0 (Appendix 1). Thus,

$$\left(\frac{s_T}{T}\right)^2 = \left(\frac{s_P}{P}\right)^2 + \left(\frac{s_{P_0}}{P_0}\right)^2$$

But shot noise is proportional to the square root of the measured quantity (p. 71). Thus,

$$s_P = k\sqrt{P}$$

where k is a proportionality constant. Also,

$$s_{P_0} = k\sqrt{P_0}$$

Substitution of these quantities into the first equation and rearrangement gives

$$\left(\frac{s_T}{T}\right)^2 = k^2\left(\frac{1}{P} + \frac{1}{P_0}\right) = \frac{k^2}{P_0}\left(\frac{P_0}{P} + 1\right)$$

or

$$s_T = \frac{Tk}{\sqrt{P_0}}\sqrt{\frac{1}{T} + 1}$$

Ordinarily, P_0 will not change greatly during a series of measurements of T. When this situation prevails, we may write

$$s_T = Tk'\sqrt{\frac{1}{T} + 1} = k'\sqrt{T + T^2} \quad (6\text{-}15)$$

where $k' = k\sqrt{P_0}$. Substitution of Equation 6-15 into 6-14 yields

$$\frac{s_c}{c} = \frac{0.434}{\log T} \times \frac{k'\sqrt{T + T^2}}{T} = \frac{0.434k'}{\log T}\sqrt{\frac{1}{T} + 1} \quad (6\text{-}16)$$

The data in column 4 of Table 6-3 were obtained with the aid of Equation 6-16. Figure 6-6, curve B is a plot of such data. Note the much larger range of absorbances that can be encompassed without serious loss of accuracy when shot noise, rather than Johnson or 0% T noise, limits the precision. This increased range represents a major advantage of photon type detectors over thermal types. As with Johnson or 0% T noise-limited instruments, shot-limited instruments do not give very reliable concentration data at transmittances greater than 95% (or $A < 0.02$).

Signal shot noise sometimes limits the accuracy of transmittance and absorbance measurements with ultraviolet and visible photometers and spectrophotometers. It is usually of little importance with infrared instruments.

Flicker Noise. Flicker noise in photometers and spectrophotometers is largely associated with the radiation source. As was pointed out in Chapter 3, the magnitude of flicker noise cannot be predicted from theory. It has been shown, however,[7] that uncertainties from the source can be described by the relationship

$$s_T = k''T \quad (6\text{-}17)$$

---

[7] See: L. D. Rothman, S. R. Crouch, and J. D. Ingle, Jr., *Anal. Chem.*, **47**, 1226 (1975).

and Equation 6-14 becomes

$$\frac{s_c}{c} = -\frac{0.434}{\log T} k'' \qquad (6\text{-}18)$$

As shown by the data in column 5 of Table 6-3 and by Figure 6-6, curve C, the concentration error due to flicker noise decreases rapidly and approaches zero at high absorbances.

The overall uncertainty in a spectrophotometric measurement at low absorbance frequently results from a combination of flicker noise and the other types mentioned earlier; occasionally, all three may contribute.

Cell Positioning Uncertainty. All cells have minor imperfections. As a consequence, reflection and scattering losses vary as different sections of the cell are exposed to the source; corresponding variations in measured transmittances must then be expected. For high-quality cells, such imperfections will be minimal. If cells become scratched or dirty, however, the dependence of transmittance upon cell position may become significant.

Rothman, Crouch, and Ingle have shown that uncertainties due to cell positioning variations behave in the same way as source flicker noise; that is, s_T is proportional to transmittance and the concentration uncertainty is given by Equation 6-18. Figure 6-7A is a plot of some of their data. Even though painstaking care was followed in cleaning and positioning the cells, significant errors

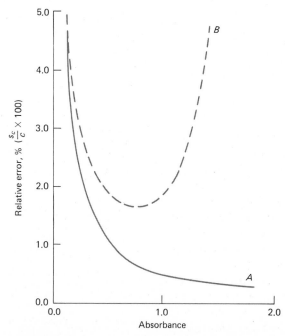

FIGURE 6-7 Curve A: Uncertainty due to uncertainties in cell positioning. (Data adapted. Reprinted with permission from L. D. Rothman, S. R. Crouch, and J. D. Ingle, Jr., *Anal. Chem.*, **47,** 1231 (1975). Copyright by the American Chemical Society.) Curve B: Combined effect of cell positioning uncertainty and Johnson noise.

resulted from this source. Indeed, it seems probable that imprecision in cell positioning is often the most important limitation to the accuracy of spectrophotometric measurements. One method of reducing the effect of cell positioning is to leave cells in place during calibration and analysis; new standards and samples are introduced after washing and rinsing the cell in place with a syringe. Care must be taken to avoid touching or jarring the cells during this process.

Summary. It is apparent that the random errors associated with a spectrophotometric analysis depend upon several variables, including instrument design, the wavelength region, the source intensity and stability, the sensitivity of the transducer, the slit widths, and the concentration of the analyte. The situation is made even more complicated with recording instruments, where several of these variables may change as the spectrum is recorded.

Infrared instruments are most commonly limited by the Johnson noise associated with the detector and by the cell position uncertainty. The latter is due to the fact that narrow cells must often be used and the width of the windows, and thus the relative sample thickness undoubtedly fluctuates considerably from one part of the cell to another. Furthermore, infrared windows are readily scratched, abraded, and attacked by the atmosphere; transmission is thus highly variable across their surfaces. Source flicker may also contribute to the uncertainty of infrared measurements, but is probably less important than Johnson noise and cell uncertainty. Figure 6-7, curve B is an error plot assuming the same cell positioning uncertainty as in Figure 6-7, curve A and a Johnson noise uncertainty equal to that in Figure 6-6, curve A. For most infrared measurements, the errors would probably be larger than those indicated by the plot.

Considerably greater variation in behavior is encountered among ultraviolet and visible photometers and spectrophotometers. For the best quality instruments, the limiting source of random error is probably cell-positioning uncertainty. It is evident that the fullest potential of such instruments can only be realized by filling and rinsing the cells in place; even with this precaution, cell uncertainty often appears to limit the performance of these instruments.

The performance of inexpensive spectrophotometers tends to be limited, at least in part, by their readout devices; it has been shown that significant improvements can be achieved by replacing these components with more sensitive ones.[8]

Effect of Slit Width on Absorption Measurements

The ability of a spectrophotometer to resolve a pair of adjacent absorption peaks depends upon the slit width employed (p. 133). Narrow slits are required to obtain the maximum detail from a complex spectrum.[9] Figure 6-8 illustrates the loss of spectral detail that accompanies the use of wider slits. In this example, the transmittance spectrum of a didymium glass was obtained at slit settings that provided bandwidths of 0.5, 9, and 20 nm. The progressive loss of spectral detail is clear. For qualitative studies, such losses often loom important.

Figure 6-9 illustrates another effect of slit width on spectra. Here, the spectrum of a praseodymium chloride solution was obtained at slit settings of 1.0, 0.5, and 0.1 mm. Note that the peak absorbance values increase significantly (by as much as 70% in one instance) as the slit width decreases. At slit settings less than about 0.14 mm, absorbances were found to become independent of

---

[8] For example, see: J. D. Ingle, Jr., *Anal. Chim. Acta*, **88**, 131 (1977).

[9] For a discussion of the effects of slit width on spectra, see: *Optimum Spectrophotometer Parameters, Application Report AR* **14-2**, Monrovia, CA: Cary Instruments; and F. C. Strong III, *Anal. Chem.*, **48**, 2155 (1976).

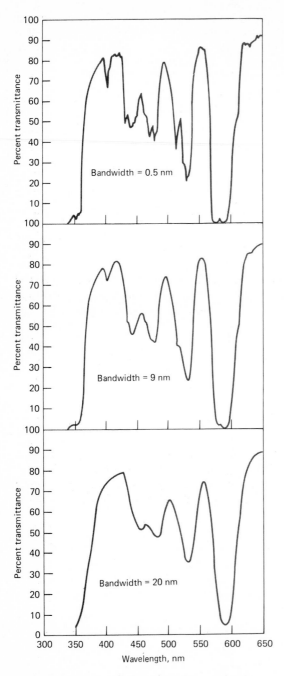

slit width. Careful inspection of Figure 6-8 reveals the same type of effect. In both sets of spectra, the areas under the individual peaks are the same, but wide slit widths result in broader, lower peaks.

It is evident from both of these illustrations that quantitative measurement of narrow absorption bands demands the use of narrow slit widths or, alternatively, very reproducible slit-width settings.

Unfortunately, a decrease in slit width is accompanied by a second-order power reduction in the emitted radiant energy; at very narrow settings, spectral detail may be lost owing to an increase in the signal-to-noise ratio. The situation becomes particularly serious in spectral regions where the output of the source or the sensitivity of the detector is low. Under such circumstances, noise in either of these components or their associated electronic circuits may result in partial or total loss of spectral fine structure.

In general, it is good practice to narrow slits no more than is necessary for resolution of the spectrum at hand. With a variable slit spectrophotometer, proper slit adjustment can be determined by obtaining spectra at progressively narrower slits until peak heights become constant.

Effect of Scattered Radiation at Wavelength Extremes of a Spectrophotometer

We have already noted that scattered radiation may cause instrumental deviations from Beer's law. When measurements are attempted at the wavelength extremes of an instrument, the effects of stray radiation may be even more serious, and on occasion lead to the appearance of false absorption peaks. For example, consider a spectrophotometer for the visible region equipped with glass optics, a tungsten source, and a photocell detector. At wavelengths below about 380 nm, the windows, cells, and prism begin to absorb radiation, thus reducing the radiant energy

FIGURE 6-8 Effect of band width on spectral detail. The sample was a didymium glass. [Spectra provided through the courtesy of Amsco Instrument Company (formerly Turner Associates) Carpinteria, California.]

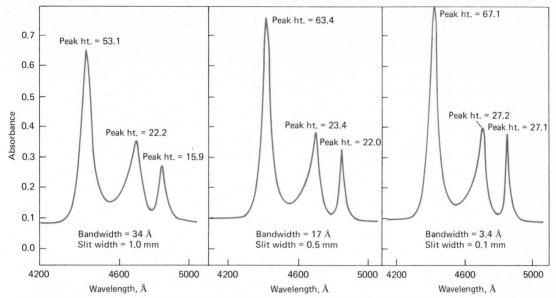

FIGURE 6-9 Effect of slit width and bandwidth on peak heights. Here, the sample was a solution of praseodymium chloride. (From *Optimum Spectrophotometer Parameters*, *Application Report AR* **14-2**, Monrovia, California. Cary Instruments. With permission.)

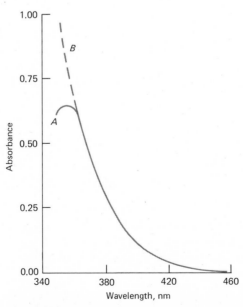

FIGURE 6-10 Spectra of cerium (IV) obtained with a spectrophotometer having glass optics (*A*) and quartz optics (*B*). The false peak in *A* arises from transmission of stray radiation of longer wavelengths.

reaching the transducer. The output of the source falls off rapidly in this region as well; so also does the sensitivity of the photo-electric device. Thus, the radiant power for the 100% T adjustment may be as low as 1 to 2% of that in the region between 500 and 650 nm.

The scattered radiation, however, is often made up of wavelengths to which the instrument is highly sensitive. Thus, its effects can be enormously magnified. Indeed, in some instances the output signal produced by the stray radiation may exceed that from the monochromator beam; under these circumstances, the measured absorbance is as much that for the stray radiation as for the radiation to which the instrument is set.

An example of a false peak appearing at the wavelength extremes of a visible-region spectrophotometer is shown in Figure 6-10. The spectrum of a solution of cerium(IV) obtained with an ultraviolet-visible spectrophotometer, sensitive in the region of 200 to 750 nm, is shown by curve B. Curve A is a spectrum of the same solution obtained with a simple visible spectrophotometer. The apparent maximum shown in curve A arises from the instrument responding to stray wavelengths longer than 400 nm, which (as can be seen from the spectra) are not absorbed by the cerium(IV) ions.

This same effect is sometimes observed with ultraviolet-visible instruments when attempts are made to measure absorbances at wavelengths lower than about 200 nm.

PROBLEMS

1. Use the data provided to evaluate the missing quantities. Wherever necessary, assume that the molecular weight of the absorbing species is 280.

	A	$\% T$	ε	b, cm	c, M	c	a
(a)	0.296			2.50	4.41×10^{-4}	mg/liter	
(b)		19.6		1.50	6.91×10^{-3}	ppm	
(c)	0.877			0.500		g/liter	0.250
(d)		84.2	7.85×10^{2}		3.00×10^{-4}	mg/100 ml	
(e)			1.15×10^{4}	1.25		3.33 ppm	
(f)			6.42×10^{3}	1.10	7.84×10^{-5}	mg/liter	
(g)		43.2		0.100		4.11 ppm	
(h)	1.025		2.86×10^{3}	1.00		g/liter	
(i)		87.2		0.980		mg/liter	0.631
(j)		6.74	9.16×10^{2}		1.86×10^{-4}	mg/liter	

2. A solution that was $1.54 \times 10^{-4}M$ in X had a transmittance of 0.0874 when measured in a 2.00-cm cell. What concentration of X would be required for the transmittance to be increased by a factor of 3 when a 1.00-cm cell was used?

3. A compound had a molar absorptivity of 6.74×10^{3} liter cm^{-1} mol^{-1}. What concentration of the compound would be required to produce a solution having a transmittance of 7.77% in a 2.50-cm cell?

4. At 580 nm, the wavelength of its maximum absorption, the complex $FeSCN^{2+}$ has a molar absorptivity of 7.00×10^{3} liter cm^{-1} mol^{-1}. Calculate

(a) the absorbance of a $3.77 \times 10^{-4}M$ solution of the complex at 580 nm when measured in a 0.750-cm cell.

(b) the transmittance of a $2.85 \times 10^{-4}M$ solution employing the same cell as in (a).

(c) the absorbance (0.750-cm cell) of a solution that has half the transmittance of that described in (a).

5. A solution containing the thiourea complex of bismuth(III) has a molar absorptivity of 9.3×10^3 liter cm^{-1} mol^{-1} at 470 nm.

(a) What will be the absorbance of a $5.0 \times 10^{-5}M$ solution of the complex when measured in a 0.500-cm cell?

(b) What will be the percent transmittance of the solution described in (a)?

(c) What concentration of bismuth could be determined by means of this complex if 1.00-cm cells are to be used and the absorbance is to be kept between 0.100 and 1.500?

6. In ethanol, acetone has a molar absorptivity of 2.75×10^3 liter cm^{-1} mol^{-1} at 366 nm. What range of acetone concentrations could be determined if the percent transmittances of the solutions are to be limited to a range of 10 to 90% and a 1.0-cm cell is to be used?

7. In neutral aqueous solution, it is found that log ε for phenol at 211 nm is 3.79. What range of phenol concentration could be determined spectrophotometrically if absorbance values in a 2.00-cm cell are to be limited to a range of 0.100 to 1.50?

8. The equilibrium constant for the reaction

$$2CrO_4^{2-} + 2H^+ \rightleftharpoons Cr_2O_7^{2-} + H_2O$$

has a value of 4.2×10^{14}. The molar absorptivities for the two principal species in a solution of $K_2Cr_2O_7$ are

λ	$\varepsilon_1(CrO_4^{2-})$	$\varepsilon_2(Cr_2O_7^{2-})$
345	1.84×10^3	10.7×10^2
370	4.81×10^3	7.28×10^2
400	1.88×10^3	1.89×10^2

Four solutions were prepared by dissolving the following number of moles of $K_2Cr_2O_7$ in water and diluting to 1.00 liter with a pH 5.40 buffer: 4.00×10^{-4}, 3.00×10^{-4}, 2.00×10^{-4}, and 1.00×10^{-4}. Derive theoretical absorbance values (1.00-cm cells) for each and plot the data for (a) 345 nm, (b) 370 nm, (c) 400 nm.

9. A species Y has a molar absorptivity of 2400. Derive absorbance data (1.00-cm cells) for solutions of Y that are 10×10^{-4}, 6.00×10^{-4}, 3.00×10^{-4}, and $1.00 \times 10^{-4}M$ in Y, assuming that the radiation employed was contaminated with the following percent of nonadsorbed radiation: (a) 0.000; (b) 0.300; (c) 2.00; (d) 6.00. Plot the data.

7

APPLICATIONS OF ULTRAVIOLET AND VISIBLE ABSORPTION MEASUREMENTS

Absorption measurements employing ultraviolet or visible radiation find widespread application to quantitative and qualitative analysis. Some of these applications are considered in this chapter.[1]

ABSORBING SPECIES

The absorption of ultraviolet or visible radiation by some species M can be considered to be a two-step process, the first of which involves excitation as shown by the equation

$$M + h\nu \rightarrow M^*$$

where M^* represents the atomic or molecular particle in the electronically excited state resulting from absorption of the photon $h\nu$. The lifetime of the excited state is brief (10^{-8} to 10^{-9} s), its existence being terminated by any of several *relaxation* processes (p. 109). The most common type of relaxation involves conversion of the excitation energy to heat; that is,

$$M^* \rightarrow M + \text{heat}$$

Relaxation may also occur by decomposition of M^* to form new species; such a process is called a *photochemical reaction*. Alternatively, relaxation may involve fluorescent or phosphorescent reemission of radiation. It is important to note that the lifetime of M^* is usually so very short that its concentration at any instant is ordinarily negligible. Further-more, the amount of thermal energy evolved is usually not detectable. Thus, absorption measurements have the advantage of creating minimal disturbance of the system under study (except when photochemical decomposition occurs).

The absorption of ultraviolet or visible radiation generally results from excitation of bonding electrons; as a consequence, the wavelengths of absorption peaks can be correlated with the types of bonds that exist in the species under study. Absorption spectroscopy is, therefore, valuable for identifying functional groups in a molecule; it also provides a somewhat selective method of quantitative analysis for compounds containing absorbing bonds.

For purposes of this discussion, it is useful to recognize three types of electronic transitions and to categorize absorbing species on this basis. The three include transitions, involving: (1) π, σ, and n electrons, (2) d and f electrons, and (3) charge-transfer electrons.

Absorbing Species Containing π, σ, and n Electrons

Absorbing species of this type include organic molecules and ions as well as a number of inorganic anions. Our discussion will deal largely with the former, although brief mention will be made of absorption by certain inorganic systems as well.

All organic compounds are capable of absorbing electromagnetic radiation because all contain valence electrons that can be excited to higher energy levels. The excitation energies associated with electrons forming most single bonds are sufficiently high that absorption by them is restricted to the so-called vacuum ultraviolet region ($\lambda < 185$ nm), where components of the atmosphere also absorb strongly. The experimental difficulties associated with the vacuum ultraviolet are formidable; as a result, most spectrophotometric investigations of organic compounds have involved the wavelength region greater

---

[1] Some useful references on absorption methods include: R. P. Bauman, *Absorption Spectroscopy*. New York: Wiley, 1962; J. R. Edisbury, *Practical Hints on Absorption Spectrometry*. New York: Plenum Press, 1968; F. Grum, *Physical Methods of Chemistry*, eds. A. Weissberger and B. W. Rossiter. New York: Wiley-Interscience, 1972, Volume I, Part III B, Chapter 3; H. H. Jaffe and M. Orchin, *Theory and Applications of Ultraviolet Spectroscopy*. New York: Wiley, 1962; G. F. Lothian, *Absorption Spectrophotometry*, 3d ed. London: Adam Hilger Ltd., 1969; and J. E. Crooks, *The Spectrum in Chemistry*. London: Academic Press, 1978.

than 185 nm. Absorption of longer-wavelength ultraviolet and visible radiation is restricted to a limited number of functional groups (called *chromophores*) that contain valence electrons with relatively low excitation energies.

The electronic spectra of organic molecules containing chromophores are usually complex, because the superposition of vibrational transitions on the electronic transitions leads to an intricate combination of overlapping lines; the result is a broad band of continuous absorption. The complex nature of the spectra makes detailed theoretical analysis difficult or impossible. Nevertheless, qualitative or semi-quantitative statements concerning the types of electronic transitions responsible for a given absorption spectrum can be deduced from molecular orbital considerations.

Types of Absorbing Electrons.[2] The electrons that contribute to absorption by an organic molecule are: (1) those that participate directly in bond formation between atoms and are thus associated with more than one atom; (2) nonbonding or unshared outer electrons that are largely localized about such atoms as oxygen, the halogens, sulfur, and nitrogen.

Covalent bonding occurs because the electrons forming the bond move in the field about two atomic centers in such a manner as to minimize the repulsive coulombic forces between these centers. The nonlocalized fields between atoms that are occupied by bonding electrons are called *molecular orbitals* and can be considered to result from the overlap of atomic orbitals. When two atomic orbitals combine, either a low-energy *bonding molecular orbital* or a high-energy *antibonding molecular orbital* results. The electrons of a molecule occupy the former in the ground state.

The molecular orbitals associated with single bonds in organic molecules are designated as *sigma* (σ) *orbitals*, and the corresponding electrons are σ electrons. As shown in Figure 7-1a, the distribution of charge density of a sigma orbital is rotationally symmetric around the axis of the bond. Here, the average negative charge density arising from the motion of the two electrons around the two positive nuclei is indicated by the degree of shading.

The double bond in an organic molecule contains two types of molecular orbitals: a *sigma* (σ) orbital corresponding to one pair of the bonding electrons; and a *pi* (π) *molecular orbital* associated with the other pair. Pi orbitals are formed by the parallel overlap of atomic p orbitals. Their charge distribution is characterized by a *nodal plane* (a region of low-charge density) along the axis of the bond and a maximum density in regions above and below the plane (see Figure 7-1b).

Also shown in Figure 7-1c and 7-1d are the charge-density distributions for antibonding sigma and pi orbitals; these orbitals are designated by σ^* and π^*.

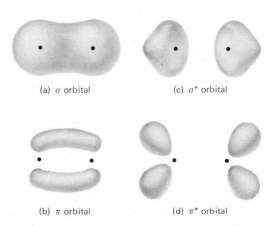

(a) σ orbital   (c) σ^* orbital

(b) π orbital   (d) π^* orbital

---

[2] For further details, see: C. N. R. Rao, *Ultra-Violet and Visible Spectroscopy*, 3d ed. London: Butterworths, 1975; and R. P. Bauman, *Absorption Spectroscopy*. New York: Wiley, 1962, Chapters 6 and 8.

FIGURE 7-1   Electron distribution in sigma and pi molecular orbitals.

In addition to σ and π electrons, many organic compounds contain nonbonding electrons. These unshared electrons are designated by the symbol n. An example showing the three types of electrons in a simple organic molecule is shown in Figure 7-2.

As shown in Figure 7-3, the energies for the various types of molecular orbitals differ significantly. Quite generally, the energy level of a nonbonding electron lies between those of the bonding and the antibonding π and σ orbitals.

Electronic transitions among certain of the energy levels can be brought about by the absorption of radiation. As shown in Figure 7-3, there are four of these: $\sigma \rightarrow \sigma^*$, $n \rightarrow \sigma^*$, $n \rightarrow \pi^*$, and $\pi \rightarrow \pi^*$.

$\sigma \rightarrow \sigma^*$ **Transitions.** Here, an electron in a bonding σ orbital of a molecule is excited to the corresponding antibonding orbital by the absorption of radiation. The molecule is then described as being in the σ,σ^* excited state. Relative to other possible transitions, the energy required to induce a $\sigma \rightarrow \sigma^*$ transition is large (see Figure 7-3), corresponding to radiant frequencies in the vacuum ultraviolet region. Methane, for example, which contains only single C—H bonds and can thus undergo only $\sigma \rightarrow \sigma^*$ transitions, exhibits an absorption maximum at 125 nm. Ethane has an absorption peak at 135 nm, which must also arise from the same type of transition, but here, electrons of the C—C bond appear to be involved. Because the strength of the C—C bond is less than that of the C—H bond, less energy is required for excitation; thus, the absorption peak occurs at a longer wavelength.

Absorption maxima due to $\sigma \rightarrow \sigma^*$ transitions are never observed in the ordinary, accessible ultraviolet region; for this reason, they will not be considered further.

$n \rightarrow \sigma^*$ **Transitions.** Saturated compounds containing atoms with unshared electron pairs (nonbonding electrons) are capable of $n \rightarrow \sigma^*$ transitions. In general, these transitions require less energy than the $\sigma \rightarrow \sigma^*$ type

FIGURE 7-2 Types of molecular orbitals in formaldehyde.

and can be brought about by radiation in the region of between 150 and 250 nm, with most absorption peaks appearing below 200 nm. Table 7-1 shows absorption data for some typical $n \rightarrow \sigma^*$ transitions. It will be seen that the energy requirements for such transitions depend primarily upon the kind of atomic bond and to a lesser extent upon the structure of the molecule. The molar absorptivities (ε) associated with this type of absorption are intermediate in magnitude and usually range between 100 and 3000 liter cm^{-1} mol^{-1}.

Absorption maxima for formation of the n,σ^* state tend to shift to shorter wavelengths in the presence of polar solvents such as water or ethanol. The number of organic functional groups with $n \rightarrow \sigma^*$ peaks in the readily accessible ultraviolet region is relatively small.

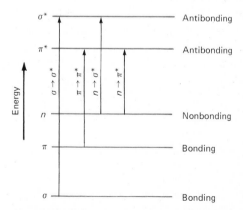

FIGURE 7-3 Electronic molecular energy levels.

$n \rightarrow \pi^*$ **and** $\pi \rightarrow \pi^*$ **Transitions.** Most applications of absorption spectroscopy to organic compounds are based upon transitions for n or π electrons to the π^* excited state, because the energies required for these processes bring the absorption peaks into an experimentally convenient spectral region (200 to 700 nm). Both transitions require the presence of an unsaturated functional group to provide the π orbitals. Strictly speaking, it is to these unsaturated absorbing centers that the term chromophore applies.

The molar absorptivities for peaks associated with excitation to the n,π^* state are generally low and ordinarily range from 10 to 100 liter cm^{-1} mol^{-1}; values for $\pi \rightarrow \pi^*$ transitions, on the other hand, normally fall in the range between 1000 and 10,000. Another characteristic difference between the two types of absorption is the effect exerted by the solvent on the wavelength of the peaks. Peaks associated with $n \rightarrow \pi^*$ transitions are generally shifted to shorter wavelengths (a *hypsochromic* or *blue shift*) with increasing polarity of the solvent. Usually, but not always, the reverse trend (a *bathochromic* or *red shift*) is observed for $\pi \rightarrow \pi^*$ transitions. The hypsochromic effect apparently arises from the increased solvation of the unbonded electron pair, which lowers the energy of the n orbital. The most dramatic effects of this kind (blue shifts of 30 nm or more) are seen with polar hydrolytic solvents, such as water or

alcohols, in which hydrogen-bond formation between the solvent protons and the non-bonded electron pair is extensive. Here, the energy of the n orbital is lowered by an amount approximately equal to the energy of the hydrogen bond. When an $n \rightarrow \pi^*$ transition occurs, however, the remaining single n electron cannot maintain the hydrogen bond; thus, the energy of the n,π^* *excited* state is not affected by this type of solvent interaction. A blue shift, also roughly corresponding to the energy of the hydrogen bond, is therefore observed.

A second solvent effect that undoubtedly influences both $\pi \rightarrow \pi^*$ and $n \rightarrow \pi^*$ transitions leads to a bathochromic shift with increased solvent polarity. This effect is small (usually less than 5 nm), and as a result is completely overshadowed in $n \rightarrow \pi^*$ transitions by the hypsochromic effect just discussed. Here, attractive polarization forces between the solvent and the absorber tend to lower the energy levels of both the unexcited and the excited states. The effect on the excited state is greater, however, and the energy differences thus become smaller with increased solvent polarity; small bathochromic shifts result.

Organic Chromophores. Table 7-2 lists common organic chromophores and the approximate location of their absorption maxima. These data can serve only as rough guides for the identification of functional

---

TABLE 7-1   SOME EXAMPLES OF ABSORPTION DUE TO $n \rightarrow \sigma^*$ **TRANSITIONS**[a]

Compound	λ_{max}(nm)	ε_{max}	Compound	λ_{max}(nm)	ε_{max}
H_2O	167	1480	$(CH_3)_2S$[b]	229	140
CH_3OH	184	150	$(CH_3)_2O$	184	2520
CH_3Cl	173	200	CH_3NH_2	215	600
CH_3I	258	365	$(CH_3)_3N$	227	900

[a] Samples in vapor state.
[b] In ethanol solvent.

groups, since the positions of maxima are also affected by solvent and structural details. Furthermore, the peaks are ordinarily broad because of vibrational effects; the precise determination of the position of a maximum is thus difficult.

Effect of Conjugation of Chromophores. In the molecular-orbital treatment, π electrons are considered to be further delocalized by the conjugation process; the orbitals thus involve four (or more) atomic centers. The effect of this delocalization is to lower the energy level of the π^* orbital and give it less antibonding character. Absorption maxima are shifted to longer wavelengths as a consequence.

As seen from the data in Table 7-3, the absorptions of multichromophores in a single organic molecule are approximately additive, provided the chromophores are separated from one another by more than one single bond. Conjugation of chromophores, however, has a profound effect on spectral properties. For example, it is seen in Table 7-3 that 1,3-butadiene, $CH=CHCH=CH_2$, has a strong absorption band that is displaced to a longer wavelength by 20 nm as compared with the corresponding peak for an unconjugated diene. When three double bonds are conjugated, the bathochromic effect is even larger.

Conjugation between the doubly bonded oxygen of aldehydes, ketones, and carboxylic acids and an olefinic double bond gives rise

TABLE 7-2   ABSORPTION CHARACTERISTICS OF SOME COMMON CHROMOPHORES

Chromophore	Example	Solvent	λ_{max}(nm)	ε_{max}	Type of transition
Alkene	$C_6H_{13}CH=CH_2$	n-Heptane	177	13,000	$\pi \to \pi^*$
Alkyne	$C_5H_{11}C\equiv C-CH_3$	n-Heptane	178	10,000	$\pi \to \pi^*$
			196	2000	—
			225	160	—
Carbonyl	$CH_3\overset{O}{\underset{\|}{C}}CH_3$	n-Hexane	186	1000	$n \to \sigma^*$
			280	16	$n \to \pi^*$
	$CH_3\overset{O}{\underset{\|}{C}}H$	n-Hexane	180	large	$n \to \sigma^*$
			293	12	$n \to \pi^*$
Carboxyl	$CH_3\overset{O}{\underset{\|}{C}}OH$	Ethanol	204	41	$n \to \pi^*$
Amido	$CH_3\overset{O}{\underset{\|}{C}}NH_2$	Water	214	60	$n \to \pi^*$
Azo	$CH_3N=NCH_3$	Ethanol	339	5	$n \to \pi^*$
Nitro	CH_3NO_2	Isooctane	280	22	$n \to \pi^*$
Nitroso	C_4H_9NO	Ethyl ether	300	100	—
			665	20	$n \to \pi^*$
Nitrate	$C_2H_5ONO_2$	Dioxane	270	12	$n \to \pi^*$

to similar behavior (see Table 7-3). Analogous effects are also observed when two carbonyl or carboxylate groups are conjugated with one another. For α-β unsaturated aldehydes and ketones, the weak absorption peak due to $n \to \pi^*$ transitions is shifted to longer wavelengths by 40 nm or more. In addition, a strong absorption peak corresponding to a $\pi \to \pi^*$ transition appears. This latter peak occurs only in the vacuum ultraviolet if the carbonyl group is not conjugated.

The wavelengths of absorption peaks for conjugated systems are sensitive to the types of groups attached to the doubly bonded atoms. Various empirical rules have been developed for predicting the effect of such substitutions upon absorption maxima and have proved useful for structural determinations.[3]

---

[3] For a summary of these rules, see: R. M. Silverstein, G. C. Bassler, and T. C. Morrill, *Spectrometric Identification of Organic Compounds*, 3d ed. New York: Wiley, 1974, pp. 241–255.

Absorption by Aromatic Systems. The ultraviolet spectra of aromatic hydrocarbons are characterized by three sets of bands that originate from $\pi \to \pi^*$ transitions. For example, benzene has a strong absorption peak at 184 nm ($\varepsilon_{max} \sim 60{,}000$); a weaker band, called the E_2 band, at 204 nm ($\varepsilon_{max} = 7900$); and a still weaker peak, termed the B band, at 256 nm ($\varepsilon_{max} = 200$). The long-wavelength band of benzene, and many other aromatics, contains a series of sharp peaks (see Figure 4-11b, p. 106) due to the superposition of vibrational transitions upon the basic electronic transitions. Polar solvents tend to eliminate this fine structure, as do certain types of substitution.

All three of the characteristic bands for benzene are strongly affected by ring substitution; the effects on the two longer-wavelength bands are of particular interest because they can be readily studied with ordinary spectrophotometric equipment. Table 7-4 illustrates the effects of some common ring substituents.

By definition, an *auxochrome* is a func-

TABLE 7-3 EFFECT OF MULTICHROMOPHORES ON ABSORPTION

Compound	Type	λ_{max}(nm)	ε_{max}
$CH_3CH_2CH_2CH{=}CH_2$	Olefin	184	$\sim 10{,}000$
$CH_2{=}CHCH_2CH_2CH{=}CH_2$	Diolefin (unconjugated)	185	$\sim 20{,}000$
$H_2C{=}CHCH{=}CH_2$	Diolefin (conjugated)	217	21,000
$H_2C{=}CHCH{=}CHCH{=}CH_2$	Triolefin (conjugated)	250	—
$CH_3CH_2CH_2CH_2\overset{\overset{\displaystyle O}{\|}}{C}CH_3$	Ketone	282	27
$CH_2{=}CHCH_2CH_2\overset{\overset{\displaystyle O}{\|}}{C}CH_3$	Unsaturated ketone (unconjugated)	278	30
$CH_2{=}CH\overset{\overset{\displaystyle O}{\|}}{C}CH_3$	Unsaturated ketone (conjugated)	324	24
		219	3600

tional group that does not itself absorb in the ultraviolet region, but has the effect of shifting chromophore peaks to longer wavelengths as well as increasing their intensities. It is seen in Table 7-4 that —OH and —NH_2 have an auxochromic effect on the benzene chromophore, particularly with respect to the B band. Auxochromic substituents have at least one pair of n electrons capable of interacting with the π electrons of the ring. This interaction apparently has the effect of stabilizing the π^* state, thereby lowering its energy; a bathochromic shift results. Note that the auxochromic effect is more pronounced for the phenolate anion than for phenol itself, probably because the anion has an extra pair of unshared electrons to contribute to the interaction. With aniline, on the other hand, the nonbonding electrons are lost by formation of the anilinium cation, and the auxochromic effect disappears as a consequence.

Absorption by Inorganic Anions. A number of inorganic anions exhibit ultraviolet absorption peaks that are a consequence of $n \to \pi^*$ transitions. Examples include nitrate (313 nm),

carbonate (217 nm), nitrite (360 and 280 nm), azido (230 nm), and trithiocarbonate (500 nm) ions.

Absorption Involving d and f Electrons

Most transition-metal ions absorb in the ultraviolet or visible region of the spectrum. For the lanthanide and actinide series, the absorption process results from electronic transitions of $4f$ and $5f$ electrons; for elements of the first and second transition-metal series, the $3d$ and $4d$ electrons are responsible for absorption.

Absorption by Lanthanide and Actinide Ions. The ions of most lanthanide and actinide elements absorb in the ultraviolet and visible regions. In distinct contrast to the behavior of most inorganic and organic absorbers, their spectra consist of narrow, well-defined, and characteristic absorption peaks, which are little affected by the type of ligand associated with the metal ion. Portions of a typical spectrum are shown in Figure 7-4.

The transitions responsible for absorption

TABLE 7-4 ABSORPTION CHARACTERISTICS OF AROMATIC COMPOUNDS

Compound		E_2 **Band**		B **Band**	
		λ_{max}(nm)	ε_{max}	λ_{max}(nm)	ε_{max}
Benzene	C_6H_6	204	7900	256	200
Toluene	$C_6H_5CH_3$	207	7000	261	300
m-Xylene	$C_6H_4(CH_3)_2$	—	—	263	300
Chlorobenzene	C_6H_5Cl	210	7600	265	240
Phenol	C_6H_5OH	211	6200	270	1450
Phenolate ion	$C_6H_5O^-$	235	9400	287	2600
Aniline	$C_6H_5NH_2$	230	8600	280	1430
Anilinium ion	$C_6H_5NH_3^+$	203	7500	254	160
Thiophenol	C_6H_5SH	236	10,000	269	700
Naphthalene	$C_{10}H_8$	286	9300	312	289
Styrene	$C_6H_5CH{=}CH_2$	244	12,000	282	450

by elements of the lanthanide series appear to involve the various energy levels of 4f electrons, while it is the 5f electrons of the actinide series that interact with radiation. These inner orbitals are largely screened from external influences by electrons occupying orbitals with higher principal quantum numbers. As a consequence, the bands are narrow and relatively unaffected by the nature of the solvent or the species bonded by the outer electrons.

Absorption by Elements of the First and Second Transition-Metal Series. The ions and complexes of the 18 elements in the first two transition series tend to absorb visible radiation in one if not all of their oxidation states. In contrast to the lanthanide and actinide elements, however, the absorption bands are often broad (Figure 7-5) and are strongly influenced by chemical environmental factors. An example of the environmental effect is found in the pale blue color of the aquo

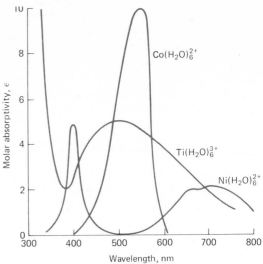

FIGURE 7-5   Absorption spectra of some transition-metal ions.

copper(II) ion and the much darker blue of the copper complex with ammonia.

Metals of the transition series are characterized by having five partially occupied d orbitals (3d in the first series and 4d in the second), each capable of accommodating a pair of electrons. The electrons in these orbitals do not generally participate in bond formation; nevertheless, it is clear that the spectral characteristics of transition metals involve electronic transitions among the various energy levels of these d orbitals.

Two theories have been advanced to rationalize the colors of transition-metal ions and the profound influence of chemical environment on these colors. The *crystal-field theory*, which we shall discuss briefly, is the simpler of the two and is adequate for a qualitative understanding. The more complex molecular-orbital treatment, however, provides a better quantitative treatment of the phenomenon.[4]

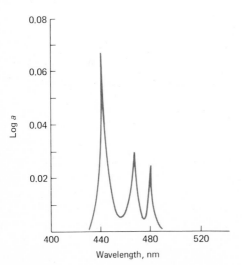

FIGURE 7-4   The absorption spectrum of a praseodymium chloride solution; a = absorptivity in liter cm^{-1} g^{-1}. (Reprinted with permission from T. Moeller and J. C. Brantley, *Anal. Chem.*, **22**, 433 (1950). Copyright by the American Chemical Society.)

---

[4] For a nonmathematical discussion of these theories, see: L. E. Orgel, *An Introduction to Transition Metal Chemistry, Ligand-Field Theory.* New York: Wiley, 1960.

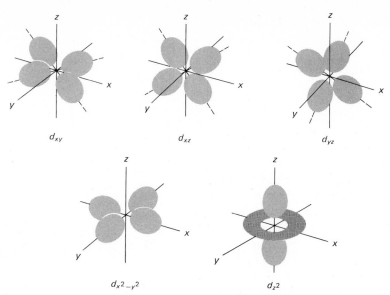

FIGURE 7-6 Electron density distribution in various d orbitals.

Both theories are based upon the premise that the energies of d orbitals of the transition-metal ions in solution are not identical, and that absorption involves the transition of electrons from a d orbital of lower energy to one of higher energy. In the absence of an external electric or magnetic field (as in the dilute gaseous state), the energies of the five d orbitals are identical, and absorption of radiation is not required for an electron to move from one orbital to another. On the other hand, complex formation in solution occurs between the metal ion and water or some other ligand. Splitting of the d-orbital energies then results, owing to the differential forces of electrostatic repulsion between the electron pair of the donor and the electrons in the various d orbitals of the central metal ion. In order to understand this effect, we must first consider the spatial distribution of electrons in the various d orbitals.

The electron-density distribution of the five d orbitals around the nucleus is shown in Figure 7-6. Three of the orbitals, termed d_{xy}, d_{xz}, and d_{yz}, are similar in every regard

except for their spatial orientation. Note that these orbitals occupy spaces *between* the three axes; consequently, they have minimum electron densities along the axes and maximum densities on the diagonals between axes. In contrast, the electron densities of the $d_{x^2-y^2}$ and the d_{z^2} orbitals are directed along the axes.

Let us now consider a transition-metal ion that is coordinated to six molecules of water (or some other ligand). These ligand molecules or ions can be imagined as being symmetrically distributed around the central atom, one ligand being located at each end of the three axes shown in Figure 7-6; the resulting octahedral structure is the most common orientation for transition-metal complexes. The negative ends of the water dipoles are pointed toward the metal ion, and the electrical fields from these dipoles tend to exert a repulsive effect on all of the d orbitals, *thus increasing their energy*; the orbitals are then said to have become *destabilized*. The maximum charge density of the d_{z^2} orbital lies along the bonding axis. The negative field of a

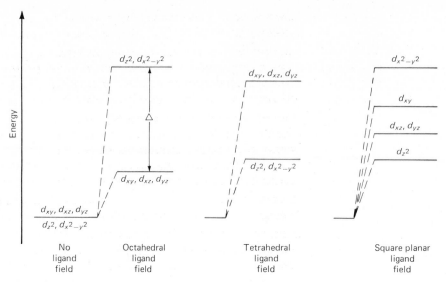

FIGURE 7-7 Effect of ligand field on *d*-orbital energies.

bonding ligand therefore has a greater effect on this orbital than upon the d_{xy}, d_{xz}, and d_{yz} orbitals, whose charge densities do not coincide with the bonding axes. These latter orbitals will be destabilized equally, inasmuch as they differ from one another only in the matter of orientation. The effect of the electrical field on the $d_{x^2-y^2}$ orbital is less obvious, but quantum calculations have shown that it

is destabilized to the same extent as the d_{z^2} orbital. Thus, the energy-level diagram for the octahedral configuration (Figure 7-7) shows that the energies of all of the *d* orbitals rise in the presence of a ligand field but, in addition, that the *d* orbitals are split into levels differing in energy by Δ. Also shown are energy diagrams for complexes involving four coordinated bonds. Two configurations are en-

TABLE 7-5   EFFECT OF LIGANDS ON ABSORPTION MAXIMA ASSOCIATED WITH *d–d* **TRANSITIONS**

	λ_{max}(nm) for the indicated ligands				
	Increasing ligand field strength				
Central ion	$6Cl^-$	$6H_2O$	$6NH_3$	$3en^a$	$6CN^-$
Cr(III)	736	573	462	456	380
Co(III)	—	538	435	428	294
Co(II)	—	1345	980	909	—
Ni(II)	1370	1279	925	863	—
Cu(II)	—	794	663	610	—

^{a} en = ethylenediamine, a bidentate ligand

countered: the *tetrahedral*, in which the four groups are symmetrically distributed around the metal ion; and the *square planar*, in which the four ligands and the metal ion lie in a single plane. Unique *d*-orbital splitting patterns for each configuration can be deduced by arguments similar to those used for the octahedral structure.

The magnitude of Δ (Figure 7-7) depends upon a number of factors, including the valence state of the metal ion and the position of the parent element in the periodic table. An important variable attributable to the ligand is the *ligand field strength*, which is a measure of the extent to which a complexing group will split the energies of the *d* electrons; that is, a complexing agent with a high ligand field strength will cause Δ to be large.

It is possible to arrange the common ligands in the order of increasing ligand field strengths: $I^- < Br^- < Cl^- < F^- < OH^- < C_2O_4^{2-} \sim H_2O < SCN^- < NH_3 <$ ethylenediamine $< o$-phenanthroline $< NO_2^- < CN^-$. With only minor exceptions, this order applies to all transition-metal ions and permits qualitative predictions about the relative positions of absorption peaks for the various complexes of a given transition-metal ion. Since Δ increases with increasing field strength, the wavelength of the absorption maxima decreases. This effect is demonstrated by the data in Table 7-5.

Charge-Transfer Absorption

For analytical purposes, species that exhibit *charge-transfer absorption*[5] are of particular importance because molar absorptivities are very large ($\varepsilon_{max} > 10,000$). Thus, these complexes provide a highly sensitive means for detecting and determining absorbing species. Many inorganic complexes exhibit charge-

transfer absorption and are therefore called, *charge-transfer complexes*. Common examples of such complexes include the thiocyanate and phenolic complexes of iron(III), the *o*-phenanthroline complex of iron(II), the iodide complex of molecular iodine, and the ferro-ferricyanide complex responsible for the color of Prussian blue.

In order for a complex to exhibit a charge-transfer spectrum, it is necessary for one of its components to have electron-donor characteristics and for the other component to have electron-acceptor properties. Absorption of radiation then involves transfer of an electron from the donor to an orbital that is largely associated with the acceptor. As a consequence, the excited state is the product of an internal oxidation-reduction process. This behavior differs from that of an organic chromophore, where the electron in the excited state is in the *molecular* orbital formed by two or more atoms.

A well-known example of charge-transfer absorption is observed in the iron(III)/thiocyanate complex. Absorption of a photon results in the transfer of an electron from the thiocyanate ion to an orbital associated with the iron(III) ion. The product is thus an excited species involving predominantly (iron(II) and the neutral thiocyanate radical SCN. As with other types of electronic excitation, the electron, under ordinary circumstances, returns to its original state after a brief period. Occasionally, however, dissociation of the excited complex may occur, producing photochemical oxidation-reduction products.

As the tendency for electron transfer increases, less radiant energy is required for the charge-transfer process, and the resulting complexes absorb at longer wavelengths. For example, thiocyanate ion is a better electron donor (reducing agent) than is chloride ion; thus, the absorption of the iron(III)/thiocyanate complex occurs in the visible region, whereas the absorption maximum for the corresponding yellow chloride complex is in the ultraviolet region. Presumably, the iodide

---

[5] For a brief discussion of this type of absorption, see: C. N. R. Rao, *Ultra-Violet and Visible Spectroscopy, Chemical Applications*, 3d ed. London: Butterworths, 1975, Chapter 11.

complex of iron(III) would absorb at still longer wavelengths. Such an absorption peak has not been observed, however, because the electron-transfer process is complete, giving iron(II) and iodine as products.

In most charge-transfer complexes involving a metal ion, the metal serves as the electron acceptor. An exception is the o-phenanthroline complex of iron(II) or copper(I), where the ligand is the acceptor and the metal ion is the donor. Other examples of this type are known.

Organic compounds form many interesting charge-transfer complexes. An example is quinhydrone (a 1:1 complex of quinone and hydroquinone), which exhibits strong absorption in the visible region. Other examples include iodine complexes with amines, aromatics, and sulfides, among others.

SOME TYPICAL INSTRUMENTS

A host of instruments are available commercially for absorbance measurements in the ultraviolet and visible spectral regions. The simplest of these are colorimeters, in which the eye serves as the transducer. The most complex are double-beam, recording instruments that cover the entire spectral range from about 185 to 3000 nm. A few typical instruments are considered in this section.

Photometers

The photometer provides a simple, relatively inexpensive tool for performing absorption analyses. Convenience, ease of maintenance, and ruggedness are properties of a filter photometer that may not be found in the more sophisticated spectrophotometer. Moreover, where high spectral purity is not important to a method (and often it is not), analyses can be performed as accurately with a photometer as with more complex instrumentation.

Visible Photometers. Figure 7-8 presents schematic diagrams for two photometers. The first is a single-beam, direct-reading instrument consisting of a tungsten-filament lamp, a lens to provide a parallel beam of light, a filter, and a photovoltaic cell. The current produced is indicated with a microammeter, the face of which is ordinarily scribed with a linear scale from 0 to 100. In some instruments, adjustment to obtain a full-scale response with the solvent in the light path involves changing the voltage applied to the lamp. In others, the aperture size of a diaphragm located in the light path is altered. Since the signal from the photovoltaic cell is linear with respect to the radiation it receives, the scale reading with the sample in the light path will be the percent transmittance (that is, the percent of full scale). Clearly, a logarithmic scale could be substituted to give the absorbance of the solution directly.

Also shown in Figure 7-8 is a schematic representation of a double-beam, null-type photometer. Here, the light beam is split by a mirror. One part passes through the sample, and thence to a photovoltaic cell; the other part passes through the solvent to a similar detector. The currents from the two photovoltaic cells are passed through variable resistances; one of these is calibrated as a transmittance scale in linear units from 0 to 100. A sensitive galvanometer, which serves as a null indicator, is connected across the two resistances. When the potential drop across AB is equal to that across CD, no electricity passes through the galvanometer; under all other circumstances, a current is indicated. At the outset, the solvent is placed in both cells, and contact A is set at 100; contact C is then adjusted until no current is indicated. Introduction of the sample into the one cell results in a decrease in radiant power that reaches the working phototube and a corresponding decrease in the potential drop across CD; this lack of balance is compensated for by moving A to a lower value. At balance, the percent transmittance is read directly from the scale.

Commercial photometers usually cost a few hundred dollars. The majority employ the

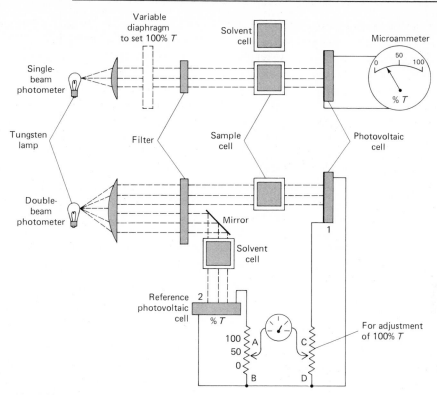

FIGURE 7-8 Schematic diagram for a single-beam and a double-beam photometer.

double-beam principle because this design largely compensates for fluctuations in the source intensity due to voltage variations.

Filter Selection for Photometric Analysis. Photometers are generally supplied with several filters, each of which transmits a different portion of the spectrum. Selection of the proper filter for a given application is important inasmuch as the sensitivity of the measurement is directly dependent upon this choice. The color of the light absorbed is the complement of the color of the solution itself. For example, a liquid appears red because it transmits the red portion of the spectrum but absorbs the green. It is the intensity of green radiation that varies with concentration; a green filter should thus be employed. In general, then, the most suitable filter for a photometric analysis will be the color com-

plement of the solution being analyzed. If several filters possessing the same general hue are available, the one that causes the sample to exhibit the greatest absorbance (or least transmittance) should be used.

Ultraviolet Photometers. The most important application of ultraviolet photometry is as a detector in high-performance liquid chromatography. In this application, a mercury-vapor lamp serves as a source and the emission line at 254 nm is isolated by filters. This type of detector is discussed in more detail in Chapter 25.

Spectrophotometers

A large number of spectrophotometers are available from commercial sources. A few of these are considered in the paragraphs that follow.

Instruments for the Visible Range. Several spectrophotometers are available that are designed to operate within the general region of 380 to 800 nm; within these limits, the range varies somewhat from instrument to instrument.

Spectrophotometers designed for the visible region are frequently simple, single-beam, grating instruments that are inexpensive (less than $1000), rugged, and readily portable. At least one is battery operated and light enough to be hand-held. The most common application of these instruments is to quantitative analysis, although several produce surprisingly good absorption spectra as well.

Figure 7-9 depicts two typical visible spectrophotometers.

The Bausch and Lomb Spectronic 20 spectrophotometer shown in Figure 7-9a employs a reference phototube, which serves to compensate for fluctuations in the output of the tungsten-filament light source; this design eliminates the need for a stabilized lamp power supply. The amplified difference signal from the two phototubes powers a meter with a 5 1/2-in. scale calibrated in transmittance and absorbance.

The Spectronic 20 is equipped with an occluder, which is a vane that automatically falls between the beam and the detector whenever the cuvette is removed from its holder; the 0% T adjustment can then be made. The light control device shown in Figure 7-9a consists of a V-shaped slot that

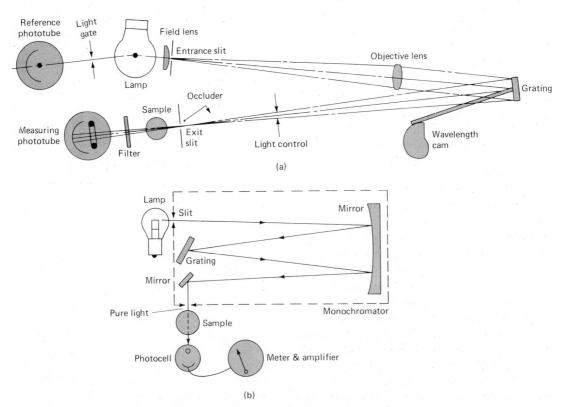

FIGURE 7-9 Two examples of simple spectrophotometers. (a) The Spectronic 20; diagram courtesy of Bausch and Lomb, Rochester, New York. (b) The Turner 350; diagram courtesy of Amsco Instrument Company (formerly Turner Associates) Carpinteria, California.

can be moved in or out of the beam in order to set the meter to 100% T.

The range of the Spectronic 20 is from 340 to 625 nm; an accessory phototube extends this range to 950 nm. Other specifications for the instrument include a bandwidth of 20 nm and a wavelength accuracy of ± 2.5 nm.

The Turner instrument, shown schematically in Figure 7-9b, makes use of a tungsten-filament bulb as a source, a plane reflection grating in an *Ebert mounting* for dispersion, and a phototube detector that is sensitive in the range between 210 and 710 nm. The readout device is a meter calibrated both in transmittance and absorbance; instruments with 4- or 7-in. scales are offered. The transmittance is first set to zero by adjustment of the amplifier output while a shutter screens the lamp from the detector. With the solvent in the light path, the 100% T adjustment is then accomplished by varying the output of the stabilized lamp power supply. Finally, the transmittance or absorbance is read with the sample in the beam. The instrument specifications include a bandwidth of 9 nm, a wavelength accuracy of ± 2 nm, and a photometric accuracy of 0.5% A.

Several accessories are offered with the Turner instrument. One, which includes a deuterium lamp, extends the range of the instrument to 210 nm; another provides an additional phototube that permits measurements to 1000 nm.

Single-Beam Instruments for the Ultraviolet-Visible Region. Several instrument manufacturers offer single-beam instruments which can be used for both ultraviolet and visible measurements. The lower wavelength extremes for these instruments vary from 190 to 210 nm and the upper from 800 to 1000 nm. All are equipped with interchangeable tungsten and hydrogen or deuterium lamps. Most employ photomultiplier tubes as detectors and gratings for dispersion. Some are equipped with digital readout devices; others employ large meters. The prices for these instruments range from $2000 to $7000.

As might be expected, performance specifications vary considerably among instruments and are related, at least to some degree, to instrument price. Typically, bandwidths vary from 2 to 8 nm; wavelength accuracies of ± 0.5 to ± 2 nm are reported.

The optical designs for the various grating instruments do not differ greatly from those of the two instruments shown in Figure 7-9. One manufacturer, however, employs a concave rather than a plane grating; a simpler and more compact design results.

Figure 7-10 is a schematic diagram of the Beckman DU-2 spectrophotometer, a high-quality, single-beam instrument for the ultraviolet-visible region. This instrument, which employs a Littrow prism rather than a grating, originally appeared on the market in 1941, at which time it was the first ultraviolet-visible spectrophotometer to become commercially available to chemists. The optics were of remarkably high quality; many of the early instruments are still in use today, although most have been refitted with more modern electronic systems. The optical design of the instrument is still being employed by one instrument manufacturer.[6]

The DU-2 spectrophotometer is equipped with quartz optics, which permits its use in both the ultraviolet and visible regions of the spectrum. Interchangeable radiation sources are provided, including a deuterium or hydrogen discharge tube for the lower wavelengths and a tungsten-filament lamp for the visible region. A pair of mirrors reflects radiation through an adjustable slit into the monochromator compartment. After traversing the length of the instrument, the radiation is reflected into a Littrow prism; by adjusting the position of the prism, light of the desired wavelength can be focused on the exit slit. The optics are so arranged that the entrance and exit beams are displaced from one another on the vertical axis; thus, the exit

---

[6] Gilford Instrument, Oberlin, Ohio 44074.

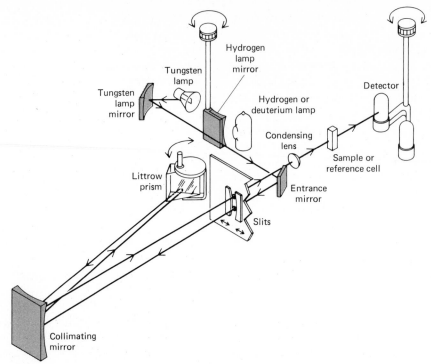

FIGURE 7-10   Schematic diagram of the Beckman DU-2® Spectrophotometer. (By permission, Beckman Instruments, Inc., Fullerton, California.)

beam passes above the entrance mirror as it leaves the monochromator.

The cell compartment accommodates as many as four rectangular 1-cm cells, any one of which can be positioned in the path of the beam by movement of a carriage arrangement. Compartments are also available that will hold both cylindrical cells and cells up to 10 cm in length.

The original DU-2 employed interchangable detectors—one for the range of 190 to 625 nm and the other for 625 to 800 nm. Gilford Instrument now offers a photometer modernization kit for the DU-2 in which the phototubes are replaced by a single photomultiplier with the same range as the original two. An amplifier circuit is then employed to convert the transducer signal to a voltage that

is directly proportional to absorbance; the readout is digital. Gilford also offers an accessory that oscillates the cuvette carriage at a rate of five cycles per second, alternately placing the sample and reference cells in the light path. With this addition, the readout of the instrument is automatic in absorbance. Addition of wavelength and slit-drive mechanisms and a recorder, makes possible the direct recording of spectra.

The modified DU-2 design achieves photometric accuracies as great as $\pm 0.5\% \, A$ by employing high-quality electronic components that are operated well below their rated capacities. Narrow effective bandwidths (less than 0.5 nm) can be obtained throughout the spectrum region by suitable variation of the slit adjustment (note that grating instruments discussed earlier are operated at fixed slit widths because of the linear characteristics of their dispersing elements).

The manual DU-2 is particularly well suited for quantitative analytical measurements that require absorbance data at a limited number of wavelengths.

Double-Beam Instruments for the Ultraviolet-Visible Region. Numerous double-beam spectrophotometers for the ultraviolet-visible region of the spectrum are now available. Generally, these instruments are more expensive than their single-beam counterparts, with the nonrecording variety ranging in cost from about $4000 to $10,000. The most sophisticated instrument may cost as much as $30,000 or more when all accessories are included. The latter will generally provide spectra in the wavelength range of 185 to 3000 nm.

Figure 7-11 shows construction details of two commercially available, ultraviolet-visible, double-beam instruments. The upper diagram is for a relatively inexpensive ($\sim$$5000), manual instrument (the Hitachi Model 100-60), which employs a concave grating for dispersing and focusing the radiation on the entrance slit. A motor-driven sector mirror divides the beam; after passage through the sample or solvent, the beams are recombined to give an ac signal (unless their powers are identical) which is transduced, amplified, and converted to a ratio of P and P_0. Either a meter or a digital readout is available; a recorder can also be purchased as an accessory.

Figure 7-11b shows details of a high-performance, recording spectrophotometer (Varian Cary Model 219). Its cost is significantly greater than the simpler instrument ($\sim$$12,000). A unique feature of the Cary instrument is the double-pass optics by means of which radiation is reflected and diffracted by the grating twice before being split for passage through the sample and reference cells. The effect is the same as with a double monochromator—that is, enhanced dispersion and reduction in scattered radiation.

The performance characteristics of the two instruments differ significantly. The stray light at 220 nm, the wavelength accuracy, and the wavelength repeatability for the Hitachi instrument are $< 0.07\%$, ± 0.4 nm, and ± 0.2 nm, respectively; the corresponding figures for the Cary instrument are $< 0.002\%$, ± 0.2 nm, and ± 0.1 nm. The photometric linearity of the Hitachi instrument is 0.005 to 0.01 A compared with 0.0016 to 0.003 A for the Cary. In addition, the Cary instrument has eight recorder ranges (compared with three for the Hitachi) and a number of automated features such as an automatic slit control to provide either constant resolution or constant noise, automatic 0% T set, and digital display of concentrations.

For many applications, simpler and less expensive instruments, such as that shown in Figure 7-11a, provide just as satisfactory results as do the more expensive ones.

Comparison of Single- and Double-Beam Instruments. Because the measurement of P and P_0 is made simultaneously or nearly simultaneously, a double-beam instrument compensates for all but the most short-term electrical fluctuations, as well as other time-dependent irregularities in the source, the detector, and the amplifier. Therefore, the electrical components of a double-beam photometer need not be of such high quality as those for a single-beam instrument. Offsetting this advantage, however, is the greater number and complexity of components associated with double-beam instrumentation.

Single-beam instruments are particularly well adapted to the quantitative analysis that involves an absorbance measurement at a single wavelength. Here, the simplicity of the instrument and the concomitant ease of maintenance offer real advantages. The greater speed and convenience of measurement, on the other hand, make the double-beam instrument particularly useful for qualitative analyses, where absorbance measurements must be made at many wavelengths.

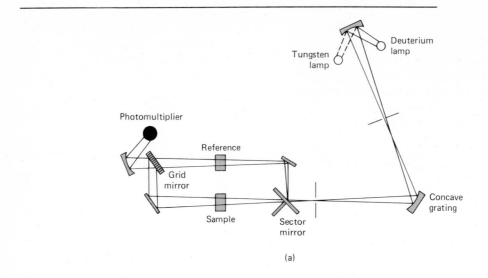

(a)

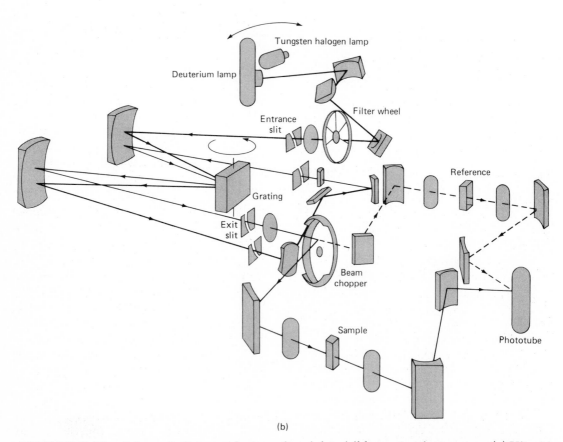

(b)

FIGURE 7-11   Schematic diagrams for two ultraviolet-visible spectrophotometers. (a) Hitachi Model 100-60 (courtesy of Hitachi Scientific Instruments, Sunnyvale, California); (b) Varian Cary Model 219 (courtesy of Varian Instrument Division, Palo Alto, California).

APPLICATION OF ABSORPTION MEASUREMENT TO QUALITATIVE ANALYSIS

Ultraviolet and visible spectrophotometry have somewhat limited application for qualitative analysis because the number of absorption maxima and minima are relatively few. Thus, unambiguous identification is frequently impossible.

Qualitative Techniques

Solvents. In choosing a solvent, consideration must be given not only to its transparency, but also to its possible effects upon the absorbing system. Quite generally, polar solvents such as water, alcohols, esters, and ketones tend to obliterate spectral fine structure arising from vibrational effects; spectra that approach those of the gas phase (see Figure 7-12) are more likely to be observed in nonpolar solvents such as hydrocarbons. In addition, the positions of absorption maxima are also influenced by the nature of the solvent. Clearly, the same solvent must be used when comparing absorption spectra for identification purposes.

Table 7-6 lists some common solvents and the approximate wavelength below which they cannot use because of absorption. These minima depend strongly upon the purity of the solvent.[7]

Methods for Plotting Spectra. Figure 7-13 shows absorption spectra for solutions of permanganate plotted with three different ordinates. Note that the greatest differences among transmittance plots occur when the transmittance ranges between 20 to 60%. In contrast, greater differentiation of absorbance plots is

---

[7] Most major suppliers of reagent chemicals in the United States offer spectrochemical grades of solvents; these meet or exceed the requirements set forth in *Reagent Chemicals, American Chemical Society Specifications*, 5th ed. Washington, D.C.: American Chemical Society, 1974.

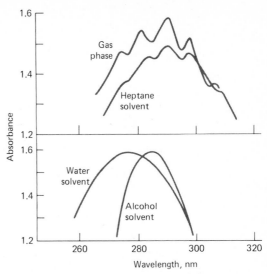

FIGURE 7-12 Effect of solvent on the absorption spectrum of acetaldehyde.

found at low transmittance or high absorbance (0.8 to 1.6).

The log absorbance plots shown in Figure 7-13c are advantageous for the comparison of spectra. If Beer's law is written in the form

$$\log A = \log \varepsilon + \log bc$$

it is seen that of the terms on the right-hand side of this equation, only ε varies with wavelength. Thus, if the concentrations or cell lengths for the spectra being matched are not identical, $\log A$ is displaced by exactly the same absolute amount at each wavelength. This effect is shown in Figure 7-13c. When the plots of $\log A$ are compared with those of A (Figure 7-13b), the advantage of the former for qualitative analysis becomes evident. It should also be noted that plots of $\log \varepsilon$ are also readily compared for the same reason.

Detection of Functional Groups

Even though they may not provide the unambiguous identification of an organic compound, absorption measurements in the

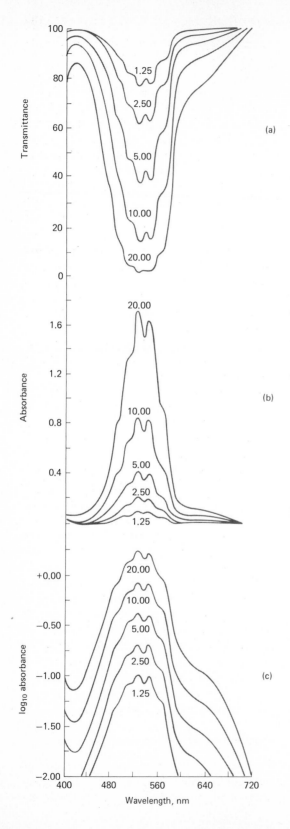

visible and the ultraviolet regions are nevertheless useful for detecting the presence of certain functional groups that act as chromophores.[8] For example, a weak absorption band in the region of 280 to 290 nm, which is displaced toward shorter wavelengths with increased solvent polarity, strongly indicates the presence of the carbonyl group. A weak absorption band at about 260 nm with indications of vibrational fine structure constitutes evidence for the existence of an aromatic ring. Confirmation of the presence of an aromatic amine or a phenolic structure may be obtained by comparing the effects of pH on the spectra of solutions containing the sample with those shown in Table 7-4 for phenol and aniline.

QUANTITATIVE ANALYSIS BY ABSORPTION MEASUREMENTS

Absorption spectroscopy is one of the most useful tools available to the chemist for quantitative analysis. Important characteristics of spectrophotometric and photometric methods include:

1. *Wide applicability.* As we have pointed out, numerous inorganic and organic species absorb in the ultraviolet and visible ranges and are thus susceptible to quantitative deter-

[8] See: R. M. Silverstein, G. C. Bassler, and T. C. Morrill, *Spectrometric Identification of Organic Compounds*, 3d ed. New York: Wiley, 1974, Chapter 5.

FIGURE 7-13 Three methods for plotting absorption spectra. The numbers beneath the curves indicate the ppm $KMnO_4$ in the solution; 2.0-cm cells were employed. (Plots from M. G. Mellon, *Analytical Absorption Spectroscopy.* New York: Wiley, 1950, pp. 104–106.)

mination. In addition, many nonabsorbing species can be analyzed after conversion to absorbing species by suitable chemical treatment.

2. *High sensitivity.* Molar absorptivities in the range of 10,000 to 40,000 are common, particularly among the charge-transfer complexes of inorganic species. Thus, analyses for concentrations in the range of 10^{-4} to $10^{-5}M$ are ordinary; the range can often be extended to 10^{-6} or even $10^{-7}M$ with suitable procedural modifications.

3. *Moderate to high selectivity.* It may be possible to locate a wavelength region in which the only absorbing component in a sample is the substance being determined. Furthermore, where overlapping absorption bands do occur, corrections based on additional measurements at other wavelengths are sometimes possible. As a consequence, the separation step can be omitted.

4. *Good accuracy.* For the typical spectrophotometric or photometric procedure, the relative error in concentration measurements lies in the range of 1 to 3%. Such errors can often be reduced to a few tenths of a percent with special precautions.

5. *Ease and convenience.* Spectrophotometric or photometric measurements are easily and rapidly performed with modern instruments.

Scope

The applications of quantitative absorption methods are not only numerous, but also touch upon every field in which quantitative chemical information is required. The reader can obtain a notion of the scope of spectrophotometry by consulting a series of review articles published periodically in *Analytical Chemistry*[9] and from monographs on the subject.[10]

Applications to Absorbing Species. Tables 7-2, 7-3, and 7-4 list many common organic chromophoric groups. Spectrophotometric analysis for any organic compound containing one or more of these groups is potentially feasible; many examples of this type of analysis are found in the literature.

A number of inorganic species also absorb

---

[9] See, for example, D. F. Boltz and M. G. Mellon, *Anal. Chem.*, **48**, 216R (1976); and J. A. Howell, and L. G. Hargis, *Anal. Chem.*, **50**, 243R (1978).

[10] See, for example: E. B. Sandell, *Colorimetric Determination of Traces of Metals*, 4th ed. New York: Interscience, 1978; *Colorimetric Determination of Nonmetals*, 2d ed., eds. D. F. Boltz and J. A. Howell. New York: Wiley, 1978; Z. Marczenko, *Spectrophotometric Determination of Elements.* New York: Halsted Press, 1975; and M. Pisez and J. Bartos, *Colorimetric and Fluorometric Analysis of Organic Compounds and Drugs.* New York: Marcel Dekker, 1974.

TABLE 7-6   SOLVENTS FOR THE ULTRAVIOLET AND THE VISIBLE REGIONS

Solvent	Approximate transparency minimum (nm)	Solvent	Approximate transparency minimum (nm)
Water	180	Carbon tetrachloride	260
Ethanol	220	Diethyl ether	210
Hexane	200	Acetone	330
Cyclohexane	200	Dioxane	320
Benzene	280	Cellosolve	320

and are thus susceptible to direct determination; we have already mentioned the various transition metals. In addition, a number of other species also show characteristic absorption. Examples include nitrite, nitrate, and chromate ions; osmium and ruthenium tetroxides; molecular iodine; and ozone.

Applications to Nonabsorbing Species. Numerous reagents react with nonabsorbing species to yield products that absorb strongly in the ultraviolet or visible regions. The successful application of such reagents to quantitative analysis usually requires that the color-forming reaction be forced to near completion. It should be noted that these reagents are frequently employed as well for the determination of an absorbing species, such as a transition-metal ion; the molar absorptivity of the product will frequently be orders of magnitude greater than that of the uncombined species.

A host of complexing agents have been employed for the determination of inorganic species. Typical inorganic reagents include thiocyanate ion for iron, cobalt, and molybdenum; the anion of hydrogen peroxide for titanium, vanadium, and chromium; and iodide ion for bismuth, palladium, and tellurium. Of even more importance are organic chelating agents which form stable, colored complexes with cations. Examples include o-phenanthroline for the determination of iron, dimethylglyoxime for nickel, diethyldithiocarbamate for copper, and diphenyldithiocarbazone for lead.

Procedural Details

The first steps in a photometric or spectrophotometric analysis involve the establishment of working conditions and the preparation of a calibration curve relating concentration to absorbance.

Selection of Wavelength. Spectrophotometric absorbance measurements are ordinarily made at a wavelength corresponding to an absorption peak because the change in absorbance per unit of concentration is greatest at this point; the maximum sensitivity is thus realized. In addition, the absorption curve is often flat in this region; under these circumstances, good adherence to Beer's law can be expected (p. 156). Finally, the measurements are less sensitive to uncertainties arising from failure to reproduce precisely the wavelength setting of the instrument.

The absorption spectrum, if available, aids in choosing the most suitable filter for a photometric analysis; if this information is lacking, the alternative method for selection, given on page 181, may be used.

Variables that Influence Absorbance. Common variables that influence the absorption spectrum of a substance include the nature of the solvent, the pH of the solution, the temperature, high electrolyte concentrations, and the presence of interfering substances. The effects of these variables must be known; conditions for the analysis must be chosen such that the absorbance will not be materially influenced by small, uncontrolled variations in their magnitudes.

Determination of the Relationship Between Absorbance and Concentration. After deciding upon the conditions for the analysis, it is necessary to prepare a calibration curve from a series of standard solutions. These standards should approximate the overall composition of the actual samples and should cover a reasonable concentration range of the analyte. Seldom, if ever, is it safe to assume adherence to Beer's law and use only a single standard to determine the molar absorptivity. The results of an analysis should *never* be based on a literature value for the molar absorptivity.

Cleaning and Handling of Cells. It is apparent that accurate spectrophotometric analysis requires the use of good-quality, matched cells. These should be regularly calibrated against one another to detect differences that can arise from scratches, etching, and wear. Equally important is the use of proper cell cleaning and drying techniques. Erickson and Surles[11] recommend the following cleaning sequence for the outside windows of cells. Prior to measurement, the cell surfaces are cleaned with a lens paper soaked in spectrograde methanol. The paper is held with a hemostat; after wiping, the methanol is allowed to evaporate, leaving the cell surfaces free of contaminants. The authors showed that this method was far superior to the usual procedure of wiping the cell surfaces with a dry lens paper, which apparently leaves lint and films on the surface.

Analysis of Mixtures of Absorbing Substances. The total absorbance of a solution at a given wavelength is equal to the sum of the absorbances of the individual components present (p. 153). This relationship makes possible the analysis of the individual components of a mixture, even if their spectra overlap. Consider, for example, the spectra of M and N, shown in Figure 7-14. Obviously, no wavelength exists at which the absorbance of this mixture is due simply to one of the components; thus, an analysis for either M or N is impossible by a single measurement. However, the absorbances of the mixture at the two wavelengths λ' and λ'' may be expressed as follows:

$$A' = \varepsilon_M' b c_M + \varepsilon_N' b c_N \qquad \text{(at } \lambda')$$

$$A'' = \varepsilon_M'' b c_M + \varepsilon_N'' b c_N \qquad \text{(at } \lambda'')$$

The four molar absorptivities ε_M', ε_N', ε_M'', and ε_N'' can be evaluated from individual standard solutions of M and of N, or better, from the slopes of their Beer's law plots. The absorbances of the mixture, A' and A'', are experimentally determinable, as is b, the cell thickness. Thus, from these two equations, the concentration of the individual components

[11] J. O. Erickson and T. Surles, *American Laboratory*, **8** (6), 50 (1976).

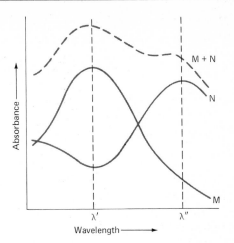

FIGURE 7-14 Absorption spectrum of a two-component mixture.

in the mixture, c_M and c_N, can be readily calculated. These relationships are valid only if Beer's law is followed. The greatest accuracy in an analysis of this sort is attained by choosing wavelengths at which the differences in molar absorptivities are large.

Mixtures containing more than two absorbing species can be analyzed, in principle at least, if a further absorbance measurement is made for each added component. The uncertainties in the resulting data become greater, however, as the number of measurements increases.

EXAMPLE

A 0.246-g sample of steel was analyzed for Cr and Mn by solution of the sample in acid and dilution to exactly 250 ml. A 50.0-ml aliquot was treated with potassium persulfate, which, in the presence of a catalytic amount of Ag^+, converted the Cr and Mn to $Cr_2O_7^{2-}$ and MnO_4^-. The resulting solution was then diluted to 100 ml. In a 1.00-cm cell, the absorbance of the solution at 440 and 545 nm was found to be 0.932 and 0.778, respectively. Calculate the percent Mn and Cr employing the following previously determined data:

λ, nm	ε, $Cr_2O_7^{2-}$	ε, MnO_4^-
440	369	95
545	11	2350

For the absorbance at 440 nm, we may write

$$0.932 = 369 \times 1.00 \times M_{Cr}$$
$$+ 95 \times 1.00 \times M_{Mn}$$

and at 545 nm,

$$0.778 = 11 \times 1.00 \times M_{Cr}$$
$$+ 2350 \times 1.00 \times M_{Mn}$$

The first equation can be rewritten as

$$M_{Cr} = (0.932 - 95 M_{Mn})/369$$

Substituting into the second equation gives

$$0.778 \times 369 = 11 \times 0.932 - 11 \times 95 M_{Mn}$$
$$+ 2350 \times 369 M_{Mn}$$
$$M_{Mn} = 3.20 \times 10^{-4}$$

Substitution of this into the first equation yields

$$M_{Cr} = 2.44 \times 10^{-3}$$

Employing 54.9 as the atomic weight of manganese gives

$$\% \ Mn = \frac{\begin{array}{c} 3.20 \times 10^{-4} \times 100(50.0/250) \\ \times 0.0549 \times 100 \end{array}}{0.246}$$

$$= 0.143$$

Similarly,

$$\% \ Cr = \frac{\begin{array}{c} 2.44 \times 10^{-3} \times 100(50.0/250) \\ \times 0.0520 \times 100 \end{array}}{0.246}$$

$$= 1.03$$

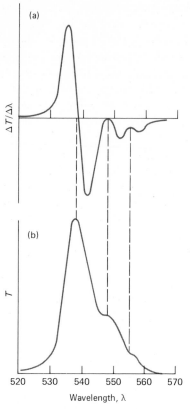

FIGURE 7-15 Comparison of a derivative curve (a) with a standard transmittance curve (b).

Derivative Spectrophotometry[12]

In derivative spectrophotometry, the first or second derivative of absorbance or transmittance with respect to wavelength is employed as the ordinate in plotting spectral data. As shown by Figure 7-15, such plots often reveal spectral detail that is less apparent in an ordinary absorption spectral plot.

Derivative spectra are obtained in two ways. The output signal from the transducer

[12] For additional information, see: T. C. O'Haver and G. L. Green, *American Laboratory*, **7** (3), 15 (1975); and T. J. Porro, *Anal. Chem.*, **44** (4), 93A (1972).

of a recording spectrophotometer can be differentiated electronically (p. 66) or mechanically and recorded as a function of wavelength. Alternatively, use is made of a *dual-wavelength* spectrophotometer such as that shown in Figure 7-16. Note that this instrument is designed to operate either in a dual-wavelength mode, as shown, or in an ordinary double-beam mode instrument. In the dual mode, the reference cell is not required and the sample cell is alternately exposed to radiation from each of the gratings. In the double-beam mode, only monochromator 1 is used and the beam is directed first through the sample and then through the reference.

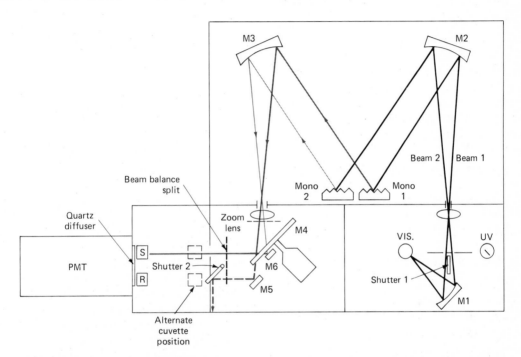

FIGURE 7-16 Schematic diagram of a dual-wavelength spectrophotometer. Radiation from the source is focused by lamp mirror M1 on the slit of the duochromator. This mirror and the aperture define two independent beams used in the dual wavelength mode. After collimation by mirror M2 and diffraction by gratings MONO 1 and MONO 2, the beams are focused by mirror M3 on the exit slit. After passing through the exit slit and the zoom lens, the beams alternately pass through, or are deflected by the rotating chopper mirror M4. During one half of the cycle, light from MONO 1 is reflected by mirror M5 toward a light trap, and light from MONO 2 is reflected by mirror M6, located behind the chopper mirror, toward the sample cuvette. During the second half of the cycle, light from MONO 1 is reflected by the chopper mirror toward the sample cuvette while light from MONO 2 is trapped. For ordinary double-beam spectra, shutter 1 is turned so that it deflects beam 1 away from the entrance aperture, shutter 2 is also moved so that the second half of the radiation passes through the solvent contained in the alternate cuvette. Twin detectors are then employed to compare the two beams. (Courtesy of American Instruments Company, Division of Travenol Laboratories, Silver Springs, Md.)

A spectrum such as that shown in Figure 7-15 is obtained with a dual-wavelength instrument by offsetting one monochromator from the other by one or two nanometers and recording the resulting derivative spectrum.

Dual-wavelength spectrometers are also useful for quantitative analysis of one species in the presence of a second species, when the spectra of the two are alike. Here, the instrument is operated in the nonscanning mode with one monochromator set at a wavelength at which the two species have identical molar absorptivities. The other monochromator is then set to a wavelength where the analyte absorbance exhibits a large change as a function of concentration. The resulting signal is then proportional to the concentration of the analyte under most circumstances.

Dual-wavelength spectrophotometers are also useful for monitoring the concentration of two species in kinetic studies, and for deriving absorption spectra of a species in a turbid medium.

Differential Absorption Methods[13]

In the previous chapter (p. 160), it was pointed out that the accuracy of many photometers and spectrophotometers is limited by the sensitivities of their readout devices. With such instruments, the uncertainty s_T in the measurement of T is *constant*, and the resulting relative concentration errors are given by Equation 6-14.

Curve O in Figure 7-17 gives a plot of the relative concentration errors to be expected from a *readout-limited* uncertainty of $s_T = \pm 0.005$. This uncertainty would be typical for an instrument equipped with a small D'Arsonval-type meter. Clearly, serious analytical errors are encountered when the analyte concentration is such that $\%\, T$ is smaller than

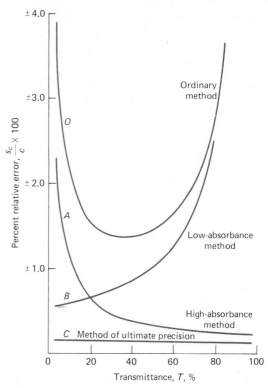

Figure 7-17 Relative errors in spectrophotometric analysis. Curve O: ordinary method; reference solution has transmittance of 100%. Curve A: high-absorbance method; reference solution has transmittance of 10%. Curve B: low-absorbance method; reference solution has transmittance of 90%. Curve C: method of ultimate precision; reference solutions have transmittances of 45% and 55%, respectively. s_T is 0.5% for each curve. See Figure 7-18 for the type of scale expansion corresponding to curves A, B, and C.

10% ($A = 1.0$) or greater than 70% ($A = 0.15$).

Differential methods provide a means of expanding the scale of a readout-limited instrument and thus reducing this type of error significantly. These methods employ solutions of the analyte for adjusting the zero and/or the 100% transmittance setting of the

[13] For a complete analysis of these methods, see: C. N. Reilley and C. M. Crawford, *Anal. Chem.*, **27**, 716 (1955); C. F. Hiskey, *Anal. Chem.*, **21**, 1440 (1949).

photometer or spectrophotometer rather than the shutter and the solvent. The three types of differential methods are compared with the ordinary method in Table 7-7.

High-Absorbance Method. In the high-absorbance procedure, the zero adjustment is carried out in the usual way, with the shutter imposed between the source and the detector. The 100% transmittance adjustment, however, is made with a standard solution of the analyte, which is less concentrated than the sample, in the light path. Finally, the standard is replaced by the sample, and a relative transmittance is read directly. As shown in Figure 7-18a, the effect of this modification is to expand a small portion of the transmittance scale to a full 100%; thus, as shown by the upper scale of Figure 7-18a, the transmittance of the reference standard employed is 10% *when compared with pure solvent*, while the sample exhibits a 4% transmittance against the same solvent. That is, the sample transmittance is four-tenths that of the standard when compared with a common solvent. When the standard is substituted as the reference, however, its transmittance becomes

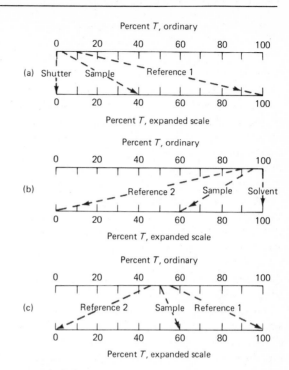

FIGURE 7-18   Scale expansion by various methods: (a) high-absorbance method; (b) low-absorbance method; and (c) method of ultimate precision.

TABLE 7-7   COMPARISON OF METHODS FOR ADSORPTION MEASUREMENTS

| Method | Designation[a] | Imposed in Beam for Indicator Setting of: | |
		0% T	100% T
Ordinary	O	Shutter	Solvent
High absorbance	A or a	Shutter	Standard solution less concentrated than sample
Low absorbance	B or b	Standard solution more concentrated than sample	Solvent
Ultimate precision	C or c	Standard solution more concentrated than sample	Standard solution less concentrated than sample

[a] See Figures 7-17 and 7-18.

100% (lower scale of Figure 7-18a). The transmittance of the sample remains four-tenths that of the reference standard; here, then, it is 40%.

If the instrumental uncertainty in the measurement of absorbance is not affected by the modification, use of the standard as a reference brings the transmittance into the middle of the scale, where instrumental uncertainties have a minimal effect on the relative concentration error.

It can be shown that a linear relationship exists between concentration and the relative absorbance measured by the high-absorbance technique.[14] The method is particularly useful for the analysis of samples with absorbances greater than unity.

Low-Absorbance Method. In the low-absorbance procedure, a standard solution somewhat more concentrated than the sample is employed in lieu of the shutter for zeroing the indicator scale; the full-scale adjustment is made in the usual way, with the solvent in the light path. The sample transmittance is then obtained by replacing the solvent with the sample. The effect of this procedure is shown in Figure 7-18b; note that, again, a small segment of the scale is expanded to 100% transmittance, and that the transmittance of the sample now lies near the center of the expanded scale.

The low-absorbance method is particularly applicable to samples having absorbances of less than 0.1. Here a nonlinear relationship exists between relative absorbance and concentration. As a consequence, a number of standard solutions must be employed to establish accurately the calibration curve for the analysis.

The Method of Ultimate Precision. The two techniques just described can be combined to give the method of ultimate precision. As shown in Figure 7-18c, a scale expansion is again achieved; here, two reference solutions, one having a smaller and one having a greater transmittance than the sample, are employed to adjust the 100% and the 0% scale readings. A nonlinear relationship between relative absorbance and concentration is again observed.

Precision Gain by Differential Methods. It is possible to derive relationships analogous to Equation 6-14 for the three differential methods.[15] Figure 7-17 compares the corresponding error curves with that of the ordinary method. From curve A, it is apparent that a significant gain in precision results when the high-absorbance method elevates the transmittances of solutions to ten percent or greater. The plot was based upon the reference standard employed in Figure 7-18a; that is, for a reference having a normal transmittance of 10%. The error curve approaches that for the ordinary procedure (see curve O, Figure 7-17) as the percent transmittance of the reference increases and approaches 100. Error curves for the low-absorbance method (curve B) and the method of ultimate precision (curve C) are based on the corresponding data in the caption accompanying Figure 7-18.

Instrumental Requirements for Precision Methods. For the low-absorbance method, it is clearly necessary that the spectrophotometer possess a dark-current compensating circuit capable of offsetting larger currents than are normally produced when no radiation strikes the photoelectric detector. For the high-absorbance method, on the other hand, the instrument must have a sufficient reserve capacity to permit setting the indicator at 100% transmittance when an absorbing solution is placed in the light path. Here, the full-scale reading may be realized either by increasing the radiation intensity (most often by widening the slits) or by increasing the

[14] See: C. N. Reilley and C. M. Crawford, *Anal. Chem.*, **27**, 716 (1955).

[15] See: C. N. Reilley and C. M. Crawford, *Anal. Chem.*, **27**, 716 (1955).

amplification of the photoelectric current. The method of ultimate precision requires instruments with both of these qualities.

The capability of a spectrophotometer to be set to full scale with an absorbing solution in the radiation path will depend both upon the quality of its monochromator and the stability of its electronic circuit. Furthermore, this capacity will be wavelength-dependent, since the intensity of the source and the sensitivity of the detector change as the wavelength is varied. In regions where the intensity and sensitivity are low, an increase in slit width may be necessary to realize a full-scale setting; under these circumstances, scattered radiation may lead to errors unless the quality of the monochromator is high. Alternatively, a very high current amplification may be necessary; again, unless the electronic stability is good, significant photometric error may result.

PHOTOMETRIC TITRATIONS

Photometric or spectrophotometric measurements can be employed to advantage in locating the equivalence point of a titration.[16] The end point in a direct photometric titration is the result of a change in the concentration of a reactant or a product, or both; clearly, at least one of these species must absorb radiation at the wavelength selected. In the indirect method, the absorbance of an indicator is observed as a function of titrant volume.

Titration Curves

A photometric titration curve is a plot of corrected absorbance as a function of the volume of titrant. If conditions are chosen properly, the curve will consist of two straight-line por-

tions with differing slopes, one occurring at the outset of the titration and the other located well beyond the equivalence-point region; the end point is taken as the intersection of extrapolated linear portions. Figure 7-19 shows some typical titration curves. Titration of a nonabsorbing species with a colored titrant that is decolorized by the reaction produces a horizontal line in the initial stages, followed by a rapid rise in absorbance beyond the equivalence point (Figure 7-19a). The formation of a colored product from colorless reactants, on the other hand, initially produces a linear rise in the absorbance, followed by a region in which the absorbance becomes independent of reagent volume (Figure 7-19b). Depending upon the absorption characteristics of the reactants and the products, the other curve forms shown in Figure 7-19 are also possible.

In order to obtain a satisfactory photometric end point, it is necessary that the absorbing system(s) obey Beer's law; otherwise, the titration curve will lack the linear portions needed for end-point extrapolation. Further, it is necessary to correct the absorbance for volume changes. The observed

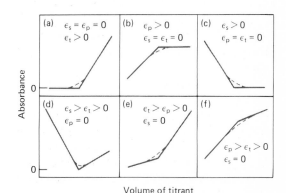

FIGURE 7-19 Typical photometric titration curves. Molar absorptivities of the substance titrated, the product, and the titrant are given by ϵ_s, ϵ_p, ϵ_t, respectively.

[16] For further information concerning this technique, see: J. B. Headridge, *Photometric Titrations*. New York: Pergamon Press, 1961.

values are multiplied by $(V + v)/V$, where V is the original volume of the solution and v is the volume of added titrant.

Instrumentation

Photometric titrations are ordinarily performed with a spectrophotometer or a photometer that has been modified to permit insertion of the titration vessel in the light path.[17] After the zero adjustment of the meter scale has been made, radiation is allowed to pass through the solution of the analyte, and the instrument is adjusted by varying the source intensity or the detector sensitivity until a convenient absorbance reading is obtained. Ordinarily, no attempt is made to measure the true absorbance, since relative values are perfectly adequate for the purpose of end-point detection. Data for the titration are then collected without alteration of the instrument setting.

The power of the radiation source and the response of the detector must be reasonably constant during the period required for a photometric titration. Cylindrical containers are ordinarily used, and care must be taken to avoid any movement of the vessel that might alter the length of the radiation path.

Both filter photometers and spectrophotometers have been employed for photometric titrations. The latter are preferred, however, because their narrower bandwidths enhance the probability of adherence to Beer's law.

Application of Photometric Titrations

Photometric titrations often provide more accurate results than a direct photometric analysis because the data from several measurements are pooled in determining the end point. Furthermore, the presence of other absorbing species may not interfere, since only a change in absorbance is being measured.

The photometric end point possesses the advantage over many other commonly used end points in that the experimental data are taken well away from the equivalence-point region. Thus, the titration reactions need not have such favorable equilibrium constants as those required for a titration that depends upon observations near the equivalence point (for example, potentiometric or indicator end points). For the same reason, more dilute solutions may be titrated.

The photometric end point has been applied to all types of reactions.[18] Most of the reagents used in oxidation-reduction titrations have characteristic absorption spectra and thus produce photometrically detectable

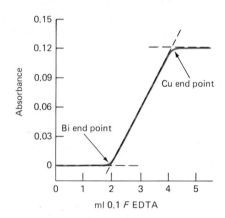

FIGURE 7-20 Photometric titration curve of 100 ml of a solution that was 2.0×10^{-3} F in Bi^{3+} and Cu^{2+}. Wavelength: 745 nm. (Reprinted with permission from A. L. Underwood, *Anal. Chem.*, **26**, 1322 (1954). Copyright by the American Chemical Society.)

---

[17] Titration flasks and cells for use in a Spectronic 20 spectrophotometer are available from the Kontes Manufacturing Corp., Vineland, N.J.

[18] See, for example, the review: A. L. Underwod, *Advances in Analytical Chemistry and Instrumentation*, ed. C. N. Reilley. New York: Interscience, 1964, vol. 3, pp. 31–104.

end points. Acid-base indicators have been employed for photometric neutralization titrations. The photometric end point has also been used to great advantage in titrations with EDTA and other complexing agents. Figure 7-20 illustrates the application of this end point to the successive titration of bismuth(III) and copper(II). At 745 nm, neither cation nor the reagent absorbs, nor does the more stable bismuth complex, which is formed in the first part of the titration; the copper complex, however, does absorb. Thus, the solution exhibits no absorbance until essentially all of the bismuth has been titrated. With the first formation of the copper complex, an increase in absorbance occurs. The increase continues until the copper equivalence point is reached. Further reagent additions cause no further absorbance change. Clearly, two well-defined end points result.

The photometric end point has also been adapted to precipitation titrations; here, the suspended solid product has the effect of diminishing the radiant power by scattering; titrations are carried to a condition of constant turbidity.

AUTOMATIC PHOTOMETRIC AND SPECTROPHOTOMETRIC ANALYSIS

One of the major instrumental developments during the last two decades has been the appearance, from commercial sources, of automatic analysis systems, which provide analytical data with a minimum of operator intervention. The need for these systems has been greatest in clinical laboratories, where perhaps thirty or more analyses are routinely used in large number for diagnostic and screening purposes. The number of such analyses required by modern medicine is enormous; the need to keep their costs at a reasonable level is obvious. These two considerations have led to a focus of effort toward using automatic instruments for clinical laboratories.[19] As an outgrowth, automatic instruments are now beginning to find application in such diverse fields as analyses for the control of industrial processes and the routine analyses of air, water, soil, and agricultural products.

Most clinical analyses are based upon photometric or spectrophotometric measurements employing ultraviolet or visible radiation. Thus, an important component of most automatic instruments is a photometer or spectrophotometer. As will be pointed out later, however, automatic systems may also employ atomic spectroscopy and electroanalytical methods to measure analyte concentrations.

Two examples of automatic systems are discussed briefly in the paragraphs that follow.

Continuous-Flow Analyzer

In a continuous analyzer, successive samples pass through the same system of tubes and chambers. Thus, samples must be isolated from one another to avoid cross contamination, and means must be provided for rinsing the system between samples. In the Technicon AutoAnalyzer®, the earliest of the commercial instruments, movement of sample, reagents, and diluent through plastic tubing is accomplished by means of a peristaltic pump. Successive samples are isolated from one another by introducing bubbles of air into the tubing.

Figure 7-21 is a schematic diagram of a single-channel autoanalyzer used for the analysis of one of the constituents of blood. Here, samples are removed automatically and successively from containers held in the rotating table sampler and are mixed with diluent and air bubbles. The latter promote mixing and serve to separate the sample from earlier

---

[19] For a description of some commercially available instruments, see: R. H. Laessig, *Anal. Chem.*, **43** (8), 18A (1971); and J. K. Foreman and P. B. Stockwell, *Automatic Chemical Analysis.* New York: Wiley, 1975, Chapter 4.

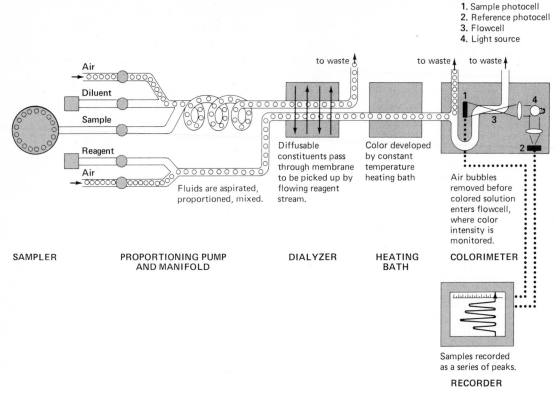

1. Sample photocell
2. Reference photocell
3. Flowcell
4. Light source

Air

Diluent

Sample

Reagent

Air

Fluids are aspirated, proportioned, mixed.

to waste

Diffusable constituents pass through membrane to be picked up by flowing reagent stream.

to waste   to waste

Color developed by constant temperature heating bath

Air bubbles removed before colored solution enters flowcell, where color intensity is monitored.

SAMPLER

PROPORTIONING PUMP AND MANIFOLD

DIALYZER

HEATING BATH

COLORIMETER

Samples recorded as a series of peaks.

RECORDER

FIGURE 7-21   A single-channel Technicon AutoAnalyzer System®. (Courtesy of Technicon Instruments Corporation, Tarrytown, New York.)

and later ones. The diluted sample then passes into the dialyzer, which contains membranes through which the small analyte ions or molecules are free to diffuse into the reagent stream. The residual large-protein molecules of the blood remain in the diluent stream and pass from the system to waste. The remainder of the system shown in Figure 7-21 is self-explanatory.

A system for routine blood analysis includes a module for diluting and partitioning the sample into several aliquots, each of which passes through a separate channel similar to the one shown in Figure 7-21. The final output from a typical system is shown in Figure 7-22. The shaded areas show the range of concentrations which are considered to be normal for the population.

Centrifugal Fast Analyzer

A type of batch analyzer that is capable of analyzing as many as 16 samples simultaneously for a single constituent is based upon the use of a centrifuge to mix the sample with the reagent and to transfer the mixture to a cell for photometric or spectrophotometric analysis.[20] The system is such that conversion from one type of reagent to another is usually easy.

The principle of the instrument is seen in

[20] For a brief description of a typical system, see: R. L. Coleman, W. D. Shults, M. T. Kelley, and J. A. Dean, *Amer. Lab.*, **3** (7), 26 (1971); and C. D. Scott and C. A. Burtis, *Anal. Chem.*, **45** (3), 327A (1973).

Figure 7-23, which is a cross-sectional view of the circular, plastic rotor of a centrifuge. The rotor has 17 dual compartments arranged radially around the axis of rotation. Samples and reagents are pipetted automatically into 16 of the compartments as shown; solvent and reagent are measured into the seven- teenth. When the rotor reaches a rotation rate of about 350 rpm, reagent and samples in the 17 compartments are mixed simultaneously and carried into individual cells located at the outer edge of the rotor; these cells are equipped with horizontal quartz windows. Mixing is hastened by drawing air through

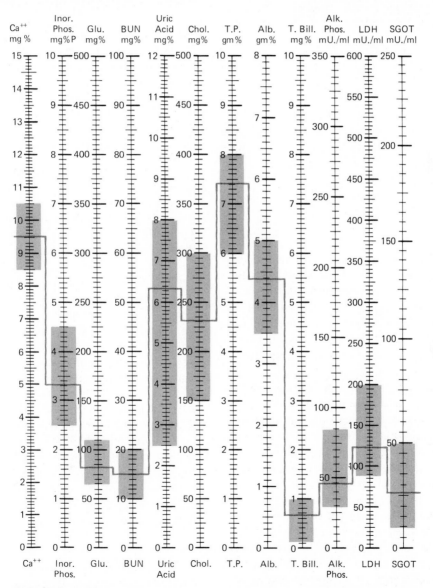

FIGURE 7-22 Readout from a multichannel Technicon AutoAnalyzer System®. (Courtesy of Technicon Instruments Corporation, Tarrytown, New York.)

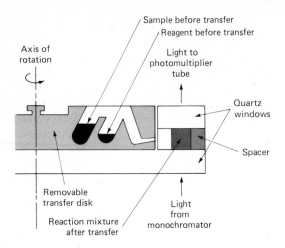

FIGURE 7-23   Rotor for a centrifugal fast analyzer. (From R. L. Coleman, W. D. Shults, M. T. Kelley, and J. A. Dean, *Amer. Lab.*, **3**(7), 26 (1971). With permission. Copyright 1971 by International Scientific Communications, Inc.)

the mixtures. Radiation from an interference-filter photometer or a spectrophotometer passes through the cells and falls upon a photomultiplier tube. For each rotation, a series of electrical pulses are produced, 16 for the samples and 1 for the blank. Between each of these pulses is a signal corresponding to the dark current.

The successive signals from the instrument are collected in the memory of a dedicated minicomputer for subsequent manipulation. Signal averaging can be employed to optimize the signal-to-noise ratio.

One of the most important applications of the centrifugal fast analyzer is for the determination of enzymes. Ordinarily, enzyme analyses are based upon the catalytic effect of the analyte upon a reaction that involves formation or consumption of an absorbing species. Here, a calibration curve relating the rate of appearance or disappearance of the absorbing species as a function of enzyme concentration serves as the basis for

the analysis. The centrifugal fast analyzer permits the simultaneous determination of the rates of 16 reactions under exactly the same conditions; thus, 16 simultaneous enzyme analyses are feasible.

PHOTOACOUSTIC SPECTROSCOPY

Photoacoustic spectroscopy is a recently developed method for determining the ultraviolet and visible absorption spectra of solids or semisolids.[21] Acquisition of spectra for this type of sample by ordinary methods is usually difficult at best because of light scattering.

The Photoacoustic Effect

Photoacoustic spectroscopy is based upon an absorption effect which was first investigated in the 1880s by Alexander Graham Bell and others. This effect is observed when a gas in a closed cell is irradiated with a chopped beam of radiation of a wavelength that is absorbed by the gas. The absorbed radiation is ultimately converted partially or completely to kinetic energy of the gas molecules, owing to nonradiative relaxation of the excited absorbing species. The kinetic energy of the gas molecules gives rise to pressure fluctuations within the cell, which, because of the chopping rate, are of acoustical frequencies and can thus be detected by a sensitive microphone. The photoacoustic effect has been used since the turn of the century for gas analysis, and has recently taken on new importance for this purpose with the advent of tunable infrared lasers as sources.

In photoacoustic studies of solids, the sample is placed in a closed cell containing air or some other *nonabsorbing* gas and a sen-

[21] For a review on applications, see: A. Rosencwaig, *Anal. Chem.*, **47** (6), 592A (1975); and *Optoacoustic Spectroscopy and Detection*, ed. Y. Pao, New York: Academic Press, 1977, Chapter 8.

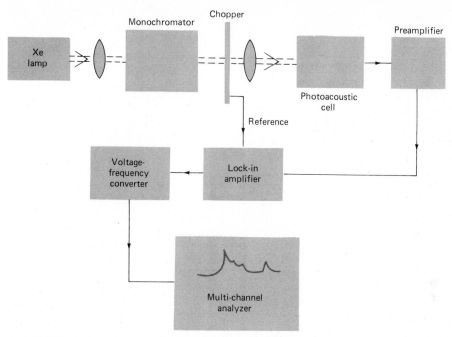

FIGURE 7-24   Block diagram of a single-beam photoacoustic spectrometer with digital data processing. (Reprinted with permission from A. Rosencwaig, *Anal. Chem.*, **47**, 592A (1975). Copyright by the American Chemical Society.)

sitive microphone. The solid is then irradiated with a chopped beam from a monochromator. The photoacoustic effect is observed *provided the radiation is absorbed by the solid*; the power of the resulting sound is directly related to the extent of absorption. Radiation reflected or scattered by the sample has no effect on the microphone and thus does not interfere. This latter property is perhaps the most important characteristic of the method.

The source of the photoacoustic effect in solids appears to be similar to that in gases. That is, nonradiative relaxation of the absorbing solid causes a periodic heat flow from the solid to the surrounding gas; the resulting pressure fluctuations in the gas are then detected by the microphone.

Figure 7-24 is a block diagram showing the components of a single-beam photoacoustic spectrometer. Figure 7-25 shows a typical application of the instrument. Here, photoacoustic spectra of smears of whole blood, blood cells freed of plasma, and hemoglobin extracted from the cells are shown. Conventional spectroscopy, even on dilute solutions of whole blood, does not yield satisfactory results because of the strong light-scattering properties of the protein and lipid molecules present. Photoacoustic spectroscopy clearly permits spectroscopic studies of blood without the necessity of a preliminary separation of large molecules.

Other applications which suggest themselves include the study of minerals, semiconductors, and the spots on thin-layer or paper chromatograms.

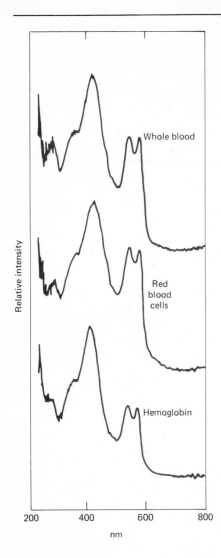

FIGURE 7-25 Photoacoustic spectra of smears of blood and blood components. (Reprinted with permission from A. Rosencwaig, *Anal. Chem.*, **47**, 592A (1975). Copyright by the American Chemical Society.)

PROBLEMS

1. The following data were obtained for solutions of substance X and Y (1.20-cm cells):

	C, M	A at 320 nm	A at 460 nm
X	4.60×10^{-3}	0.965	0.106
Y	5.11×10^{-3}	0.094	1.213

(a) Calculate the molar absorptivity of X and Y at each wavelength.

(b) A solution containing both X and Y was found to have an absorbance of 0.431 at 320 nm and 0.686 at 460 nm when measured in 0.500-cm cells. Calculate the molar concentration of each species in the solution assuming both species obey Beer's law.

2. A. J. Mukhedkar and N. V. Deshpande [*Anal. Chem.*, **35**, 47 (1963)] report on a simultaneous determination for cobalt and nickel based upon absorption by their respective 8-quinolinol complexes. Molar absorptivities corresponding to absorption maxima are:

	Wavelength	
	365 nm	700 nm
ε_{Co}	3529	428.9
ε_{Ni}	3228	0.00

Calculate the concentration of nickel and cobalt in each of the following solutions on the basis of the accompanying data:

	Absorbance, 1.00-cm Cell	
Solution	365 nm	700 nm
1	0.816	0.074
2	0.516	0.033

3. Solutions of P and Q individually obey Beer's law over a large concentration range. Spectral data for these species, measured in 1.00-cm cells, are tabulated below.

	Absorbance, A	
λ, nm	$8.55 \times 10^{-5} M$ P	$2.37 \times 10^{-4} M$ Q
400	0.078	0.550
420	0.087	0.592
440	0.096	0.599
460	0.102	0.590
480	0.106	0.564
500	0.110	0.515
520	0.113	0.433
540	0.116	0.343
560	0.126	0.255
580	0.170	0.170
600	0.264	0.100
620	0.326	0.055
640	0.359	0.030
660	0.373	0.030
680	0.370	0.035
700	0.346	0.063

(a) Plot an absorption spectrum (1.00-cm cells) for a solution that is $7.76 \times 10^{-5} M$ with respect to P and $3.14 \times 10^{-4} M$ with respect to Q.

(b) Calculate the absorbance (1.00-cm cells) at 440 nm of a solution that is $200 \times 10^{-5} M$ in P and $400 \times 10^{-4} M$ in Q.

4. Use the data in the previous problem to calculate the quantities sought (1.50-cm cells are used for all measurements).

	Concentration, mol/liter		Absorbance	
	P	Q	440 nm	620 nm
(a)			0.862	0.450
(b)			0.313	0.799
(c)			0.276	0.347
(d)			0.897	0.350
(e)			0.495	0.816
(f)			0.202	0.297

5. Absorptivity data for the cobalt and nickel complexes with 2,3-quinoxalinedithiol at the respective absorption peaks are as follows:

	Wavelength	
	510 nm	656 nm
ε_{Co}	36,400	1,240
ε_{Ni}	5,520	17,500

A 0.376-g soil sample was dissolved and subsequently diluted to 50.0 ml. A 25.0-ml aliquot was treated to eliminate interferences; after addition of 2,3-quinoxalinedithiol, the volume was adjusted to 50.0 ml. This solution had an absorbance of 0.467 at 510 nm and 0.347 at 656 nm in a 1.00-cm cell. Calculate the respective percentages of cobalt and nickel in the soil.

6. The indicator HIn has an acid-dissociation constant of 5.20×10^{-6} at ordinary temperatures. The accompanying absorbance data are for $7.50 \times 10^{-5} F$ solutions of the indicator measured in 1.00-cm cells in strongly acidic and strongly alkaline media.

	Absorbance				Absorbance	
λ, nm	pH = 1.00	pH = 13.00		λ, nm	pH = 1.00	pH = 13.00
420	0.535	0.050		550	0.119	0.324
445	0.657	0.068		570	0.068	0.352
450	0.658	0.076		585	0.044	0.360
455	0.656	0.085		595	0.032	0.361
470	0.614	0.116		610	0.019	0.355
510	0.353	0.223		650	0.014	0.284

Estimate the wavelength at which absorption by the indicator becomes independent of the pH (that is, the isosbestic point).

7. Calculate the absorbance (1.00-cm cells) at 450 nm of a solution in which the total formal concentration of the indicator described in Problem 6 is 7.50×10^{-5} and the pH is (a) 4.670, (b) 5.217, and (c) 5.866.

8. What will be the absorbance at 595 nm (1.00-cm cells) of a solution that is $1.25 \times 10^{-4}F$ with respect to the indicator in Problem 6 and whose pH is (a) 5.451, (b) 6.213, and (c) 5.158?

9. Several buffer solutions were made $1.25 \times 10^{-4}F$ with respect to the indicator in Problem 6; absorbance data (1.00-cm cells) at 450 nm and 595 nm were as follows:

	Absorbance	
Solution	450 nm	595 nm
A	0.333	0.307
B	0.497	0.193
C	0.666	0.121
D	0.210	0.375

Calculate the pH of each solution.

10. Construct an absorption spectrum for an $8.00 \times 10^{-5}F$ solution of the indicator in Problem 6 when measurements are made with 1.00-cm cells and

(a) $\dfrac{[\text{HIn}]}{[\text{In}^-]} = 3$

(b) $\dfrac{[\text{HIn}]}{[\text{In}^-]} = 1$

(c) $\dfrac{[\text{HIn}]}{[\text{In}^-]} = \dfrac{1}{3}$

11. The chelate ZnQ_2^{2-} exhibits a maximum absorption at 480 nm. When the chelating agent is present in at least a fivefold excess, the absorbance depends only upon the formal concentration of Zn(II) and obeys Beer's law over a large range. Neither Zn^{2+} nor Q^{2-} absorbs at 480 nm.

A solution that is $2.30 \times 10^{-4}F$ in Zn^{2+} and $8.60 \times 10^{-3}F$ in Q has an absorbance of 0.690 when measured in a 1.00-cm cell at 480 nm. Under the same conditions, a solution that is $2.30 \times 10^{-4}F$ in Zn^{2+} and $5.00 \times 10^{-4}F$ in Q^{2-} has an absorbance of 0.540. Calculate the numerical value of K for the process.

$$\text{Zn}^{2+} + 2\text{Q}^{2-} \rightleftharpoons \text{ZnQ}_2^{2-}$$

12. What colored glass filter should be employed for a photometric analysis based upon
    (a) the blue color of the ammonia complex of Cu(II)?
    (b) the red SCN^- complex of Fe^{3+}?
    (c) the yellow complex formed when H_2O_2 is added to a solution containing Ti(IV)?

13. Derive titration curves for 50.0 ml of $2.00 \times 10^{-4}F$ A with $0.00100F$ M if the reaction is

$$M + A \rightleftharpoons MA$$

Assume that the path length of the cell is 5.5 cm and that the equilibrium is forced far to the right after the following additions of reagent: 2.00, 4.00, 6.00, 14.00, 16.00, and 18.00 ml. Assume the following molar absorptivities for the three species:

	ε_M	ε_A	ε_{MA}
(a)	0.0	18.6	6530
(b)	762	0.0	6530
(c)	8.2	6530	0.0
(d)	6680	0.0	36.5
(e)	6680	0.0	1430

8

INFRARED ABSORPTION SPECTROSCOPY

The infrared region of the spectrum encompasses radiation with wavenumbers ranging from about 12,800 to 10 cm^{-1} or wavelengths from 0.78 to 1000 μm.[1] From the standpoint of both application and instrumentation, it is convenient to subdivide the spectrum into *near*, *middle*, and *far*-infrared radiation; the limits of each are shown in Table 8-1. The large majority of analytical applications are confined to a portion of the middle region extending from 4000 to 670 cm^{-1} or 2.5 to 15 μm.

Infrared spectroscopy finds widespread application to qualitative and quantitative analyses.[2] Its most important use has been for the identification of organic compounds, because their spectra are generally complex and provide numerous maxima and minima that can be employed for comparison purposes (see Figure 8-1). Indeed, the infrared spectrum of an organic compound represents one of its unique physical properties; with the exception of optical isomers, no two compounds, in theory, absorb in exactly the same way.

In addition to its application as a qualitative-analytical tool, infrared spectrophotometry is finding increasing use for quantitative analysis as well. Here, its great advantage lies in its selectivity, which often makes possible the quantitative estimation of an analyte in a complex mixture with little or no prior separation steps. The most important analyses of this type have been of atmospheric pollutants from industrial processes.

THEORY OF INFRARED ABSORPTION

Introduction

A typical infrared spectrum, obtained with a double-beam recording spectrophotometer, is shown in Figure 8-1. In contrast to most ultraviolet and visible spectra, a bewildering array of maxima and minima are observed.

Spectral Plots. The plot shown in Figure 8-1 is a reproduction of the recorder output of a widely used commercial infrared spectrophotometer. As is ordinarily the case, the ordinate is linear in transmittance; it should be noted, however, that the chart paper of some manufacturers also contains a nonlinear absorbance scale for reference. Note also that

---

[1] Until recently, the unit of wavelength that is 10^{-6} m was called the *micron*, μ; it is now more properly termed the micrometer, μm.

[2] For detailed discussion of infrared spectroscopy, see: N. B. Colthup, L. N. Daly, and S. E. Wiberley, *Introduction to Infra Red and Raman Spectroscopy*, 2d ed. New York: Academic Press, 1974; K. Nakanishi, *Infrared Absorption Spectroscopy*. San Francisco: Holden-Day, 1977; N. L. Alpert, W. E. Keiser, and H. A. Szymanski, *Theory and Practice of Infrared Spectroscopy*, 2d ed. New York: Plenum Press, 1970; R. T. Conley, *Infrared Spectroscopy*, 2d ed. Boston: Allyn and Bacon, 1972; and *Infrared and Raman Spectroscopy*, eds. E. G. Brame and J. G. Grasselli. New York: Marcel Dekker, 1977.

TABLE 8-1   INFRARED SPECTRAL REGIONS

Region	Wavelength (λ) Range, μm	Wavenumber (σ) Range, cm^{-1}	Frequency (v) Range, Hz
Near	0.78 to 2.5	12800 to 4000	3.8×10^{14} to 1.2×10^{14}
Middle	2.5 to 50	4000 to 200	1.2×10^{14} to 6.0×10^{12}
Far	50 to 1000	200 to 10	6.0×10^{12} to 3.0×10^{11}
Most used	2.5 to 15	4000 to 670	1.2×10^{14} to 2.0×10^{13}

the abscissa in this chart is linear in units of reciprocal centimeters; other instruments, particularly older ones, employ a scale that is linear in wavelength (an easy way to remember how to convert from one scale to the other is the relationship $cm^{-1} \times \mu m = 10,000$). For comparison, a wavelength scale has been added on the upper axis of the original chart paper shown in Figure 8-1. Several brands of chart paper contain both wavelength and wavenumber scales; obviously, only one can be linear.

The preference for a linear wavenumber scale is based upon the direct proportionality between this quantity and both energy and frequency; the frequency of the absorbed radiation is, in turn, the molecular vibrational frequency actually responsible for the absorption process. Frequency, however, is seldom if ever employed as the abscissa, probably because of the inconvenient size of the unit (for example, the frequency scale of the plot in Figure 8-1 would extend from 1.2×10^{14} to 2.0×10^{13} Hz). A scale in terms of cm^{-1} is often referred to as a frequency scale; the student should bear in mind that this terminology is not strictly correct, the wavenumber being only proportional to frequency.

Finally, it should be noted that the horizontal scale of Figure 8-1 changes at 2000 cm^{-1}, with the units at higher wavenumbers being represented by half the linear distance of those at lower. This discontinuity is introduced for convenience, since much useful qualitative infrared data appear at wavenumbers smaller than 2000 cm^{-1}.

Dipole Changes During Vibrations and Rotations. Most electronic transitions require energies in the ultraviolet or visible regions; absorption of infrared radiation is thus confined largely to molecular species for which small energy differences exist between various vibrational and rotational states.

In order to absorb infrared radiation, a molecule must undergo a net change in dipole moment as a consequence of its vibrational or

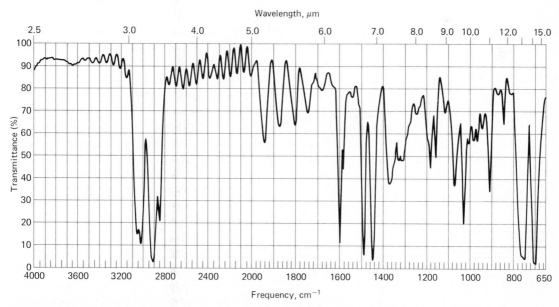

FIGURE 8-1 Infrared absorption spectrum of a thin polystyrene film recorded with a modern IR spectrophotometer. Note that the abscissa scale changes at 2000 cm^{-1}.

rotational motion. Only under these circumstances can the alternating electrical field of the radiation interact with the molecule and cause changes in the amplitude of one of its motions. For example, the charge distribution around a molecule such as hydrogen chloride is not symmetric, the chlorine having a higher electron density than the hydrogen. Thus, hydrogen chloride has a significant dipole moment and is said to be polar. The dipole moment is determined by the magnitude of the charge difference and the distance between the two centers of charge. As a hydrogen chloride molecule vibrates longitudinally, a regular fluctuation in dipole moment occurs, and a field is established which can interact with the electrical field associated with radiation. If the frequency of the radiation matches a natural vibrational frequency of the molecule, there occurs a net transfer of energy that results in a change in the *amplitude* of the molecular vibration: absorption of the radiation is the consequence. Similarly, the rotation of asymmetric molecules around their centers of mass results in a periodic dipole fluctuation; again, interaction with radiation is possible.

No net change in dipole moment occurs during the vibration or rotation of homonuclear species such as O_2, N_2, or Cl_2; consequently, such compounds cannot absorb in the infrared.

Rotational Transitions. The energy required to cause a change in rotational level is minute and corresponds to radiation of 100 μm and greater (< 100 cm^{-1}). Because rotational levels are quantized, absorption by gases in this far-infrared region is characterized by discrete, well-defined lines. In liquids or solids, intramolecular collisions and interactions cause broadening of the lines into a continuum.

Vibrational-Rotational Transitions. Vibrational energy levels are also quantized, and the energy differences between quantum states correspond to the readily accessible regions of the infrared from about 13,000 to 675 cm^{-1} (0.75 to 15 μm). The infrared spectrum

of a gas usually consists of a series of closely spaced lines, because there are several rotational energy states for each vibrational state. On the other hand, rotation is highly restricted in liquids and solids; in such samples, discrete vibrational-rotational lines disappear, leaving only somewhat broadened vibrational peaks. Our concern is primarily with the spectra of solutions, liquids, and solids, in which rotational effects are minimal.

Types of Molecular Vibrations. The relative positions of atoms in a molecule are not exactly fixed but instead fluctuate continuously as a consequence of a multitude of different types of vibrations. For a simple diatomic or triatomic molecule, it is easy to define the number and nature of such vibrations and relate these to energies of absorption. An analysis of this kind becomes difficult if not impossible for polyatomic molecules, not only because of the large number of vibrating centers, but also because interactions among several centers occur and must be taken into account.

Vibrations fall into the basic categories of *stretching* and *bending*. A stretching vibration involves a continuous change in the interatomic distance along the axis of the bond between two atoms. Bending vibrations are characterized by a change in the angle between two bonds and are of four types: *scissoring*, *rocking*, *wagging*, and *twisting*. The various types of vibrations are shown schematically in Figure 8-2.

All of the vibration types shown in Figure 8-2 may be possible in a molecule containing more than two atoms. In addition, interaction or *coupling* of vibrations can occur if the vibrations involve bonds to a single central atom. The result of coupling is a change in the characteristics of the vibrations involved.

In the treatment that follows, we shall first consider isolated vibrations with a simple mechanical model called the *harmonic oscillator*. Modifications to the theory of the harmonic oscillator, which are needed to describe a molecular system, will be taken

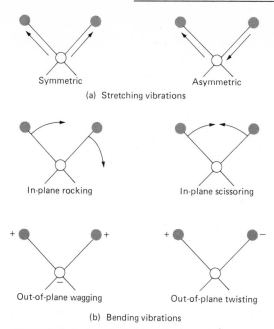

Symmetric   Asymmetric

(a) Stretching vibrations

In-plane rocking   In-plane scissoring

Out-of-plane wagging   Out-of-plane twisting

(b) Bending vibrations

FIGURE 8-2 Types of molecular vibrations. Note: + indicates motion from plane toward reader; − indicates motion from plane away from reader.

up next. Finally, the effects of vibrational interactions in molecular systems will be discussed.

Mechanical Model of Stretching Vibrations

The characteristics of an atomic stretching vibration can be approximated by a mechanical model consisting of two masses connected by a spring. A disturbance of one of these masses along the axis of the spring results in a vibration called a *simple harmonic motion*.

Let us first consider the vibration of a mass attached to a spring that is hung from an immovable object (see Figure 8-3a). If the mass is displaced a distance y from its equilibrium position by application of a force along the axis of the spring, the restoring force is proportional to the displacement (Hooke's law). That is,

$$F = -ky \qquad (8-1)$$

where F is the restoring force and k is the *force constant*, which depends upon the stiffness of the spring. The negative sign indicates that F is a restoring force.

Potential Energy of a Harmonic Oscillator. The potential energy of the mass and spring can be considered to be zero when the mass is in its rest or equilibrium position. As the spring is compressed or stretched, however, the potential energy of this system increases by an amount equal to the work required to displace the mass. If, for example, the mass is moved from some position y to $(y + dy)$, the work and hence the change in potential energy E is equal to the force F times the distance dy. Thus,

$$dE = -F dy \qquad (8-2)$$

Combining Equations 8-2 and 8-1 yields

$$dE = ky dy$$

Integrating between the equilibrium position $(y = 0)$ and y gives

$$\int_0^E dE = k \int_0^y y dy$$

$$E = \tfrac{1}{2}ky^2 \qquad (8-3)$$

The potential-energy curve for a simple harmonic oscillation, derived from Equation 8-3, is plotted in Figure 8-3a. It is seen that the potential energy is a maximum when the spring is stretched or compressed to its maximum amplitude A, and decreases parabolically to zero at the equilibrium position.

Vibrational Frequency. The motion of the mass as a function of time t can be deduced as follows. Newton's law states that

$$F = ma$$

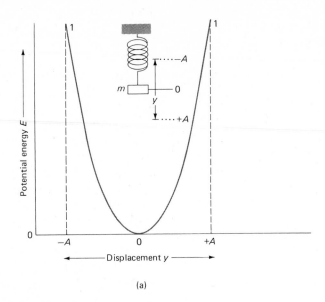

(a)

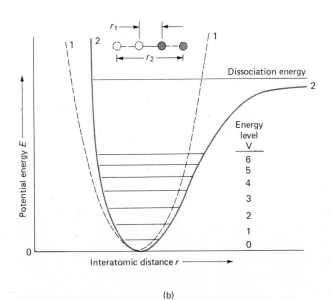

(b)

FIGURE 8-3 Potential energy diagrams. Curve 1, harmonic oscillator. Curve 2, anharmonic oscillator.

where m is the mass and a is its acceleration. But acceleration is the second derivative of distance with respect to time:

$$a = \frac{d^2y}{dt^2}$$

Substituting these expressions into 8-1 gives

$$m\frac{d^2y}{dt^2} = -ky \qquad (8\text{-}4)$$

One solution (but not the only one) to Equation 8-4 can be shown by substitution to be

$$y = A \sin\left(\sqrt{\frac{k}{m}}\,t\right) \qquad (8\text{-}5)$$

where A is the amplitude of the vibration, a constant that is equal to the maximum value of y.

Equation 8-5 is analogous to the sinusoidal function described by Equation 2-16, page 12. That is,

$$y = A \sin 2\pi v t$$

Comparison of the two equations reveals that

$$\sqrt{\frac{k}{m}}\,t = 2\pi v t \qquad (8\text{-}6)$$

This equation can be rearranged to give

$$v_m = \frac{1}{2\pi}\sqrt{\frac{k}{m}} \qquad (8\text{-}7)$$

where v_m is the *natural frequency* of the mechanical oscillator. While it is dependent upon the force constant of the spring and the mass of the attached body, the natural frequency is *independent* of the energy imparted to the system; changes in energy merely result in a change in the amplitude A of the vibration.

The equations just developed can be readily modified to describe the behavior of a system consisting of two masses m_1 and m_2 connected by a spring. Here, it is only necessary to substitute the *reduced mass* μ for the single mass m where

$$\mu = \frac{m_1 m_2}{m_1 + m_2} \qquad (8\text{-}8)$$

Thus, the vibrational frequency for such a system is given by

$$v_m = \frac{1}{2\pi}\sqrt{\frac{k}{\mu}} = \frac{1}{2\pi}\sqrt{\frac{k(m_1 + m_2)}{m_1 m_2}} \qquad (8\text{-}9)$$

Molecular Vibrations. The approximation is ordinarily made that the behavior of a molecular vibration is analogous to the mechanical model just described. Thus, the frequency of the molecular vibration is calculated from Equation 8-9 by substituting the masses of the two atoms for m_1 and m_2; the quantity k becomes the force constant for the chemical bond, which is a measure of its stiffness (but not necessarily its strength).

Quantum Treatment of Vibrations

Harmonic Oscillators. The equations of ordinary mechanics, such as we have used thus far, do not completely describe the behavior of particles with atomic dimensions. For example, the quantized nature of molecular vibrational energies (and of course other atomic and molecular energies as well) does not appear in these equations. It is possible, however, to employ the concept of the simple harmonic oscillator for the development of the wave equations of quantum mechanics. Solutions of these equations for potential energies are found when

$$E = \left(v + \frac{1}{2}\right)\frac{h}{2\pi}\sqrt{\frac{k}{\mu}} \qquad (8\text{-}10)$$

where v, *the vibrational quantum number,* can take only positive integer values (including zero). Thus, in contrast to ordinary mechanics where vibrators can have any positive potential energy, quantum mechanics requires that only certain discrete energies be assumed by a vibrator.

It is of interest to note that the term $(\sqrt{k/\mu})/2\pi$ appears in both the mechanical and the quantum equations; by substituting Equation 8-9 into 8-10, we find

$$E = (v + \tfrac{1}{2})hv_m \qquad (8\text{-}11)$$

where v_m is the vibrational frequency of the mechanical model.

We now assume that transitions in vibrational energy levels can be brought about by radiation, provided the energy of the radiation exactly matches the difference in energy levels ΔE between the vibrational quantum states (and provided also that the vibration causes a fluctuation in dipole). *This difference is identical between any pair of adjacent levels,* since v in Equations 8-10 and 8-11 can assume only whole numbers; that is,

$$\Delta E = hv_m = \frac{h}{2\pi}\sqrt{\frac{k}{\mu}} \qquad (8\text{-}12)$$

At room temperature, the majority of molecules are in the ground state ($v = 0$); thus, from Equation 8-11,

$$E_0 = \tfrac{1}{2}hv_m$$

Promotion to the first excited state ($v = 1$) with energy

$$E_1 = \tfrac{3}{2}hv_m$$

requires radiation of energy

$$(\tfrac{3}{2}hv_m - \tfrac{1}{2}hv_m) = hv_m$$

The frequency of radiation v that will bring about this change is *identical to the classical vibration frequency of the bond* v_m. That is,

$$E_{\text{radiation}} = hv = \Delta E = hv_m = \frac{h}{2\pi}\sqrt{\frac{k}{\mu}}$$

or

$$v = v_m = \frac{1}{2\pi}\sqrt{\frac{k}{\mu}} \qquad (8\text{-}13)$$

If we wish to express the radiation in wavenumbers,

$$\sigma = \frac{1}{2\pi c}\sqrt{\frac{k}{\mu}} = \frac{1}{2\pi c}\sqrt{\frac{k(m_1 + m_2)}{m_1 m_2}} \qquad (8\text{-}14)$$

where σ is the wavenumber of an absorption peak in cm^{-1}, k is the force constant for the bond in dynes/cm, c is the velocity of light in

cm/s, and m_1 and m_2 are the masses of atoms 1 and 2 in g.

Equation 8-14 and infrared measurements have been employed to evaluate the force constants for various types of chemical bonds. Generally, k has been found to lie in the range between 3×10^5 and 8×10^5 dynes/cm for most single bonds, with 5×10^5 serving as a reasonable average value. Double and triple bonds are found by this same means to have force constants of about two and three times this value, respectively. With these average experimental values, Equation 8-14 can be used to estimate the wavenumber of the fundamental absorption peak (the absorption peak due to the transition from the ground state to the first excited state) for a variety of bond types. The following example demonstrates such a calculation.

EXAMPLE

Calculate the approximate wavenumber and wavelength of the fundamental absorption peak due to the stretching vibration of a carbonyl group   C=O.

The force constant for a double bond has an approximate value of 1×10^6 dynes/cm. The masses of the carbon and oxygen atoms are approximately $12/6.0 \times 10^{23}$ and $16/6.0 \times 10^{23}$, or 2.0×10^{-23} and 2.7×10^{-23} g per atom. Substituting into Equation 8-14, we obtain

$$\sigma = \frac{1}{2 \times 3.14 \times 3 \times 10^{10}}$$

$$\times \sqrt{\frac{1 \times 10^6(2.0 + 2.7) \times 10^{-23}}{2.0 \times 2.7 \times 10^{-46}}}$$

$$= 1.6 \times 10^3 \text{ cm}^{-1}$$

$$\lambda = \frac{10^4}{1.6 \times 10^3} = 6.3 \ \mu\text{m}$$

The carbonyl stretching band is found experimentally to be in the region of 5.3 to 6.7 μm or 1500 to 1900 cm^{-1}.

Selection Rules. As given by Equations 8-11 and 8-12, the energy for a transition from energy level 1 to 2 or from level 2 to 3 should be identical to that for the $0 \rightarrow 1$ transition. Furthermore, quantum theory indicates that the only transitions that can take place are those in which the vibrational quantum number changes by unity; that is, the so-called selection rule states that $\Delta v = \pm 1$. Since the vibrational levels are equally spaced, only a single absorption peak should be observed for a given transition.

Anharmonic Oscillator. Thus far, we have considered the classical and quantum mechanical treatments of the harmonic oscillator. The potential energy of such a vibrator changes periodically as the distance between masses fluctuates (Figure 8-3a). From qualitative considerations, however, it is apparent that this description of a molecular vibration is imperfect. For example, as the two atoms approach one another, coulombic repulsion between the two nuclei produces a force that acts in the same direction as the restoring force of the bond; thus, the potential energy can be expected to rise more rapidly than the harmonic approximation predicts. At the other extreme of oscillation, a decrease in the restoring force, and thus the potential energy, occurs as the interatomic distance approaches that at which dissociation of atoms takes place.

In theory, the wave equations of quantum mechanics permit the derivation of more nearly correct potential-energy curves for molecular vibrations. Unfortunately, however, the mathematical complexity of these equations precludes their quantitative application to all but the very simplest of systems. It is qualitatively apparent, however, that the curves must take the *anharmonic* form shown in Figure 8-3b. These curves depart from harmonic behavior by varying degrees, depending upon the nature of the bond and the atoms involved. Note, however, that the harmonic and anharmonic curves are nearly alike at low potential energies. This fact accounts for the success of the approximate methods described.

Anharmonicity leads to deviations of two kinds. At higher quantum numbers, ΔE becomes smaller (see Figure 8-3b), and the selection rule is not rigorously followed; as a result, transitions of $\Delta v = \pm 2$ or ± 3 are observed. Such transitions are responsible for the appearance of *overtone lines* at frequencies approximately twice or three times that of the fundamental line; the intensity of overtone absorption is frequently low, and the peaks may not be observed.

Vibrational spectra are further complicated by the fact that two different vibrations in a molecule can interact to give absorption peaks with frequencies that are approximately the sums or differences of their fundamental frequencies. Again, the intensities of combination and difference peaks are generally low.

Vibrational Modes

It is ordinarily possible to deduce the number and kinds of vibrations in simple diatomic and triatomic molecules and whether these vibrations will lead to absorption. Complex molecules may contain several types of atoms as well as bonds; for these, the multitude of possible vibrations gives rise to infrared spectra that are difficult, if not impossible, to analyze.

The number of possible vibrations in a polyatomic molecule can be calculated as follows. Three coordinates are needed to locate a point in space; to fix N points requires a set of three coordinates for each for a total of $3N$. Each coordinate corresponds to one degree of freedom for one of the atoms in a polyatomic molecule; for this reason, a molecule containing N atoms is said to have $3N$ *degrees of freedom*.

In defining the motion of a molecule, we need to consider: (1) the motion of the entire molecule through space (that is, the translational motion of its center of gravity); (2)

the rotational motion of the entire molecule around its center of gravity; and (3) the motion of each of its atoms relative to the other atoms (in other words, its individual vibrations). Definition of translational motion requires three coordinates and uses up three degrees of freedom. Another three degrees of freedom are needed to describe the rotation of the molecule as a whole. The remaining $(3N - 6)$ degrees of freedom involve inter-atomic motion, and hence represent the number of possible vibrations within the molecule. A linear molecule is a special case since, by definition, all of the atoms lie on a single, straight line. Rotation about the bond axis is not possible, and two degrees of freedom suffice to describe rotational motion. Thus, the number of vibrations for a linear molecule is given by $(3N - 5)$. Each of the $(3N - 6)$ or $(3N - 5)$ vibrations is called a *normal mode*.

For each normal mode of vibration, there exists a potential-energy relationship such as that shown by the solid line in Figure 8-3b. The same selection rules discussed earlier apply for each of these. In addition, to the extent that a vibration approximates harmonic behavior, the differences between the energy levels of a given vibration are the same; that is, a single absorption peak should appear for each vibration in which there is a change in dipole.

In fact, however, the number of normal modes does not necessarily correspond exactly to the number of observed absorption peaks. The number of peaks is frequently less because: (1) the symmetry of the molecules is such that no change in dipole results from a particular vibration; (2) the energies of two or more vibrations are identical or nearly identical; (3) the absorption intensity is so low that as to be undetectable by ordinary means; or (4) the vibrational energy is in a wavelength region beyond the range of the instrument. As we have pointed out, additional peaks arise from overtones as well as from combination or difference frequencies.

Vibrational Coupling

The energy of a vibration, and thus the wavelength of its absorption peak, may be influenced (or coupled) by other vibrators in the molecule. A number of factors that influence the extent of such coupling can be identified.

1. Strong coupling between stretching vibrations occurs only when there is an atom common to the two vibrations.

2. Interaction between bending vibrations requires a common bond between the vibrating groups.

3. Coupling between a stretching and a bending vibration can occur if the stretching bond forms one side of the angle that varies in the bending vibration.

4. Interaction is greatest when the coupled groups have individual energies that are approximately equal.

5. Little or no interaction is observed between groups separated by two or more bonds.

6. Coupling requires that the vibrations be of the same symmetry species.[3]

As an example of coupling effects, let us consider the infrared spectrum of carbon dioxide. If no coupling occurred between the two $C{=}O$ bonds, an absorption peak would be expected at the same wavenumber as the peak for the $C{=}O$ stretching vibration in an aliphatic ketone (about 1700 cm^{-1}, or 6 μm; see example, p. 216). Experimentally, carbon dioxide exhibits two absorption peaks, the one at 2330 cm^{-1} (4.3 μm) and the other at 667 cm^{-1} (15 μm).

Carbon dioxide is a linear molecule and thus has four normal modes ($3 \times 3 - 5$). Two stretching vibrations are possible; furthermore, interaction between the two can occur since the bonds involved are associated with a common carbon atom. As may be seen, one

---

[3] For a discussion of symmetry operations and symmetry species, see: R. P. Bauman, *Absorption Spectroscopy*. New York: Wiley, 1962, Chapter 10.

of the coupled vibrations is symmetric and the other is asymmetric.

Symmetric   Asymmetric

The symmetric vibration causes no change in dipole, since the two oxygen atoms simultaneously move away from or toward the central carbon atom. Thus, the symmetric vibration is infrared-inactive. One oxygen approaches the carbon atom as the other moves away during asymmetric vibration. As a consequence, a net change in charge distribution occurs periodically; absorption at 2330 cm^{-1} results.

The remaining two vibrational modes of carbon dioxide involve scissoring, as shown below.

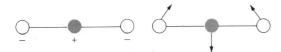

The two bending vibrations are the resolved components (at 90 deg to one another) of the bending motion in all possible planes around the bond axis. The two vibrations are identical in energy and thus produce but one peak at 667 cm^{-1}. (Quantum energy differences that are identical, as these are, are said to be *degenerate*.)

It is of interest to compare the spectrum of carbon dioxide with that of a nonlinear, triatomic molecule such as water, sulfur dioxide, or nitric oxide. These molecules have $(3 \times 3 - 6)$, or 3, vibrational modes which take the following forms:

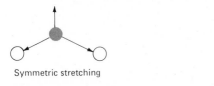

Symmetric stretching

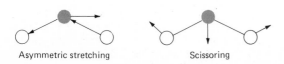

Asymmetric stretching   Scissoring

Since the central atom is not in line with the other two, a symmetric stretching vibration will produce a change in dipole and will thus be responsible for infrared absorption. For example, stretching peaks at 3650 and 3760 cm^{-1} (2.74 and 2.66 μm) are observed for the symmetric and asymmetric vibrations of the water molecule. Only one component to the scissoring vibration exists for this nonlinear molecule, since motion in the plane of the molecule constitutes a rotational degree of freedom. For water, the bending vibration causes absorption at 1595 cm^{-1} (6.27 μm).

The difference in behavior of linear and nonlinear triatomic molecules (two and three absorption peaks, respectively) illustrates how infrared-absorption spectroscopy can sometimes be used to deduce molecular shapes.

Coupling of vibrations is a common phenomenon; as a result, the position of an absorption peak corresponding to a given organic functional group cannot be specified exactly. For example, the C—O stretching frequency in methanol is 1034 cm^{-1} (9.67 μm); in ethanol it is 1053 cm^{-1} (9.50 μm); and in methylethylcarbinol it is 1105 cm^{-1} (9.05 μm). These variations result from a coupling of the C—O stretching with adjacent C—C or C—H vibrations.

While interaction effects may lead to uncertainties in the identification of functional groups contained in a compound, it is this very effect that provides the unique features of an infrared-absorption spectrum that are so important for the positive identification of a specific compound.

INFRARED INSTRUMENT COMPONENTS

Infrared spectrophotometers have the same basic components as do instruments used for the study of absorption in the ultraviolet and visible regions of the spectrum (see Figure 5-1b, p. 15).

Wavelength Selection

Infrared wavelength selection can be accomplished by means of interference filters, prisms, or gratings.

Interference Filters. Interference filters (p. 135) for quantitative infrared analysis of specific compounds are available commercially.[4] For example, a filter with a transmission maximum of 9.0 μm is offered for the determination of acetaldehyde; one with a maximum at 13.4 μm is suggested for o-dichlorobenzene; and a filter exhibiting a transmission peak at 4.5 μm is employed for the determination of nitrous oxide. In general, these filters have effective bandwidths of about 1.5% of their peak wavelength.

Filter Wedges. Wilks Scientific Corporation also offers an instrument that provides narrow and continuously variable bands of infrared radiation by means of three filter wedges (p. 136) formed as segments of a circle. The ranges of the three segments are 2.5 to 4.5, 4.5 to 8, and 8 to 14.5 μm. The wedges are mounted on a wheel which can be rotated to provide the desired wavelength at the slit. The effective bandwidth of the device is about 1.5% of the wavelength that is transmitted to the slit.

Prisms. Several materials have been used for prism construction. Quartz is employed for the near-infrared region (0.8 to 3 μm), even though its dispersion characteristics for this region are far from ideal. It absorbs strongly beyond about 4 μm (2500 cm^{-1}). Crystalline sodium chloride is the most common prism material; its dispersion is high in the region between 5 and 15 μm (2000 and 670 cm^{-1}) and is adequate to 2.5 μm (4000 cm^{-1}). Beyond 20 μm (500 cm^{-1}), sodium chloride absorbs strongly and cannot be used. Crystalline potassium bromide and cesium bromide prisms are suitable for the far-infra-

red region (15 to 40 μm or 670 to 250 cm^{-1}), while lithium fluoride is useful in the near-infrared region (1 to 5 μm or 10,000 to 2000 cm^{-1}). Many spectrophotometers are designed for convenient prism interchange. Unfortunately, all common infrared-transmitting materials except quartz are easily scratched and are water-soluble. Desiccants or heat are required to protect them from the condensation of moisture.

Reflection Gratings. Reflection gratings offer a number of advantages as dispersing elements for the infrared region and are replacing prisms as a consequence. Inherently better resolution is possible because there is less loss of radiant energy than in a prism system; thus, narrower slit widths can be realized. Other advantages include more nearly linear dispersion and resistance to attack by water. An infrared grating is usually constructed from glass or plastic that has been coated with aluminum.

The disadvantage of a grating lies in the greater amounts of scattered radiation and the appearance of radiation of other spectral orders. In order to minimize these effects, gratings are blazed to concentrate the radiation into a single order. Filters (and occasionally prisms) are used in conjunction with the grating to minimize these problems.

Beam Attenuation

Infrared instruments are generally of a null type, in which the power of the reference beam is reduced, or *attenuated*, to match that of the beam passing through the sample. Attenuation is accomplished by imposing a device that removes a variable fraction of the reference beam. The attenuator commonly takes the form of a fine-toothed comb, the teeth of which are tapered so that a linear relationship exists between the lateral movement of the comb and the decrease in power of the beam. Movement of the comb occurs when a difference in power of the two beams is sensed by the detector. This movement is

---

[4] Wilks Scientific Corporation, South Norwalk, CT. 06856.

synchronized with the recorder pen so that its position gives a measure of the relative power of the two beams and thus the transmittance of the sample.

Many instruments also employ a beam attenuator in conjunction with the sample beam for the purpose of making the 100% T adjustment.

Infrared Sources, Sample Containers, and Detectors

Details regarding the nature and properties of infrared sources and detectors are found in Chapter 5. Discussion of infrared cells is deferred to the section in this chapter that describes sample handling.

SOME TYPICAL INSTRUMENTS

A large number of infrared instruments for qualitative applications are available commercially. In addition, photometers and spectrophotometers designed specifically for quantitative analysis have recently become available from instrument manufacturers.

Infrared Spectrophotometers for Qualitative Analysis

Because of the complexity of infrared spectra, a recording instrument is required for qualitative work. As a consequence, all commercially available instruments of this type are of double-beam design. An added reason for employing two beams is that this design is less demanding in terms of performance of source and detector than the single-beam arrangement (see p. 185). This property is important because of the low energy of infrared radiation, the low stability of sources and detectors, and the need for large signal amplification.

All commercial infrared spectrophotometers incorporate a low-frequency chopper (5 to 13 cycles per minute) to modulate the source output. This feature permits the detector to discriminate between the signal from the source and signals from extraneous radiation, such as infrared emission from various bodies surrounding the detector. Low chopping rates are demanded by the slow response times of most infrared detectors.

Infrared instruments are generally of the null type with the beam being attenuated by a comb or an absorbing wedge.

Several dozen infrared instruments, ranging in cost from $3000 to more than $30,000, are available from various instrument manufacturers. In general, the optical design of these instruments does not differ greatly from the ultraviolet-visible spectrophotometers discussed in the previous chapter except that the sample and reference compartment is always located between the source and the monochromator in infrared instruments. This arrangement is desirable because the monochromator then removes most of the stray radiation generated by the sample, the reference, and the cells. In ultraviolet-visible spectroscopy, stray radiation from these sources is seldom serious. Location of the cell compartment between the monochromator and the detector of such instruments, however, has the advantage of protecting the sample from the full power and wavelength range of the ultraviolet source; thus, photochemical decomposition is less likely.

Typical Instrument Design. Figure 8-4 shows schematically the arrangement of components in a typical infrared spectrophotometer. Note that three types of systems link the components: (1) a radiation linkage indicated by dashed lines; (2) a mechanical linkage shown by thick dark lines; and (3) an electrical linkage shown by narrow solid lines.

Radiation from the source is split into two beams, half passing into the sample-cell compartment and the other half into the reference area. The reference beam then passes through an attenuator and on to a chopper. The chopper consists of a motor-driven disk that alternately reflects the reference beam or transmits

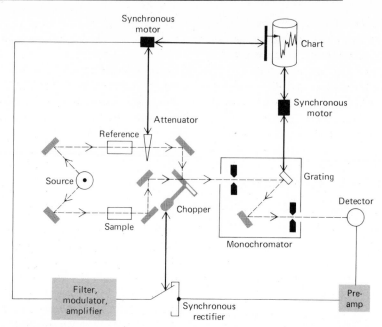

FIGURE 8-4 Schematic diagram of a double-beam spectrophotometer. Heavy dark line indicates mechanical linkage; light line indicates electrical linkage; dashed line indicates radiation path.

the sample beam into the monochromator. After dispersion by a prism or grating, the alternating beams fall on a detector and are converted to an electrical signal. The signal is amplified and passed to the synchronous rectifier, a device that is mechanically or electrically coupled to the chopper to cause the rectifier switch and the beam leaving the chopper to change simultaneously. If the two beams are identical in power, the signal from the rectifier is an unfluctuating direct current. If, on the other hand, the two beams differ in power, a fluctuating or ac current is produced, the polarity of which is determined by which beam is the more intense. The current from the rectifier is filtered and further amplified to drive a synchronous motor in one direction or the other, depending upon the polarity of the input current. The synchronous motor is mechanically linked to both

the attenuator and the pen drive of the recorder and causes both to move until a null is achieved. A second synchronous motor drives the chart and varies the wavelength simultaneously. There is frequently a mechanical linkage between the wavelength and slit drives so that the radiant power reaching the detector is kept approximately constant by variations in the slit width.

Figure 8-5 shows the optics of a simple, inexpensive commercial spectrophotometer manufactured by Beckman Instruments, Inc. Radiation from the hot nichrome-wire source is split and passes through the sample compartment. Both beams have attenuators, one for the 100% T adjustment and the other for reducing the reference beam to null for the %T measurement. The beams are recombined by means of the segmented chopper, dispersed by the monochromator, and detected by a thermocouple. The ac output of the latter, after amplification, drives the reference attenuator to a null position; the pen drive is coupled mechanically to this movement.

Instrument Automation. As with ultraviolet and visible instruments, newer infrared instruments usually contain microprocessor systems for partial automation of the measurement process. In the less-expensive instruments, automation includes slit programs, scan-time adjustments, calibration modes for checking the frequency scale, and gain adjustment for survey scans. More sophisticated instruments offer push-button control of the normal scanning parameters, a visual display of selected nonstandard parameters, automatic means for superimposing spectra, and computer storage of spectra for subsequent signal averaging.

Computer Search Systems. Two companies have recently offered computer search systems. Both are floppy-disk data storage systems that attach to existing spectrophotometers. Spectral data are processed to identify the characteristic absorption peaks for the sample. The resulting peak profile is then compared with profiles of pure compounds stored in the computer memory to provide positive or tentative identification of the analyte. The memory bank of one of these instruments contains 2000 compounds; the other holds 4800. The systems cost about $25,000.

Instruments for Quantitative Work

For two reasons double-beam recording spectrophotometers are, in principle, less satisfactory for infrared quantitative work than single-beam instruments. First, the double-beam instrument, with its more complex electronic and switching devices, is inherently more noisy. Second, the zero-transmittance calibration and low-transmittance measurements with most double-beam instruments are subject to an error arising from their optical null design. The zero-percent transmittance setting for such an instrument is obtained by blocking the sample beam from the detector. The reference-beam attenuator is then driven in such a way as to decrease the

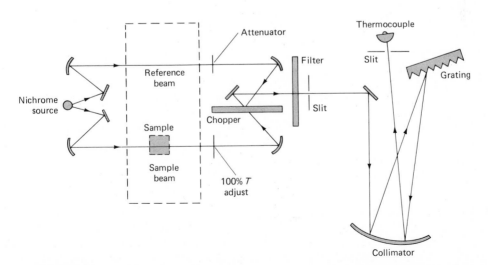

FIGURE 8-5 Schematic diagram of the Beckman AccuLab® Infrared Spectrophotometer. (Courtesy, Beckman Instruments, Inc., Fullerton, Ca. With permission.)

power of the reference beam to zero also. Under these circumstances, essentially no energy reaches the detector and the exact null position cannot be located with precision. In practice, the intensity of the sample beam is slowly diminished by the gradual introduction of a shutter across its path; in this way, the tendency for the pen drive to overshoot the zero is avoided, and a more accurate calibration is achieved. Figure 8-14 illustrates the tendency of a typical recording instrument to overshoot when measuring zero or low transmittances; again, the cause is the optical null design. Fortunately, the inevitable small error arising from the uncertainty in the zero in low transmittance measurements is not serious for qualitative work, provided the transmittances being measured are greater than 5 to 10%.

Because of these limitations inherent in typical infrared-recording spectrophotometers, a number of simple, rugged (and usually single-beam) instruments have been designed for quantitative work. Some are simple filter or nondispersive photometers; others are spectrophotometers that employ filter wedges as the dispersing element.

Filter Photometers. Figure 8-6 is a schematic diagram of a portable (weight = 18 lb), infrared filter photometer designed for quantitative analysis of various organic substances in the atmosphere. The source is a nichrome, wire-wound, ceramic rod; the transducer is a pyroelectric detector. A variety of interference filters, which transmit in the range between about 3 and 14 μm (3000 to 750 cm^{-1}), are available; each is designed for the analysis of a specific compound. The filters are readily interchangeable.

The gaseous sample is introduced into the cell by means of a battery-operated pump. The path length of the cell as shown is 0.5 m; a series of reflecting mirrors (not shown in Figure 8-6) permit increases in cell length to 20 m in increments of 1.5 m. This feature greatly enhances the concentration range of the instrument.

The photometer is reported to be sensitive to a few tenths parts per million of such substances as acrylonitrile, chlorinated hydrocarbons, carbon monoxide, phosgene, and hydrogen cyanide.

Nondispersive Photometers. Nondispersive photometers, which employ no wavelength-restricting device, are widely employed to

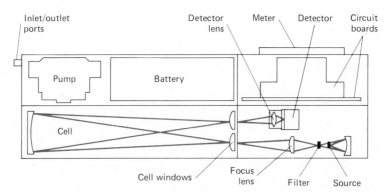

FIGURE 8-6   A portable, infrared photometer designed for gas analysis. (Courtesy of Foxboro Analytical, Wilks Infrared Center, South Norwalk, CT. With permission.)

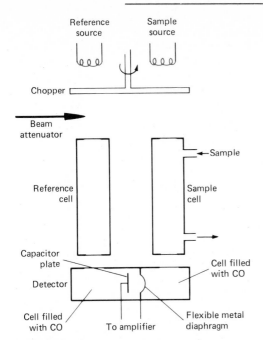

FIGURE 8-7 A nondispersive infrared photometer for monitoring carbon monoxide.

monitor gas streams for a single component. Figure 8-7 shows a typical nondispersive instrument designed to determine carbon monoxide in a gaseous mixture. The reference cell is a sealed container filled with a nonabsorbing gas; as shown in the figure, the sample flows through a second cell that is of similar length. The chopper blade is so arranged that the beams from identical sources are chopped simultaneously at the rate of about five times per second. Selectivity is obtained by filling both compartments of the detector cell with the gas being analyzed, here, carbon monoxide. The two chambers of the detector are separated by a thin, flexible, metal diaphragm that serves as one plate of a capacitor; the second plate is contained in the detector compartment on the left.

In the absence of carbon monoxide in the sample cell, the two detector chambers are

heated equally by infrared radiation from the two sources. If the sample contains carbon monoxide, however, the right-hand beam is attenuated somewhat and the corresponding detector chamber becomes cooler with respect to its reference counterpart; the consequence is a movement of the diaphragm to the right and a change in capacitance of the capacitor. This change in capacitance is sensed by the amplifier system, the output of which drives a servomotor that moves the beam attenuator into the reference beam until the two compartments are again at the same temperature. The instrument thus operates as a null device. The chopper serves to provide a dynamic, ac-type signal which is needed because the electrical system responds more reproducibly to an ac signal than to a slow dc drift.

The instrument is highly selective because heating of the detector gas occurs only from that narrow portion of the spectrum that is absorbed by the carbon monoxide in the sample. Clearly, the device can be adapted to the analysis of any infrared-absorbing gas.

Automated Instruments for Quantitative Analysis. Figure 8-8 is a schematic diagram of a computer-controlled instrument designed specifically for quantitative infrared analysis. The dispersing element, which consists of three filter wedges (p. 136) mounted in the form of a segmented circle, is shown in Figure 8-8b. The motor drive and potentiometric control permit rapid computer-controlled wavelength selection in the region between 4000 and 690 cm^{-1} (or 2.5 to 14.5 μm) with an accuracy of 0.4 cm^{-1}. The source and detector are similar to those described in the earlier section on filter photometers; note that a beam chopper is used here. The sample area can be readily adapted to solid, liquid, or gaseous samples. The instrument can be programmed to determine the absorbance of a multicomponent sample at several wavelengths and then compute the concentration of each component.

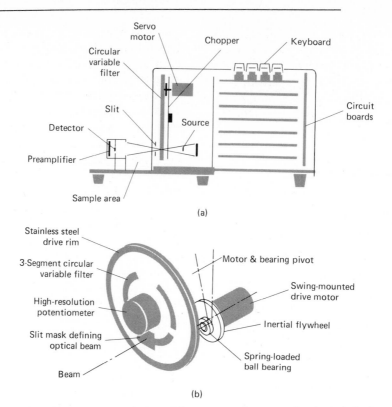

FIGURE 8-8   An infrared instrument for quantitative analysis. (a) Schematic of the instrument; (b) circular variable filter wheel. (Courtesy of Foxboro Analytical, Wilks Infrared Center, South Norwalk, CT. With permission.)

SAMPLE HANDLING TECHNIQUES[5]

As we have seen, ultraviolet and visible spectra are most conveniently obtained from dilute solutions of the analyte. Absorbance measurements in the optimum range are obtained by suitably adjusting either the con-

centration or the cell length. Unfortunately, this approach is not generally applicable for infrared spectroscopy because no good solvents exist that are transparent throughout the region. As a consequence, techniques must often be employed for liquid and solid samples that make the accurate determination of molar absorptivities difficult if not impossible. Some of these techniques are discussed in the paragraphs that follow.

Gas Samples

The spectrum of a low-boiling liquid or gas can be obtained by permitting the sample to expand into an evacuated cell. For this

[5] For a more complete discussion, see: N. B. Colthup, L. H. Daly, and S. E. Wiberley, *Introduction to Infra Red and Raman Spectroscopy*, 2d ed. New York: Academic Press, 1975, p. 85; R. P. Bauman, *Absorption Spectroscopy*. New York: Wiley, 1962, p. 184; *Treatise on Analytical Chemistry*, eds. I. M. Kolthoff and P. J. Elving. New York: Interscience, 1965, Part I, vol. 6, p. 3582.

purpose, a variety of cells are available with path lengths that range from a few centimeters to several meters. The longer path lengths are obtained in compact cells by providing reflecting internal surfaces, so that the beam makes numerous passes through the sample before exiting from the cell.

Solutions

Solvents. Figure 8-9 lists the more common solvents employed for infrared studies of organic compounds. It is apparent that no single solvent is transparent throughout the entire middle-infrared region (Table 8-1).

Water and the alcohols are seldom employed, not only because they absorb strongly, but also because they attack alkali-metal halides, the most common materials used for cell windows. For these reasons also, care must be taken to dry the solvents shown in Figure 8-9 before use.

Cells. Because of the tendency for solvents to absorb, infrared cells are ordinarily much narrower (0.1 to 1 mm) than those employed in the ultraviolet and visible regions. Light paths in the infrared range normally require sample concentrations from 0.1 to 10%. The cells are frequently demountable, with spacers

to allow variation in path length. Fixed path-length cells can be filled or emptied with a hypodermic syringe.

Sodium chloride windows are most commonly employed; even with care, however, their surfaces eventually become fogged due to absorption of moisture. Polishing with a buffing powder returns them to their original condition.

The thickness of a narrow infrared cell can be determined by measuring the transmittance of the empty cell with air as the reference. Radiation reflected off the two walls of the cell interacts with the transmitted radiation to produce an interference pattern such as that shown in Figure 8-10. The thickness b of the cell in centimeters is obtained from the relationship

$$b = \frac{N}{2(\sigma_1 - \sigma_2)}$$

where N is the number of peaks between two wavenumbers σ_1 and σ_2.

It should be noted that interference fringes are ordinarily not seen when a cell is filled with liquid because the refractive index of most liquids approaches that of the window material; reflection is thus reduced (Equation 4-14, p. 103). On the other hand, interference

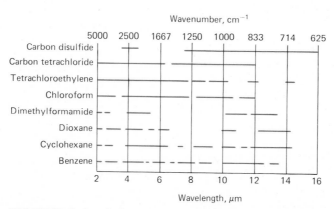

FIGURE 8-9 Infrared solvents. Black lines indicate useful regions.

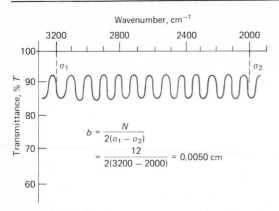

FIGURE 8-10 Determination of path length b from interference fringe produced by an empty cell.

can be observed between 2800 and 2000 cm^{-1} in Figure 8-1 (p. 211). Here, the sample is a sheet of polyethylene, which has a refractive index considerably different from air; significant reflection occurs at the two interfaces of the sheet.

Pure Liquids

When the amount of sample is small or when a suitable solvent is unavailable, it is common practice to obtain spectra on the pure (neat) liquid. Here, only a very thin film has a sufficiently short path length to produce satisfactory spectra. Commonly, a drop of the neat liquid is squeezed between two rock-salt plates to give a layer that has a thickness of 0.01 mm or less. The two plates, held together by capillarity, are then mounted in the beam path. Clearly, such a technique does not give particularly reproducible transmittance data, but the resulting spectra are usually satisfactory for qualitative investigations.

Solids

Solids that cannot be dissolved in an infrared-transparent solvent can be suspended in a suitable nonabsorbing liquid medium to form

a two-phase mixture called a *mull*. An essential condition for the acquisition of satisfactory spectra is that the particle size of the suspended solid be smaller than the wavelength of the infrared beam; if this condition is not realized, a significant portion of the radiation is lost to scattering.

Two techniques are employed. In one, 2 to 5 mg of the finely ground sample (particle size < 2 μm) is further ground in the presence of one or two drops of a heavy hydrocarbon oil (Nujol). If hydrocarbon bands are likely to interfere, Fluorolube, a halogenated polymer, can be used. In either case, the resulting mull is then examined as a film between flat salt plates.

In the second technique, a milligram or less of the finely ground sample is intimately mixed with about 100 mg of dried potassium bromide powder. Mixing can be carried out with a mortar and pestle; a small ball mill is more satisfactory, however. The mixture is then pressed in a special die at 10,000 to 15,000 pounds per square inch to yield a transparent disk. Best results are obtained if the disk is formed in a vacuum to eliminate occluded air. The disk is then held in the instrument beam for spectroscopic examination. The resulting spectra frequently exhibit bands at 2.9 and 6.1 μm (3400 and 1600 cm^{-1}) due to absorbed moisture.

Internal-Reflection Spectroscopy[6]

When a beam of radiation passes from a more dense to a less dense medium, reflection occurs; beyond a certain critical angle, reflection becomes complete. It has been shown both theoretically and experimentally that during the reflection process the beam acts as if, in fact, it penetrates a small distance into the less dense medium before reflection

---

[6] For a review of this technique, see: P. A. Wilks, Jr., *Amer. Lab.*, **4** (11), 42 (1972).

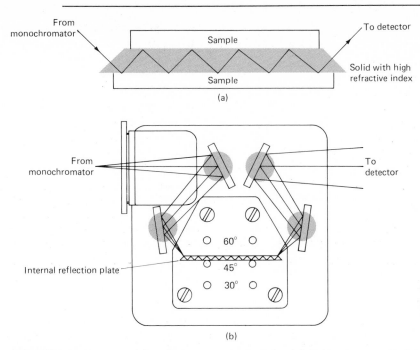

FIGURE 8-11   Internal reflectance apparatus. (a) Sample mounted on reflection plate; (b) internal reflection adapter. (Courtesy of Foxboro Analytical, Wilks Infrared Center, South Norwalk, CT. With permission.)

occurs.[7] The depth of penetration, which varies from a fraction of a wavelength up to several wavelengths, depends upon the wavelength, the index of refraction of the two materials, and the angle of the beam with respect to the interface. If the less dense medium absorbs the radiation, attenuation of the beam also occurs.

Figure 8-11 shows an apparatus that takes advantage of the internal-reflectance phenomenon for infrared-absorption measurements. As will be seen from the upper figure, the sample (here, a solid) is placed on opposite sides of a transparent crystalline material of high refractive index; a mixed crystal of thallium bromide/thallium iodide is frequently employed. By proper adjustment of the incident angle, the radiation undergoes multiple internal reflections before passing from the crystal to the detector. Absorption and attenuation takes place at each of these reflections.

Figure 8-11b is an optical diagram of a commercially available adapter that will fit into the cell area of most infrared spectrometers and will permit internal-reflectance measurements. Note that an incident angle of 30, 45, or 60 deg can be chosen. Cells for liquid samples are also available.

Internal-reflectance spectra are similar to, but not identical with, ordinary absorption spectra. In general, while the same peaks are observed, their relative intensities differ. The absorbances, while dependent upon the angle of incidence, are independent of sample thickness, since the radiation penetrates only a few micrometers into the sample.

[7] J. Fahrenfort, *Spectrochem. Acta*, **17**, 698 (1961).

Internal-reflectance spectroscopy has been applied to many substances such as polymers, rubbers, and other solids. It is of interest to note that the resulting spectra are free from the interference fringes mentioned in the previous section.

QUALITATIVE APPLICATIONS OF INFRARED ABSORPTION

We have noted that the approximate frequency at which an organic functional group, such as $C=O$, $C=C$, CH_3, and $C\equiv C$, will absorb infrared radiation can be calculated from the masses of the atoms and the force constant of the bond between them (Equation 8-14). These frequencies, called *group frequencies*, are seldom totally invariant because of interactions with other vibrations associated with one or both of the atoms comprising the group. On the other hand, such interaction effects are frequently small; as a result, a range of frequencies can be assigned within which it is highly probable that the absorption peak for a given functional group will be found.

Group frequencies often make it possible to establish the probable presence or absence of a given functional group in a molecule.

Correlation Charts

Over the years, a mass of empirical information has been accumulated concerning the frequency range within which various functional groups can be expected to absorb. *Correlation charts* provide a concise means for summarizing this information in a form that is useful for identification purposes. A number of correlation charts have been developed,[8] one of which is shown in Figure 8-12.

Correlation charts permit intelligent guesses to be made as to what functional groups are likely to be present or absent in a molecule. Ordinarily, it is impossible to identify unambiguously either the sources of all of the peaks in a given spectrum or the exact identity of the molecule. Instead, correlation charts serve as a starting point in the identification process.

Important Spectral Regions in the Infrared

The chemist interested in identifying an organic compound by the infrared technique usually examines certain regions of the spectrum in a systematic way in order to obtain clues as to the presence or absence of certain group frequencies. Some of the important regions are considered briefly.

Hydrogen Stretching Region 3700 to 2700 cm^{-1} (2.7 to 3.7 μm). The appearance of strong absorption peaks in this region usually results from a stretching vibration between hydrogen and some other atom. The motion is largely that of the hydrogen atom since it is so much lighter than the species with which it bonds; as a consequence, the absorption is not greatly affected by the rest of the molecule. Furthermore, the hydrogen stretching frequency is much higher than that for other chemical bonds, with the result that interaction of this vibration with others is usually small.

Absorption peaks in the region of 3700 to 3100 cm^{-1} (2.7 to 3.2 μm) are ordinarily due to various O—H and N—H stretching vibrations, with the former tending to appear at higher wavenumbers. The O—H bands are often broader than N—H bands and appear

[8] N. B. Colthup, *J. Opt. Soc. Am.*, **40**, 397–400 (1950); A. D. Cross, *Introduction to Practical Infra-Red Spectroscopy*, 2d ed. Washington: Butterworths, 1964, pp. 56–62; R. N. Jones, *Infrared Spectra of Organic Compounds:*

Summary Charts of Principal Group Frequencies. Ottawa: National Research Council of Canada, 1959; K. Nakanishi and P. H. Solomon, *Infrared Absorption Spectroscopy*, 2d ed. San Francisco: Holden-Day, 1977, pp. 10–56; and R. M. Silverstein, G. C. Bassler, and T. C. Morrill, *Spectrometric Identification of Organic Compounds*, 3d ed. New York: Wiley, 1974, pp. 135–152.

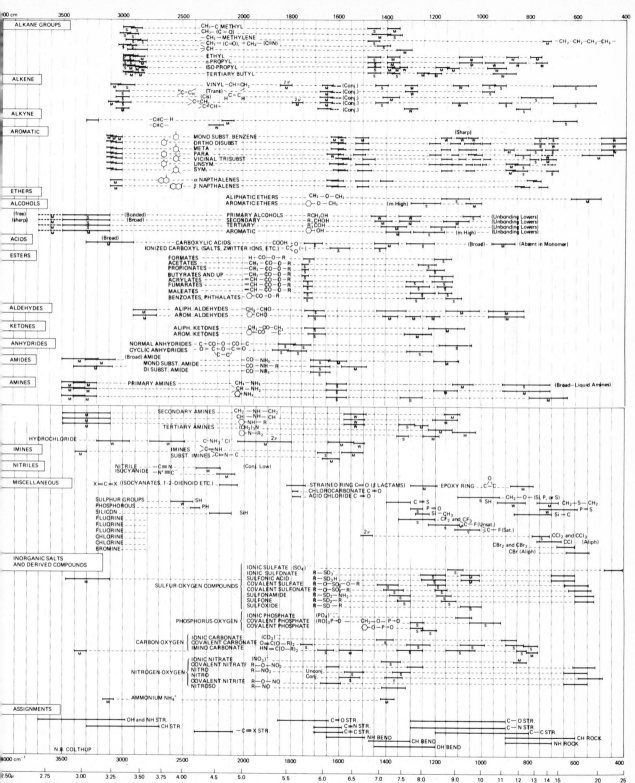

FIGURE 8-12 Correlation Chart.

only in dilute, nonpolar solvents. Hydrogen bonding tends to broaden the peaks and move them toward lower wavenumbers.

Aliphatic C—H vibrations fall in the region between 3000 and 2850 cm^{-1} (3.3 to 3.5 μm). Most aliphatic compounds have a sufficient number of C—H bonds to make this a prominent peak. Any structural variation that affects the C—H bond strength will cause a shift in the maximum. For example, the band for Cl—C—H lies just above 3000 cm^{-1} (< 3.3 μm), as do the bands for olefinic and aromatic hydrogen. The acetylenic C—H bond is strong and occurs at about 3300 cm^{-1} (3.0 μm). The hydrogen on the carbonyl group of an aldehyde usually produces a distinct peak in the region of 2745 to 2710 cm^{-1} (3.64 to 3.69 μm). Substitution of deuterium for hydrogen causes a shift to lower wavenumbers by the factor of approximately $1/\sqrt{2}$, as would be predicted from Equation 8-14; this effect has been employed to identify C—H stretching peaks.

The Triple-Bond Region Between 2700 and 1850 cm^{-1} (3.7 to 5.4 μm). A limited number of groups absorb in this spectral region; their presence is thus readily apparent. Triple-bond stretching results in a peak at 2250 to 2225 cm^{-1} (4.44 to 4.49 μm) for —C≡N, at 2180 to 2120 cm^{-1} (4.59 to 4.72 μm) for —N$^+$≡C$^-$, and at 2260 to 2190 cm^{-1} (4.42 to 4.57 μm) for —C≡C—. Also present in this region are peaks for S—H at 2600 to 2550 cm^{-1} (3.85 to 3.92 μm), P—H at 2440 to 2350 cm^{-1} (4.10 to 4.26 μm), and Si—H at 2260 to 2090 cm^{-1} (4.42 to 4.78 μm).

The Double-Bond Region Between 1950 and 1550 cm^{-1} (5.1 to 6.5 μm). The carbonyl stretching vibration is characterized by absorption throughout this region. Ketones, aldehydes, acids, amides, and carbonates all have absorption peaks around 1700 cm^{-1} (5.9 μm). Esters, acid chlorides, and acid anhydrides tend to absorb at slightly higher wavenumbers; that is, 1770 to 1725 cm^{-1} (5.65 to 5.80 μm). Conjugation tends to lower the absorption peak by about 20 cm^{-1}. It is

frequently impossible to determine the type of carbonyl that is present solely on the basis of absorption in this region; however, examination of additional spectral regions may provide the supporting evidence needed for clear-cut identification. For example, esters have a strong C—O—R stretching peak at about 1200 cm^{-1} (8.3 μm), while aldehydes have a distinctive hydrogen stretching peak just above 2700 cm^{-1} (3.7 μm), as noted previously.

Absorption peaks arising from C=C and C=N stretching vibrations are located in the 1690 to 1600 cm^{-1} (5.9 to 6.2 μm) range. Valuable information concerning the structure of olefins can be obtained from the exact position of such a peak.

The region between 1650 and 1450 cm^{-1} (6.1 to 6.9 μm) provides important information about aromatic rings. Aromatic compounds with a low degree of substitution exhibit four peaks near 1600, 1580, 1500, and 1460 cm^{-1} (6.25, 6.33, 6.67, and 6.85 μm). Variations of the spectra in this region with the number and arrangement of substituent groups are usually consistent but independent of the type of substituent; considerable structural information can thus be gleaned from careful study of aromatic absorption in the infrared region.

The "Fingerprint" Region Between 1500 and 700 cm^{-1} (6.7 to 14 μm). Small differences in the structure and constitution of a molecule result in significant changes in the distribution of absorption peaks in this region of the spectrum. As a consequence, a close match between two spectra in the fingerprint region (as well as others) constitutes strong evidence for the identity of the compounds yielding the spectra. Most single bonds give rise to absorption bands at these frequencies; because their energies are about the same, strong interaction occurs between neighboring bonds. The absorption bands are thus composites of these various interactions and depend upon the overall skeletal structure of the molecule. Exact interpretation of spectra in this region is seldom possible because of their complexity; on the other hand, it is this complexity that leads to uniqueness and the consequent usefulness of the region for identification purposes.

A few important group frequencies are to be found in the fingerprint region. These include the C—O—C stretching vibration in ethers and esters at about 1200 cm^{-1} (8.3 μm) and the C—Cl stretching vibration at 700 to 800 cm^{-1} (14.3 to 12.5 μm). A number of inorganic groups such as sulfate, phosphate, nitrate, and carbonate also absorb at wavenumbers below 1200 cm^{-1}.

Limitations to the Use of Correlation Charts. The unambiguous establishment of the identity or the structure of a compound is seldom possible from correlation charts alone. Uncertainties frequently arise from overlapping group frequencies, spectral variations as a function of the physical state of the sample (that is, whether it is a solution, a mull, in a pelleted form, and so forth), and instrumental limitations.

In employing group frequencies, it is essential that the entire spectrum, rather than a small isolated portion, be considered and interrelated. Interpretation based on one part of the spectrum should be confirmed or rejected by study of other regions.

To summarize, then, correlation charts serve only as a guide for further and more careful study. Several excellent monographs describe the absorption characteristics of functional groups in detail.[9] A study of these characteristics, as well as the other physical properties of the sample, may permit unambiguous identification. Infrared spectroscopy, when used in conjunction with other methods such as mass spectroscopy, nuclear magnetic resonance, and elemental analysis, usually makes possible the positive identification of a species.

The examples that follow illustrate how infrared spectra are employed in the identification of pure compounds. In every case, confirmatory tests would be desirable; a comparison of the experimental spectrum with that of the pure compound would suffice for this purpose.

EXAMPLE

The spectrum in Figure 8-13 was obtained for a pure colorless liquid in a 0.01-mm cell; the liquid boiled at 190°C. Suggest a structure for the sample.

The four absorption peaks in the region of 1450 to 1600 cm^{-1} are characteristic of an aromatic system, and one quickly learns to recognize this grouping. The peak at 3100 cm^{-1} corresponds to a hydrogen-stretching vibration, which for an aromatic system is usually above 3000 cm^{-1}. For aliphatic hydrogens, this type of vibration causes absorption at or below 3000 cm^{-1} (see Figure 8-12). Thus, it would appear that the compound is predominately, if not exclusively, aromatic.

We then note the sharp peak at 2250 cm^{-1} and recall that only a few groups absorb in this region. From Figure 8-12, we see that these include —C≡C—, —C≡CH, —C≡N, and Si—H. The —C≡CH group can be eliminated, however, since it has a hydrogen stretching frequency of about 3250 cm^{-1}.

The pair of strong peaks at 680 and 760 cm^{-1} suggests that the aromatic ring may be singly substituted, although the pattern of peaks in the regions of 1000 to 1300 cm^{-1} is confusing and may lead to some doubt about this conclusion.

Two likely structures that fit the infrared data would appear to be C$_6$H$_5$C≡N and C$_6$H$_5$C≡C—C$_6$H$_5$. The latter, however, is a solid at room temperature; in contrast, benzonitrile has a boiling point (191°C), similar

[9] N. B. Colthup, L. N. Daly, and S. E. Wilberley, *Introduction to Infra Red and Raman Spectroscopy*, 2d ed. New York: Academic Press, 1975; and H. A. Szymanski, *Theory and Practice of Infrared Spectroscopy*. New York: Plenum Press, 1964.

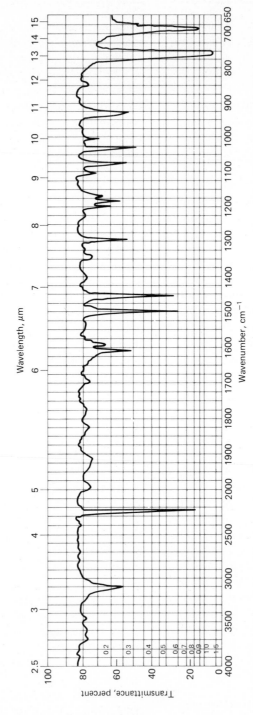

FIGURE 8-13 Spectrum of a colorless liquid. (From "Catalog of Selected Ultraviolet Spectral Data," Thermodynamics Research Center Data Project, Thermodynamics Research Center, Texas A&M University, College Station, Texas. Loose-leaf data sheets, extant 1975. With permission.)

to that of the sample. Thus, we tentatively conclude that the compound under investigation is $C_6H_5C\equiv N$.

EXAMPLE

A colorless liquid was found to have the empirical formula $C_6H_{12}O$ and a boiling point of 130°C. Its infrared spectrum (neat in a 0.025-mm cell) is given in Figure 8-14. Suggest a structure.

The presence of an aliphatic structure is suggested in Figure 8-14 by both the strong band at 3000 cm^{-1} and the empirical formula. The intense band at 1720 cm^{-1} and the single oxygen in the formula strongly indicate the likelihood of an aldehyde or a ketone. The band at about 2800 cm^{-1} could be attributed to the shift in stretching vibration associated with hydrogen that is bonded directly to a carbonyl carbon atom; if this interpretation is correct, the substance is an aldehyde. The peak at 3450 cm^{-1} is puzzling,

for absorption in this region is usually the result of an N—H or an O—H stretching vibration; the possibility that the substance is an aliphatic alcohol must thus be considered. The empirical formula, however, is inconsistent with a structure that contains both an aldehyde and a hydroxyl group. Examination of the lower-frequency region fails to yield further information. This dilemma can be resolved by attributing absorption at 3450 cm^{-1} to the strong O—H stretching band of water, which is often present as a contaminant. Thus, we conclude that the sample is probably an aliphatic aldehyde. We are unable to determine the extent of chain branching from the spectrum; to be sure, the peak at 730 cm^{-1} suggests the presence of four or more methylene groups in a row. The boiling point of *n*-hexanal is 131°C; we are thus inclined to conclude that the sample is this compound. Confirmatory tests would be needed, however.

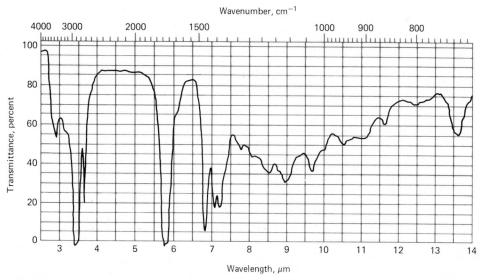

FIGURE 8-14 Infrared spectrum of an unknown. (From "Catalog of Selected Ultraviolet Spectral Data," Thermodynamics Research Center Data Project, Thermodynamics Research Center, Texas A&M University, College Station, Texas. Loose-leaf data sheets, extant 1975. With permission.)

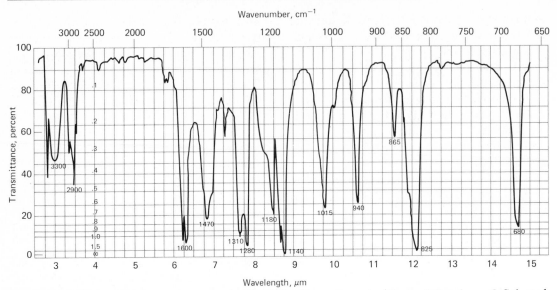

FIGURE 8-15   Spectrum of a sample in CCl_4 (2 to 8 μm). (From "Catalog of Selected Ultraviolet Spectral Data," Thermodynamics Research Center Data Project, Thermodynamics Research Center, Texas A&M University, College Station, Texas. Loose-leaf data sheets, extant 1975. With permission.)

EXAMPLE

The spectrum shown in Figure 8-15 was obtained from a 0.5% solution of a liquid sample in a 0.5-mm cell. For the region of 2 to 8 μm, the solvent was CCl_4, and for the higher wavelengths, CS_2 was employed. The empirical formula of the compound was found to be $C_8H_{10}O$. Suggest a probable structure.

The empirical formula indicates that the compound is probably aromatic, and this supposition is borne out by the characteristic pattern of peaks in the 1450 to 1600 cm^{-1} range. The C—H stretching bands, however, are somewhat lower in wavenumber than those for a purely aromatic system, which suggests the existence of some aliphatic groups as well. The broad band at 3300 cm^{-1} appears important, and suggests an O—H or an N—H stretching vibration. The formula permits us to conclude that the molecule must contain a phenolic or an alcoholic O—H group. The spectrum in the region of

1100 to 1400 cm^{-1} is compatible with the postulated O—H group, but we are unable to decide from the spectral data whether the sample is a phenol or an alcohol.

From our observations thus far, and from the empirical formula, likely structures for the unknown appear to be:

We now turn to the low-frequency end of the spectrum, where correlations of peaks with aromatic substitution patterns can often be found. The two strong peaks at 680 and 825 cm^{-1} and the weaker one at 865 cm^{-1} appear to fit best with the pattern for a symmetrically trisubstituted benzene. Thus, 3,5-di-methylphenol seems the most logical choice. Confirmatory data would be needed, however, to establish the identity unambiguously.

Collections of Spectra

As may be seen from the foregoing examples, correlation charts seldom suffice for the positive identification of an organic compound from its infrared spectrum. There are available, however, several catalogs of infrared spectra that assist in qualitative identification by providing comparison spectra for a large number of pure compounds. These collections have become so extensive as to require edge-punched cards, IBM cards, or magnetic tapes for efficient retrieval. Data presentation takes several forms including replica of spectra on notebook-size paper, cards, microfiches and microfilms, as well as magnetic tapes or disks. In some instances, digitalized peak positions are employed in lieu of replicas of spectra. A list of sources of these collections is found in a paper by Gevantman.[10]

QUANTITATIVE APPLICATIONS

Quantitative infrared-absorption methods differ somewhat from those discussed in the previous chapter because of the greater complexity of the spectra, the narrowness of the absorption bands, and the instrumental limitations of infrared instruments.

---

[10] L. H. Gevantman, *Anal. Chem.*, **44** (7), 31A (1972).

Deviations from Beer's Law

With infrared radiation, instrumental deviations from Beer's law are more common than with ultraviolet and visible wavelengths because infrared absorption bands are relatively narrow. Furthermore, the low intensity of sources and low sensitivities of detectors in this region require the use of relatively wide monochromator slit widths; thus, the bandwidths employed are frequently of the same order of magnitude as the widths of absorption peaks. We have pointed out (see Figure 6-4) that this combination of circumstances usually leads to a nonlinear relationship between absorbance and concentration. Calibration curves, determined empirically, are therefore often required for quantitative work.

Absorbance Measurement

Matched absorption cells for solvent and solution are ordinarily employed in the ultraviolet and visible regions, and the measured absorbance is then found from the relation

$$A \cong \log \frac{P_{solvent}}{P_{solution}}$$

The use of the solvent in a matched cell as a reference absorber has the advantage of largely canceling out the effects of radiation losses due to reflection at the various interfaces, scattering and absorption by the solvent, and absorption by the container windows (p. 150). This technique is seldom practical for measurements in the infrared region because of the difficulty in obtaining cells whose transmission characteristics are identical. Most infrared cells have very short path lengths that are difficult to duplicate exactly. In addition, the cell windows are readily attacked by contaminants in the atmosphere and the solvent; thus, their transmission characteristics change continually with use. For these reasons, a reference absorber is often dispensed with entirely in infrared work, and

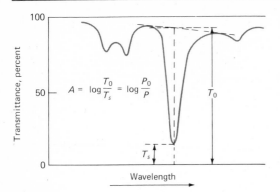

$$A = \log \frac{T_0}{T_s} = \log \frac{P_0}{P}$$

FIGURE 8-16   Base-line method for determination of absorbance.

the intensity of the radiation passing through the sample is simply compared with that of the unobstructed beam; alternatively, a salt plate may be placed in the reference beam. In either case, the resulting transmittance is ordinarily less than 100%, even in regions of the spectrum where no absorption by the sample occurs; this effect is readily seen by examining the several spectra that have appeared earlier in this chapter.

For quantitative work, it is necessary to correct for the scattering and absorption by the solvent and the cell. Two methods are employed. In the so-called *cell in-cell out* procedure, spectra of the solvent and sample are obtained successively with respect to the unobstructed reference beam. The same cell is used for both measurements. The transmittance of each solution versus the reference beam is then determined at an absorption maximum of the analyte. These transmittances can be written as

$$T_0 = P_0/P_r$$

and

$$T_s = P/P_r$$

where P_r is the power of the reference beam and T_0 and T_s are the transmittances of the solvent and sample, respectively, against this reference. If P_r remains constant during the two measurements, then the transmittance of the sample with respect to the solvent can be obtained by division of the two equations. That is,

$$T = T_s/T_0 = P/P_0$$

An alternative way of obtaining P_0 and T is the *base-line* method, in which the solvent transmittance is assumed to be constant or at least to change linearly between the shoulders of the absorption peak. This technique is demonstrated in Figure 8-16.

Applications of Quantitative Infrared Spectroscopy

With the exception of homonuclear molecules, all organic and inorganic molecular species absorb in the infrared region; thus, infrared spectrophotometry offers the potential for determining an unusually large number of substances. Moreover, the uniqueness of an infrared spectrum leads to a degree of specificity that is matched or exceeded by relatively few other analytical methods. This specificity has found particular application to analysis of mixtures of closely related organic compounds. Two examples that typify these applications follow.

Analysis of a Mixture of Aromatic Hydrocarbons. A typical application of quantitative infrared spectroscopy involves the resolution of C_8H_{10} isomers in a mixture which includes o-xylene, m-xylene, p-xylene, and ethylbenzene. The infrared absorption spectra of the individual components in the 12 to 15 μm range is shown in Figure 8-17; cyclohexane is the solvent. Useful absorption peaks for determination of the individual compounds occur at 13.47, 13.01, 12.58, and 14.36 μm, respectively. Unfortunately, however, the absorbance of a mixture at any one of these wavelengths is not entirely determined by the concentration of just one component, because of overlapping absorption bands. Thus, molar absorptivities for each of the four compounds must be determined at the four wavelengths.

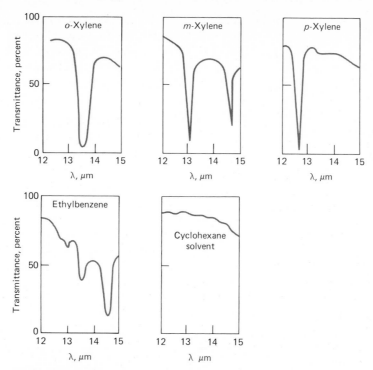

FIGURE 8-17   Spectra of C_8H_{10} isomers in cyclohexane. (From R. P. Bauman, *Absorption Spectroscopy*. New York: John Wiley & Sons, Inc., 1962, p. 406. With permission.)

Then four simultaneous equations can be written which permit the calculation of the concentration of each species from four absorbance measurements (see p. 191). Such calculations are most easily performed with a computer.

When the relationship between absorbance and concentration is nonlinear (as frequently occurs in the infrared region), the algebraic manipulations associated with an analysis of several components having overlapping absorption peaks are considerably more complex.[11]

---

[11] The treatment of infrared data for various types of mixtures is discussed in: R. P. Bauman, *Absorption Spectroscopy*. New York: Wiley, 1962, pp. 403–419.

Analysis of Air Contaminants. The recent proliferation of government regulations with respect to atmospheric contaminants has demanded the development of sensitive, rapid, and highly specific methods for a variety of chemical compounds. Infrared absorption procedures appear to meet this need better than any other single analytical tool.

Table 8-2 demonstrates the potential of infrared spectroscopy for the analysis of mixtures of gases. The standard sample of air containing five species in known concentration was analyzed with the computerized instrument shown in Figure 8-8; a 20-m gas cell was employed. The data were printed out within a minute or two after sample injection.

Table 8-3 shows potential applications of infrared filter photometers (such as that shown in Figure 8-6) for the quantitative determination of various chemicals in the atmosphere for the purpose of assuring compliance with OSHA regulations.

Of the more than 400 chemicals for which maximum tolerable limits have been set by the Occupational Safety and Health Administration, more than half appear to have absorption characteristics suitable for determination by means of infrared filter photometers or spectrophotometers. Obviously, among all of these absorbing compounds, peak overlaps are to be expected; yet the method should provide a moderately high degree of selectivity.

TABLE 8-2 AN EXAMPLE OF INFRARED ANALYSIS OF AIR CONTAMINANTS[a]

Contaminants	Concn, ppm	Found, ppm	Relative Error, %
Carbon monoxide	50	49.1	1.8
Methylethyl ketone	100	98.3	1.7
Methyl alcohol	100	99.0	1.0
Ethylene oxide	50	49.9	0.2
Chloroform	100	99.5	0.5

[a] (Courtesy of Foxboro/Wilks, Inc., South Norwalk, CT.)

TABLE 8-3 SOME EXAMPLES OF INFRARED VAPOR ANALYSIS FOR OSHA COMPLIANCE[a]

Compound	Allowable Exposure, ppm[b]	λ, μm	Minimum Detectable Concentration, ppm[c]
Carbon disulfide	20	4.54	0.5
Chloroprene	25	11.4	4
Diborane	0.1	3.9	0.05
Ethylenediamine	10	13.0	0.4
Hydrogen cyanide	10	3.04	0.4
Methyl mercaptan	10	3.38	0.4
Nitrobenzene	1	11.8	0.2
Pyridine	5	14.2	0.2
Sulfur dioxide	5	8.6	0.5
Vinyl chloride	1	10.9	0.3

[a] (Courtesy of Foxboro/Wilks, Inc., South Norwalk, CT.)
[b] 1977 OSHA exposure limits for 8-hour weighed average.
[c] For 20.25-m cell.

Disadvantages and Limitations to Quantitative Infrared Methods

Several disadvantages attend the application of infrared methods to quantitative analysis. Among these are the frequent nonadherence to Beer's law and the complexity of spectra; the latter enhances the probability of the overlap of absorption peaks. In addition, the narrowness of peaks and the effects of stray radiation make absorbance measurements critically dependent upon the slit width and the wavelength setting. Finally, the narrow cells required for many analyses are inconvenient to use and may lead to significant analytical uncertainties.

The analytical errors associated with a quantitative infrared analysis often cannot be reduced to the level associated with ultraviolet and visible methods, even with considerable care and effort.

INFRARED FOURIER TRANSFORM SPECTROSCOPY

The sensitivity of several spectroscopic methods can be enhanced significantly by application of the Fourier transform mentioned in Chapter 4 (p. 94). Use of the transform requires a dedicated high-speed computer and often, spectroscopic equipment that differs considerably from the conventional dispersive type. The Fourier transform technique has been applied advantageously to infrared, nuclear magnetic resonance, ion cyclotron resonance, atomic absorption, mass, and microwave spectroscopy. In this section, we will discuss its use for infrared studies.

Fourier transform infrared spectroscopy was first developed by astronomers in the early 1950s in order to study the spectra of distant stars; only by the Fourier technique could the very weak signals from these sources be isolated from environmental noise. The first chemical applications of Fourier transform spectroscopy, which were reported approximately a decade later, were to the energy-starved far-infrared region; by the late 1960s, instruments for chemical studies in both the far- (10 to 400 cm^{-1}) and the ordinary-infrared regions were available commercially.[12]

Before considering details of Fourier transform spectroscopy, it is worthwhile to describe the general principle of the procedure and its inherent advantage in recovering weak signals from noisy backgrounds.

The Inherent Advantage of Fourier Transform Spectroscopy

For purposes of this discussion, it is convenient to think of an experimentally derived spectrum as being made up of m individual transmittance measurements at equally spaced frequency or wavelength intervals called *resolution elements*. The quality of the spectrum—that is, the amount of spectral detail—increases as the number of resolution elements becomes larger or as the frequency intervals between measurements become smaller.[13] Thus, in order to increase spectral quality, m must be made larger; clearly, increasing the number of resolution elements must also increase the time required for obtaining a spectrum.

Consider, for example, the derivation of an infrared spectra from 500 to 5000 cm^{-1}. If resolution elements of 3 cm^{-1} were chosen, m

---

[12] For a more complete discussion of Fourier transform infrared spectroscopy, consult the following references: P. R. Griffiths, *Chemical Infrared Fourier Transform Spectroscopy*. New York: Wiley, 1975; *Fourier Transform Infrared Spectroscopy: Applications to Chemical Systems*, eds. J. R. Ferraro and L. J. Basile. New York: Academic Press, Vol. I, 1978 and Vol. II, 1979.

[13] With a recording spectrophotometer, of course, individual point-by-point measurements are not made; nevertheless, the idea of a resolution element is useful, and the ideas generated from it apply to recording instruments as well.

would be 1500; if one-half second were required for recording the transmittance of each resolution element, 750 s or 12.5 min would be needed to obtain the spectrum. Reducing the width of the resolution element to 1.5 cm^{-1} would be expected to provide significantly greater spectral detail; it would also double the number of resolution elements as well as the time required for their measurement.

The quality of an infrared spectrum depends not only upon the number of resolution elements, but also upon the signal-to-noise ratio for the instrument. Unfortunately, an increase in the number of resolution elements is usually accompanied by a decrease in the signal-to-noise ratio, because narrower slits are required. Thus, the signal reaching the transducer is weaker; the noise, however, ordinarily remains undiminished.

As was pointed out in Chapter 3 (p. 76), an effective method of extracting a weak signal from a noisy environment is by signal averaging. This technique is effective because in summing replicate measurements, the signals contained therein are additive; the random noise, on the other hand, increases with the square root of the number of measurements only. The improvement in signal-to-noise ratio brought about by signal averaging is thus proportional to $\sqrt{n}$, where n is the number of replications.

The application of signal averaging to conventional infrared spectroscopy is, unfortunately, costly in terms of time. Thus, in the example just considered, 12.5 min were required to obtain a spectrum of 1500 resolution elements. Improvement of the signal-to-noise ratio by a factor of two, would require 4×1500 measurements, or 50 min.

Fourier transform spectroscopy differs from conventional spectroscopy in that all of the resolution elements for a spectrum are measured *simultaneously*. Significant gains in signal-to-noise ratio accrue as a consequence. For example, if we assume that a single Fourier transform measurement requires the same amount of time as the measurement of a single resolution element in conventional spectroscopy (one-half second in our example), then in the 12.5 min required to record a conventional spectrum, 1500 $(12.5 \times 60/0.5)$ replicate Fourier transform measurements could be averaged. The resulting improvement in signal-to-noise ratio would be $\sqrt{1500}$ or about 39. This advantage of Fourier transform spectroscopy was first suggested by P. Fellgett in 1958 and has become known as the *Fellgett advantage*.

Spectroscopic methods, such as the Fourier transform technique, in which all resolution elements are measured simultaneously are often called *multiplex methods*. Measurement of ultraviolet-visible spectra with a fixed monochromator and a vidicon tube (p. 142) is another example of a multiplex method.

Time Domain Spectroscopy

Conventional spectroscopy can be termed *frequency domain* spectroscopy, in that radiant power data are recorded as a function of radiant frequency (or the inversely related wavelength). *Time domain* spectroscopy, in contrast, is concerned with changes in radiant power with time. Figure 8-18 illustrates the difference.

The curves in (c) and (d) are conventional spectra of two monochromatic sources with frequencies v_1 and v_2 Hz. The curve in (e) is the spectrum of a source containing both frequencies. In each case, some measure of the radiant power, $P(v)$ is plotted with respect to the frequency in hertz. The symbol in parentheses is added to emphasize the frequency dependence of the power; time domain power will be indicated by $P(t)$.

The curve in Figure 8-18a shows the time domain spectra for each of the monochromatic sources. The two have been plotted together in order to make the small frequency difference between them more obvious. Here, the instantaneous power $P(t)$ is plotted as a function of time. The curve in Figure 8-18b is

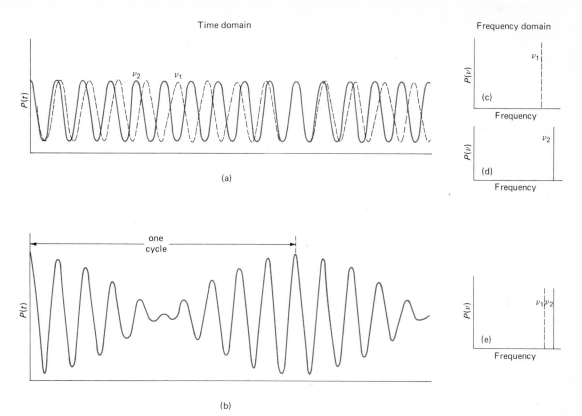

Time domain

Frequency domain

(a)

(b)

(c)

(d)

(e)

FIGURE 8-18   Illustration of time domain (a) and (b) and frequency domain (c), (d), and (e) plots.

the time domain spectrum of the source containing the two frequencies. As is shown by the horizontal arrow, the plot exhibits a periodicity or *beat* as the two waves go in and out of phase.

Examination of Figure 8-19 reveals that the time domain spectrum for a source containing several wavelengths is considerably more complex than those shown in Figure 8-18. Because a large number of wavelengths are involved, a full cycle is never realized in the time shown. To be sure, a pattern of beats can be observed as certain wavelengths pass in and out of phase. In general, the signal power decreases with time as a consequence of the various closely spaced wavelengths becoming more and more out of phase.

It is important to appreciate that a time domain spectrum contains the same information as does a spectrum in the frequency domain, and in fact, one can be converted to the other by mathematical manipulations. Thus, Figure 8-18b was derived from Figure 8-18e by means of the equation

$$P(t) = k(\cos 2\pi v_1 t + \cos 2\pi v_2 t)$$

where k is a constant and t is the time. The difference in frequency between the two lines was approximately 10% of v_2.

The interconversion of time and frequency domain spectra becomes exceedingly complex and mathematically tedious when more than a few lines are involved; the operation is only practical with a high-speed computer.

Methods of
Obtaining Time Domain Spectra

Time domain spectra, such as those shown in Figures 8-18 and 8-19, cannot be acquired experimentally with radiation of the frequency range of absorption spectroscopy (10^{15} Hz for ultraviolet to 10^7 Hz for nuclear magnetic resonance) because transducers with sufficiently rapid response times are not available. Thus, a typical transducer responds only to the average power of a high-frequency signal and not to its periodic variation. To obtain time domain spectra requires, therefore, a method of converting (or *modulating*) a high-frequency signal to one of measurable frequency without distorting the time relationships carried in the signal; that is, the frequencies in the modulated signal must be directly proportional to those in the original. Different signal-modulation procedures are employed for the various wavelength regions of the spectrum. The Michelson interferometer is used extensively for measurements in the infrared region.

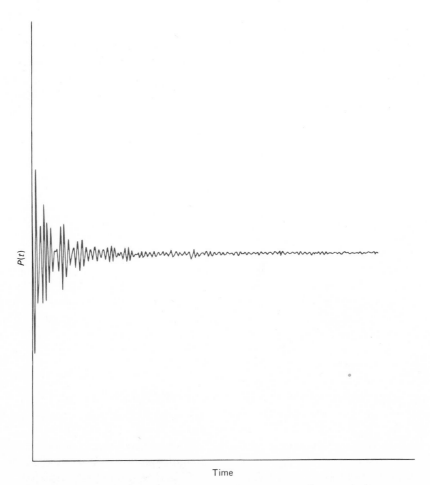

FIGURE 8-19 Time domain spectrum of a source made up of several wavelengths.

The Michelson Interferometer

The device used for modulating infrared radiation is an interferometer similar in design to one first described by Michelson in 1891. The Michelson interferometer is a device that splits a beam of radiation into two beams of nearly equal power and then recombines them in such a way that intensity variations of the combined beam can be measured as a function of differences in the lengths of the paths of the two halves. Figure 8-20 is a schematic diagram of such a device as it is used for infrared Fourier transform spectroscopy.

As shown in the figure, a beam of radiation from a source is collimated and impinges on a beam splitter, which transmits approximately half of the radiation and reflects the other half. The resulting twin beams are then reflected from mirrors, one of which is fixed and the other of which is movable. The beams then meet again at the beam splitter, with half of each beam being directed toward the sample and detector and the other two halves being directed back toward the source. Only the two halves passing through the sample to the detector are employed for analytical purposes, although the other halves contain the same information about the source.

Horizontal motion of the movable mirror will cause the power of the radiation reaching the detector to fluctuate in a predictable manner. When the two mirrors are equidistant from the splitter (position 0 in Figure 8-20), the two parts of the recombined beam will be totally in phase and the power will be a maximum. For a monochromatic source, motion of the movable mirror in either direction by a distance equal to exactly one-quarter wavelength (position B or C in the

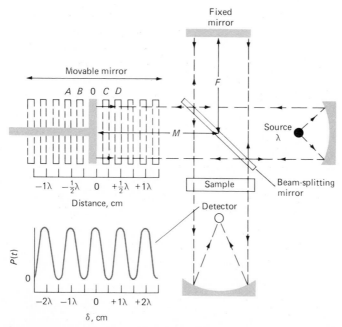

FIGURE 8-20 Schematic diagram of a Michelson interferometer illuminated by a monochromatic source.

figure) will change the path length of the corresponding reflected beam by one-half wavelength (one-quarter wavelength for each direction). Under this circumstance, destructive interference will reduce the radiant power of the recombined beams to zero. Further motion to A or D will bring the two halves back in phase so that constructive interference can again occur.

The difference in path lengths for the two beams, $2(M - F)$ in the figure is termed the *retardation* δ. A plot of the output power from the detector versus δ is called an *interferogram*; for monochromatic radiation, the interferogram takes the form of a cosine curve such as that shown in the lower part of Figure 8-20 (cosine rather than sine because the power is always a maximum when δ is zero and the two paths are identical).

In the typical infrared scanning interferometer, the mirror is motor-driven at a constant velocity of v_M cm/s. Since peaks occur at higher multiples of the wavelength of the incident radiation, the frequency f of the *interferogram* is given by

$$f = \frac{2v_M \text{ cm/s}}{\lambda \text{ cm}} \qquad (8\text{-}15)$$

Here, λ is the wavelength of the incident beam (in cm) and the factor 2 is required because the reflection process doubles the effect of the retardation.

In some instances, it is convenient to express Equation 8-15 in terms of wavenumber of the incident radiation σ rather than in wavelength. Thus,

$$f = 2v_M \sigma \qquad (8\text{-}16)$$

The relationship between the *optical frequency* of the radiation and the frequency of the interferogram is readily obtained by substitution of $\lambda = c/v$ into Equation 8-15. Thus,

$$f = \frac{2v_M}{c} v \qquad (8\text{-}17)$$

where v is the frequency of the radiation and c is the velocity of light $(3 \times 10^{10}$ cm/s). When v_M is constant, it is evident that the *interferogram frequency f is directly proportional to the optical frequency v*. Furthermore, the proportionality constant will generally be a very small number. For example, if the mirror is driven at a rate of 1.5 cm/s,

$$\frac{2v_M}{c} = \frac{2 \times 1.5 \text{ cm/s}}{3 \times 10^{10} \text{ cm/s}} = 10^{-10}$$

and

$$f = 10^{-10} v$$

As shown by the following example, the frequency of infrared radiation is readily modulated into the audio range by a Michelson interferometer.

EXAMPLE

Calculate the frequency range of a modulated signal from a Michelson interferometer with a mirror velocity of 0.50 cm/s, for infrared radiation of 2.5 and 16 μm $(1.2 \times 10^{14}$ to 1.9×10^{13} Hz).

Employing Equation 8-15, we find

$$f_1 = \frac{2 \times 0.5 \text{ cm/s}}{2.5 \ \mu\text{m} \times 10^{-4} \text{ cm/}\mu\text{m}} = 4000 \text{ Hz}$$

$$f_2 = \frac{2 \times 0.5}{16 \times 10^{-4}} = 625 \text{ Hz}$$

Certain types of infrared transducers are capable of following fluctuations in signal power that fall into the audio-frequency range. Thus, it becomes possible to record a modulated time domain spectrum which reflects exactly the appearance of the very high-frequency time domain spectra of an infrared source. Figure 8-21 shows three examples of such time domain interferograms on the left and their frequency domain counterparts on the right.

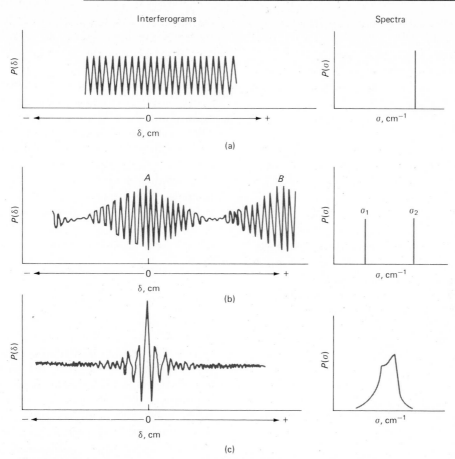

Interferograms   Spectra

(a)

(b)

(c)

FIGURE 8-21 Comparison of interferograms and optical spectra.

Fourier Transformation of Interferograms

The cosine wave of the interferogram shown in Figure 8-21a (and also in Figure 8-20) can be described in theory by the equation[14]

$$P(\delta) = \tfrac{1}{2}P(\sigma) \cos 2\pi ft \qquad (8\text{-}18)$$

where $P(\sigma)$ is the radiant power of the infrared beam incident upon the interferometer and $P(\delta)$ is the amplitude or power of the interferogram signal. The parenthetical symbols emphasize that one power (σ) is in the frequency domain and the other (δ) in the time domain. In practice, the foregoing equation is modified to take into account the fact that the interferometer ordinarily will not split the source exactly in half and that the detector response and the amplifier behavior are frequency-dependent. Thus, it is useful to introduce a new variable $B(\sigma)$ which depends upon $P(\sigma)$ but also takes these other factors into account. Therefore, we rewrite the equation in the form

$$P(\delta) = B(\sigma) \cos 2\pi ft \qquad (8\text{-}19)$$

---

[14] This equation can be derived in the same way as Equation 2-16 (p. 12).

Substitution of Equation 8-16 into Equation 8-19 leads to

$$P(\delta) = B(\sigma) \cos 2\pi 2 v_M \sigma t \qquad (8\text{-}20)$$

But the mirror velocity can be expressed in terms of retardation or

$$v_M = \frac{\delta}{2t}$$

Substitution of this relationship into Equation 8-20 gives

$$P(\delta) = B(\sigma) \cos 2\pi\delta\sigma$$

which expresses the magnitude of the interferogram signal as a function of the retardation factor and the wavenumber of the optical input signal.

The interferograms shown in part (b) of Figure 8-21 would be described by

$$P(\delta) = B_1(\sigma) \cos 2\pi\delta\sigma_1 + B_2(\sigma) \cos 2\pi\delta\sigma_2$$

$$(8\text{-}21)$$

For a continuous source, such as in part (c) of Figure 8-21, the interferogram can be represented as a sum of an infinite number of cosine terms. That is,

$$P(\delta) = \int_{n=-\infty}^{+\infty} B(\sigma) \cos 2\pi\delta\sigma_n \cdot d\delta \qquad (8\text{-}22)$$

The Fourier transform of this integral is

$$P(\sigma) = \int_{n=-\infty}^{+\infty} B(\delta) \cos 2\pi\sigma\delta_n \cdot d\sigma \qquad (8\text{-}23)$$

Infrared Fourier transform spectroscopy consists of recording $P(\delta)$ as a function of δ (Equation 8-22) and then mathematically transforming this relation to one that gives $P(\sigma)$ as a function of σ (the spectrum) as shown by Equation 8-23.

Equations 8-22 and 8-23 cannot be employed as written because they assume that the beam contains radiation from zero to infinite wavenumbers and a mirror drive of infinite length. Furthermore, Fourier transformations with a computer require that the detector output be digitalized; that is, the

output must be sampled periodically and stored in digital form. Equation 8-22, however, demands that the sampling intervals $d\delta$ be infinitely small; that is, $d\delta \rightarrow 0$.

From a practical standpoint, only a finite-sized sampling interval can be summed over a finite retardation range (a few centimeters). These limitations have the effect of limiting the resolution of a Fourier transform instrument and restricting its frequency range.

Resolution of a Fourier Transform Spectrometer. The resolution of a Fourier transform spectrometer can be described in terms of the difference in wavenumber between two lines that can be just separated by the instrument. That is,

$$\Delta\sigma = \sigma_1 - \sigma_2 \qquad (8\text{-}24)$$

where σ_1 and σ_2 are wavenumbers for a pair of barely resolvable infrared lines.

It is possible to show that in order to resolve two lines, it is necessary to scan the time domain spectrum long enough so that one complete cycle or beat for the two lines is completed; only then will all of the information contained in the spectra have been recorded. For example, resolution of the two lines σ_1 and σ_2 in Figure 8-21b would require recording the interferogram from the maximum A at zero retardation to the maximum B where the two waves are again in phase. The maximum at B occurs, however, when $\delta\sigma_2$ is larger than $\delta\sigma_1$ by 1 in Equation 8-21. That is, when

$$\delta\sigma_2 - \delta\sigma_1 = 1$$

or

$$\sigma_2 - \sigma_1 = \frac{1}{\delta}$$

Substitution into Equation 8-24 reveals that the resolution is given by

$$\Delta\sigma = \sigma_2 - \sigma_1 = \frac{1}{\delta} \qquad (8\text{-}25)$$

EXAMPLE

What length of mirror drive will provide a resolution of 0.1 cm^{-1}?

Substituting into Equation 8-25 gives

$$0.1 = \frac{1}{\delta}$$

$$\delta = 10 \text{ cm}$$

The mirror motion required is one-half the retardation, or 5 cm.

In addition to the length of mirror sweep, resolution is also influenced by the planarity of the mirrors and the coplanarity of the beam splitter. Resolution is also lost if the radiation entering the interferometer converges or diverges significantly.

Sampling the Interferogram. In order to compute the Fourier transform of an interferogram, it is necessary to digitalize and store the radiant power at equal intervals of retardation. That is, the interferogram must be broken into resolution elements and the power of each of these measured, digitalized, and stored in the memory of a computer. Information theory permits the calculation of the minimum number of resolution elements X that will permit reconstruction of a spectrum from an interferogram without a loss of detail. The number is given by

$$X = \frac{2(\sigma_{max} - \sigma_{min})}{\Delta \sigma} \qquad (8\text{-}26)$$

EXAMPLE

Calculate the number of resolution elements that would be needed to reconstruct an infrared spectrum from 4000 to 625 cm^{-1} if a resolution of 1.5 cm^{-1} is desired.

$$X = \frac{2(4000 - 625)}{1.5} = 4500$$

Thus, the power for 4500 points would need to be measured, digitalized, and stored.

Instrumentation

Several Fourier transform spectrometers are available commercially. Their prices lie in the $35,000 to $120,000 range (including the dedicated computer for performing the Fourier transformation).

Drive Mechanism. A requirement for satisfactory interferograms (and thus satisfactory spectra) is that the speed of the moving mirror be relatively constant and its position exactly known at any instant. The planarity of the mirror must also remain constant during its entire sweep of 10 cm or more.

In the far-infrared region, where the wavelengths are in the micrometer range, displacement of the mirror by a fraction of a wavelength, and accurate measurement of its position, can be accomplished by means of a motor-driven micrometer screw. A more precise and sophisticated mechanism is required for the mid- and near-infrared regions, however. Here, the mirror mount is generally floated on air cushions held within close-fitting stainless steel sleeves (see Figure 8-22). The mount is driven by an electromagnetic coil similar to the voice coil in a loudspeaker; a slowly increasing current in the coil drives the mirror at constant velocity. After reaching its terminus, the mirror is returned rapidly to the starting point for the next sweep by a rapid reversal of the current. The length of travel varies from 2 to about 18 cm; the scan rates range from 0.05 cm/s to 4 cm/s.

Two additional features of the mirror system are necessary for successful operation in the infrared regions. The first of these is a means of sampling the interferogram at precisely spaced retardation intervals. The second is a method for determining exactly the zero-retardation point to permit signal averaging. If this point is not known precisely, the signals from repetitive sweeps would not be fully in phase; averaging would tend to degrade rather than improve the signal.

The problem of precise signal sampling

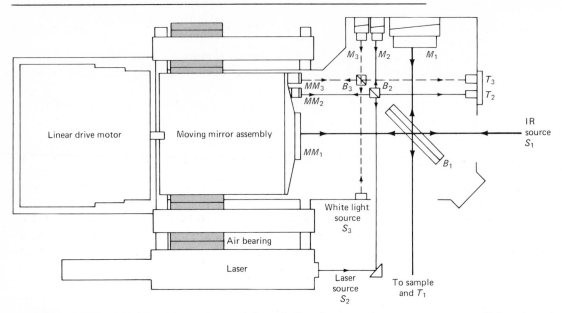

FIGURE 8-22   Interferometers in an infrared Fourier transform spectrometer. Subscripts 1 define the infrared interferometer; subscripts 2 and 3 refer to the laser and white-light interferometers, respectively. (Courtesy of Nicolet Instrument Corp., Madison, Wisconsin.)

and signal averaging is accomplished in modern instruments by using three interferometers rather than one, with a single mirror mount holding the three movable mirrors. Figure 8-22 is a schematic diagram showing the arrangement employed by one manufacturer. The components and radiation paths for each of the three interferometer systems are indicated by the subscripts 1, 2, and 3, respectively. System 1 is the infrared system that ultimately provides an interferogram similar to that shown as curve A in Figure 8-23. System 2 is a so-called *laser-fringe reference* system, which provides sampling-interval information. It consists of a helium neon laser S_2, an interferometric system including mirrors MM_2 and M_2, a beam splitter B_2, and a transducer T_2. The output from this system is sinusoidal, as shown in C of Figure 8-23. This signal is converted electronically to the square-wave form shown in D; sampling begins or terminates at

each successive zero crossing. The laser-fringe reference system gives a highly reproducible and regularly spaced sampling interval. In most instruments, the laser signal is also employed to control the speed of the mirror-drive system at a constant level.

The third interferometer system, sometimes called the "white-light" system, employs a tungsten source S_3 and transducer T_3 sensitive to visible radiation. Its mirror system is fixed to give a zero retardation that is displaced to the left from that for the analytical signal (see interferogram B, Figure 8-23). Because the source is polychromatic, its power at zero retardation is much larger than any signal before and after that point. Thus, this maximum can be employed to trigger the start of data sampling for each sweep at a highly reproducible point.

Beam Splitters. Beam splitters are constructed of transparent materials with refractive indices such that approximately 50%

of the radiation is reflected. A widely used material for the far-infrared region is a thin film of Mylar sandwiched between two plates of a low refractive-index solid. Thin films of germanium or silicon deposited on cesium iodide or bromide, sodium chloride, or potassium bromide are satisfactory for the mid-infrared region. A film of iron(III) oxide is deposited on calcium fluoride for work in the near-infrared.

Sources and Detectors. The sources for Fourier transform infrared instruments are similar to those discussed earlier in this chapter. Generally, pyroelectric detectors must be employed because their response times are shorter than those of the other infrared detectors.

Double-Beam Design. Instruments for the far-infrared region are often of single-beam design. Most instruments for the higher-frequency range are double-beam; after exiting from the interferometer, the beam is alternated between a sample compartment and a reference compartment by means of a moving mirror. The beams are then recombined and pass on to the detector.

Performance Characteristics of Commercial Instruments. Several instrument makers offer Fourier transform spectrometers which cover various frequency ranges and have var-

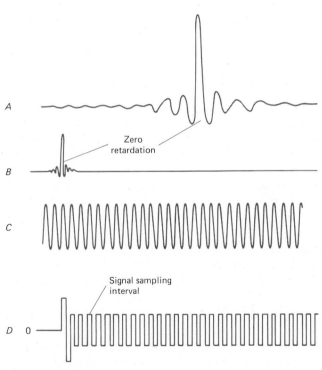

FIGURE 8-23 Time domain signals for the three interferometers contained in Fourier transform infrared instruments. Curve *A*: Infrared signal. Curve *B*: White light signal. Curve *C*: Laser-fringe reference signal. Curve *D*: Square-wave electrical signal formed from the laser signal. (From P. G. Griffiths, *Chemical Infrared Fourier Transform Spectroscopy.* New York: John Wiley & Sons, Inc., 1975, p. 102. With permission.)

ious resolutions. Some instruments are designed exclusively for the far-infrared region ($\sim$ 10 to 500 cm^{-1}); the best resolution of these instruments is about 0.1 cm^{-1}. Their cost ranges from \$40,000 to \$60,000, including the computer.

Several instruments are available for the mid-infrared region; most have interchangeable sources and beam splitters that permit coverage of several ranges of wavenumbers. When interchangeable beam splitters, detectors, and sources are used, one instrument is reported to produce spectra from the visible to the far-infrared (16,000 cm^{-1} to 10 cm^{-1} or 0.6 to 1000 μm).

Resolutions for commercial instruments vary from about 2 cm^{-1} to a low of 0.06 cm^{-1} for the more expensive instruments. Several minutes are required to obtain a complete spectrum with the highest resolution. On the other hand, a complete spectrum can be displayed on a television screen in less than one second if a resolution of several cm^{-1} is tolerable.

Applications of FT Infrared Spectroscopy[15]

As noted early in this discussion, most of the applications of Fourier transform spectroscopy have been in areas where low radiant energy is a serious problem. Here, the Fellgett advantage of the interferometric system offers significant improvements over the signal-to-noise ratios found with prism or grating instruments. Another advantage is that the optics of the interferometer provide a

much larger energy throughput (approximately 50%) than do dispersive optics, which are limited by the necessity of narrow slit widths. The higher energy reaching the detector with Fourier transform instruments often results in an increase in sensitivity. Finally, it should be noted that the interferometer is free from the problem of stray radiation because each IR frequency is, in effect, chopped at a different frequency.

The Fourier transform method offers little or no advantage over a good-quality grating spectrophotometer for routine qualitative applications in the region between 650 and 4000 cm^{-1} (15 and 2.5 μm). Furthermore, it suffers by comparison in terms of high initial cost and substantial maintenance problems. The latter arise because the quality of a Fourier transform spectrum degrades much more rapidly with instrument maladjustment than does a grating spectrum.

The paragraphs that follow outline briefly some of the important applications of Fourier transform spectroscopy.

Far-infrared Spectroscopy. The earliest chemical applications of the Fourier transform technique involved absorption studies in the range between 400 and 10 cm^{-1} (25 and 1000 μm). The energy advantage of the interferometric system over a dispersive one generally results in a significant improvement in spectral quality. In addition, the use of grating spectrometers in this wavelength region is complicated by the overlapping of several orders of diffracted radiation.

The far-infrared region is particularly useful for inorganic studies because absorption due to stretching and bending vibrations of bonds between metal atoms and both inorganic and organic ligands generally occur at frequencies lower than 600 cm^{-1} ($>$ 17 μm). For example, heavy-metal iodides generally absorb in the region below 100 cm^{-1}, while the bromides and chlorides have bands at higher frequencies. Absorption frequencies for metal-organic bonds are ordinarily dependent

---

[15] For an excellent summary of applications of FT infrared spectroscopy, see: J. R. Ferraro and L. J. Basile, *Fourier Transform Infrared Spectroscopy: Applications to Chemical Systems.* New York: Academic Press, vol. I, 1978 and vol. II, 1979.

upon both the metal atom and the organic portion of the species.

Far-infrared studies of inorganic solids have also provided useful information about lattice energies of crystals and transition energies of semiconducting materials.

Molecules composed only of light atoms absorb in the far-infrared if they have skeletal bending modes that involve more than two atoms other than hydrogen. Important examples are substituted benzene derivatives, which generally show several absorption peaks. The spectra are frequently quite specific and useful for identifying a particular compound; to be sure, characteristic group frequencies also exist in the far-infrared region.

Pure rotational absorption by gases is observed in the far-infrared region, provided the molecules have permanent dipole moments. Examples include H_2O, O_3, HCl, and AsH_3. Absorption by water is troublesome; elimination of its interference requires evacuation or at least purging of the spectrometer.

Mid-Infrared Spectroscopy. Application of Fourier transform spectroscopy in the 650 to 4000 cm^{-1} range has been largely confined to special problems where some sort of energy limitation exists. For example, it has proved useful for studying micro samples where absorption is confined to a very limited area; spectra for particles as small as 100 μm can be acquired.

The method has also been employed for studying transient species that would otherwise require rapid scanning. The gain here results from the fact that the entire spectrum is observed simultaneously.

Single-beam Fourier transform spectroscopy provides a useful method for the study of dilute solutions. Here, the interferograms for the solvent and sample are obtained individually and stored in the computer. The spectrum of the pure solute is then obtained by subtraction. For dilute solutions, the difference will often be so small as to be in-

distinguishable with a dispersive-type spectrometer. A particularly important application has been to studies of aqueous solutions.

Identification of Species After Gas Chromatographic Separations. Perhaps the most important application of the infrared Fourier technique in the mid-infrared region has been to the on-line identification of species eluted from a gas chromatographic column. An analysis of this sort must be completed in a few seconds; thus, rapid data acquisition, such as that provided by a scanning interferometer, is essential. Further discussion of this application will be found in the chapter devoted to gas chromatography.

Emission Spectroscopy. Upon being heated, molecules that absorb infrared radiation are also capable of emitting characteristic infrared wavelengths. The principal deterrent to the analytical application of this phenomenon has been the poor signal-to-noise ratio characteristic of the infrared emission signal, particularly when the sample is at a temperature only slightly higher than its surroundings. With the interferometric method, interesting and useful applications are now appearing.

An early example of the application of infrared emission spectroscopy is found in a paper[16] which describes the use of a Fourier transform spectrometer for the identification of microgram quantities of pesticides. Samples were prepared by solution in a suitable solvent and evaporation on a NaCl or KBr plate. The plate was then heated electrically near the spectrometer entrance. Pesticides such as DDT, malathion, and dieldrin were identified in amounts as low as 1 to 10 μg.

Equally interesting has been the use of the interferometric technique for the remote detection of components emitted from in-

---

[16] I. Coleman and M. J. D. Low, *Spectrochim. Acta*, **22**, 1293 (1966).

dustrial stacks. In one of these applications,[17] an interferometer was mounted on an 8-inch reflecting telescope. With the telescope focused on the plume from an industrial plant, CO_2 and SO_2 were readily detected at a distance of several hundred feet.

HADAMARD TRANSFORM SPECTROSCOPY

Another multiplexing technique that provides the Fellgett advantage is Hadamard transform spectroscopy.[18] This method employs a modified-grating infrared spectrometer in which the dispersed radiation is focused on a mask located on the focal plane of the monochromator. The mask is somewhat longer than twice the length of the focal plane and is covered with a continuous random arrangement of opaque and transparent slots, each of which has the dimensions of the entrance slit. The mask in its various positions has the effect of dividing the dispersed radiation into m resolution elements, each of which has a bandwidth determined by the width of the slots (or the entrance slit). The radiation that passes through the mask is reflected back to the grating, where it is recombined and then passes out of the entrance slit; the total power of the exiting beam is measured with a radiation detector.

The measurement process consists of stepping the mask across the focal plane one slot width at a time and measuring the total power at each step. The power data are digitalized and stored in a computer. After m steps, a sufficient number of linear simultaneous equations can be written to permit the calculation of the power P_m corresponding to each of the resolution elements. That is, a spectrum can be mathematically derived by the computer. The array of equations used takes the form

$$(P_T)_1 = a_{1,1}P_1 + a_{1,2}P_2 + \cdots + a_{1,m}P_m$$
$$(P_T)_2 = a_{2,1}P_1 + a_{2,2}P_2 + \cdots + a_{2,m}P_m$$
$$\vdots$$
$$(P_T)_n = a_{n,1}P_1 + a_{n,2}P_2 + \cdots + a_{n,m}P_m$$

The quantity $(P_T)_n$ corresponds to the power at step n, while $a_{n,m}$ is a coefficient that has a value of 0 for an opaque slot and 1 for a transparent slot. When $n = m$, sufficient data are available to solve the set of equations for P_1 through P_m.

As in the interferometric experiment, the spectrum is viewed continuously. Thus, the signal-to-noise ratio is in theory improved by the factor of $\sqrt{m}$. The Hadamard transform method does not, however, offer the advantage of greater energy throughout.

To date, the Hadamard method has not been widely used, despite the fact that the equipment is simpler and should be less expensive and easier to maintain.

---

[17] M. J. D. Low and F. K. Clancy, *Env. Sci. Technol.*, **1**, 73 (1967).

[18] For a discussion of this method, see: J. A. Decker, *App. Opt.*, **10**, 510 (1971) and *Amer. Lab.*, **4** (1), 29 (1972).

PROBLEMS

1. For $CHCl_3$, the C—H stretch occurs at 3030 cm^{-1} and the C—Cl stretch occurs at 758 cm^{-1}.
    (a) Calculate the position of the C—D stretch for $CDCl_3$.
    (b) Calculate the expected frequency of the C—Br stretch of bromoform, $CHBr_3$.

2. Most organic compounds contain C—H bonds, whose IR absorptions appear at approximately 3000 cm^{-1}. Calculate the force constants for the following types of C—H bonds:

(a) aromatic C—H, $\sigma = 3030$ cm^{-1}
(b) alkyne C—H, $\sigma = 3300$ cm^{-1}
(c) aldehydic C—H, $\sigma = 2750$ cm^{-1}
(d) alkane C—H, $\sigma = 2900$ cm^{-1}

3. Carbon disulfide is a linear molecule.
   (a) Draw schematic diagrams for the different modes of vibration of CS_2.
   (b) Indicate which of the normal vibrational modes are infrared active.
   (c) Assuming a standard double bond force constant (about 10×10^5 dynes/cm), calculate the expected absorption for the stretching vibrations. Compare your answer with the standard correlation charts. Explain the large difference.

4. Shown below are some of the normal modes of vibration for ethylene, $H_2C{=}CH_2$. Indicate whether the vibration will be active or inactive in the infrared spectrum.

   (a) C—H stretch

   (b) C—H stretch

   (c) CH$_2$ wag

   (d) CH$_2$ twist

   (e) C=C stretch

   (f) C—H bend

5. The force constant for the bond in HF is about 9×10^5 dynes/cm.
   (a) Calculate the vibrational absorption peak for HF.
   (b) Repeat the calculation for DF.

6. A compound isolated from the kernels of almonds was determined to have the molecular formula C_7H_6O. From the infrared spectrum given in Figure 8-24, determine the structure of this unknown.

7. Acetone reacts with hydrocyanic acid to give a moderately stable adduct whose infrared spectrum is shown in Figure 8-25. Interpret the major peaks in this spectrum, and deduce the structure of the adduct.

8. From the infrared spectrum given in Figure 8-26, determine the structure of an unknown having the formula C_7H_9N. By examining the fingerprint region, determine which isomer corresponds to this spectrum.

9. For the infrared spectrum shown in Figure 8-27, assign probable functional groups to all the peaks. What class of molecules would give this spectrum?

10. From the infrared spectrum given in Figure 8-28, determine what functional groups are present in this molecule. What gives rise to the peak at 1250 cm^{-1}? This compound is often used as a solvent for glue and has a strong odor of bananas. Given the molecular formula, $C_7H_{14}O_2$, suggest the structure of this unknown.

11. Deduce the structure of the compound C_3H_4O whose infrared spectrum is given in Figure 8-29. Explain the double peak at 3300 cm^{-1}.

12. Deduce the functional groups present in the unknown whose IR spectrum is shown in Figure 8-30.

13. The compound, whose spectrum is shown in Figure 8-31, was used as an antiseptic for early surgical work and is still found in some popular face creams. From the infrared spectrum determine its structure. The molecular weight of this unknown has been determined to be 94 ± 1.

14. The compound, whose infrared spectrum is shown in Figure 8-32, can be found in small quantities in black pepper. It has a characteristic fishy odor. What functional groups are indicated by the infrared spectrum? Given its molecular weight of 85, suggest possible structures for this compound.

15. From the infrared spectrum given in Figure 8-33, determine what functional groups are present in this molecule. Explain the peaks at 1700 cm^{-1}. The structural features of this unknown are all represented in its infrared spectrum. Determine the structure of this compound.

16. An empty cell showed 14 interference peaks in the wavelength range of 6.0 to 12.0 μm. Calculate the path length of the cell.

17. An empty cell exhibited 8.5 interference peaks in the region of 1000 to 1250 cm^{-1}. What was the path length of the cell?

18. What length of mirror drive in a Fourier transform spectrometer would be required to provide a resolution of (a) 0.05 cm^{-1} and (b) 0.01 cm^{-1}?

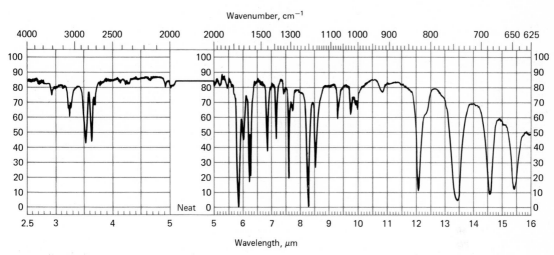

FIGURE 8-24 See Problem 6. (From C. J. Pouchert, *The Aldrich Library of Infrared Spectra*, 2d ed., 1975, and published with permission of the Aldrich Chemical Company, Inc., Milwaukee, Wisconsin.)

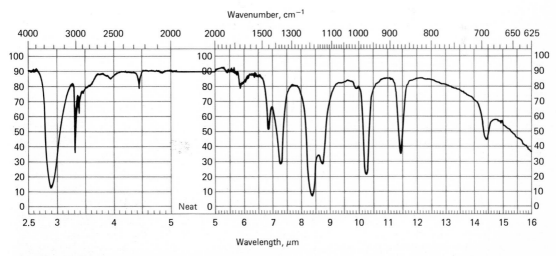

FIGURE 8-25 See Problem 7. (From C. J. Pouchert, *The Aldrich Library of Infrared Spectra*, 2d ed., 1975 and published with permission of the Aldrich Chemical Company, Inc., Milwaukee, Wisconsin.)

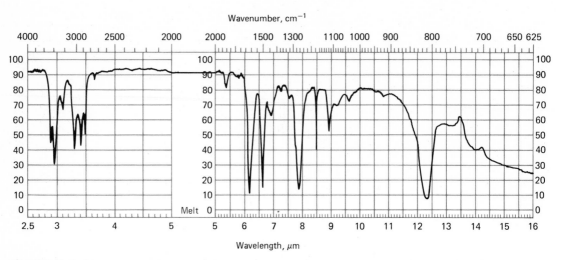

FIGURE 8-26   See Problem 8. (From C. J. Pouchert, *The Aldrich Library of Infrared Spectra*, 2d ed., 1975, and published with permission of the Aldrich Chemical Company, Inc., Milwaukee, Wisconsin.)

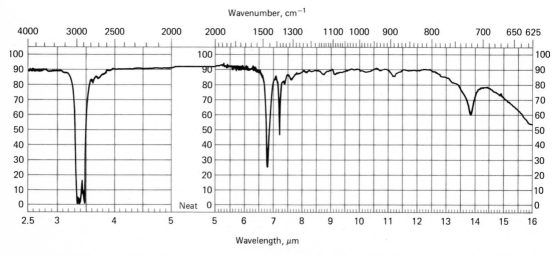

FIGURE 8-27   See Problem 9. (From C. J. Pouchert, *The Aldrich Library of Infrared Spectra*, 2d ed., 1975, and published with permission of the Aldrich Chemical Company, Inc., Milwaukee, Wisconsin.)

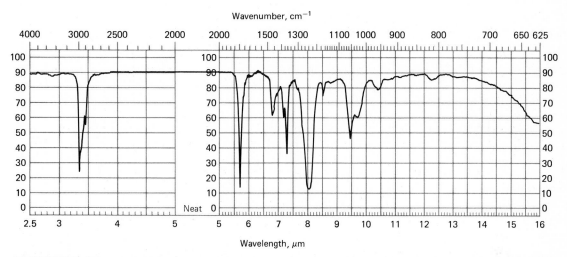

FIGURE 8-28   See Problem 10. (From C. J. Pouchert, *The Aldrich Library of Infrared Spectra*, 2d ed., 1975, and published with permission of the Aldrich Chemical Company, Inc., Milwaukee, Wisconsin.)

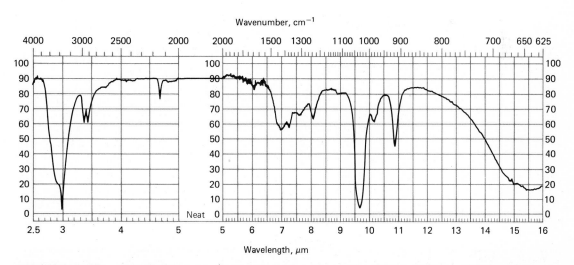

FIGURE 8-29   See Problem 11. (From C. J. Pouchert, *The Aldrich Library of Infrared Spectra*, 2d ed., 1975, and published with permission of the Aldrich Chemical Company, Inc., Milwaukee, Wisconsin.)

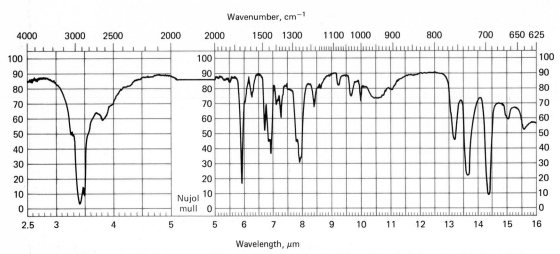

FIGURE 8-30   See Problem 12. (From C. J. Pouchert, *The Aldrich Library of Infrared Spectra*, 2d ed., 1975, and published with permission of the Aldrich Chemical Company, Inc., Milwaukee, Wisconsin.)

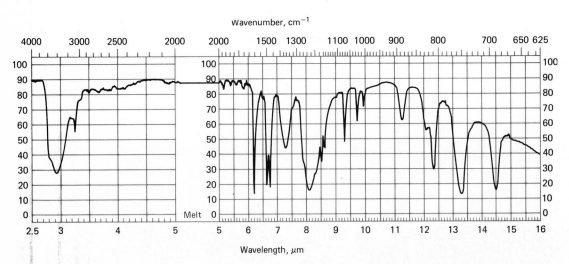

FIGURE 8-31   See Problem 13. (From C. J. Pouchert, *The Aldrich Library of Infrared Spectra*, 2d ed., 1975, and published with permission of the Aldrich Chemical Company, Inc., Milwaukee, Wisconsin.)

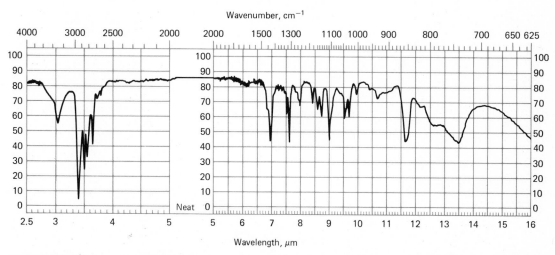

FIGURE 8-32   See Problem 14. (From C. J. Pouchert, *The Aldrich Library of Infrared Spectra*, 2d ed., 1975, and published with permission of the Aldrich Chemical Company, Inc., Milwaukee, Wisconsin.)

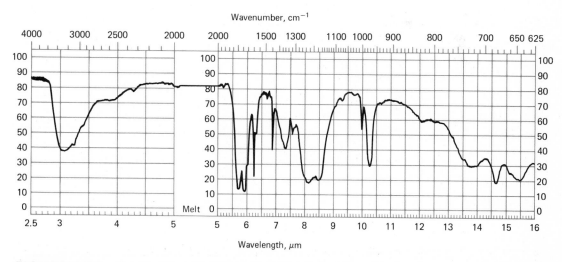

FIGURE 8-33   See Problem 15. (From C. J. Pouchert, *The Aldrich Library of Infrared Spectra*, 2d ed., 1975, and published with permission of the Aldrich Chemical Company, Inc., Milwaukee, Wisconsin.)

9

RAMAN SPECTROSCOPY

When radiation passes through a transparent medium, a fraction of the radiant power is scattered at all angles from the direction of the beam by the molecules or aggregates of molecules present (p. 104). The fraction of the beam that is scattered is undetectable to the eye when the particles are of molecular dimensions; this type of scattering is termed *Rayleigh scattering*. If the medium contains aggregates of particles with dimensions approximating that of the wavelength of the radiation, scattering becomes evident as the Tyndall effect or as a turbidity of the solution.

In 1928, the Indian physicist C. V. Raman discovered that the wavelength of a small fraction of the radiation scattered by certain molecules differs from that of the incident beam and furthermore that the shifts in wavelength depend upon the chemical structure of the molecules responsible for the scattering. He was awarded the 1931 Nobel prize in physics for this discovery and his systematic exploration of it.

The theory of Raman scattering, which by now is well understood, shows that the phenomenon results from the same type of quantized vibrational changes that are associated with infrared absorption. Thus, the *difference* in wavelength between the incident and scattered radiation corresponds to wavelengths in the mid-infrared region. Indeed, the Raman scattering spectrum and infrared absorption spectrum for a given species often resemble one another quite closely. There are, however, enough differences between the kinds of groups that are infrared and Raman active to make the techniques complementary rather than competitive. For some problems, the infrared method is the superior tool; for others, the Raman procedure offers advantages. Because of this close relationship, we have chosen to discuss Raman spectroscopy at this point.[1]

An important advantage of Raman spectra over infrared lies in the fact that water does not cause interference; indeed, Raman spectra can be obtained from aqueous solutions. In addition, glass or quartz cells can be employed, thus avoiding the inconvenience of working with sodium chloride or other atmospherically unstable windows.

THEORY OF RAMAN SPECTROSCOPY

Raman spectra are obtained by irradiating a sample with a powerful source of visible monochromatic radiation. A mercury arc was employed in early investigations; by now, this source has been superseded by high-intensity gas or solid lasers. During irradiation, the spectrum of the scattered radiation is measured at some angle (usually 90 deg) with a suitable visible-region spectrophotometer. At the very most, the intensities of Raman lines are 0.01 % of the source; as a consequence, their detection and measurement creates some experimental problems.

Excitation of Raman Spectra

Figure 9-1b shows a photograph of the Raman spectrum for carbon tetrachloride excited by an intense mercury arc. For reference, the spectrum of the source is shown in part (a) of the figure. Note that the two mercury lines appear unchanged in wavelength on the lower spectrum as a consequence of normal Rayleigh scattering. In addition, however, two sets of five lines appear on the

---

[1] For more complete discussions of the theory and prac-

tice of Raman spectroscopy, see: T. R. Gilson and P. J. Hendra, *Laser Raman Spectroscopy.* New York: Wiley-Interscience, 1970; *Raman Spectroscopy, Theory and Practice*, 2 vols., ed. H. A. Szymanski. New York: Plenum Press, 1967 and 1970; J. R. Durig and W. C. Harris, *Physical Methods of Chemistry.* New York: Wiley-Interscience, 1972, Vol. 1, Part III B, Chapter 2; and *Infrared and Raman Spectroscopy*, eds. E. G. Brame and J. G. Grasselli. New York: Dekker, 1977.

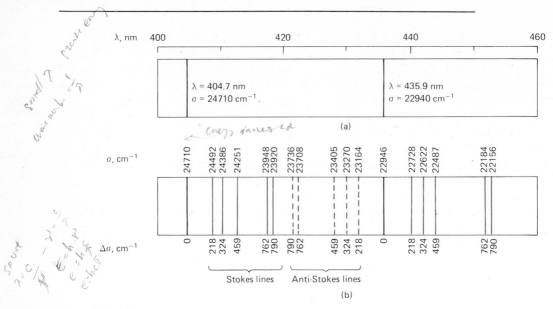

FIGURE 9-1 (a) Intense lines of a mercury arc spectrum and (b) Raman spectrum for CCl_4.

longer wavelength side of the two intense mercury lines at 404.7 and 435.8 nm. If it is assumed that each set of Raman lines is excited by the nearer of the two intense mercury lines, it is found that the pattern of displacement in wavenumbers (cm^{-1}) is identical for the two sets. Thus, the 404.7-nm line with a wavenumber of 24,710 cm^{-1} exhibits Raman lines at 24,492, 24,386, and 24,251. As shown by the lower scale of Figure 9-1b, these lines correspond to Raman shifts ($\Delta\sigma$) of 218, 324, and 459 cm^{-1}. Note that lines having identical shifts appear to the right of the 435.8 nm (22,946 cm^{-1}) line. Additional lines appear at shifts of 762 and 790 cm^{-1}.

Another set of less intense lines appear to the left of both of the mercury lines (the set associated with the 404.7-nm line is not shown for lack of space). These lines are displaced by amounts identical to the displacements to the right.

Superficially, the appearance of Raman spectral lines at lower energies (longer wavelengths) is analogous to what is observed in a fluorescence experiment; thus, Raman shifts in this direction are sometimes called *Stokes shifts* (Chapter 10). We shall see, however, that the two phenomena arise from fundamentally different processes; for this reason, the application of the same nomenclature to both fluorescent and Raman spectra is unfortunate.

Shifts toward higher energies are termed *anti-Stokes*; quite generally, anti-Stokes lines have appreciably weaker intensities than the corresponding Stokes lines.

Raman spectra are characteristic of the molecules responsible for the scattering and are thus useful for analytical purposes. Most commonly, the spectra are plotted in terms of the wavenumber *shift* $\Delta\sigma$ with respect to the source line. Since Stokes peaks are the more intense, $\Delta\sigma$ (the shift in units of cm^{-1}) is usually evaluated on the basis of the shift to lower wavenumbers; that is,

$$\Delta\sigma = \sigma_s - \sigma \qquad (9\text{-}1)$$

where σ_s is the wavenumber (in cm^{-1}) of the

particular source line and σ is the wave-number for the Raman peak. Thus, the spectrum for carbon tetrachloride in Figure 9-1 reveals Raman peaks at 218, 324, 459, 762, and 790 cm^{-1}. It should be noted that anti-Stokes shifts may be more convenient to measure in samples that also fluoresce.

Transitions Responsible for Raman Peaks. The discrete energy shifts that characterize Raman spectra indicate the presence of quantized energy transitions. These shifts can be rationalized by assuming that the electrical field of the radiation interacts with the electrons of the sample and causes periodic polarization and depolarization. As a consequence, the energy of the radiation is momentarily retained in a so-called *virtual* state as a distorted polarized species. The energetics of this process is shown by the two vertical arrows on the left in Figure 9-2; note that the interactions, in contrast to absorption, *do not* involve transition to a higher electronic energy level. Note also that the second arrow represents the interaction between photons and a molecule that has been thermally excited to the

first vibrational level. After a retention time of perhaps 10^{-15} to 10^{-14} s, the species ordinarily return to their ground state, as shown by the dashed arrows, with the release, in all directions, of radiation with exactly the same energy as the source. The small fraction of this radiation that is transmitted at an angle to the beam is Rayleigh scattering. Under some circumstances, however, the molecule may return from the distorted or virtual state to the *first excited vibrational level of the ground state* (broad arrow in Figure 9-2). The frequency emitted in this transition will be less by a quantized amount corresponding to the difference in energies between the ground state and the first vibrational level, ΔE. That is,

$$h\nu = h\nu_s - \Delta E$$

or

$$\Delta \nu = \Delta E/h$$

where ν and ν_s are the frequencies of the Raman peak and the source peak, respectively, and h is Planck's constant.

As shown to the right in Figure 9-2, anti-Stokes scattering results from interaction of the radiation with a molecule that is initially in its first excited vibrational level. The resulting emission is *increased* in frequency by the amount ΔE if the distorted species returns to the ground vibrational state. Since the fraction of molecules in the first excited state is smaller than in the ground state at room temperature, anti-Stokes radiation is weaker than Stokes.

Vibrational Modes Associated with the Raman Effect

The transitions responsible for the Raman effect are of the same type as those described in Chapter 8 for infrared absorption; that is, transitions between the ground state of a molecule and its first vibrational states are involved in both phenomena. Therefore, for a given molecule the *energy shifts* observed in a

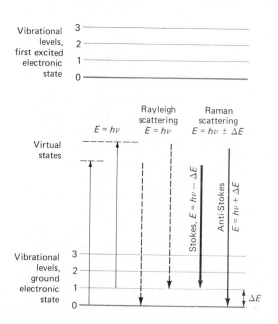

FIGURE 9-2   Origin of Raman spectra.

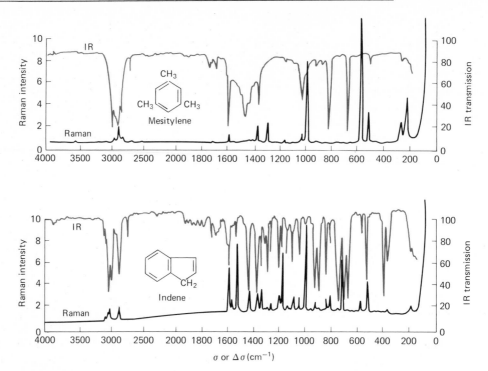

FIGURE 9-3 Comparison of Raman and infrared spectra. (Courtesy Perkin-Elmer Corp., Norwalk, Connecticut.)

Raman experiment should be identical to the *energies* of its infrared absorption bands, provided that the vibrational modes involved are active toward both infrared absorption and Raman scattering. Figure 9-3 illustrates the similarity of the two types of spectra; it is seen that several peaks with identical σ and $\Delta\sigma$ values exist for the two compounds. It is also noteworthy, however, that the relative size of the corresponding peaks is frequently quite different; moreover, certain peaks that occur in one spectrum are absent in the other.

The differences between a Raman and an infrared spectrum are not surprising when it is considered that the basic mechanisms, although dependent upon the same vibrational modes, arise from processes that are mechanistically different. Infrared absorption requires that a vibrational mode of the

molecule have a change in dipole or charge distribution associated with it. Only then can radiation of the same frequency interact with the molecule and promote it to an excited vibrational state. In contrast, scattering involves a momentary elastic distortion of the electrons distributed around a bond in a molecule, followed by reemission of the radiation in all directions as the bond returns to its normal state. In its distorted form, the molecule is temporarily polarized; that is, it develops, momentarily, an induced dipole which disappears upon relaxation and reemission. The effectiveness of a bond toward scattering thus depends directly upon the ease with which the electrons of the bond can be distorted from their normal positions (that is, the *polarizability* of the bonds); polarizability decreases with increasing electron density, in-

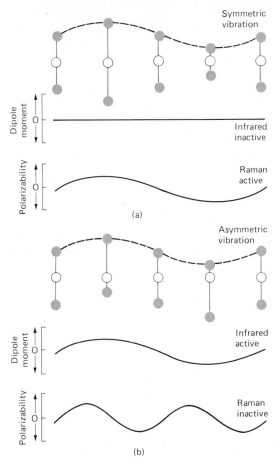

Symmetric vibration

Infrared inactive

Raman active

(a)

Asymmetric vibration

Infrared active

Raman inactive

(b)

FIGURE 9-4 Raman and infrared activity of two vibrational modes of carbon dioxide: (a) symmetric, (b) asymmetric.

creasing bond strength, and decreasing bond length. The Raman shift in scattered radiation, then, requires that there be a *change in polarizability*—rather than a change in dipole—associated with the vibrational mode of the molecule; as a consequence, the Raman activity of a given mode may differ markedly from its infrared activity. For example, a homonuclear molecule such as nitrogen, chlorine, or hydrogen has no dipole moment either in its equilibrium position or when a stretching vibration causes a change in the distance

between the two nuclei. Thus, absorption of radiation of the vibration frequency cannot occur. On the other hand, the polarizability of the bond between the two atoms of such a molecule varies periodically in phase with the stretching vibrations, reaching a maximum at the greatest separation and a minimum at the closest approach. A Raman shift corresponding in frequency to that of the vibrational mode thus results.

It is of interest to compare the infrared and the Raman activities of coupled vibrational modes such as those described earlier (p. 218) for carbon dioxide. In the symmetric mode, shown in Figure 9-4a, it is seen that no change in dipole occurs as the two oxygen atoms move away from or toward the central carbon atom; thus, this mode is infrared-inactive. The polarizability, however, fluctuates in phase with the vibration since distortion of bonds becomes easier as they lengthen and more difficult as they shorten; Raman activity is associated with this mode.

As shown in Figure 9-4b, the dipole moment of carbon dioxide fluctuates in phase with the asymmetric vibrational mode; thus, an infrared absorption peak arises from this mode. The polarizability also changes during the vibration, reaching a maximum at the equilibrium position and a minimum at either vibrational extreme. Note, however, that the fluctuations in polarizability are not in phase with the vibrational changes but occur at twice the frequency. As a consequence, Raman scattering is not associated with these asymmetric vibrations.

As shown earlier (p. 219), a dipole change, and thus infrared absorption, is also associated with the two degenerate scissoring vibrational modes of carbon dioxide. Here again, the change in polarizability does not occur at the vibrational frequency but at twice the frequency; these modes do not, therefore, produce Raman scattering.

Often, as in the foregoing examples, parts of Raman and infrared spectra are complementary, each being associated with a different set of vibrational modes within a molecule. Other vibrational modes may be both Raman- and infrared-active. Here, the two spectra resemble one another with peaks involving the same energies. The relative intensities of corresponding peaks may differ, however, because the probability for the transition may be different for the two mechanisms.

Intensity of Raman Peaks

The intensity or power of a Raman peak depends in a complex way upon the polarizability of the molecule, the intensity of the source, and the concentration of the active group, as well as other factors. In the absence of absorption, the power of Raman emission increases with the fourth power of the frequency of the source; however, advantage can seldom be taken of this relationship because of the likelihood that ultraviolet irradiation will cause photodecomposition.

Raman intensities are usually directly proportional to the concentration of the active species. In this regard, Raman spectroscopy more closely resembles fluorescence than absorption, where the concentration-intensity relationship is logarithmic.

INSTRUMENTATION

Instrumentation for modern Raman spectroscopy consists of three components, namely, an intense source, a sample illumination system, and a suitable spectrophotometer.[2]

Sources

The most widely used Raman source is probably the helium/neon laser, which operates continuously at a power of 50 mW. Laser radiation is produced at 632.8 nm; several other lower-intensity nonlasing lines accompany the principal line and must be removed by suitable narrow band filters. Alternatively, the effect of these lines can be eliminated by taking advantage of the fact that nonlasing lines diverge much more rapidly than the lasing line; thus, by making the distance between the source and the entrance slit large, the intensities of the former can be made to approach zero.

Argon-ion lasers, with lines at 488.0 and 514.5 nm, are also employed, particularly when higher sensitivity is required. Because the intensity of Raman scattering varies as the fourth power of the frequency of the exciting source, the argon line at 488 nm provides Raman lines that are nearly three times as intense as those excited by the helium/neon source, given the same input power.

A variety of other laser sources are available; undoubtedly new and improved sources will appear in the future. The need exists for several sources, inasmuch as one must be chosen which is not absorbed by the sample or solvent. Furthermore, care must be taken to avoid wavelengths which will cause the sample to fluoresce or to photodecompose.

Sample-Illumination Systems

Sample handling for Raman studies tends to be simpler than for infrared because the measured wavenumber differences are between two *visible* frequencies. Thus, glass can be employed for windows, lenses, and other optical components. In addition, the source is readily focused on a small area and the emitted radiation can be efficiently focused on a slit. Very small samples can be examined as a consequence. In fact, a common sample holder for liquid samples is an ordinary glass melting-point capillary.

Figure 9-5 shows two of many configurations for the handling of liquids. The size of the tube in (b) has been enlarged to show

---

[2] For a description of commercially available instruments up to 1969, see: B. J. Bulkin, *J. Chem. Educ.*, **46**, A781, A859 (1969).

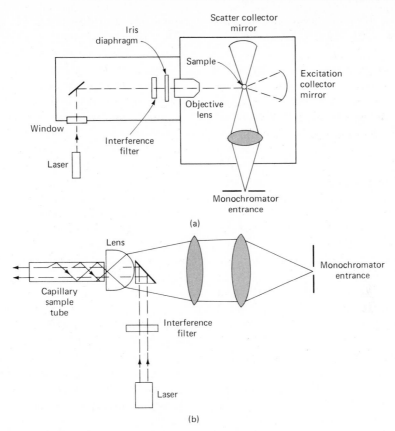

FIGURE 9-5   Two sample excitation systems.

details of the reflection of the Raman radiation off the walls; in fact, the holder is a 1-mm o.d. glass capillary that is about 5 cm long.

For unusually weak signals such as might occur with a dilute gas sample, the cell is often placed between the mirrors of the laser source; enhanced excitation power results.

Raman Spectrophotometers

All early Raman studies were performed with prism spectrographs, with the spectra being recorded on photographic films or plates. Intensities were determined by measuring the blackness of the lines.

Several Raman spectrophotometers that incorporate photomultipliers to record peak intensities are now available. Their designs are not significantly different from the recording spectrophotometers discussed in Chapter 7. Most employ double monochromators to minimize spurious radiation reaching the detector. In addition, a split-beam design is employed to compensate for the effects of fluctuations in the intensity of the source.

Comparison of Raman and Infrared Sample-Handling Techniques

In terms of ease of sample preparation and handling, Raman spectroscopy possesses the great advantage of permitting use of glass cells instead of the more fragile and atmospherically less stable crystalline halides required for work in the infrared.

Substances of limited solubility can often be finely ground and tapped into an open-ended cavity for Raman examination; the more troublesome mull technique is required for infrared studies. Polymers can often be examined for Raman activity directly with no preliminary sample treatment; comparable infrared analyses may require that the polymer be compressed, molded, or cast into a thin film prior to examination.

Another important difference is that water is a weak scatterer of Raman radiation but a strong absorber of infrared. Thus, aqueous solutions of samples can be employed for Raman studies. This advantage is particularly pronounced for the study of biological systems, inorganic substances, and water pollution problems.

Liquid samples containing colloidal or suspended particles ordinarily scatter sufficient amounts of the laser beam to make observation of the Raman effect difficult or impossible. Such samples must be treated to remove solids before a Raman spectrum can be obtained.

Comparison of Raman and Infrared Instrumentation

H. J. Sloane has compared Raman and infrared spectroscopy on the basis of instrumentation, sample handling, and applicability.[3]

Optics. Raman spectrophotometers are somewhat less complicated from the standpoint that glass or quartz optics can be used throughout. Furthermore, Raman spectra in both the mid- and the far-infrared regions (4000 to 25 cm^{-1}) can be examined with a single optical system; in contrast, infrared studies require several gratings to cover the same range. The grating for a Raman instrument, however, must be of higher quality because spectral artifacts arising from imperfections have more serious consequences.

Detectors. Raman instruments use photomultiplier detectors which offer distinct signal-to-noise advantages over the thermal detectors employed in infrared instruments (see p. 143).

Resolution. The resolution of the best infrared and Raman spectrometers is about the same (~ 0.2 cm^{-1}).

Cost. There are no low-cost Raman spectrophotometers equivalent to bench-top infrared spectrometers. The prices of Raman instruments have decreased significantly recently; it is to be hoped that instruments for routine applications will appear shortly at a cost of $5000 to $10,000.

APPLICATIONS OF RAMAN SPECTROSCOPY

Raman spectroscopy has been applied to the qualitative and quantitative analysis of organic, inorganic, and biological systems.

Raman Spectra of Inorganic Species [4]

The Raman technique is often superior to the infrared for investigating inorganic systems because aqueous solutions can be employed. In addition, the vibrational energies of metal-ligand bonds are generally in the range of 100 to 700 cm^{-1}, a region of the infrared that is experimentally difficult to study. These vibrations are frequently Raman-active, however, and peaks with $\Delta\sigma$ values in this range are readily observed. Raman studies are potentially useful sources of information concerning the composition, structure, and stability

[3] H. J. Sloane, *Appl. Spec.*, **25**, 430 (1971).

[4] For a review of inorganic applications, see: K. Nakamoto, *Infrared and Raman Spectra of Inorganic and Coordination Compounds*, 3d ed. New York: Wiley, 1978; D. J. Gardiner, *Anal. Chem.*, **50**, 131R (1978); and W. E. L. Grossman, *Anal. Chem.*, **48**, 261R (1976); **46**, 345R (1974).

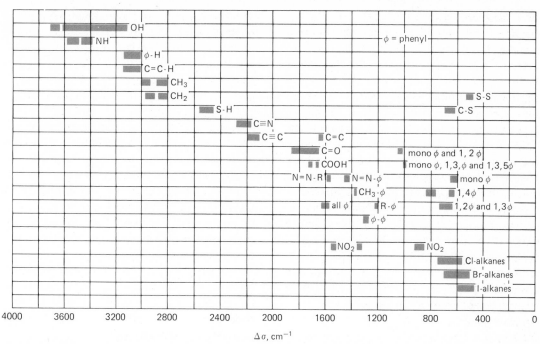

FIGURE 9-6 Correlation chart showing group frequencies for various organic functional groups. (Cary Instruments, Monrovia, California, with permission.)

of coordination compounds. For example, numerous halogen and halogenoid complexes produce Raman spectra and thus are susceptible to investigation by this means. Metal-oxygen bonds are also active. Spectra for such species as VO_4^{3-}, $Al(OH)_4^-$, $Si(OH)_6^{2-}$, and $Sn(OH)_6^{2-}$ have been obtained; Raman studies have permitted conclusions regarding the probable nature of such species. For example, in perchloric acid solutions, vanadium(IV) appears to be present as $VO^{2+}(aq)$ rather than as $V(OH)_2^{2+}(aq)$; studies of boric acid solutions show that the anion formed by acid dissociation is the tetrahedral $B(OH)_4^-$ rather than $H_2BO_3^-$. Dissociation constants for strong acids such as H_2SO_4, HNO_3, H_2SeO_4 and H_5IO_6 have been obtained by Raman measurement.

It seems probable that the future will see even wider use of Raman spectroscopy for theoretical studies of inorganic systems.

Raman Spectra of Organic Species

Raman spectra are similar to infrared spectra in that they have regions that are useful for functional group detection and fingerprint regions that permit the identification of specific compounds. Figure 9-6 shows a correlation chart that can be employed for functional group recognition. Dollish[5] has published a comprehensive treatment of functional group frequencies. Catalogs of Raman spectra for organic compounds are also available.[6]

Raman spectra yield more information about certain types of organic compounds

[5] F. R. Dollish, W. G. Fateley, and F. F. Bentley, *Characteristic Raman Frequencies of Organic Compounds*. New York: Wiley-Interscience, 1971.
[6] Samuel P. Sadtler and Sons, Inc., 2100 Arch Street, Philadelphia, PA.

than do their infrared counterparts. For example, the double-bond stretching vibration for olefins results in weak and sometimes undetected infrared absorption. On the other hand, the Raman band, (which, like the infrared band, occurs at about 1600 cm^{-1}) is intense, and its position is sensitive to the nature of substituents as well as to their geometry. Thus, Raman studies are likely to yield useful information about the olefinic functional group that may not be revealed by infrared spectra. This statement applies to cycloparaffin derivatives as well; these compounds have a characteristic Raman peak in the region of 700 to 1200 cm^{-1}. This peak has been attributed to a "breathing" vibration in which the nuclei move in and out symmetrically with respect to the center of the ring. The position of the peak decreases continuously from 1190 cm^{-1} for cyclopropane to 700 cm^{-1} for cyclooctane; Raman spectroscopy thus appears to be an excellent diagnostic tool for the estimation of ring size in paraffins. The infrared peak associated with this vibration is weak or nonexistent.

Biological Applications of Raman Spectroscopy

Raman spectroscopy has been applied advantageously for the study of biological systems.[7] The advantages of this technique include the small sample requirement, the minimal sensitivity toward interference by water, the spectral detail, and the conformational and environmental sensitivity.

Quantitative Applications

Raman spectra tend to be less cluttered with peaks than infrared spectra. As a consequence, peak overlap in mixtures is less likely and quantitative measurements are simpler. In addition, Raman instrumentation is not subject

to attack by moisture, and small amounts of water in a sample do not interfere. Despite these advantages, Raman spectroscopy has not yet been exploited widely for quantitative analysis.

An example of the potentialities of the method for the analysis of mixtures is provided in a paper by Nicholson,[8] in which a procedure for the determination of the constituents in an eight-component mixture is described. The components included benzene, isopropyl benzene, three diisopropyl benzenes, two triisopropyl derivatives, and 1,2,4,5-tetraisopropyl benzene. The power of the characteristic peaks, compared with the power of a reference peak (CCl_4), was assumed to vary linearly with volume percent of each component. Analysis of synthetic mixtures of all the components by the procedure produced results that agreed with the preparatory data to about 1% (absolute).

Application of Other Types of Raman Spectroscopy

With the development of tunable lasers, two new Raman spectroscopic methods were developed in the early 1970s. Neither has as yet become widely employed. Thus, only brief mention will be made of the techniques.

Resonance Raman Spectroscopy.[9] Resonance Raman spectroscopy is based upon the observation that a laser source with a frequency near or coincident with an electronic absorption peak of a substance causes a marked enhancement of the Raman scattering associated with the chromophore. Significant increases in sensitivity and selectivity result. Because local heating is likely to bring about sample decomposition when this technique is employed, it is often necessary to rotate or flow the sample

---

[7] For a review of biological applications, see: B. P. Gaber, *Amer. Lab.*, **9** (3), 15 (1977).

[8] D. E. Nicholson, *Anal. Chem.*, **32**, 1634 (1960).

[9] For a brief review of this topic, see: D. P. Strommen and K. Nakamoto, *J. Chem. Educ.*, **54**, 474 (1977).

past the laser beam or to use a pulsed beam.

Coherent Anti-Stokes Raman Spectroscopy (CARS).[10] This technique has been employed to overcome some of the drawbacks of conventional Raman spectroscopy, namely, its low efficiency, its limitation to the visible and near-ultraviolet regions, and its susceptibility to interference from fluorescence.

One way of carrying out a CARS experiment is to employ a laser to pump a tunable dye laser. The two lasers are arranged in such a way that part of the beam from the pumping laser of frequency v_p is employed for pumping and part for sample excitation. The dye laser is also employed for sample excitation, and its frequency v_d is varied until the difference between the two excitation frequencies is equal to the resonance frequency v_r for one of the Raman lines. That is,

$$v_r = v_p - v_d$$

Under these circumstances, a beam of radiation is generated having a frequency v_a that is given by

$$v_a = 2v_p - v_d$$

This radiation differs from normal Raman radiation in that it is *coherent* rather than scattered in all directions. Furthermore, it is emitted at an angle such that it can be readily separated from the excitation beam without a monochromator. Finally, because it occurs in the anti-Stokes region (a similar beam occurs at $v_a = 2v_p + v_d$ in the Stokes region), interference from Stokes fluorescence by the sample is avoided.

The efficiency of coherent anti-Stokes emission is high, with as much as 1 % of the excitation radiation being converted to the new frequency. Furthermore, because the beam is coherent, high detection efficiencies are realized. As a consequence of these two properties, sensitivity is enhanced.

[10] For brief reviews, see: R. F. Begley, A. B. Harvey, R. L. Beyer, and B. S. Hudson, *Amer. Lab.*, **6** (11), 11 (1974); and A. B. Harvey, *Anal. Chem.*, **50** (9), 905A (1978).

PROBLEMS

1. For two energy levels separated by the energy ΔE, the population of the higher state compared with that of the lower state at thermal equilibrium is given by the Boltzmann distribution:

$$\frac{\text{population of higher state}}{\text{population of lower state}} = e^{-\Delta E/kT}$$

where k is Boltzmann's constant and T is the temperature in °K. The intensity of a Raman line is proportional to the population of molecules from which the transition originates. If all other factors are equal, calculate the ratio of anti-Stokes intensity to that of Stokes intensity for the CCl_4 lines at 218 and 459 cm^{-1} at 300°K.

2. Two reasonable structures for the $Cu(CN)_3^{2-}$ ion are planar or bent.

Each of these species is expected to have three types of $C\equiv N$ stretching modes, two of which have the same energy. Vibrations with the same energy are said to be *degenerate* and can give only a single IR or Raman band. The three $C\equiv N$ stretching modes are therefore expected to give, at most, two bands in the $C\equiv N$ stretching region (near 2100 cm^{-1}) of the IR or Raman spectrum. The symmetric and asymmetric stretching modes of the planar species are as follows:

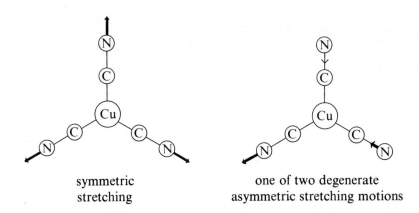

<table>
<tr><td>symmetric
stretching</td><td>one of two degenerate
asymmetric stretching motions</td></tr>
</table>

The asymmetric stretching mode is allowed by both the IR and Raman selection rules for both the planar and bent structures. Using reasoning analogous to that for Figure 9-4, decide whether the symmetric stretching is allowed in either the IR or Raman spectrum, or in both. Use this reasoning for both the planar and bent structures.

The IR spectrum of $Cu(CN)_3^{2-}$ in aqueous solution exhibits one band in the $C\equiv N$ stretching region at 2094 cm^{-1}. The Raman spectrum exhibits bands at 2094 and 2108 cm^{-1}. Is $Cu(CN)_3^{2-}$ planar or bent?

3. Suggest structures for the compounds whose *IR* and Raman spectra are shown in Figures 9-7 through 9-14.

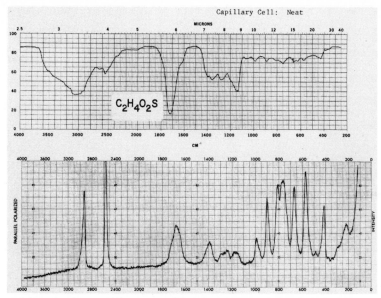

FIGURE 9-7 IR and Raman spectra of a compound with the formula $C_2H_4O_2S$. (Reproduced with permission of the Sadtler Research Laboratories, Philadelphia, Pennsylvania.)

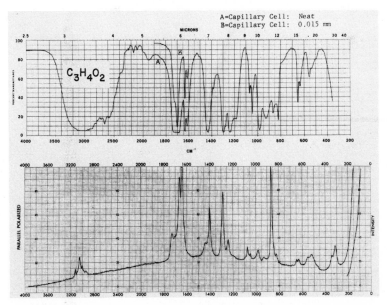

FIGURE 9-8 IR and Raman spectra of a compound with the formula $C_3H_4O_2$. (Reproduced with permission of the Sadtler Research Laboratories, Philadelphia, Pennsylvania.)

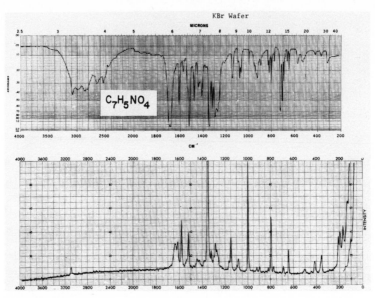

FIGURE 9-9 IR and Raman spectra of a compound with the formula $C_7H_5NO_4$. (Reproduced with permission of the Sadtler Research Laboratories, Philadelphia, Pennsylvania.)

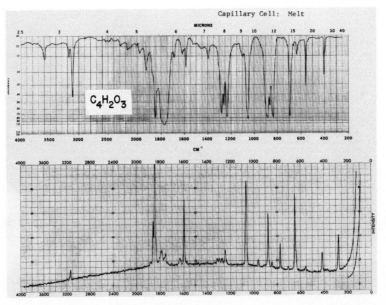

FIGURE 9-10 IR and Raman spectra of a compound with the formula $C_4H_2O_3$. (Reproduced with permission of the Sadtler Research Laboratories, Philadelphia, Pennsylvania.)

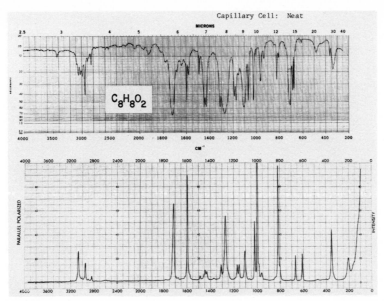

FIGURE 9-11 IR and Raman spectra of a compound with the formula $C_8H_8O_2$. (Reproduced with permission of the Sadtler Research Laboratories, Philadelphia, Pennsylvania.)

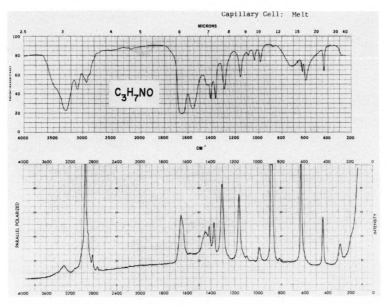

FIGURE 9-12 IR and Raman spectra of a compound with the formula C_3H_7NO. (Reproduced with permission of the Sadtler Research Laboratories, Philadelphia, Pennsylvania.)

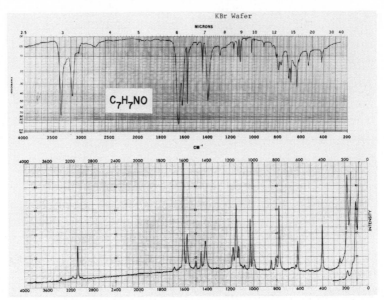

FIGURE 9-13 IR and Raman spectra of a compound with the formula C_7H_7NO. (Reproduced with permission of the Sadtler Research Laboratories, Philadelphia, Pennsylvania.)

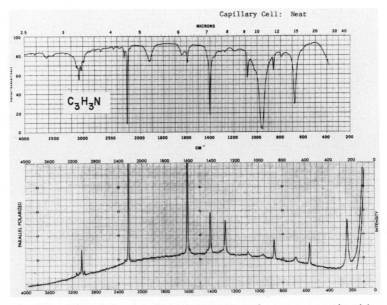

FIGURE 9-14 IR and Raman spectra of a compound with the formula C_3H_3N. (Reproduced with permission of the Sadtler Research Laboratories, Philadelphia, Pennsylvania.)

10

MOLECULAR FLUORESCENCE SPECTROSCOPY

Many chemical systems are photoluminescent; that is, they can be excited by electromagnetic radiation and, as a consequence, reemit radiation either of the same or longer wavelengths. The two most common manifestations of photoluminescence are *fluorescence* and *phosphorescence* which, as we shall see, are produced by somewhat different mechanisms. The two phenomena can be distinguished experimentally by observing the lifetime of the excited state. With fluorescence, the luminescent process ceases almost immediately ($<10^{-6}$ s) after irradiation is discontinued; phosphorescence usually endures for an easily detectable length of time. Fluorescence is more important than phosphorescence from an analytical standpoint and will be emphasized in the discussion that follows.

Measurement of fluorescent intensity permits the quantitative determination of many inorganic and organic species in trace amounts; many useful fluorometric methods exist, particularly for biological systems.

One of the most attractive features of fluorometry is its inherent sensitivity. The lower limits for the method frequently are less than those for an absorption method by a factor of ten or better and are in the range between a few thousandths to perhaps one-tenth part per million. In addition, selectivity is at least as good and may be better than that of other methods. Fluorometry, however, is less widely applicable than absorption methods because of the relatively limited number of chemical systems that can be made to fluoresce.[1]

---

[1] For further discussion of fluorescence and phosphorescence theory and applications, see: G. G. Guilbault, *Practical Fluorescence: Theory, Methods, and Technique.* New York: Marcel Dekker, 1973; S. G. Schulman, *Fluorescence and Phosphorescence Spectroscopy: Physiochemical Principles and Practice.* New York: Pergamon Press, 1977; *Modern Fluorescence Spectroscopy,* ed. E. L. Wehry. New York: Plenum Press, 1976; and J. D. Winefordner, S. G. Schulman, and T. C. O'Haver, *Luminescence Spectrometry in Analytical Chemistry.* New York: Wiley-Interscience, 1972.

THEORY OF FLUORESCENCE

Fluorescent behavior occurs in simple as well as in complex gaseous, liquid, and solid chemical systems. The simplest kind of fluorescence is that exhibited by dilute atomic vapors. For example, the $3s$ electrons of vaporized sodium atoms can be excited to the $3p$ state by absorption of radiation of 5896 and 5890 Å. After approximately 10^{-8} s, the electrons return to the ground state, and in so doing emit radiation of the same two wavelengths in all directions. This type of fluorescence, in which the absorbed radiation is reemitted without alteration, is known as *resonance radiation* or *resonance fluorescence*.

Polyatomic molecules or ions also exhibit resonance radiation; in addition, characteristic radiation of longer wavelengths is emitted. This phenomenon is called the *Stokes shift*.

Excited States

To understand the characteristics of fluorescence and phosphorescence phenomena, recall that (p. 170) a bond between two atoms consists of one or more molecular orbitals formed from the overlap of the atomic orbitals of the electron pair that forms the bond. Combination of two atomic orbitals gives rise to both a bonding and an antibonding molecular orbital; the former has the lower energy and is thus occupied by the electrons in the ground state. Superimposed upon the electronic energy level of each molecular orbital is a series of closely spaced vibrational energy levels. As a consequence, each electronic absorption band contains a series of closely spaced vibrational peaks corresponding to transitions from the ground state to the several vibrational levels of an excited electronic state.

Most molecules contain an even number of electrons; in the ground state, these electrons exist as pairs in the various atomic or molecular orbitals. The Pauli exclusion principle demands that the two electrons in a given orbital have opposing spins (the spins are then said to be paired). As a consequence of spin pairing, most molecules have no net electron spin and are, therefore, diamagnetic.[2] A molecular electronic state in which all electron spins are paired is called a *singlet* state, and no splitting of the energy level occurs when the molecule is exposed to a magnetic field (we are here neglecting the effects of nuclear spin). The ground state for a free radical, on the other hand, is a *doublet* state; here, the odd electron can assume two orientations in a magnetic field and thus gives rise to a splitting of the energy level.

When one of the electrons of a molecule is excited to a higher energy level, a singlet or a *triplet* state can result. In the excited singlet state, the spin of the promoted electron is still paired with the ground-state electron; in the triplet state, however, the spins of the two electrons have become unpaired and are thus parallel. These states can be represented as follows:

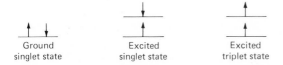

| Ground singlet state | Excited singlet state | Excited triplet state |

The nomenclature of singlet, doublet, and triplet comes from spectroscopic multiplicity considerations which need not concern us here.

The properties of a molecule in the excited triplet state differ significantly from those of the corresponding singlet state. For example, a molecule is paramagnetic in the former and diamagnetic in the latter. More important, however, is the fact that a singlet-triplet transition (or the reverse) that also involves a change in electronic state is a significantly less probable event than the corresponding singlet-

singlet transition. As a consequence, the average lifetime of an excited triplet state may be as long as a second or more, as compared with an average lifetime of about 10^{-8} s for an excited singlet state. Furthermore, radiation-induced excitation of a ground-state molecule to an excited triplet state does not occur readily, and absorption peaks due to this process are several orders of magnitude less intense than the analogous singlet-singlet transition. We shall see, however, that an excited triplet state can be populated from an *excited* singlet state of certain molecules; the consequence of this process is phosphorescent behavior.

Figure 10-1 is a partial energy-level diagram for a typical photoluminescent molecule. The lowest heavy horizontal line represents the ground-state energy of the molecule, which is normally a singlet state and is labeled S_0. At room temperature, this state represents the energies of essentially all of the molecules in a solution.

The upper heavy lines are energy levels for the ground vibrational states of three excited electronic states. The two lines on the left represent the first (S_1) and second (S_2) electronic *singlet* states. The one on the right (T_1) represents the energy of the first electronic *triplet* state. As is normally the case, the energy of the first excited triplet state is lower than the energy of the corresponding singlet state.

Numerous vibrational energy levels are associated with each of the four electronic states, as suggested by the lighter horizontal lines.

As shown in Figure 10-1, excitation of this molecule can be brought about by absorption of two bands of radiation, one centered about the wavelength λ_1 ($S_0 \rightarrow S_1$) and the second around the shorter wavelength λ_2 ($S_0 \rightarrow S_2$). Note that the excitation process can result in conversion of the molecule to any of the several excited vibrational states. Note also that direct excitation to the triplet state does

---

[2] Diamagnetic species tend to move out from magnetic fields, whereas paramagnetic species tend to move into a field.

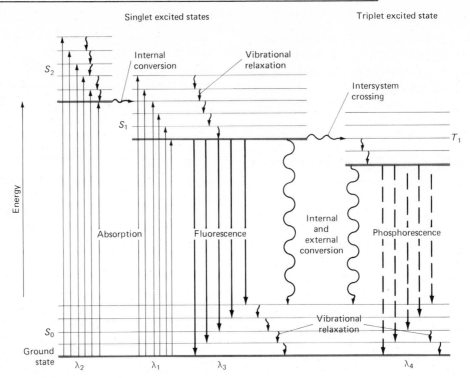

FIGURE 10-1 Partial energy diagram for a photoluminescent system.

not occur to any significant extent because this process involves a change in multiplicity, an event which, as we have mentioned, has a low probability of occurrence.

Deactivation Processes

An excited molecule can return to its ground state by a combination of several mechanistic steps. As shown by the straight vertical arrows in Figure 10-1, two of these steps, fluorescence and phosphorescence, involve the release of a photon of radiation. The other deactivation steps, indicated by wavy arrows, are radiationless processes. The favored route to the ground state is the one that minimizes the lifetime of the excited state. Thus, if deactivation by fluorescence is rapid with respect to the radiationless processes, such emission is observed. On the other hand, if a radia-

tionless path has a more favorable rate constant, fluorescence is either absent or less intense.

The fluorescence phenomenon is limited to a relatively small number of systems incorporating structural and environmental features that cause the rate of radiationless relaxation or deactivation processes to be slowed to a point where the emission reaction can compete kinetically. Information concerning the emission process is sufficiently complete to permit a quantitative accounting of its rate. Understanding of other deactivation routes, however, is rudimentary at best; for these, only qualitative statements or speculations can be put forth. Nevertheless, the interpretation of fluorescence requires consideration of these other processes.

Emission Rate. Since fluorescent emission is the reverse of the excitation process, it is perhaps not surprising that a simple inverse relationship exists between the lifetime of an excited state and the molar absorptivity of the absorption peak corresponding to the excitation process. This relationship (and experiment as well) shows that typical lifetimes of excited states that are deactivated by emission are 10^{-7} to 10^{-9} s, when the molar absorptivities range between 10^3 and 10^5; for weakly absorbing systems, where the probability of the transition process is low, the lifetimes may be as long as 10^{-6} to 10^{-5} s. Any deactivation process that occurs in a shorter time thus reduces the fluorescent intensity.

Vibrational Relaxation. As shown in Figure 10-1, a molecule may be promoted to any of several vibrational levels during the electronic excitation process. In solution, however, the excess vibrational energy is immediately lost as a consequence of collisions between the molecules of the excited species and those of the solvent; the result is an energy transfer and a minuscule increase in temperature of the solvent. This relaxation process is so efficient that the average lifetime of a vibrationally excited molecule is 10^{-12} s or less, a period significantly shorter than the average lifetime of an electronically excited state. As a consequence, fluorescence from solution, when it occurs, always involves a transition *from the lowest vibrational level of an excited state*. Several closely spaced peaks are produced, however, since the electron can return *to any one of the vibrational levels of the ground state* (Figure 10-1), whereupon it will rapidly fall to the lowest ground state by further vibrational relaxation.

A consequence of the efficiency of vibrational relaxation is that the fluorescent band for a given electronic transition is displaced toward lower frequencies or longer wavelengths from the absorption band; overlap occurs only for the resonance peak involving transitions between the lowest vibrational level of the ground state and the corresponding level of the excited state.

Internal Conversion. The term *internal conversion* is employed to describe intermolecular processes by which a molecule passes to a lower energy *electronic* state without emission of radiation. These processes are neither well defined nor well understood, but it is apparent that they are often highly efficient, since relatively few compounds exhibit fluorescence.

Internal conversion appears to be particularly efficient when two electronic energy levels are sufficiently close for the existence of an overlap in vibrational levels. This situation is depicted for the two excited singlet states in Figure 10-1. At the overlaps shown, the potential energies of the two excited states are identical; this equality apparently permits an efficient transition. Internal conversion through overlapping vibrational levels is usually more probable than the loss of energy by fluorescence from a higher excited state. Thus, referring again to Figure 10-1, excitation by radiation of λ_2 frequently produces fluorescence of wavelength λ_3 to the exclusion of a band that would result from a transition between S_2 and S_0. Here, the excited molecule proceeds from the higher electronic state to the lowest vibrational state of the lower electronic excited state via a series of vibrational relaxations, an internal conversion, and then further relaxations. Under these circumstances, the fluorescence would be of λ_3 *only*, regardless of whether radiation of wavelength λ_1 or λ_2 was responsible for the excitation.

The mechanisms of the internal conversion processes $S_1 \rightarrow S_0$ and $T_1 \rightarrow S_0$ shown in Figure 10-1 are not well understood. The vibrational levels of the ground state may overlap those of the first excited electronic state; under such circumstances, deactivation will occur rapidly by the mechanism just described. This situation prevails with aliphatic compounds, for example, and accounts for the fact that these species seldom fluoresce; that is, deactivation by energy transfer through overlapping vibrational levels occurs so rapidly that fluorescence is prevented.

Internal conversion may also result in the

phenomenon of *predissociation*. Here, the electron moves from a higher electronic state to an upper vibrational level of a lower electronic state in which the vibrational energy is great enough to cause rupture of a bond. In a large molecule, there is an appreciable probability for the existence of bonds with strengths less than the electronic excitation energy of the chromophores. Rupture of these bonds can occur as a consequence of absorption by the chromophore followed by internal conversion of the electronic energy to vibrational energy associated with the weak bond.

A predissociation should be differentiated from a *dissociation*, in which the absorbed radiation excites the electron of a chromophore directly to a sufficiently high vibrational level to cause rupture of the chromophoric bond; no internal conversion is involved. Dissociation processes also compete with the fluorescence process.

External Conversion. Deactivation of an excited electronic state may involve interaction and energy transfer between the excited molecule and the solvent or other solutes. These processes are called *external conversions*. Evidence for external conversion includes the marked effect upon fluorescent intensity exerted by the solvent; furthermore, those conditions that tend to reduce the number of collisions between particles (low temperature and high viscosity) generally lead to enhanced fluorescence. The details of external conversion processes are not well understood.

Radiationless transitions to the ground state from the lowest excited singlet and triplet states (Figure 10-1) probably involve external conversions, as well as internal conversions.

Intersystem Crossing. *Intersystem crossing* is a process in which the spin of an excited electron is reversed and a change in multiplicity of the molecule results. As with internal conversion, the probability of this transition is enhanced if the vibrational levels of the two states overlap. The singlet-triplet transition shown in Figure 10-1 is an example; here, the lowest singlet vibrational state overlaps one of the upper triplet vibrational levels and a change in spin state is thus more probable.

Intersystem crossings are most common in molecules that contain heavy atoms, such as iodine or bromine. Apparently interaction between the spin and orbital motions becomes large in the presence of such atoms, and a change in spin is thus more favored. The presence of paramagnetic species such as molecular oxygen in solution also enhances intersystem crossing and a consequent decrease in fluorescence.

Phosphorescence. Deactivation may also involve phosphorescence. After intersystem crossing to an excited triplet state, further deactivation can occur either by internal or external conversion or by phosphorescence. A triplet-singlet transition is much less probable than a singlet-singlet conversion, and the average lifetime of the excited triplet state with respect to emission ranges from 10^{-4} to several seconds. Thus, emission from such a transition may persist for some time after irradiation has been discontinued.

External and internal conversions compete so successfully with phosphorescence that this kind of emission is ordinarily observed only at very low temperatures or in viscous media.

Variables that Affect Fluorescence and Phosphorescence

Both molecular structure and chemical environment are influential in determining whether a substance will or will not fluoresce (or phosphoresce), as well as the intensity of the emission that might occur. The effects of some of these variables are considered briefly in this section.

Quantum Yield. The *quantum yield*, or *quantum efficiency*, for a fluorescent process is simply the ratio of the number of molecules that fluoresce to the total number of excited molecules (the quantum yield for phosphorescence can be defined in an analogous way). For a highly fluorescent molecule such as fluorescein, the quantum efficiency under some conditions approaches unity. Chemical species that do not fluoresce appreciably have an efficiency that approaches zero.

From a consideration of Figure 10-1 and our discussion of deactivation processes, it is apparent that the fluorescent quantum yield ϕ for a compound must be determined by the relative rates for the processes by which the lowest excited singlet state is deactivated—namely, fluorescence, intersystem crossing, external and internal conversion, predissociation, and dissociation. We may express these relationships by the equation

$$\phi = \frac{k_f}{k_f + k_i + k_{ec} + k_{ic} + k_{pd} + k_d} \quad (10\text{-}1)$$

where the k terms are the respective rate constants for the several processes enumerated above.

Equation 10-1 permits a qualitative interpretation of many of the structural and environmental factors that influence fluorescent intensity. Clearly, those variables that lead to high values for the fluorescence rate constant k_f and low values for the other k terms enhance fluorescence. The magnitude of k_f, the predissociation rate constant k_{pd}, and the dissociation rate constant k_d are mainly dependent upon chemical structure; the remaining constants are strongly influenced by environment and to a somewhat lesser extent by structure.

Transition Types in Fluorescence. It is important to note that fluorescence seldom results from absorption of ultraviolet radiation of wavelengths lower than 250 nm, because such radiation is sufficiently energetic to cause deactivation of the excited states by predissociation or dissociation. For example, 200-nm radiation corresponds to about 140 kcal/mol; most molecules have at least some bonds that can be ruptured by energies of this magnitude. As a consequence, fluorescence due to $\sigma^* \rightarrow \sigma$ transitions is seldom observed; instead, such emission is confined to the less energetic $\pi^* \rightarrow \pi$ and $\pi^* \rightarrow n$ processes (see Figure 7-3, p. 171 for the relative energies associated with these transitions).

As we have noted, an electronically excited molecule ordinarily returns to its *lowest excited state* by a series of rapid vibrational relaxations and internal conversions that produce no emission of radiation. Thus, any fluorescence observed most commonly arises from a transition from the first excited state to the ground state. For the majority of fluorescent compounds then, radiation is produced by deactivation of either the n,π^* or the π,π^* excited state, depending upon which of these is the less energetic.

Quantum Efficiency and Transition Type. It is observed empirically that fluorescent behavior is more commonly found in compounds in which the lowest energy excited state is of a π,π^* type than in those with a lowest energy n,π^* state; that is, the quantum efficiency is greater for $\pi^* \rightarrow \pi$ transitions.

The greater quantum efficiency associated with the π,π^* state can be rationalized in two ways. First, the molar absorptivity of a $\pi \rightarrow \pi^*$ transition is ordinarily 100- to 1000-fold greater than for $n \rightarrow \pi^*$ process, and this quantity represents a measure of transition probability in either direction; thus, the inherent lifetime associated with a $\pi \rightarrow \pi^*$ transition is shorter (10^{-7} to 10^{-9} s compared with 10^{-5} to 10^{-7} s for an n,π^* state) and k_f in Equation 10-1 is larger.

It is also believed that the rate constant for intersystem crossing k_i is smaller for π,π^* excited states because the energy difference between the singlet-triplet states is larger; that is, more energy is required to unpair the electrons of the π,π^* excited state. As a consequence, overlap of triplet vibrational levels with those of the singlet state is less, and the probability of an intersystem crossing is smaller.

In summary, then, fluorescence is more

commonly associated with π,π^* states than with n,π^* states because the former possess shorter average lifetimes (k_f is larger) and because the deactivation processes that compete with fluorescence are less likely to occur.

Fluorescence and Structure. The most intense and most useful fluorescent behavior is found in compounds containing aromatic functional groups with low-energy $\pi \rightarrow \pi^*$ transition levels. Compounds containing aliphatic and alicyclic carbonyl structures or highly conjugated double-bond structures may also exhibit fluorescence, but the number of these is small compared with the number in the aromatic systems.

Most unsubstituted aromatic hydrocarbons fluoresce in solution, the quantum efficiency usually increasing with the number of rings and their degree of condensation. The simplest heterocyclics, such as pyridine, furan, thio-phene, and pyrrole, do not exhibit fluorescent behavior; on the other hand, fused-ring structures ordinarily do. With nitrogen heterocyclics, the lowest-energy electronic transition is believed to involve an $n \rightarrow \pi^*$ system which rapidly converts to the triplet state and prevents fluorescence. Fusion of benzene rings to a heterocyclic nucleus, however, results in an increase in the molar absorptivity of the absorption peak. The lifetime of an excited state is shorter in such structures; fluorescence is thus observed for compounds such as quinoline, isoquinoline, and indole.

Substitution on the benzene ring causes shifts in the wavelength of absorption maxima and corresponding changes in the fluorescence peaks. In addition, substitution frequently affects the fluorescent efficiency; some of these effects are illustrated by the data for benzene derivatives in Table 10-1.

TABLE 10-1 EFFECT OF SUBSTITUTION ON THE FLUORESCENCE OF BENZENE[a,b]

Compound	Formula	Wavelength of Fluorescence, nm	Relative Intensity of Fluorescence
Benzene	C_6H_6	270–310	10
Toluene	$C_6H_5CH_3$	270–320	17
Propylbenzene	$C_6H_5C_3H_7$	270–320	17
Fluorobenzene	C_6H_5F	270–320	10
Chlorobenzene	C_6H_5Cl	275–345	7
Bromobenzene	C_6H_5Br	290–380	5
Iodobenzene	C_6H_5I	—	0
Phenol	C_6H_5OH	285–365	18
Phenolate ion	$C_6H_5O^-$	310–400	10
Anisole	$C_6H_5OCH_3$	285–345	20
Aniline	$C_6H_5NH_2$	310–405	20
Anilinium ion	$C_6H_5NH_3^+$	—	0
Benzoic acid	C_6H_5COOH	310–390	3
Benzonitrile	C_6H_5CN	280–360	20
Nitrobenzene	$C_6H_5NO_2$	—	0

[a] In ethanol solution.
[b] Taken from W. West, *Chemical Applications of Spectroscopy* (*Techniques of Organic Chemistry*, Vol. IX). New York: Interscience Publishers, Inc., 1956, p. 730. With permission.

The influence of halogen substitution is striking; the decrease in fluorescence with increasing atomic number of the halogen is thought to be due in part to the heavy atom effect (p. 284), which increases the probability for intersystem crossing to the triplet state. Predissociation is thought to play an important role in iodobenzene and in nitro derivatives as well; these compounds have easily ruptured bonds that can absorb the excitation energy following internal conversion.

Substitution of a carboxylic acid or carbonyl group on an aromatic ring generally inhibits fluorescence. In these compounds, the energy of the n,π^* system is less than in the π,π^* system; as we have pointed out earlier, the fluorescent yield from the former type of system is ordinarily low.

Effect of Structural Rigidity. It is found experimentally that fluorescence is particularly favored in molecules that possess rigid structures. For example, the quantum efficiencies for fluorene and biphenyl are nearly 1.0 and 0.2, respectively, under similar conditions of measurement.

Fluorene   Biphenyl

The difference in behavior appears to be largely a result of the increased rigidity furnished by the bridging methylene group in fluorene. Many similar examples can be cited. In addition, enhanced emission frequently results when fluorescing dyes are adsorbed on a solid surface; here again, the added rigidity provided by the solid surface may account for the observed effect.

The influence of rigidity has also been invoked to account for the increase in fluorescence of certain organic chelating agents when they are complexed with a metal ion. For example, the fluorescent intensity of 8-hydroxyquinoline is much less than that of the zinc complex:

Lack of rigidity in a molecule probably causes an enhanced internal conversion rate (k_{ic} in Equation 10-1) and a consequent increase in the likelihood for radiationless deactivation. One part of a nonrigid molecule can undergo low-frequency vibrations with respect to its other parts; such motions undoubtedly account for some energy loss.

Temperature and Solvent Effects. The quantum efficiency of fluorescence by most molecules decreases with increasing temperature because the increased frequency of collisions at elevated temperatures improves the probability for deactivation by external conversion. A decrease in solvent viscosity also increases the likelihood of external conversion and leads to the same result.

The polarity of the solvent may also have an important influence. In Chapter 7, we pointed out that the energy for $n \to \pi^*$ transitions is often increased in polar solvents, while that for a $\pi \to \pi^*$ transition suffers the opposite effect. Such shifts may occasionally be great enough to lower the energy of the $\pi \to \pi^*$ process below that of the $n \to \pi^*$ transition; enhanced fluorescence results.

The fluorescence of a molecule is decreased by solvents containing heavy atoms or other solutes with such atoms in their structure; carbon tetrabromide and ethyl iodide are examples. The effect is similar to that which occurs when heavy atoms are substituted into fluorescent compounds; orbital spin interactions result in an increase in the rate of triplet formation and a corresponding decrease in fluorescence. Compounds containing heavy atoms are frequently incorporated into solvents when enhanced phosphorescence is desired.

Effect of pH on Fluorescence. The fluorescence of an aromatic compound with acidic or basic ring substituents is usually pH-dependent. Both the wavelength and the emission intensity are likely to be different for the ionized and nonionized forms of the compound. The data for phenol and aniline shown in Table 10-1 illustrate this effect. The changes in emission exhibited by these compounds are analogous to the absorption changes that occur with acid-base indicators; indeed, fluorescent indicators have been proposed for acid-base titrations in highly colored solutions. For example, fluorescence of the phenolic form of 1-naphthol-4-sulfonic acid is not detectable by the eye since it occurs in the ultraviolet region. When the compound is converted to the phenolate ion by the addition of base, the emission peak shifts to visible wavelengths where it can readily be detected. It is of interest that this change occurs at a different pH than would be predicted from the acid dissociation constant for the phenol. The explanation of this discrepancy is that the acid dissociation constant for the *excited* molecule differs from that for the same species in its ground state. Changes in acid or base dissociation constants with excitation are common and are occasionally as large as four or five orders of magnitude.

It is clear from these observations that analytical procedures based on fluorescence frequently require close control of pH.

Effect of Dissolved Oxygen. The presence of dissolved oxygen often reduces the emission intensity of a fluorescent solution. This effect may be the result of a photochemically induced oxidation of the fluorescent species. More commonly, however, the quenching takes place as a consequence of the paramagnetic properties of molecular oxygen that can be expected to promote intersystem crossing and conversion of excited molecules to the triplet state. Other paramagnetic species also tend to quench fluorescence.

Effect of Concentration on Fluorescent Intensity. The power of fluorescent radiation F is proportional to the radiant power of the excitation beam that is absorbed by the system. That is,

$$F = K'(P_0 - P) \qquad (10\text{-}2)$$

where P_0 is the power of the beam incident upon the solution and P is its power after traversing a length b of the medium. The constant K' depends upon the quantum efficiency of the fluorescence process. In order to relate F to the concentration c of the fluorescing particle, we write Beer's law in the form

$$\frac{P}{P_0} = 10^{-\varepsilon bc} \qquad (10\text{-}3)$$

where ε is the molar absorptivity of the fluorescent molecules and εbc is the absorbance A. By substitution of Equation 10-3 into Equation 10-2, we obtain

$$F = K'P_0(1 - 10^{-\varepsilon bc}) \qquad (10\text{-}4)$$

The exponential term in Equation 10-4 can be expanded to

$$F = K'P_0 \qquad (10\text{-}5)$$
$$\times \left[2.3\varepsilon bc - \frac{(-2.3\varepsilon bc)^2}{2!} - \frac{(-2.3\varepsilon bc)^3}{3!} - \cdots \right]$$

Provided $\varepsilon bc = A < 0.05$, all of the subsequent terms in the brackets become small with respect to the first; under these conditions, we may write

$$F = 2.3K'\varepsilon bcP_0 \qquad (10\text{-}6)$$

or at constant P_0,

$$F = Kc \qquad (10\text{-}7)$$

Thus, a plot of the fluorescent power of a solution versus concentration of the emitting species should be linear at low concentrations c. When c becomes great enough so that the absorbance is larger than about 0.05, linearity is lost and F lies below an extrapolation of the straight-line plot.

Two other factors responsible for further negative departures from linearity at high concentration are *self-quenching* and *self-absorption*. The former is the result of collisions between excited molecules. Radiationless transfer of energy occurs, perhaps in a fashion analogous to the transfer to solvent molecules that occurs in an external conversion. Self-quenching can be expected to increase with concentration.

Self-absorption occurs when the wavelength of emission overlaps an absorption peak; fluorescence is then decreased as the emitted beam traverses the solution.

The effects of these phenomena are such that a plot relating fluorescent power to concentration often exhibits a maximum.

INSTRUMENTS FOR FLUORESCENCE ANALYSIS

The various components of fluorescence instruments are similar to those found in ultraviolet-visible photometers or spectrophotometers. Figure 10-2 shows a typical configuration for these components in a *fluorometer* or a *spectrofluorometer*. Nearly all fluorescence instruments employ double-beam optics as shown in order to compensate for fluctuations in the power of the source. The sample beam first passes through an excitation filter or monochromator which serves to transmit that part of the beam which will excite fluorescence but exclude wavelengths which are subsequently produced by the irradiated sample. Fluorescent radiation is emitted by the sample in all directions but is most conveniently observed at right angles to the excitation beam; at other angles, increased scattering from the solution and the cell walls can be the cause of large errors in the measurement of fluorescent intensity. The emitted radiation reaches a photoelectric detector after passing through a second filter or monochromator that isolates a fluorescent peak for measurement.

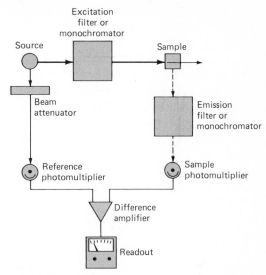

FIGURE 10-2 Components of a fluorometer or a spectrofluorometer.

The reference beam passes through an attenuator that reduces its power to approximately that of the fluorescent radiation (the power reduction is usually by a factor of 100 or more). The outputs from the reference and sample phototubes are then fed into a difference amplifier and the output into a meter or recorder. Many fluorescent instruments are of the null type, this state being achieved by optical or electrical attenuators.

The sophistication, performance characteristics, and costs of fluorometers and spectrofluorometers differ as widely as do the corresponding instruments for absorption measurements. Fluorometers are analogous to absorption photometers in that filters are employed to restrict the wavelengths of the excitation and emission beams. Spectrofluorometers are of two types. The first employs a suitable filter to limit the excitation radiation and a grating or prism monochromator to isolate a peak of the fluorescent emission spectrum. Several commercial spectrophotometers can be purchased with adapters that permit their use as spectrofluorometers.

True spectrofluorometers are specialized instruments equipped with two monochromators. One of these restricts the excitation radiation to a narrow band; the other permits the isolation of a particular fluorescent wavelength. Such instruments permit measurement of *fluorescence*, *excitation*, and *absorption* spectra. An excitation spectrum is obtained with the emission monochromator fixed at a wavelength of strong fluorescence; a plot of fluorescence output as a function of the excitation wavelength is then obtained. A fluorescence spectrum, on the other hand, involves a fixed excitation wavelength while the fluorescent wavelengths are scanned. With suitable corrections for variations in source output as a function of wavelength, an absolute excitation spectrum for a substance is obtained which is similar in appearance to the absorption spectrum.

The selectivity provided by spectrofluorometers is of prime importance to investigations concerned with the electronic and structural characteristics of molecules, and is of value in both qualitative and quantitative analytical work as well. For quantitative purposes, however, the information provided by simpler instruments is entirely satisfactory. Indeed, relatively inexpensive fluorometers that have been designed specifically to meet the measurement problems peculiar to fluorescent analysis are frequently as specific and selective as modified spectrophotometers.

The discussion that follows is largely focussed on the simpler instruments for fluorescence analysis.

Components of Fluorometers and Spectrofluorometers

The components of fluorometers and spectrofluorometers differ only in detail from those of photometers and spectrophotometers; we need to consider only these differences.

Sources. In most applications, a more intense source is needed than the tungsten or hydrogen lamp employed for the measurement of absorption. A mercury or xenon arc lamp is commonly employed.

The xenon arc lamp produces intense radiation by the passage of current through an atmosphere of xenon. The spectrum is continuous over the range between about 250 and 600 nm, with the peak intensity occurring at about 470 nm (see Figure 4–13, p. 110). In some instruments, regularly spaced flashes are obtained by discharging a capacitor through the lamp; high intensities result. In addition, the output of the phototubes is then ac, which can be readily amplified and processed.

Mercury arc lamps produce an intense line spectrum. High-pressure lamps (~ 8 atmospheres) give lines at 366, 405, 436, 546, 577, 691, and 773 nm. Low-pressure lamps, equipped with silica windows, additionally provide intense radiation at 254 nm. Inasmuch as fluorescent behavior can be induced in most fluorescing compounds by a variety of wavelengths, at least one of the mercury lines ordinarily proves suitable.

A recent development has been the use of various types of lasers as excitation sources for fluorometry. Of particular interest is a tunable dye laser employing a pulsed nitrogen laser as the primary source. Radiation in the region between 360 and 650 nm is produced. Such a device eliminates the need for an excitation monochromator.

Filters and Monochromators. Both interference and absorption filters have been employed in fluorometers. Most spectrofluorometers are equipped with grating monochromators.

Detectors. The typical fluorescent signal is of low intensity; large amplification factors are thus required for its measurement. Photomultiplier tubes have come into widespread use as detectors in sensitive fluorescence instruments.

Cells and Cell Compartments. Both cylindrical and rectangular cells fabricated of glass or silica are employed for fluorescence measurements. Care must be taken in the design of the cell compartment to reduce the amount of scattered radiation reaching the detector. Baffles are often introduced into the compartment for this purpose.

Instrument Designs

Fluorometers. Figure 10-3 is a schematic diagram of a double-beam fluorometer that employs a mercury lamp and a single photomultiplier tube as a detector. Part of the radiation from the lamp passes through a filter to the sample. The fluorescent radiation then passes through a second filter to the detector. A reference beam is reflected off the mirrored surface of the light cam to a lucite light pipe which directs it to the photomultiplier tube. The rotating light interrupter causes this reference beam and the fluorescent beam to strike the detector surface alternately, thus producing an ac signal whenever the power of one differs from the other; the phase of the ac signal, however, will depend upon which is the stronger. A phase-sensitive device is employed to translate this difference and its sign into a deflection of a meter needle (not shown). The power of the reference beam can then be varied by rotation of the light cam, which mechanically increases or decreases the fraction of the reference beam that reaches the detector. The cam is equipped with a linear dial, each increment of which corresponds to an equal fraction of light.

Accurate adjustment of any null device requires that the condition of null be approachable from both directions. For a totally nonfluorescent sample, however, no light would reach the detector in one of the phases and

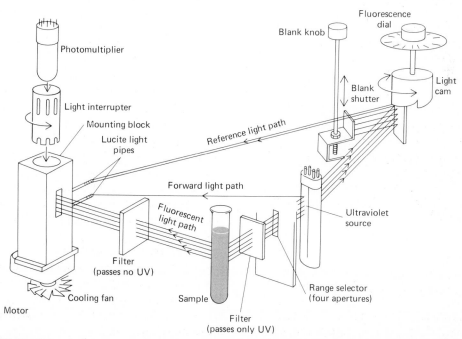

FIGURE 10-3   Optical design of the Turner Model 110 Fluorometer. (Courtesy of Amsco Instrument Company (formerly G. K. Turner Associates), Carpinteria, CA.)

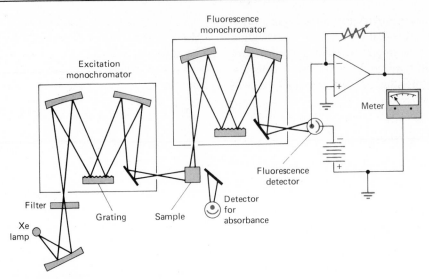

FIGURE 10-4   A spectrofluorometer. Note that the instrument can also be employed for absorbance measurements.

the null point could then be approached from one direction only. To avoid the resultant error, a third beam of constant intensity (the forward light path) is directed to the detector *in phase* with the fluorescent beam so that, under all conditions, some radiation strikes the photomultiplier. Correction for the effect of the third beam on the measured fluorescence is accomplished by setting the fluorescence dial to zero with a solvent blank or a nonfluorescing dummy cuvette in the cell compartment; the intensity of the reference beam can then be varied by means of the blank shutter until an optical null is indicated; this operation must be carried out frequently during a set of analyses.

The single-detector aspect of this double-beam instrument imparts high reproducibility, even with long-term changes in detector sensitivity and output of the source. Thus, only occasional checks of calibration curves with a standard are required.

The instrument just described is representative of the dozen or more fluorometers available commercially. Some of these are simpler single-beam instruments. The cost of such fluorometers ranges from a few hundred dollars to perhaps $2000.

Spectrofluorometers. Several companies offer spectrofluorometers. An example of a typical simple instrument is shown in Figure 10-4. Two grating monochromators are employed. Radiation from the xenon lamp is dispersed by the first and serves to excite the sample. The resulting fluorescent radiation, after dispersion by the second monochromator, is detected by a photocell. Readout is by means of a meter or recorder.

Note that the instrument can be employed for absorbance measurements by use of the first monochromator only.

An instrument such as that shown in Figure 10-4 provides perfectly satisfactory spectra for quantitative analysis. The spectra obtained will, however, not necessarily compare well with spectra from other instruments because the output depends not only upon the fluorescence but also upon the characteristics of the lamp, detector, and monochromators. All of these instrument characteristics vary with wavelength and differ from instrument to instrument.

A number of methods have been developed for obtaining a *corrected* spectrum, which is the true fluorescent spectrum freed from instrumental effects; many of the newer and more sophisticated commercial instruments provide a means for obtaining corrected spectra directly.[3]

APPLICATIONS OF FLUOROMETRY

Fluorometric methods are inherently applicable to lower concentration ranges than are spectrophotometric determinations and are thus among the most sensitive analytical techniques available to the scientist. The basic difference in the sensitivity between the two methods arises from the fact that the concentration-related parameter for fluorometry F can be measured independently of the power of the source P_0. In contrast, a spectrophotometric measurement requires evaluation of both P_0 and P, because the concentration-dependent parameter A is dependent upon the ratio between these two quantities. The sensitivity of a fluorometric method can be improved by increasing P_0 or by further amplifying the fluorescent signal. In spectrophotometry, an increase in P_0 results in a proportionate change in P and therefore fails to affect A; thus, no improvement in sensitivity results. Similarly, amplification of the detector signal has the same effect on both P and P_0 and results in no net gain with respect to A. Thus, fluorometric methods generally have sensitivities that are two to four orders of magnitude better than the corresponding spectrophotometric procedures.

Inorganic Analysis

Inorganic fluorometric methods are of two types. Direct methods involve the formation of a fluorescent chelate and the measurement of its emission. A second group is based upon the diminution of fluorescence resulting from the quenching action of the substance being determined. The latter technique has been most widely used for anion analysis.

Cations that Form Fluorescent Chelates. Two factors greatly limit the number of transition-metal ions that form fluorescent chelates. First, many of these ions are paramagnetic; this property increases the rate of intersystem crossing to the triplet state. Deactivation by fluorescence is thus unlikely, although phosphorescent behavior may be observed. A second reason is that transition-metal complexes are characterized by many closely spaced energy levels, which enhance the likelihood of deactivation by internal conversion. Nontransition-metal ions are less susceptible to the foregoing deactivation processes; it is for these elements that the principal applications of fluorometry are to be found. It is noteworthy that nontransition-metal cations are generally colorless and tend to form chelates which are also without color. Thus, fluorometry often complements spectrophotometry.

Fluorometric Reagents.[4] The most successful fluorometric reagents for cation analyses have aromatic structures with two or more donor functional groups that permit chelate

---

[3] For a summary of correction methods, see: N. Wotherspoon, G. K. Oster, and G. Oster, in *Physical Methods of Chemistry*, eds. A. Weissberger and B. R. Rossiter. New York: Wiley-Interscience, 1972, vol. 1, Part III B, pp. 460–462 and pp. 473–478.

[4] For a more detailed discussion of fluorometric reagents, see: T. S. West, in *Trace Characterization, Chemical and Physical*, eds. W. W. Meinke and B. E. Scribner. Washington, D.C.: National Bureau of Standards Monograph 100, 1967, pp. 237–266; and C. E. White, in *Trace Analysis*, eds. J. H. Yoe and H. J. Koch, New York: Wiley 1957, Chapter 7.

formation with the metal ion. The structures of four common reagents follow:

8-Hydroxyquinoline
(Reagent for Al, Be, and other metal ions)

Alizarin garnet R
(Reagent for Al, F^-)

Flavanol
(Reagent for Zr and Sn)

Benzoin
(Reagent for B, Zn, Ge, and Si)

Selected fluorometric reagents and their applications are presented in Table 10-2. For a more complete summary see Meites,[5] St. John,[6] or the review articles in Analytical Chemistry.[7]

---

[5] L. Meites, *Handbook of Analytical Chemistry.* New York: McGraw-Hill, 1963, pp. **6**-178 to **6**-181.

[6] P. A. St. John, in *Trace Analysis*, ed. J. D. Winefordner. New York: Wiley, 1976, pp. 263–271.

[7] C. M. O'Donnell and T. N. Solle, *Anal. Chem.,* **50,** 189R (1978); **48,** 175R (1976).

Organic Species

The number of applications of fluorometric analysis to organic problems is impressive. Weissler and White have summarized the most important of these in several tables.[8] Under a heading of *Organic and General Biochemical Substances* are found over one hundred entries that include such diverse substances as adenine, anthranilic acid, aromatic polycyclic hydrocarbons, cysteine, guanidine, indole, naphthols, certain nerve gases, proteins, salicylic acid, skatole, tryptophan, uric acid, and warfarin. Some fifty medicinal agents that can be determined fluorometrically are listed. Included among these are adrenaline, alkylmorphine, chloroquin, digitalis principles, lysergic acid diethylamide (LSD), penicillin, phenobarbital, procaine, and reserpine. Methods for the analysis of ten steroids and an equal number of enzymes and coenzymes are also found in these tables. Some of the plant products listed include chlorophyll, ergot alkaloids, rauwolfia serpentina alkaloids, flavonoids, and rotenone. Some eighteen listings for vitamins and vitamin products are also included; among these are ascorbic acid, folic acid, nicotinamide, pyridoxal, riboflavin, thiamin, vitamin A, and vitamin B_{12}.

Without question, the most important applications of fluorometry are in the analyses of food products, pharmaceuticals, clinical samples, and natural products. The sensitivity and selectivity of the method make it a particularly valuable tool in these fields.

NEPHELOMETRY AND TURBIDIMETRY

Nephelometry and turbidimetry are closely related analytical methods which are based upon the scattering of radiation by particu-

---

[8] A. Weissler and C. E. White, in *Handbook of Analytical Chemistry*, ed. L. Meites. New York: McGraw-Hill, 1963, pp. **6**-182 to **6**-196.

late matter.[9] The instrumentation required for nephelometry is similar to the fluorometers we have just considered, whereas a filter photometer can be used for turbidimetric measurements. It is convenient to consider these two methods briefly at this point.

When light passes through a transparent medium in which solid particles are dispersed, part of the radiation is scattered in all directions, giving a turbid appearance to the mixture. The diminution of power of a collimated beam as a result of scattering by particles is the basis of *turbidimetric* methods. *Nephelometric* methods, on the other hand, are based upon the measurement of the scat-

tered radiation, usually at a right angle to the incident beam. Nephelometry is generally more sensitive than turbidimetry for the same reasons that fluorometry is more sensitive than photometry.

The choice between a nephelometric and a turbidimetric measurement depends upon the fraction of light scattered. When scattering is extensive, owing to the presence of many particles, a turbidimetric measurement is the more satisfactory. If scattering is minimal and the diminution in power of the incident beam is small, nephelometric measurements provide more satisfactory results.

Theory of Nephelometry and Turbidimetry

It is important to appreciate that the scattering associated with nephelometry and turbidimetry (in contrast to Raman spectroscopy)

---

[9] For a more complete discussion, see: F. P. Hochgesang, in *Treatise on Analytical Chemistry*, eds. I. M. Kolthoff and P. J. Elving. New York: Interscience, 1964, Part I, vol. 5, Chapter 63.

TABLE 10-2   SELECTED FLUOROMETRIC METHODS FOR INORGANIC SPECIES[a]

| Ion | Reagent | Wavelength, nm | | Sensitivity μg/ml | Interference |
		Absorption	Fluorescence		
Al^{3+}	Alizarin garnet R	470	500	0.007	Be, Co, Cr, Cu, F^-, NO_3^-, Ni, PO_4^{3-}, Th, Zr
F^-	Al complex of Alizarin garnet R (quenching)	470	500	0.001	Be, Co, Cr, Cu, Fe, Ni, PO_4^{3-}, Th, Zr
$B_4O_7^{2-}$	Benzoin	370	450	0.04	Be, Sb
Cd^{2+}	2-(o-Hydroxyphenyl)-benzoxazole	365	Blue	2	NH_3
Li^+	8-Hydroxyquinoline	370	580	0.2	Mg
Sn^{4+}	Flavanol	400	470	0.1	F^-, PO_4^{3-}, Zr
Zn^{2+}	Benzoin	—	Green	10	B, Be, Sb, Colored ions

[a] L. Meites, *Handbook of Analytical Chemistry*. New York: McGraw-Hill Book Company, Inc., 1963, pp. **6**-178 to **6**-181. With permission.

involves no net loss in radiant power; only the direction of propagation is affected. The intensity of radiation appearing at any angle depends upon the number of particles, their size and shape, the relative refractive indexes of the particles and the medium, and the wavelength of the radiation. The relationship among these variables is complex. A theoretical treatment is feasible but seldom applied to specific analytical problems because of its complexity. In fact, most nephelometric and turbidimetric procedures are highly empirical.

Effect of Concentration on Scattering. The attenuation of a parallel beam of radiation by scattering in a dilute suspension is given by the relationship

$$P = P_0 e^{-\tau b} \qquad (10\text{-}8)$$

where P_0 and P are the power of the beam before and after passing through the length b of the turbid medium. The quantity τ is called the *turbidity coefficient*, or the turbidity; its value is often found to be linearly related to the concentration c of the scattering particles. As a consequence, a relationship analogous to Beer's law results; that is,

$$\log_{10} \frac{P_0}{P} = kbc \qquad (10\text{-}9)$$

where $k = 2.3\ \tau/c$.

Equation 10-9 is employed in turbidimetric analysis in exactly the same way as Beer's law is used in photometric analysis. The relationship between $\log_{10} P_0/P$ and c is established with standard samples, the solvent being used as a reference to determine P_0. The resulting calibration curve is then used to determine the concentration of samples.

For nephelometric measurements, the power of the beam scattered at right angles to the incident beam is normally plotted against concentration; a linear relationship is frequently obtained. The procedure here is entirely analogous to a fluorometric method.

Effect of Particle Size on Scattering. The fraction of radiation scattered at any angle depends upon the size and shape of the particles responsible for the scattering; the effect is large. Since most analytical applications of scattering involve the generation of a colloidally dispersed phase in a solution, those variables that influence particle size during precipitation also affect both turbidimetric and nephelometric measurements. Thus, such factors as concentration of reagents, rate and order of mixing, length of standing, temperature, pH, and ionic strength are important experimental variables. Care must be exercised to reproduce all conditions likely to affect particle size during calibration and analysis.

Effect of Wavelength on Scattering. It has been shown experimentally that the turbidity coefficient varies with wavelength as given by the equation

$$\tau = s\lambda^{-t}$$

where s is a constant for a given system. The quantity t is dependent on particle size and has a value of 4 when the scattering particles are significantly smaller than the wavelength of radiation (Rayleigh scattering); for particles with dimensions similar to the wavelength (the usual situation in a turbidimetric analysis) t is found to be about 2.

For purposes of analysis, ordinary white light is employed. If the solution is colored, it is necessary to select a portion of the spectrum in which absorption by the medium is minimized.

Instruments

Nephelometric and turbidimetric measurements can be readily made with the various fluorometers and photometers discussed earlier. Rectangular cells are ordinarily employed. With the exception of those areas through which the radiation is transmitted, the cell walls are given a dull black coating to eliminate reflection of unwanted radiation to the detector.

Figure 10-5 shows a simple visual turbidimeter. The viewing tube is adjusted in the suspension until the special S-shaped lamp filament just disappears. The length of solution is then related to concentration by calibration. This very elementary device yields remarkably accurate data for the analysis of sulfate in low concentrations. Here, a $BaSO_4$ suspension is formed by the addition of $BaCl_2$.

Applications of Scattering Methods

Turbidimetric or nephelometric methods are widely used in the analysis of water, for the determination of clarity, and for the control of treatment processes. In addition, the concentration of a variety of ions can be determined by using suitable precipitating reagents. Conditions must be chosen so that the solid phase forms as a stable colloidal suspension. Surface-active agents (such as gelatin) are frequently added to prevent coagulation of the colloid. As noted earlier, reliable analytical data are obtained only when care is taken to control scrupulously all of the variables that affect particle size.

Table 10-3 lists some of the species that have been determined by turbidimetric or nephelometric methods. Perhaps most widely employed is the method for sulfate ion. Nephelometric methods permit the determination of concentrations as low as a few parts per million with a precision of 1 to 5%. Turbidimetric methods are reported to give the same degree of reproducibility with more concentrated solutions.

TABLE 10-3   SOME TURBIDIMETRIC AND NEPHELOMETRIC METHODS[a]

Element	Method[b]	Suspensions	Reagent	Interferences
Ag	T, N	AgCl	NaCl	—
As	T	As	KH_2PO_2	Se, Te
Au	T	Au	$SnCl_2$	Ag, Hg, Pd, Pt, Ru, Se, Te
Ca	T	CaC_2O_4	$H_2C_2O_4$	Mg, Na, SO_4^{2-} (in high concentration)
Cl^-	T, N	AgCl	$AgNO_3$	Br^-, I^-
K	T	$K_2NaCo(NO_2)_6$	$Na_3Co(NO_2)_6$	SO_4^{2-}
Na	T, N	$NaZn(UO_2)_3(OAc)_9$	$Zn(OAc)_2$ and $UO_2(OAc)_2$	Li
SO_4^{2-}	T, N	$BaSO_4$	$BaCl_2$	Pb
Se	T	Se	$SnCl_2$	Te
Te	T	Te	NaH_2PO_2	Se, As

[a] Data taken from L. Meites, *Handbook of Analytical Chemistry*. New York: McGraw-Hill Book Company, Inc., 1963, p. **6**-175. With permission.

[b] T = turbidimetric; N = nephelometric.

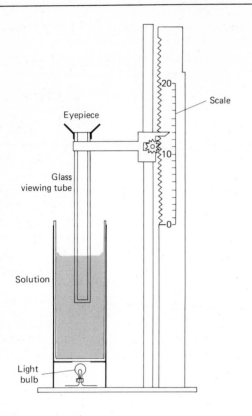

Turbidimetric measurements have also been employed for determining the end point in precipitation titrations. The apparatus can be very simple, consisting of a light source and a photocell located on opposite sides of the titration vessel. The photocurrent is then plotted as a function of volume of reagent. Ideally, the turbidity increases linearly with volume of reagent until the end point is reached, whereupon it remains essentially constant.[10]

[10] For an analysis of the factors affecting turbidity titration curves, see: F. J. Meehan and G. Chiu, *Anal. Chem.*, **36**, 536 (1964).

FIGURE 10-5   A simple turbidimeter.

PROBLEMS

1. Define the following terms: (a) fluorescence, (b) phosphorescence, (c) resonance fluorescence, (d) singlet state, (e) triplet state, (f) vibrational relaxation, (g) internal conversion, (h) external conversion, (i) intersystem crossing, (j) predissociation, (k) dissociation, and (l) quantum yield.

2. Explain the difference between a fluorescence emission spectrum and a fluorescence excitation spectrum. Which most closely resembles an absorption spectrum?

3. Why is nephelometry more sensitive than turbidimetry?

4. Which compound below is expected to have a greater fluorescence quantum yield? Explain.

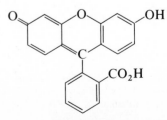

Phenolphthalein   Fluorescein

5. In which solvent would the fluorescence of naphthalene be expected to be greatest: 1-chloropropane, 1-bromopropane, or 1-iodopropane? Why?

6. Would the fluorescence of aniline ($C_6H_5NH_2$) be greater at pH 3 or at pH 10? Explain.

7. (a) Assume that the rate constants for various electronic processes of a certain molecule are: k(fluorescence) $= 2 \times 10^8$ s^{-1}, k(internal conversion, $S_1 \rightarrow S_0$) $= 5 \times 10^7$ s^{-1}, k(external conversion, $S_1 \rightarrow S_0$) $= 5 \times 10^7$ s^{-1}, k(dissociation) $= 3 \times 10^7$ s^{-1}, k(predissociation) $= 1 \times 10^5$ s^{-1}. Referring to Figure 10-1, calculate the quantum yield for fluorescence.

    (b) Assume that in addition to the processes in part (a), the following rate constants also apply: k(intersystem crossing, $S_1 \rightarrow T_1$) $= 2 \times 10^8$ s^{-1}, k(internal conversion, $T_1 \rightarrow S_0$) $= 0.1$ s^{-1}, k(external conversion, $T_1 \rightarrow S_0$) $= 0.2$ s^{-1}, k(phosphorescence) $= 0.7$ s^{-1}. Calculate the quantum yield for fluorescence and the quantum yield for phosphorescence. What fraction of molecules which reach the T_1 state phosphoresces?

8. The reduced form of nicotinamide adenine dinucleotide (NADH) is an important and highly fluorescent coenzyme. It has an absorption maximum at 340 nm and an emission maximum at 465 nm. Standard solutions of NADH gave the following fluorescence intensities:

Concn NADH, μM	Relative Intensity	Concn NADH, μM	Relative Intensity
0.100	13.0	0.500	59.7
0.200	24.6	0.600	71.2
0.300	37.9	0.700	83.5
0.400	49.0	0.800	95.1

Construct a calibration curve and estimate the concentration of NADH in an unknown with a relative fluorescence intensity of 42.3 units.

9. In the absence of self-absorption, the fluorescence intensity of a sample is proportional to concentration only at low concentrations. Use Equation 10-4 to calculate the relative fluorescence intensities of 2.5×10^{-5}, 2.5×10^{-4}, and $1.0 \times 10^{-3} M$ solutions of a compound with $\varepsilon = 4.0 \times 10^3 M^{-1}$ cm^{-1} if $b = 0.20$ cm. Suppose that a fluorometer is calibrated with a $2.5 \times 10^{-5} M$ solution of the compound and the response is assumed to be linear. What would be the percentage of error when a $2.5 \times 10^{-4} M$ solution is analyzed?

10. When a fluorescing sample absorbs appreciably at the exciting and emitting wavelengths (λ_{ex} and λ_{em}, respectively) the observed emission intensity does not follow Beer's law. Consider the cell below in which

the observed fluorescence originates from the central region with approximate dimensions $l_2 \times l_2$:

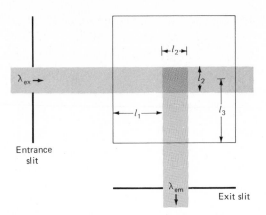

Incoming radiation is diminished by sample absorbance of λ_{ex} over the distance l_1. Outgoing radiation is diminished by absorbance of λ_{em} over the distance l_3. The fluorescence intensity depends upon absorbance of λ_{ex} over the distance l_2. It can be shown that the net observed emission intensity, F, is given by:[11]

$$F = K[10^{-\varepsilon_{ex}l_1 C}(1 - 10^{-\varepsilon_{ex}l_2 C})10^{-\varepsilon_{em}l_3 C}]$$

where ε_{ex} is the molar absorptivity at λ_{ex}, ε_{em} is the molar absorptivity at λ_{em}, C is the sample concentration, and K is a constant which incorporates various instrumental parameters and the fluorescence quantum yield. Construct a graph showing the relative fluorescence intensity of 2,3-butanedione $(CH_3COCOCH_3)$ at the following concentrations: 0, 0.01, 0.02, 0.03, 0.04, 0.05, 0.06, 0.08 and 0.12M. Use the parameters $\lambda_{ex} = 422$ nm, $\varepsilon_{ex} = 19.2M^{-1}$ cm^{-1}, $\lambda_{em} = 464$ nm, $\varepsilon_{em} = 2.3M^{-1}$ cm^{-1}, $l_1 = 0.400$ cm, $l_2 = 0.200$ cm, and $l_3 = 0.500$ cm.

[11] G. Henderson, *J. Chem. Educ.*, **54**, 57 (1977).

11

ATOMIC SPECTROSCOPY

Atomic spectroscopy is based upon the absorption, emission, or fluorescence of electromagnetic radiation by atomic particles. Two regions of the spectrum yield atomic spectral data—the ultraviolet-visible and the X-ray. This and the following chapter will be concerned with the former; a discussion of X-ray spectroscopy will be found in Chapter 15.

In order to obtain ultraviolet and visible atomic spectra, it is necessary to *atomize* the sample. In this process, the constituent molecules are decomposed and converted to gaseous elementary particles. The emission, absorption, and fluorescent spectrum of an atomized element consists of a relatively limited number of discrete lines at wavelengths that are characteristic of the element. To the extent that molecules or complex ions are absent, band spectra are not observed because vibrational and rotational quantum states cannot exist; thus, the number of possible transitions is relatively small.

Table 11-1 outlines the various analytical methods based upon atomic spectroscopy. These procedures frequently offer the advantages of high specificity, wide applicability, excellent sensitivity, speed, and convenience; they are among the most selective of all analytical methods. Perhaps 70 elements can be determined. The sensitivities are typically

TABLE 11-1   CLASSIFICATION OF ATOMIC SPECTRAL METHODS

	Type of Spectroscopy	Method of Atomization	Radiation Source
Emission	Arc	Sample heated in an electric arc	Sample
	Spark	Sample excited in a high voltage spark	Sample
	Argon plasma	Sample heated in an argon plasma	Sample
	Atomic or flame emission	Sample solution aspirated into a flame	Sample
	X-Ray emission or electron probe X-ray	None required; sample bombarded with electrons	Sample
Absorption	Atomic absorption (flame)	Sample solution aspirated into a flame	Hollow cathode tube
	Atomic absorption (nonflame)	Sample solution evaporated and ignited on a hot surface	Hollow cathode tube
	X-Ray absorption	None required	X-Ray tube
Fluorescence	Atomic fluorescence (flame)	Sample solution aspirated into a flame	Sample (excited by radiation from a pulsed lamp)
	Atomic fluorescence (nonflame)	Sample solution evaporated and ignited on a hot surface	Same as above
	X-Ray fluorescence	None required	Sample (excited by X-ray radiation)

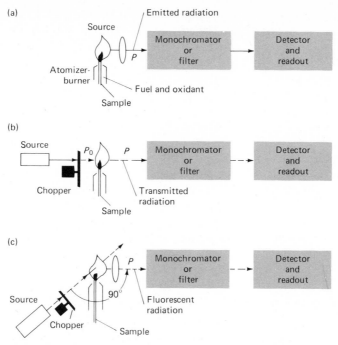

FIGURE 11-1 Three types of flame spectroscopy: (a) atomic emission, (b) atomic absorption, and (c) atomic fluorescence.

in the parts-per-million to parts-per-billion range. An atomic spectral analysis can frequently be completed in a few minutes.

This chapter is concerned principally with ultraviolet and visible atomic emission, absorption, and fluorescence spectroscopy that employ flames or other relatively low-temperature sources.[1] Arc, spark, and argon plasma atomic spectroscopy are considered in Chapter 12.

---

[1] For a more detailed discussion of these topics, see: G. F. Kirkbright and M. Sargent, *Atomic Absorption and Fluorescence Spectroscopy*. New York: Academic Press, 1974; *Flame Emission and Atomic Absorption Spectrometry*, eds. J. A. Dean and T. C. Rains. New York: Marcel Dekker, vol. 1: *Theory*, 1969; vol. 2: *Components and Techniques*, 1971; vol. 3: *Elements and Matrices*, 1974; W. G. Schrenk, *Analytical Atomic Spectroscopy*. New York: Plenum Press, 1975; and E. L. Grove, *Applied Atomic Absorption Spectroscopy*. New York: Plenum Press, 1978.

THEORY OF FLAME SPECTROSCOPY

When an aqueous solution of inorganic salts is aspirated into the hot flame of a burner, a substantial fraction of the metallic constituents are reduced to the elemental state; to a much lesser extent, monatomic ions are also formed. Thus, within the flame is produced a gaseous solution or *plasma* containing a significant concentration of elementary particles. Flame spectroscopy employs the emission, absorption, or fluorescence by these particles as a basis for analyses.

Types of Flame Spectroscopy

Figure 11-1 illustrates the three types of flame spectroscopy. The first, called *atomic emission* or *flame emission*, is based upon the measurement of the radiant power of a characteristic line emitted by the analyte when aspirated into a hot flame. A monochromator or filter is employed to isolate the line from the radiation produced by the flame and the other constituents of the sample.

Figure 11-1b is a schematic diagram showing the components of an *atomic absorption* or *flame absorption* apparatus. Here, the flame serves the same purpose as the cell or cuvette in an ordinary photometer or spectrophotometer; that is, the flame can be considered to be a dilute gaseous solution of the atomized sample held in place by the aspirator-burner. Radiation from a suitable source is passed through the atomized sample and into the slit of a photometer or spectrophotometer. In order to discriminate between radiation from the source and that which is emitted by the flame, it is common practice to chop the beam from the source before it reaches the flame. The detector circuitry is then designed to reject the dc output from flame emission and measure the ac absorption signal from the source and sample.

Atomic fluorescence or *flame fluorescence* is illustrated in Figure 11-1c. Here again, the flame serves to form atomic particles and hold them in the light path for excitation by a suitable source. The resulting fluorescence is then directed to a photometer or spectrophotometer.

Flame Spectra

The emission, absorption, and fluorescence spectra of gaseous, atomic particles consist of well-defined narrow lines arising from electronic transitions of the outermost electrons. For metals, the energies of many of these transitions appear in the ultraviolet and visible regions.

Energy Level Diagrams. The energy level diagrams for the outer electrons of an element provide a convenient basis for describing the processes upon which the various types of atomic spectroscopy are based. The diagram for sodium shown in Figure 11-2a is typical.

Note that the energy scale is linear in units of electron volts (eV), with the 3s orbital being assigned a value of zero. The scale extends to about 5.2 eV, the energy necessary to remove the single 3s electron from the influence of the central atom, thus producing a sodium ion.

The energies of several atomic orbitals are indicated on the diagram by horizontal lines. Note that the p orbitals are split into two levels that differ but slightly in energy. This difference is rationalized by assuming that an electron spins about its own axis and that the direction of this motion may either be the same as or opposed to its orbital motion. Both the spin and the orbital motions create magnetic fields owing to the rotation of the charge carried by the electron. The two fields interact in an attractive sense if these two motions are in the opposite direction; a repulsive force is generated when the motions are parallel. As a consequence, the energy of the electron whose spin opposes its orbital motion is slightly smaller than one in which the motions are alike. Similar differences exist in the d and f orbitals, but their magnitudes are ordinarily so slight as to be undetectable; thus, only a single energy level is indicated for d orbitals in Figure 11-2a.

The splitting of higher energy p, d, and f orbitals into two states is characteristic of all species containing a single external electron. Thus, the energy level diagram for the singly charged magnesium ion, shown in Figure 11-2b, has much the same general appearance as that for the uncharged sodium atom. So also do the diagrams for the dipositive aluminum ion and the remainder of the alkali-metal atoms. It is important to note, however, that the energy difference between the 3p and 3s states is approximately twice as great for the magnesium ion as for the sodium atom because of the larger nuclear charge of the former.

A comparison of Figure 11-2b with Figure 11-3 shows that the energy levels, and thus the spectrum, of an ion is significantly different from that of its parent atom. For atomic magnesium, with two outer electrons, excited singlet and triplet states with different energies exist (Figure 11-3). In the singlet state, the spins of the two electrons are opposed (or paired); in the triplet state the spins are parallel (p. 281). The p, d, and f orbitals of the triplet state are split into three levels that differ slightly in energy. These phenomena can be also rationalized by taking into account the interaction between the fields associated with the spins of the two outer electrons and the net field arising from the orbital motions of all the electrons. In the singlet state, the two spins are paired, and their respective magnetic effects cancel; thus, no energy splitting is observed. In the triplet state, however, the two spins are unpaired (that is, their spin

moments lie in the same direction). The effect of the orbital magnetic moment on the magnetic field of the combined spins produces a splitting of the p level into a triplet. This behavior is characteristic of all of the alkaline-earth atoms, singly charged aluminum and beryllium ions, and so forth.

As the number of electrons outside the closed shell increases, the energy level diagrams become more and more complex. Thus, with three outer electrons, a splitting of energy levels into two and four states occurs; with four outer electrons, singlet, triplet, and quintet states exist.

Although correlation of atomic spectra with energy level diagrams for elements such as sodium and magnesium is relatively straight-

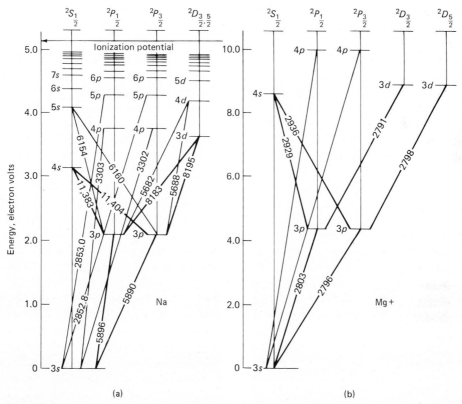

FIGURE 11-2 Energy level diagrams for (a) atomic sodium and (b) magnesium(I) ion.

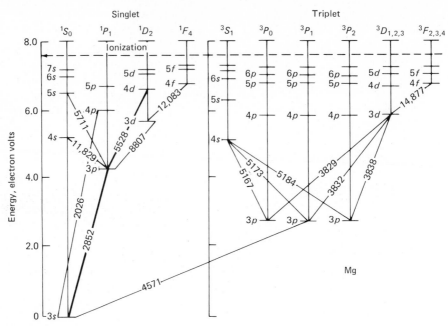

FIGURE 11-3 Energy level diagram for atomic magnesium.

forward and amenable to theoretical interpretation, the same cannot be said for the heavier elements, and particularly the transition metals. These species have larger numbers of closely spaced energy levels; as a consequence, the number of absorption or emission lines can be enormous. For example, Harvey[2] has listed the number of lines observed in the arc and spark spectra of neutral and singly ionized atoms for a variety of elements. For the alkali metals, this number ranges from 30 for lithium to 645 for cesium; for the alkaline earths, magnesium has 173, calcium 662, and barium 472. Typical of the transition series, on the other hand, are chromium, iron, and cerium with 2277, 4757, and 5755 lines, respectively. Fewer lines are excited in flames because of their lower temperatures; still, the flame spectra of the transition metals are considerably more complex that the spectra of species with low atomic numbers.

Atomic Emission Spectra. At room temperature, essentially all of the atoms of a sample of matter are in the ground state. For example, the single outer electron of metallic sodium occupies the $3s$ orbital under these circumstances. Excitation of this electron to higher orbitals can be brought about by the heat of a flame or an electric arc or spark. The lifetime of the excited atom is brief, however, and its return to the ground state is accompanied by the emission of a quantum of radiation. The vertical lines in Figure 11-2a indicate some of the common electronic transitions that follow excitation of sodium atoms; the wavelength of the resulting radiation is also indicated. The two lines at 5890 and 5896 Å are the most intense under usual excitation conditions and are, therefore, the ones most commonly employed for analytical purposes. A slit width that is wide enough to permit

[2] C. E. Harvey, *Spectrochemical Procedures*. Glendale, CA: Applied Research Laboratories, 1950, Chapter 4.

simultaneous measurement of the two lines is frequently used.

Atomic Absorption Spectra. In a flame plasma, sodium atoms are capable of *absorbing* radiation of wavelengths characteristic of electronic transitions from the 3s state to higher excited states. For example, sharp absorption peaks at 5890, 5896, 3302, and 3303 Å are observed experimentally; referring again to Figure 11-2a, it is apparent that each adjacent pair of these peaks corresponds to transitions from the 3s level to the 3p and the 4p levels, respectively. It should be mentioned that absorption due to the 3p to 5s transition is so weak as to go undetected because the number of sodium atoms in the 3p state is so very small in a flame. Thus, typically, an atomic absorption spectrum from a flame consists predominately of *resonance lines*, which are the result of transitions from the ground state to upper levels.

Atomic Fluorescence Spectra. Atoms in a flame can be made to fluoresce by irradiation with an intense source containing wavelengths that are absorbed by the element. As in molecular fluorescence, the fluorescent spectra is most conveniently observed at an angle of 90 deg to the light path. The observed radiation is frequently the result of resonance fluorescence. For example, when magnesium atoms are exposed to an ultraviolet source, radiation of 2852 Å is absorbed as electrons are promoted from the 3s to the 3p level (see Figure 11-3); the resonance fluorescence emitted at this same wavelength may then be used for analysis. In contrast, when sodium atoms absorb radiation of wavelength 3302 Å, electrons are promoted to the 4p state (see Figure 11-2a). A radiationless transition to the two 3p states takes place more rapidly than resonance fluorescence. As a consequence, the observed fluorescence occurs at 5890 and 5896 Å. Figure 11-4 illustrates yet a third type of mechanism for atomic fluorescence. Here, some of the thallium atoms excited in a flame return to the ground state

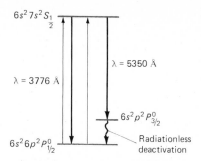

FIGURE 11-4 Energy level diagram for thallium showing the source of two fluorescence lines.

in two steps, including a fluorescent step producing a line at 5350 Å; radiationless deactivation to the ground state quickly follows. Resonance fluorescence at 3776 Å is also observed.

Line Widths. Atomic emission and absorption peaks are much narrower than the bands resulting from emission or absorption by molecules. The natural width of atomic lines can be shown to be about 10^{-4} Å. Two effects, however, combine to broaden the lines to a range from 0.02 to 0.05 Å.

Doppler broadening arises from the rapid motion of the atomic particles in the flame plasma. Atoms that are moving toward the monochromator emit lower wavelengths owing to the well-known Doppler shift; the effect is reversed for atoms moving away from the monochromator.

Doppler broadening is also observed for absorption lines. Those atoms traveling toward the source absorb radiation of slightly shorter wavelength than that absorbed by particles traveling perpendicular to the incident beam. The reverse is true of atoms moving away from the source.

Pressure broadening also occurs; here collisions among atoms cause small changes in the ground-state energy levels and a consequent broadening of peaks.

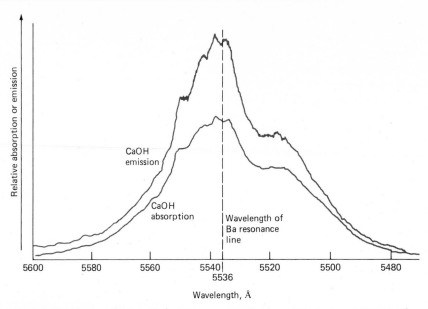

FIGURE 11-5 Molecular flame and flame absorption spectra for CaOH and Ba. (Adapted from L. Capacho-Delgado and S. Sprague, *Atomic Absorption Newsletter*, **4**, 363 (1965). Courtesy of Perkin-Elmer Corporation, Norwalk, Connecticut.)

Molecular Spectra in Flames. Hydrogen or hydrocarbon fuels produce absorption and emission bands over certain wavelength ranges, owing to the presence of such species as OH and CN radicals and C_2 molecules. Some alkaline-earth and rare-earth metals also form volatile oxides or hydroxides that absorb and emit over broad spectral ranges. One example is shown in Figure 11-5 where the emission and absorption spectra of CaOH are shown. The dotted line in the figure shows the wavelength of the barium resonance line. The potential interference of calcium in the atomic absorption determination of barium can be avoided by employing a hotter flame which decomposes the CaOH molecule, thus causing the molecular absorption band shown in the figure to disappear.

A band spectrum is useful for the determination of perhaps one-third of the elements that are amenable to emission analysis. For both emission and absorption spectroscopy, the presence of such bands represents a potential source of interference, which must be dealt with by proper choice of wavelength, by background correction, or by a change in combustion conditions.

FLAME CHARACTERISTICS

The greatest source of uncertainty in flame spectroscopic methods arises from variations in the behavior of the flame. Thus, it is important to understand the important characteristics of flames and the variables that affect these characteristics.

Flame Temperature

Table 11-2 lists the common fuels and oxidants employed in flame spectroscopy and the maximum observed temperatures when the reactants are present in stoichiometric proportions.

The temperatures provided by the burning of natural or manufactured gas in air are so low that only the alkali and alkaline-earth metals, with very low excitation energies, produce useful spectra. Acetylene/air mixtures give a somewhat higher temperature. Oxygen or nitrous oxide must be employed as the oxidant in order to excite the spectra of many metals; with the common fuels, temperatures of 2500 to 3100°C are obtained. The hottest practical flame results from the combustion of cyanogen in oxygen:

$$C_2N_2 + O_2 \rightarrow 2CO + N_2$$

Temperature Profiles. Figure 11-6 is a temperature profile of a typical flame for atomic spectroscopy. The maximum temperature is located somewhat above the inner cone. Clearly, it is important—particularly for emission methods—to focus the same part of the flame on the entrance slit for all calibrations and analytical measurements.

Effect of Temperature on the Types of Spectra. The temperature of the flame plasma determines the nature of the emission, absorption, or fluorescence spectrum that is ob-served for an element. For example, when a sample containing magnesium is aspirated into a low-temperature flame (2000 to 2500°K), the element is converted almost completely to the atomic state; the spectrum, therefore, consists of the lines shown in Figure 11-3. On the other hand, more than 50% of the element may be present as Mg^+ in a cyanogen flame; under these circumstances, the spectrum will contain wavelengths indicated in Figure 11-2b as well as those in Figure 11-3.

Effect of Temperature on Emission, Absorption, and Fluorescence. The flame temperature also determines the fraction of species that exists in excited states, and thus influences emission intensities. The Boltzmann equation permits the calculation of this fraction. If N_j and N_0 are the number of atoms in an excited state and the ground state respectively, their ratio is given by

$$\frac{N_j}{N_0} = \frac{P_j}{P_0} \exp\left(-\frac{E_j}{kT}\right) \qquad (11\text{-}1)$$

where k is the Boltzmann constant (1.38×10^{-16} erg/deg), T is the temperature in degrees Kelvin, and E_j is the energy difference in ergs between the excited state and the ground state.

TABLE 11-2 MAXIMUM FLAME TEMPERATURE FOR VARIOUS FUELS AND OXIDANTS[a]

Fuel	Oxidant	Measured Temperatures, °C
Natural gas	Air	1700–1900
Natural gas	Oxygen	2740
Hydrogen	Air	2000–2050
Hydrogen	Oxygen	2550–2700
Acetylene	Air	2125–2400
Acetylene	Oxygen	3060–3135
Acetylene	Nitrous oxide	2600–2800
Cyanogen	Oxygen	4500

[a] Data from R. N. Kniseley in *Flame Emission and Atomic Absorption Spectroscopy*, eds. J. A. Dean and T. C. Rains, London: Marcel Dekker, 1969, vol. 1, p. 191. With permission.

The quantities P_j and P_0 are statistical factors that are determined by the number of states having equal energy at each quantum level. As shown by the following example, the fraction of excited atoms in a typical gas flame ($T = 2500°K$) is very small.

EXAMPLE

Calculate the ratio of sodium atoms in the $3p$ excited states to the number in the ground state at 2500 and 2510°K.

In order to calculate E_j in Equation 11-1, we employ an average wavelength of 5893 Å for the two sodium emission lines involving the $3p \to 3s$ transitions. To obtain the energy in ergs, we employ the conversion factors found inside the front cover.

$$\text{Wavenumber} = \frac{1}{5893 \text{ Å} \times 10^{-8} \text{ cm/Å}}$$

$$= 1.697 \times 10^4 \text{ cm}^{-1}$$

$$E_j = 1.697 \times 10^4 \text{ cm}^{-1} \times 1.99$$
$$\times 10^{-16} \text{ erg/cm}^{-1}$$

$$= 3.38 \times 10^{-12} \text{ erg}$$

There are two quantum states in the $3s$ level and six in the $3p$. Thus,

$$\frac{P_j}{P_0} = \frac{6}{2} = 3$$

Substituting into Equation 11-1 yields

$$\frac{N_j}{N_0} = 3 \exp\left(-\frac{3.38 \times 10^{-12}}{1.38 \times 10^{-16} \times 2500}\right)$$

which can be rewritten as

$$\log_{10} \frac{N_j}{3N_0}$$

$$= -\frac{3.38 \times 10^{-12}}{2.303 \times 1.38 \times 10^{-16} \times 2500} = -4.254$$

or

$$\frac{N_j}{N_0} = 1.67 \times 10^{-4}$$

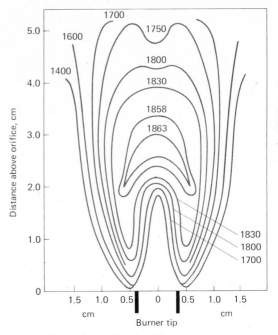

FIGURE 11-6   Temperature profiles (in °C) for a natural gas-air flame. (From B. Lewis and G. van Elbe, *J. Chem. Phys.*, **11**, 75 (1943). With permission.)

Replacing 2500 with 2510 in the foregoing equations yields

$$\frac{N_j}{N_0} = 1.74 \times 10^{-4}$$

The foregoing example demonstrates that a temperature fluctuation of only 10°K results in a 4% increase in the number of excited sodium atoms. A corresponding increase in emitted power by the two lines would result. Thus, an analytical method based on the measurement of emission requires close control of flame temperature.

Absorption and fluorescence methods are theoretically less dependent upon temperature because both measurements are based upon initially *unexcited* atoms rather than thermally excited ones. In the example just considered, only about 0.017% of the sodium atoms were thermally excited at the temperature of a hydrogen/oxygen flame. An emission method is based upon this small fraction of the analyte. In contrast, absorption and fluorescence measurements use the 99.8% of the analyte present as unexcited sodium atoms to produce the analytical signals. Note also that while a 10°K temperature change causes a 4% increase in sodium ions, the corresponding change in percent of sodium atoms is inconsequential.

Temperature fluctuations actually do exert an indirect influence on flame absorption and fluorescence measurements in several ways. An increase in temperature usually increases the efficiency of the atomization process and hence the total number of atoms in the flame. In addition, line broadening and a consequent decrease in peak height occurs because the atomic particles travel at greater rates which enhance the Doppler effect. Increased concentrations of gaseous atoms at higher temperatures also cause pressure broadening of the absorption lines. Because of these indirect effects, a reasonable control of the flame temperature is also required for quantitative absorption and fluorescence measurements.

The large ratio of unexcited to excited atoms in flames leads to another interesting comparison of the three atomic flame methods. Because atomic absorption and fluorescence methods are based upon a much larger population of particles, both procedures might be expected to be more sensitive than the emission procedure. This apparent advantage is offset in the absorption method, however, by the fact that an absorbance measurement involves evaluation of a difference $(\log P_0 - \log P)$; when the two numbers are nearly alike, larger errors must be expected in the difference. As a consequence, atomic emission and atomic absorption procedures tend to be complementary in sensitivity, the one being advantageous for one group of elements and the other for a different group.

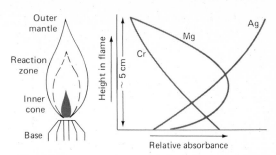

FIGURE 11-7 Flame absorption profiles for three elements. (Reprinted from J. W. Robinson, *Atomic Absorption Spectroscopy*, 2d ed. New York: p. 69. Courtesy of Marcel Dekker, 1975, p. 56.)

Flame Profiles

Important regions of the flame, from bottom to top, include the *base*, the *inner cone*, the *reaction zone*, and the *outer mantle* (see Figure 11-7). The sample enters the base of the flame in the form of minute droplets. Within this region, water evaporates from a substantial fraction of the droplets; some of the sample thus enters the inner cone as solid particles. Here, vaporization and decomposition to the atomic state occurs; it is here, also, that the excitation and absorption processes commence. Upon entering the reaction zone, the atoms are converted to oxides; these then pass into the outer mantle and are subsequently ejected from the flame. This sequence is not necessarily suffered by every drop that is aspirated into the flame; indeed, depending upon the droplet size and the flow rate, a significant portion of the sample may pass essentially unaltered through the flame.

A flame profile provides useful information about the processes that go on in different parts of a flame; it is a plot that reveals regions of the flame that have similar values for a parameter of interest. Some of these parameters include temperature (Figure 11-6), chemical composition, absorbance, and radiant or fluorescent intensity.

Emission Profiles. Figure 11-8 is a three-dimensional profile showing the emission in-

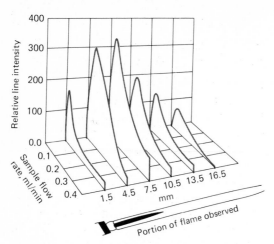

FIGURE 11-8 Flame profile for calcium line in a cyanogen-oxygen flame for different sample flow rates. (Reprinted with permission from K. Fuwa, R. E. Thiers, B. L. Vallee, and M. R. Baker, *Anal. Chem.*, **31**, 2039 (1959). Copyright by the American Chemical Society.)

tensity for a calcium line produced in a cyanogen flame. Note that the emission maximum is found just above the inner cone.

Figure 11-8 also demonstrates that the intensity of emission is critically dependent upon the rate at which the sample is introduced into the flame. Initially, the line intensity rises rapidly with increasing flow rate as a consequence of the increasing number of calcium particles. A rather sharp maximum in intensity occurs, beyond which the water, which is aspirated with the sample, lowers the flame temperature.

Where molecular band spectra form the basis for emission analysis, the peaks often appear lower in the flame than line maxima. For example, calcium produces a useful band in the region of 540 to 560 nm, probably due to the presence of CaOH in the flame (see Figure 11-5). The maximum intensity of this emission band occurs in the inner cone and decreases rapidly in the mantle as the molecules

responsible for the emission dissociate at the higher temperatures of the latter region.

The more sophisticated instruments for flame emission spectroscopy are equipped with monochromators that sample the radiation from a relatively small part of the flame; adjustment of the position of the flame with respect to the entrance slit is thus critical. Filter photometers, on the other hand, scan a much larger portion of the flame; here, control of flame position is less important.

Flame Absorbance Profiles. Figure 11-7 shows typical absorbance profiles for three elements. Magnesium exhibits a maximum in absorbance at about the middle of the flame because of two opposing effects. The initial increase in absorbance as the distance from the base becomes larger results from an increased number of atomic magnesium particles produced with longer exposure to the heat of the flame. In the reaction zone, however, appreciable oxidation of the magnesium starts to occur. This process leads to an eventual decrease in absorbance because the oxide particles formed are nonabsorbing at the wavelength used. To obtain maximum analytical sensitivity, then, the flame must be adjusted with respect to the beam until a maximum absorbance reading is obtained.

The behavior of silver, which is not readily oxidized, is quite different; here, a continuous increase in atoms is observed from the base to the periphery of the flame. In contrast, chromium, which forms very stable oxides, shows a continuous decrease in absorbance beginning at the base; this observation suggests that oxide formation predominates from the start. Clearly, a different portion of the flame should be used for the analysis of each of these elements.

ATOMIZERS FOR ATOMIC SPECTROSCOPY

Two types of atomizers are encountered in atomic spectroscopy. Flame atomizers are employed for atomic emission, absorption, and fluorescence measurements. Nonflame atomizers have been confined to the latter two procedures.

Flame Atomizers for Atomic Spectroscopy

The most common atomization device for atomic spectroscopy consists of a *nebulizer* and a burner.[3] The nebulizer produces a fine spray or aerosol from the liquid sample, which is then fed into the flame. Both *total consumption* (*turbulent flow*) and *premixed* (*laminar flow*) *burners* are encountered.

Turbulent Flow Burner. Figure 11-9 is a schematic diagram of a commercially available turbulent flow burner. Here, the nebulizer and burner are combined into a single unit. The sample is drawn up the capillary and nebulized by venturi action caused by the flow of gases around the capillary tip. Typical sample flow rates are 1 to 3 ml/min.

Turbulent flow burners offer the advantage of letting a relatively large and representative sample reach the flame. Furthermore, no possibility of flashback and explosion exists. The disadvantages of such burners include a relatively short flame path length and problems with clogging of the tip. In addition, turbulent flow burners are noisy both from the electronic and auditory standpoint. Although sometimes used for emission and fluorescence analyses, turbulent flow burners find little use in present-day absorption work.

Laminar Flow Burner. Figure 11-10 is a diagram of a typical commercial laminar flow burner. The sample is nebulized by the flow of oxidant past a capillary tip. The resulting aerosol is then mixed with fuel and flows past a series of baffles that remove all but the finest droplets. As a result of the baffles, the majority of the sample collects in the bottom

---

[3] For a detailed discussion of flame atomizers, see: R. D. Dresser, R. A. Mooney, E. M. Heithmar, and F. W. Pankey, *J. Chem. Educ.*, **52**, A403 (1975).

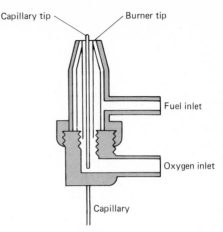

FIGURE 11-9 A turbulent flow burner. (Courtesy of Beckman Instrument, Inc., Fullerton, California.)

of the mixing chamber where it is drained to a waste container. The aerosol, oxidant, and fuel are then burned in a slotted burner that provides a flame which is usually 5 or 10 cm in length.

Laminar flow burners provide a relatively quiet flame and a significantly longer path length. These properties tend to enhance sensitivity and reproducibility. Furthermore, clogging is seldom a problem. Disadvantages include a lower rate of sample introduction (which may offset the longer path length advantage) and the possibility of selective evaporation of mixed solvents in the mixing chamber, which may lead to analytical uncertainties. Furthermore, the mixing chamber contains a potentially explosive mixture which can be ignited by a flashback. Note that the burner in Figure 11-10 is equipped with pressure relief vents for this reason.

Fuel and Oxidant Regulators. An important variable that requires close control in flame spectroscopy is the flow rate of both oxidant and fuel. It is desirable to be able to vary each over a considerable range so that ideal atomization conditions can be found experimentally. Fuel and oxidant are ordinarily combined in approximately stoichiometric amounts. For the analysis of metals that form stable oxides, however, a reducing flame, which contains an excess of fuel, may prove more desirable. Flow rates are ordinarily controlled by means of double-diaphragm pressure regulators followed by needle valves in the instrument housing.

A requirement for reproducible analytical conditions is that both the fuel and oxidant systems be equipped with some type of flowmeter. The most widely used of these is the rotameter, which consists of a tapered transparent tube that is mounted vertically with the smaller end down. A light-weight conical or spherical float is lifted by the gas flow; its vertical position is determined by the flow rate.

Nonflame Atomizers

In terms of reproducible behavior, flame atomization appears to be superior to all other methods that have been thus far developed. In terms of sampling efficiency (and thus sensitivity), however, other atomization methods are markedly better. Two reasons for the lower sampling efficiency of the flame can be cited. First, a large portion of the sample flows down

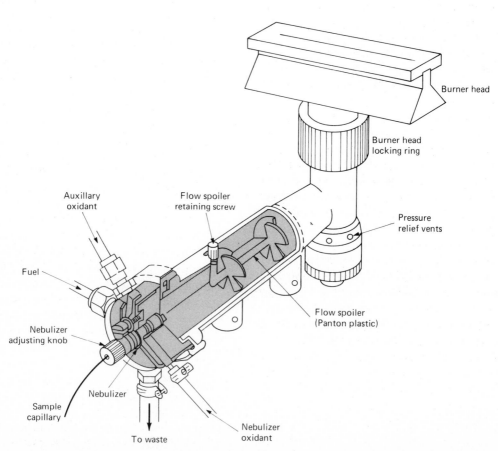

FIGURE 11-10   A laminar flow burner. (Courtesy of Perkin-Elmer Corporation, Norwalk, Connecticut.)

the drain (laminar burner) or is not completely atomized (turbulent burner). Second, the residence time of individual atoms in the optical path in the flame is brief ($\sim 10^{-4}$ s).

Since 1970, nonflame atomizers have appeared on the market.[4] These devices generally provide enhanced sensitivity because the entire sample is atomized in a short period and the average residence time of the atoms in the optical path is a second or more.

In a nonflame atomizer, a few microliters of sample are evaporated and ashed at low temperatures on an electrically heated surface of carbon, tantalum, or other conducting material. The conductor can be a hollow tube, a strip or rod, a boat, or a trough. After ashing, the current is increased to 100 A or more, which causes the temperature to soar to perhaps 2000 to 3000°C; atomization of the sample occurs in a period of a few seconds. The absorption or fluorescence of the atomized particles can then be measured in the region immediately above the heated conductor. At a wavelength where absorption or fluorescence takes place, the signal is observed to rise to a maximum after a few seconds and then decay to zero, corresponding to the atomization and subsequent escape of the volatilized sample; analyses are based upon peak height or area. Atomization is ordinarily performed in an inert gas atmosphere to prevent oxidation of the conductor.

Nonflame atomizers offer the advantage of unusually high sensitivity for small volumes of sample. Typically, sample volumes between 0.5 and 10 μl are employed; under these circumstances, absolute limits lie in the range of 10^{-10} to 10^{-13} g of analyte. These sensitivities are up to 1000 times greater than those obtained with flames.

The relative precision of nonflame methods is generally in the range of 5 to 10 % compared with the 1 to 2 % that can be expected for flame atomization.

ATOMIC ABSORPTION SPECTROSCOPY[5]

Atomic absorption spectroscopy finds considerably wider use than either of the other two atomic spectroscopic methods because it is best suited to routine analyses in the hands of relatively unskilled operators.

Sources for Atomic Absorption Methods

Analytical methods based on atomic absorption are potentially highly specific because atomic absorption lines are remarkably narrow and because electronic transition energies are unique for each element. On the other hand, the limited line widths create a measurement problem not encountered in molecular absorption. Recall that Beer's law applies only for monochromatic radiation; a linear relationship between absorbance and concentration, however, can be expected only if the bandwidth of the source is narrow with respect to the width of the absorption peak (p. 155). No ordinary monochromator is capable of yielding a band of radiation as narrow as the peak width of an atomic absorption line (0.002 to 0.005 nm). Thus, when a continuous source is employed with a monochromator for atomic absorption, only a minute fraction of the radiation is of a wavelength that is absorbed,

---

[4] See: R. D. Dresser, R. A. Mooney, E. M. Heithmar, and F. W. Plankey, *J. Chem. Educ.*, **52**, A451, A503 (1975); R. E. Sturgeon, *Anal. Chem.*, **49**, 1255A (1977); and C. W. Fuller, *Electrothermal Atomization for Atomic Absorption Spectroscopy.* London: The Chemical Society, 1978.

[5] Reference books on atomic absorption spectroscopy include: G. F. Kirkbright and M. Sargent, *Atomic Absorption and Fluorescence Spectroscopy.* New York: Academic Press, 1974; J. W. Robinson, *Atomic Absorption Spectroscopy*, 2d ed. New York: Marcel Dekker, 1975; and W. Slavin, *Atomic Absorption Spectroscopy*, 2d ed. New York: Interscience, 1978.

and the relative change in intensity of the emergent band is small in comparison to the change suffered by radiation that actually corresponds to the absorption peak. Under these conditions, Beer's law is not followed; in addition, the sensitivity of the method is lessened significantly.

This problem has been overcome by employing a source of radiation that emits a line of the same wavelength as the one to be used for the absorption analysis. For example, if the 589.6-nm line of sodium is chosen for the absorption analysis of that element, a sodium vapor lamp is useful as a source. Gaseous sodium atoms are excited by electrical discharge in such a lamp; the excited atoms then emit characteristic radiation as they return to lower energy levels. An emitted line will thus have the same wavelength as the resonance absorption line. With a properly designed source (one that operates at a lower temperature than the flame to minimize Doppler broadening), the emission lines will have bandwidths significantly narrower than the absorption bandwidths. It is only necessary, then, for the monochromator to have the capability of isolating a suitable emission line for the absorption measurement (see Figure 11-11). The radiation employed in the analysis is thus sufficiently limited in bandwidth to permit measurements at the absorption peak. Greater sensitivity and better adherence to Beer's law result.

A separate lamp source is needed for each element (or sometimes, group of elements). To avoid this inconvenience, attempts have been made to employ a continuous source with a very high resolution monochromator or, alternatively, to produce a line source by introducing a compound of the element to be determined into a high-temperature flame. Neither of these alternatives is as satisfactory as individual lamps.

Hollow Cathode Lamps. The most common source for atomic absorption measurements is the *hollow cathode lamp*, which consists of a tungsten anode and a cylindrical cathode

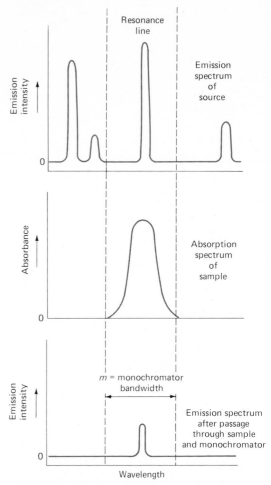

FIGURE 11-11   Absorption of a resonance line by atoms.

sealed in a glass tube that is filled with neon or argon at a pressure of 1 to 5 torr (see Figure 11-12). The cathode is constructed of the metal whose spectrum is desired or serves to support a layer of that metal.

Ionization of the gas occurs when a potential is applied across the electrodes, and a current of about 5 to 10 mA is generated as the ions migrate to the electrodes. If the potential is sufficiently large, the gaseous cations acquire enough kinetic energy to dislodge some of the metal atoms from the cathode surface and produce an atomic cloud; this process is called *sputtering*. A portion of the sputtered metal atoms are in excited states and thus emit their characteristic radiation in the usual way. Eventually, the metal atoms diffuse back to the cathode surface or the glass walls of the tube and are redeposited.

The cylindrical configuration of the cathode tends to concentrate the radiation in a limited region of the tube; this design also enhances the probability that redeposition will occur at the cathode rather than on the glass walls.

The efficiency of the hollow cathode lamp depends upon its geometry and the operating potential. High potentials, and thus high currents, lead to greater intensities. This advantage is offset somewhat by an increase in Doppler broadening of the emission lines. Furthermore, the greater currents result in an increase in the number of unexcited atoms in the cloud; the unexcited atoms, in turn, are capable of absorbing the radiation emitted by the excited ones. This *self-absorption* leads to lowered intensities, particularly at the center of the emission band.

A variety of hollow cathode tubes are available commercially. The cathodes of some consist of a mixture of several metals; such lamps permit the analysis of more than a single element.

Gaseous Discharge Lamps. Gas discharge lamps produce a line spectrum as a consequence of the passage of an electrical current through a vapor of metal atoms; the familiar sodium and mercury vapor lamps are examples. Sources of this kind are particularly useful for producing spectra of the alkali metals.

Source Modulation. In the typical atomic absorption instrument, it is necessary to eliminate interferences caused by emission of radiation by the flame. Most of the emitted radiation can be removed by locating the monochromator between the flame and the detector; nevertheless, this arrangement does not remove the flame radiation corresponding to the wavelength selected for the analysis. The flame will contain such radiation, since excitation and radiant emission by some atoms of the analyte can occur. This difficulty is overcome by modulating the output of the source so that its intensity fluctuates at a constant frequency. The detector then receives two types of signal, an alternating one from the source and a continuous one from the flame. These signals are converted to the corresponding types of electrical response. A simple high-pass, RC filter (p. 23) is then employed to remove the unmodulated dc signal and pass the ac signal for amplification.

A simple and entirely satisfactory way of modulating radiation from the source is to interpose a circular disk in the beam between the source and the flame. Alternate quadrants of this disk are removed to permit passage of light. Rotation of the disk at constant speed provides a beam that is chopped to the desired frequency. As an alternative, the power supply for the source can be designed for intermittent or ac operation.

Instruments for Atomic Absorption Spectroscopy

Instruments for atomic absorption work are offered by numerous manufacturers. The range of sophistication and cost (upward

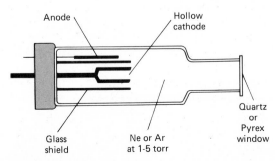

Anode

Hollow cathode

Glass shield

Ne or Ar at 1-5 torr

Quartz or Pyrex window

FIGURE 11-12 Schematic cross section of a hollow cathode lamp.

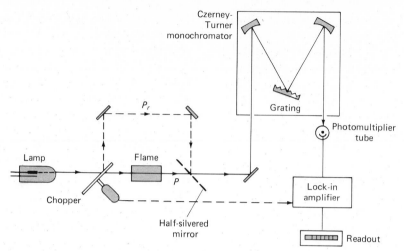

FIGURE 11-13 A typical double-beam atomic absorption spectrophotometer.

from a few thousand dollars) is substantial; as always, the potential user must select the most appropriate design for the intended use.

Single-Beam Spectrophotometers. A typical single-beam instrument for multielement analyses consists of several hollow cathode sources, a chopper, an atomizer, and a simple grating spectrophotometer with a photomultiplier transducer. It is used in the same way as a single-beam instrument for molecular absorption work. Thus, the dark current is nulled with a shutter in front of the transducer. The 100% T adjustment is then made while a blank is aspirated into the flame (or ignited in a nonflame atomizer). Finally, the transmittance is obtained with the sample replacing the blank.

Single-beam atomic absorption instruments have the same advantages and disadvantages as their molecular absorption counterparts, which were discussed on page 185.

Double-Beam Spectrophotometers. Figure 11-13 is a schematic diagram of a typical double-beam instrument. The beam from the hollow cathode source is split by a mirrored chopper, one half passing through the flame and the other half around it. The two beams are then recombined by a half-silvered mirror

and passed into a Czerney-Turner grating monochromator; a photomultiplier tube serves as the transducer. The output from the latter is fed to a lock-in amplifier which is synchronized with the chopper drive. The ratio between the reference and sample signal is then amplified and fed to the readout which may be a meter, digital device, or recorder. Alternatively, the amplified signal from the reference beam may be attenuated to match the sample signal by means of a potentiometer; the transmittance or absorbance is then read from the position of the slide wire contact.

It should be noted that the reference beam in atomic absorption instruments does not pass through the flame and thus does not correct for loss of radiant power due to absorption or scattering by the flame itself.

Applications of Atomic Absorption Spectroscopy

Atomic absorption spectroscopy is a sensitive means for the determination of more than 60 elements. Details concerning the methods for sample preparation and quantitative determination of individual elements are available in the several reference works listed in footnote 5 (p. 315).

Sensitivity and Detection Limits. Two terms are employed in characterizing atomic absorption methods. The *sensitivity* is defined as the concentration of an element in $\mu g/ml$ (or ppm) which produces a transmittance signal of 0.99 or a corresponding absorbance signal of 0.0044. Modern atomic instruments have adequate precision to discriminate between absorbance signals that differ by less than 0.0044. For this reason, the term *detection limit* has been introduced, which is defined as the concentration of the element that produces an analytical signal equal to twice the standard deviation of the background signal. (For flame atomization, the standard deviation of the background signal is obtained by observing the signal variation when a blank is sprayed into the flame.) Both the sensitivity and the detection limits vary widely with such variables as flame temperature, spectral bandwidth, detector sensitivity, and type of signal processing. Small differences among quoted values of the two parameters are not significant; for example, whereas an order of magnitude difference is certainly meaningful, a factor of 2 or 3 is probably not.

Detection limits ranging from about 3×10^{-4} ppm to 20 ppm are observed for the various metallic elements when flame atomization is employed in atomic absorption. Nonflame atomization often enhances this limit by a factor of 10 to 1000.

Columns three and four of Table 11-3 list detection limits for several common elements by flame and nonflame atomization procedures.

TABLE 11-3   DETECTION LIMITS FOR THE ANALYSIS OF SELECTED ELEMENTS BY ATOMIC ABSORPTION AND FLAME EMISSION SPECTROMETRY

| Element | Wavelength, nm | Detection Limit, $\mu g/ml$ | | |
		Nonflame Absorption[a]	Flame Absorption[b]	Flame Emission[b]
Aluminum	396.2	0.03		0.005 (N_2O)
	309.3		0.1 (N_2O)	
Calcium	422.7	0.0003	0.002 (air)	0.005 (air)
Cadmium	326.1	0.0001		2 (N_2O)
	228.8		0.005 (air)	
Chromium	425.4	0.005		0.005 (N_2O)
	357.9		0.005 (air)	
Iron	372.0	0.003		0.05 (N_2O)
	248.3		0.005 (air)	
Lithium	670.8	0.005	0.005 (air)	0.00003 (N_2O)
Magnesium	285.2	0.00006	0.00003 (air)	0.005 (N_2O)
Potassium	766.5	0.0009	0.005 (air)	0.0005 (air)
Sodium	589.0	0.0001	0.002 (air)	0.0005 (air)

[a] Data from J. W. Robinson and P. J. Slevin, *American Laboratory*, **4** (8), 14 (1972). With permission, International Scientific Communications, Inc., Fairfield, CT © 1972; wavelength not reported. Copyright 1972 by International Scientific Communications, Inc.

[b] Taken from data compiled by E. E. Pickett and S. R. Koirtyohann, *Anal. Chem.*, **41** (14), 28A (1969) and reprinted with permission. Copyright by the American Chemical Society. Data are for an acetylene flame with the oxidant shown in parentheses.

Accuracy

Under usual conditions, the relative error associated with a flame absorption analysis is of the order of 1 to 2%. With special precautions, this figure can be lowered to a few tenths of 1%.

Spectral Interferences

Interferences of two types are encountered in atomic absorption methods. *Spectral interferences* arise when the absorption of an interfering species either overlaps or lies so close to the analyte absorption that resolution by the monochromator becomes impossible. *Chemical interferences* result from various chemical processes occurring during atomization that alter the absorption characteristics of the analyte. A brief discussion of spectral interferences follows; sources of chemical interference are considered in the next section.

Because the emission lines of hollow cathode sources are so very narrow, interference due to overlap of atomic spectral lines is rare. For such an interference to occur, the separation between the two lines would have to be less than perhaps 0.1 Å. For example, a vanadium line at 3082.11 Å interferes in an analysis based upon the aluminum absorption line at 3082.15 Å. The interference is readily avoided, however, by employing the aluminum line at 3092.7 Å instead.

Spectral interferences also result from the presence of combustion products that exhibit broad band absorption or particulate products that scatter radiation. Both diminish the power of the transmitted beam and lead to positive analytical errors. Where the source of these products is the fuel and oxidant mixture alone, corrections are readily obtained from absorbance measurements while a blank is aspirated into the flame. Note that this correction must be employed with a double-beam as well as a single-beam instrument because the reference beam of the former does not pass through the flame (see Figure 11-13).

A much more troublesome problem is encountered when the source of absorption or scattering originates in the sample matrix; here, the power of the transmitted beam, P, is reduced by the nonanalyte components of the sample matrix, but the incident beam power, P_0 is not; a positive error in absorbance and thus concentration results. An example of a potential matrix interference due to absorption occurs in the determination of barium in alkaline-earth mixtures. As shown by the dotted line in Figure 11-5, the wavelength of the barium line used for atomic absorption analysis appears in the center of an absorption band for CaOH; clearly, interference of calcium in a barium analysis is to be expected. In this particular situation, the effect is readily eliminated by substituting nitrous oxide for air as the oxidant for the acetylene; the higher temperature decomposes the CaOH and eliminates the absorption band.

Spectral interference due to scattering by products of atomization is most often encountered when concentrated solutions containing elements such as Ti, Zr, and W—which form refractory oxides—are aspirated into the flame. Metal oxide particles with diameters greater than the wavelength of light appear to be formed; significant scattering of the incident beam results.

Fortunately, spectral interferences by matrix products are not widely encountered and often can be avoided by variations in the analytical parameters, such as temperature and fuel-to-oxidant ratio. Alternatively, if the source of interference is known, an excess of the interfering substance can be added to both sample and standards; provided the excess is large with respect to the concentration from the sample matrix, the contribution of the latter will become insignificant. The added substance is sometimes called a *radiation buffer*.

The foregoing technique is not effective with complex samples because the source of the interference may not be known; a background correction must, therefore, be employed.[6] A description of some of the most commonly used techniques follows.

The Two-Line Correction Method. The two-line correction procedure requires the presence of a reference line from the source; this line should lie as close as possible to the analyte line but *must not be absorbed by the analyte*. If these conditions are met, it is assumed that any decrease in power of the reference line from that observed during calibration arises from absorption or scattering by the matrix products of the sample; this decrease is then used to correct the power of the analyte line.

The reference line may be from an impurity in the hollow cathode lamp, a neon or argon line from the gas contained in the lamp, or a nonresonant emission line of the element being determined.

Unfortunately, a suitable reference line is often not available.

The Continuous-Source Correction Method. A second method for background correction is available with suitably equipped double-beam instruments. Here, a hydrogen or deuterium lamp is employed as a source of continuous radiation throughout the ultraviolet region. The configuration of the chopper is then rearranged so that the reference beam shown in Figure 11-13 is dispensed with, and instead, radiation from the continuous source and the hollow cathode lamp are passed alternately through the flame. The power of the beams from the two sources is then compared rather than the power of a sample and reference beam from the hollow cathode source. The slit width is kept sufficiently wide so that the fraction of the continuous source that is absorbed by the atoms of the sample is negligible. Thus, the attenuation of its power during passage through the flame reflects only the broad band absorption or scattering by the flame components. A background correction is thus achieved.

The Zeeman Effect Correction Method. One commercial atomic absorption instrument achieves a background correction by taking advantage of the Zeeman effect to split the analyte line into two components, one of which is displaced from the other by a small wavelength increment (about 0.01 nm). The two components are also polarized at 90 deg to one another and can be monitored alternatively by inserting a rotating polarizer into the beam path. Here, the displaced component is sufficiently separated from the absorption peak to provide a means of correction for background absorption or scattering.

Zeeman splitting can be accomplished by exposing the atomizer device or the light source to a strong magnetic field.[7]

Chemical Interferences

Chemical interferences are more common than spectral ones. Their effects can frequently be minimized by suitable choice of conditions.

Both theoretical and experimental evidence suggest that many of the processes occurring in the mantle of a flame are in approximate equilibrium. As a consequence, it becomes possible to regard the burned gases of the flame as a solvent medium to which thermodynamic calculations can be applied. The processes of principal interest include formation of compounds of low volatility, dissociation reactions, and ionizations.

Formation of Compounds of Low Volatility. Perhaps the most common type of interference is by anions which form compounds

---

[6] For a critical discussion of the various methods for background correction, see: A. T. Zander, *Amer. Lab.*, **8** (11), 11 (1976).

[7] For a detailed discussion of the application of the Zeeman effect to atomic absorption, see: T. Hadeishi and R. D. McLaughlin, *Science*, **174**, 404 (1974); T. Hadeishi and R. D. McLaughlin, *Anal. Chem.*, **48**, 1009 (1976); H. Koizuma and K. Yasuda, *Anal. Chem.*, **48**, 1178 (1976); and S. D. Brown, *Anal. Chem.*, **49** (14), 1269A (1977).

of low volatility with the analyte and thus reduce the rate at which it is atomized. Low results are the consequence. An example is the decrease in calcium absorbance that is observed with increasing concentrations of sulfate or phosphate. For a fixed calcium concentration, the absorbance is found to fall off nearly linearly with increasing sulfate or phosphate concentration until the anion to calcium ratio is about 0.5; the absorbance then levels off at about 30 to 50% of its original value and becomes independent of anion concentration.

Examples of cation interference have also been recognized. Thus, aluminum is found to cause low results in the determination of magnesium, apparently as a result of the formation of a heat-stable aluminum/magnesium compound (perhaps an oxide).

Interferences due to formation of species of low volatility can often be eliminated or moderated by use of higher-temperature flames. Alternatively, *releasing agents*, which are cations that react preferentially with the interference and prevent its interaction with the analyte, can be employed. For example, addition of an excess of strontium or lanthanum ion minimizes the interference of phosphate in the determination of calcium. The same two species have also been employed as releasing agents for the determination of magnesium in the presence of aluminum. In both instances, the strontium or lanthanum replaces the analyte in the compound formed with the interfering species.

Protective agents prevent interferences by forming stable but volatile species with the analyte. Three common reagents for this purpose are EDTA, 8-hydroxyquinoline, and APDC (the ammonium salt of 1-pyrrolidine-carbodithioic acid). The presence of EDTA has been shown to eliminate the interference of aluminum, silicon, phosphate, and sulfate in the determination of calcium. Similarly, 8-hydroxyquinoline suppresses the interference of aluminum in the determination of calcium and magnesium.

Dissociation Equilibria. In the hot, gaseous environment of a flame, numerous dissociation and association reactions lead to conversion of the metallic constituents to the elemental state. It seems probable that at least some of these reactions are reversible and can be treated by the laws of thermodynamics. Thus, in theory, it should be possible to formulate equilibria such as

$$MO \rightleftharpoons M + O$$

$$M(OH)_2 \rightleftharpoons M + 2OH$$

or, more generally,

$$MA \rightleftharpoons M + A$$

In practice, not enough is known about the nature of the chemical reactions in a flame to permit a quantitative treatment such as that for an aqueous solution. Instead, reliance must be placed on empirical observations.

Dissociation reactions involving metal oxides and hydroxides clearly play an important part in determining the nature of the emission or absorption spectra for an element. For example, the alkaline-earth oxides are relatively stable, with dissociation energies in excess of 5 eV. Molecular bands arising from the presence of metal oxides or hydroxides in the flame thus constitute a prominent feature of their spectra. Except at very high temperatures, these bands are more intense than the lines for the atoms or ions. In contrast, the oxides and hydroxides of the alkali metals are much more readily dissociated so that line intensities for these elements are high, even at relatively low temperatures.

It seems probable that dissociation equilibria involving anions other than oxygen may also influence flame emission. For example, the line intensity for sodium is markedly decreased by the presence of HCl. A likely explanation is the mass-action effect on the equilibrium

$$NaCl \rightleftharpoons Na + Cl$$

Chlorine atoms formed from the added HCl decrease the atomic sodium concentration and thereby lower the line intensity.

Another example of this type of interference involves the enhancement of the absorption by vanadium when aluminum or titanium is present. The interference is significantly more pronounced in fuel-rich flames than in lean. These effects are readily explained by assuming that the three metals interact with such species as O and OH, which are always present in flames. If the oxygen-bearing species are given the general formula Ox, a series of equilibrium reactions can be postulated. Thus,

$$VOx \rightleftharpoons V + Ox$$

$$AlOx \rightleftharpoons Al + Ox$$

$$TiOx \rightleftharpoons Ti + Ox$$

In fuel-rich combustion mixtures, the concentration of Ox is sufficiently small that its concentration is lowered significantly when aluminum or titanium is present in the sample. This decrease causes the first equilibrium to shift to the right with an accompanying increase in metal concentration as well as absorbance. In lean mixtures, on the other hand, the concentration of Ox is apparently high relative to the total concentration of metal atoms. Thus, addition of aluminum or titanium scarcely changes the concentration of Ox. Therefore, the position of the first equilibrium is not disturbed significantly.

Ionization in Flames. Ionization of atoms and molecules is small in combustion mixtures that involve air as the oxidant, and generally can be neglected. At the higher temperatures of flames employing oxygen or nitrous oxide, however, ionization becomes important, and a significant concentration of free electrons exists as a consequence of the equilibrium

$$M \rightleftharpoons M^+ + e^- \qquad (11\text{-}2)$$

where M represents a neutral atom or molecule and M^+ is its ion. We will focus upon equilibria in which M is a metal atom.

The equilibrium constant K for this reaction may take the form

$$K = \frac{[M^+][e^-]}{[M]} = \left(\frac{x^2}{1-x}\right)p \qquad (11\text{-}3)$$

where the bracketed terms are activities, x is the fraction of M that is ionized, and p is the partial pressure of the metal in the gaseous solvent before ionization.

The effect of temperature on K is given by the *Saha equation*; namely,

$$\log K = \frac{-5041E_i}{T} + \tfrac{5}{2}\log T$$

$$- 6.49 + \log \frac{g_{M^+}g_{e^-}}{g_M} \qquad (11\text{-}4)$$

where E_i is the ionization potential for the metal in electron volts, T is the absolute temperature of the medium, and g is a statistical weight for each of the three species indicated by the subscripts. For the alkali metals, the final term has a value of zero while for the alkaline earths it is equal to 0.6.

Table 11-4 shows the calculated degree of ionization for several common metals under conditions that approximate those used in flame emission spectroscopy. The temperatures correspond roughly to conditions that exist in air/fuel, oxygen/acetylene, and oxygen/cyanogen flames, respectively.

It is important to appreciate that treatment of the ionization process as an equilibrium—with free electrons as one of the products—immediately implies that the degree of ionization of a metal will be strongly influenced by the presence of other ionizable metals in the flame. Thus, if the medium contains not only species M, but species B as well, and if B ionizes according to the equation

$$B \rightleftharpoons B^+ + e^-$$

then the degree of ionization of M will be decreased by the mass-action effect of the electrons formed from B. Determination of the degree of ionization under these conditions requires a calculation involving the dissociation constant for B and the mass-balance expression

$$[e^-] = [B^+] + [M^+]$$

The presence of atom-ion equilibria in flames has a number of important consequences in flame spectroscopy. For example, the intensity of atomic emission or absorption lines for the alkali metals, particularly potassium, rubidium, and cesium, is affected by temperature in a complex way. Increased temperatures cause an increase in the population of excited atoms, according to the Boltzmann relationship (p. 309); counteracting this effect, however, is a decrease in concentration of atoms as a result of ionization. Thus, under some circumstances a decrease in emission or absorption may be observed in hotter flames. It is for this reason that lower excitation temperatures are usually specified for the analysis of alkali metals.

The effects of shifts in ionization equilibria can frequently be eliminated by addition of an *ionization suppressor*, which provides a relatively high concentration of electrons to the flame; suppression of ionization of the analyte results. The effect of a suppressor is demonstrated by the calibration curves for strontium shown in Figure 11-14. Note the marked steepening of these curves as strontium ionization is repressed by the increasing concentration of potassium ions and electrons. Note also the enhanced sensitivity that results from the use of nitrous oxide instead of air as the oxidant; the higher temperature achieved with nitrous oxide undoubtedly enhances the rate of decomposition and volatilization of the strontium compounds in the plasma.

Analytical Techniques

Both calibration curves and the standard addition method are suitable for atomic absorption spectroscopy.

Calibration Curves. While in theory, ab-

TABLE 11-4 DEGREE OF IONIZATION OF METALS AT FLAME TEMPERATURES[a]

Element	Ionization Potential, eV	Fraction Ionized at the Indicated Pressure and Temperature					
		$p = 10^{-4}$ atm			$p = 10^{-6}$ atm		
		2000°K	3500°K	5000°K	2000°K	3500°K	5000°K
Cs	3.893	0.01	0.86	> 0.99	0.11	> 0.99	> 0.99
Rb	4.176	0.004	0.74	> 0.99	0.04	> 0.99	> 0.99
K	4.339	0.003	0.66	0.99	0.03	0.99	> 0.99
Na	5.138	0.0003	0.26	0.98	0.003	0.90	> 0.99
Li	5.390	0.0001	0.18	0.95	0.001	0.82	> 0.99
Ba	5.210	6×10^{-4}	0.41	0.99	6×10^{-3}	0.95	> 0.99
Sr	5.692	1×10^{-4}	0.21	0.97	1×10^{-3}	0.87	> 0.99
Ca	6.111	3×10^{-5}	0.11	0.94	3×10^{-4}	0.67	0.99
Mg	7.644	4×10^{-7}	0.01	0.83	4×10^{-6}	0.09	0.75

[a] Data calculated from Equation 11-4; from B. L. Vallee and R. R. Thiers, in *Treatise on Analytical Chemistry*, eds. I. M. Kolthoff and P. J. Elving. New York: Interscience, 1965, Part I, vol. 6, p. 3500. With permission.

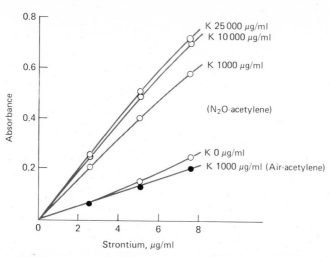

FIGURE 11-14 Effect of potassium concentration on the calibration curve for strontium. (Reprinted with permission from J. A. Bowman and J. B. Willis, *Anal. Chem.*, **39**, 1210 (1967). Copyright by the American Chemical Society.)

sorbance should be proportional to concentration, deviations from linearity often occur. Thus, empirical calibration curves must be prepared. In addition, there are sufficient uncontrollable variables in the production of an atomic vapor to warrant measuring the absorbance of at least one standard solution each time an analysis is performed. Any deviation of the standard from the original calibration curve can then be employed to correct the analytical results.

Standard Addition Method. The standard addition method is widely used in atomic absorption spectroscopy. Here, two or more aliquots of the sample are transferred to volumetric flasks. One is diluted to volume, and the absorbance of the solution is obtained. A known amount of analyte is added to the second, and its absorbance is measured after dilution to the same volume. Data for other standard additions may also be obtained. If a linear relationship between absorbance and concentration exists (and this should be established by several standard additions), the following relationships apply:

$$A_x = kC_x$$

$$A_T = k(C_s + C_x)$$

where C_x is the analyte concentration in the diluted sample and C_s is the contribution of the added standard to the concentration; A_x and A_T are the two measured absorbances. Combination of the two equations yields

$$C_x = C_s \frac{A_x}{(A_T - A_x)} \qquad (11\text{-}5)$$

If several additions are made, A_T can be plotted against C_s. The resulting straight line can be extrapolated to $A_T = 0$. Substituting this value into Equation 11-5 reveals that at the intercept, $C_x = -C_s$.

The standard addition method has the advantage that it tends to compensate for variations caused by physical and chemical interferences in the sample solution.

Use of Organic Solvents. It has been found that the heights of flame absorption and emission peaks are generally enhanced in the presence of lower-molecular-weight alcohols, ketones, and esters, both in the presence and in the absence of water. The effect is largely attributable to an increased nebulizer efficiency; the lower surface tension of such solutions results in finer drop sizes and a consequent increase in the amount of sample that reaches the flame.

Leaner fuel-oxidant ratios must be employed with organic solvents in order to offset the effect of the added organic material. Unfortunately, however, the leaner mixture results in lower flame temperatures and a consequent increase in the possibility for chemical interferences.

An important analytical application of organic solvents to flame spectroscopy is encountered when it is necessary to concentrate the analyte by solvent extraction with a water-immiscible organic compound. Often in these applications a chelating agent, such as dithizone or 8-hydroxyquinoline, converts the analyte ion to a species that is soluble in organic solvents. Ordinarily, the organic extract is nebulized directly into the flame for atomic absorption, emission, or fluorescence measurement.

ATOMIC EMISSION SPECTROSCOPY

Atomic emission spectroscopy (also called flame emission spectroscopy or flame photometry) has found widespread application to elemental analysis. Its most important uses have been in the analysis of sodium, potassium, lithium, and calcium, particularly in biological fluids and tissues. For reasons of convenience, speed, and relative freedom from interferences, flame emission spectroscopy has become the method of choice for these otherwise difficult-to-determine elements. The method has also been applied, with varying degrees of success, to the determination of perhaps half the elements in the periodic table. Thus, flame emission spectroscopy must be con-

sidered to be one of the important tools for analysis.[8]

Instrumentation

Instruments for flame emission work are similar in construction to the flame absorption instruments except that the flame now acts as the radiation source; the hollow cathode lamp and chopper are, therefore, unnecessary. Many modern instruments are adaptable to either emission or absorption analysis.

Much of the early work in atomic emission analyses was accomplished with turbulent flow burners. Laminar flow burners, however, are becoming more and more widely used.

Spectrophotometers. For nonroutine analysis, a recording, ultraviolet-visible spectrophotometer with a resolution of perhaps 0.5 Å is desirable. The recording feature provides a simple means for making background corrections (see Figure 11-17).

Photometers. Simple filter photometers often suffice for routine analyses of the alkali and alkaline-earth metals. A low-temperature flame is employed to eliminate excitation of most other metals. As a consequence, the spectra are simple, and glass or interference filters can be used to isolate the desired emission line.

Several instrument manufacturers supply flame photometers designed specifically for the analysis of sodium, potassium, and lithium in blood serum and other biological samples. In these instruments, the radiation from the flame is split into three beams of approximately equal power. Each then passes

---

[8] For a more complete discussion of the theory and applications of flame emission spectroscopy, see: *Flame Emission and Atomic Absorption Spectroscopy*, vol. 1: *Theory*; vol. 2: *Components and Techniques*; vol. 3: *Elements and Matrices*, eds. J. A. Dean and T. C. Rains. New York: Marcel Dekker, 1974; J. A. Dean, *Flame Photometry.* New York: McGraw-Hill, 1960; and B. L. Vallee and R. E. Thiers, in *Treatise on Analytical Chemistry*, eds. I. M. Kolthoff and P. J. Elving. New York: Interscience, 1965, Part I, vol. 6, Chapter 65.

into a separate photometric system consisting of an interference filter (which transmits an emission line of one of the elements while absorbing those of the other two), a phototube, and an amplifier. The outputs can then be measured separately if desired. Ordinarily, however, lithium serves as an *internal standard* for the analysis. For this purpose, a fixed amount of lithium is introduced into each standard and sample. The ratios of outputs of the sodium and lithium transducer and the potassium and lithium transducer then serve as analytical parameters. This system provides improved accuracy because the intensities of the three lines are affected in the same way by most analytical variables, such as flame temperature, fuel flow rates, and background radiation. Clearly, lithium must be absent in the sample.

Automated Flame Photometers. Fully automated photometers now exist for the determination of sodium and potassium in clinical samples. In one of these, the samples are withdrawn sequentially from a sample turntable, dialyzed to remove protein and particulates, diluted with the lithium internal standard, and aspirated into a flame. Sample and reagent transport is accomplished with a roller-type pump. Air bubbles serve to separate samples. Results are printed out on a paper tape. Calibration is performed automatically after every nine samples. The instrument costs about $9000.

Instruments for Simultaneous Multielement Analyses

During the past decade, considerable effort has been made toward the development of instruments for rapid sequential or simultaneous flame determination of several elements in a single sample.[9] We have already considered one example, the simple photometer for the simultaneous determination of sodium and potassium. Some of these efforts have resulted in the development of computer-controlled monochromators that permit rapid sequential measurement of radiant power at several wavelengths corresponding to peaks for various elements. With such instruments, two to three seconds are required to move from one peak to the next. The detector photocurrent is then measured for one or two seconds. Thus, it is possible to determine the concentration of as many as ten elements per minute. Instruments of this type have been employed with all three types of flame methods, although the emission method has the distinct advantage of not requiring several sources.

Simultaneous, multielement, flame emission analyses have been made possible by the use of the optical multichannel analyzer that was described earlier (p. 142). As an example,[10] a silicon diode vidicon tube was mounted on the optical plane originally occupied by the slit of an ordinary grating monochromator. The diameter of the tube surface was such that a 20-nm band of radiation was continuously monitored; by adjustment of the tube along the focal plane of the monochromator, various 20-nm bands of the spectrum could be observed. The resolution within the 20-nm band was such that lines 0.14 nm apart could be resolved.

Figure 11-15 shows a spectrum obtained simultaneously for eight elements that have emission peaks in the wavelength range of 388.6 to 408.6 nm. A nitrous oxide/acetylene flame was employed for excitation. Slightly more than half a minute was required to accumulate data for the eight analyses. A relative precision better than 5% was obtained.

---

[9] For a review of this topic, see: K. W. Busch and G. H. Morrison, *Anal. Chem.*, **45** (8), 712A (1973).

[10] K. W. Busch, N. G. Howell, and G. H. Morrison, *Anal. Chem.*, **46**, 575 (1974).

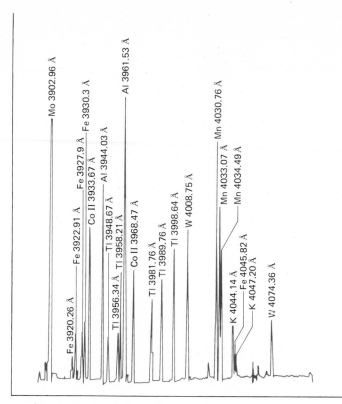

FIGURE 11-15 Multielement flame emission spectrum from 388.6 to 408.6 nm. [K. W. Busch, N. G. Howell, and G. H. Morrison, *Anal. Chem.*, **46,** 578 (1974). With permission of the American Chemical Society.]

Interferences

The interferences encountered in flame emission spectroscopy arise from the same sources as those in atomic absorption methods (see p. 320); the severity of any given interference will often differ for the two procedures, however.

Spectral Line Interference. Interference between two overlapping atomic absorption peaks occurs only in the occasional situation where the peaks are within about 0.1 Å of one another. That is, the high degree of spectral specificity is more the result of the narrow line properties of the source than the high resolution of the monochromator. Atomic emission spectroscopy, in contrast, depends entirely upon the monochromator for selectivity; the probability of spectral interference due to line overlap is consequently greater. Figure 11-16 shows an emission spectrum for three transition elements, iron, nickel, and chromium. Note that several unresolved peaks exist and that care would have to be taken to avoid spectral interference in the analysis for any one of these elements.

Band Interference; Background Correction. Emission lines are often superimposed on bands emitted by oxides and other molecular species from the sample, the fuel, or the oxidant. An example appears in Figure 11-17. As shown in the figure, a background correction for band emission is readily made by scanning for a few ångström units on either side of the analyte peak. For nonrecording instruments, a measurement on either side of the peak suffices. The average of the two measurements is then subtracted from the total peak height.

Chemical Interferences. Chemical interferences in flame emission studies are essentially the same as those encountered in flame absorption methods. They are dealt with by judicious choice of flame temperature and the use of protective agents, releasing agents, and ionization suppressors.

Self-Absorption. The center of a flame is hotter than the exterior; thus, atoms that emit in the center are surrounded by a cooler region which contains a higher concentration of unexcited atoms; *self-absorption* of the resonance wavelengths by the atoms in the cooler layer will occur. Doppler broadening of the emission line is greater than the corresponding broadening of the resonance absorption line, however, because the particles are moving more rapidly in the hotter-emission zone. Thus, self-absorption tends to alter the center of a line more than its edges. In the extreme, the center may become less intense than the edges, or it may even disappear; the result is division of the emission maximum into what appears

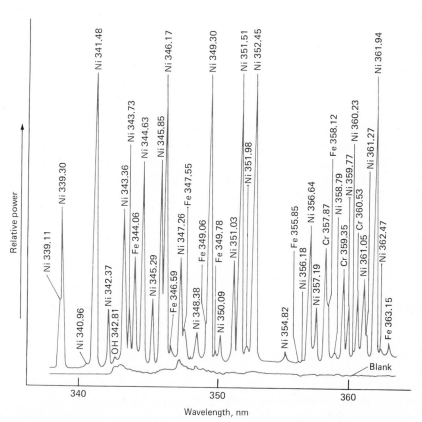

FIGURE 11-16   Partial oxyhydrogen flame-emission spectrum for a sample containing 600 ppm Fe, 600 ppm Ni, and 200 ppm Cr. (Taken from R. Herrmann and C. T. J. Alkemade, *Chemical Analysis by Flame Photometry*, 2d ed. New York: Interscience, 1963, p. 527. With permission.)

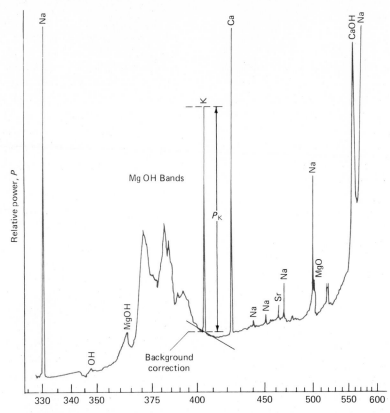

FIGURE 11-17 Flame emission spectrum for a natural brine showing the method used for correcting for background radiation. (Taken from R. Herrmann and C. T. J. Alkemade, *Chemical Analysis by Flame Photometry*, 2d ed. New York: Interscience, 1963, p. 484. With permission.)

to be two peaks by *self-reversal*. Figure 11-18 shows an example of severe self-absorption and self-reversal.

Self-absorption often becomes troublesome when the analyte is present in high concentration. Under these circumstances, a nonresonance line, which cannot undergo self-absorption, may be preferable for an analysis.

Self-absorption and ionization sometimes result in S-shaped emission calibration curves with three distinct segments. At intermediate concentrations of potassium, for example, a

linear relationship between intensity and concentration is observed (Figure 11-19). At low concentrations, curvature is due to the increased degree of ionization in the flame. Self-absorption, on the other hand, causes negative departures from a straight line at higher concentrations.

Analytical Techniques

The analytical techniques for flame emission spectroscopy are similar to those described earlier for atomic absorption spectroscopy

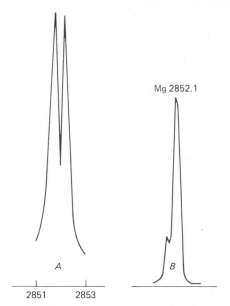

FIGURE 11-18 Curve *A* illustrates the self reversal that occurs with high concentration of Mg (2000 μg/ml). Curve *B* shows the normal spectrum of 100 μg/ml of Mg.

(p. 324). Both calibration curves and the standard addition method are employed. In addition, internal standards may be used to compensate for flame variables.

Comparison of Atomic Emission and Atomic Absorption Methods

For purposes of comparison, the main advantages and disadvantages of the two widely used flame methods are listed in the paragraphs that follow.[11] The comparisons apply to versatile spectrophotometers that are readily adapted to the determination of numerous elements.

[11] For an excellent comparison of the two methods, see: E. E. Pickett and S. R. Koirtyohann, *Anal. Chem.*, **41** (14), 28A (1969).

1. *Instruments*. A major advantage of the emission procedure is that the flame serves as the source. In contrast, absorption methods require an individual lamp for each element (or sometimes, for a limited group of elements). On the other hand, the quality of the monochromator for an absorption instrument does not have to be so great to achieve the same degree of selectivity because of the narrow lines emitted by the hollow cathode lamp.

2. *Operator Skill*. Emission methods generally require a higher degree of operator skill because of the critical nature of such adjustments as wavelength, flame zone sampled, and fuel-to-oxidant ratio.

3. *Background Correction*. Correction for band spectra arising from sample constituents is more easily, and often more exactly, carried out for emission methods.

4. *Precision and Accuracy*. In the hands of a skilled operator, uncertainties are about the same for the two procedures (± 0.5 to 1% relative). With less skilled personnel, atomic absorption methods have an advantage.

5. *Interferences*. The two methods suffer from similar chemical interferences. Atomic

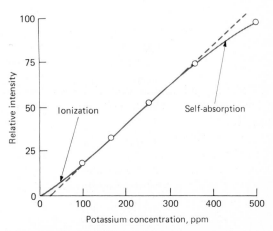

FIGURE 11-19 Effects of ionization and self-absorption on a calibration curve for potassium.

absorption procedures are less subject to spectral line interferences, although such interferences are usually easily recognized and avoided in emission methods. Spectral band interferences were considered under background correction.

6. *Detection Limits.* The data in Table 11-5 provide a comparison of detection limits and emphasizes the complementary nature of the two procedures.

ATOMIC FLUORESCENCE SPECTROSCOPY

Since 1966, considerable research has been carried out on the principles and applications of atomic fluorescence spectroscopy.[12] It is evident from such work that this procedure is somewhat more sensitive for perhaps five to ten elements than either of the two atomic methods we have just considered, particularly

when nonflame atomization is employed. On the other hand, for many elements, the method is less sensitive and appears to have a smaller useful concentration range. Furthermore, for comparable performance, fluorescence instruments appear to be more complex and potentially more expensive to purchase and maintain.[13]

Instrumentation

An instrument for atomic fluorescence measurements contains a modulated source, an atomizer (flame or nonflame), a monochromator or an interference-filter system, a detector, and a signal-processing system. With the exception of the source, most of these components are similar to those discussed in earlier parts of this chapter.

A continuous source would clearly be desirable for atomic fluorescence measurements. Unfortunately, however, the output power of a continuous source over a region as narrow

---

[12] For further information on atomic fluorescence spectroscopy, see: C. Veillon, in *Trace Analysis*, ed. J. D. Winefordner, New York: Wiley, 1976, Chapter VI; and J. D. Winefordner, *J. Chem. Educ.*, **55**, 72 (1978).

[13] See: W. B. Barnett and H. L. Kahn, *Anal. Chem.*, **44**, 935 (1972).

TABLE 11-5 COMPARISON OF DETECTION LIMITS FOR VARIOUS ELEMENTS BY FLAME ABSORPTION AND FLAME EMISSION METHODS[a]

Flame Emission More Sensitive	Sensitivity About the Same	Flame Absorption More Sensitive
Al, Ba, Ca, Eu, Ga, Ho, In, K, La, Li, Lu, Na, Nd, Pr, Rb, Re, Ru, Sm, Sr, Tb, Tl, Tm, W, Yb	Cr, Cu, Dy, Er, Gd, Ge, Mn, Mo, Nb, Pd, Rh, Sc, Ta, Ti, V, Y, Zr	Ag, As, Au, B, Be, Bi, Cd, Co, Fe, Hg, Ir, Mg, Ni, Pb, Pt, Sb, Se, Si, Sn, Te, Zn

[a] Adapted from E. E. Pickett and S. R. Koirtyohann, *Anal. Chem.*, **41** (14), 42A (1969) and reprinted with permission. Copyright by the American Chemical Society.

as an atomic line width is so low as to restrict the sensitivity of the method.

Conventional hollow cathode lamps operated continuously do not provide enough radiant power for fluorescence analysis. However, very intense bursts of radiant energy can be achieved by pulsing the lamps with large currents for brief periods. Provided the pulse width and period are chosen to give a low average current, destruction of the lamp is avoided. The detector must be gated to observe the fluorescence only during the pulse.

Gaseous discharge lamps with sufficient radiant power for fluorescence measurement are available for several of the more volatile elements such as the alkalis, mercury, cadmium, zinc, thallium, and gallium.

Electrodeless discharge lamps provide high-intensity line sources for the various elements. Typically, these lamps consist of a sealed quartz tube containing an inert gas at perhaps 1 torr and the element or a salt of the analyte element. Excitation is obtained by placing the tube in a microwave field generated by an antenna or a resonant cavity.

Tunable dye lasers are beginning to find use as sources for atomic fluorescence studies.

Interferences

Interferences encountered in atomic fluorescence spectroscopy appear to be of the same type and of about the same magnitude as those found in atomic absorption spectroscopy.[14]

Applications

Atomic fluorescence methods have been applied to the analysis of metals in such materials as lubricating oils, seawater biological material, graphite, and agricultural samples.[15]

---

[14] See: J. D. Winefordner and R. C. Elser, *Anal. Chem.*, **43** (4), 25A (1971).

[15] For a summary of applications, see: J. D. Winefordner, *J. Chem. Educ.*, **55**, 72 (1978).

PROBLEMS

1. Describe the basic differences between atomic emission and atomic fluorescence spectroscopy.

2. Define the following terms: (a) atomization, ((b) pressure broadening, (c) Doppler broadening, (d) turbulent flow nebulizer, (e) laminar flow nebulizer, (f) hollow cathode lamp, (g) sputtering, (h) self-absorption, (i) sensitivity, (j) detection limit, (k) spectral interference, (l) chemical interference, (m) radiation buffer, (n) releasing agent, (o) protective agent, and (p) ionization suppressor.

3. Why is the CaOH spectrum in Figure 11-5 so much broader than the Ba resonance line?

4. Why is atomic emission more sensitive to flame instability than atomic absorption or fluorescence?

5. Describe the effects which are responsible for the three different absorbance profiles in Figure 11-7.

6. Why is a nonflame atomizer more sensitive than a flame atomizer?

7. Why is source modulation employed in atomic absorption spectroscopy?

8. Describe how a deuterium lamp can be employed to provide a background correction for an atomic absorption spectrum.

9. Why do organic solvents enhance atomic absorption and emission peaks?

10. Explain why curve A in Figure 11-18 is split into a doublet.

11. The cobalt in an aqueous sample was determined by pipetting 10.0 ml of the unknown into each of four 50.0-ml volumetric flasks. Various volumes of a standard containing 6.23 ppm Co were added to the flasks following which the solutions were diluted to volume. Calculate the parts per million of cobalt in the sample from the data that follow.

Sample	Unknown, ml	Standard, ml	Absorbance
blank	0.0	0.0	0.042
A	10.0	0.0	0.201
B	10.0	10.0	0.292
C	10.0	20.0	0.378
D	10.0	30.0	0.467
E	10.0	40.0	0.554

12. What is the equilibrium ratio of excited Tl atoms to those in the ground state when radiation of 3776 Å is absorbed and the temperature is (see Figure 11-4)
    (a) 2200°K?
    (b) 2800°K?

13. For the flame shown in Figure 11-6, calculate the percent decrease in sodium emission intensity (5893 Å) when the flame is observed at 4.0 cm rather than 3.0 cm above the orifice?

14. Calculate the equilibrium constant for the reaction $Ca \rightleftharpoons Ca^+ + e^-$ at 2800°K.

15. Calculate the equilibrium constant for the reaction $Ca^+ + Rb \rightleftharpoons Ca + Rb^+$ at 2800°K.

16. The Doppler effect is one of the sources of line broadening in atomic absorption spectroscopy. Atoms moving toward the light source see a higher frequency than do atoms moving away from the source. The difference in wavelength, $\Delta\lambda$, experienced by an atom moving at speed v (compared to one at rest) is $\Delta\lambda/\lambda = v/c$, where c is the speed of light. Estimate the line width (in Å) of the sodium D line (5893 Å) when the absorbing atoms are at a temperature of 2000°K. The average speed of an atom is given by $v = \sqrt{8kT/\pi m}$, where k is Boltzmann's constant, T is temperature, and m is the mass.

12

EMISSION SPECTROSCOPY

The term *emission spectroscopy* usually refers to a type of atomic spectroscopy that employs more energetic excitation sources than the flames and ovens described in Chapter 11. Until recently, two types of sources were employed in emission methods, the electric arc and the electric spark.[1] During the last decade, however, argon *plasma sources* have been developed that combine many of the best features of flame sources with the attributes of the classical arc and spark. As a consequence, there has been a recent resurgence of interest in emission methods, and several new instruments employing argon plasma sources have appeared on the market since 1977.

Arc and spark emission methods have several advantages that account for their widespread use since the 1930s or earlier. Among these is their wide applicability (70 or more metal or metalloid elements), their high degree of specificity, and their sensitivity, which often permits analyses in the parts-per-million or parts-per-billion range. Arc and spark emission methods offer certain advantages over flame methods, which account for their continued use even after the remarkable growth of the latter beginning in the mid-1960s. One of these advantages is the minimal sample preparation required for emission methods, because excitation can usually be carried out directly on liquids, powders, metals, and glasses. Second, the higher energies employed tend to reduce interelement interference. Third, good spectra for most elements can be obtained

under a single set of excitation conditions; as a consequence, spectra for dozens of elements can be recorded *simultaneously*. This property is of particular importance for the multielement analysis of very small samples. In contrast, optimum flame excitation conditions vary widely from element to element; high temperatures are needed for excitation of some elements and low temperatures for others. Reducing conditions work best in some analyses and oxidizing conditions in others. Finally, the region of the flame that gives the optimum line intensity varies from element to element. This sensitivity to operating variables limits the applicability of flame methods for simultaneous multielement analyses.

Several disadvantages associated with arc and spark spectroscopy account for its displacement by flame methods for many analyses. One disadvantage arises from the fact that except for the most expensive instruments, arc and spark spectra must be recorded photographically. The time needed to develop and read a plate or film is substantially greater than that involved in the direct acquisition of spectral data with a photoelectric detector. To be sure, direct-reading photoelectric arc and spark instruments, which permit the determination of two dozen or more elements in a matter of minutes, are available, but their costs are prohibitive ($\sim$ \$150,000) for many laboratories. In addition, the quantitative applications of arc and spark source emission spectroscopy are limited, primarily because of difficulties encountered in reproducing radiation intensities. Only with the greatest care can relative errors be reduced to 1 to 2%; uncertainties on the order of 10 to 20% or greater are more common. Where traces are being determined, errors of such magnitude are often tolerable and not significantly greater than those associated with other methods. In the determination of a major constituent, however, arc and spark emission spectroscopy suffers by comparison with flame methods.

Plasma methods appear to offer all of the

[1] For a more complete discussion of arc and spark emission spectroscopy, see: B. F. Scribner and M. Margoshes, in *Treatise on Analytical Chemistry*, eds. I. M. Kolthoff and P. J. Elving. New York: Interscience, 1965, Part I, vol. 6, Chapter 64; J. Mika and T. Török, *Analytical Emission Spectroscopy*. New York: Crane, Russak & Co., 1974; M. Slavin, *Emission Spectrochemical Analysis*. New York: Wiley-Interscience, 1971; and E. L. Grove, *Analytical Emission Spectroscopy*. New York: Marcel Dekker, Part I, 1971, Part II, 1972.

advantages of the classical emission methods. An exception is encountered in the analysis of powdered samples such as minerals or ores; no practical method has yet been developed for direct plasma excitation of such materials, and solution in a suitable solvent is required. Plasma excitation also appears to have all of the advantages of flame excitation with few if any of the disadvantages. Thus, it is to be expected that the applications of plasma spectroscopy will continue to expand in the near future.

EMISSION SPECTRA

The central problem in analytical emission spectroscopy is the large influence of the source upon both the pattern and the intensity of lines produced by a species. It is apparent that the source serves two functions. First, it must provide sufficient energy to vaporize the sample; in this process, it is essential that the distribution of the elements in the vapor be reproducibly related to their concentration or distribution in the sample. The second function is to supply sufficient energy to cause electronic excitation of the elementary particles in the gas, again in a reproducible manner.

An examination of the emission produced by an electrical arc or spark or by an argon plasma reveals three types of superimposed spectra. First, there is *continuous background radiation*. For arc and spark sources, this radiation is emitted by the heated electrodes and perhaps also by hot particulate matter detached from the electrode surface. The frequency distribution of this radiation depends upon the temperature and approximates that of a black body (see p. 109). As will be pointed out later, the continuum found with plasma sources apparently arises from a recombination of thermally produced electrons and ions.

Band spectra, made up of a series of closely spaced lines, are also frequently observed in certain wavelength regions. This type of emission is due to molecular species in the vapor state, which produce bands resulting from the superposition of vibrational energy levels upon electronic levels. The cyanogen band, caused by the presence of CN radicals, is always observed when carbon electrodes are employed in an atmosphere containing nitrogen. Samples containing a high silicon content may yield an additional molecular band spectrum due to SiO. Another common source of bands is the OH radical. If these bands obscure line spectra of interest, precautions must be taken to eliminate them.

Emission spectroscopy is based upon the line spectra produced by excited atoms. The source and nature of these spectra were discussed in the previous chapter (p. 306).

Arc, spark, and plasma sources are generally richer in lines than the sources described in Chapter 11 because of their greater energies. Arc sources ordinarily provide lower energies than do spark sources; as a result, lines of neutral atoms tend to predominate in an arc spectrum. On the other hand, spectra excited by a spark typically contain lines associated with excited ions. Similarly, many of the lines found in plasma spectra also arise from ions rather than atoms.

ARC AND SPARK SOURCES

In arc and spark sources, sample excitation occurs in the gap between a pair of electrodes. Passage of electricity through the electrodes and the gap provides the necessary energy to atomize the sample and excite the resulting atoms to higher electronic states.

Sample Introduction

Samples for arc and spark source spectroscopy may be solid or liquid; they must be distributed more or less regularly upon the surface of at least one of the electrodes that serves as the source.

Metal Samples. If the sample is a metal or an alloy, one or both electrodes can be formed from the sample by milling, by turning, or by casting the molten metal in a mold. Ideally, the electrode will be shaped as a cylindrical rod that is one-eighth inch to one-quarter inch in diameter and tapered at one end. For some samples, it is more convenient to employ a polished, flat surface of a large piece of the metal as one electrode and a graphite or metal rod as the other. Regardless of the ultimate shape of the sample, care must be taken to avoid contamination of the surface while it is being formed into an electrode.

Electrodes for Nonconducting Samples. For nonmetallic materials, the sample is supported on an electrode whose emission spectrum will not interfere with the analysis. Carbon is an ideal electrode material for many applications. It can be obtained in a highly pure form, is a good conductor, has good heat resistance, and is readily shaped. Manufacturers offer carbon electrodes in many sizes, shapes, and forms. Frequently, one of the electrodes is shaped as a cylinder with a small crater drilled into one end; the sample is then packed into this cavity. The other electrode is commonly a tapered carbon rod with a slightly rounded tip. This configuration appears to produce the most stable and reproducible arc or spark.

Silver or copper rods are also employed to hold samples when these elements are not of analytical interest. The surfaces of these electrodes can be cleaned and reshaped after each analysis.

Excitation of the Constituents of Solutions. Several techniques are employed to excite the components of solutions or liquid samples. One common method is to evaporate a measured quantity of the solution in a small cup into the surface of a graphite or a metal electrode. Alternatively, a porous graphite electrode may be saturated by immersion in the solution; it is then dried before use.

Arc Sources and Arc Spectra

The usual arc source for a spectrochemical analysis is formed with a pair of graphite or metal electrodes spaced about 1 to 20 mm apart. The arc is initially ignited by a low current spark that causes momentary formation of ions for current conduction in the gap. Once the arc is struck, thermal ionization maintains the current. Alternatively, the arc can be started by bringing the electrodes together to provide the heat for ionization; they are then separated to the desired distance.

In the typical arc, currents lie in the range of 1 to 30 A. A dc source usually has an open circuit voltage of about 200 V; ac source voltages range from 2200 to 4400 V.

Electricity is carried in an arc by the motion of the electrons and ions formed by thermal ionization; the high temperature that develops is a result of the resistance to this motion by the atoms in the arc gap. Thus, the arc temperature depends upon the composition of the plasma, which in turn depends upon the rate of formation of atomic particles from the sample and the electrodes. Little is known of the mechanisms by which a sample is dissociated into atoms and then volatilized in an arc. It can be shown experimentally, however, that the rates at which various species are volatilized differ widely. The spectra of some species appear early and then disappear; those for other species reach their maximum intensities at a later time. Thus, the composition of the plasma, and therefore the temperature, may undergo variation with time. Typically, the plasma temperature is 4000 to 5000°K.

The precision obtainable with an arc is generally poorer than that with a spark and much poorer than that with a plasma or flame. On the other hand, an arc source is more sensitive to traces of an element in a sample than is a spark source. In addition, because of its high temperature, chemical interferences, such as those found in flame spectroscopy, are much less common. For these reasons, an arc source is often preferred for qualitative analysis as well as for the quantitative determination of trace constituents.

Spark Sources

Formation of a Spark. Figure 12-1 shows a typical electrical circuit for driving an alternating current across a spark gap. The transformer converts line power to 15,000 to 40,000 V, which then charges the capacitor. When the potential becomes large enough to break down the two air gaps shown on the right, a series of oscillating discharges follows. The voltage then drops until the capacitor is incapable of forcing further current across the gaps, following which the cycle is repeated. The frequency and current of the discharge are determined by the magnitude of the capacitance, inductance, and resistance of the circuit in Figure 12-1; these parameters also affect the relative intensities of the neutral atom and ion lines as well as the background intensity.

Most spark sources employ a pair of spark gaps arranged in series (Figure 12-1). The analytical gap is formed by a pair of carbon or metal electrodes that contain the sample. The control gap may be formed from smoothly rounded and carefully spaced electrodes over which is blown a stream of air. Conditions

at the analytical gap are subject to continual change during operation. Accordingly, electrode spacing is arranged so that the voltage breakdown is determined by the more reproducible control gap. The employment of air-cooling tends to maintain constant operating conditions for the control gap; a higher degree of reproducibility is thus imparted to the analytical results than would otherwise be attained.

The *average* current with a high-voltage spark is usually significantly less than that of the typical arc, being on the order of a few tenths of an ampere. On the other hand, during the initial phase of the discharge the *instantaneous* current may exceed 1000 A; here, electricity is carried by a narrow streamer that involves but a minuscule fraction of the total space in the spark gap. The temperature within this streamer is estimated to be as great as 40,000°K. Thus, while the average electrode temperature of a spark source is much lower than that of an arc, the *energy* in the small volume of the streamer may be several times greater. As a consequence, ionic spectra are more pronounced in a high voltage spark than in an arc.

Laser Excitation

A pulsed ruby laser provides sufficient energy to a small area on the surface of a sample to cause atomization and excitation of emission lines. This source is far from ideal, however, because of high background intensity, self-absorption, and weak line intensities. One technique that has shown promise is the laser microprobe, which is available commercially. Here, the laser is employed to vaporize the sample into a gap between two graphite electrodes, which serves as a spark excitation source. The device is applicable even to nonconducting materials and provides a surface analysis of a spot no greater than 50 μm in diameter.

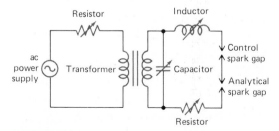

FIGURE 12-1   Power supply for a high-voltage spark source.

ARGON PLASMA SOURCES

By definition, a plasma is a gaseous mixture in which a significant fraction of the atomic or molecular species present are in the form of ions. The plasma employed for emission analyses is generally made up of a mixture of argon ions and atoms. When a sample is injected into this medium, atomization occurs as a consequence of the high temperature, which may be as great as 10,000°K. Argon ions, once formed in a plasma, are capable of absorbing sufficient power from an external source to maintain a sufficiently high temperature to continue the ionization process and thus sustain the plasma indefinitely. Three power sources have been employed in argon plasma spectroscopy. One is a dc electrical source capable of maintaining a current of several amperes between electrodes immersed in a stream of argon. The second and third are powerful radio frequency and microwave frequency generators through which the argon is flowed. Of the three, a radio-frequency source appears to offer the greatest advantage in terms of sensitivity and freedom from interference. Thus, the discussion will focus on these, which are often called *inductively coupled plasma* sources (sometimes abbreviated ICP).

The Inductively Coupled Plasma Source[2]

Figure 12-2 is a schematic drawing of an inductively coupled plasma source. It consists of three concentric quartz tubes through which streams of argon flow at a total rate between 11 and 17 liters/min. The diameter of the largest tube is about 2.5 cm. Surrounding the top of this tube is a water-cooled induc-

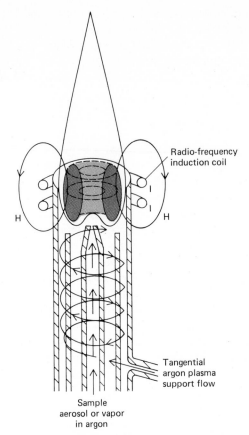

Radio-frequency induction coil

H

H

Tangential argon plasma support flow

Sample aerosol or vapor in argon

FIGURE 12-2   A typical inductively coupled plasma source. (From V. A. Fassel, *Science*, **202**, 185 (1978). With permission. Copyright 1978 by the American Association for the Advancement of Science.)

tion coil that is powered by a radio frequency generator, which produces 2 kW of energy at about 27 MHz. Ionization of the flowing argon is initiated by a spark from a Tesla coil. The resulting ions, and their associated electrons, then interact with the fluctuating magnetic field (labeled H in Figure 12-2) produced by the induction coil. This interaction causes the ions and electrons within the coil to flow in the closed annular paths depicted in the figure; ohmic heating is the consequence of their resistance to this movement.

---

[2] For a more complete discussion, see: V. A. Fassel, *Science*, **202**, 183 (1978); A. F. Ward, *Amer. Lab.*, **10** (11), 79 (1978); and R. M. Ajhar, P. D. Dalager, and A. L. Davison, *Amer. Lab.*, **8** (3), 71 (1976).

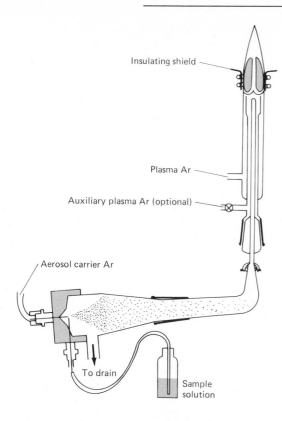

Insulating shield

Plasma Ar

Auxiliary plasma Ar (optional)

Aerosol carrier Ar

To drain

Sample solution

FIGURE 12-3 A typical nebulizer for sample injection into a plasma source. [From V. A. Fassel, *Science*, **202**, 186 (1978). With permission. Copyright 1978 by the American Association for the Advancement of Science.]

The temperature of the plasma formed in this way is high enough to require thermal isolation from the outer quartz cylinder. This isolation is achieved by flowing argon tangentially around the walls of the tube as indicated by the arrows in Figure 12-2; the flow rate of this stream is 10 to 15 liters/min. The tangential flow cools the inside walls of the center tube and centers the plasma radially.

Sample Injection. The sample is carried into the hot plasma at the head of the tubes by argon flowing at about 1 liter/min through the central quartz tube. The sample may be

an aerosol, a thermally generated vapor, or a fine powder.

The most widely used apparatus for sample injection is similar in construction to the nebulizer employed for flame methods. Figure 12-3 shows a typical arrangement. Here, the sample is nebulized by the flow of argon, and the resulting finely divided droplets are carried into the plasma. Aerosols have also been produced from liquids and solids by means of an ultrasonic nebulizer. Still another method of sample introduction involves deposition of the sample on a tantalum strip and vaporization by passage of a large current. The vapors are then swept into the plasma by the argon stream. No practical method has yet been devised for the introduction of powder samples such as those encountered in minerals and ores.

Plasma Appearance and Spectra. The typical plasma has a very intense, brilliant white, nontransparent core topped by a flamelike tail. The core, which extends a few millimeters above the tube, is made up of a continuum upon which is superimposed the atomic spectrum for argon. The source of the continuum apparently arises from recombination of argon and other ions with electrons. In the region 10 to 30 mm above the core, the continuum fades, and the plasma is optically transparent. Spectral observations are generally made at a height of 15 to 20 mm above the induction coil. Here, the background radiation is remarkably free of argon lines and is well suited for analysis. Many of the most sensitive analyte lines in this region of the plasma are from ions such as Ca(I), Ca(II), Cd(I), Cr(II), and Mn(II).

Analyte Atomization and Ionization. Figure 12-4 shows temperatures at various parts of the plasma. By the time the sample atoms have reached the observation point, they will have had a residence time of about 2 ms at temperatures ranging from 6000 to 8000°K. These times and temperatures are roughly twice as great as those found in the hotter combustion flames (acetylene/nitrous oxide) employed in flame methods. As a consequence, atomization is more complete, and there are

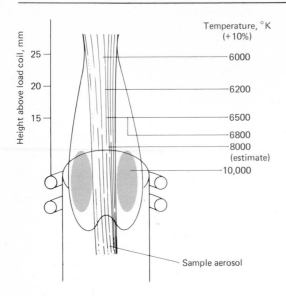

FIGURE 12-4 Temperature in a typical inductively coupled plasma source. (From V. A. Fassel, *Science*, **202**, 187 (1978). With permission. Copyright 1978 by the American Association for the Advancement of Science.)

fewer interference problems due to dissociation reactions (p. 322). Surprisingly, ionization interference effects (p. 323) are small or non-existent, perhaps because the electron concentration from ionization of the argon is large compared with that resulting from ionization of sample components.

Several other advantages are associated with the plasma source. First, atomization occurs in a chemically inert environment, which should also enhance the lifetime of the analyte. In addition, and in contrast to arc, spark, and flame, the temperature cross section of the plasma is relatively uniform; as a consequence, self-absorption and self-reversal effects (p. 329) are not encountered. Consequently, linear calibration curves over several orders of magnitude of concentration are often observed.

INSTRUMENTS FOR EMISSION SPECTROSCOPY

Instruments for emission spectroscopy are of two types, simultaneous multielement spectrometers and sequential spectrophotometers. The former are capable of detecting and measuring emission lines for a large number of elements (sometimes as many as 50 or 60) *simultaneously*. Sequential instruments, on the other hand, measure line intensities for various elements on a one-by-one basis. Here, a wavelength change is required after the signal for each element has been monitored. Clearly, the total excitation time required for a multi-element analysis will be significantly greater with the sequential instrument; as a consequence, such instruments are considerably more costly in terms of elapsed time and sample consumption.

Arc and spark sources demand the use of a simultaneous multielement instrument because of their instability. In order to obtain reproducible line intensities with these sources, it is necessary to integrate and average a signal for at least 20 s and often for a minute or more. When several elements must be determined, the time required and the sample consumed by a sequential procedure is usually unacceptable.

Simultaneous multielement instruments are of three types. One is a spectrograph, which employs a photographic film or plate as a detector and integrator. The second is a photoelectric spectrometer that employs an array of photomultiplier tubes (one for each element) located at appropriate positions on the focal plane of a spectrometer; a suitable integrating circuit is required for each detector. The third type employs a vidicon tube as a detector (see p. 143). Spectrographs are relatively inexpensive. On the other hand, processing of films or plates is somewhat time consuming and generally inconvenient when compared with photoelectric detection. Photoelectric multielement spectrometers are capable of providing for the determination of 20 or 30

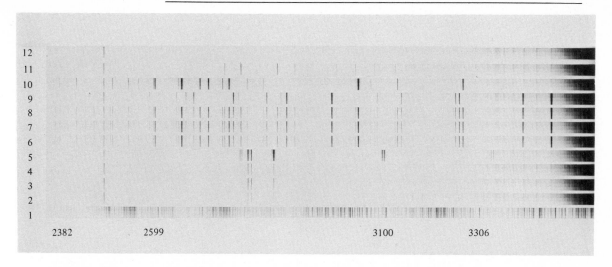

FIGURE 12-5 Typical spectra obtained with a 3.4-meter grating spectrograph. Numbers on horizontal axis are wavelengths in Å. Spectra: (1) iron standard; (2)-(5) casein samples; (6)-(8) Cd-Ge arsenide samples; (9)-(11) pure Cd, Ge, and As, respectively; (12) pure graphite electrode.

elements in a few minutes. They are, however, expensive (~$150,000) and not very versatile; they cannot ordinarily be switched from the determination of one set of elements to another. Their sum can only be justified for a situation demanding a large number of routine analyses for a certain set of elements.

Plasma sources are sufficiently stable to permit completion of line-intensity measurements in one or two seconds. As a consequence, sequential instruments have become a reasonable alternative to simultaneous instruments. While not as rapid as the photoelectric multielement spectrometer, sequential instruments often compete favorably in time consumption with spectrographs and provide the convenience of photoelectric detection. In addition, the cost of sequential spectrophotometers is competitive with spectrographs.

Spectrographs

To permit simultaneous intensity recording

by photographic means, the exit slit of a monochromator (see Figure 5-7) is replaced by a photographic plate (or film) located along the focal plane of the instrument.[3] After exposure and development of the emulsion, the various spectral lines of the source appear as a series of black images of the entrance slit distributed along the length of the plate (see Figure 12-5). The location of the lines provides qualitative information concerning the sample; the darkness of the images can be related to line intensities and hence to concentrations.

Types of Monochromators

The dispersing element in a spectrograph may be either a prism or a grating. High-

---

[3] The focal points of a prism spectrometer lie on a very slightly curved surface. To compensate for this curvature, spectrographic plate holders are designed to force a slight bend in the photographic plate.

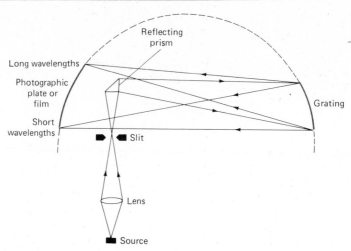

FIGURE 12-6 The Eagle mounting for a grating spectro-graph.

dispersion prism instruments employ a Lit-trow-mount monochromator; a Cornu-type is often used in lower dispersion spectrographs. Prism spectrographs suffer from the disad-vantage of nonlinear dispersion, which compli-cates the identification of lines and also crowds the longer wavelength lines together.

Reflection gratings are widely used in the manufacture of spectrographs. One of the common arrangements is shown in Figure 12-6.[4]

The spectral region of greatest importance for emission analysis lies between 2500 and 4000 Å, although both longer and shorter wavelengths are occasionally useful. Thus, the optical components through which radiation is transmitted must be of quartz or fused silica, and the detector must be sensitive to ultraviolet radiation.

The Hartmann Diaphragm. Spectrographs

are generally provided with a masking device called a *Hartmann diaphragm*, by means of which only a fraction of the vertical length of the slit is illuminated (Figure 12-7). Adjust-ment of the diaphragm permits the successive recording of several spectra on a single plate or film (see Figure 12-5).

Photographic Detection. A photographic emulsion can serve as a detector, an amplifier, and an integrating device for evaluating the intensity-time integral for a spectral line. When a photon of radiation is absorbed by a silver halide particle of the emulsion, the radiant energy is stored in the form of a latent image.

---

[4] Here also, the spectral lines are focused on a curved surface. With smaller instruments, the curvature may be so great as to preclude the use of glass photographic plates; flexible film must thus be employed.

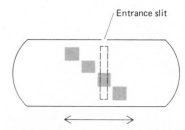

FIGURE 12-7 The Hart-mann diaphragm.

Treatment with a reducing agent results in the formation of a large number of silver atoms for each absorbed photon. This process is an example of chemical amplification; the number of black silver particles, and thus the darkness of the exposed area, is a function of the *exposure E*, which is defined by

$$E = I_\lambda t \qquad (12\text{-}1)$$

where I_λ is the intensity of radiation and t is the exposure time.[5] In obtaining spectra for quantitative purposes, t remains constant for both sample and standards. Under these circumstances, exposure is directly proportional to the line intensity and thus to the concentration of the emitting species.

The blackness of a line on a photographic plate is expressed in terms of its *optical density D*, which is defined as

$$D = \log \frac{I_0}{I}$$

where I_0 is the intensity (power) of a beam of radiation after it has passed through an unexposed portion of the emulsion, and I is its intensity after being attenuated by the line. Note the similarity between optical density and absorbance; indeed, the former can be measured with an instrument that is similar to the photometer shown in Figure 7-8.

In order to convert the optical density of a line to its original intensity (or exposure), it is necessary to obtain empirically a plate calibration curve, which is a plot of the optical density as a function of the logarithm of relative exposure. Such a plot is obtained experimentally by exposing portions of a plate to radiation of constant intensity for various lengths of time; generally, the absolute intensity of the source will not be known, so that

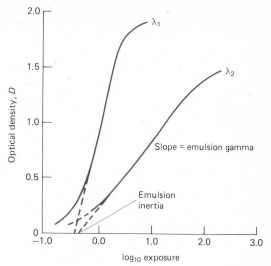

FIGURE 12-8 Typical plate calibration curves at two wavelengths.

only relative exposures can be calculated from Equation 12-1.

Figure 12-8 shows typical calibration curves for an emulsion obtained at two wavelengths. With the aid of such curves, the measured optical densities of various analytical lines can be related to their *relative intensities*. These relative intensities are the concentration-dependent parameter employed in quantitative spectrographic analysis. Several calibration curves are needed if a spectrum covers a large wavelength range because of the dependence of the slope on wavelength region (Figure 12-8).

Simultaneous Multielement Spectrometers

For routine emission analyses, photoelectric detection of integrated line intensities can be employed. Instruments for this purpose are equipped with as many as 48 photomultiplier tubes, which are located behind fixed slits along the focal curve of the spectrometer (see Figure 12-9). The position of each slit must

---

[5] Here, we follow the convention of emission spectroscopists and employ the term intensity I, rather than the power P. The difference is small (p. 93), and for practical purposes, the terms can be considered synonomous.

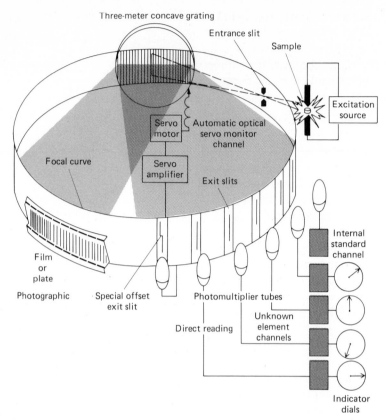

FIGURE 12-9 Optical system of a direct-reading grating spectrometer. (Courtesy of Baird Corporation, Bedford, Massachusetts.)

correspond exactly to the position of a line of interest.

The output of each photomultiplier is fed into a suitable capacitor-resistor circuit which integrates the output over a period of 10 to 40 s. The capacitor voltage at the end of the exposure is a function of its total charge which, in turn, is the product of the detector current multiplied by time; thus, this voltage provides a measure of the time integral of the line intensity.

Photoelectric multielement instruments are highly automated, complex, expensive, and permit measurement of only certain elements, which are chosen at the time of instrument manufacture. For rapid routine analysis, however, such spectrometers are often ideal. For example, in the production of alloys, quantitative analyses for 5 to 20 or more elements can' be accomplished within five minutes of receipt of a sample; close control over the composition of the final product thus becomes possible.

In addition to speed, photoelectric spectrometers offer the advantage of better analytical precision. Under ideal conditions reproducibilities of the order of 1 % relative of the amount present have been demonstrated.

Sequential Spectrometers

It is practical to employ automated sequential spectrometers for emission analysis with an inductively coupled plasma source because the relative stability of this source permits intensity measurements to be completed in a few seconds. Instruments of this kind are similar to the automated flame emission instrument described earlier (p. 327).

APPLICATIONS OF EMISSION SPECTROSCOPY

Spectroscopic instruments that rely upon photographic detection are particularly useful for qualitative or semiquantitative analyses. Photoelectric instruments find wide use for the routine quantitative analysis of a limited number of elements.

Qualitative Analysis

Photographic emission spectroscopy permits the detection of some seventy elements by brief excitation of a few milligrams of sample. Furthermore, from a subjective judgment of line blackening, the chemist with a little experience can rapidly estimate the concentration of an element with an accuracy of perhaps one order of magnitude. In this application, photographic emission spectroscopy is unsurpassed.

The sensitivity of spectrographic methods depends upon the nature and amount of sample, the type of excitation, and the instrument employed. Table 12-1 lists some representative data for lower limits of detection that can be achieved with dc arc excitation.

Excitation Technique. For qualitative work, a dc arc is most often employed. The current and exposure times are adjusted so that com-

TABLE 12-1   DETECTION LIMITS WITH A DC ARC[a]

Element	Wavelength, Å	Lower Limit of Detection[b]	
		Percent	Micrograms
Ag	3280.68	0.0001	0.01
As	3288.12	0.002	0.2
B	2497.73	0.0004	0.04
Ca	3933.66	0.0001	0.01
Cd	2288.01	0.001	0.1
Cu	3247.54	0.00008	0.008
K	3446.72	0.3	30
Mg	2852.12	0.00004	0.004
Na	5895.92	0.0001	0.01
P	2535.65	0.002	0.2
Pb	4057.82	0.0003	0.03
Si	2516.12	0.0002	0.02
Sr	3464.45	0.02	2
Ti	3372.80	0.001	0.1
Zn	3345.02	0.003	0.3

[a] Data taken from N. W. H. Addink, *Spectrochim. Acta*, **11**, 168 (1957). With permission.
[b] Values are based upon complete volatilization of a 10-mg sample in a dc arc.

plete volatilization of the sample occurs; currents of 5 to 30 A for 20 to 100 s are common. Generally, 2 to 50 mg of sample are packed into the cavity of a graphite electrode or evaporated upon its surface; a second (or counter) electrode, also of graphite, then completes the arc. The sample electrode is ordinarily made the anode because of its higher temperature. On the other hand, an enhancement of spectral intensities is sometimes observed in the vicinity of a sample electrode that is operated as a cathode, as a result of retarded diffusion by positive ions in the arc plasma. Advantage is sometimes taken of this phenomenon by using radiation adjacent to the cathode as the source.

Identification of Elements. For a qualitative analysis, a standard iron spectrum is usually recorded as a reference on each plate or film. Identification of unknown lines is greatly facilitated with a projector that permits the simultaneous comparison of this plate with a master iron spectrum upon which the wavelengths of the more intense iron lines have been identified. Also indicated on the master spectrum are the positions of several of the most sensitive lines for each of the elements. The analytical plate and the comparison plate are then aligned so that the two iron spectra coincide exactly. Identification of three or four of the prominent lines of a particular element constitutes nearly certain evidence for its presence in the sample.

Quantitative Applications

Quantitative emission analyses demand precise control of the many variables involved in sample preparation and excitation (and also in film processing with spectrographs). In addition, quantitative measurements require a set of carefully prepared standards for calibration; these standards should approximate as closely as possible the composition and physical properties of the samples to be analyzed.

Internal Standards and Data Treatment. As we have pointed out, the central problem of quantitative emission analysis is the very large number of variables that affect the blackness of the image of a spectral line on a photographic plate or the intensity of a line reaching the photoelectric detector. Most variables that are associated with the excitation and the photographic processes are difficult or impossible to control completely. In order to compensate for their effects, an *internal standard* is generally employed.

An internal standard is an element incorporated in a fixed concentration into each sample and each calibration standard. The relative intensity of a line of the internal standard is then determined for each sample or standard at the same time the relative intensities of the elements of interest are measured. A ratio of relative intensity of the analyte to that of standard is then employed for the concentration determination. Experimentally, a direct proportionality often exists between this ratio and concentration. Occasionally, however, the relationship is nonlinear; the resulting curve is then employed for concentration determinations.

Criteria in the Choice of an Internal Standard. The ideal internal standard has the following properties.

1. Its concentration in samples and standards is always the same.

2. Its chemical and physical properties are as similar as possible to those of the element being determined; only under these circumstances will the internal standard provide adequate compensation for the variables associated with volatilization.

3. It should have an emission line that has about the same excitation energy as one for the element being determined, so that the two lines are similarly affected by temperature fluctuations in the source.

4. The ionization energies of the internal standard and the element of interest should be similar to assure that both have the same distribution ratio of atoms to ions in the source.

5. The lines of the standard and the analyte should be similar in intensity and should be in the same spectral region so as to provide adequate compensation for emulsion variables (or differences in detector response with photoelectric spectrometers).

It is seldom possible to find an internal standard that will meet all of these criteria, and compromises must be made, particularly where the same internal standard is used for the determination of several elements.

If the samples to be analyzed are in solution, considerable leeway is available in the choice of internal standards, since a fixed amount can be introduced volumetrically. Here, an element must be chosen whose concentration in the sample is small relative to the amount to be added as the internal standard.

The introduction of a measured amount of an internal standard is seldom possible with metallic samples. Instead, the major element in the sample is chosen, and the assumption is made that its concentration is essentially invariant. For example, in the quantitative analysis of the minor constituents in a brass, either zinc or copper might be employed as the internal standard.

For powdered samples, the internal standard is sometimes introduced as a solid. Weighed quantities of the finely ground sample and the internal standard are thoroughly mixed prior to excitation.

The foregoing criteria provide theoretical guidelines for the selection of an internal standard; nevertheless, experimental verification of the effectiveness of a particular element and the line chosen is necessary. These experiments involve determining the effects of variation in excitation times, source temperatures, and development procedures on the relative intensities of the lines of the internal standard and the analytes.

Standard Samples. In addition to the choice of an internal standard, a most critical phase in the development of a quantitative emission method involves the preparation or acquisition of a set of standard samples from which calibration curves are prepared. For the ultimate in accuracy, the standards must closely approximate the samples both in chemical composition and physical form; their preparation often requires a large expenditure of time and effort—an expenditure that can be justified economically only if a large number of analyses is anticipated.

In some instances, standards can be synthesized from pure chemicals; solution samples are most readily prepared by this method. Standards for an alloy analysis might be prepared by melting together weighed amounts of the pure elements.

Another common method involves chemical analysis of a series of typical samples encompassing the expected concentration range of the elements of interest. A set of standards is then chosen on the basis of these results.

The National Bureau of Standards has available a large number of carefully analyzed metals, alloys, and mineral materials; occasionally, suitable standards can be found among these. Standard samples are also available from commercial sources.[6]

The Method of Standard Additions. When the number of samples is too small to justify extensive preliminary work, the method of standard additions described on page 325 may be employed.

Semiquantitative Methods. Numerous semiquantitative spectrographic methods have been described which provide concentration data reliable to within 30 to 200% of the true amount of an element present in the sample.[7]

---

[6] For a listing of sources, see: R. E. Michaelis, *Report on Available Standard Samples and Related Materials for Spectrochemical Analysis, Am. Soc. Testing Materials Spec. Tech. Publ. No. 58-E.* Philadelphia: American Society for Testing and Materials, 1963.

[7] For references to a more complete description of some semiquantitative procedures, see: C. E. Harvey, *Spectrochemical Procedures.* Glendale, CA: Applied Research Laboratories, 1950, Chapter 7.

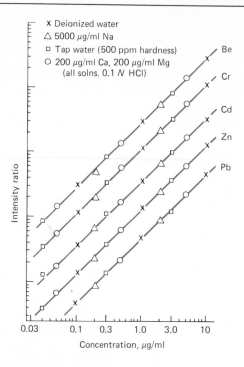

FIGURE 12-10 Calibration curves with an inductively coupled plasma source. Here, an yttrium line at 242.2 nm served as an internal standard. Notice the lack of interelement interference. (From V. A. Fassel, *Science*, **202**, 187 (1978). With permission. Copyright 1978 by the American Association for the Advancement of Science.)

These methods are useful where the preparation of good standards is not economical. Several such procedures are based upon the total vaporization of a measured quantity (1 to 10 mg) of sample in the arc. The concentration estimate may then be based on a knowledge of the minimum amount of an element required to cause the appearance of each of a series of lines. In other methods, optical densities of elemental lines are measured and compared with the line of an added matrix material, or with the background. Concentration calculations are then based on the assumption that the line intensity is independent of the state in which the element occurs. The effects of other elements on the line intensity can be minimized by mixing a large amount of a suitable matrix material (a *spectroscopic buffer*) with the sample.

It is sometimes possible to estimate the concentration of an element that occurs in small amount by comparing the blackness of several of its lines with a number of lines of a major constituent. Matching densities are then used to establish the concentration of the minor constituent.

Quantitative Applications with the Inductively Coupled Plasma Source. As we have noted earlier, the inductively coupled plasma source appears to offer all the advantages of the flame methods for quantitative measure-

TABLE 12-2 COMPARISON OF DETECTION LIMITS FOR SEVERAL ATOMIC SPECTRAL METHODS[a]

Method	Number of Elements Detected at Concentrations of				
	< 1 ppb	1–10 ppb	11–100 ppb	101–500 ppb	> 500 ppb
Inductively coupled plasma emission	9	32	14	6	0
Atomic emission	4	12	19	6	19
Atomic fluorescence	4	14	16	4	6
Atomic absorption	1	14	25	3	14

[a] Data from V. A. Fassel and R. N. Kniseley, *Anal. Chem.*, **48** (13), 1111A (1974). Reprinted with permission. Copyright by the American Chemical Society. Detection limits correspond to a signal that is twice as great as the deviation for the background noise.

ments, along with fewer chemical interference problems. For example, Figure 12-10 shows calibration curves for several elements. Note that some of the data were obtained from solutions prepared by dissolving a compound of the element in pure water; others contained high concentrations of different species, indicating a freedom from interelement interferences. Note also that the curves cover a concentration range of nearly three orders of magnitude.

In general, the detection limits with the inductively coupled plasma source appear comparable to or better than other atomic spectral procedures. Table 12-2 compares the sensitivity of several of these methods. Note that more elements can be detected in the ten-parts-per-billion (or less) range with plasma excitation than with any other of the methods.

PROBLEMS

1. State the differences between emission spectroscopy described in this chapter and atomic emission spectroscopy described in Chapter 11.

2. Explain how an inductively coupled plasma source produces high intensity emission spectra for elements.

3. What is an internal standard and how is it used?

4. Explain the statement that "a photographic emulsion can serve as a detector, an amplifier, and an integrating device for evaluation of the intensity-time integral for a spectral line."

5. Element 1 exhibited an emission line at wavelength λ_1 in Figure 12-8. A standard mixture containing this element at a concentration of 20.0 ppb exhibited optical density of 1.10 at λ_1. An alloy containing this element produced an optical density of 0.68 at this same wavelength. Use Figure 12-8 to estimate the concentration (in ppb) of the element in the alloy.

6. Element 2, which emitted at λ_2 (Figure 12-8) was used as an internal standard for element 1 (with emission at λ_1). A standard containing 11.8 ppb of element 1 and 94.7 ppb of element 2 yielded optical densities of 1.00 at λ_1 and 0.92 at λ_2. Element 2 was mixed at a concentration of 85.0 ppb with an unknown containing element 1. This sample produced optical densities of 0.75 at λ_1 and 0.95 at λ_2. Assuming that the emission from each element has been affected by the matrix equally, calculate the concentration of element 1 in the unknown.

13

MISCELLANEOUS OPTICAL METHODS

This chapter contains brief discussions of some optical methods which, while important, find less general use by the average practicing chemist.

REFRACTOMETRY

When radiation passes through a transparent medium, interaction occurs between the electric field of the radiation and the bound electrons of the medium; as a consequence, the rate of propagation of the beam is less than in a vacuum (see p. 102). The refractive index of a substance n_i at a wavelength i is given by the relationship

$$n_i = \frac{c}{v_i} \qquad (13\text{-}1)$$

where v_i is the velocity of propagation of radiation in the medium and c is the velocity in vacuum (a constant under all conditions). The refractive index of most liquids is in the range between 1.3 and 1.8; it is 1.3 to 2.5 or higher for solids.[1]

The Measurement of Refractive Index

The refractive index of a substance is ordinarily determined by measuring the change in direction (refraction) of collimated radiation as it passes from one medium to another. As was shown on page 103,

$$\frac{n_2}{n_1} = \frac{v_1}{v_2} = \frac{\sin \theta_1}{\sin \theta_2} \qquad (13\text{-}2)$$

where v_1 is the velocity of propagation in the less dense medium M_1 and v_2 is the velocity in medium M_2; n_1 and n_2 are the correspond-

ing refractive indexes and θ_1 and θ_2 are the angles of incidence and refraction, respectively (see Figure 4-9). When M_1 is a vacuum, n_1 is unity because v_1 becomes equal to c in Equation 13-1. Thus,

$$n_2 = n_{\text{vac}} = \frac{c}{v_2} = \frac{\sin \theta_1}{\sin \theta_2} \qquad (13\text{-}3)$$

where n_{vac} is the *absolute refractive index* of M_2. Thus, n_{vac} can be obtained by measuring the two angles θ_1 and θ_2.

It is much more convenient to measure the refractive index with respect to some medium other than vacuum, and air is commonly employed as a standard for this purpose. Most compilations of n for liquids and solids in the literature are with reference to air at laboratory temperatures and pressures. Fortunately, the change in the refractive index of air with respect to temperature and pressure is small enough so that a correction from ambient laboratory conditions to standard conditions is needed for only the most precise work. A refractive index n_D measured with respect to air with radiation from the D line of sodium can be converted to n_{vac} with the equation

$$n_{\text{vac}} = 1.00027 n_D \qquad (13\text{-}4)$$

This conversion is seldom required.

It is usually necessary to measure a refractive index to an accuracy of at least 2×10^{-4}. Accuracies on the order of 6 to 7×10^{-5} may be required for the routine analysis of solutions. For the detection of impurities, a difference in refractive index between the sample and a pure standard is measured; here, the capability of detecting a difference on the order of 1×10^{-6} or better is needed.

Variables That Affect Refractive Index Measurements

Temperature, wavelength, and pressure are the most common experimentally controllable variables that affect a refractive index measurement.

---

[1] For a more complete discussion of refractometry, see: S. Z. Lewin and N. Bauer, in *Treatise on Analytical Chemistry*, eds. I. M. Kolthoff and P. J. Elving. New York: Interscience, 1965, Part I, vol. 6, Chapter 70.

Temperature. Temperature influences the refractive index of a medium primarily because of the accompanying change in density. For many liquids, the temperature coefficient lies in the range of -4 to -6×10^{-4} deg^{-1}. Water is an important exception, with a coefficient of about -1×10^{-4}; aqueous solutions behave similarly. Solids have temperature coefficients that are roughly an order of magnitude smaller than that of the typical liquid.

It is apparent from the foregoing that the temperature must be controlled closely for accurate refractive index measurements. For the average liquid, temperature fluctuations should be less than $\pm 0.2°C$ if fourth place accuracy is required, and $\pm 0.02°C$ for measurements to the fifth place.

Wavelength of Radiation. As noted on page 102, the refractive index of a transparent medium gradually decreases with increasing wavelength; this effect is referred to as *normal dispersion*. In the vicinity of absorption bands, rapid changes in refractive index occur; here, the dispersion is referred to as *anomalous* (see Figure 4-8, p. 102).

Dispersion phenomena make it essential that the wavelength employed be specified in quoting a refractive index. The D line from a sodium vapor lamp ($\lambda = 589$ nm) is most commonly used as a source in refractometry, and the corresponding refractive index is designated as n_D (often the temperature in °C is also indicated by a superscript; for example n_D^{20}). Other lines commonly employed for refractive index measurements include the C and F lines from a hydrogen source ($\lambda = 656$ nm and 486 nm, respectively) and the G line of mercury ($\lambda = 436$ nm).

Pressure. The refractive index of a substance increases with pressure because of the accompanying rise in density. The effect is most pronounced in gases, where the change in n amounts to about 3×10^{-4} per atmosphere; the figure is less by a factor of 10 for liquids, and it is yet smaller for solids. Thus,

only for precise work with gases and for the most exacting work with liquids and solids is the variation in atmospheric pressure important.

Instruments for Measuring Refractive Index

Two types of instruments for measuring refractive index are available from commercial sources. *Refractometers* are based upon measurement of the so-called *critical angle* or upon the determination of the displacement of an image. *Interferometers* utilize the interference phenomenon to obtain differential refractive indexes with very high precision. We shall consider only refractometers.

Critical Angle Refractometers. The most widely used instruments for the measurement of refractive index are of the critical angle type. The *critical angle* is defined as the angle of refraction in a medium when the angle of the incident radiation is 90 deg (the *grazing angle*); that is, when θ_1 in Equation 13-2 is 90 deg, θ_2 becomes the critical angle θ_c. Thus,

$$\frac{n_2}{n_1} = \frac{\sin 90}{\sin \theta_c} = \frac{1}{\sin \theta_c} \qquad (13\text{-}5)$$

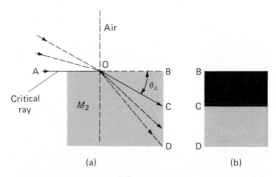

(a)   (b)

FIGURE 13-1   (a) Illustration of the critical angle, θ_c, and the critical ray AOC; (b) end-on view showing sharp boundary between the dark and light fields formed at the critical angle.

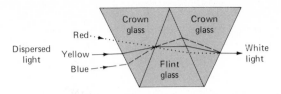

FIGURE 13-2 Amici prism for compensation of dispersion by sample. Note that the yellow radiation (sodium D line) suffers no net deviation from passage through the prism.

Figure 13-1a illustrates the critical angle that is formed when the critical ray approaches the surface of the medium M_2 at 90 deg to the normal and is then refracted at some point 0 on the surface. Note that if the medium could be viewed end-on, as in Figure 13-1b, the critical ray would appear as the boundary between a dark and a light field. It should be noted, however, that the illustration is unrealistic in that the rays are shown as entering the medium at but one point 0; in fact, they would be expected to enter at all points along the surface and thus create an entire family of critical rays with the same angle θ_c. A condensing or focusing lens is needed to produce a single dark-light boundary such as shown in Figure 13-1b.

It is important to realize that the critical angle depends upon wavelength. Thus, if polychromatic radiation is employed, no single sharp boundary such as that in Figure 13-1b is observed. Instead, a diffuse chromatic region between the light and dark areas develops; the precise establishment of the critical angle is thus impossible. This difficulty is often overcome in refractometers by the use of monochromatic radiation. As a convenient alternative, many critical angle refractometers are equipped with a compensator that permits the use of radiation from a tungsten source, but compensates for the resulting dispersion in such a way as to give a refractive index in

terms of the sodium D line. The compensator consists of one or two *Amici prisms*, as shown in Figure 13-2. The properties of this complex prism are such that the dispersed radiation is converged to give a beam of white light that travels in the path of the yellow sodium D line.

Abbé Refractometer. The Abbé instrument is undoubtedly the most convenient and widely used refractometer; Figure 13-3 shows a schematic diagram of its optical system. The sample is contained as a thin layer (~0.1 mm) between two prisms. The upper prism is firmly mounted on a bearing that permits its rotation by means of the side arm shown in dotted lines. The lower prism is hinged to the upper to allow separation for cleaning and for intro-

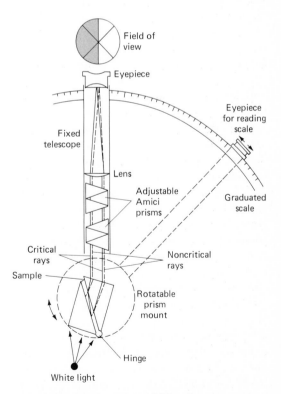

FIGURE 13-3 The Abbé refractometer.

duction of the sample. The lower prism face is rough-ground; when light is reflected into the prism, this surface effectively becomes the source for an infinite number of rays that pass through the sample at all angles. The radiation is refracted at the interface of the sample and the smooth-ground face of the upper prism, whereupon it passes into the fixed telescope. Two Amici prisms, which can be rotated with respect to one another, serve to collect the divergent critical angle rays of different colors into a single white beam, which corresponds in path to that of the sodium D ray. The eyepiece of the telescope is equipped with crosshairs; in making a measurement, the prism angle is changed until the light-dark interface just coincides with the crosshairs. The position of the prism is then established from the fixed scale (which is normally graduated in units of n_D). Thermostating is accomplished by circulation of water through jackets surrounding the prisms.

The Abbé refractometer owes its popularity to its convenience, its wide range ($n_D = 1.3$ to 1.7), and to the minimal sample required. The accuracy of the instrument is about ± 0.0002; its precision is half this figure. The most serious error in the Abbé instrument is caused by the fact that the nearly grazing rays are cut off by the arrangement of the two prisms; the boundary is thus less sharp than is desirable.

A *precision* Abbé refractometer, which diminishes the uncertainties of the ordinary instrument by a factor of about three, is also available; the improvement in accuracy is obtained by replacing the compensator with a monochromatic source and by using larger and more precise prism mounts. The former provides a much sharper critical boundary, and the latter permits a more accurate determination of the prism position.

Image Displacement Refractometer. The most straightforward method for determining refractive index involves measurement of the angles of incidence and refraction by means of a spectrometer arrangement somewhat like that shown in Figure 5-7a (p. 124). Liquid samples are contained in a prism-shaped container mounted at the center of a large circular metal table; solid samples are cut into the shape of a prism and are similarly mounted. A light source, a slit, and a collimator are employed to direct a parallel beam of radiation onto one surface of the prism. The refracted image of the slit is then viewed with a telescope mounted on the circle. Since the slit image can be made very sharp, the accuracy of the determination depends only upon the accuracy of the angular measurements and the control of temperature; uncertainties of 1×10^{-6} or smaller can be attained. An instrument of this type is often used as a detector in liquid chromatography (see Figure 25-6).

Applications of Refractometry

In common with density, melting point, and boiling point, the refractive index is one of the classical physical constants that can be used to describe a chemical species. While it is a nonspecific property, few substances have identical refractive indexes at a given temperature and wavelength. Thus, this constant is useful for confirming the identity of a compound and measuring its purity. In addition, refractive index determinations are useful for the quantitative analysis of binary mixtures. Finally, combined with other measurements, the refractive index provides structural and molecular weight information about a substance.

For all applications, periodic calibration of refractometers is necessary. Standards for this purpose include purified liquids such as water ($n_D^{20} = 1.3330$), toluene ($n_D^{20} = 1.4969$), and methylcyclohexane ($n_D^{20} = 1.4231$). The latter two compounds can be obtained from the National Bureau of Standards as certified samples with five-decimal indexes at 20, 25, and 30°C and for each of seven wavelengths. A glass test piece, supplied with most refractometers, can also be employed as a reference. The difference between the refractive index of the standard and the instrument scale reading is applied as an arithmetic correction to subsequent determinations. With the Abbé refractometer, the objective of the telescope can be adjusted mechanically so that the instrument indicates the proper refractive index for the standard.

POLARIMETRY

Optical activity is a measure of the ability of certain substances to rotate plane-polarized light. This phenomenon, first reported for quartz in 1811, has been studied intensely since that time. By the mid-nineteenth century, many of the laws relating to optical activity had been formulated; these in turn played a direct part in the development of the ideas of organic stereochemistry and structure later in the same century. Some of the early concepts of optical activity have stood the test of time and remain essentially unaltered today. It is of interest, however, that despite this long history, the interactions of radiation with matter that cause the rotation of polarized light are less clearly understood than the processes responsible for absorption, emission, or refraction.

The term *polarimetry*, as it is used by most chemists, can be defined as the study of the rotation of polarized light by transparent substances. The direction and the extent of rotation (the *optical rotatory power*) is useful for both qualitative and quantitative analysis and for the elucidation of chemical structure as well.[2]

---

[2] For a more complete discussion of the various aspects of polarimetry, see: *Physical Methods of Chemistry*, eds. A. Weissberger and B. W. Rossiter. New York: Wiley, 1972, vol. 1, Part III C, Chapters 1 and 2; and W. A. Struck and E. C. Olson, in *Treatise on Analytical Chemistry*, eds. I. M. Kolthoff and P. J. Elving. New York: Interscience, 1965, Part I, vol. 6, Chapter 71.

Transmission and Refraction of Radiation in Optically Anisotropic Media

Optically *isotropic* substances transmit radiation at equal velocities in all directions regardless of the polarization of the radiation. Examples of isotropic material include homogeneous gases and liquids, solids that crystallize in the cubic form, and noncrystalline solids such as glasses and many polymers. Noncubic crystals, on the other hand, are *anisotropic* and may transmit polarized radiation at different velocities depending upon the angular relationship between the plane of polarization and a given axis of the crystal.

Transmission of Polarized Radiation Through Anistropic Crystal. Recall (p. 102) that radiation is slowed as it passes through a medium containing atoms, ions, or molecules because its electrical vector interacts momentarily with the electrons of these particles, causing temporary polarization. Reemission of the radiation occurs after 10^{-14} to 10^{-15} s as the polarized particles return to their original state. A beam of radiation traversing an isotropic medium encounters a symmetrical distribution of particles around its path of travel. Thus, the slowing of a beam of polarized radiation is the same regardless of the angle of the plane of polarization around the direction of travel. In contrast, the distribution of atomic or molecular particles around most paths through an anisotropic crystal is not symmetrical. Thus, radiation vibrating in one plane along this path encounters a different particle environment from that in another; a difference in rate of transmission results.

All anisotropic crystals have at least one axis, called the *optic axis*, around which there exists a symmetrical distribution of the particles making up the crystal. Polarized radiation travels along the optic axis at a constant rate regardless of its angle of polarization with respect to that axis.

Figure 13-4a depicts the electrical vectors at maximum amplitude for two beams of radiation that are in phase and polarized at 90 deg to one another. The arrows are the vectors for the beam which is vibrating in the plane of the paper. The dots represent the vector that is fluctuating in a plane perpendicular to the page. Figure 13-4a shows that the wavelengths of both beams are decreased equally as they travel along the optic axis of the crystal; their velocities remain the same as a consequence (see Figure 4-2, p. 94). Thus, the two beams remain in phase both within the crystal and as they emerge from it.

Figure 13-4b contrasts the behavior of the two beams when they strike the crystal at an angle that is 90 deg to the optic axis. Here, the beam that is vibrating in the plane of the paper encounters an atomic or molecular environment similar to that depicted in Figure 13-4a; its wavelength and velocity behavior are thus similar to the two beams moving along the optic axis. The beam vibrating in the plane perpendicular to the page, however, encounters a less dense environment; thus, its velocity and wavelength are not decreased as

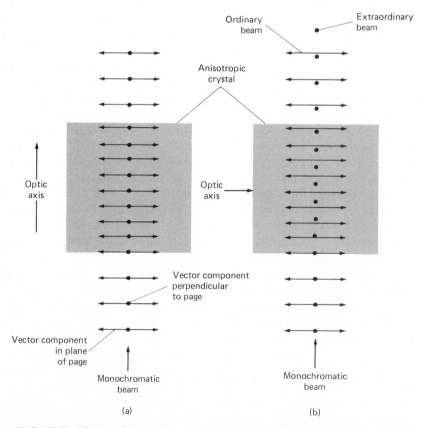

(a) (b)

FIGURE 13-4 Schematic representation of passage of a monochromatic beam through two axes of an anisotropic crystal of calcite. Arrows represent the electric vector component in the plane of the page. Dots represent the vector component in a plane perpendicular to the page.

much, and the two beams become out of phase not only within the crystal but also after emerging from it.

The polarized beam that travels at the same velocity along the optic axis and perpendicular to it is called the *ordinary beam*; the faster moving beam is the *extraordinary beam*. (In some instances, the relative velocities of the two types of beams are reversed.) The velocity of the extraordinary beam varies continuously as the angle of its travel with respect to the optic axis is varied from 0 to 90 deg, and reaches a maximum (or minimum) at the latter angle.

Transmission of Unpolarized Radiation Through Anisotropic Crystals. When a beam of *unpolarized* monochromatic radiation passes through an anisotropic crystal at an angle to the optic axis, separation into an ordinary and extraordinary beam occurs. To understand this behavior, recall that the various electrical vectors of unpolarized monochromatic radiation can be resolved into two mutually perpendicular vectors, as shown in Figure 4-10 (p. 105); thus, a beam of ordinary radiation can be considered to be made up of two plane-polarized beams of equal amplitude whose planes are oriented at 90 deg to one another. Therefore, Figure 13-4 applies to an unpolarized beam as well as to two plane-polarized beams; here, it is only necessary to specify that the amplitudes of the latter beams are identical.

Double Refraction by Anisotropic Crystal. From the foregoing discussion, it is evident that the velocity, and thus the index of refraction of the extraordinary ray, in an anisotropic crystal is dependent upon direction, being identical with that for the ordinary ray along the optic axis and changing continuously to a maximum or a minimum along a perpendicular axis. Refractive indexes for the extraordinary ray are normally reported in terms of this perpendicular axis. Table 13-1 compares the refractive index for the ordinary ray n_o with that for the extraordinary ray n_e in some common anisotropic crystals. In calcite,

the extraordinary ray is clearly propagated at a greater rate than the ordinary ray; in quartz, the reverse is the case. Because anisotropic crystals have two characteristic refractive indexes, they can be made to refract the ordinary and the extraordinary rays at different angles; anisotropic crystals are thus *double-refracting*. This property provides a convenient means for separating an unpolarized beam into two beams that are plane-polarized at 90 deg to one another.

The Nicol Prism. Figure 13-5 depicts a Nicol prism, a device that exploits the double-refracting properties of crystalline calcite $(CaCO_3)$ to produce plane-polarized radiation. The end faces of a natural crystal are trimmed slightly to give an angle of 68 deg, as shown; the crystal is then cut across its short diagonal. A layer of Canada balsam, a transparent substance with a refractive index intermediate between the two refractive indexes of calcite, is placed between the two crystal halves. This layer is totally reflecting for the ordinary ray with its greater refractive index but, as shown in the figure, transmits the extraordinary ray almost unchanged.

Pairs of Nicol prisms are employed in measurements involving the rotation of the plane of polarized light. One prism serves to produce a polarized beam that is passed into the medium under study. A second analyzer prism then determines the extent of rotation caused by the medium. If the two Nicol prisms are identically oriented with respect to

TABLE 13-1   REFRACTIVE INDEX DATA FOR SELECTED ANISOTROPIC CRYSTALS

Crystal	n_o	n_e
Calcite	1.6583	1.4864
Quartz	1.544	1.553
Ice	1.306	1.307

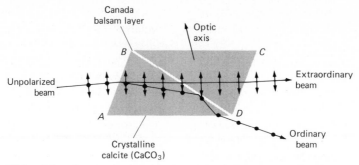

FIGURE 13-5   A Nicol prism for resolving an unpolarized beam into two beams plane-polarized at 90 deg to one another.

the beam, and if the medium has no effect, the extraordinary ray, comprising nearly 50% of the original intensity, is emitted from the analyzer (see Figure 13-6a). If the polarizer is rotated (Figure 13-6b), only the vertical component $m'n'$ of the beam mn emitted from the polarizer is transmitted through the analyzer, the horizontal component being now reflected from the Canada balsam layer. Here, less than 50% of the incident beam appears at the face

of the analyzer. Rotation of the polarizer by 90 deg results in a beam from the analyzer that has no vertical component. Thus, no radiation is observed at the face of the analyzer. If a medium affecting the rotation of light is interposed between the two Nicol prisms, the relative orientations needed to achieve a maximum or a minimum in transmission are changed by an amount corresponding to the rotatory power of the medium.

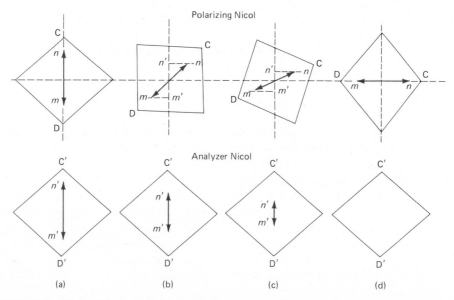

FIGURE 13-6   End view of a polarizer and analyzer Nicol. mn = electrical vector of beam transmitted by polarizer. $m'n'$ = vertical component of beam transmitted by polarizer and analyzer.

Interference Effects with Polarized Radiation

To account for many of the experimental observations regarding the interactions of polarized radiation, it is necessary to assume that interference between polarized beams can occur, provided the beams are *coherent* (p. 99). The effect of interference can then be visualized by vector addition of the electromagnetic components of the individual beams.

Figure 13-7a illustrates the interference between two plane-polarized beams of equal amplitude that are in phase but oriented 90 deg to one another. Addition of the electrical vectors of the two beams is shown schematically in the end-on view to the right. Note that the numbered points along the axis MN have been projected onto the plane ABCD which is perpendicular to MN, thus giving a two-dimensional representation of the resultant and its components. It will be seen that the resultant is a vector that oscillates in a plane oriented at 45 deg to the planes of the two component beams; *interference thus produces a single, plane-polarized beam.*

Figure 13-7b demonstrates interference when the phase relationship between the two plane-polarized beams differs by one-half wavelength (180 deg). Here, the waves are again in phase, and the resultant is a plane-polarized beam which, however, is perpendicular with respect to the resultant of Figure 13-7a.

Circularly and Elliptically Polarized Radiation. It is of interest now to examine the behavior of a plane-polarized beam of monochromatic radiation as it passes through an anisotropic crystal. In Figure 13-8a, the path of the incident beam is normal to the optic axis of the crystal, with the plane of the polarized radiation oriented at 45 deg to that axis (the angle of the plane of polarization is given by the arrow MP). As shown by Figure 13-7, however, the plane-polarized beam can be considered to consist of two *coherent* components lying in perpendicular planes oriented along MB and MA. These components are also indicated in Figure 13-8. Upon entering the crystal, the component lying in the MA direction travels at the rate of an ordinary ray since this direction lies along the optic axis of the crystal; the orientation of component MB, on the other hand, corresponds to that for an extraordinary ray, and its rate of propagation is thus different. As a result of the velocity difference, *the two components are no longer coherent* and thus cannot interfere. That is, *within the crystal, the beam can be considered to consist of two components having different velocities*; because of their incoherence; interference cannot occur.

When the two rays leave the crystal, their velocities again become equal in the isotropic air medium; thus, they can again interfere since they are once more coherent. The nature of the resultant will, however, depend upon the phase relationship between the two that exists at the instant they emerge from the crystal surface. This phase relationship is determined by the relative velocities of the two rays in the medium as well as by the length of traverse. If, for example, the path in the crystal is such that the two rays are completely in phase upon exiting, then constructive interference similar to that shown in Figure 13-7a occurs. That is, the resultant beam will be polarized at the same angle as the entering beam. If, on the other hand, the crystal thickness is such that the phase relationship between the two rays at the face is shifted exactly one-half wavelength, interference such as shown in Figure 13-7b results. Here, the plane of the exit beam is oriented 90 deg with respect to the entering beam.

In Figure 13-8a, we have shown the emerging waves as one-quarter wavelength out of phase and have indicated their relationship to one another, assuming that no interference takes place. In fact, however, interaction does occur as the two rays enter the air medium, and the path of the electrical vector for the resulting wave can be obtained by adding the two vectors. The resulting vector quantity is seen to travel in a helical pathway around the direction of travel (Figure 13-8b). If the vector sum is plotted in two-dimensional form (Figure 13-8c), a circle is obtained. This condition is in distinct contrast to the original, linearly polarized radiation in which

the electrical vector lies in a single plane. A helical beam of this type is called *circularly polarized light*. Note that if the two waves had been out of phase one-quarter wavelength in the other sense, the direction of travel of the vector would have been clockwise rather than

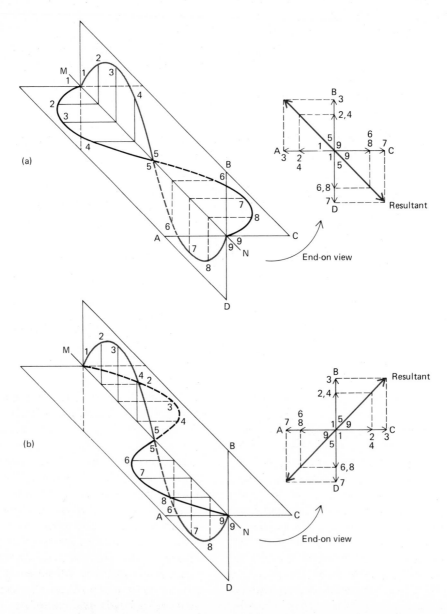

FIGURE 13-7   Interference between two in-phase plane-polarized beams.

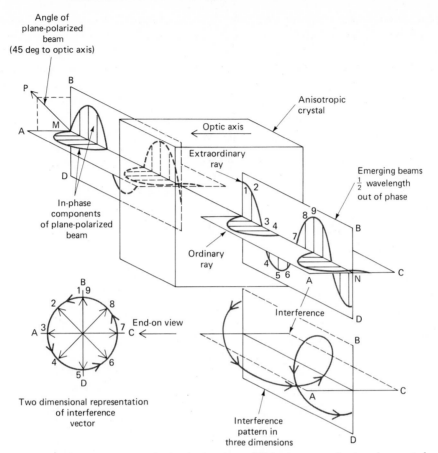

FIGURE 13-8 Circular polarization of light by an anisotropic crystal.

counterclockwise.

Thus far, we have considered the nature of the exit beam from the anisotropic crystal when the phase difference created was 0, $\frac{1}{4}$, $\frac{1}{2}$, or some multiple of these fractions. If the light path in the crystal is such as to produce phase differences other than these, the path traced by the resultant electrical vector is an ellipse, and the radiation is called *elliptically polarized light*. Figure 13-9 summarizes the states that result when the components of a plane-polarized beam are emitted from an anisotropic crystal with various phase differences

Anisotropic crystals of suitable length are employed experimentally to produce circularly polarized radiation. Such crystals are called *quarter-wave plates* and find use in circular dichroism studies.

Relationship Between Plane Polarized and Circularly Polarized Radiation. In the preceding section, we have seen that the behavior of plane-polarized radiation upon passage through an anisotropic crystal can be rationalized by considering the beam to be the resultant of two plane-polarized rays that are in phase and oriented at 90 deg to one another. It is of equal importance to understand that plane-polarized radiation can also be treated as the interference product of *two coherent* *circular rays of equal amplitude that rotate in opposite directions.* Figure 13-10 shows how the vectors for the d and l circular components are added to produce the equivalent vectors of a plane-polarized beam. From the middle figure, it can be seen that each of the two rotating vectors describes a helical path around the axis of travel of the beam.

The rationalization of many phenomena to be considered in this chapter is based upon the idea that plane-polarized radiation consists of a d and an l circular component. Here, d (*dextrorotatory*) refers to the clockwise rotation as the beam approaches the observer; l (*levorotatory*) is the counterclockwise component.

Circular Double Refraction

The rotation of plane-polarized light by an optically active species can be explained if one assumes that the rates of propagation of the d and the l circular components of a plane-polarized beam are different in the presence of such a species; that is, the refractive index of the substance with respect to d radiation (n_d) is different from that for l (n_l). Thus, optically active substances are anisotropic with respect to circularly polarized light and show *circular double refraction*. Note that the two circular components are no longer coherent in the anisotropic medium and cannot interfere until they again reach an isotropic medium.

The rotation of a beam of light by an optically active medium is shown schematically in Figure 13-11. Initially in Figure 13-11a, the beam is polarized in a vertical plane with the circular d and l components rotating at equal velocities. Upon entering the anisotropic medium, the rate of propagation of the d component is slowed more than that of the l because $n_d > n_l$. Thus, at some location (Figure 13-11b) in the medium, the d vector will lag behind the l; if at this point the two rays could interfere, the resultant would still be a plane, but one that was rotated from the

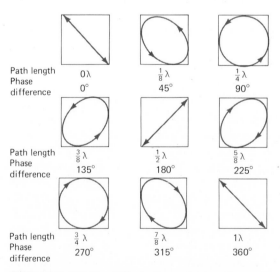

Path length Phase difference	0λ $0°$	$\frac{1}{8}\lambda$ $45°$	$\frac{1}{4}\lambda$ $90°$
Path length Phase difference	$\frac{3}{8}\lambda$ $135°$	$\frac{1}{2}\lambda$ $180°$	$\frac{5}{8}\lambda$ $225°$
Path length Phase difference	$\frac{3}{4}\lambda$ $270°$	$\frac{7}{8}\lambda$ $315°$	1λ $360°$

FIGURE 13-9 Effect of an anisotropic crystal on plane-polarized radiation. Each diagram corresponds to a different path length in the crystal.

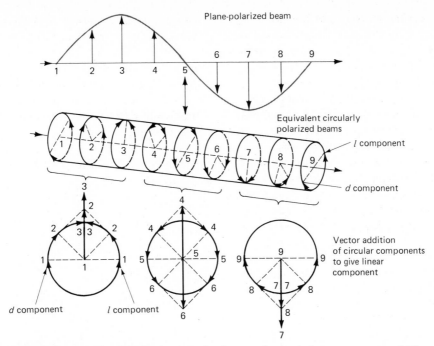

FIGURE 13-10 Equivalency of plane-polarized beam to two (d,l) circularly polarized beams.

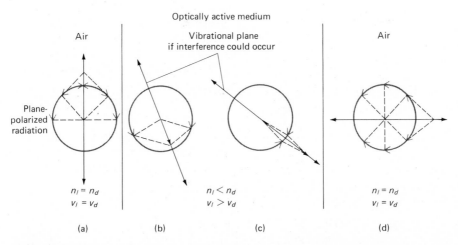

FIGURE 13-11 Rotation of plane-polarized light in a medium in which $n_l < n_d$.

vertical. At a further point (Figure 13-11c), the *d* component would be still further retarded and a greater rotation would result. Figure 13-11 has been drawn to show that, upon emerging from the medium, retardation has been such that the resultant plane is now horizontal. As shown in Figure 13-11c, the *d* and *l* components are again propagated at identical rates in the isotropic air medium, and interference of the coherent rays occurs. Passage through the anisotropic medium has had the effect, however, of shifting the phase relationship such that the observed plane of polarization is now 90 deg with respect to the original.

Quantitative Relationships. It is readily shown that the rotation (α_λ) in degrees caused by an optically active substance is given by the relationship

$$\alpha_\lambda = \frac{180\,l}{\lambda}\left(n_l - n_d\right) \qquad (13\text{-}6)$$

where l is the path length of the medium in centimeters and λ is the wavelength of the plane-polarized radiation (in vacuum), also in centimeters. The quantity $(n_l - n_d)$ is termed the *circular birefringence*. The following calculation demonstrates the magnitude of the circular birefringence required to bring about a typical rotation.

EXAMPLE

A solution contained in a 10-cm cell is found to rotate the plane-polarized radiation of the sodium D line by 100 deg. What is the difference in refractive index of the medium for the l and d circularly polarized components?

$$\lambda = 589 \text{ nm} \qquad \text{or} \qquad 5.89 \times 10^{-5} \text{ cm}$$

Substituting into Equation 13-6 yields

$$100 \text{ deg} = \frac{180 \text{ deg} \times 10 \text{ cm}}{5.89 \times 10^{-5} \text{ cm}}\left(n_l - n_d\right)$$

$$\left(n_l - n_d\right) = 3.3 \times 10^{-6}$$

It is apparent from this calculation that a relatively small difference in refractive index has a large effect in terms of optical rotatory power. Note that if the refractive index for the sodium D line is about 1.5 (a typical value), then a 2.2 ppm difference between n_d and n_l is responsible for the 100-deg rotation shown in the example.

Optically Active Compounds

Optical activity is associated with two types of species: (1) crystalline compounds, which lose their activity when the crystal is converted to a liquid, a gas, or a solution (quartz is the classic example of this manifestation of optical activity); (2) compounds in which the optical activity is inherent in the molecule itself and is observed regardless of the physical state of the compound. It is with this second type that we are concerned.

The structural requirements for an optically active molecule are well understood and treated in detail in organic chemistry textbooks. It is sufficient to note that the optically active forms of a molecule, called *enantiomers*, are mirror images which cannot be superimposed, regardless of the orientation of one with respect to the other. Enantiomers thus bear the same relationship to one another as the left to the right hand. The two isomers then rotate polarized light equally but in opposite directions; if one is in excess in a mixture or is isolated from the other, rotation is observed. The most common form of optical activity results from the presence of four different substituents on a tetrahedral carbon atom in an organic compound; this feature leads to two nonsuperimposable *chiral* molecules. Other types of asymmetric centers can also occur, both in organic and inorganic structures.[3]

---

[3] For a discussion of optical activity in inorganic systems, see: J. C. Bailar and D. H. Busch, *The Chemistry of the Coordination Compounds*. New York: Reinhold, 1956.

It is important to appreciate that the inter-actions which lead to rotation of polarized light are not peculiar to molecules with asymmetric centers but instead are characteristic of most molecules. However, rotation is not *observed* in noncrystalline samples that lack asymmetric character because the molecules are randomly oriented; rotation caused by a molecule in one orientation is canceled by the equal and opposite rotation of another oriented as its mirror image. For the same reason, samples containing *d* and *l* isomers in equal concentrations exhibit to net rotation because their individual effects cancel. It is only when one form of a chiral molecule is present in excess that a net rotation can be observed.

Variables that Affect Optical Rotation

The rotation of plane-polarized radiation by optically active compounds can range from several hundred to a few hundredths of a degree. Experimental variables that influence the observed rotation include the wavelength of radiation, the optical path length, the temperature, the density of the substance if undiluted, and its concentration if in solution. For solutions, the rotation may also vary with the kind of solvent.

The *specific rotation* or the *specific rotatory power* $[\alpha]_\lambda^T$ is widely employed to describe the rotatory characteristics of a liquid. It is defined as

$$[\alpha]_\lambda^T = \frac{\alpha}{lc} \qquad (13\text{-}7)$$

where α is the observed rotation in degrees, l is the path length in decimeters, and c is the grams of solute in 100 cc of solution. The wavelength λ and temperature T are usually specified with a subscript and a superscript, as shown. Most specific rotations are measured at 20°C with sodium D line and are thus reported as $[\alpha]_D^{20}$. For a pure liquid, c is replaced by its density. By convention, counter-

clockwise, or l rotation as the observer faces the beam, is given the negative sign. Clockwise (*d*) rotation is positive.

The term *molecular rotation* $[M]$ is also encountered; it is defined as

$$[M] = M[\alpha]/100 \qquad (13\text{-}8)$$

where M is the molecular weight.

It is frequently necessary to measure the optical rotatory power of a substance as a solute. Unfortunately, the specific rotation of a compound is nearly always found to vary with the nature of the solvent. Because of solubility considerations, no single standard solvent can be designated. In addition, the specific rotation in a given solvent may not be entirely independent of concentration, although variations in dilute solutions are usually small. Because of these effects, it is common practice to designate both the kind of solvent and the solute concentration in reporting a specific rotation.

The variation in specific rotation with temperature is approximately linear, but the temperature coefficient differs widely from substance to substance. For example, the specific rotation of tartaric acid solutions may vary as much as 10% per degree; on the other hand, the variation for sucrose is less than 0.1% per degree.[4]

The effect of wavelength on rotation is considered in a later section on optical rotatory dispersion.

Mechanism of Optical Rotation

While the structural requirements for optical activity can be precisely defined, the mechanism by which a beam of circularly polarized radiation interacts with matter and is thus

---

[4] It is important to remember that most organic solvents have temperature coefficients of expansion of about 0.1% per deg C. Therefore, accurate measurements of $[\alpha]_\lambda^0$ for solutions require close control of temperature both during the measurement and *during the solution preparation*.

retarded is much less obvious. That is, we can predict with certainty that a compound such as 2-iodobutane is chiral because it possesses an asymmetric center. On the other hand, it is difficult to account for the specific rotations ($[\alpha]_D^{20} = \pm 32$ deg) for these isomers, compared with those of 2-butanol ($[\alpha]_D^{20} = \pm 13.5$ deg).

Several theories concerning the mechanism of optical rotation have been developed. Generally, these are couched in terms of quantum mechanics and are of sufficient complexity to evade all but the most mathematically oriented chemist. Furthermore, while in principle it may be possible to predict optical rotatory power for certain types of compounds with the aid of these theories, such calculations are not practical in terms of effort; nor has it been demonstrated that the values obtained from such calculations can be sufficiently precise to be useful.

Polarimeters

The basic components of a polarimeter include a monochromatic light source, a polarizing prism, a sample tube, an analyzer prism with a circular scale, and a detector (see Figure 13-12). The eye serves as the detector for most polarimeters although photoelectric polarimeters are becoming more common.

Sources. Because optical rotation varies with wavelength, monochromatic radiation is employed. Historically, the sodium D line was obtained by introducing a sodium salt into a gas flame. Suitable filters then removed other lines and background radiation. Sodium vapor lamps with a filter to remove all but the D line are now employed. Mercury vapor lamps are also useful, the line at 546 nm being isolated by a suitable filter system.

Polarizer and Analyzer. Nicol prisms are most commonly employed to produce plane-polarized light and to determine the angle through which the light has been rotated by the sample. In principle, the measurement could be made by first adjusting the two prisms to a crossed position that yields a minimum in light intensity in the absence of sample. With a sample in place, rotation of the beam would cause an increase in light intensity which could then be offset by rotation of the analyzer prism. The angular change required to minimize the intensity would correspond to the rotatory power of the sample. Unfortunately, however, the position of minimum intensity cannot be determined accurately with the eye (nor with a photoelectric detector, for that matter) because the rate of change in intensity per degree of rotation is at a minimum in this region. Therefore, polarimeters are equipped with *half-shadow* devices which permit the determination to be made by matching two halves of a field at a radiation intensity greater than the minimum.

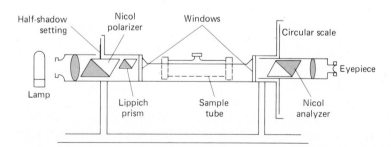

FIGURE 13-12   Typical visual polarimeter.

Half-Shadow Devices. Figure 13-12 shows a typical half-shadow device consisting of a small Nicol prism (called a *Lippich prism*) that intercepts about half the beam emerging from the polarizer. The position of the Lippich prism is adjusted to alter the plane of polarization by a few degrees; thus, in the absence of sample and with the analyzer Nicol prism at 90 deg with respect to the polarizer, a split, light-dark field is observed. The light portion, of course, corresponds to that half of the beam that has been rotated by the Lippich prism and the dark part of the field corresponds to the unobstructed beam. The intensity of the two halves is then balanced by rotation of the analyzer. The analyzer scale is adjusted to read zero at this point. With the sample in place, the analyzer is rotated until the same balance is obtained. The rotation of the sample can then be read directly from the circular analyzer scale.

Other end point devices, which operate upon the same principle as the Lippich prism, permit the determination of optical rotatory power with a precision of 0.005 to 0.01 deg under ideal conditions. Photoelectric detectors yield a precision of about 0.001 deg.

Sample Tubes. The sample for polarimetry is contained in a cylindrical tube, usually 5, 10, or 20 cm in length. The ends are plane-parallel glass disks that are either fused to the tube walls or held in place with screw-cap holders. For precise measurements, the tubes are surrounded by a jacket for temperature control. Tubes can be calibrated for length by measuring the rotation of a liquid of known rotatory power; nicotine/alcohol or sucrose/water mixtures are often used for this purpose.

Applications of Polarimetry

Qualitative Analysis. The optical rotation of a pure compound under a specified set of conditions provides a basic physical constant that is useful for identification purposes in the same way as its melting point, boiling point, or refractive index. Optical activity is characteristic of many naturally occurring substances such as amino acids, steroids, alkaloids, and carbohydrates; polarimetry represents a valuable tool for identifying such compounds.

Structural Determination. In this application, the change in optical rotation resulting from a chemical transformation is measured. Empirical correlations obtained from the study of known structures are then employed to deduce information about the unknown compound. Details of steroid structures, in particular, have been acquired from polarimetric measurements; similar information has been obtained for carbohydrates, amino acids, and other organic compounds.

Quantitative Analysis. Polarimetric measurements are readily adapted to the quantitative analysis of optically active compounds. Empirical calibration curves are used to relate optical rotation to concentration; these plots may be linear, parabolic, or hyperbolic.

The most extensive use of optical rotation for quantitative analysis is in the sugar industry. For example, if the only optically active constituent is sucrose, its concentration can be determined from a simple polarimetric analysis of an aqueous solution of the sample. The concentration is directly proportional to the measured rotation. If other optically active materials are present, a more complex procedure is required; here, the change in rotation resulting from the hydrolysis of the sucrose is determined. The basis for this analysis is shown by the equation

$$C_{12}H_{22}O_{11} + H_2O \xrightarrow{\text{acid}} C_6H_{12}O_6 + C_6H_{12}O_6$$

Sucrose	Glucose	Fructose
$[\alpha]_D^{20} = +66.5°$	$+52.7°$	$-92.4°$

This reaction is termed an *inversion* because of the change in sign of the rotation that occurs. The concentration of sucrose is directly proportional to the difference in rotation before and after inversion.

OPTICAL ROTATORY DISPERSION AND CIRCULAR DICHROISM

Optical rotatory dispersion and circular dichroism are two closely related physical methods that are based upon the interaction of circulatory polarized radiation with an optically active species.[5] The former method measures the wavelength dependence of the molecular rotation of a compound. As we have already indicated, optical rotation at any wavelength depends upon the difference in refractive index of a substance toward d and l circularly polarized radiation—that is, upon the *circular birefringence* $(n_l - n_d)$; this quantity is found to vary in a characteristic way as a function of wavelength. In contrast, circular dichroism depends upon the fact that the *molar absorptivity* of an optically active compound is different for the two types of circularly polarized radiation. Here, the wavelength dependence of $(\varepsilon_l - \varepsilon_d)$ is studied, where ε_l and ε_d are the respective molar absorptivities.

The inequality of molar absorptivities was first reported by A. Cotton in 1895, and the whole complex relationship between absorptivity and refractive index differences is now termed the *Cotton effect*. The application of the Cotton effect to chemical investigations was delayed until convenient instruments for its study became available.

General Principles

Cotton employed solutions of potassium chromium tartrate, which absorb in the visible region, to demonstrate that right circularly polarized radiation was not only refracted but also absorbed to a different extent than the left circularly polarized beam; that is, $\varepsilon_d \neq \varepsilon_l$.

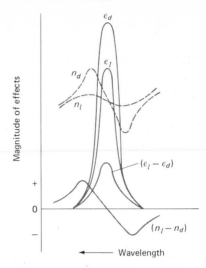

FIGURE 13-13 The Cotton effect. (Adapted from W. Heller and H. G. Curmé in *Physical Methods of Chemistry*, eds. A. Weissberger and B. W. Rossiter. New York: Wiley-Interscience, 1972, Vol. 1, Part III C, p. 66. With permission.)

At the same time, Cotton observed that dramatic changes occurred in the optical rotation $(n_l - n_d)$, as well as in the difference in the absorptivities $(\varepsilon_l - \varepsilon_d)$ of the beams in the region of an absorption maximum. These effects are shown in igure 13-13. It is important to appreciate that the $(n_l - n_d)$ curve for a substance is similar in shape to curves index as a function of the wavelength of unpolarized light (see Figure 4-8, p. 102); here, too, marked changes in the refractive index (anomalous dispersion) occur in the region of absorption.

---

[5] For a more complete discussion of optical rotatory dispersion and circular dichroism, see: K. P. Wong, *J. Chem., Educ.*, **52**, A9, A83, A573 (1975); C. Djerassi, *Optical Rotatory Dispersion: Application to Organic Chemistry*. New York: McGraw-Hill, 1960; P. Crabbé, *Optical Rotatory Dispersion and Circular Dichroism*. San Francisco: Holden-Day, 1965; and A. Abu-Shumays and J. J. Duffield, *Anal. Chem.*, **38** (7), 29A (1966).

Optical Rotatory Dispersion Curves

An optical rotatory dispersion curve consists of a plot of specific or molecular rotation as a function of wavelength. Two types of curvature can be discerned. The first is the normal dispersion range wherein $[\alpha]$ changes only gradually with wavelength. The second, the region of anomalous dispersion, occurs near an absorption peak. If one peak is isolated from others, the anomalous part of the dispersion curve will have the appearance of the curve labeled $(n_l - n_d)$ in Figure 13-13. That is, the rotation undergoes rapid change to some maximum (or minimum) value, alters direction to a minimum (or maximum), and then finally reverts to values corresponding to normal dis-

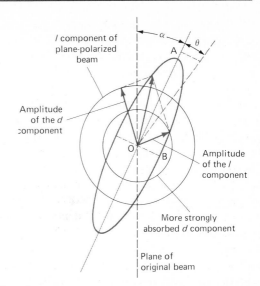

FIGURE 13-15   Elliptically polarized radiation after leaving a medium in which $\varepsilon_d > \varepsilon_l$ and $n_l > n_d$.

persion. As indicated in Figure 13-13, a change in the sign of the rotation may accompany these changes.

If molecules have multiple absorption peaks, as is usually the case, overlapping regions of anomalous dispersion lead to optical rotatory dispersion curves that are complex, as shown in Figure 13-14. Note that the ultraviolet absorption spectrum for the compound is also included for reference.

Circular Dichroism Curves

In circular dichroism, one of the circular components of a plane-polarized beam is more strongly absorbed than the other. The effect of this differential absorption is to convert the plane-polarized radiation to an elliptically polarized beam. Figure 13-15 illustrates how two circular components of unequal amplitude, which result from the differential absorption by a medium, are combined to give a resultant that travels in an elliptical path. The l component of the original beam is shown as retarded more than the d compo-

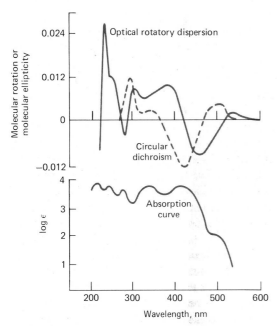

FIGURE 13-14   Optical rotatory dispersion, circular dichroism, and absorption curves for $(+)$-camphor trithione. (Adapted from H. Wolf, E. Bunnenberg, C. Djerassi, A. Lüttring-haus, and A. Stockhausen, *Ann. Chem.*, **674**, 62 (1964). With permission.)

nent because $n_l > n_d$; on the other hand, the amplitude of the d component is less than that of the l component because we have assumed that its molar absorptivity is greater; that is, $\varepsilon_d > \varepsilon_l$.

The angle of rotation α is taken as the angle between the major axis of the emergent elliptical beam and the plane of polarization of the incident beam. The *ellipticity* is given by the angle θ; the tangent of θ is clearly equal to the ratio of the minor axis of the elliptical path to the major (that is, OB/OA).

It can be shown that the ellipticity is approximated by the relation

$$\theta = \tfrac{1}{4}(k_l - k_d) \qquad (13-9)$$

where k_l and k_d are the absorption coefficients[6] of the circularly polarized l and d radiation and θ is expressed in radians. The quantity $(k_l - k_d)$ is termed the *circular dichroism*. The molecular ellipticity $[\theta]$ can be shown to be equal to

$$[\theta] = 3305(\varepsilon_l - \varepsilon_d) \qquad (13-10)$$

where $[\theta]$ has the units of degree-cm² per decimole and the ε's are the molar absorptivities of the respective circular components.

Circular dichroism curves consist of a plot of $[\theta]$ versus wavelength. Note that $[\theta]$ can be either negative or positive depending on the relative magnitudes of ε_l and ε_d. The dashed line in Figure 13-14 is a typical curve.

Instrumentation

Optical Rotatory Dispersion. A number of recording spectropolarimeters are now manufactured that directly provide optical rotatory dispersion curves in the ultraviolet and visible regions. In these instruments, radiation from a conventional monochromator is passed through a polarizer, the sample, and an ana-

lyzer and is detected with a photomultiplier tube. The signal from the detector is amplified; it is then employed to adjust the analyzer position to compensate for rotation caused by the sample and to position a recorder pen as well. As with a visual polarimeter, the half-shade method is the most efficient way of determining the null position of the analyzer. In one instrument, the polarizer is mechanically rocked through a small angle at low frequency. The amplifier system of the detector responds to the resulting ac signal and adjusts the analyzer until the signal is symmetric around the null point. Another spectropolarimeter employs an ordinary double-beam spectrometer with two sets of polarizer-analyzer prisms. The two analyzers are offset from one another by a few degrees, and both beams are passed through the sample. The ratio of the power of the two beams is compared electronically and gives a measure of the optical rotation of the sample.

Circular Dichroism. A conventional spectrometer can be adapted to measure molecular ellipticity. From Beer's law, Equation 13-10 can be written in the form

$$[\theta] = \frac{3305}{bc}\left(\log\frac{P_{l0}}{P_l} - \log\frac{P_{d0}}{P_d}\right)$$

where P_{l0} and P_l represent the power of the circularly polarized l beam before and after it has passed through a solution of length b and containing a molar concentration c of the sample. The terms P_d and P_{d0} have equivalent meanings for the d radiation. If now $P_{d0} = P_{l0}$, then

$$[\theta] = \frac{3305}{bc}\log\frac{P_d}{P_l} \qquad (13-11)$$

Thus, the molecular ellipticity can be obtained directly by comparing the power of the two transmitted beams, provided the intensities of the incident l and d circularly polarized beams are identical.

In order to employ Equation 13-11 with an ordinary spectrophotometer, a device for producing d and l circularly polarized radiation

---

[6] The absorption coefficient k is related to the more common molar absorptivity ε by the equation $k = 2.303\varepsilon c$, where c is the concentration in moles per liter.

must be provided. We have noted (p. 361) that circularly polarized radiation can be obtained by passing plane-polarized radiation through an anisotropic crystal which has a thickness such that the extraordinary and ordinary rays are one-quarter wavelength out of phase. Rotation of the optic axis of the quarter-wave plate by 90 deg yields either d or l circularly polarized radiation. For the measurement of circular dichroism with a single-beam spectrophotometer, a polarizer followed by a quarter-wave plate is inserted in the cell compartment of the instrument, provision being made to allow rotation of the plate by ± 45 deg. The sample cell is then placed between the plate and the detector; with the plate set to produce d circular radiation, the instrument is set on 100% transmittance, or zero absorbance. The plate is then rotated 90 deg; the new absorbance reading then corresponds to $\log (P_d/P_l)$ for the sample. In order to cover a very wide spectral range, several plates of different thicknesses must be employed.

Other methods exist for producing circularly polarized radiation. One of these, a *Fresnel rhomb*, is incorporated as an adapter for a double-beam spectrophotometer; see Figure 13-16. When a polarized beam undergoes internal reflection in this device, one of the perpendicular components is retarded with respect to the other. The retardation

depends upon the refractive index of the medium, the angle of incidence of the reflected beam, and the number of reflections. Quarter-wave retardation, and hence circular polarization, can be attained through proper adjustment of these variables.

The unit shown in Figure 13-16 is fitted into the sample compartment of a double-beam spectrophotometer, with a similar unit in the reference beam. The units are so adjusted that d radiation is emitted from one and l radiation from the other. The two beams then pass through identical cells containing the sample and their relative powers are compared photometrically.

Applications of Optical Rotatory Dispersion and Circular Dichroism

Optical rotatory dispersion and circular dichroism studies often provide spectral details for optically active compounds that are absent in their ultraviolet spectra. Thus, in the lower plot in Figure 13-14, the absorption spectrum is seen to consist of a group of overlapping peaks that would be difficult to interpret. On the other hand, the molecular rotation and ellipticity curves for the optically active groups are much more clearly defined and lend themselves to detailed analysis.

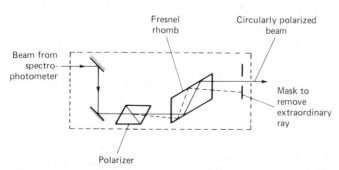

FIGURE 13-16 Spectrophotometer adapter for production of circularly polarized radiation.

Optical rotatory dispersion curves have been mainly applied to structural determinations in two major areas: (1) amino acids, polypeptides, and proteins; and (2) complex natural products such as steroids, terpenes, and antibiotics. Most of the structural conclusions from this work are empirical, being based upon spectral observations of known structures. The curves can provide information concerning the configuration of angular substituents at ring junctures, the location of ketone groups, conformational analysis of substituents exerting a vicinal action on an optically active chromophore, the degree of coiling of protein helices, and the type of substitution in amino acids.

The applications of circular dichroism are less developed than optical rotatory dispersion; it appears, however, that the technique will also provide much useful structural information regarding organic and biological systems as well as metal-ligand complexes.

PROBLEMS

1. In binary mixtures of liquids that approach ideal behavior, the refractive index varies linearly with volume percent. Mesitylene and pentane exhibit this behavior. For mesitylene, $n_D^{20} = 1.4998$, the molecular weight is 120.20, and the density is 0.8642 g/ml (all at 20°C). For pentane, $n_D^{20} = 1.3579$, the molecular weight is 72.15 and the density is 0.6262 g/ml. Find the molarity of pentane in a mixture of pentane and mesitylene having a refractive index of 1.4522.

2. Binary liquid mixtures that do not form ideal solutions exhibit refractive indexes which are complicated functions of the composition. One of the simplest functions applies to an acetone/carbon tetrachloride solution whose refractive index is given by $n_D^{20} = ac + b$, where c is the molar concentration of acetone and a and b are constants. For acetone, $n_D^{20} = 1.3588$, the molecular weight is 58.08 and the density is 0.7908 g/ml (all at 20°C). For carbon tetrachloride, $n_D^{20} = 1.4664$, the molecular weight is 153.82 and the density is 1.5942 g/ml. Calculate the refractive index of 3.00M acetone in CCl_4.

3. The molar refraction, R, of a compound is defined as

$$R = \left(\frac{n^2 - 1}{n^2 + 2}\right)\frac{M}{d}$$

where n is the index of refraction, M is the molecular weight, and d is the density. To a good approximation, the molar refraction of a compound is the sum of refractions contributed by each atom in the compound. Tables of atomic contributions to molar refraction can be found in such reference books as the *Handbook of Chemistry and Physics*. For example, carbon contributes a refraction of 2.591, hydrogen contributes 1.028, oxygen (doubly bonded) contributes 2.122, and chlorine contributes 5.844.

(a) Calculate the molar refraction of acetone from the data in Problem 2 and compare it to the value calculated from atomic contributions.

(b) Calculate the molar refraction of CCl_4 from atomic contributions.

(c) Using $n_D^{20} = 1.4664$ and $M = 153.82$, calculate the density of CCl_4. Compare your calculated value to the observed value of 1.5942 g/ml.

4. What is the difference between unpolarized, plane polarized, circularly polarized, and elliptically polarized light?

5. Distinguish between ordinary and extraordinary beams of radiation.

6. Define the following terms: (a) optic axis, (b) Nicol prism, (c) quarter wave plate, (d) double refraction, (e) circular double refraction, (f) circular birefringence, (g) ellipticity, (h) enantiomers, (i) dextrorotatory, and (j) levorotatory.

7. What is the difference between optical rotatory dispersion and circular dichroism?

8. Describe two methods of producing circularly polarized radiation.

9. How can it be determined whether a sample is rotating the plane of polarization of light by $+20$ deg or -160 deg?

10. If the circular birefringence of a solution at 360 nm is 1.0×10^{-5}, what will be the angle of rotation, α_{360}, for a 5-cm path length? If the circular birefringence is 1.0×10^{-5} at 720 nm, what will be α_{720} for a 5-cm path length?

11. For sucrose, $C_{12}H_{22}O_{11}$, $[\alpha]_D^{20} = +66.5$ deg.
    (a) Calculate the molecular rotation of sucrose.
    (b) Calculate the angle of rotation expected for a solution containing sucrose at a concentration of 5.0 g/liter in a 10-cm cell.
    (c) A solution originally containing 5.0 g of sucrose per liter exhibits $\alpha = +18.6$ deg in a 10-cm cell after a period of heating with acid (reaction on page 369). Calculate the fraction of sucrose which has been hydrolyzed.

14

NUCLEAR MAGNETIC RESONANCE SPECTROSCOPY

A strong magnetic field causes the energies of certain nuclei to be split into two or more quantized levels, owing to the magnetic properties of these particles. Transitions among the resulting magnetically induced energy levels can be brought about by the absorption of electromagnetic radiation of suitable frequency, just as electronic transitions are caused by the absorption of ultraviolet or visible radiation. The energy differences between magnetic quantum levels for atomic nuclei correspond to radiation energies in the frequency range of 0.1 to 100 MHz[1] (wavelengths between 3000 and 3 m), which is in the radio-frequency portion of the electromagnetic spectrum (see Figure 4-7, p. 101).

The study of absorption of radio-frequency radiation by nuclei is called *nuclear magnetic resonance*[2] (often abbreviated NMR or nmr); it has proved to be one of the most powerful tools available for determining the structure of both organic and inorganic species.

This chapter is mainly concerned with the theory, instrumentation, and applications of NMR spectroscopy; a somewhat analogous method called *electron spin resonance* is briefly discussed at the end.

THEORY OF NUCLEAR MAGNETIC RESONANCE

As early as 1924, Pauli suggested that certain atomic nuclei might have the properties of spin and magnetic moment and that as a con-

sequence, exposure to a magnetic field would lead to splitting of their energy levels. During the next decade, experimental verification of these postulates was obtained. It was not until 1946, however, that Bloch at Stanford and Purcell at Harvard, working independently, were able to demonstrate that nuclei absorbed electromagnetic radiation in a strong magnetic field as a consequence of the energy level splitting induced by the magnetic force. The two physicists shared the 1952 Nobel prize for their work. In the first five years following the discovery of nuclear magnetic resonance, chemists became aware that molecular environment influences the absorption by a nucleus in a magnetic field and that this effect can be correlated with molecular structure. Since then, the growth of NMR spectroscopy has been explosive, and the technique has had profound effect on the development of organic, inorganic, and biochemistry. It is doubtful that there has ever been as short a delay between an initial discovery and its widespread application and acceptance.

In common with optical spectroscopy, both classical and quantum mechanics are useful to provide an explanation of the nuclear magnetic resonance phenomenon. The two treatments yield identical relationships. Quantum mechanics, however, is more useful for relating absorption frequencies to energy states of molecules, while classical mechanics is more helpful in providing a physical picture of the absorption process and how it is measured.

Quantum Description of NMR

To account for some properties of nuclei, it is necessary to assume that they rotate about an axis and thus have the property of *spin*. Furthermore, it is necessary to postulate that the angular momentum associated with the spin of the particle is an integral or a half-integral multiple of $h/2\pi$, where h is Planck's constant. The maximum spin component for

---

[1] MHz = 10^6 Hz = 10^6 cycles per second.

[2] The following references are recommended for additional study: E. D. Becker, *High Resolution NMR*. New York: Academic Press, 1969; A. Ault and G. O. Dudek, *An Introduction to Proton Magnetic Resonance Spectroscopy*. San Francisco: Holden-Day, 1976; L. M. Jackman and S. Sternhall, *Nuclear Magnetic Resonance Spectroscopy*, 2d ed. New York: Pergamon Press, 1969; D. E. Leyden and R. H. Cox, *Analytical Applications of NMR*. New York: Wiley, 1977; and D. W. Mathieson, *Nuclear Magnetic Resonance for Organic Chemists*. New York: Academic Press, 1967.

a particular nucleus is its *spin quantum number* I; it is found that a nucleus will then have $(2I + 1)$ discrete states. The component of angular momentum for these states in any chosen direction will have values of I, $I - 1$, $I - 2$, ..., $-I$. In the absence of an external field, the various states have *identical* energies.

The spin number for the proton is $\frac{1}{2}$; thus two spin states exist, corresponding to $I = +\frac{1}{2}$ and $I = -\frac{1}{2}$. Heavier nuclei, being assemblages of various elementary particles, have spin numbers that range from zero (no net spin component) to at least $\frac{9}{2}$. As shown in Table 14-1, the spin number of a nucleus is related to the relative number of protons and neutrons it contains.

Magnetic Properties of Elementary Particles. Since a nucleus bears a charge, its spin gives rise to a magnetic field that is analogous to the field produced when electricity flows through a coil of wire. The resulting magnetic dipole μ is oriented along the axis of spin and has a value that is characteristic for each type of nucleus.

The interrelation between particle spin and magnetic moment leads to a set of observable magnetic quantum states m given by

$$m = I, I - 1, I - 2, \ldots, -I \quad (14\text{-}1)$$

Energy Levels in a Magnetic Field. When brought into the influence of an external magnetic field, a particle possessing a magnetic moment tends to become oriented such that its magnetic dipole—and hence its spin axis—is parallel to the field. The behavior of the particle is somewhat like that of a small bar magnet when introduced into such a field, the potential energy of either depending upon the orientation of the dipole with respect to the field. The energy of the bar magnet can assume an infinite number of values depending upon its alignment; in contrast, however, the energy of the nucleus is limited to $(2I + 1)$ discrete values (that is, the alignment is limited to $2I + 1$ positions). Whether quantized or not,

the potential energy of a magnet in a field is given by the relationship

$$E = -\mu_z H_0 \quad (14\text{-}2)$$

where μ_z is the *component* of magnetic moment in the direction of an external field having a strength of H_0.

The quantum character of nuclei limits the number of possible energy levels to a few. Thus, for a particle with a spin number of I and a magnetic quantum number of m, the energy of a quantum level is given by

$$E = -\frac{m\mu}{I} \beta H_0 \quad (14\text{-}3)$$

where H_0 is the strength of the external field in gauss (G) and β is a constant called the *nuclear magneton* (5.051×10^{-24} erg G^{-1}); μ is the magnetic moment of the particle expressed in units of nuclear magnetons. The value of μ for the proton is 2.7927 nuclear magnetons.

Turning now to the proton, for which $I = \frac{1}{2}$, we see from Equation 14-1 that this particle has magnetic quantum numbers of $+\frac{1}{2}$ and $-\frac{1}{2}$. The energies of these states in a magnetic field (Equation 14-3) take the following values:

$$m = +\tfrac{1}{2}, \quad E = -\frac{\tfrac{1}{2}(\mu\beta H_0)}{\tfrac{1}{2}}$$

$$= -\mu\beta H_0$$

$$m = -\tfrac{1}{2}, \quad E = -\frac{-\tfrac{1}{2}(\mu\beta H_0)}{\tfrac{1}{2}}$$

$$= +\mu\beta H_0$$

These two quantum energies correspond to the two possible orientations of the spin axis with respect to the magnetic field; as shown in Figure 14-1, for the lower energy state ($m = \frac{1}{2}$) the vector of the magnetic moment is aligned with the field, and for the higher energy state ($m = -\frac{1}{2}$) the alignment is reversed. The

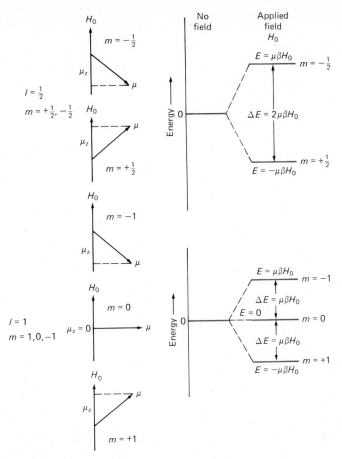

FIGURE 14-1 Orientations of magnetic moments and energy levels for nuclei in a magnetic field H_0.

TABLE 14-1 SPIN QUANTUM NUMBER FOR VARIOUS NUCLEI

Number of Protons	Number of Neutrons	Spin Quantum Number I	Examples
Even	Even	0	^{12}C, ^{16}O, ^{32}S
Odd	Even	$\frac{1}{2}$ $\frac{3}{2}$	^{1}H, ^{19}F, ^{31}P ^{11}B, ^{79}Br
Even	Odd	$\frac{1}{2}$ $\frac{3}{2}$	^{13}C ^{127}I
Odd	Odd	1	^{2}H, ^{14}N

energy difference between the two levels is given by

$$\Delta E = 2\mu\beta H_0$$

Also shown in Figure 14-1 are the orientations and energy levels for a nucleus such as ^{14}N, which has a spin number of 1. Here, three energy levels ($m = 1$, 0, and -1) are found, and the difference in energy between each is $\mu\beta H_0$. In general, the energy differences are given by

$$\Delta E = \mu\beta \frac{H_0}{I} \qquad (14\text{-}4)$$

As with other types of quantum states, excitation to a higher nuclear magnetic quantum level can be brought about by absorption of a photon with energy $h\nu$ that just equals ΔE. Thus, Equation 14-4 can be written as

$$h\nu = \mu\beta \frac{H_0}{I} \qquad (14\text{-}5)$$

EXAMPLE

Many NMR instruments employ a magnet that provides a field strength of 14,092 G. At what frequency would the proton nucleus absorb in such a field?

Substituting into Equation 14-5 we find

$$\nu = \frac{2.7927 \times 5.051 \times 10^{-24} \times 14,092}{6.6256 \times 10^{-27} \times \frac{1}{2}}$$

$$= 60.0 \times 10^6 \text{ Hz}$$

$$= 60.0 \text{ MHz}$$

The foregoing example reveals that radio-frequency radiation of 60.0 MHz will bring about a change in alignment of the magnetic moment of the proton from a direction that parallels the field to a direction that opposes it.

Distribution of Particles Between Magnetic Quantum States. In the absence of a magnetic field, the energies of the magnetic quantum states are identical. Consequently, a large as-semblage of protons will contain an identical number of nuclei with $m = +\frac{1}{2}$ and $m = -\frac{1}{2}$. When placed in a field, however, the nuclei tend to orient themselves so that the lower energy state ($m = +\frac{1}{2}$) predominates. Because thermal energies at room temperatures are several orders of magnitude greater than these magnetic energy differences, thermal agitation tends to offset the magnetic effects, and only a minute excess (<10 ppm) of nuclei persists in the lower-energy state. The success of nuclear magnetic resonance, however, depends upon this slight excess. If the number of protons in the two states were identical, the probability for absorption of radiation would be the same as the probability for reemission by particles passing from the higher energy state to the lower; under these circumstances, the net absorption would be nil.

Classical Description of NMR

To understand the absorption process, and in particular the measurement of absorption, a more classical picture of the behavior of a charged particle in a magnetic field is helpful.

Precession of Particles in a Field. Let us first consider the behavior of a nonrotating magnetic body, such as a compass needle, in an external magnetic field. If momentarily displaced from alignment with the field, the needle will swing in a plane about its pivot, as a consequence of the force exerted by the field on its two ends; in the absence of friction, the ends of the needle will fluctuate back and forth indefinitely about the axis of the field. A quite different motion occurs, however, if the magnet is spinning rapidly around its north-south axis. Because of the gyroscopic effect, the force applied by the field to the axis of rotation causes movement not in the plane of the force but perpendicular to this plane; the axis of the rotating particle, therefore, moves in a circular path (or *precesses*) around the magnetic field. This motion, illustrated in Figure 14-2, is similar to the motion of a

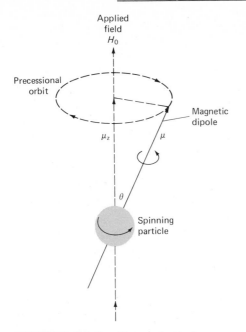

FIGURE 14-2 Precession of a rotating particle in a magnetic field.

gyroscope when it is displaced from the vertical by application of a force.

From classical mechanics, it is known that the angular velocity of precession is directly proportional to the applied force and inversely proportional to the angular momentum of the spinning body to which the force is applied. The force on a spinning nucleus in a magnetic field is the product of the field strength H_0 and the magnetic moment $\mu\beta$ of the particle, or $\mu\beta H_0$; as noted earlier, the angular momentum is given by $I(h/2\pi)$. Therefore, the precessional velocity is

$$\omega_0 = \frac{2\pi\mu\beta}{Ih} \cdot H_0 = \gamma H_0 \qquad (14\text{-}6)$$

where γ is a constant called the *magnetogyric ratio* (or, less appropriately, the *gyromagnetic ratio*). The magnetogyric ratio expresses the relationship between the magnetic moment

and the angular momentum of a rotating particle; that is,

$$\gamma = \frac{\mu\beta}{I(h/2\pi)} \qquad (14\text{-}7)$$

The magnetogyric ratio has a characteristic value for each type of nucleus.

Equation 14-6 can be converted to a frequency of precession v_0 (the *Larmor frequency*) by division by 2π. Thus,

$$v_0 = \frac{\omega_0}{2\pi} = \frac{\gamma H_0}{2\pi} \qquad (14\text{-}8)$$

Equations 14-8 and 14-7 can also be combined to give

$$hv_0 = \frac{\mu\beta}{I} H_0 \qquad (14\text{-}9)$$

A comparison of Equations 14-9 and 14-5 suggests that the precessional frequency of the particle derived from classical mechanics is identical to the quantum mechanical frequency of radiant energy required to bring about the transition of a rotating particle from one spin state to another; that is, $v_0 = v$. Substituting this equality in Equation 14-8 gives a useful relationship between the frequency of absorbed radiation and the strength of the magnetic field:

$$v = \frac{\gamma H_0}{2\pi} \qquad (14\text{-}10)$$

Absorption Process. The potential energy E of the precessing particle shown in Figure 14-2 is given by

$$E = -\mu_z H_0 = -\mu H_0 \cos\theta$$

Thus, when radio-frequency energy is absorbed by a nucleus, its angle of precession must change. Hence, we imagine that absorption involves a flipping of the magnetic moment that is oriented in the field direction to a state in which the moment is in the opposite direction. The process is pictured in Figure 14-3. In order for the dipole to flip, there must be present a magnetic force at

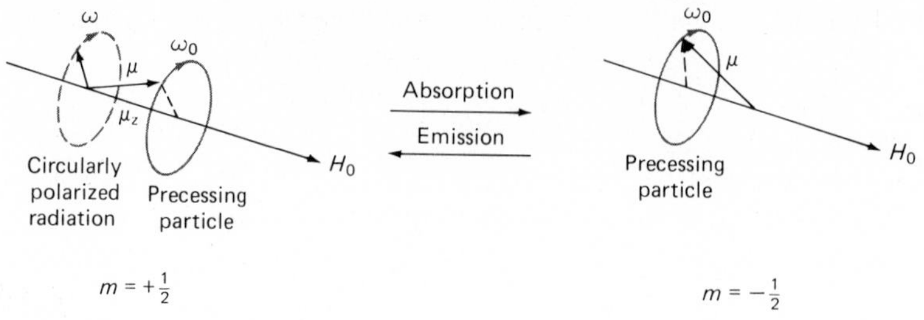

FIGURE 14-3 Model for the absorption of radiation by a precessing particle.

right angles to the fixed field and one with a circular component that can move in phase with the precessing dipole. Circularly polarized radiation (p. 361) of a suitable frequency has these necessary properties; that is, its magnetic vector has a circular component, as represented by the dotted line in Figure 14-3. If the rotational frequency of the magnetic vector of the radiation is the same as the precessing frequency, absorption and flipping can occur. The process is reversible, and the excited particle can thus return to the ground state by reemission of the radiation.

As shown in Figure 13-10 (p. 365), plane-polarized radiation can be considered to be composed of two circularly polarized beams rotating in opposite directions, in phase, and in a plane perpendicular to the plane of linear polarization. Thus, by irradiating nuclear particles with a beam polarized at 90 deg to the direction of the fixed magnetic field, circularly polarized radiation is introduced in the proper plane for absorption. Only that component of the beam that rotates in the precessional direction is absorbed; the other half of the beam, being out of phase, passes through the sample unchanged. The process is depicted in Figure 14-4.

Relaxation Processes

We must now consider mechanisms by which a nucleus in an upper energy or excited spin state can return to its lower energy state. One obvious path would involve emission of radiation of a frequency corresponding to the energy difference between the states. Radiation theory, however, predicts a low probability of occurrence for this process; it is necessary, therefore, to postulate radiationless pathways by which energy is lost by nuclei in higher spin states. The various mechanisms for radiationless energy transfers are termed nuclear *relaxation processes*.

The rates at which relaxation processes occur affect the nature and quality of an

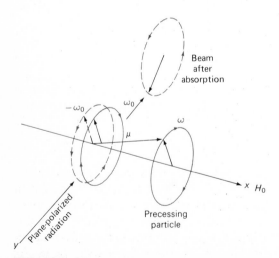

FIGURE 14-4 Absorption of one circular component of a beam that is polarized in the *xy* plane.

NMR signal; to some extent, these rates are subject to experimental control because they depend upon the physical state of the sample.

It is clear that relaxation processes are needed for the observation of a steady NMR absorption signal. Recall that such a signal depends upon the few parts-per-million excess of lower energy nuclei that exist in a strong magnetic field. Since absorption depletes this excess, the signal would rapidly fall to zero if additional low energy particles were not being produced at a sufficient rate by some radiationless energy-transfer processes. These processes must also be responsible for the creation of the slight excess of lower energy particles when the sample is first introduced into the magnetic field.

To produce a readily detectable absorption signal, then, the relaxation process should be as rapid as possible; that is, the lifetime of the excited state should be small. A second factor—the inverse relationship between the lifetime of an excited state and the width of its absorption line—negates the advantage of very short lifetimes. Thus, when relaxation rates are high, or the lifetimes low, line broadening is observed, which prevents high-resolution measurements. These two opposing factors cause the optimum half-life for an excited species to range from perhaps 0.1 to 1 s.

Two types of nuclear relaxation are recognized. The first is called *longitudinal* or *spin-lattice* relaxation; the second is termed *transverse* or *spin-spin* relaxation.

Spin-Lattice Relaxation. The absorbing nuclei in an NMR experiment are part of the larger assemblage of atoms that constitutes the sample. The entire assemblage is termed the lattice, regardless of whether the sample is a solid, a liquid, or a gas. In the latter two states particularly, the various nuclei comprising the lattice are in violent vibrational and rotational motion, which creates a complex field about each magnetic nucleus. The resulting lattice field thus contains an infinite number of magnetic components, at least some of which must correspond in frequency and phase with the precessional frequency of the magnetic nuclei of interest. These vibrationally and rotationally developed components interact with and convert nuclei from a higher to a lower spin state; the absorbed energy then simply increases the amplitude of the thermal vibrations or rotations. This change corresponds to a minuscule temperature rise for the sample.

Spin-lattice relaxation is a first-order process that can be characterized by a time T_1, which is a measure of the average lifetime of the nuclei in the higher energy state. In addition to depending upon the magnetogyric ratio of the absorbing nuclei, T_1 is strongly affected by the mobility of the lattice. In crystalline solids and viscous liquids, where mobilities are low, T_1 is large. As the mobility increases (at higher temperatures, for example), the vibrational and rotational frequencies increase, thus enhancing the probability for existence of a magnetic fluctuation of the proper magnitude for a relaxation transition; T_1 becomes shorter as a consequence. At very high mobilities, on the other hand, the fluctuation frequencies are further increased and spread over such a broad range that the probability of a suitable frequency for a spin-lattice transition again decreases. The result is a minimum in the relationship between T_1 and lattice mobility.

The spin-lattice relaxation time is greatly shortened in the presence of an element with an unpaired electron which, because of its spin, creates strong fluctuating magnetic fields. A similar effect is caused by nuclei that have spin numbers greater than one-half. These particles are characterized by a nonsymmetrical charge distribution; their rotation also produces a strong fluctuating field that provides yet another pathway for an excited nucleus to give up its energy to the lattice. The marked shortening of T_1 causes line broadening in the presence of such species. An example is found in the NMR spectrum for the proton attached to a nitrogen atom (for ^{14}N, $I = 1$).

Spin-Spin Relaxation and Line Broadening.
Several other effects tend to diminish relaxation times and thereby broaden NMR lines. These effects are normally lumped together and described by a spin-spin relaxation time T_2. Values for T_2 are generally so small for crystalline solids or viscous liquids (as low as 10^{-4} s) as to prohibit the use of samples of these kinds for high-resolution spectra.

When two neighboring nuclei of the same kind have identical precession rates, but are in different magnetic quantum states, the magnetic fields of each can interact to cause an interchange of states. That is, a nucleus in the lower spin state can be excited while the excited nucleus relaxes to the lower energy state. Clearly, no net change in the relative spin state population results, but the average lifetime of a particular excited nucleus is shortened. Line broadening is the result.

Two other causes of line broadening should be noted. Both arise if H_0 in Equation 14-10 differs slightly from nucleus to nucleus; under these circumstances, a band of frequencies, rather than a single frequency, is absorbed. One cause for such a variation in the static field is the presence in the sample of other magnetic nuclei whose spins create local fields that may act to enhance or diminish the external field acting on the nucleus of interest. In a mobile lattice, these local fields tend to cancel because the nuclei causing them are in rapid and random motion. In a solid or a viscous liquid, however, the local fields may persist long enough to produce a range of field strengths and thus a range of absorption frequencies. Variations in the static field also result from small inhomogeneities in the field source itself. This effect can be largely offset by rapidly spinning the entire sample in the magnetic field.

Measurement of Absorption

Absorption Signal. The analytical measurement in all of the types of absorption spectroscopy that we have discussed thus far has

consisted of determining the decrease in the power (the attenuation) of radiation caused by the absorbing sample. While this same technique has been applied in NMR spectroscopy, it suffers from the disadvantage that the excess of absorbing particles is so small that the resulting attenuation is difficult to measure accurately. Consequently, nearly all NMR spectrometers employ an arrangement by which the magnitude of a positive absorption-related signal is determined.

Figure 14-5b is a schematic representation of the major components of an NMR spectrometer; these are described in greater detail in a subsequent section. The radiation source is a coil that is part of a radio-frequency oscillator circuit (p. 24 to 27). Electromagnetic radiation from such a coil is plane-polarized (in the xz plane in the figure). The detector is a second coil located at right angles to the source (on the y axis in the figure) and is part of a radio-receiver circuit. The magnetic field used in NMR experiments has a direction along the z axis, perpendicular with respect to both the source and the detector.

As shown in Figure 14-5c, the plane-polarized radiation can be resolved into two circularly polarized vectors that rotate in opposite directions to one another (in the xy plane). Vector addition of these components, regardless of their angular position, indicates that there is no net component *along the y axis*. Thus, no signal is received by a detector located on that axis.

Figure 14-5b and 14-5d shows the effect of a sample positioned at the origin of the three axes. If the source has a frequency that is absorbed by a particular type of nucleus in the sample, the power of one of the two circular components of the beam is diminished. Vector addition of the resulting components indicates that the radiation now has a fluctuating component in the y direction, which causes the detector to respond. Thus, the sample serves to couple the generator to the receiver, provided the frequency of radiation corresponds to the precessing frequency

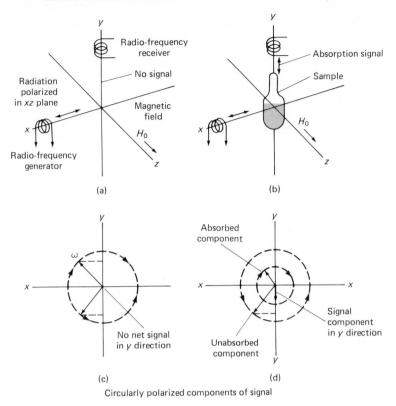

FIGURE 14-5 The absorption signal in NMR.

of the nuclei in the magnetic field. The extent of coupling, and thus the signal strength, depends upon the number of absorbing nuclei.

Absorption Spectra. Two methods are available for producing NMR spectra. The first is analogous to the method for obtaining optical spectra; it involves measuring an absorption signal as the electromagnetic frequency is varied. No dispersing elements such as prisms or gratings exist for radio frequencies, however. Thus, a variable-frequency oscillator with a linear sweep is required to produce radio frequencies that can be varied continuously over a range of 1 kHz for hydrogen and as great as 10 kHz for such nuclei as ^{13}C and ^{19}F.

An alternative way of collecting NMR

data is to employ a constant-frequency radio oscillator and sweep the magnetic field H_0 continuously. Since for a given nucleus, frequency and field strength are directly proportional (Equation 14-10), H_0 can serve equally well as the abscissa for an absorption spectrum. Figure 14-6 shows such a spectrum.

Early commercial NMR instruments employed the field sweep method of producing spectra because the electronic equipment required to produce a linear variation in magnetic field is considerably simpler and less expensive than a linear sweep oscillator. The latter, however, has the advantage of being more efficient in producing spin-decoupled spectra (see p. 402). As a consequence, several more recent instruments make use of frequency sweep; some are capable of both sweep modes.

Table 14-2 provides spectral data for several common nuclei employed for NMR studies.

EXPERIMENTAL METHODS OF NMR SPECTROSCOPY

Nuclear magnetic resonance instruments are either *high-resolution* or *wide-line* instruments. Only the former can resolve the fine structures associated with absorption peaks; the nature of this fine structure is determined by the chemical environment of the nucleus. High resolution requires the use of magnetic fields greater than 7000 G. Wide-line instruments are useful for quantitative elemental analysis and for studying the physical environment of a nucleus. Figure 14-6 is an example of a low-resolution spectrum obtained with a wide-line instrument. Wide-line spectrometers, which employ magnets with field strengths of a few thousand gauss, are

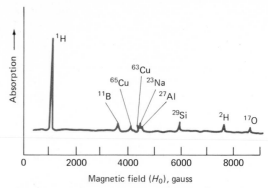

FIGURE 14-6 A low-resolution NMR spectrum of water in glass container. Frequency = 5 MHz. (Courtesy of Varian Associates, Palo Alto, California.)

much simpler and less expensive than their high-resolution counterparts.

From an instrumental standpoint, the equipment required for high-resolution NMR

TABLE 14-2   SPECTRAL AND MAGNETIC PROPERTIES OF SOME COMMON NUCLEI

Nucleus	Spin	Absorption Frequency[a]	Isotopic Abundance, %	Sensitivity[b]
^1H	$\frac{1}{2}$	60.0	99.98	1.000
^7Li	$\frac{3}{2}$	23.3	92.57	0.294
^{13}C	$\frac{1}{2}$	15.1	1.11	0.00018
^{14}N	1	4.3	99.63	0.001
^{17}O	$\frac{5}{2}$	8.1	0.04	0.00001
^{19}F	$\frac{1}{2}$	56.5	100	0.833
^{23}Na	$\frac{3}{2}$	15.9	100	0.093
^{25}Mg	$\frac{5}{2}$	3.8	10.05	0.027
^{27}Al	$\frac{5}{2}$	15.6	100	0.206
^{29}Si	$\frac{1}{2}$	11.9	4.70	0.00037
^{31}P	$\frac{1}{2}$	24.3	100	0.066
^{33}S	$\frac{3}{2}$	4.6	7.67	0.002
^{109}Ag	$\frac{1}{2}$	2.8	48.65	0.0001

[a] MHz in a 14092-G magnetic field.
[b] Sensitivity relative to an equal number of protons at a constant field.

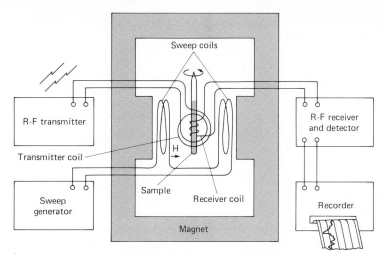

FIGURE 14-7   Schematic diagram of an NMR spectrometer. (Courtesy of Varian Associates, Palo Alto, California.)

spectroscopy is the most elaborate needed for any absorption method. Thus, most chemists are forced to accept only a general understanding of the operating principles of the NMR spectrometer, and leave its design and maintenance to those skilled in electronics. On the other hand, the sample-handling techniques, the interpretation of spectra, and an appreciation of the effects of variables are no more complex than for other types of absorption spectroscopy; these areas are of greatest interest to the chemist.

Instrumentation[3]

Figure 14-7 is a schematic diagram showing the important components of an NMR spectrometer. A brief description of each follows.

   The Magnet. The sensitivity and resolution of an NMR spectrometer are critically dependent upon the strength and quality of its magnet. Both sensitivity and resolution in-

crease with increases in field strength; in addition, however, the field must be highly homogeneous and reproducible. These requirements make the magnet by far the most expensive component of an NMR spectrometer.

   Spectrometric magnets are of three types: permanent magnets, conventional electromagnets, and superconducting solenoids. Permanent magnets with field strengths of 7046 or 14,092 G are used in several commercial instruments; corresponding oscillator frequencies for proton studies are 30 and 60 MHz. Permanent magnets are highly temperature-sensitive and require extensive thermostating and shielding as a consequence.

   Electromagnets are relatively insensitive to temperature fluctuations but require cooling systems to remove the heat generated by the large currents that are required. Elaborate power supplies are also necessary to provide the required stability. Commercial electromagnets generate fields of 14,092, 21,140, and 23,490 G, corresponding to proton absorption frequencies of 60, 90, and 100 MHz.

   Superconducting magnets are used in the highest resolution and most expensive instru-

[3] For a concise description of NMR spectrometers, see: D. G. Howery, *J. Chem. Educ.*, **48**, A327, A389 (1971).

ments. Here, fields as great as 110,390 G are attained, corresponding to a proton frequency of 470 MHz.

The performance specifications for a spectrometer magnet are stringent. The field produced must be homogeneous to a few parts per billion within the sample area and must be stable to a similar degree for short periods of time. Unfortunately, the inherent stability of most magnets is considerably lower than this; variations as large as one part in 10^7 are often observed over a period of 1 hr. In order to offset the effect of field fluctuations, a *frequency lock system* is employed in all commercial NMR instruments. Here, a reference nucleus is continuously irradiated and monitored at a frequency corresponding to its resonance maximum at the rated field strength of the magnet. Changes in the intensity of the reference absorption signal, resulting from drifts in the magnetic field, control a feedback circuit, the output of which is fed into coils in the magnetic gap in such a way as to correct for the drift.

Recall that the ratio between the field strengths and resonance frequencies is a constant *regardless of the nucleus involved* (Equation 14-10). Thus, the drift correction for the reference signal is applicable to the signals for all nuclei in the sample area.

Two types of lock systems are encountered. The *external lock* employs a separate sample container for the reference substance; this container must be located as close as possible to the analyte container. In an *internal lock* system, the reference substance is dissolved in the solution containing the sample. Here, the reference compound is usually tetramethylsilane (TMS), which also serves as an internal standard (see p. 392).

The external lock is simpler to operate, and control is not lost during sample changes. It does not, however, provide as good control over field changes encountered by the sample because of its displacement from the sample. Thus, the external system provides control of the order of parts in 10^9, while the internal

system gives control of the order of parts in 10^{10}.

The Field Sweep Generator. A pair of coils located parallel to the magnet faces (see Figure 14-7) permits alteration of the applied field over a small range. By varying a direct current through these coils, the effective field can be changed by a few hundred milligauss without loss of field homogeneity.

Ordinarily, the field strength is changed automatically and linearly with time, and this change is synchronized with the linear drive of a chart recorder. For a 60-MHz proton instrument, the sweep range is 1000 Hz (235 milligauss) or some integral fraction thereof. For nuclei such as ^{19}F and ^{13}C, sweeps as great as 10 kHz are required.

As was noted earlier, several more recent instruments employ a frequency sweep system in lieu of or in addition to magnetic field sweep.

The Radio-Frequency Source. The signal from a radio-frequency oscillator (transmitter) is fed into a pair of coils mounted at 90 deg to the path of the field. A fixed oscillator of exactly 60, 90, or 100 MHz is ordinarily employed; for high-resolution work, the frequency must be constant to about 1 part in 10^9. The power output of this source is less than 1 W and should be constant to perhaps 1% over a period of several minutes. The output is plane polarized.

The Signal Detector and Recorder System. The radio-frequency signal produced by the resonating nuclei is detected by means of a coil that surrounds the sample and is at right angles to the source coil. The electrical signal generated in the coils is small and must be amplified by a factor of 10^5 or more before it can be recorded.

The abscissa drive of an NMR recorder is synchronized with the spectrum sweep; often, in fact, the recorder controls the sweep rate. The response of the recorder must be rapid; it is also desirable that the sweep rate be variable.

All modern NMR recorders are equipped

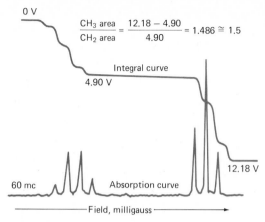

$$\frac{CH_3 \text{ area}}{CH_2 \text{ area}} = \frac{12.18 - 4.90}{4.90} = 1.486 \cong 1.5$$

FIGURE 14-8 Absorption and integral curve for a dilute ethylbenzene solution (aliphatic region). (Courtesy of Varian Associates, Palo Alto, California.)

with electronic integrators to provide information regarding areas under absorption peaks (see Figure 3-26, p. 66 for a simple integration circuit). Usually, the integral data appear as step functions superimposed on the NMR spectrum (see Figure 14-8). Generally, the area data are reproducible to a few percent relative.

Peak areas are important because they per-

mit estimation of the relative number of absorbing nuclei in each chemical environment (Figure 14-8). This information is vital to the deduction of chemical structure. In addition, peak areas are useful for quantitative analytical work.

Peak heights are not particularly satisfactory measures of concentration because a number of variables that are difficult to control can alter the widths and thus the heights of the peaks. Peak areas, in contrast, are not affected by these variables.

The Sample Holder and Sample Probe. The usual NMR sample cell consists of a 5-mm O.D. glass tube containing about 0.4 ml of liquid. Microtubes for smaller sample volumes are also available.

The sample probe is a device for holding the sample tube in a fixed spot in the field. As will be seen from Figure 14-9, the probe contains not only the sample holder but the sweep source and detector coils as well, to assure reproducible positioning of the sample with respect to those components; it may also contain the reference cell and liquid used as an external lock. The detector and receiver coils are oriented at right angles to one another to minimize the transfer of power in the absence of sample. Even with this arrange-

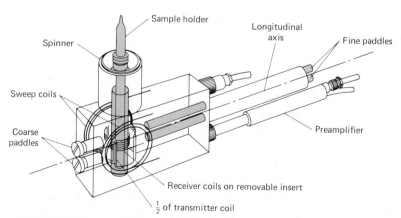

FIGURE 14-9 NMR probe. (Courtesy of Varian Associates, Palo Alto, California.)

ment, some leakage of power between the source and receiver does occur; the so-called paddles, shown in Figure 14-9, reduce this leakage to a tolerable level.

The sample probe is also provided with an air-driven turbine for rotating the sample tube at several hundred rpm along its longitudinal axis. This rotation serves to average out the effects of inhomogeneities in the field; sharper lines and better resolution are obtained as a consequence.

Cost of NMR Instruments. The foregoing brief discussion suggests that NMR instruments are complex. As might be expected, their cost is high, ranging from about $6000 to $200,000. The typical 60-MHz high-resolution instrument employed for routine work in many laboratories is priced from approximately $30,000 to $40,000.

Sample Handling

For high-resolution work, samples must be in a nonviscous liquid state. Most commonly, solutions of the sample (2 to 15%) are employed. The sample can also be examined neat if it has suitable physical properties.

The best solvents for proton NMR spectroscopy contain no protons; from this standpoint, carbon tetrachloride is ideal. The low solubility of many compounds in carbon tetrachloride limits its value, however, and a variety of deuterated solvents are used instead. Deuterated chloroform ($CDCl_3$) and deuterated benzene are commonly encountered.

ENVIRONMENTAL EFFECTS ON PROTON NMR SPECTRA

Figure 14-6 is an example of a low-resolution NMR spectrum. Each kind of nucleus is characterized by a single absorption peak, the location of which appears to be independent of the chemical state of the atom. If, however, the spectral region around one of the nuclear

absorption peaks is examined in detail with an instrument that permits the determination of absorption over very much smaller increments of the abscissa (H_0 or v), the single peak is usually found to be composed of several peaks (for example, see the proton spectra in Figure 14-10); moreover, the position and intensity of the component peaks depend critically upon the chemical environment of the nucleus responsible for the absorption. It is this dependence of the fine-structure spectrum upon the environment of a nucleus that makes NMR spectroscopy such a valuable tool.

The effect of chemical environment on nuclear magnetic absorption spectra (Figure 14-10) is general; as a consequence, NMR rep-

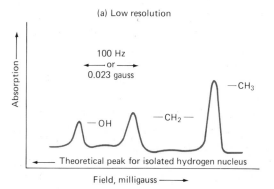

(a) Low resolution

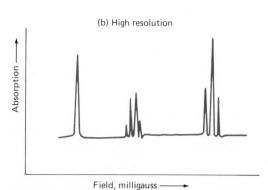

(b) High resolution

FIGURE 14-10 NMR spectra of ethanol at a frequency of 60 MHz. (a) Resolution $\sim 1/10^6$; (b) resolution $\sim 1/10^7$.

resents a profitable approach of the study of nuclei that possess magnetic properties. The fact that the proton appears in most organic compounds, and in many inorganic ones as well, has resulted in a concentration of effort upon this particular nucleus; our discussion follows this trend.

Types of Environmental Effects

The spectra for ethyl alcohol, shown in Figure 14-10, illustrate two types of environmental effects. The curve in Figure 14-10a, obtained with a lower resolution instrument, shows three proton peaks with areas in the ratio 1 : 2 : 3 (left to right). On the basis of this ratio, it appears logical to attribute the peaks to the hydroxyl, the methylene, and the methyl protons, respectively. Other evidence confirms this conclusion; for example, if the hydrogen atom of the hydroxyl group is replaced by deuterium, the first peak disappears from this part of the spectrum. Thus, small differences occur in the absorption frequency of the proton; such differences depend upon the group to which the hydrogen atom is bonded. This effect is called the *chemical shift.*

The higher resolution spectrum of ethanol, shown in Figure 14-10b, reveals that two of the three proton peaks are split into additional peaks. This secondary environmental effect, which is superimposed upon the chemical shift, has a different cause; it is termed *spin-spin splitting.*

Both the chemical shift and spin-spin splitting are important in structural analysis.

Experimentally, the two types of NMR peaks are readily distinguished, for it is found that the peak separations (in units of v or H_0) resulting from a chemical shift are directly proportional to the field strength or to the oscillator frequency. Thus, if the spectrum in Figure 14-10a were to be obtained at 100 MHz rather than at 60 MHz, the horizontal distance between any pair of the peaks would be increased by 100/60 (see Figure 14-11). In contrast, the distance between the fine-structure peaks within a group (lower spectrum) would not be altered by this frequency change.

Measurement of the Chemical Shift and Spin-Spin Splitting

Source of the Chemical Shift. The chemical shift arises from a circulation of the electrons surrounding the nucleus under the influence of the applied magnetic field. This phenomenon is discussed later in greater detail; for the present, it suffices to say that this movement of electrons creates a small magnetic field that ordinarily opposes the applied field. As a consequence, the nucleus is exposed to an effective field that is somewhat smaller (but in some instances, larger) than the external field. The magnitude of the field developed internally is directly proportional to the applied external field, so that we may write

$$H_0 = H_{\text{appl}} - \sigma H_{\text{appl}} \qquad (14\text{-}11)$$
$$= H_{\text{appl}}(1 - \sigma)$$

where H_{appl} is the applied field and H_0 *is the resultant field which determines the resonance behavior of the nucleus.* The quantity σ is the *shielding parameter,* which is determined by the electron density around the nucleus and which depends upon the structure of the compound containing the nucleus.

The shielding parameter for protons in a methyl group is larger than σ for methylene protons, and the parameter is even smaller for the proton in an —OH group. It is, of course, zero for an isolated hydrogen nucleus. Thus, in order to bring any of the protons in ethanol into resonance at a given oscillator frequency v, it is necessary to employ a field H_{appl} that is greater than H_0 (Equation 14-11), the resonance value for the isolated proton. Since σ differs for protons in various functional groups, the required applied field differs from group to group. This effect is shown in the spectrum of Figure 14-10a, where the hydroxyl proton appears at the lowest applied field, the methylene protons next, and finally the methyl protons. Note that all of these peaks occur at an applied field greater than the theoretical one for the isolated hydrogen nucleus, which would lie far to the left in Figure 14-10a. Note also that if the applied field is held constant at a level necessary to excite the methyl proton, an increase in frequency would be needed to bring the methylene protons into resonance.

Source of Spin-Spin Splitting. The splitting of chemical shift peaks can be explained by assuming that the effective field around one nucleus is further enhanced or reduced by local fields generated by *the hydrogen nuclei bonded to an adjacent atom.* Thus, the fine structure of the methylene peak shown in Figure 14-10b can be attributed to the effect of the local fields associated with the adjacent methyl protons. Conversely, the three methyl peaks are caused by the adjacent methylene protons. These effects are independent of the applied field and are superimposed on the effects of the chemical shift.

Abscissa Scales for NMR Spectra. The determination of the absolute field strength with the accuracy required for high resolution is difficult or impossible; on the other hand, it is entirely feasible to determine within a few milligauss the *change* in field strength caused by the subsidiary sweep coils. Thus, it is expedient to report the position of resonance absorption peaks relative to the resonance peak for a standard substance that can be measured at essentially the same time. In this way, the effect of fluctuations in the fixed magnetic field is minimized. The use of an internal standard is also advantageous in that chemical shifts can be reported in terms that are independent of the oscillator frequency.

A variety of internal standards have been proposed, but the compound now most generally accepted is tetramethylsilane (TMS), $(CH_3)_4Si$. All of the protons in this compound are identical, and for reasons to be

considered later, the shielding parameter for TMS is larger than for most other protons. Thus, the compound provides, at a high applied field, a single sharp peak that is isolated from most of the peaks of interest in a spectrum. In addition, TMS is inert, readily soluble in most organic liquids, and easily removed from samples by distillation (bp = 27°C). Unfortunately, TMS is not water soluble; in aqueous media, the sodium salt of 2,2-dimethyl-2-silapentane-5-sulfonate,

$$(CH_3)_3SiCH_2CH_2CH_2SO_3Na$$

may be used in its stead. The methyl protons of this compound produce a peak analogous to that of TMS; the methylene protons give a series of small peaks that are readily identified and can thus be ignored.

The field strength H_{ref} required to produce the TMS resonance line at frequency v is given by Equation 14-11

$$H_0 = H_{ref}(1 - \sigma_{ref})$$

which can be rewritten as

$$\sigma_{ref} = \frac{H_{ref} - H_0}{H_{ref}} \qquad (14\text{-}12)$$

Similarly, for a given absorption peak of the sample, we may write

$$\sigma_{sple} = \frac{H_{sple} - H_0}{H_{sple}} \qquad (14\text{-}13)$$

where H_{sple} is the field necessary to produce the peak. We then define a *chemical shift parameter* δ as

$$\delta = (\sigma_{ref} - \sigma_{sple}) \times 10^6 \qquad (14\text{-}14)$$

and substitute Equations 14-12 and 14-13 into this expression to obtain

$$\delta = \frac{H_0(H_{ref} - H_{sple})}{H_{sple}H_{ref}} \times 10^6$$

But H_{sple} is very nearly the same as H_0, so that the ratio H_0/H_{sple} is very nearly unity. Thus,

$$\delta \cong \frac{H_{\text{ref}} - H_{\text{sple}}}{H_{\text{ref}}} \times 10^6 \qquad (14\text{-}15)$$

The quantity δ is dimensionless and expresses the relative shift in parts per million; for a given peak, δ will be the same regardless of whether a 60-, 90-, or 100-MHz instrument is employed. Most proton peaks lie in the δ range of 1 to 12.

Another chemical shift parameter τ is defined as

$$\tau = 10 - \delta \qquad (14\text{-}16)$$

Generally, NMR plots have linear scales in δ (and sometimes τ), and the data are plotted with the field increasing from left to right (see Figure 14-11). Thus, if TMS is employed as the reference, its peak will appear on the far right-hand side of the plot, since σ for TMS is

large. As shown, the zero value for the δ scale corresponds to the TMS peak and the value of δ increases from right to left. The τ scale, of course, changes in the opposite way. Referring again to Figure 14-11, note that the various peaks appear at the same values of δ and τ in spite of the fact that the two spectra were obtained with instruments having different fixed fields.

For the purpose of reporting spin-spin splitting, it is desirable to utilize scalar units of milligauss or hertz. The latter scale has now come into more general use and is the one shown in Figure 14-11. The position of the reference TMS peak is arbitrarily taken as zero, with frequencies increasing from right to left. The effect of this choice is to make the frequency of a given peak identical with the increase in oscillator frequency that would be needed to bring that proton into resonance if the field were maintained constant at the level required to produce the TMS peak.

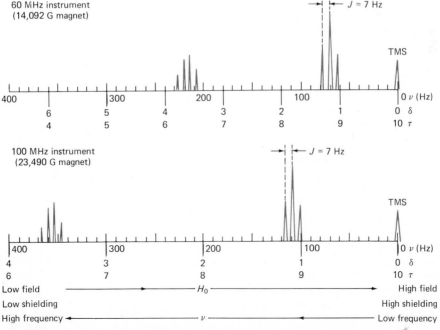

FIGURE 14-11 Abscissa scales for NMR spectra.

It can be seen in Figure 14-11 that the spin-spin splitting in frequency units (J) is the same for the 60-MHz and the 100-MHz instruments. Note, however, that the chemical shift *in frequency units* is enhanced with the higher frequency instrument.

The Chemical Shift

As noted earlier, chemical shifts arise from the secondary magnetic fields produced by the circulation of electrons in the molecule. These electronic currents (local *diamagnetic currents*[4]) are induced by the fixed magnetic field and result in secondary fields that may either decrease or enhance the field to which a given proton responds. The effects are complex, and we consider only the major aspects of the phenomenon here. More complete treatments can be found in the several reference works listed under footnote 2, page 377.

Under the influence of the magnetic field, electrons bonding the proton tend to precess around the nucleus in a plane perpendicular to the magnetic field (see Figure 14-12). A consequence of this motion is the development of a secondary field, which opposes the primary field; the behavior here is analogous to the passage of electrons through a wire loop. The nucleus then experiences a resultant field, which is smaller (the nucleus is said to be *shielded* from the full effect of the primary field); as a consequence, the external field must be increased to cause nuclear resonance. The frequency of the precession, and thus the magnitude of the secondary field, is a direct function of the external field.

The shielding experienced by a given nucleus is directly related to the electron density sur-

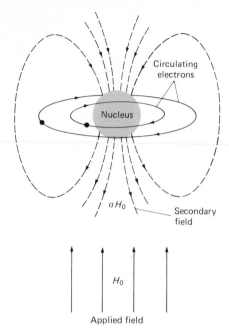

FIGURE 14-12   Diamagnetic shielding of a nucleus.

rounding it. Thus, in the absence of the other influences, shielding would be expected to decrease with increasing electronegativity of adjacent groups. This effect is illustrated by the δ values for the protons in the methyl halides, CH_3X, which lie in the order I (2.16), Br (2.68), Cl (3.05), and F (4.26). Here, iodine (the least electronegative) is the least effective of the halogens in withdrawing electrons from the protons; thus, the electrons of iodine provide the largest shielding effect. Similarly, electron density around the methyl protons of methanol is greater than around the proton associated with oxygen because oxygen is more electronegative than carbon. Thus, the methyl peaks are upfield from the hydroxyl peak. The position of the proton peaks in TMS is also explained by this model, since silicon is relatively electropositive. Finally, acidic protons have very low electron densities, and the peak for the proton in RSO_3H or $RCOOH$ lies far downfield ($\delta > 10$).

---

[4] The intensity of magnetization induced in a *diamagnetic* substance is smaller than that produced in a vacuum with the same field. Diamagnetism is the result of motion induced in bonding electrons by the applied field; this motion (a *diamagnetic current*) creates a secondary field that opposes the applied field. *Paramagnetism* (and the resulting *paramagnetic currents*) operates in just the opposite sense.

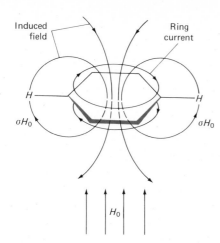

FIGURE 14-13 Deshielding of aromatic protons brought about by ring current.

Effect of Magnetic Anisotropy. It is apparent from an examination of the spectra of compounds containing double or triple bonds that local diamagnetic effects do not suffice to explain the position of certain proton peaks. Consider, for example, the irregular change in δ values for protons in the following hydrocarbons, arranged in order of increasing acidity (or increased electronegativity of the groups to which the protons are bonded): CH_3-CH_3 ($\delta = 0.9$), $CH_2=CH_2$ ($\delta = 5.8$), and $HC\equiv CH$ ($\delta = 2.9$). Furthermore, the aldehydic proton RCHO ($\delta \sim 10$) and the protons on benzene ($\delta \sim 7.3$) appear considerably farther downfield than might be expected on the basis of the electronegativity of the groups to which they are attached.

The effects of multiple bonds upon the chemical shift can be explained by taking into account the anisotropic magnetic properties of these compounds. For example, the magnetic susceptibilities[5] of crystalline aromatic

compounds have been found to differ appreciably, depending upon the orientation of the ring with respect to the applied field. This anisotropy is readily understood from the model shown in Figure 14-13. Here, the plane of the ring is perpendicular to the magnetic field; in this position, the field can induce a flow of the π electrons around the ring (a ring current). The consequence is similar to that of a current in a wire loop; namely, a secondary field is produced that acts in opposition to the applied field. This secondary field, however, exerts a magnetic effect on the protons attached to the ring; as shown in Figure 14-13, this effect is in the direction of the field. Thus, the aromatic protons require a lower external field to bring them into resonance. This effect is either absent or self-canceling in other orientations of the ring.

A somewhat analogous model can be envisioned for the ethylenic or carbonyl double bonds. Here, one can imagine circulation of the π electrons in a plane along the axis of the bond when the molecule is oriented to the field as shown in Figure 14-14a. Again, the secondary field produced acts upon the proton to reinforce the applied field. Thus, deshielding

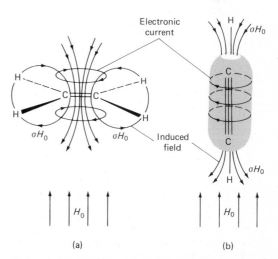

(a)   (b)

FIGURE 14-14 Deshielding of ethylene and shielding of acetylene brought about by electronic currents.

---

[5] The magnetic susceptibility of a substance can be thought of as the extent to which it is susceptible to induced magnetization by an external field.

shifts the peak to larger values of δ. With an aldehyde, this effect combines with the deshielding brought about by the electronegative nature of the carbonyl group; a very large value of δ results.

In an acetylenic bond, the symmetrical distribution of π electrons about the bond axis permits electron circulation around the bond (in contrast, such circulation is prohibited by the nodal plane in the electron distribution of a double bond). From Figure 14-14b, it can be seen that in this orientation the protons are shielded. This effect is apparently large enough to offset the deshielding resulting from the acidity of the protons and from the electronic currents at perpendicular orientations of the bond.

Correlation of Chemical Shift with Structure. The chemical shift is employed for the identification of functional groups and as an aid in determining structural arrangements of groups. These applications are based upon empirical correlations between structure and shift. A number of correlation charts[6] and tables[7] have been published. Two of these are shown in Figure 14-15 and Table 14-3. It

---

[6] N. F. Chamberlain, *Anal. Chem.*, **31**, 56 (1959).

[7] R. M. Silverstein, G. C. Bassler, and T. C. Morrill, *Spectrometric Identification of Organic Compounds*, 3d ed. New York: Wiley, 1974, Chapter 4; and L. M. Jackman and S. Sternhall, *Nuclear Magnetic Resonance Spectroscopy*, 2d ed. New York: Pergamon Press, 1969.

TABLE 14-3   APPROXIMATE CHEMICAL SHIFTS FOR CERTAIN METHYL, METHYLENE, AND METHINE PROTONS

Structure	δ, ppm		
	$M = CH_3$	$M = CH_2$	$M = CH$
Aliphatic β substituents			
M—C—Cl	1.5	1.8	2.0
M—C—Br	1.8	1.8	1.9
M—C—NO$_2$	1.6	2.1	2.5
M—C—OH (or OR)	1.2	1.5	1.8
M—C—OC(=O)R	1.3	1.6	1.8
M—C—C(=O)H	1.1	1.7	—
M—C—C(=O)R	1.1	1.6	2.0
M—C—C(=O)OR	1.1	1.7	1.9
M—C—ϕ	1.1	1.6	1.8
Aliphatic α substituents			
M—Cl	3.0	3.5	4.0
M—Br	2.7	3.4	4.1
M—NO$_2$	4.3	4.4	4.6
M—OH (or OR)	3.2	3.4	3.6
M—O—ϕ	3.8	4.0	4.6
M—OC(=O)R	3.6	4.1	5.0
M—C=C	1.6	1.9	—
M—C≡C	1.7	2.2	2.8
M—C(=O)H	2.2	2.4	—
M—C(=O)R	2.1	2.4	2.6
M—C(=O)ϕ	2.4	2.7	3.4
M—C(=O)OR	2.2	2.2	2.5
M—ϕ	2.2	2.6	2.8

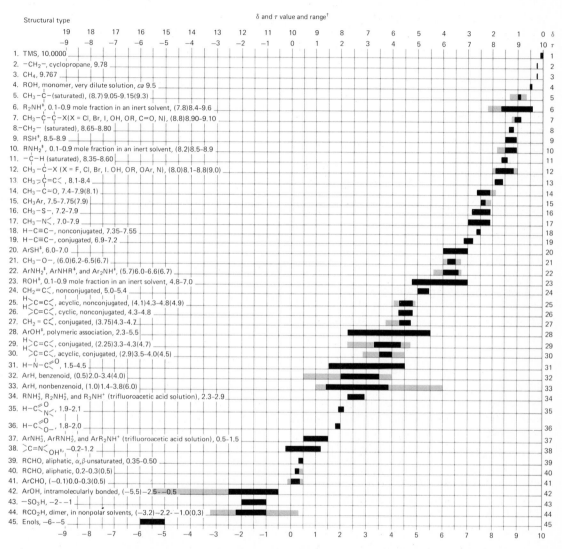

δ and τ value and range†

Structural type

1. TMS, 10.0000
2. −CH₂−, cyclopropane, 9.78
3. CH₄, 9.767
4. ROH, monomer, very dilute solution, ca 9.5
5. CH₃−Ċ−(saturated), (8.7)9.05-9.15(9.3)
6. R₂NH‡, 0.1-0.9 mole fraction in an inert solvent, (7.8)8.4-9.6
7. CH₃−Ċ−Ċ−X(X = Cl, Br, I, OH, OR, C=O, N), (8.8)8.90-9.10
8. −CH₂− (saturated), 8.65-8.80
9. RSH‡, 8.5-8.9
10. RNH₂‡, 0.1-0.9 mole fraction in an inert solvent, (8.2)8.5-8.9
11. −Ċ−H (saturated), 8.35-8.60
12. CH₃−Ċ−X (X = F, Cl, Br, I, OH, OR, OAr, N), (8.0)8.1-8.8(9.0)
13. CH₃−Ċ=C, 8.1-8.4
14. CH₃−C≈O, 7.4-7.9(8.1)
15. CH₃Ar, 7.5-7.75(7.9)
16. CH₃−S−, 7.2-7.9
17. CH₃−N, 7.0-7.9
18. H−C≡C−, nonconjugated, 7.35-7.55
19. H−C≡C−, conjugated, 6.9-7.2
20. ArSH‡, 6.0-7.0
21. CH₃−O−, (6.0)6.2-6.5(6.7)
22. ArNH₂‡, ArNHR‡, and Ar₂NH‡, (5.7)6.0-6.6(6.7)
23. ROH‡, 0.1-0.9 mole fraction in an inert solvent, 4.8-7.0
24. CH₂=C, nonconjugated, 5.0-5.4
25. H₂C=C, acyclic, nonconjugated, (4.1)4.3-4.8(4.9)
26. C=C, cyclic, nonconjugated, 4.3-4.8
27. CH₂ = C, conjugated, (3.75)4.3-4.7
28. ArOH‡, polymeric association, 2.3-5.5
29. H₂C=C, conjugated, (2.25)3.3-4.3(4.7)
30. C=C, acyclic, conjugated, (2.9)3.5-4.0(4.5)
31. H−N−C≈O, 1.5-4.5
32. ArH, benzenoid, (0.5)2.0-3.4(4.0)
33. ArH, nonbenzenoid, (1.0)1.4-3.8(6.0)
34. RNH₃⁺, R₂NH₂⁺, and R₃NH⁺ (trifluoroacetic acid solution), 2.3-2.9
35. H−C≈O/N, 1.9-2.1
36. H−C≈O/O, 1.8-2.0
37. ArNH₃⁺, ArRNH₂⁺, and ArR₂NH⁺ (trifluoroacetic acid solution), 0.5-1.5
38. C=N−OH‡, −0.2-1.2
39. RCHO, aliphatic, α,β-unsaturated, 0.35-0.50
40. RCHO, aliphatic, 0.2-0.3(0.5)
41. ArCHO, (−0.1)0.0-0.3(0.5)
42. ArOH, intramolecularly bonded, (−5.5)−2.5−−0.5
43. −SO₃H, −2−−1
44. RCO₂H, dimer, in nonpolar solvents, (−3.2)−2.2−−1.0(0.3)
45. Enols, −6−−5

†Normally, absorptions for the functional groups indicated will be found within the range shown. Occasionally, a functional group will absorb outside this range. Approximate limits for this are indicated by absorption values in parentheses and by shading in the figure.

‡The absorption positions of these groups are concentration-dependent and are shifted to higher τ values in more dilute solutions.

FIGURE 14-15 Absorption positions of protons in various structural environments. (Table taken from J. R. Dyer, *Applications of Absorption Spectroscopy of Organic Compounds*. Englewood Cliffs, N. J.: Prentice-Hall, Inc., © 1965, p. 85. With permission.)

should be noted that the exact values for δ may depend upon the nature of the solvent as well as upon the concentration of the solute. These effects are particularly pronounced for protons involved in hydrogen bonding; an example is the hydrogen atom in the alcoholic functional group.

Spin-Spin Splitting

As may be seen in Figure 14-10, the absorption bands for the methyl and methylene protons in ethanol consist of several narrow peaks that can be separated only with a high-resolution instrument. Careful examination of these peaks shows that the spacing for the three components of the methyl band is identical to that for the four peaks of the methylene band; this spacing in frequency units is called the *coupling constant* for the interaction and is given the symbol J. Moreover, the areas of the peaks in a multiplet approximate an integral ratio to one another. Thus, for the methyl triplet, the ratio of areas is $1:2:1$; for the quartet of methylene peaks, it is $1:3:3:1$.

Origin. It seems plausible to attribute these observations to the effect that the spins of one set of nuclei exert upon the resonance behavior of another. That is to say, a small interaction or coupling exists between the two groups of protons. This explanation presupposes that such coupling takes place via interactions between the nuclei and the bonding electrons rather than through free space. For our purpose, however, the details of this mechanism are not important.

Let us first consider the effect of the methylene protons in ethanol on the resonance of the methyl protons. We must first remember that the ratio of protons in the two possible spin states is very nearly one, even in a strong magnetic field. We can imagine, then, that the two methylene protons in a molecule can have four possible combinations of spin states and that in an entire sample each of these combinations will be approximately equally represented. If we represent the spin orienta-

tion of each nucleus with a small arrow, the four states are

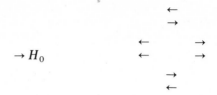

Field direction — Possible spin orientations of methylene protons

In one combination, the spins of the two methylene protons are paired and aligned against the field, while in a second, the paired spins are reversed; there are also two combinations in which the spins are opposed to one another. The magnetic effect that is transmitted to the methyl protons on the adjacent carbon atoms is determined by the spin combinations that exist in the methylene group at any instant. If the spins are paired and opposed to the external field, the effective applied field on the methyl protons is slightly lessened; thus, a somewhat higher field is needed to bring them into resonance, and upfield shift results. Spins paired and aligned with the field result in a downfield shift. Neither of the combinations of opposed spin has an effect on the resonance of the methyl protons. Thus, splitting into three peaks results. The area under the middle peak is twice that of either of the other two, since two spin combinations are involved.

Let us now consider the effect of the three methyl protons upon the methylene peak. Possible spin combinations for the methyl protons are

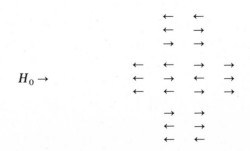

Here, we have eight possible spin combinations; however, among these are two groups containing three combinations that have equivalent magnetic effects. The methylene peak is thus split into four peaks having areas in the ratio $1:3:3:1$.

The interpretation of spin-spin splitting pattern is relatively simple and straightforward for *first-order* spectra. First-order spectra are those in which the chemical shift between interacting groups of nuclei is large with respect to their coupling constant J. Rigorous first-order behavior requires that $\Delta v/J$ be greater than 20; frequently, however, analysis of spectra by first-order techniques can be accomplished down to value of $\Delta v/J$ of somewhat less than 10. The ethanol spectrum shown in Figure 14-10 is an example of a pure first-order spectrum with J for the methyl and methylene peaks being 7 Hz and the separation between the centers of the two multiplets being about 140 Hz.

Interpretation of second-order NMR spectra is difficult and complex and will not be dealt with in this text. It is noteworthy, however, that because Δv increases with increases in the magnetic field while J does not, spectra obtained with an instrument having a high magnetic field are much more readily interpreted than those produced by a spectrometer with a weaker magnet.

Rules Governing the Interpretation of First-Order Spectra. The following rules govern the appearance of first-order spin-spin spectra.

1. Equivalent nuclei do not interact with one another to give multiple absorption peaks. The three protons in the methyl group in ethanol give rise to splitting of the adjacent methylene protons only and not to splitting among themselves.

2. Coupling constants decrease with separation of groups, and coupling is seldom observed at distances greater than three bond lengths.

3. The multiplicity of a band is determined by the number n of magnetically equivalent protons on the neighboring atoms and is given by $(n + 1)$. Thus, the multiplicity for the methylene band in ethanol is determined by the number of protons in the adjacent methyl groups and is equal to $(3 + 1)$.

For nuclei having spins (I) other than 1/2, the multiplicity is given by $(2nI + 1)$.

4. If the protons on atom B are affected by protons on atoms A and C that are nonequivalent, the multiplicity of B is equal to $(n_A + 1)(n_C + 1)$, where n_A and n_C are the number of equivalent protons on A and C, respectively.

5. The approximate relative areas of a multiplet are symmetric around the midpoint of the band and are proportional to the coefficients of the terms in the expansion $(x + 1)^n$. The application of this rule is demonstrated in Table 14-4 and in the examples that follow.

6. The coupling constant is independent of the applied field; thus, multiplets are readily distinguished from closely spaced chemical shift peaks.

EXAMPLE

For each of the following compounds, calculate the number of multiplets for each band and their relative areas.

(a) $ClCH_2CH_2CH_2Cl$. The multiplicity of the band associated with the four equivalent protons on the two ends of the molecule would be determined by the number of protons on the central carbon; thus, the multiplicity is $(2 + 1) = 3$ and the areas would be $1:2:1$. The multiplicity of two central methylene protons would be determined by the four equivalent protons at the ends and would thus be $(4 + 1) = 5$. Expansion of $(x + 1)^4$ gives the following coefficients (Table 14-4), which are proportional to the areas of the peaks $1:4:6:4:1$.

(b) $CH_3CHBrCH_3$. The band for the six methyl protons will be made up of $(1 + 1) = 2$ peaks having relative areas of $1:1$; the proton on the central carbon atom has a multiplicity of $(6 + 1) = 7$. These peaks will have areas (Table 14-4) in the ratio of

$$1:6:15:20:15:6:1$$

(c) $CH_3CH_2OCH_3$. The methyl protons on the right are separated from the other protons by more than three bonds so that only a single peak will be observed for them. The protons of the central methylene group will have a multiplicity of $(3 + 1) = 4$ and a ratio of $1 : 3 : 3 : 1$. The methyl protons on the left have a multiplicity of $(2 + 1) = 3$ and an area ratio of $1 : 2 : 1$.

The foregoing examples are relatively simple because all of the protons influencing the multiplicity of any single peak are magnetically equivalent. A more complex splitting pattern results when a set of protons is affected by two or more nonequivalent protons. As an example, consider the spectrum of 1-iodopropane, $CH_3CH_2CH_2I$. If we label the three carbon atoms (a), (b), and (c) from left to right, the chemical shift bands are found at $\delta_{(a)} = 1.02$, $\delta_{(b)} = 1.86$, and $\delta_{(c)} = 3.17$. The band at $\delta_{(a)} = 1.02$ will be split by the two methylene protons on (b) into $(2 + 1) = 3$ peaks having relative areas of $1 : 2 : 1$. A similar splitting of the band $\delta_{(c)} = 3.17$ will also be observed. The experimental coupling constants for the two shifts are $J_{(ab)} = 7.3$ and

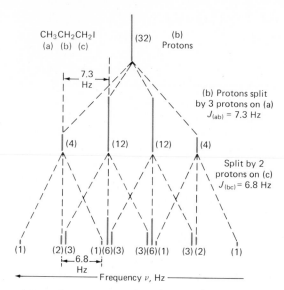

FIGURE 14-16 Splitting pattern for methylene (b) protons in $CH_3CH_2CH_2I$. Figures in parentheses are relative areas under peaks.

$J_{(bc)} = 6.8$. The band for the methylene protons (b) is affected by two groups of protons which are not magnetically equivalent, as is

TABLE 14-4   RELATIVE INTENSITIES OF FIRST-ORDER MULTIPLETS
$(I = \frac{1}{2})$

Number of Equivalent Protons, n	Multiplicity, $(n + 1)$	Relative Peak Areas								
0	1					1				
1	2				1		1			
2	3			1		2		1		
3	4		1		3		3		1	
4	5	1		4		6		4		1
5	6	1	5		10		10		5	1
6	7	1	6	15		20		15	6	1
7	8	1	7	21	35		35	21	7	1

evident from the difference between $J_{(ab)}$ and $J_{(bc)}$. Thus, invoking rule 4, the number of peaks will be $(3 + 1)(2 + 1) = 12$. In cases such as this, derivation of a splitting pattern, as shown in Figure 14-16, is helpful. Here, the effect of the (a) proton is first shown and leads to four peaks of relative areas $1 : 3 : 3 : 1$ spaced at 7.3 Hz. Each of these is then split into three new peaks spaced at 6.8 Hz, having relative areas of $1 : 2 : 1$. (The same final pattern is produced if the original band is first split into a triplet.) At very high resolution, the spectrum for 1-iodopropane exhibits a series of peaks that approximates the series shown at the bottom of Figure 14-16. At lower resolution (so low that the instrument does not detect the difference between $J_{(ab)}$ and $J_{(bc)}$, only six peaks are observed with relative areas of

$$1 : 5 : 10 : 10 : 5 : 1.$$

Second-Order Spectra. Coupling constants are usually smaller than 20, whereas chemical shifts may be as high as 1000. Therefore, the splitting behavior described by the rules in the previous section is common. However, when $\Delta v/J$ becomes less than perhaps 7, these rules no longer apply. Generally, as Δv approaches J, the peaks on the inner side of two multiplets tend to be enhanced at the expense of the peaks on the outer side, and the symmetry of each multiplet is thus destroyed, as noted earlier. Analysis of a spectrum under these circumstances is difficult.

Effect of Chemical Exchange on Spectra. Turning again to the spectrum of ethanol (Figure 14-10), it is interesting to consider why the OH proton appears as a singlet rather than a triplet. The methylene protons and the OH proton are separated by only three bonds; coupling should occur to increase the multiplicity of both OH and the methylene peaks. Actually, as shown in Figure 14-17, the expected multiplicity is observed by employing a highly purified sample of the

alcohol. Note the triplet OH peaks and the eight methylene peaks in this spectrum. If, now, a trace of acid or base is added to the pure sample, the spectrum reverts to the form shown in Figure 14-10.

The exchange of OH protons among alcohol molecules is known to be catalyzed by both acids and bases, as well as by the impurities that commonly occur in alcohol. It is thus plausible to associate the decoupling observed in the presence of these catalysts to an exchange process. If exchange is rapid, each OH group will have several protons associated with it during any brief period; within this interval, all of the OH protons will experience the effects of the three spin arrangements of the methylene protons. Thus, the magnetic effects on the alcoholic proton are averaged and a single sharp peak is observed. Spin decoupling always occurs when the exchange frequency is greater than the separation (in frequency units) between the interacting components.

Chemical exchange can affect not only spin-

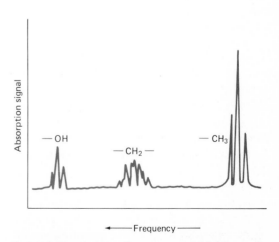

FIGURE 14-17 Spectrum of highly purified ethanol showing additional splitting of OH and CH$_2$ peaks (compare with Figure 14-10).

spin spectra but also chemical shift spectra. Purified alcohol/water mixtures have two well-defined and easily separated OH proton peaks. Upon addition of an acid or base as a catalyst, however, the two peaks coalesce to form a single sharp line. Here, the catalyst enhances the rate of proton exchange between the alcohol and the water and thus averages the shielding effect. A single sharp line is obtained when the exchange rate is significantly greater than the separation frequency of the individual lines of alcohol and water. On the other hand, if the exchange frequency is about the same as this frequency difference, shielding is only partially averaged and a broad line results. The correlation of line breadth with exchange rates has provided a direct means for investigating the kinetics of such processes and represents an important application of the NMR experiment.

Aids in the Analysis of Complex Spectra

Several methods are available for the simplification of spectra that are too complex for ready analysis. The more common of these are discussed briefly in this section.

Increases in the Magnetic Field. As mentioned earlier, coupling constants are unaffected by increases in the magnetic field, whereas chemical shifts are enhanced. Thus, the use of an instrument with a stronger magnet may convert an uninterpretable second-order spectra to one that is susceptible to first-order analysis.

Isotopic Substitution. Substitution of deuterium for one or more of the protons in a molecule simplifies the spectrum by removal of the absorption peaks corresponding to the substituted protons. Further simplification may also be observed because the coupling between a deuterium and a proton is significantly less strong than between two protons.

Spin Decoupling. Figure 14-18 illustrates the spectral simplification that may accom-

pany spin decoupling. Spectrum B shows the absorption associated with the four protons on the pyridine ring of nicotine. Spectrum C was obtained by sweeping the same portion of the spectrum while simultaneously irradiating the sample with a strong, second, radiofrequency signal having a frequency corresponding to the absorption peaks of protons (d) and (c) (about 8.6 ppm). The consequence is a decoupling of the interaction between these two protons and protons (a) and (b). Here, the complex absorption spectra for (a) and (b) collapse to two doublet peaks which arise from coupling between these protons. Similarly, the spectra for (d) and (c) could be simplified by decoupling with a beam having a frequency corresponding to the peaks for protons (a) or (b).

The interaction between dissimilar nuclei can also be decoupled. An important example is the decoupling of ^{14}N nuclei and protons, which can be accomplished by irradiation of the sample with a beam having a frequency of about 4.3 MHz when the magnetic field is 14,092 G. Absorption of this radiation by the ^{14}N nuclei results in a complete decoupling of their interaction with protons and often produces spectral simplifications.

The theory of spin-spin decoupling is complex and beyond the scope of this book.

Chemical Shift Reagents.[8] Important aids in the interpretation of proton NMR spectra are *shift reagents*, which have the effect of dispersing the absorption peaks for certain types of compounds over a much larger frequency range. This dispersion will frequently separate otherwise overlapping peaks and permit easier interpretation.

Shift reagents are generally complexes of europium or praseodymium. A typical example is the dipivalomethanato complex of

---

[8] For a brief summary of shift reagents, see: *Shift Reagents in NMR, Perkin-Elmer NMR Quarterly*, Number 7, August 1977, The Perkin-Elmer Corporation, Norwalk, CT 06852.

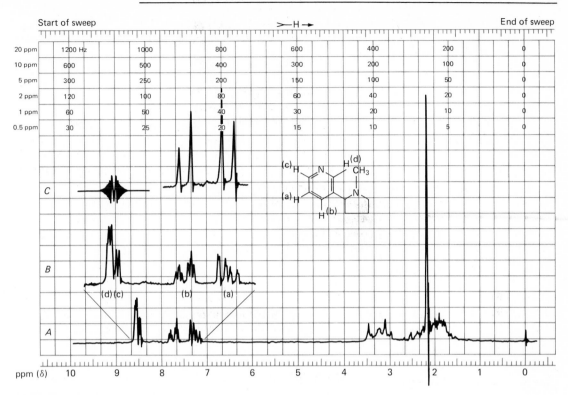

FIGURE 14-18 Effect of spin decoupling on the NMR spectrum of nicotine dissolved in $CDCl_3$. Curve A: the entire spectrum. Curve B: expanded spectrum for the four protons on the pyridine ring. Curve C: spectrum for protons (a) and (b) when decoupled from (d) and (c) by irradiation with a second beam that has a frequency corresponding to about 8.6 ppm. (Courtesy Varian Instrument Division, Palo Alto, California 94303.)

praseodymium(III) [usually abbreviated as $Pr(DPM)_3$].

The praseodymium ion in this neutral complex is capable of increasing its coordination by interaction with lone electron pairs. Therefore, reactions can take place between the com-

plex and molecules containing oxygen, nitrogen, or other atoms that contain free electron pairs.

The DPM complexes of europium and praseodymium are generally employed in nonpolar solvents such as CCl_4, $CDCl_3$, and C_6D_6 to avoid solvent competition with the analyte for electron receiver sites on the metal ion.

Figure 14-19 illustrates the dramatic effect that $Pr(DPM)_3$ has on the complex spectrum for styrene oxide. Here, the hydrogens closest to the oxygen binding site are shifted to higher fields and actually above that for the TMS reference. Note also that the ortho hydrogen in the ring is shifted to a greater extent than the meta or para hydrogens, again as a con-

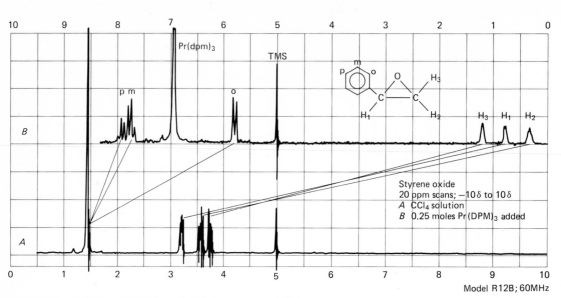

FIGURE 14-19   Effect of Pr(DPM)$_3$ on the NMR spectrum of styrene oxide. Spectrum *A*: in the absence of the reagent. Spectrum *B*: in the presence of the reagent. (Spectra courtesy of Perkin-Elmer, Norwalk, Connecticut.)

sequence of its being closer to the metal ion than the other two.

Similar effects are observed with Eu(DPM)$_3$ except that the shifts are to lower fields. Again, the protons closest to the metal ion are affected most.

The primary source of the chemical shift arises from the secondary magnetic field generated by the large magnetic moments of the paramagnetic praseodymium or europium ions. If the geometry of the complex between the analyte and the shift reagent is known, reasonably good estimates can be made as to the extent of shift for various protons in the analyte.

APPLICATIONS OF PROTON NMR

Unquestionably, the most important applications of proton NMR spectroscopy have been to the identification and structural elucidation of organic, metal-organic, and biochemi-

cal molecules. In addition, however, the method often proves useful for quantitative determination of absorbing species.

Identification of Compounds

An NMR spectrum, like an infrared spectrum, seldom suffices by itself for the identification of an organic compound. However, in conjunction with other observations such as elemental analysis, as well as ultraviolet, infrared, and mass spectra, NMR is an important tool for the characterization of a pure compound. The simple examples that follow give some idea of the kinds of information that can be extracted from NMR studies.

EXAMPLE

The NMR spectrum shown in Figure 14-20 is for an organic compound having the empirical formula $C_5H_{10}O_2$. Identify the compound.

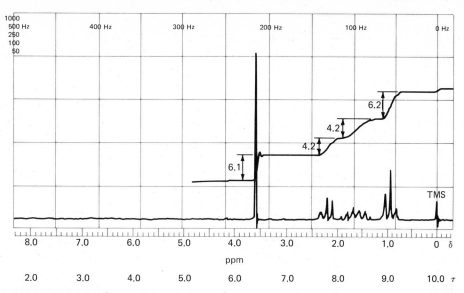

FIGURE 14-20   NMR spectrum and peak integral curve for the organic compound $C_5H_{10}O_2$ in CCl_4. (From R. M. Silverstein, G. C. Bassler, and T. C. Morrill, *Spectrometric Identification of Organic Compounds*, 3d ed. New York: John Wiley & Sons, Inc., 1974, p. 296. With permission.)

We obtain relative areas from left to right of about 6.1, 4.2, 4.2, and 6.2 from integral plot. These figures suggest a distribution of the 10 protons of 3, 2, 2, and 3. The single peak at $\delta = 3.6$ must be due to an isolated methyl group; upon inspection of Figure 14-15 and Table 14-3, the functional group $CH_3OC(=O)-$ is suggested. The empirical formula and the $2:2:3$ distribution of the remaining protons indicate the presence of an *n*-propyl group as well. The structure $CH_3OC(=O)CH_2CH_2CH_3$ is consistent with all of these observations. In addition, the positions and the splitting patterns of the three remaining peaks are entirely compatible with this hypothesis. The triplet at $\delta = 0.9$ is typical of a methyl group adjacent to a methylene. From Table 14-3, the two protons of the methylene adjacent to the carboxylate peak should yield the observed triplet peak at about $\delta = 2.2$. The other methylene group would be expected to produce a pattern of 12 peaks

(3×4) at about $\delta = 1.7$. Only six are observed, presumably because the resolution of the instrument is insufficient.

EXAMPLE

The spectra shown in Figure 14-21 are for colorless, isomeric, liquids containing only carbon and hydrogen. Identify the two compounds.

The single peak at about $\delta = 7.2$ in the upper figure suggests an aromatic structure; the relative area of this peak corresponds to 5 protons; from this we conclude that we may have a monosubstituted derivative of benzene. The seven peaks for the single proton appearing at $\delta = 2.9$ and the six-proton doublet at $\delta = 1.2$ can only be explained by the structure

$$\begin{array}{c} CH_3 \\ | \\ -C-CH_3 \\ | \\ H \end{array}$$

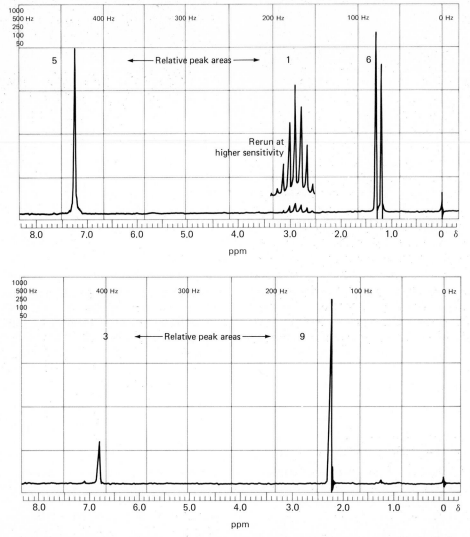

FIGURE 14-21 NMR spectra for two organic isomers in CDCl$_3$ solution. (Courtesy of Varian Associates, Palo Alto, California.)

Thus, we conclude that this compound is cumene.

The isomeric compound has an aromatic peak at $\delta = 6.8$; its relative area suggests a trisubstituted benzene, which can only mean

that the compound is $C_6H_3(CH_3)_3$. The relative peak areas confirm this diagnosis. We cannot, however, decide which of the three trimethylbenzene derivatives we have from the NMR data.

EXAMPLE

The spectrum shown in Figure 14-22 is for an organic compound having a molecular weight of 72 and containing carbon, hydrogen, and oxygen only. Identify the compound.

The triplet peak at $\delta = 9.8$ appears (Figure 14-15) to be that of an aliphatic aldehyde, RCHO. If this is the case, R has a molecular weight of 43, which corresponds to a C_3H_7 fragment. The triplet nature of the peak at $\delta = 9.8$ requires that there be a methylene group adjacent to the carbonyl. Thus, the compound would appear to be

n-butyraldehyde, $CH_3CH_2CH_2CHO$

The triplet peak at $\delta = 0.97$ appears to be that of the terminal methyl. The protons on the adjacent methylene would be expected to show a complicated splitting pattern of 12 peaks (4×3); the grouping of peaks around $\delta = 1.7$ is compatible with this prediction. Finally, the peak for the protons on the methylene group adjacent to the carbonyl should appear as a sextet downfield from the other methylene proton peaks. The group at $\delta = 2.4$ is consistent with this conclusion.

APPLICATION OF PROTON NMR TO QUANTITATIVE ANALYSIS

Quantitative Analysis

A unique aspect of NMR spectra is the direct proportionality between peak areas and the number of nuclei responsible for the peak. As a consequence, a quantitative determination of a specific compound does not require pure samples for calibration. Thus, if an identifiable peak for one of the constituents of a sample does not overlap the peaks of the other constituents, the area of this peak can be employed to establish the concentration of the species directly, provided only that the signal area per proton is known. This latter parameter can be obtained conveniently from a known concentration of an internal standard. For example, if the solvent present in a known amount were benzene, cyclohexane, or water, the areas of the single proton peak for these compounds could be used to give the desired information; of course, the peak of the internal standard should not overlap with

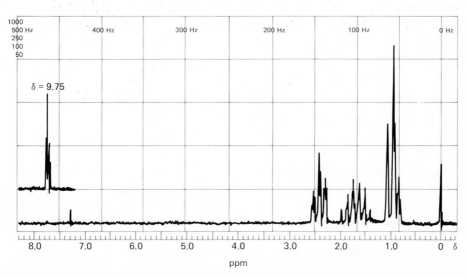

FIGURE 14-22 NMR spectrum of a pure organic compound containing C, H, and O only. (Courtesy of Varian Associates, Palo Alto, California.)

any of the sample peaks. Organic silicon derivatives are uniquely attractive for calibration purposes, owing to the high upfield location of their proton peaks.

The widespread use of NMR spectroscopy for quantitative work has been inhibited by the cost of the instruments. In addition, the probability that resonance peaks will overlap becomes greater as the complexity of the sample increases. Often, too, analyses that are possible by the NMR method can be as conveniently accomplished by other techniques.

One of the main problems in quantitative NMR methods is the result of the saturation effect. As we have pointed out, the NMR absorption signal depends upon a very minute excess of nuclei in the lower magnetic energy state and that the absorption process tends to depopulate this excess. Whether or not depopulation has a significant effect on the absorption intensity depends upon the relaxation time for the species, the power of the source, and the rate at which the spectrum is scanned. Errors arising from saturation can usually be avoided through control of these variables.

Analysis of Multicomponent Mixtures. Methods for the analysis of many multicomponent mixtures have been reported. For example, Hollis[9] has described a method for the determination of aspirin, phenacetin, and caffeine in commercial analgesic preparations. The procedure requires about 20 min, and the relative errors are in the range of 1 to 3%. Chamberlain[10] describes a procedure for the rapid analysis of benzene, heptane, ethylene glycol, and water in mixtures. A wide range of these mixtures was analyzed with a precision of 0.5%.

An important quantitative application of the NMR technique has been to the determination of water in food products, pulp and paper, and agricultural materials. The water in these substances is sufficiently mobile to give a narrow peak suitable for quantitative measurement.[11]

Elemental Analysis. NMR spectroscopy can be employed to determine the total concentration of a given kind of magnetic nucleus in a sample. For example, Jungnickel and Forbes[12] have investigated the integrated NMR intensities of the proton peaks for numerous organic compounds and have concluded that accurate quantitative determinations of total hydrogen in organic mixtures is possible. Paulsen and Cooke[13] have shown that the resonance of fluorine-19 can be used for the quantitative analysis of that element in an organic compound—an analysis that is difficult to carry out by classical methods. For quantitative work, a low-resolution or *wide-line* spectrometer can be employed.

Study of Isotopes Other Than the Proton

Table 14-2 lists several nuclei which, in addition to the proton, have magnetic moments and can thus be studied by the magnetic resonance technique. More than one hundred other isotopes also possess magnetic moments[14]; the resonance behavior of only a few of these has been investigated to date.

[9] D. P. Hollis, *Anal. Chem.*, **35**, 1682 (1963).

[10] N. F. Chamberlain, in *Treatise on Analytical Chemistry*, eds. I. M. Kolthoff and P. J. Elving. New York: Interscience, 1963, Part I, vol. 4, p. 1932.

[11] T. M. Shaw and R. H. Elsken, *J. Chem. Phys.*, **18**, 1113 (1950); *J. Appl. Physics*, **26**, 313 (1955); and T. M. Shaw, R. H. Elsken, and C. H. Kunsman, *J. Assoc. Offic. Agr. Chemists*, **36**, 1070 (1953).

[12] J. L. Jungnickel and J. W. Forbes, *Anal. Chem.*, **35**, 938 (1963).

[13] P. J. Paulsen and W. D. Cooke, *Anal. Chem.*, **36**, 1721 (1964).

[14] See, for example: J. A. Pople, W. G. Schneider, and H. J. Bernstein, *High-resolution Nuclear Magnetic Resonance*. New York: McGraw-Hill, 1959, pp. 480–485.

Fluorine. Fluorine, with an atomic number of 19, has a spin quantum number of $\frac{1}{2}$ and a magnetic moment of 2.6285 nuclear magnetons. Thus, the resonance frequency of fluorine in similar fields is only slightly lower than the proton (56.5 MHz, as compared with 60.0 MHz at 14,092 gauss). Therefore, with relatively minor changes, a proton NMR spectrometer can be adapted to the study of fluorine resonance.

It is found experimentally that fluorine absorption is also sensitive to the environment; the resulting chemical shifts, however, extend over a range of about 300 ppm compared with a maximum of 20 ppm for the proton. In addition the solvent plays a much more important role in determining fluorine peak positions than with the proton.

Empirical correlations of the fluorine shift with structure are relatively sparse when compared with information concerning proton behavior. It seems probable, however, that the future will see further developments in this field, particularly for structural investigation of organic fluorine compounds.

Phosphorus. Phosphorus-31, with spin number $\frac{1}{2}$, also exhibits sharp NMR peaks with chemical shifts extending over a range of 700 ppm. The resonance frequency of ^{31}P at 14,092 gauss is 24.3 MHz. Several investigations correlating the chemical shift of the phosphorus nucleus with structure have been reported.

Other Nuclei. Other nuclei that offer considerable potential for NMR studies include ^{13}C, ^{17}O, ^{2}H, ^{11}B, ^{109}Ag, and ^{29}Si; an increasing amount of work is being reported for each of these. Carbon-13 NMR spectroscopy will be considered in the next section, which deals with the application of Fourier transform methods to NMR spectroscopy.

FOURIER TRANSFORM NMR[15]

Conventional NMR spectroscopy is not very sensitive. As a consequence, the generation of good proton spectra for materials available in microgram quantities has been difficult, time consuming, and sometimes impossible. This lack of sensitivity also seriously inhibited the growth of carbon-13 NMR spectroscopy. Here, the low natural abundance of this isotope and its relatively small magnetogyric ratio provides an NMR signal that is less by a factor of about 6000 than that for the proton (see Table 14-2). Thus, producing a useful spectrum for carbon in unenriched samples required repeated scans and signal averaging for periods of 24 hr or more.

The commercial development of pulsed, Fourier transform NMR spectrometers by several companies since 1970 has resulted in dramatic increases in the sensitivity of NMR measurements; as a result, the routine application of this technique to naturally occurring carbon-13 and to protons in microgram quantities of chemical or biological materials has become widespread. The basis of the increased sensitivity is the same as that discussed in the section on Fourier transform infrared spectroscopy (p. 241). In both, all of the resolution elements of a spectrum are observed in a very brief period by measurements that yield a time-domain rather than a frequency-domain spectrum. A time-domain spectrum can be obtained in a few seconds or less; thus, it becomes practical to replicate spectra hundreds of thousands of times and average the measurements to give a vastly improved signal-to-noise ratio. The frequency-domain spectrum can then be obtained by a Fourier transform employing a digital computer.

---

[15] For a more complete discussion of Fourier transform NMR spectroscopy, see: E. D. Becker and T. C. Farrar, *Science*, **178**, 361 (1972); D. Shaw, *Fourier Transform NMR Spectroscopy*. New York: Elsevier, 1976; R. J. Abraham and P. Loftus, *Proton and Carbon-13 NMR Spectroscopy*. Philadelphia: Heyden and Sons, 1978; and T. C. Farrar and E. D. Becker, *Pulse and Fourier Transform NMR*. New York: Academic Press, 1971,

Pulsed NMR Spectra

In Fourier transform or pulse NMR studies, the sample is irradiated periodically with brief, highly intense pulses of radio-frequency radiation, following which the *free induction decay signal*—a characteristic radio-frequency emission signal stimulated by the irradiation—is recorded as a function of time. Typically, pulses of 1 to 10 μs are employed

and the observation time between pulses is approximately 1 s. Figure 14-23a depicts the time relations for the input signal.

Time-Domain Decay Spectra. The periodic pulses serve the same function as the interferometer in infrared Fourier transform spectroscopy in the respect that they produce a time-domain spectrum during the period between pulses. Two such spectra are shown in Figure 14-23b and 14-23c. These consist of

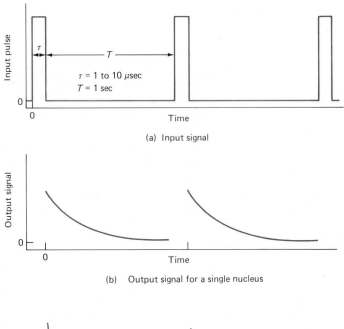

(a) Input signal

(b) Output signal for a single nucleus

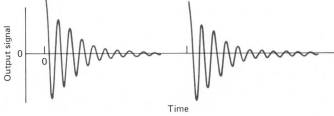

(c) Output signal for several types of nuclei

FIGURE 14-23 (a) Input signal for pulsed NMR. (b) Free induction decay signal when a single type of nucleus is present. (c) Free induction decay signal when several types of nuclei are present.

the free induction decay signals emitted by nuclei as they return to their ground state from the excited state that was induced by the pulse of radiation. The frequencies of the emitted radiation are, of course, identical to the absorption frequencies that would appear in a conventional NMR spectrum for the various types of nuclei present.

Figure 14-23b shows the decay signal when the sample contains but a single type of nucleus and the excitation frequency corresponds exactly to the resonance frequency for that nucleus. Figure 14-23c depicts the decay signal when more than one type of nucleus is excited. Here, a more complex time-domain response is observed because of interference among the radiations of differing frequencies.

Because the free induction decay curves owe their characteristic appearance to interference among various emitted radiations, it is to be expected that these time-domain spectra will be unique and characteristic for every conceivable combination of frequencies. Thus, just as an infrared time-domain spectrum, obtained with a Michelson interferometer (p. 244), contains all of the information necessary to derive a conventional infrared spectrum, so also does the free induction decay spectrum contain the information needed to yield a normal frequency-domain NMR spectrum. Conversion of the data from the time domain to the frequency domain can be realized with the aid of a digital computer programmed to perform a fast Fourier transform. Figure 14-24 demonstrates the relationship between the two types of spectra. The four peaks shown in the frequency-domain spectrum for carbon-13 arise from the coupling between the carbon nucleus and the three protons. Figure 14-24a indicates the parts of the time-domain spectrum that contain information about the chemical shift, δ, and the coupling constant, J.

Frequency Range of a Pulse. Radio-frequency generators, when operated continuously, produce radiation of a single frequency. The NMR experiments we have just described, however, require a range of frequencies suffi-

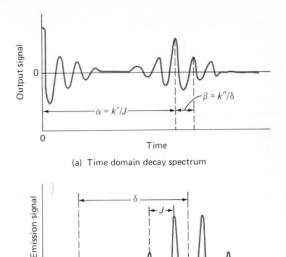

FIGURE 14-24 Spectra for $H_3^{13}CCl$. (a) Free-induction decay spectrum. (b) Conventional spectrum for ^{13}C showing spin-spin coupling between ^{13}C and protons. [Adapted from T. C. Farrar, *Anal. Chem.*, **42** (4), 109A, (1970). Reprinted with permission. Copyright by the American Chemical Society.]

ciently great to excite nuclei with different resonance frequencies. Fortunately, a sufficiently short pulse of radiation, such as that shown in Figure 14-23a, provides a band of frequencies from a monochromatic source. The frequency range of this band is about $\frac{1}{4}\tau$ Hz, where τ is the length in seconds of each pulse. Thus, by employing a pulse of 1 μs with a 15-MHz transmitter, a frequency range of 15 MHz $\pm$ 120 kHz would result. This production of *side bands* by pulsing can be understood by reference to Figure 4-4 (p. 96), where it is shown that a rectangular wave form is made up of a series of sine or cosine functions differing from one another by small frequency increments. Thus, a pulse generated by rapid on and off switching of a radio-frequency oscillator will consist of an envelope of power having a shape somewhat similar to the positive half of the solid line in Figure 4-4b.

Carbon-13 NMR[16]

Carbon-13 NMR has several advantages over proton NMR in terms of its power to elucidate organic and biochemical structures. First, there is the obvious advantage that carbon-13 NMR provides information about the backbone of molecules rather than about the periphery. In addition, the chemical shifts for carbon-13 in a majority of organic compounds is about 200 ppm, compared with approximately 10 to 20 ppm for the proton; less overlap of peaks is the consequence. Thus, for example, it is often possible to observe individual carbon resonance peaks for compounds ranging in molecular weight from 200 to 400. Also, homonuclear, spin-spin coupling between carbon atoms is not encountered, because in unenriched samples, the probability of two carbon-13 atoms occurring in the same molecule is small. Furthermore, heteronuclear spin coupling between carbon-13 and carbon-12 does not occur because the spin quantum number of the latter is zero. Finally, good methods exist for decoupling the interaction between carbon-13 atoms and protons. Thus, generally, the spectrum for a particular type of carbon consists of but a single line.

Figure 14-25 demonstrates some of the chemical shifts which are observed for carbon-13 in various chemical environments. As with proton spectra, these shifts are relative to tetramethylsilane.

---

[16] For a thorough discussion of carbon-13 NMR spectroscopy, see: J. B. Stothers, *Carbon-13 NMR Spectroscopy.* New York: Academic Press, 1972; D. E. Leyden and R. H. Cox, *Analytical Applications of NMR.* New York: Wiley, 1977, Chapter 5; and E. Breitmaier, and W. Voelters, *^{13}C NMR Spectroscopy*, 2d ed. New York: Verlag Chemie, 1978.

ELECTRON SPIN RESONANCE SPECTROSCOPY

Electron spin resonance spectroscopy or ESR (also called electron paramagnetic resonance spectroscopy or EPR) is based upon the absorption of microwave radiation by an unpaired electron when it is exposed to a strong magnetic field.[17] Species that contain unpaired electrons, and can therefore be detected by ESR spectroscopy, include free radicals, odd-electron molecules, transition-metal complexes, rare-earth ions, and triplet-state molecules.

Principles of ESR

The principles of ESR spectroscopy are similar to those of NMR spectroscopy, which were discussed in the previous sections. The electron, like the proton, has a spin quantum number of one-half and thus has two energy levels that differ slightly in energy under the influence of a strong magnetic field. In contrast to the proton, however, the lower energy level corresponds to $m = -\frac{1}{2}$ and the higher to $m = +\frac{1}{2}$; this difference results from the negative charge of the electron.

Applying Equations 14-4 and 14-5 to an unpaired electron yields

$$\Delta E = h\nu = \mu\beta_N \frac{H_0}{I} = g\beta_N H_0 \quad (14\text{-}17)$$

where g is the *splitting factor* and β_N is the *Bohr magneton*, which has a value of 9.27×10^{-21} erg G^{-1}. The value for g varies with the electron's environment. For a free electron, its value is 2.0023; for an unpaired electron in a molecule or ion, g lies within a few percent of this number.

---

[17] For further details on ESR spectroscopy, see: M. C. R. Symons, *Chemical and Biochemical Aspects of Electron-Spin Resonance Spectroscopy.* New York: Wiley, 1978; and J. E. Wertz and J. R. Bolton, *Electron Spin Resonance: Elementary Theory and Practical Applications.* New York: McGraw-Hill, 1972.

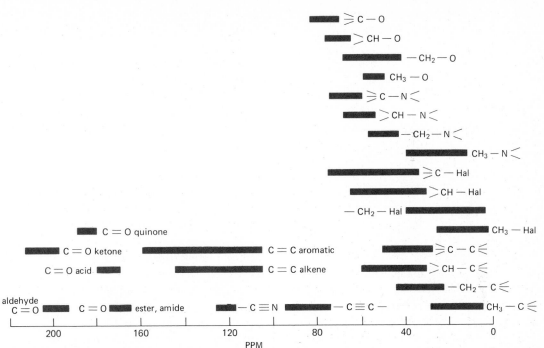

FIGURE 14-25 Chemical shifts for carbon-13. (From D. E. Leyden and R. H. Cox, *Analytical Applications of NMR*. New York: Wiley, 1977, p. 196. With permission.)

ESR spectrometers often employ a field of 3400 G. Substituting this value into Equation 14-17 gives

$$v = \frac{g\beta H_0}{h} = \frac{2.0 \times 9.27 \times 10^{-21} \times 3400}{6.63 \times 10^{-27}}$$

$$= 9.51 \times 10^9 \text{ Hz} \quad \text{or} \quad 9500 \text{ MHz}$$

Thus, the resonance frequency for an unpaired electron is about 9500 MHz, which lies in the microwave region (see Figure 4-7, p. 101).

Instrumentation

The source of microwave radiation is a klystron tube, which is operated to produce monochromatic radiation having a frequency of about 9500 MHz. The klystron tube is an electronic oscillator in which a beam of electrons is pulsed between a cathode and a reflector. The oscillating output of the klystron is transmitted to a *wave guide* by a loop of wire, which sets

up a fluctuating magnetic field (electromagnetic radiation) in the guide. The wave guide, which is a rectangular metal tube, transmits the microwave radiation to the sample, which is generally held in a small quartz tube positioned between the poles of the permanent magnet. Helmholtz coils provide a means for varying the field over the small range in which resonance occurs.

Generally, ESR spectra are recorded in derivative form to enhance sensitivity and resolution. Figure 14-26 contrasts two derivative spectra with their ordinary absorption counterparts.

ESR Spectra

Most molecules fail to exhibit an ESR spectrum because they contain an even number of electrons, and the number in the two spin states is identical (that is, the spins are paired); as a consequence, the magnetic effects of elec-

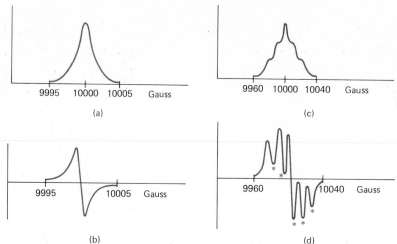

FIGURE 14-26   Comparison of spectral presentations as absorption (a and c) and the corresponding derivative (b and d) curves. Note that the shoulders in (c) never reach a maximum; consequently the corresponding derivative peaks do not pass through the abscissa. The number of peaks can be obtained by a count of the number of maxima or minima (as shown by the asterisks). (From R. S. Drago, *Physical Methods in Chemistry*, 2d ed. Philadelphia: Saunders, 1977, p. 323. With permission.)

tron spin are canceled. Such substances are diamagnetic because of the small fields induced from the *orbital precession* of the electrons around the nuclei; these fields act in opposition to the applied field. In contrast, the spin of an unpaired electron induces a field that reinforces the applied field. This paramagnetic effect is much larger than the diamagnetic behavior. Splitting of the energy levels of an electron occurs in a magnetic field; this splitting can be observed, as noted earlier, by microwave absorption studies.

The spin of an unpaired electron can couple with the spin of nuclei in the species to give splitting patterns analogous to those observed for nuclear spin-spin coupling. When an electron interacts with n equivalent nuclei, its resonance peak is split into $(2nI + 1)$ peaks, where I is the spin quantum number of the nuclei and n is the number of equivalent nuclei. This process is termed *hyperfine splitting*. Figure 14-27 shows the spectrum that results from hyperfine splitting that occurs in the

simplest of all free radicals, the hydrogen atom. Here, the spin of the hydrogen nucleus couples with that of the unpaired electron to produce $(2 \times \frac{1}{2} \times 1 + 1)$ or 2 peaks.

Figure 14-28 illustrates the application of ESR to the detection of a free radical intermediate in a chemical process. This spectrum was obtained after the addition of base to a solution of hydroquinone in the presence of air. The reaction involves oxidation of the anion of hydroquinone to give quinone. That is,

Hydroquinone ion   Semiquinone radical   Quinone

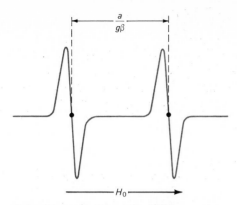

FIGURE 14-27 The ESR spectrum for the hydrogen atom. The hyperfine splitting is $a/g\beta$ gauss. (From R. S. Drago, *Physical Methods in Chemistry*, 2d ed. Philadelphia: Saunders, 1977, p. 324. With permission.)

tative determination of paramagnetic ions in biological systems in concentrations as low as a few parts per billion.

Another important biological application involves the use of *spin-label reagents*. These compounds are stable molecules that contain an odd electron and, in addition, react selectively with certain amino acids or functional groups in a biological system. The ESR spectra of the product then provide information about the environment, such as structural features, polarity, viscosity, phase changes, and chemical reactivity.

Hydroquinone and quinone are not ESR active because neither contains an odd electron. The postulated intermediate does, however, and the five-peak hyperfine pattern is consistent with a semiquinone radical. Here, the four ring protons are assumed to be equivalent with respect to the interactions of their spins with that of the odd electron; five peaks result. This picture is reasonable since the position of the odd electron is not confined to the oxygen atom as shown. Indeed, several resonance structures can be written in which the odd electron is associated with any of several other atoms in the molecule.

Electron spin has been widely applied to the study of chemical, photochemical, and electrochemical reactions which proceed via free radical mechanisms. Often, as in the foregoing example, the technique has provided useful structural information about intermediate radicals.

ESR studies are also used to obtain structural information about the transition metals and their complexes. In addition, they have been used for the detection and semiquanti-

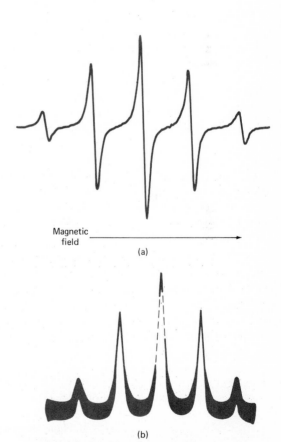

FIGURE 14-28 The ESR spectrum for the semiquinone radical. (a) Derivative spectrum. (b) Absorption spectrum. [From B. Venkataraman and G. K. Fraenkel, *J. Amer. Chem. Soc.*, **77**, 2707 (1955). With permission of the American Chemical Society.]

PROBLEMS

1. Diagram the high-resolution NMR spectrum expected for the following compounds. Give chemical shifts, splitting patterns, and relative intensities of each peak.
   (a) Toluene, $C_6H_5CH_3$
   (b) Diethyl ether, $C_2H_5OC_2H_5$
   (c) Propionaldehyde, CH_3CH_2CHO
   (d) Isopropyl chloride, $(CH_3)_2CHCl$
   (e) 2-Phenylethanol, $C_6H_5CH_2CH_2OH$

2. Diagram the NMR spectrum expected for each of the following compounds. Give chemical shifts, splitting patterns, and relative intensities of each peak.
   (a) 1,2-Dimethoxyethane, $CH_3OCH_2CH_2OCH_3$
   (b) 2,3-Dibromobutane, $CH_3CHBrCHBrCH_3$

   (c) Cyclohexane, $\overline{CH_2CH_2CH_2CH_2CH_2CH_2}$

   (d) Tetrahydrofuran, $\begin{array}{c} CH_2\text{———}CH_2 \\ CH_2 \qquad CH_2 \\ O \end{array}$

   (e) p-Dimethylaminobenzaldehyde, $(CH_3)_2N\!-\!\!\langle\ \rangle\!-\!CHO$

3. From the spectrum given in Figure 14-29, deduce the structure of this strong-smelling compound with an empirical formula of $C_3H_6O_2$.

4. From the spectrum given in Figure 14-30, deduce the structure of this hydrocarbon.

5. From the spectrum given in Figure 14-31, determine the structure of this compound, which is a commonly used pain killer; its empirical formula is $C_{10}H_{13}NO_2$.

6. From the NMR spectrum shown in Figure 14-32, determine the structure of this pleasant-smelling compound, $C_8H_8O_3$.

7. From the NMR spectrum shown in Figure 14-33, determine the structure of this easily oxidized compound.

8. Determine the structure of the basic compound whose NMR spectrum is shown in Figure 14-34; its empirical formula is $C_8H_{11}N$.

9. Figures 14-35 and 14-36 show the NMR spectra of two isomers of $C_4H_8O_2$. Both of these compounds show a strong absorption at 1735 cm^{-1} in their IR spectrum. Identify the two compounds.

10. The NMR spectra of two isomers of $C_5H_{10}O$ are shown in Figures 14-37 and 14-38. Identify these isomers. They both show a carbonyl peak in their IR spectra.

11. Determine the possible structure(s) for the compound giving the spectrum in Figure 14-39. Can any conclusions be drawn about the different possible isomers?

12. From the 60-MHz spectrum given in Figure 14-40, deduce the structure of this compound, whose empirical formula is $C_4H_{11}NO$. Draw the spectrum as it would appear if obtained with a 100-MHz instrument.

13. From the 60-MHz spectrum given in Figure 14-41, deduce the structure of this chlorine containing compound. Draw the spectrum as it would appear if obtained with a 100-MHz instrument.

14. Shown in Figure 14-42 is the NMR spectrum of a common aprotic solvent having the empirical formula C_3H_7NO. When the spectrum of this compound is obtained at a higher temperature, the two peaks at 2.9 δ coalesce into one sharp singlet. What is this compound? Explain the high-temperature behavior.

15. It is apparent from the spectra in Figure 14-43 that the sample gives a very different NMR spectrum when DCl is added to it. What is this bromine-containing compound? Explain the effect of the DCl.

16. From the spectrum shown in Figure 14-44, deduce the structure of this biologically interesting molecule with the empirical formula $C_5H_{11}NO_2$. Notice that the spectrum has been taken in deuterium oxide (D_2O).

17. The ^1H NMR spectrum of a pleasant-smelling liquid consists of two singlets. Shown in Figure 14-45, is a proton decoupled ^{13}C NMR spectrum of the compound. Deduce the structure of this unknown.

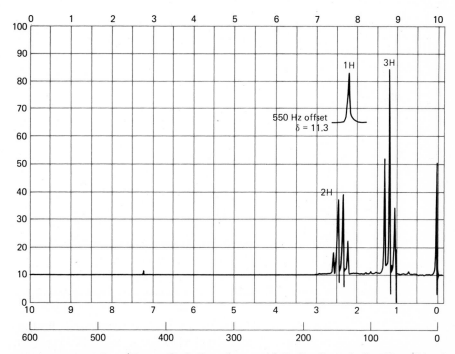

FIGURE 14-29 (From C. J. Pouchert and J. R. Campbell, *The Aldrich Library of NMR Spectra*. Milwaukee, WI: The Aldrich Chemical Company, 1974. With permission.) See Problem 3.

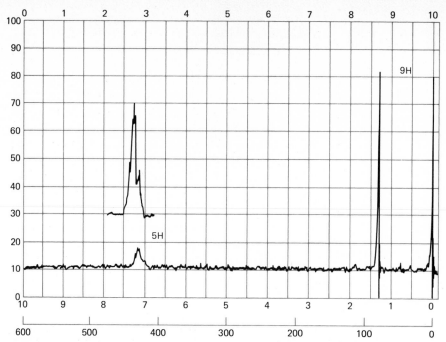

FIGURE 14-30 (From C. J. Pouchert and J. R. Campbell, *The Aldrich Library of NMR Spectra.* Milwaukee, WI: The Aldrich Chemical Company, 1974. With permission.) See Problem 4.

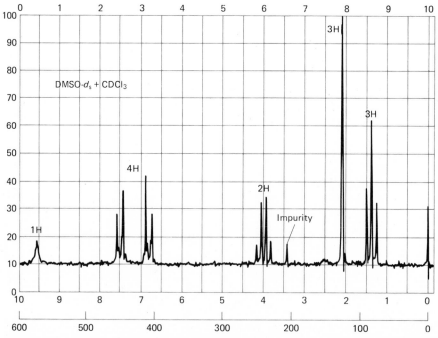

FIGURE 14-31 (From C. J. Pouchert and J. R. Campbell, *The Aldrich Library of NMR Spectra.* Milwaukee, WI: The Aldrich Chemical Company, 1974. With permission). See Problem 5.

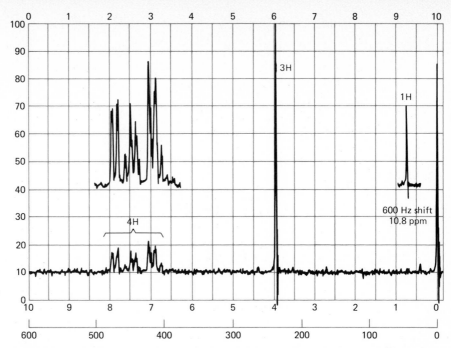

FIGURE 14-32 (From C. J. Pouchert and J. R. Campbell, *The Aldrich Library of NMR Spectra.* Milwaukee, WI: The Aldrich Chemical Company, 1974. With permission.) See Problem 6.

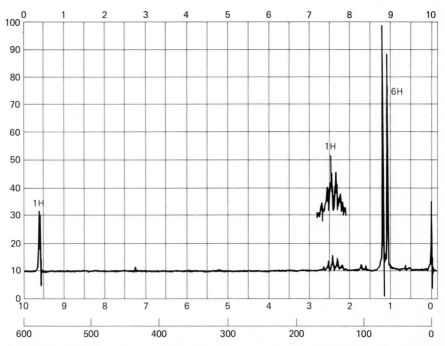

FIGURE 14-33 (From C. J. Pouchert and J. R. Campbell, *The Aldrich Library of NMR Spectra.* Milwaukee, WI: The Aldrich Chemical Company, 1974. With permission.) See Problem 7.

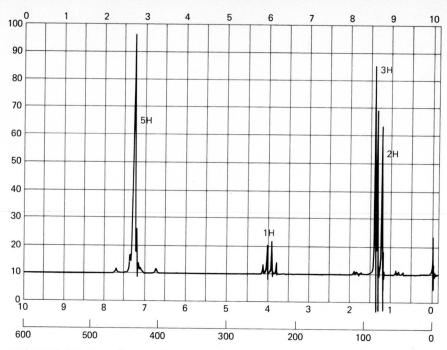

FIGURE 14-34 (From C. J. Pouchert and J. R. Campbell, *The Aldrich Library of NMR Spectra.* Milwaukee, WI: The Aldrich Chemical Company, 1974. With permission.) See Problem 8.

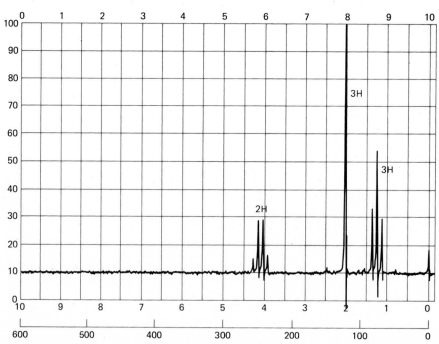

FIGURE 14-35 (From C. J. Pouchert and J. R. Campbell, *The Aldrich Library of NMR Spectra.* Milwaukee, WI: The Aldrich Chemical Company, 1974. With permission.) See Problem 9.

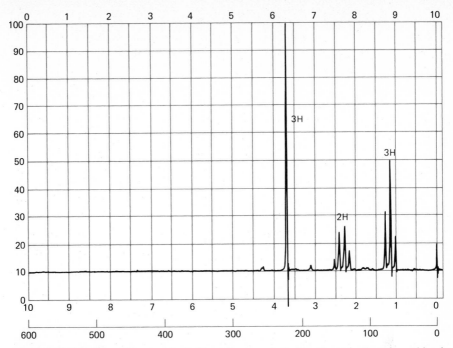

FIGURE 14-36 (From C. J. Pouchert and J. R. Campbell, *The Aldrich Library of NMR Spectra*. Milwaukee, WI: The Aldrich Chemical Company, 1974. With permission.) See Problem 9.

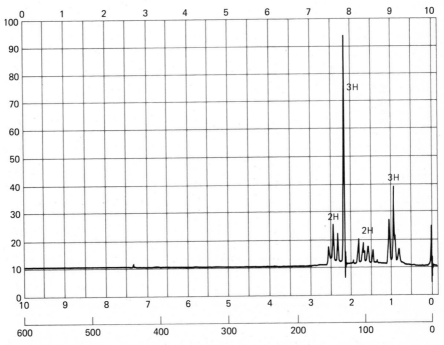

FIGURE 14-37 (From C. J. Pouchert and J. R. Campbell, *The Aldrich Library of NMR Spectra*. Milwaukee, WI: The Aldrich Chemical Company, 1974. With permission.) See Problem 10.

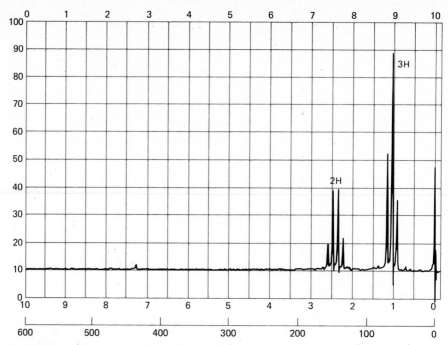

FIGURE 14-38 (From C. J. Pouchert and J. R. Campbell, *The Aldrich Library of NMR Spectra.* Milwaukee, WI: The Aldrich Chemical Company, 1974. With permission.) See Problem 10.

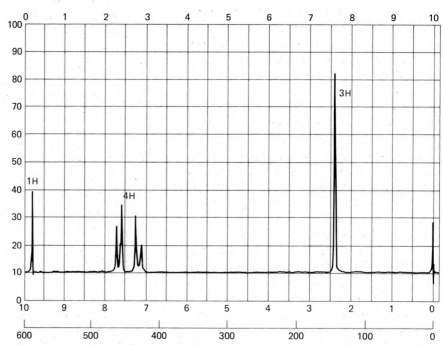

FIGURE 14-39 (From C. J. Pouchert and J. R. Campbell, *The Aldrich Library of NMR Spectra.* Milwaukee, WI: The Aldrich Chemical Company, 1974. With permission.) See Problem 11.

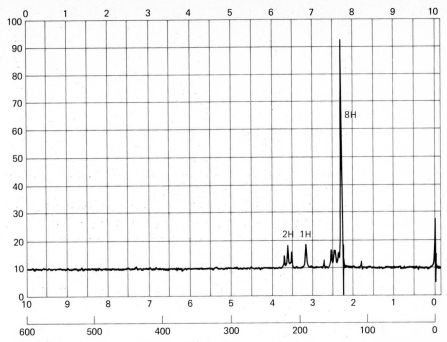

FIGURE 14-40 (From C. J. Pouchert and J. R. Campbell, *The Aldrich Library of NMR Spectra*. Milwaukee, WI: The Aldrich Chemical Company, 1974. With permission.) See Problem 12.

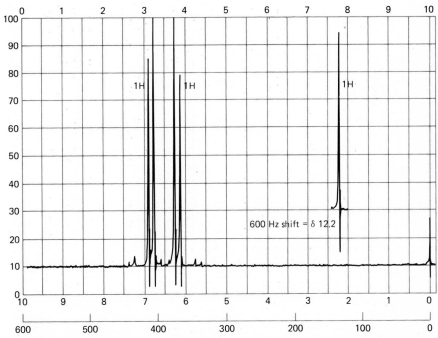

FIGURE 14-41 (From C. J. Pouchert and J. R. Campbell, *The Aldrich Library of NMR Spectra*. Milwaukee, WI: The Aldrich Chemical Company, 1974. With permission.) See Problem 13.

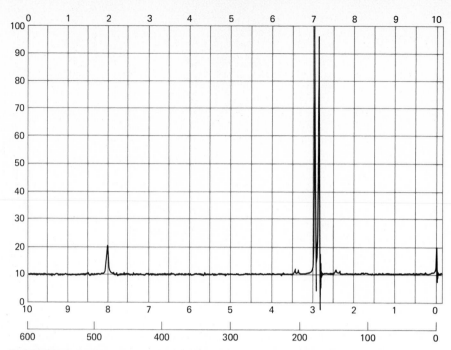

FIGURE 14-42 (From C. J. Pouchert and J. R. Campbell, *The Aldrich Library of NMR Spectra*. Milwaukee, WI: The Aldrich Chemical Company, 1974. With permission.) See Problem 14.

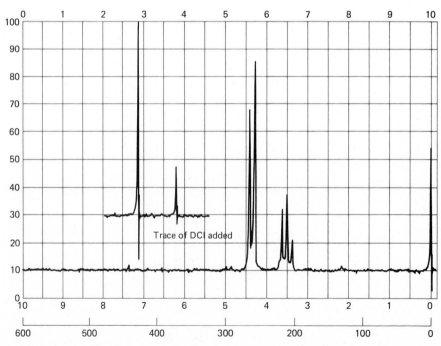

FIGURE 14-43 (From C. J. Pouchert and J. R. Campbell, *The Aldrich Library of NMR Spectra*. Milwaukee, WI: The Aldrich Chemical Company, 1974. With permission.) See Problem 15.

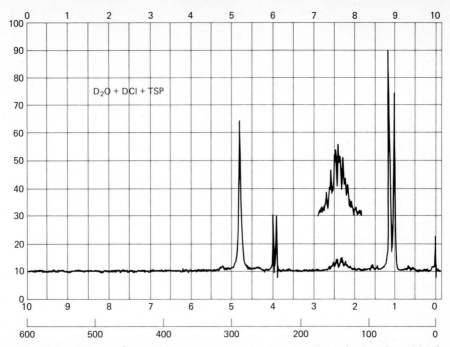

FIGURE 14-44 (From C. J. Pouchert and J. R. Campbell, *The Aldrich Library of NMR Spectra.* Milwaukee, WI: The Aldrich Chemical Company, 1974. With permission.) See Problem 16.

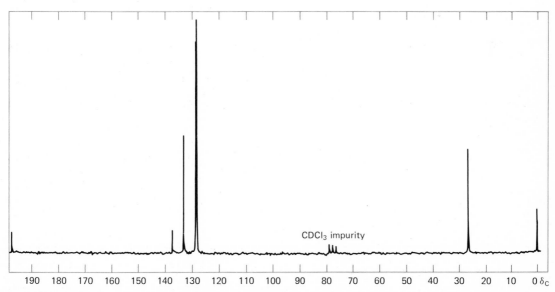

FIGURE 14-45 (From L. F. Johnson and W. C. Jankowski, *Carbon-13 NMR Spectra.* New York: Wiley-Interscience, 1972. With permission.) See Problem 17.

15

X-RAY METHODS

Several types of X-ray spectroscopic methods exist that are similar in many respects to those encountered in optical spectroscopy. Thus, methods based upon emission (including fluorescent emission), absorption, and diffraction of X-rays now find wide use for qualitative and quantitative analysis.[1]

FUNDAMENTAL PRINCIPLES

Before describing the applications of X-rays to analytical problems, it is desirable to consider certain theoretical aspects that relate to the emission, absorption, and diffraction of radiation in the wavelength range between about 0.1 and 25 Å. (The ångström unit Å, which is 0.1 nm, was first used because it provided a convenient way of expressing the wavelengths of X-rays.)

Emission of X-Rays

For analytical purposes, X-rays are obtained in three ways, namely: (1) by bombardment of a metal target with a beam of high-energy electrons, (2) by exposure of a substance to a primary beam of X-rays in order to generate a secondary beam of fluorescent X-rays, and (3) by employment of a radioactive source whose decay process results in X-ray emission.

X-Ray sources, like ultraviolet and visible emitters, often produce both a continuous and a discontinuous (line) spectrum; both types are of importance in analysis. Continuous radiation is also called *white* radiation

[1] For a more extensive discussion of the theory and analytical applications of X-rays, see: E. P. Bertin, *Introduction to X-Ray Spectrometric Analysis*. New York: Plenum Press, 1978; L. S. Birks, *X-Ray Spectrochemical Analysis*. New York: Interscience, 1969; H. A. Liebhafsky, H. G. Pfeiffer, E. H. Winslow, and P. D. Zemany, *X-Rays, Electrons and Analytical Chemistry*. New York: Wiley-Interscience, 1972; and R. O. Muller, *Spectrochemical Analysis by X-Ray Fluorescence*. New York: Plenum Press, 1972.

or *Bremsstrahlung* (meaning radiation that arises from retardation by particles; such radiation is generally continuous).

Continuous Spectra from Electron Beam Sources. In an X-ray tube, electrons produced at a heated cathode are accelerated toward an anode (the *target*) by a potential as great as 100 kV; upon collision, part of the energy of the electron beam is converted to X-rays. Under some conditions, only a continuous spectrum such as that shown in Figure 15-1 results; under others, a line spectrum is superimposed upon the continuum (see Figure 15-2).

The continuous X-ray spectrum shown in the two figures is characterized by a well-defined, short-wavelength limit (λ_0), which is dependent upon the accelerating voltage V but independent of the target material. Thus, λ_0 for the spectrum produced with a molybdenum target at 35 kV (Figure 15-2) is identical to λ_0 for a tungsten target at the same voltage (Figure 15-1).

The continuous radiation from an electron beam source results from collisions between the electrons of the beam and the atoms of

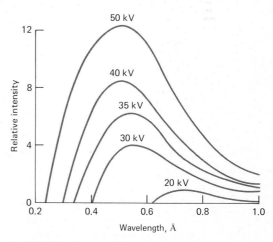

FIGURE 15-1   Distribution of continuous radiation from an X-ray tube with a tungsten target. The numbers above the curves indicate the accelerating voltages.

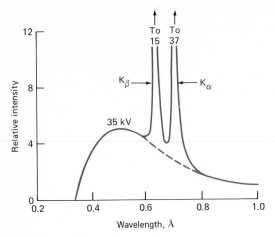

FIGURE 15-2   Line spectrum for a tube with a molybdenum target.

the target material. At each collision, an electron is decelerated and a photon of X-ray energy is produced. The energy of the photon will be equal to the difference in kinetic energies of the electron before and after the collision. Generally, the electrons in a beam are decelerated in a series of collisions; the resulting loss of kinetic energy differs from collision to collision. Thus, the energies of the emitted X-ray photons vary continuously over a considerable range. The maximum photon energy generated corresponds to the instantaneous deceleration of the electron to zero kinetic energy in a single collision. For such an event, we may write

$$h\nu_0 = \frac{hc}{\lambda_0} = Ve \qquad (15\text{-}1)$$

where Ve, the product of the accelerating voltage and the charge on the electron, is the kinetic energy of all of the electrons in the beam, h is Planck's constant, and c is the velocity of light. The quantity ν_0 is the maximum frequency of radiation that can be produced at voltage V, while λ_0 is the low-wavelength limit for the radiation. This relationship is known as the *Duane-Hunt law*.

Upon substituting numerical values for the constants and rearranging, Equation 15-1 becomes

$$\lambda_0 = 12{,}398/V \qquad (15\text{-}2)$$

where λ_0 and V have units of ångströms and volts, respectively. It is of interest to note that Equation 15-1 has provided a direct means for the highly accurate determination of Planck's constant.

Characteristic Line Spectra from Electron Beam Sources. As shown in Figure 15-2, bombardment of a molybdenum target produces intense emission lines at about 0.63 and 0.71 Å; an additional simple series of lines occurs in the longer wavelength range of 4 to 6 Å.

The emission behavior of molybdenum is typical of all elements having atomic numbers larger than 23; that is, the X-ray line spectra are remarkably simple when compared with ultraviolet emission and consist of two series of lines. The shorter wavelength group is called the K series and the other the L series.[2] Elements with atomic numbers smaller than 23 produce only a K series. Table 15-1 (p. 430) presents wavelength data for the emission spectra of a few elements.

A second characteristic of X-ray spectra is that the minimum acceleration voltage required for the excitation of the lines for each element increases with atomic number. Thus, the line spectrum for molybdenum (atomic number = 42) disappears if the excitation voltage drops below 20 kV. As shown in Figure 15-1, bombardment of tungsten (atomic number = 74) produces no lines in the region of

---

[2] For the heavier elements, additional series of lines (M, N, and so forth) are found at longer wavelengths. Their intensities are low, however, and little use is made of them.

The designations K and L arose from the German words kurtz and lang for short and long wavelengths. The additional alphabetical designations were then added for lines occurring at progressively longer wavelengths.

0.1 to 1.0 Å even at 50 kV. Characteristic K lines appear at 0.18 and 0.21 Å, however, if the voltage is raised to 70 kV.

Figure 15-3 illustrates the linear relationship between the square root of the frequency for a given (K or L) line and the atomic number of the element responsible for the radiation. This property was first discovered by H. G. S. Moseley in 1914.

X-Ray line spectra result from electronic transitions that involve the innermost atomic orbitals. The short-wavelength K series is produced when the high-energy electrons from the cathode remove electrons from those orbitals nearest to the nucleus of the target atom (in X-ray terminology, the orbital of principal quantum number $n = 1$ is called the K shell; the orbital of quantum number $n = 2$ is called the L shell, and so forth). The collision results in the formation of an excited *ion*, which then loses quanta of X-radiation as electrons from outer orbitals undergo transitions to the

vacated orbital. As shown in Figure 15-4, the lines in the K series involve electronic transitions between higher energy levels and the K shell. The L series of lines results when an electron is lost from the second principal quantum level, either as a consequence of ejection by an electron from the cathode or from the transition of an L electron to the K level that accompanies the production of a quantum of K radiation. It is important to appreciate that the energy scale in Figure 15-4 is logarithmic. Thus, the energy difference between the L and K levels is significantly larger than that between the M and L levels. The K lines therefore appear at shorter wavelengths. It is also important to note that the energy differences between the transitions labeled α_1 and α_2 as well as those between β_1 and β_2 are so small that only single lines are observed in all but the highest resolution spectrometers (see Figure 15-2).

The energy level diagram in Figure 15-4 would be applicable to any element with sufficient electrons to permit the number of transitions shown. The differences in energies between the levels increase regularly with atomic number because of the increasing charge on the nucleus; therefore, the radiation for the K series appears at shorter wavelengths for the heavier elements (see Table 15-1). The effect of nuclear charge is also reflected in the increase in minimum voltage required to excite the spectra of these elements.

It is important to note that for all but the lightest elements, the wavelengths of characteristic X-ray lines are independent of chemical combination because the transitions responsible for these lines involve electrons that take no part in bonding. Thus, the position of the K_α lines for molybdenum is the same regardless of whether the target is the pure metal, its sulfide, or its oxide.

Fluorescent Line Spectra. Another convenient way of producing a line spectrum is to irradiate the element or one of its compounds with the continuous radiation from an X-ray

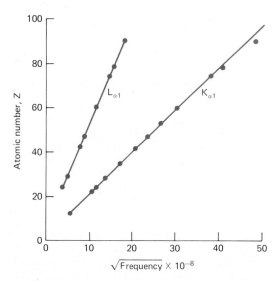

FIGURE 15-3 Relationship between X-ray emission frequency and atomic number ($K_{\alpha 1}$ and $L_{\alpha 1}$ lines).

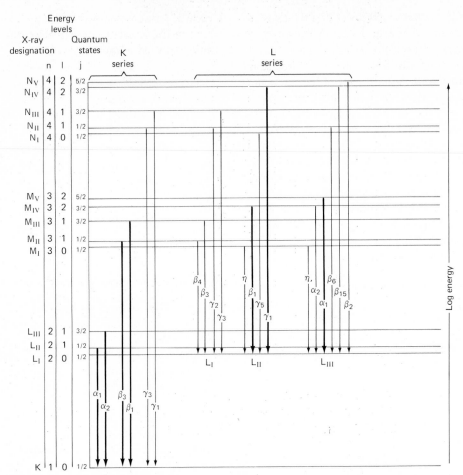

FIGURE 15-4 Partial energy level diagram showing common transitions leading to X-radiation. The most intense lines are indicated by the widest arrows.

TABLE 15-1 WAVELENGTHS IN ÅNGSTRÖM UNITS OF THE MORE INTENSE EMISSION LINES FOR SOME TYPICAL ELEMENTS

Element	Atomic Number	K Series		L Series	
		α_1	β_1	α_1	β_1
Na	11	11.909	11.617	—	—
K	19	3.742	3.454	—	—
Cr	24	2.290	2.085	21.714	21.323
Rb	37	0.926	0.829	7.318	7.075
Cs	55	0.401	0.355	2.892	2.683
W	74	0.209	0.184	1.476	1.282
U	92	0.126	0.111	0.911	0.720

tube. This process is considered further in a later section.

Radioactive Sources. X-Radiation occurs in two radioactive decay processes. *Gamma rays*, which are indistinguishable from X-rays, owe their production to intranuclear reactions. *Electron capture* or *K capture* also produces X-radiation. This process involves capture of a K electron (less commonly, an L or an M electron) by the nucleus and formation of an element of the next lower atomic number. As a result of K capture, electronic transitions to the vacated orbital occur, and the X-ray line spectrum of the newly formed element is observed. The half-lives (p. 462) of K-capture processes range from a few minutes to several thousands of years.

Artificially produced radioactive isotopes provide a very simple source of mono-energetic radiation for certain analytical applications. The best known example is iron-55, which undergoes a K-capture reaction with a half-life of 2.6 years:

$$^{55}Fe \rightarrow {}^{54}Mn + h\nu$$

The resulting manganese K_α line at about 2.1 Å has proved to be a useful source for both fluorescence and absorption methods.

Absorption of X-Rays

When a narrow beam of X-rays is passed through a thin layer of matter, its intensity or power is generally diminished as a consequence of absorption and scattering. The effect of scattering is ordinarily small and can be neglected in those wavelength regions where appreciable absorption occurs. As shown in Figure 15-5, the absorption spectrum of an element, like its emission spectrum, is simple and consists of a few well-defined absorption peaks. Here again, the wavelengths of the peaks are characteristic of the element and are largely independent of its chemical state.

A peculiarity of X-ray absorption spectra is the appearance of sharp discontinuities, called *absorption edges*, at wavelengths immediately beyond absorption maxima.

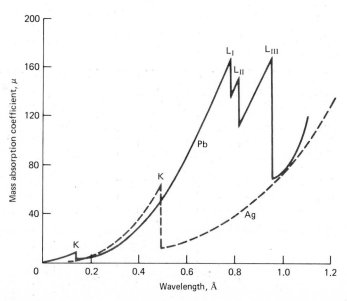

FIGURE 15-5 X-Ray absorption spectra for lead and silver.

The Absorption Process. Absorption of an X-ray quantum causes ejection of one of the innermost electrons from an atom and the consequent production of an excited ion. In this process, the entire energy hv of the radiation is partitioned between the kinetic energy of the electron (the *photoelectron*) and the potential energy of the excited ion. The highest probability for absorption arises when the energy of the quantum is exactly equal to the energy required to remove the electron just to the periphery of the atom (that is, as the kinetic energy of the ejected electron approaches zero).

The absorption spectrum for lead, shown in Figure 15-5, exhibits four peaks, the first occurring at 0.14 Å. The energy of the quantum corresponding to this wavelength exactly matches the energy required to just eject the highest energy K electron of the element; immediately beyond this wavelength, the energy of the radiation is insufficient to bring about removal of a K electron, and an abrupt decrease in absorption occurs. At wavelengths lower than 0.14 Å, the probability of interaction between the electron and the radiation diminishes and results in a smooth decrease in absorption. In this region, the kinetic energy of the ejected photoelectron increases continuously with the decrease in wavelength.

The additional peaks at longer wavelengths correspond to the removal of an electron from the L energy levels of lead. Three sets of L levels, differing slightly in energy, exist (see Figure 15-4); three peaks are, therefore, observed. Another set of peaks, arising from ejections of M electrons, will be located at still longer wavelengths.

Figure 15-5 also shows the K absorption edge for silver, which occurs at 0.485 Å. The longer wavelength for the silver peak reflects the lower atomic number of the element compared with lead.

The Mass Absorption Coefficient. Beer's law is as applicable to the absorption of X-radiation as to other types of electromagnetic radiation; thus, we may write

$$\ln \frac{P_0}{P} = \mu_1 x$$

where x is the sample thickness in centimeters and P and P_0 are the powers of the transmitted and incident beams. The constant μ_1 is called the *linear absorption coefficient* and is characteristic of the element as well as the number of its atoms in the path of the beam. A more convenient form of Beer's law is

$$\ln \frac{P_0}{P} = \mu \rho x \tag{15-3}$$

where ρ is the density of the sample and μ is the *mass absorption coefficient*, a quantity that is *independent* of the physical and chemical states of the element. Thus, the mass absorption coefficient for bromine has the same value in gaseous HBr as in solid sodium bromate.

Mass absorption coefficients are additive functions of the weight fractions of elements contained in a sample. Thus,

$$\mu_M = W_A \mu_A + W_B \mu_B + W_C \mu_C + \cdots \tag{15-4}$$

where μ_M is the mass absorption coefficient of a sample containing the weight fractions W_A, W_B, and W_C of elements A, B, and C. The terms μ_A, μ_B, and μ_C are the respective mass absorption coefficients for each of the elements.

X-Ray Fluorescence

The absorption of X-rays produces electronically excited ions that return to their ground state by transitions involving electrons from higher energy levels. Thus, an excited ion with a vacant K shell is produced when lead absorbs radiation of wavelengths shorter than 0.14 Å (Figure 15-5); after a brief period, the ion returns to its ground state via a series of electronic transitions characterized by the emission of X-radiation (fluorescence) of wavelengths identical to those that result from excitation produced by electron bombardment. The wavelengths of the fluorescent lines are always somewhat greater than the wavelength of the corresponding absorption edge, however, because absorption requires a complete removal of the electron (that is, ionization), whereas emission involves transitions of an electron from a higher energy level within the atom. For example, the K absorption edge for silver occurs at 0.485 Å, while the K emission lines for the element have wavelengths at 0.497 and 0.559 Å. When fluorescence is to be excited by radiation from an X-ray tube, the operating voltage must be sufficiently great so that the cutoff wavelength λ_0 (Equation 15-2) is shorter than the absorption edge of the element whose spectrum is to be excited. Thus, to generate the K lines for silver, the tube voltage would need to be

$$V \geq \frac{12398}{0.485} \times 10^{-3} = 25.6 \text{ kV}$$

Diffraction of X-Rays

In common with other types of electro-magnetic radiation, interaction between the electric vector of X-radiation and the electrons of the matter through which it passes results in scattering. When X-rays are scattered by the ordered environment in a crystal, interference (both constructive and destructive) takes place among the scattered rays because the distances between the scattering centers are of the same order of magnitude as the wavelength of the radiation. Diffraction is the result.

Bragg's Law. When an X-ray beam strikes a crystal surface at some angle θ, a portion is scattered by the layer of atoms at the surface. The unscattered portion of the beam penetrates to the second layer of atoms where again a fraction is scattered, and the remainder passes on to the third layer. The cumulative effect of this scattering from the regularly spaced centers of the crystal is a diffraction of the beam in much the same way as visible radiation is diffracted by a reflection grating (p. 129). The requirements for diffraction are: (1) the spacing between layers of atoms must be roughly the same as the wavelength of the radiation and (2) the scattering centers must be spatially distributed in a highly regular way.

In 1912, W. L. Bragg treated the diffraction of X-rays by crystals as shown in Figure 15-6. Here, a narrow beam strikes the crystal surface at angle θ; scattering occurs as a consequence of interaction of the radiation with atoms located at O, P, and R. If the distance

$$AP + PC = \mathbf{n}\lambda$$

where $\mathbf{n}$ is an integer, the scattered radiation will be in phase at OCD, and the crystal will

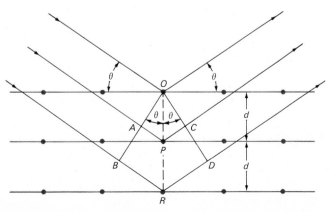

FIGURE 15-6 Diffraction of X-rays by a crystal.

appear to reflect the X-radiation. But it is readily seen that

$$AP = PC = d \sin \theta$$

where d is the interplanar distance of the crystal. Thus, we may write that the conditions for constructive interference of the beam at angle θ are

$$n\lambda = 2d \sin \theta \qquad (15\text{-}6)$$

Equation 15-6 is called the *Bragg equation* and is of fundamental importance. Note that X-rays appear to be reflected from the crystal only if the angle of incidence satisfies the condition that

$$\sin \theta = \frac{n\lambda}{2d}$$

At all other angles, destructive interference occurs.

INSTRUMENT COMPONENTS

Absorption, emission, fluorescence, and diffraction of X-rays all find applications in analytical chemistry. Instruments for these applications contain components that are analogous in function to the five components of instruments for optical spectroscopic measurement; these components include a source, a device for restricting the wavelength range to be employed, a sample holder, a radiation detector or transducer, and a signal processor and readout. These components differ considerably in detail from their optical counterparts. Their functions, however, are the same, and the ways in which they are combined to form instruments are often similar to those shown in Figure 5-1 (p. 115).

As with optical instruments, both X-ray photometers and spectrophotometers are encountered, the first employing filters and the second monochromators for restricting radiation from the source. In addition, however, a third method is available for obtaining information about isolated portions of an X-ray spectrum. Here, isolation can be achieved electronically with devices that have the power to discriminate between various parts of a spectrum based on the energy rather than the wavelength of the radiation. Thus, X-ray instruments are often described as *wavelength dispersive instruments* or *energy dispersive instruments*, depending upon the method by which they resolve spectra.

Sources

Three types of sources are encountered in X-ray instruments, namely, Coolidge tubes, radioisotopes, and secondary fluorescent sources.

The Coolidge Tube. The most common source of X-rays for analytical work is the Coolidge tube, which can take a variety of shapes and forms. Basically, however, it is a highly evacuated tube in which is mounted a tungsten filament cathode and a massive anode constructed of such metals as tungsten, copper, molybdenum, chromium, silver, nickel, cobalt, rhodium or iron (Figure 15-7). Separate circuits are used to heat the filament and to accelerate the electrons to the target. The heater circuit provides the means for controlling the intensity of the emitted X-rays while the accelerating potential determines their energy or wavelength. The Coolidge tube is normally self-rectifying, and a high-voltage ac source is connected directly to the cathode to provide the accelerating potential.

The production of X-rays by electron bombardment is a highly inefficient process. Less than one percent of the electrical power is converted to radiant power, the remainder being degraded to heat. As a consequence, water cooling of the anodes of X-ray tubes is required.

Radioisotopes. A variety of radioactive substances have been employed as sources in X-ray fluorescence and absorption methods. Generally, the radioisotope is encapsulated to prevent contamination of the laboratory and shielded to absorb radiation in all but certain directions.

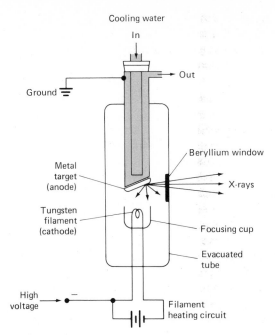

Cooling water

In

Out

Ground

Beryllium window

Metal
target
(anode)

X-rays

Tungsten
filament
(cathode)

Focusing cup

Evacuated
tube

High
voltage

Filament
heating circuit

FIGURE 15-7 Schematic diagram of the Coolidge tube.

tungsten target (Figure 15-1) could be used to excite the K_α and K_β lines of molybdenum (Figure 15-2). The resulting fluorescence spectrum would then be similar to the spectrum in Figure 15-2 except that the continuum would be removed.

Filters for X-Ray Beams

In many applications, it is desirable to employ an X-ray beam that is restricted in its wavelength range. As in the visible region, both filters and monochromators are used for this purpose.

Figure 15-8 illustrates a common technique for producing a relatively monochromatic beam by use of a filter. Here, the K_β line and most of the continuous radiation from the emission of a molybdenum target is removed by a zirconium filter having a thickness of about 0.01 cm. The pure K_α line is then available for analytical purposes. Several other target-filter combinations of this type have been developed, each of which serves to isolate

The best radioactive sources provide simple line spectra. Because of the shape of X-ray absorption curves, a given radioisotope will be suitable for excitation of fluorescence or for absorption studies for a range of elements. For example, a source producing a line in the region between 0.3 and 0.47 Å would be suitable for fluorescence or absorption studies involving the K absorption edge for silver (see Figure 15-5). Sensitivity would, of course, improve as the wavelength of the source line approaches the absorption edge. Iodine-125 with a line at 0.46 Å would be ideal from this standpoint.

Secondary Fluorescent Sources. In some applications, the fluorescence spectrum of an element that has been excited by radiation from a Coolidge tube serves as a source for absorption or fluorescence studies. This arrangement has the advantage of eliminating the continuous component emitted by a primary source. For example, a Coolidge tube with a

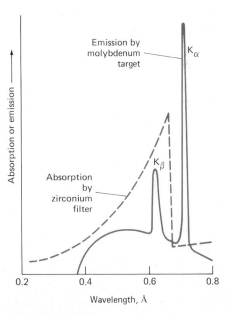

Emission by
molybdenum
target

K_α

Absorption or emission

K_β

Absorption
by
zirconium
filter

0.2   0.4   0.6   0.8

Wavelength, Å

FIGURE 15-8 Use of a filter to produce monochromatic radiation.

one of the intense lines of a target element. Monochromatic radiation produced in this way is widely used in X-ray diffraction studies. The choice of wavelengths available by this technique is limited by the relatively small number of target-filter combinations that are available.

Filtration of the continuous radiation from a Coolidge tube is also feasible with thin strips of metal. As with glass filters for visible radiation, relatively broad bands are produced with a significant loss in intensity of the desired wavelengths.

Wavelength Dispersion with Monochromators

Figure 15-9 shows the essential components of an X-ray spectrometer. The monochromator consists of a pair of beam collimators, which serve the same purpose as the slits in an optical instrument, and a dispersing element. The latter is a single crystal mounted on a *goniometer* or rotatable table that permits variation and precise determination of the angle θ between the crystal face and the

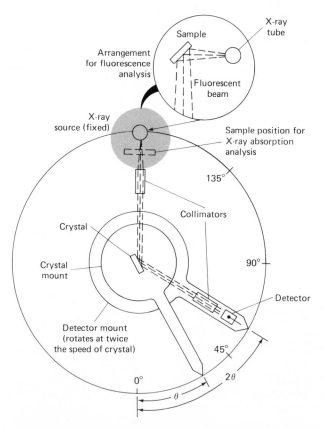

FIGURE 15-9 An X-ray monochromator and detector. Note that the angle of the detector with respect to the beam (2θ) is twice that of the crystal face. For absorption analysis, the source is an X-ray tube and the sample is located in the beam as shown. For emission work, the sample becomes a fluorescent source of X-rays as shown in the insert.

collimated incident beam. From Equation 15-6, it is evident that any given angular setting of the goniometer, only a few wavelengths are diffracted (λ, $\lambda/2$, $\lambda/3$, ..., λ/n, where $\lambda = 2d \sin \theta$). Thus, an X-ray monochromator does not disperse an entire spectrum simultaneously as does a grating or prism; instead, a particular wavelength is diffracted only when the goniometer is set at an appropriate angle.

In order to derive a spectrum, it is necessary that the exit beam collimator and the detector be mounted on a second table that rotates at twice the rate of the first; that is, as the crystal rotates through an angle θ, the detector must simultaneously move through an angle 2θ. Clearly, the interplanar spacing d for the crystal must be known precisely (Equation 15-6).

The collimators for X-ray monochromators ordinarily consist of a series of closely spaced metal plates or tubes that absorb all but the parallel beams of radiation.

X-Radiation longer than about 2 Å is absorbed by constituents of the atmosphere. Therefore, provision is usually made for a continuous flow of helium through the sample compartment and monochromator when longer wavelengths are required. Alternatively, provisions may be made to evacuate these areas by pumping.

The loss of intensity is high in a monochromator equipped with a flat crystal because as much as 99% of the radiation is sufficiently divergent to be absorbed in the collimators. Increased intensities, by as much as a factor of ten, have been realized by employing a curved crystal surface which acts not only to diffract but also to focus the divergent beam from the source upon the exit collimator.

As illustrated in Table 15-1, most analytically important X-ray lines lie in the region between about 0.1 and 10 Å. A consideration of data in Table 15-2, however, leads to the conclusion that no single crystal satisfactorily disperses radiation over this entire range. As a consequence, an X-ray monochromator must be provided with at least two (and preferably more) interchangeable crystals.

The useful wavelength range for a crystal is determined by its lattice spacing d and the problems associated with detection of the radiation when 2θ approaches zero or 180 deg. When a monochromator is set at angles of 2θ that are much less than 10 deg, the amount of polychromatic radiation scattered from the surface becomes prohibitively high. Generally, values of 2θ greater than about 160 deg can-

TABLE 15-2 PROPERTIES OF TYPICAL DIFFRACTING CRYSTALS

Crystal	Lattice Spacing d, Å	Wavelength Range[a], Å		Dispersion $d\theta/d\lambda$, deg/Å	
		λ_{max}	λ_{min}	at λ_{max}	at λ_{min}
Topaz	1.356	2.67	0.24	2.12	0.37
LiF	2.014	3.97	0.35	1.43	0.25
NaCl	2.820	5.55	0.49	1.02	0.18
EDDT[b]	4.404	8.67	0.77	0.65	0.11
ADP[c]	5.325	10.50	0.93	0.54	0.09

[a] Based on assumption that the measurable range of 2θ is from 160 deg for λ_{max} to 10 deg for λ_{min}.

[b] Ethylenediamine d-tartrate.

[c] Ammonium dihydrogen phosphate.

not be measured because the location of the source unit prohibits positioning of the detector at such an angle (see Figure 15-9). The minimum and maximum values for λ_{max} in Table 15-2 were determined from these limitations.

It will be seen from Table 15-2 that a crystal such as ammonium dihydrogen phosphate, with a large lattice spacing, has a much greater wavelength range than a crystal in which this parameter is small. The advantage of large values of d is offset, however, by the consequent lower dispersion. This effect can be seen by differentiation of Equation 15-6, which leads to

$$\frac{d\theta}{d\lambda} = \frac{n}{2d \cos \theta}$$

Here, $d\theta/d\lambda$, a measure of dispersion, is seen to be inversely proportional to d. Table 15-2 provides dispersion data for the various crystals at their maximum and minimum wavelengths. The low dispersion of ammonium dihydrogen phosphate prohibits its use in the region of low wavelengths; here, a crystal such as topaz or lithium fluoride must be substituted.

X-Ray Detectors and Signal Processors

Early X-ray equipment employed photographic emulsions for detection and measurement of radiation. For reasons of convenience, speed, and accuracy, however, modern instruments are generally equipped with detectors that convert radiant energy into an electrical signal. Three types of transducers are encountered: gas-filled detectors, scintillation counters, and semiconductor detectors. Before considering how each of these devices functions, it is worthwhile to discuss *photon counting*, a signal processing method, which is commonly employed with X-ray detectors as well as detectors of radiation from radioactive sources (Chapter 16). As was mentioned earlier (p. 146), photon

counting is also beginning to find use in ultraviolet and visible spectroscopy.

Photon Counting. In contrast to the various photoelectric detectors we have thus far considered, X-ray detectors are usually operated as *photon counters*. In this mode, the individual pulse of electricity produced as a quantum of radiation is absorbed by the transducer is counted; the power of the beam is then recorded digitally in terms of number of counts per unit of time. This type of operation requires rapid response times for the detector and signal processor with respect to the rate at which quanta are absorbed by the transducer; thus, photon counting is applicable only to beams of relatively low intensity. As the beam intensity increases, the pulse rate becomes greater than the response time of the instrument, and only a steady-state current, which represents an average number of pulses per second, can be measured.

For weak sources of radiation, photon counting generally provides more accurate intensity data than are obtainable by averaging the pulses and measuring the resulting current. The improvement results from the tendency of signal pulses to be larger than the pulses arising from background noise in the source, detector, and associated electronics; separation of the two can then be achieved with a *pulse height discriminator*, an electronic device that will be discussed in a later section.

Photon counting is used in X-ray work because the power of available sources is often low. (Photon counting has also been profitably applied to weak sources of ultraviolet and visible radiation such as those encountered in Raman spectrometry.) In addition, photon counting permits spectra to be obtained without the use of a monochromator. This property is considered in the section devoted to energy dispersive systems.

Gas-Filled Detectors. When X-radiation passes through an inert gas such as argon, xenon, or krypton, interactions occur that produce a large number of positive gaseous ions and electrons (ion pairs) for each X-ray quantum. Three types of X-radiation detectors, namely, *ionization chambers*, *proportional counters*, and *Geiger tubes*, are based upon the enhanced conductivity resulting from this phenomenon.

A typical gas-filled detector is shown schematically in Figure 15-10. Radiation enters the chamber through a transparent window of mica, beryllium, aluminum, or Mylar. Each photon of X-radiation may interact with an atom of argon, causing it to lose one of its outer electrons. This *photoelectron* has a large kinetic energy, which is equal to the difference between the X-ray photon energy and the binding energy of the electron in the argon atom. The photoelectron then loses this excess kinetic energy by ionizing several hundred additional atoms of the gas. Under the influence of an applied potential, the mobile electrons migrate toward the central wire anode while the slower moving cations are attracted toward the cylindrical metal cathode.

Figure 15-11 shows the effect of applied potential upon the number of electrons that reach the anode of a gas-filled detector for each entering X-ray photon. Several characteristic voltage regions are indicated. At potentials less than V_1, the accelerating force on the ion pairs is low, and the rate at which the positive and negative species separate is insufficient to prevent partial recombination. As a consequence, the number of electrons reaching the anode is smaller than the number produced initially by the incoming radiation.

In the region between V_1 and V_2, the number of electrons reaching the anode is reasonably constant and represents the total number formed by a single photon.

In the region between V_3 and V_4, the number of electrons increases rapidly with applied potential. This increase is the result of secondary ion-pair production caused by collisions between the accelerated electrons and gas molecules; amplification (*gas amplification*) of the ion current results.

In the range V_5 to V_6, amplification of the electrical pulse is enormous but is limited by the positive space charge created as the faster moving electrons migrate away from the slower positive ions. Because of this effect, the number of electrons reaching the anode is independent of the type and energy of incoming radiation and is governed instead by the geometry and gas pressure of the tube.

Figure 15-11 also illustrates that a larger number of electrons is produced by the more energetic, 0.6-Å radiation than the longer wavelength, 5-Å X-rays. Thus, the size of the pulse (the pulse height) is greater for the former than the latter.

The Geiger Tube. The Geiger tube is a gas-filled detector operated in the voltage region

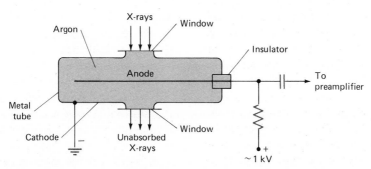

FIGURE 15-10   Cross section of a gas-filled detector.

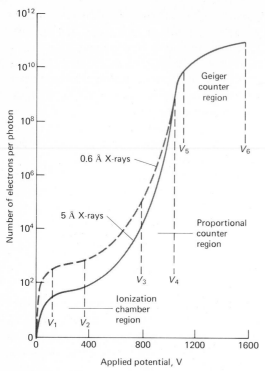

FIGURE 15-11 Gas amplification for various types of gas-filled detectors.

represents an upper limit in the response capability of the tube. Typically, the dead time of a Geiger tube is in the range from 50 to 200 μs.

Geiger tubes are usually filled with argon; a low concentration of an organic substance, often alcohol or methane (a *quench gas*), is also present to minimize the production of secondary electrons when the cations strike the chamber wall. The lifetime of a tube is limited to some 10^8 to 10^9 counts, by which time the quench gas has been depleted.

With a Geiger tube, radiation intensity is determined by a count of the pulses of current. The device is applicable to all types of nuclear and X-radiation. However, it lacks the large counting range of other detectors because of its relatively long dead time; its use in X-ray spectrometers is, therefore, limited.

Proportional Counters. The proportional counter is a gas-filled detector that is operated in the V_3 to V_4 voltage region of Figure 15-11. Here, the pulse produced by a photon is amplified by a factor of 500 to 10,000, but the number of positive ions produced is small enough so that the dead time is only about 1 μs. In general, the pulses from a proportional counter tube must be amplified before being counted.

The number of electrons per pulse (the *pulse height*) produced in the proportional region depends directly upon the energy of the incoming radiation. A proportional counter can be made sensitive to a restricted range of X-ray frequencies with a *pulse-height analyzer*, which counts a pulse only if its amplitude falls within certain limits. A pulse-height analyzer in effect permits electronic filtration of radiation; its function is analogous to that of a monochromator.

Proportional counters have been widely used as detectors in X-ray spectrometers.

Ionization Chambers. Ionization chambers are operated in the voltage range from V_1 to V_2 in Figure 15-11. Here, the currents are small (10^{-13} to 10^{-16} A typically) and relatively independent of applied voltage. Ioniza-

between V_5 and V_6 in Figure 15-11; here, a gas amplification of greater than 10^9 occurs. Each photon produces an avalanche of electrons and cations; the resulting currents are thus large and relatively easy to detect and measure.

The conduction of electricity through a chamber operated in the Geiger region (and in the proportional region as well) is not continuous because the space charge mentioned earlier, terminates the flow of electrons to the anode. The net effect is a momentary pulse of current followed by an interval during which the tube does not conduct. Before conduction can again occur, this space charge must be dissipated by migration of the cations to the walls of the chamber. During the *dead time*, when the tube is nonconducting, response to radiation is impossible; the dead time thus

tion chambers are not employed in X-ray spectrometry because of their lack of sensitivity.

Scintillation Counters. The luminescence produced when radiation strikes a phosphor represents one of the oldest methods of detecting radioactivity and X-rays, and one of the newest as well. In its earliest application, the technique involved the manual counting of flashes that resulted when individual photons or radiochemical particles struck a zinc sulfide screen. The tedium of counting individual flashes by eye led Geiger to the development of gas-filled detectors, which were not only more convenient and reliable but more responsive to radiation as well. The advent of the photomultiplier tube (p. 140) and better phosphors, however, has reversed this trend, and scintillation counting has again become one of the important methods for radiation detection.

The most widely used modern scintillation detector consists of a transparent crystal of sodium iodide that has been activated by the introduction of perhaps 1% thallium. Often, the crystal is shaped as a cylinder that is 3 to 4 in. in each dimension; one of the plane surfaces then faces the cathode of a photomultiplier tube. As the incoming radiation traverses the crystal, its energy is first lost to the scintillator; this energy is subsequently released in the form of photons of fluorescent radiation. Several thousand photons with a wavelength of about 400 nm are produced by each primary particle or photon over a period of about 0.25 μs (the *decay time*). The dead time of a scintillation counter is thus significantly smaller than the dead time of a gas-filled detector.

The flashes of light produced in the scintillator crystal are transmitted to the photocathode of the photomultiplier tube and are in turn converted to electrical pulses that can be amplified and counted. An important characteristic of scintillators is that the number of photons produced in each flash is approximately proportional to the energy of the incoming radiation. Thus, incorporation of a

pulse-height analyzer to monitor the output of a scintillation counter forms the basis of energy dispersive photometers, to be discussed later.

In addition to sodium iodide crystals, a number of organic scintillators such as stilbene, anthracene, and terphenyl have been used. In crystalline form, these compounds have decay times of 0.01 to 0.1 μs. Organic liquid scintillators have also been developed and are used to advantage because they exhibit less self-absorption of radiation than do solids. An example of a liquid scintillator is a solution of p-terphenyl in toluene.

Semiconductor Detectors. Semiconductor detectors have assumed major importance as detectors of X-radiation. These devices are sometimes called *lithium drifted silicon* or *germanium* detectors.

Figure 15-12 illustrates one form of a lithium drifted detector, which is fashioned from a wafer of crystalline silicon. Three layers exist in the crystal, a p-type semiconducting layer that faces the X-ray source, a central *intrinsic* zone, and an n-type layer. The outer surface of the p-type layer is coated with a thin layer of gold for electrical contact; often, it is also covered with a thin beryllium window which is transparent to X-rays. The signal output is taken from an aluminum layer which coats the n-type silicon and is fed into a preamplifier with an amplification factor of about 10. The preamplifier is frequently a field-effect transistor which is made an integral part of the detector.

The detector and preamplifier must be thermostated at the temperature of liquid nitrogen ($-196°C$) to decrease electronic noise to a tolerable level. Furthermore, the performance of the detector degrades badly if allowed to come to room temperature, owing to the tendency of lithium to diffuse rapidly in the silicon. Thus, the liquid nitrogen cryostat is connected to a large (~ 20 liter) Dewar flask, which must be filled every few days.

A lithium drift detector is formed by depositing lithium on the surface of a p-doped sili-

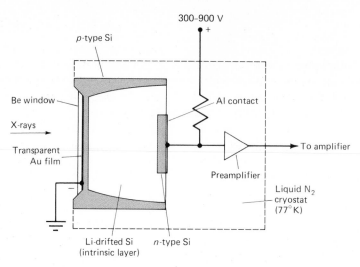

FIGURE 15-12 Vertical cross section of a lithium drifted silicon detector for X-rays and radioactive radiation.

con crystal. Upon heating to 400 to 500°C, the lithium diffuses into the crystal; because this element easily loses electrons, its presence converts the *p* region to an *n*-type. While still at an elevated temperature, a dc potential is applied across the crystal to cause withdrawal of electrons from the lithium layer and holes from the *p*-type layer. Current passage across the *np* junction requires migration (or drifting) of lithium *ions* into the *p* layer and formation of the intrinsic layer where the lithium ions replace the holes lost by conduction. Upon cooling, this central layer has a high resistance relative to the other layers because the lithium ions in this medium are less mobile than the holes they displaced.

The intrinsic layer of a silicon detector functions in a way that is analogous to argon in the gas-filled detector. Initially, absorption of a photon results in formation of a highly energetic photoelectron, which then loses its kinetic energy by elevating several thousand electrons in the silicon to a conduction band; a marked increase in conductivity results. When a potential is applied across the crystal,

a current pulse accompanies the absorption of each photon. In common with a proportional detector, the size of the pulse is directly proportional to the energy of the absorbed photons. In contrast to the proportional detector, however, secondary amplification of the pulse does not occur.

Distribution of Pulse Heights from X-Radiation Detectors. To understand the properties of energy dispersive spectrometers, it is important to appreciate that the size of current pulses resulting from absorption of successive X-ray photons of identical energy by the detector will not be exactly the same. Variations arise because the ejection of photoelectrons and their subsequent generation of conduction electrons are random processes governed by the probability law. Thus, a Gaussian distribution of pulse heights around a mean is observed. The breadth of this distribution varies from one type of detector to another, with the drift detector providing a significantly narrower band of pulse heights. It is this property that has made lithium drift detectors so important for energy dispersive X-ray spectroscopy.

Signal Processes and Readout Devices

The signal from the preamplifier of an X-ray spectrometer is fed into a linear fast response amplifier whose amplification can be varied by a factor up to 10,000. The result is a voltage pulse as large as 10 V.

Counters and Scalers. For low counting rates (500 to 1000 counts/min), a simple electromechanical counter suffices. One or more *scalers* are required for higher counting rates, however. A scaler is an electronic device arranged so that its output terminals transmit only some fixed fraction of the total number of input pulses; that is, every second, every fourth, every eighth, and so on. This device reduces the pulses to a sufficiently small number to be accommodated by the counter.

Energy Dispersive Systems

A *discriminator* is an electronic circuit that responds only to pulses having voltage heights above some preset minimum value. All modern solid-state X-ray spectrometers (wavelength dispersive as well as energy dispersive) are equipped with discriminators that reject pulses of about 0.5 V or less (after amplification). In this way, detector and amplifier noise is reduced significantly.

A *pulse height analyzer* or *selector* is an electronic circuit that rejects all pulses with heights below some predetermined minimum level and above a preset maximum level; that is, it rejects all pulses except those that lie within a limited *channel* or *window* of pulse heights. Figure 15-13 provides a schematic diagram of a pulse height analyzer and its

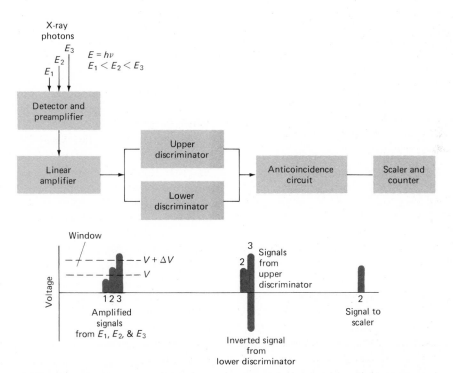

FIGURE 15-13 Schematic diagram of a signal height analyzer. Lower plot shows height of transmitted signals upon exit from various electronic components.

method of operation. The output pulses from the detector and preamplifier are further amplified and appear as voltage signals (in the 10-V range). These signals are fed into the linear pulse-height analyzer, the first stage of which consists of two discriminator circuits. Each discriminator can be set to reject any signal below a certain voltage. As shown in the lower part of Figure 15-13, the upper discriminator rejects signal 1, which is smaller than V in voltage, but transmits signals 2 and 3. The lower discriminator, on the other hand, is set to $V + \Delta V$ and thus rejects all but signal 3. In addition, the lower circuit is so arranged that its output signal is reversed in polarity and thus cancels out signal 3 from the upper circuit in the anticoincidence circuit. As a consequence, only signal 2, with a voltage in the range ΔV, reaches the counter.

Pulse height analyzers are either single- or multiple-channel devices. A single-channel analyzer typically has a voltage range of perhaps 10 V or more with a window of 0.1 to 0.5 V. The window can be manually or automatically adjusted to scan the entire voltage range, thus in effect providing data for energy dispersion.

Multichannel analyzers contain as few as two or as many as several hundred separate channels, each of which acts as a single channel that is set for a different voltage span or width. Such an arrangement permits simultaneous counting and recording of an entire spectrum.

X-RAY FLUORESCENCE METHODS

Although it is feasible to excite an X-ray emission spectrum by incorporating the sample into the target area of an X-ray tube, the inconvenience of this technique discourages its application to many types of materials. Instead, excitation is more commonly brought about by irradiation of the sample with a beam of X-rays from a Coolidge tube or a radioactive source. Under these circumstances, the elements in the sample are excited by absorption of the primary beam and emit their own characteristic fluorescent X-rays. This procedure is thus properly called an *X-ray fluorescence* or *emission* method. X-Ray fluorescence is perhaps the most widely used of all analytical methods for the qualitative identification of elements having atomic numbers greater than oxygen (>8); in addition, it is often employed for semiquantitative or quantitative determination of these elements.

Instruments

Various combinations of the instrument components discussed in the previous section lead to several recognizable types of X-ray fluorescence instruments. The three basic types are *wavelength dispersive*, *energy dispersive*, and *nondispersive*; the latter two can be further subdivided depending upon whether a Coolidge tube or a radioactive substance serves as a spectral source.

Wavelength Dispersive Instruments. Wavelength dispersive instruments always employ tubes as a source because of the large energy losses suffered when an X-ray beam is collimated and dispersed into its component wavelengths. Radioactive sources produce X-ray photons at a rate less than 10^{-4} that of a Coolidge tube; the added attenuation by a monochromator would then result in a beam that was difficult or impossible to detect and measure accurately.

Wavelength dispersive instruments are of two types, *single-channel* or *sequential*, and *multichannel* or *simultaneous*. The spectrometer shown in Figure 15-9 (p. 436) is a sequential instrument that can be readily employed for X-ray fluorescence analysis; here, the Coolidge tube and sample are arranged as shown in the circular insert at the top of the figure. Single-channel instruments may be manual or automatic. The former are entirely satisfactory for the quantitative determination of a few elements. In this application, the crystal and detector are set at the proper angles (θ and 2θ) and counting is continued until sufficient counts have accumulated for an accurate analysis. Automatic instruments are much more convenient for qualitative analysis, where an entire spectrum must be swept. Here, the electric drive for the crystal and detector is synchronized with the motor of a recorder while the detector output determines the position of the pen.

Most modern single-channel spectrometers are provided with two X-ray sources; typically, one has a chromium target for longer wavelengths and the other a tungsten target for shorter. For wavelengths longer than 2 Å, it is necessary to remove air between the source and detector by pumping or by displacement with a continuous flow of helium. A means must also be provided for ready interchange of dispersing crystals.

Recording single-channel instruments cost approximately $40,000.

Multichannel instruments are large, expensive ($\sim$ $150,000) installations that permit the simultaneous detection and determination of as many as 24 elements. Here, individual channels consisting of an appropriate crystal and a detector are arranged radially around an X-ray source and sample holder. Ordinarily, the crystals for all or most of the channels are fixed at an appropriate angle for a given analyte line; in some instruments, one or more of the crystals can be moved to permit a spectral scan.

Each detector in a multichannel instrument is provided with its own amplifier, pulse height selector, scaler, and counter or integrator. These instruments are ordinarily equipped with a computer for instrument control, data processing, and display of analytical results. An analysis for 20 or more elements can be completed in a few seconds to a few minutes.

Multichannel instruments are widely used for the determination of several components in materials of industry such as steel, other alloys, cement, ores, and petroleum products.

Both multichannel and single-channel instruments are equipped to handle samples in the form of metals, powdered solids, evaporated films, pure liquids, or solutions. Where necessary, the materials are held in a cell with a Mylar or cellophane window.

Energy Dispersive Instruments

As shown in Figure 15-14, an energy dispersive spectrometer consists of a polychromatic

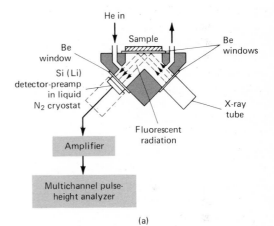

(a)

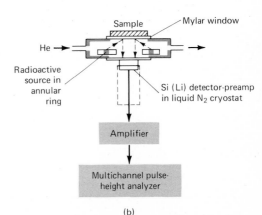

(b)

FIGURE 15-14 Energy dispersive X-ray fluorescence spectrometer. Excitation by X-rays from (a) a Coolidge tube and (b) a radioactive substance.

source, which may be a Coolidge tube or a radioactive material, a sample holder, a lithium-drifted silicon detector, and the various electronic components required for energy discrimination.

An obvious advantage of energy dispersive systems is the simplicity and lack of moving parts in the excitation and detection components of the spectrometer. Furthermore, the absence of collimators and a crystal diffractor, as well as the closeness of the detector to the sample, result in a 100-fold or more increase in energy reaching the detector. These features permit the use of weaker sources such as radioactive materials or low-power X-ray tubes, which are cheaper and less likely to cause radiation damage to the sample.

In a multichannel, energy dispersive instrument, all of the emitted X-ray lines are measured simultaneously. Increased sensitivity and improved signal-to-noise ratio result from the Fellgett advantage (see p. 242).

The principal disadvantage of energy dispersive systems, when compared with crystal spectrometers, is their lower resolutions at wavelengths longer than about 1 Å (at shorter wavelengths, energy dispersive systems exhibit superior resolution).

Nondispersive Instruments

Figure 15-15 is a cutaway view of a simple, commercial, nondispersive instrument which has been employed for the routine determination of sulfur and lead in gasoline. For a sulfur analysis, the sample is irradiated with manganese K-radiation produced by radioactive iron-55 (p. 431), which results in excitation of the sulfur line at 5.4 Å. The fluorescent radiation passes through a pair of adjacent filters and into twin proportional counters.

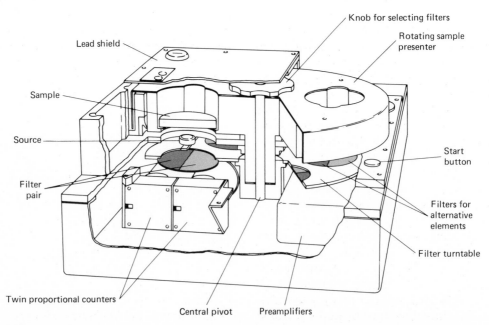

FIGURE 15-15 Cutaway view of a commercial nondispersive X-ray fluorescence instrument. (Reprinted from *Amer. Lab.* **6**(9), 62 (1974). Copyright 1974 by International Scientific Communications, Inc.)

The absorption edge of one of the filters lies just below 5.4 Å, while that of the other is just above it. The difference between the two signals is proportional to the sulfur content of the sample. A sulfur analysis with this instrument requires a counting time of about 1 min. Relative standard deviations of about 1% are obtained for replicate measurements.

Qualitative and Semiquantitative Analysis

Figure 15-16 illustrates an interesting qualitative application of the X-ray method. Here, the untreated sample, which was excited in the X-ray beam by irradiation, was subsequently recovered unchanged. Note that the abscissa for wavelength dispersive instruments is often plotted in terms of the angle 2θ, which can be readily converted to wavelength with knowledge of the crystal spacing of the monochromator (Equation 15-6). Identification of peaks is then accomplished by reference to tables of emission lines of the elements.

Figure 15-17 is a spectrum obtained with an energy dispersive instrument. With such equipment, the abscissa is generally calibrated in channel numbers or energies in keV. Each

dot represents the counts collected by one of the several hundred channels.

Qualitative information, such as that shown in Figure 15-16, can be converted to semiquantitative data by the careful measurement of peak heights. To obtain a rough estimate of concentration, the following relationship is used:

$$P_x = P_s W_x \qquad (15\text{-}7)$$

where P_x is the relative line intensity measured in terms of number of counts for a fixed period, and W_x is the weight fraction of the element in the sample. The term P_s is the relative intensity of the line that would be observed under identical counting conditions if W_x were unity. The value of P_s is determined with a sample of the pure element or a standard sample of known composition.

The use of Equation 15-7, as outlined in the previous paragraph, carries with it the assumption that the emission from the species of interest is unaffected by the presence of other elements in the sample. We shall see that this assumption may not be justified; as a consequence, a concentration estimate based upon the equation may be in error by a factor of two or more. On the other hand, this uncertainty is significantly smaller than that asso-

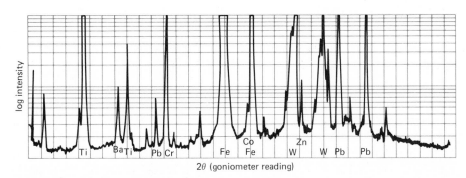

FIGURE 15-16 X-Ray fluorescence spectrum for a genuine bank note recorded with a wavelength dispersive spectrometer. (Taken from H. A. Liebhafsky, H. G. Pfeiffer, E. H. Winslow, and P. D. Zemany, *X-ray Absorption and Emission in Analytical Chemistry.* New York: Wiley, 1960, p. 163. With permission.)

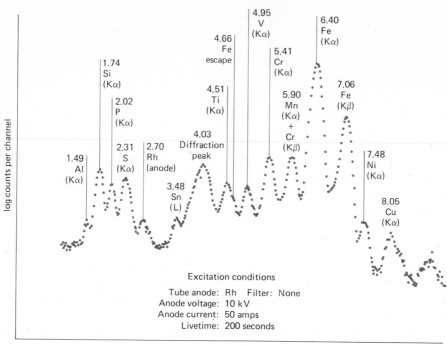

FIGURE 15-17 Spectrum of an iron sample obtained with an energy dispersive instrument with a Rh anode Coolidge tube source. The numbers above the peaks are energies in keV. (Reprinted from *Amer. Lab.* **8**(11), 44, (1976). Copyright 1976 by International Scientific Communications, Inc.)

ciated with a semiquantitative analysis by optical emission where an order of magnitude error is not uncommon.

Quantitative Analysis

Modern X-ray fluorescence instruments are capable of producing quantitative analyses of complex materials with a precision that equals or exceeds that of the classical wet chemical methods or other instrumental methods. For the accuracy of such analyses to approach this level, however, requires either the availability of calibration standards that closely approach the samples in overall chemical and physical composition or suitable methods for dealing with matrix effects.

Matrix Effects. It is important to realize that the X-rays produced in the fluorescence process are generated not only from atoms at the surface of a sample but also from atoms well below the surface. Thus, a part of both the incident beam and the resulting fluorescent beam traverse a significant thickness of sample within which absorption and scattering can occur. The extent to which either beam is attenuated depends upon the mass absorption coefficient of the medium, which in turn is determined by the coefficients of *all* of the elements in the sample. Thus, while the net intensity of a line reaching the detector in an X-ray emission analysis depends upon the concentration of the element producing the line, it is also affected by the concentration and mass absorption coefficients of the matrix elements as well.

Absorption effects by the matrix may cause results calculated by Equation 15-7 to be either high or low. If, for example, the matrix contains a significant amount of an element that absorbs either the incident or the emitted beam more strongly than the element being determined, then W_x will be low, since P_s was evaluated with a standard in which absorption was smaller. On the other hand, if the matrix elements of the sample absorb less than those in the standard, high values for W_x result.

A second matrix effect, called the *enhancement effect*, can also yield results that are greater than expected. This behavior is encountered when the sample contains an element whose characteristic emission spectrum is excited by the incident beam, and this spectrum in turn causes a secondary excitation of the analytical line.

Absorption and enhancement effects can clearly cause the intensity of an analytical line to depend not only upon the concentration of the element of interest but also, to a lesser degree, upon the concentrations of the various other elements in the sample matrix. Several techniques have been developed to compensate for such effects.

Calibration Against Standards. Here, the relationship between the analytical line intensity and the concentration is determined empirically with a set of standards that closely approximate the samples in overall composition. The assumption is then made that absorption and enhancement effects are identical for both samples and standards, and the empirical data are employed to convert emission data to concentrations. Clearly, the degree of compensation achieved in this way depends upon the closeness of the match between the samples and the standards.

Use of Internal Standards. In this procedure, an element is introduced in known and fixed concentration into both the calibration standards and the samples; the added element must be absent in the original sample. The ratio of the intensities between the element being determined and the internal standard serves as the analytical parameter. The assumption here is that absorption and enhancement effects are the same for the two lines and that use of intensity ratios compensates for these effects.

Dilution of Sample and Standards. Here, both sample and standards are diluted with a substance that absorbs X-rays only weakly (that is, a substance containing elements with low atomic numbers). Examples of such diluents include water; organic solvents containing carbon, hydrogen, oxygen, and nitrogen only; starch; lithium carbonate; alumina; and boric acid or borate glass. By employing an excess of diluent, matrix effects become essentially constant for the diluted standards and samples, and adequate compensation is achieved. This procedure has proved particularly useful for mineral analyses, where both samples and standards are dissolved in molten borax; after cooling, the fused mass is excited in the usual way.

Some Quantitative Applications of X-Ray Fluorescence. With proper correction for matrix effects, X-ray fluorescence spectrometry is perhaps the most powerful tool available to the chemist for the rapid quantitative determination of all but the lightest elements in complex samples. For example, Baird and Henke[3] have demonstrated that nine elements can be determined in samples of granitic rocks in an elapsed time, including sample preparation, of about 12 min. The precision of the method is better than wet chemical analyses and averages 0.08% relative. It is noteworthy that one of the elements analyzed is oxygen, which ordinarily can be determined by difference only.

X-Ray methods also find widespread application for quality control in the manu-

---

[3] A. K. Baird and B. L. Henke, *Anal. Chem.*, **37**, 727 (1965).

facture of metals and alloys. Here, the speed of the analysis permits correction of the composition of the alloy during its manufacture.

X-Ray emission methods are readily adapted to liquid samples. Thus, as mentioned earlier, methods have been devised for the direct quantitative determination of lead and bromine in aviation gasoline samples. Similarly, calcium, barium, and zinc have been determined in lubricating oils by excitation of fluorescence in the liquid hydrocarbon samples. The method is also convenient for the direct determination of the pigments in paint samples.

X-Ray fluorescence methods are being widely applied to the analysis of atmosphere pollutants. For example, one procedure for detecting and determining contaminants involves drawing an air sample through a stack consisting of a micropore filter for particulates and three filter paper disks impregnated with orthotolidine, silver nitrate, and sodium hydroxide, respectively. The latter three retain chlorine, sulfides, and sulfur dioxide in that order. The filters then serve as samples for X-ray fluorescence analysis.

Summary of the Advantages and Disadvantages of X-Ray Fluorescence Methods. X-Ray fluorescence offers a number of impressive advantages. The spectra are relatively simple; spectral line interference is thus unlikely. Generally, the X-ray method is nondestructive and can be used for the analysis of paintings, archeological specimens, jewelry, coins, and other valuable objects without harm to the sample. Furthermore, analyses can be performed on samples ranging from a barely visible speck to a massive object. Other advantages include the speed and convenience of the procedure, which permits multielement analyses to be completed in a few minutes. Finally, the accuracy and precision of X-ray fluorescence methods often equal or exceed those of other methods.

X-Ray fluorescence methods are generally not so sensitive as the various optical methods that have been discussed earlier in this text. In the most favorable cases, concentrations of a few parts per million can be measured. More commonly, however, the concentration range of the method will be from perhaps 0.01 to 100%. X-Ray fluorescence methods for the lighter elements are inconvenient; difficulties in detection and measurement become progressively worse as atomic numbers become smaller than 23 (vanadium), in part because a competing process, called Auger emission, reduces the fluorescent intensity (p. 454). Present commercial instruments are limited to atomic numbers of 8 (oxygen) or 9 (fluorine). Another disadvantage of the X-ray emission procedure is the high cost of instruments, which range from about $5000 for an energy dispersive system with a radioactive source, to more than $100,000 for automated and computerized wavelength dispersive systems.

X-RAY DIFFRACTION METHODS

Since its discovery in 1912 by von Laue, X-ray diffraction has provided a wealth of important information to science and industry. For example, much that is known about the arrangement and the spacing of atoms in crystalline materials has been directly deduced from diffraction studies. In addition, such studies have led to a much clearer understanding of the physical properties of metals, polymeric materials, and other solids. X-Ray diffraction is currently of prime importance in elucidating the structures of such complex natural products as steroids, vitamins, and antibiotics. Such applications are beyond the scope of this text.

X-Ray diffraction also provides a convenient and practical means for the qualitative identification of crystalline compounds. This application is based upon the fact that an X-ray diffraction pattern is unique for each crystalline substance. Thus, if an exact match can be found between the pattern of an unknown and an authentic sample, chemical identity can be assumed. In addition, diffraction data sometimes yield quantitative information concerning a crystalline compound in a mixture. The method may provide data that are difficult or impossible to obtain by other means as, for example, the percentage of graphite in a graphite-charcoal mixture.

Identification of Crystalline Compounds by X-Ray Diffraction

Sample Preparation. For analytical diffraction studies, the crystalline sample is ground to a fine homogeneous powder. In such a form, the enormous number of small crystallites are oriented in every possible direction; thus, when an X-ray beam traverses the material, a significant number of the particles can be expected to be oriented in such ways

as to fulfill the Bragg condition for reflection from every possible interplanar spacing.

Samples may be held in the beam in thin-walled glass or cellophane capillary tubes. Alternatively, a specimen may be mixed with a suitable noncrystalline binder and molded into a suitable shape.

Photographic Recording. The classical, and still widely used, method for recording powder diffraction patterns is photographic. Perhaps the most common instrument for this purpose is the *Debye-Scherrer* powder camera, which is shown schematically in Figure 15-18. Here, the beam from a Coolidge tube is filtered to produce a nearly monochromatic beam (often the copper or molybdenum K_x line), which is collimated by passage through a narrow tube. The undiffracted radiation then passes out of the camera via a narrow exit tube. The camera itself is cylindrical and equipped to hold a

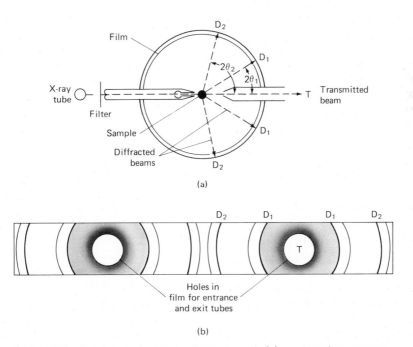

FIGURE 15-18 Schematic diagram of (a) a powder camera; (b) the film strip after development. D_2, D_1, and T indicate positions of the film in the camera.

strip of film around its inside wall. The inside diameter of the cylinder usually is 11.46 cm, so that each lineal millimeter of film is equal to 0.5 deg in θ.

The sample is held in the center of the beam by an adjustable mount.

Figure 15-18b depicts the appearance of the exposed and developed film; each set of lines (D_1, D_2, and so forth) represents diffraction from one set of crystal planes. The Bragg angle θ for each line is easily evaluated from the geometry of the camera.

Electronic Recording. Diffraction patterns can also be obtained with an instrument such as that shown in Figure 15-9. Here again, the fixed source is a filtered Coolidge tube. The sample, however, replaces the single crystal on its mount. The diffraction pattern is then obtained by automatic scanning in the same way as for an emission or an absorption spectrum.

Interpretation of Diffraction Patterns

The identification of a species from its powder diffraction pattern is based upon the position of the lines (in terms of θ or 2θ) and their relative intensities. The diffraction angle 2θ is determined by the spacing between a particular set of planes; with the aid of the Bragg equation, this distance d is readily calculated from the known wavelength of the source and the measured angle. Line intensities depend upon the number and kind of atomic reflection centers that exist in each set of planes.

Identification of crystals is empirical. The American Society for Testing Materials (ASTM) publishes file cards that provide d spacings and relative line intensities for pure compounds; data for nearly 10,000 crystalline materials have been compiled. The cards are arranged in order of the d spacing for the most intense line; cards are withdrawn from this file on the basis of a d spacing that lies within a few hundredths of an ångström of the d spacing of the most intense line for the analyte. Further elimination of possible compounds is accomplished by consideration of the spacing for the second most intense line, then the third, and so forth. Ordinarily, three or four spacings serve to identify the compound unambiguously.

If the sample contains two or more crystalline compounds, identification becomes more complex. Here, various combinations of the more intense lines are used until a match can be found.

By measuring the intensity of the diffraction lines and comparing with standards, a quantitative analysis of crystalline mixtures is also possible.

THE ELECTRON MICROPROBE METHOD

In the electron microprobe method, X-ray emission is stimulated on the surface of the sample by a narrow, focused beam of electrons. The resulting X-ray emission is detected and analyzed with a wavelength or energy dispersive spectrometer.[4]

Instruments

Figure 15-19 is a schematic diagram of an electron microprobe system. The instrument employs three integrated beams of radiation, namely, electron, light, and X-ray. In addition, it requires a vacuum system that provides a pressure of less than 10^{-5} torr and a wavelength- or energy-dispersive X-ray spectrometer (a wavelength dispersive system is shown in Figure 15-19). The electron beam is produced by a heated tungsten cathode and an accelerating anode (not shown). Two electromagnet lenses focus the beam on the specimen;

---

[4] For a detailed discussion of this method, see: L. S. Birks, *Electron Probe Microanalysis*. New York: Interscience, 1963.

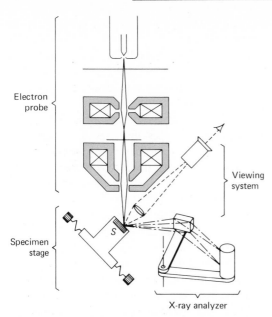

FIGURE 15-19 Schematic view of an electron-microprobe instrument. (From D. B. Wittry, in *Treatise on Analytical Chemistry*, eds. I. M. Kolthoff and P. J. Elving. New York: Interscience, 1964, part I, vol. 5, p. 3178. With permission.)

the diameter of the beam lies between 0.1 and 1 μm. An associated optical microscope is used to locate the area to be bombarded. Finally, the fluorescent X-rays produced by the electron beam are collimated, dispersed by a single crystal, and detected by a gas-filled detector. Considerable design effort is required to arrange the three systems spatially so that they do not interfere with one another.

In addition to the foregoing, the specimen stage is provided with a mechanism whereby the sample can be moved in two mutually perpendicular directions and rotated as well, thus permitting scanning of the surface.

Applications

The electron microprobe provides a wealth of information about the physical and chemical

nature of surfaces. It has had important applications to phase studies in metallurgy and ceramics, the investigation of grain boundaries in alloys, the determination of diffusion rates of impurities in semiconductors, the analysis of occluded species in crystals, and the studies of the active sites of heterogeneous catalysts. In all of these applications, both qualitative and quantitative information about surfaces is obtained.

X-RAY PHOTOELECTRON SPECTROSCOPY

X-Ray photoelectron spectroscopy, sometimes referred to as *ESCA* or electron spectroscopy for chemical analysis, is one of several types of *electron spectroscopy*[5]; others include *ultraviolet photoelectric spectroscopy*, *electron impact spectroscopy*, and *Auger* (pronounced Oh-jay) *spectroscopy*. All use an electron (rather than an X-ray) spectrometer. These instruments differ in that the former sorts electrons according to energy while the latter resolves X-radiation according to energy or wavelength. The inclusion of a brief discussion of ESCA in this chapter is convenient because the method is based upon the same processes that result in X-ray absorption and emission.

Principles of ESCA

As pointed out earlier, absorption of an X-ray photon by an atom generally produces an excited ion and an electron. The process can be described by the equation

$$A + h\nu_1 \rightarrow A^{+*} + e_1 \qquad (15\text{-}8)$$

---

[5] For reviews of electron spectroscopy, see: D. M. Hercules, *Anal. Chem.*, **42** (1), 20A (1970); T. A. Carlson, *Photoelectron and Auger Spectroscopy*. New York: Plenum Press, 1975; and *Handbook of X-Ray and Ultraviolet Photoelectron Spectroscopy*, ed. D. Briggs. Philadelphia: Heyden, 1977.

where A^{+*} represents the excited ion formed when the element A interacts with the X-ray photon $h\nu_1$. Relaxation of the excited ion can occur in two ways. That is,

$$A^{+*} \rightarrow A^+ + h\nu_2 \tag{15-9}$$

or

$$A^{+*} \rightarrow A^{++} + e_2 \tag{15-10}$$

Here, the subscripts indicate that the energies of the resulting photon or electron are different from their counterparts in the first equation.

The first relaxation process, of course, is the basis of X-ray fluorescence methods (Figure 15-20a). The second (Equation 15-10) is called Auger emission and is the basis of another type of electron spectroscopy, which will not be discussed in this text. Auger emission is a radiationless process in which energy is lost by a transfer of one electron to a lower energy state, with a simultaneous ejection of a second electron. For example, if the absorption process involves ejection of a $1s$ electron, Auger relaxation can occur by transition of one $2p$ electron to the $1s$ orbital accompanied by the simultaneous ejection of a second $2p$ electron (Figure 15-20c). Auger spectroscopy is based upon determination of the kinetic energy of the emitted electron E_k' by means of an electron spectrometer.

The two relaxation processes shown by Equations 15-9 and 15-10 are competitive, and their relative rates depend upon the atomic number of the element involved. High atomic numbers favor fluorescence, while Auger emission predominates with atoms of very low atomic numbers. As a consequence, X-ray fluorescence is not a very sensitive means for detecting elements with atomic numbers lower than about 10.

Both X-ray absorption and X-ray photoelectron spectroscopy are based upon Equation 15-8. In the former, an X-ray spectrometer is employed to study absorption as a function of wavelength or energy as the sample is irradiated with a polychromatic source. In the latter, a monochromatic source is employed, and the energy of the emitted electron is measured with an electron spectrometer (Figure 15-20b). The kinetic energy E_k'' of the emitted electron is employed to calculate the binding energy of the electron by means of the equation

$$E_b = h\nu - E_k'' \tag{15-11}$$

The binding energy is unique for each electron of each element and thus is valuable for purposes of identification.

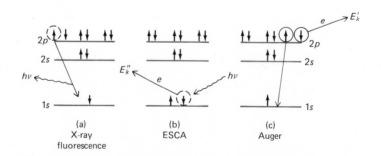

FIGURE 15-20 Mechanisms associated with three types of spectroscopy. In (a), the energy $h\nu$ of the emitted radiation is measured; in (b) and (c), the energy of the emitted electron e is measured.

The binding energies of K and L electrons are influenced to a small extent by the outer bonding electrons since the latter influence the force field of the nucleus. Thus, the binding energy varies slightly as a function of the oxidation state of an element as well as the kind of element or elements with which it is combined.

Figure 15-21 is a photoelectric spectrum for carbon in ethyl trifluoroacetate. Four peaks are observed, each corresponding to the carbon atom shown above it. It is evident from this figure that ESCA can provide useful structural information.

Figure 15-22 indicates the position of peaks for sulfur in its several oxidation states and in various types of organic compounds. The data in the top row clearly demonstrate the effect of oxidation state. Note also in the last four rows of the chart that ESCA discriminates between two sulfur atoms contained in a single ion or molecule. Thus, two peaks are observed for thiosulfate ion ($S_2O_3^{2-}$), suggesting different oxidation states for the two sulfur atoms contained therein.

It should be pointed out that the information obtained by ESCA must also be present in the absorption edge of an X-ray absorption spectrum for a compound. Most X-ray spectrometers, however, do not have sufficient resolution to permit ready extraction of the information.

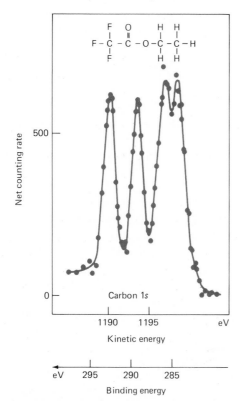

FIGURE 15-21 Carbon 1s X-ray photoelectron spectrum for ethyl trifluoroacetate. (From K. Siegbahn, *et al.*, *ESCA: Atomic, Molecular, and Solid-State Structure by Means of Electron Spectroscopy*. Upsala: Almqist and Wilcksells, 1967, p. 21. With permission.)

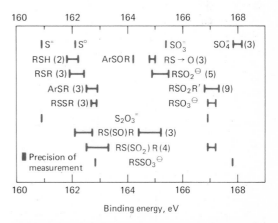

FIGURE 15-22 Correlation chart for sulfur 2s electron binding energies. The numbers in parentheses indicate the number of compounds examined. [Reprinted with permission from D. M. Hercules, *Anal. Chem.*, **42** (1), 20A (1970). Copyright by the American Chemical Society.]

Instrumentation

X-Ray photoelectron spectrometers are large and expensive pieces of equipment. The source is an ordinary Coolidge tube. Monochromatic radiation is obtained with filters (see Figure 15-8) or with a crystal monochromator. The electrons produced by irradiation of the sample are then passed into an electron spectrometer. The most common type of spectrometer utilizes one or more curved magnetic fields to focus the electron beam on a detector. The trajectory of an electron depends upon its kinetic energy and the strength of the field. Variation in the latter permits scanning of the electron energy spectrum.

Because the electron path in a spectrometer is influenced by the earth's magnetic field, elaborate shielding and Helmholtz coils must be employed to reduce the effect of this field to essentially zero. The entire instrument must be operated at a pressure of 10^{-5} to 10^{-8} torr to prevent absorption of electrons by gaseous species.

Applications

The photoelectrons produced in ESCA are incapable of passing through more than perhaps 20 to 50 Å of a solid. Thus, the most important applications of electron spectroscopy, like X-ray microprobe spectroscopy, are for the accumulation of information about surfaces. Examples of some of its uses include identification of active sites and poisons on catalytic surfaces, determination of surface contaminants on semiconductors, analysis of the composition of human skin, and study of oxide surface layers on metals and alloys.

It is evident that the method has a substantial potential in the elucidation of chemical structure (see Figures 15-21 and 15-22); the information obtained appears comparable to that from NMR or infrared spectroscopy. A noteworthy attribute is the ability of ESCA to distinguish among oxidation states of an element.

PROBLEMS

1. What is the short-wavelength limit of the continuum produced by a Coolidge tube having a chromium target and operated at 75 kV?

2. What minimum tube voltage would be required to excite the K and the L series of lines for (a) Cr, (b) Cs, (c) W, (d) U?

3. The $K_{\alpha 1}$ lines for Ca, Zn, Zr, and Sn occur at 3.36, 1.44, 0.79, and 0.49 Å, respectively. Calculate an approximate wavelength for the K_α lines of (a) Ti, (b) Fe, (c) As, (d) Ag, (e) I, (f) Br.

4. The L_α lines for Ca, Zn, Zr, and Sn are found at 36.3, 11.9, 6.07, and 3.60 Å, respectively. Estimate the wavelengths for the L_α lines for the elements listed in Problem 3.

5. The mass absorption coefficient for Ni, measured with the Cu K_α line, is 49.2 cm^2/g. Calculate the thickness of a nickel foil that was found to transmit 27.3% of the incident power of a beam of Cu K_α radiation.

6. For Mo K_α radiation (0.711 Å), the mass absorption coefficients for K, I, H, and O are 16.7, 39.2, 0.0, and 1.50 cm^2/g, respectively.

   (a) Calculate the mass absorption coefficient for a solution prepared by mixing 5.00 g of KI with 95 g of water.

   (b) The density of the solution described in (a) is 1.04 g/cm^3. What fraction of the radiation from a Mo K_α source would be transmitted by a 0.50-cm layer of the solution?

7. Aluminum windows are to be employed as windows for a cell for X-ray absorption measurements with the Ag K_α line. The mass absorption coefficient for aluminum at this wavelength is 2.74. What maximum thickness of aluminum foil could be employed to fashion the windows if no more than 1.0% of the radiation is to be absorbed by them?

8. A solution of I_2 in ethanol had a density of 0.794 g/cm^3. A 1.5-cm layer was found to transmit 27.3% of the radiation from a Mo K_α source. Mass absorption coefficients for I, C, H, and O are 39.3, 0.70, 0.00, and 1.50, respectively.

   (a) Calculate the percent I_2 present, neglecting absorption by the alcohol.

   (b) Correct the results in part (a) for the presence of alcohol.

9. Calculate the goniometer setting, in terms of 2θ, required to observe the $K_{\alpha 1}$ lines for Fe (1.76 Å), Se (0.992 Å), and Ag (0.497 Å) when the diffracting crystal is

   (a) topaz.

   (b) LiF.

   (c) NaCl.

10. Calculate the goniometer setting, in terms of 2θ, required to observe the $L_{\beta 1}$ lines for Br at 8.126 Å when the diffracting crystal is

    (a) ethylenediamine d-tartrate.

    (b) ammonium dihydrogen phosphate.

11. Calculate the minimum tube voltage required to excite the following lines. The numbers in parentheses are the wavelengths in Å for the corresponding absorption edges.

    (a) K lines for Ca (3.064)

    (b) L_α lines for As (9.370)

    (c) L_β lines for U (0.592)

    (d) K lines for Mg (0.496).

16

RADIOCHEMICAL METHODS

The discovery and production of both natural and artificial radioactive isotopes have made possible the development of analytical methods (radiochemical methods) that are both sensitive and specific. These procedures are often characterized by good accuracy and widespread applicability; in addition, some minimize or eliminate chemical separations that often precede the measurement step.[1]

Radiochemical methods are of three types. In *activation analysis*, activity is induced in one or more elements of the sample by irradiation with suitable particles (most commonly thermal neutrons from a nuclear reactor); the resulting radioactivity is then measured. In an *isotope dilution* procedure, a pure but radioactive form of the substance to be determined is mixed with the sample in known amount. After equilibration, a fraction of the component of interest is isolated by suitable means; the analysis is then based upon the activity of this isolated fraction. In a *radiometric analysis*, a

radioactive reagent is employed to separate completely the analyte from the bulk of the sample; the activity of the isolated portion is measured. Alternatively, the analyte may be titrated with a radioactive reagent; here, an end point is established by activity measurements.

THE RADIOACTIVE DECAY PROCESS

The disintegration of radioactive isotopes produces energetic particles and electromagnetic radiation. Alteration of the nucleus accompanies these processes.

Types of Radiation

Table 16-1 lists some of the types of particles and radiation that are encountered in radiochemical studies. Several of these serve as the basis for analysis and are thus discussed in the sections that follow.

Alpha Particles. Alpha particles usually result from the disintegration of isotopes possessing high atomic numbers. An example is shown by the equation

$$^{238}_{92}U \rightarrow ^{234}_{90}Th + ^{4}_{2}He$$

---

[1] For a detailed treatment of radiochemical methods, see: G. Friedlander, J. W. Kennedy, and J. M. Miller, *Nuclear and Radiochemistry*, 2d ed. New York: Wiley, 1964; H. M. Clark, in *Physical Methods of Chemistry*, eds. A. Weissberger and B. W. Rossiter. New York: Wiley-Interscience, 1972, Part IIID, vol. 1, Chapter 9.

TABLE 16-1   CHARACTERISTICS OF COMMON RADIOACTIVE DECAY PRODUCTS

Particle	Symbol	Charge	Mass Number
Alpha	α	+2	4
Electron	β^-	−1	$\dfrac{1}{1840}$
Positron	β^+	+1	$\dfrac{1}{1840}$
Gamma ray	γ	0	0
X-Ray	x	0	0
Neutron	n	0	1
Neutrino	v	0	0

The alpha particle is a helium nucleus and carries a positive charge of 2. Its formation results from the transition of a uranium-238 nucleus to a nucleus of thorium-234. Alpha radiation is generally encountered as a decay product of heavier nuclei only.

Alpha particles from a particular decay process are either monoenergetic or are distributed among relatively few discrete energies. They progressively lose their energy as a result of collisions as they pass through matter, and are ultimately converted into helium atoms through capture of two electrons from their surroundings. Their relatively large mass and charge render alpha particles highly effective in producing ion pairs within the matter through which they pass; this property makes their detection and measurement easy. Because of their high mass and charge, alpha particles have a low penetrating power in matter. The identity of an isotope that is an alpha emitter can often be established by measuring the length (or range) over which the emitted alpha particles produce ion pairs within a particular medium (often air).

Alpha particles are relatively ineffective for producing artificial isotopes because of their low penetrating power.

Beta Particles. Beta particles are produced within a nucleus by the spontaneous transformation of a neutron to a proton or a proton to a neutron. The particles consist of an electron (*negatron*) in the former case and a positive electron (a *positron*) in the latter. Examples of two reactions that yield β rays are

$$^{14}_{6}\text{C} \rightarrow {}^{14}_{7}\text{N} + e^- + v$$
$$^{65}_{30}\text{Zn} \rightarrow {}^{65}_{29}\text{Cu} + e^+ + v$$

where v represents a neutrino, a particle of no analytical significance. A third process that yields negatrons is electron capture; here, an inner electron (usually a K electron) is captured by the nucleus, leaving an excited ion with an atomic number that is one less than that of the original isotope. As shown earlier (p. 431), relaxation of the excited ion can produce negative electrons in the form of Auger electrons (see p. 454).

In contrast to alpha emission, beta decay is characterized by production of particles with a continuous spectrum of energies ranging from nearly zero to some maximum that is characteristic of each decay process. The beta particle is not nearly as effective as the alpha particle in producing ion pairs in matter because of its small mass (about 1/7000 that of an alpha particle); at the same time, its penetrating power is substantially greater. Beta ranges in air are difficult to evaluate because of the high likelihood that scattering will occur. As a result, beta energies are based upon the thickness of an absorber, ordinarily aluminum, required to stop the particle. This thickness is the *range* expressed in mg/cm^2.

Beta particles that carry a unit positive charge are called positrons. Positrons have highly transient lifetimes; their ultimate fate is annihilation by interaction with ordinary electrons to give two gamma photons.

Gamma Ray Emission. Many alpha and beta emission processes leave a nucleus in an excited state, which then returns to the ground state in one or more quantized steps with the release of gamma rays (electromagnetic radiation of very high energy). The gamma ray emission spectrum is characteristic for each nucleus and is thus useful for identifying radioisotopes.

Not surprisingly, gamma radiation is highly penetrating. Upon interaction with matter, gamma rays lose energy by three mechanisms; the one that predominates depends upon the energy of the gamma photon. With low energy gamma radiation, the *photoelectric effect* results in the ejection of a single electron from a high atomic weight target atom. With relatively energetic gamma rays, the *Compton effect* is encountered in which a gamma photon and an electron participate in an elastic collision. The electron acquires only a portion of the photon energy and recoils at a corresponding angle with respect to the projected path of the photon. The photon, now with diminished energy, suffers further energy losses; finally, it causes the photoelectric ejection of an electron from a component of the medium. If the gamma photon possesses sufficiently high energy (at least 1.02 MeV), *pair production* can occur. Here, the photon is converted into a positron and an electron in the field surrounding a nucleus.

X-Ray Emission. Two types of nuclear processes, *electron capture* and *internal conversion*, are followed by the emission of X-ray photons. In the previous chapter (p. 454), it was pointed out that the former yields an excited ion that relaxes by formation of either X-rays or Auger electrons. Which of these processes predominates depends upon the atomic number of the excited species.

Internal conversion is a process in which an excited nucleus (resulting from a decay event) loses its excitation energy by ejecting an electron from one of the orbitals near the nucleus. This alternative to gamma ray emission results in a vacated K, L, or M shell, which is then filled by an electron from a higher energy level. An X-ray photon characteristic of the element results from this transition.

It is important to note that gamma rays and X-rays differ only in their source, the former arising from nuclear events and the latter from electronic transitions outside the nucleus.

Neutrons. The neutron (n) is a particle of unit mass and zero charge. As such, it is a highly effective bombarding particle, being uninfluenced by the electrostatic charge barrier surrounding a target nucleus. In contrast to charged particles, which require high kinetic energies to surmount such barriers, slow (or thermal) neutrons are generally more reactive than high-energy neutrons. It is for this reason that the neutrons emitted from the source (ordinarily a nuclear reactor) are caused to lose much of their kinetic energy through collisions

with a moderating substance of low atomic weight; the result is a neutron flux containing low energies distributed about some average value.

Neutrons can interact with matter in any of several ways. The product (or products) depend in large measure on the energies of the bombarding neutrons. Irradiation of a stable isotope with thermal neutrons is most likely to give rise to a highly excited isotope with an atomic mass number one unit larger than that of the target. The product achieves greater stability through the prompt (within $\sim 10^{-12}$ s) emission of a gamma ray photon, γ. The process can be depicted by the sequence

$$^{A}_{Z}X + ^{1}_{0}n \rightarrow [^{(A+1)}_{Z}X]^* \xrightarrow{10^{-12}\,s} {}^{(A+1)}_{Z}X + \gamma$$

$$\text{Excited state} \qquad\qquad \text{Ground state}$$

where the superscript refers to the mass of element X having an atomic number Z. Fast neutrons interact with matter by other mechanisms, but these reactions are not pertinent to our discussion.

Units of Radioactivity

The *curie* is the fundamental unit of radioactivity; it is defined as that quantity of nuclide in which 3.7×10^{10} disintegrations occur per second. Note that the curie is an enumerative quantity only and provides no information concerning the products of the decay process or their energies.

The *millicurie* and *microcurie* are frequently more convenient units.

The Decay Law

Radioactive decay is a completely random process. Thus, while no prediction can be made concerning the lifetime of an individual nucleus, the behavior of a large ensemble of like nuclei can be described by the expression

$$-\frac{dN}{dt} = \lambda N$$

where N represents the number of radioactive nuclei in the sample at time t and λ is the characteristic *decay constant* for a particular radioisotope. Upon rearranging this equation and integrating over the interval between $t = 0$ and $t = t$ (during which the number of nuclei in the sample decreases from N_0 to N), we obtain

$$\ln \frac{N}{N_0} = -\lambda t \qquad (16\text{-}1)$$

The *half-life* of a radioactive isotope is defined as the time required for the number of atoms to decrease to one-half of its original quantity; that is, for N to become equal to $N_0/2$. Substitution of $N_0/2$ for N in Equation 16-1 gives

$$t_{1/2} = \frac{0.693}{\lambda} \qquad (16\text{-}2)$$

Counting Errors[2]

The randomness inherent in the decay process prevents an exact prediction of the number of disintegrations that will occur in any time interval. Nevertheless, given a sufficiently long counting period, the extent of decay during the interval is found to be reproducible within a predetermined level of precision.[3]

Figure 16-1 shows the deviation from the true average count that would be expected if 1000 replicate observations were made on the same sample. Curve A gives the distribution for a substance for which the true average

count r for a selected period is 5; curves B and C correspond to samples having true averages of 15 and 35. Note that the *absolute* deviations become greater with increases in r, but the *relative* deviations become smaller. Note also that for the smallest number of counts the distribution is distinctly not symmetric around the average; this lack of symmetry is a consequence of the fact that a negative count is impossible, while a finite likelihood always exists that a given count can exceed the average by a factor greater than two.

Standard Deviation of Counting Data. When the total number of counts becomes large ($r >$ 100), the distribution of deviations from the mean approaches that of a symmetrical Gaussian or normal error curve. Under these circumstances, it can be shown that

$$\sigma_N = \sqrt{N} \qquad (16\text{-}3)$$

where N is the number of counts for any given period and σ_N is the standard deviation in N (see Appendix 2). The relative standard deviation $(\sigma_N)_r$ is given by

$$(\sigma_N)_r = \frac{\sigma_N}{N} = \frac{\sqrt{N}}{N} = \frac{1}{\sqrt{N}} \qquad (16\text{-}4)$$

Thus, although the absolute standard deviation increases with the number of counts, the relative standard deviation decreases.

In normal practice, activities of samples are more conveniently expressed in terms of counting rates R (counts per minute) than in counts for some arbitrary time. Clearly,

$$R = \frac{N}{t} \qquad (16\text{-}5)$$

where t is the time in minutes required to obtain N counts. The standard deviation in rate units σ_R can be obtained by dividing both sides of Equation 16-3 by t; thus,

$$\sigma_R = \frac{\sigma_N}{t} = \frac{\sqrt{N}}{t}$$

---

[2] For a more complete discussion, see: G. Friedlander, J. W. Kennedy, and J. M. Miller, *Nuclear and Radiochemistry*, 2d ed. New York: Wiley, 1964, Chapter 8.

[3] The counting period should be short, however, with respect to the half-life so that no significant change in the number of radioactive atoms occurs. Further restrictions include a detector that responds to the decay of a single isotope only and an invariant counting geometry so that the detector responds to a constant fraction of the decay events that occur.

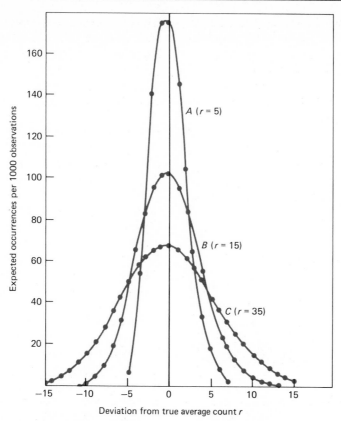

FIGURE 16-1   Distribution of counting data.

Substitution of Equation 16-5 leads to

$$\sigma_R = \sqrt{\frac{R}{t}} \qquad (16\text{-}6)$$

In relative terms,

$$(\sigma_R)_r = \frac{\sqrt{R/t}}{R} = \sqrt{\frac{1}{Rt}} \qquad (16\text{-}7)$$

Uncertainty in a Single Measurement. The standard deviation can be used to define a range about the measured count within which the true average count (or the true average rate) can be expected to be with a given degree of confidence. For a Gaussian distribution, it can be shown that (see Appendix 2)

$$r = N \pm z\sigma_N = N \pm z\sqrt{N} \qquad (16\text{-}8)$$

where r is the true average count and z is a constant that depends upon the confidence level desired. The quantity $\pm z\sigma_N = \pm z\sqrt{N}$ is clearly the absolute uncertainty of the measurement; that is,

$$\text{Absolute uncertainty} = \pm z\sqrt{N} \quad (16\text{-}9)$$

Some values of z for various confidence levels follow:

Confidence level	50%	90%	95%	99%
z	0.68	1.65	1.96	2.58

Thus, the uncertainty associated with a single count N at the 50% confidence level would be

$$z\sigma_N = \pm 0.68\sigma_N = \pm 0.68\sqrt{N}$$

The uncertainty at the 50% confidence level ($\pm 0.68\sqrt{N}$) is often termed the *probable error* of a count. The probable error defines the interval around N within which the true mean r could be expected to be found 50 times out of 100.

The uncertainty in a counting measurement can also be expressed in relative terms. Thus,

$$\text{Relative uncertainty} = \pm z(\sigma_N)_r \quad (16\text{-}10)$$

and from Equation 16-4,

$$\text{Relative uncertainty} = \pm z/\sqrt{N} \quad (16\text{-}11)$$

EXAMPLE

Calculate the absolute and relative uncertainties at the 95% confidence level in the measurement of a sample that produced 675 counts in a given period.

$$\text{Absolute uncertainty} = z\sigma_N$$
$$= \pm 1.96\sqrt{675}$$
$$= \pm 51 \text{ counts}$$

Thus, for 95 out of 100 measurements the true average count r lies in the range 624 to 726. From Equation 16-11,

$$\text{Relative uncertainty} = \pm \frac{1.96}{\sqrt{N}} \times 100 = 7.5\%$$

Figure 16-2 illustrates the relationship between total counts and tolerable levels of uncertainty as calculated from Equation 16-11. Note that the horizontal axis is logarithmic; it is clear that a tenfold decrease in the relative uncertainty requires an approximately hundredfold increase in the number of counts.

The foregoing treatment of uncertainty can be performed equally well in terms of counting rates rather than of total counts; here, σ_R and $(\sigma_R)_r$, as defined by Equations 16-6 and 16-7, are employed.

Background Corrections. The count recorded in a radiochemical analysis includes a contribution from sources other than the sample. Background activity can be traced to the existence of minute quantities of radon isotopes in the atmosphere, to the materials used in construction of the laboratory, to accidental contamination within the laboratory, to cosmic radiation, and to the release of radioactive

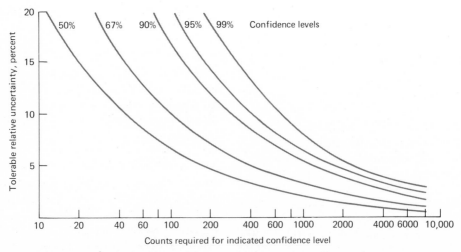

FIGURE 16-2 Relative uncertainty in counting.

materials into the earth's atmosphere. In order to obtain a true assay, then, it is necessary to correct the total count for background. The counting period required to establish the background correction frequently differs from that for the sample; as a result, it is more convenient to employ counting rates. Then,

$$R_c = R_x - R_b \qquad (16\text{-}12)$$

where R_c is the corrected counting rate and R_x and R_b are the rates for the sample and the background, respectively.

The square of the standard deviation of a sum or difference is equal to the sum of the squares of standard deviations of the individual components making up the sum or the difference (see Appendix 1). Thus, we may write that

$$\sigma_c = \sqrt{\sigma_x^2 + \sigma_b^2} \qquad (16\text{-}13)$$

where σ_c is the standard deviation in R_c, and σ_x and σ_b are the corresponding standard deviations for the sample and the background. Substituting Equation 16-6 yields

$$\sigma_c = \sqrt{\frac{R_x}{t_x} + \frac{R_b}{t_b}} \qquad (16\text{-}14)$$

The relative standard deviation $(\sigma_c)_r$ is given by

$$(\sigma_c)_r = \frac{\sqrt{R_x/t_x + R_b/t_b}}{R_x - R_b} \qquad (16\text{-}15)$$

EXAMPLE

A sample yielded 1800 counts in a 10-min period. Background was found to be 80 counts in 4 min. Calculate the relative uncertainty in the corrected counting rate at the 95% confidence level.

$$R_x = \frac{1800}{10} = 180 \text{ cpm}$$

$$R_b = \frac{80}{4} = 20 \text{ cpm}$$

Substituting into Equation 16-15 yields

$$(\sigma_c)_r = \frac{\sqrt{180/10 + 20/4}}{180 - 20} = 0.0300$$

At the 95% confidence level,

Relative uncertainty $= z(\sigma_c)_r$

$$= 1.96 \times 0.0300$$

$$= 0.059 \quad \text{or} \quad 5.9\%$$

Thus, the chances are 95 in 100 relative that the uncertainty in the corrected count is smaller than 5.9% relative.

This example suggests that the contribution of background to the standard deviation is minimized under circumstances where the background counting rate is small with respect to the sample rate. It can be shown that the optimum division of time between background and sample counting is given by

$$\frac{t_b}{t_x} = \sqrt{\frac{R_b}{R_x}} \qquad (16\text{-}16)$$

Instrumentation

Radiation from radioactive sources can be detected and measured in essentially the same way as X-radiation (Chapter 15, p. 438). Gas-filled detectors, scintillation counters, and semi-conductor detectors are all sensitive to alpha, beta, and gamma rays because absorption of these particles produces photoelectrons which can in turn produce thousands of ion pairs. A detectable electrical pulse is thus produced for each particle observed.

Measurement of Alpha Particles. Samples to be assayed for alpha activity should be as thin as possible to minimize self-absorption losses. Similarly, any window between the sample and counter must also be extremely thin. To eliminate the absorption problem, alpha sources are often sealed and counted in windowless gas flow proportional counters.

As mentioned earlier, alpha spectra consist of discrete energies, which are useful for identification. Pulse height analyzers (p. 443) can be employed for the derivation of energy spectra of alpha emitters.

Measurement of Beta Particles. For beta sources having energies greater than about 0.2 MeV, a uniform layer of the sample is ordinarily counted with thin windowed Geiger or proportional tube counters. For low energy beta emitters, such as carbon-14, sulfur-35, and tritium, liquid scintillation counters (p. 441) are preferable. Here, the sample is dissolved in a solution of the scintillating compound. A vial containing the solution is then placed between two photomultiplier tubes housed in a light-tight container. The output from the two tubes is fed into a *coincidence counter*, an electronic device that records a count only when pulses from the two detectors arrive at the same time. The coincidence counter reduces background noise from the detectors and amplifiers because

of the low probability of such noise affecting both systems simultaneously.

Because beta spectra are ordinarily continuous, pulse height analyzers are less useful.

Measurement of Gamma Radiation. Gamma radiation, which is indistinguishable from X-radiation, is detected and measured by the methods described in Chapter 15. Interference from α- and β-radiation is readily avoided by filtering the beam with a thin window of aluminum or Mylar.

Gamma ray spectrometers are similar to the pulse height analyzers described in the previous chapter. Figure 16-3 shows a typical gamma ray spectrum obtained with a 400-channel analyzer.

Figure 16-4 is a schematic diagram of a *well-type* scintillation counter. Here, the sample is contained in a small vial and placed in a cylindrical hole or well in the scintillating crystal of the counter.

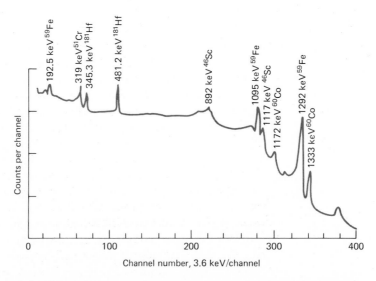

FIGURE 16-3   Gamma ray spectrum of aluminum wire after neutron activation. [Reprinted with permission and taken from S. G. Prussin, J. A. Harris and J. H. Hollander, *Anal. Chem.*, **37**, 1127 (1965). Copyright by the American Chemical Society.]

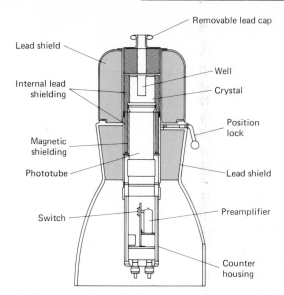

Removable lead cap
Lead shield
Well
Internal lead shielding
Crystal
Position lock
Magnetic shielding
Phototube
Lead shield
Switch
Preamplifier
Counter housing

FIGURE 16-4 A well-type scintillation counter. (Courtesy of Texas Nuclear Division, Ramsey Engineering Co., Austin, Texas. Formerly Nuclear-Chicago Corporation).

NEUTRON ACTIVATION ANALYSIS

The basis for activation analysis is the measurement of the radioactivity induced in a sample as a result of irradiation by nuclear particles (usually thermal neutrons from a reactor). The single most important advantage of activation methods is high sensitivity, which exceeds that of other methods for many elements by a factor of 100 or more; concentration determinations in the parts-per-billion range are common.[4]

Classification of Activation Methods

Activation methods can be classified in several ways. One is based on the type of radiation

employed for excitation of the sample; slow neutrons, fast neutrons, gamma rays, and various charged particles have been used. Most activation methods are based upon thermal neutrons, although gamma excitation appears to be a field of growing interest. We shall confine our discussion to irradiation by thermal neutrons.

A second variable, which is used to characterize activation methods, is the type of emission measured in the final step of the analysis. Here, both beta and gamma radiation have been monitored; the former is often more sensitive but, on the other hand, frequently suffers from being less selective since the radiation is continuous rather than discrete.

Finally, activation methods can be classified as being destructive or nondestructive of the sample. In destructive methods, the irradiated sample is dissolved, and the element of interest is counted after it has been isolated by suitable chemical or physical means; possible interferences from other species made radioactive by the irradiation are thus eliminated. In the nondestructive procedure, the activated sample is counted without preparatory treatment; here, the ability of a gamma ray spectrometer to discriminate between radiation of different energies is called upon to provide the required selectivity. The nondestructive method offers the advantage of great speed. On the other hand, the resolution of a gamma ray spectrometer may be insufficient to eliminate interferences. Also, it does not permit the use of beta emission for an analysis.

Destructive Methods

The most common activation procedure involves solution of a known amount of the irradiated sample followed by separation of the analyte from interferences. The isolated material or a fraction thereof is then counted for its beta or gamma activity.

Conventional neutron activation analysis involves irradiation of a standard containing

---

[4] Monographs on neutron activation analysis include: D. DeSoete, R. Gybels, and J. Hoste, *Neutron Activation Analysis.* New York: Wiley-Interscience, 1972; P. Kruger, *Principles of Activation Analysis.* New York: Wiley-Interscience, 1971.

a known mass w_s of the analyte simultaneously and in the same neutron flux as the sample. Insofar as the activity that results is proportional to mass, and provided also that the other components of the sample do not produce detectable radioactivity, the weight w_x of the element in the sample is given by

$$w_x = \frac{A_x}{A_s} w_s \qquad (16\text{-}17)$$

where A_x and A_s are the activities of the sample and standard. Generally, however, the neutron flux can be expected to generate activity in elements other than the analyte. Thus, chemical isolation of the species of interest from a solution of the sample ordinarily precedes radioassay. When the analyte is present as a trace (as it usually is in a neutron activation analysis), its separation from the major constituents may be difficult and the source of large error. This problem is minimized by introducing a known weight W_x of the element to the solution of the irradiated sample as a nonactive *carrier* or *collector*. Separation of the carrier plus the irradiated element ($W_x + w_x$) is then accomplished by precipitation, extraction, ion exchange, or chromatographic means. A weighted quantity w_x' of the isolated material is counted, and the resulting activity a_x is related to the total activity of the original sample A_x by the relationship

$$a_x = A_x \frac{w_x'}{W_x + w_x} \qquad (16\text{-}18)$$

Ordinarily, the amount of nonactive element added is several orders of magnitude greater than the weight from the sample; that is, $w_x \ll W_x$. Under this circumstance, Equation 16-18 simplifies to

$$a_x = \frac{A_x w_x'}{W_x} \qquad (16\text{-}19)$$

The standard sample is treated in an identical way; thus, an analogous expression can be

written

$$a_s = \frac{A_s w_s'}{W_s} \qquad (16\text{-}20)$$

Substituting these expressions into Equation 16-17 yields

$$w_x = w_s \frac{a_x W_x w_s'}{a_s W_s w_x'} \qquad (16\text{-}21)$$

Where the condition $w_x \ll W_x$ is not satisfied, a more complex equation must be employed.

The Substoichiometric Method.[5] It is experimentally feasible to impose the further conditions that, at the time of assay, $W_x = W_s$ and $w_x' = w_s'$. Equation 16-21 then simplifies to

$$w_x = w_s \frac{a_x}{a_s} \qquad (16\text{-}22)$$

which forms the basis for the substoichiometric method. Where the mass of collector greatly exceeds that of the radioisotope, W_x and W_s are essentially identical provided the same weight of collector is added to the sample and the standard. The requirement that the same total amount of species be taken for radioassay (that is, that $w_x' = w_s'$) is analytically unique, in that the same quantity must be taken from solutions of inherently dissimilar concentration. The problem is resolved by introducing a suitable reagent in an amount insufficient (*substoichiometric*) for complete removal of the species of interest from either sample or standard. If the same amount of this reagent is used for both sample and standard, w_x' and w_s' are identical because their amounts are determined by the amount of added reagent.

Nondestructive Method

In the nondestructive method, a gamma ray spectrometer is used to measure the activities

[5] For a more detailed treatment, see: J. Ruzicka and J. Stary, *Substoichiometry in Radiochemical Analysis*. New York: Pergamon Press, 1968.

of the sample and the standard immediately after irradiation. The weight of the analyte is then calculated directly from Equation 16-17.

Clearly, success of the nondestructive method requires that the spectrometer be able to isolate the gamma ray signal produced by the analyte from signals arising from the other components. Whether or not an adequate resolution is possible depends upon the complexity of the sample, the presence or absence of elements which produce gamma rays of about the same energy as that of the element of interest, and the resolving power of the spectrometer. Improvements in resolving power, which have been made in the last few years, have greatly broadened the scope of the nondestructive method. At the present time, however, the most selective and sensitive activation methods are still based upon isolation of the analyte. The great advantage of the nondestructive approach is its simplicity in terms of sample handling and speed; to be sure, the required instrumentation is more complex.

Application of Neutron Activation

Figure 16-5 illustrates that neutron activation is potentially applicable to the determination of 69 elements. In addition, four of the inert gases form active isotopes with thermal neutrons and thus can also be determined. Finally, three additional elements (oxygen, nitrogen, and yttrium) can be activated with fast neutrons. A list of types of materials to which the method has been applied is impressive and includes metals, alloys, archeological objects, semiconductors, biological specimens, rocks, minerals, and water. Acceptance of evidence developed from activation analysis by courts of law has led to its use in forensic chemistry. Most applications have involved the determination of traces of various elements.

Accuracy. The principal errors that arise in activation analyses are due to self-shielding, unequal neutron flux at sample and standard, counting uncertainties, and errors in counting due to scattering, absorption, and differences in geometry between sample and standard. The errors from these causes can usually be reduced to less than 10% relative; uncertainties in the range of 1 to 3% are frequently obtainable.

Sensitivity. The most important characteristic of the neutron activation method is its remarkable sensitivity for many elements. Note in Figure 16-5, for example, that as little as 10^{-5} μg of several elements can be detected. Note also the wide variations in sensitivities among the elements; thus, about 50 μg of iron are required for detection, in contrast to 10^{-6} μg for europium.

The sensitivity of the activation method for an element is a function of a number of variables. Some of these are associated with the properties of the particular nucleus. Others are related to the irradiation process; still others have to do with the efficiency of the counting apparatus.

The effect of a number of these variables on the activity A induced in a sample after irradiation for a time t is given by the expression

$$A = N\sigma\phi \left[1 - \exp\left(-\frac{0.693\ t}{t_{1/2}} \right) \right] \quad (16\text{-}23)$$

where A is given in counts per second. The quantity N refers to the number of target nuclei, and σ is their neutron capture cross section in cm^2 per nucleus. The neutron flux in units of neutrons per cm^2 second is given by ϕ, while t and $t_{1/2}$ are the irradiation time and half-life of the isotope formed; the two times are expressed in the same units.

The neutron capture cross section is a measure of the probability that a given kind of nucleus will capture a neutron. This quantity depends in a complex way upon the energy of the neutron; typically, one or more neutron energies will correspond to a very high probability for capture.

Figure 16-6 illustrates how the induced activity varies with neutron flux and with irradia-

Estimated sensitivities of neutron activation methods. β (upper) and γ (lower) sensitivities:

Element	β	γ
Na	5×10^{-3}	5×10^{-3}
Mg	5×10^{-1}	5×10^{-1}
Al	1×10^{-1}	1×10^{-2}
Si	5×10^{-2}	500
P	5×10^{-1}	—
S	5	200
Cl	1×10^{-2}	1×10^{-1}
F	—	1
K	5×10^{-2}	5×10^{-2}
Ca	1.0	5
Sc	1×10^{-2}	5×10^{-2}
Ti	5×10^{-1}	5×10^{-2}
V	5×10^{-3}	1×10^{-3}
Cr	—	1
Mn	5×10^{-5}	5×10^{-5}
Fe	50	200
Co	5×10^{-3}	1×10^{-1}
Ni	5×10^{-2}	5×10^{-1}
Cu	1×10^{-3}	1×10^{-1}
Zn	1×10^{-1}	1×10^{-1}
Ga	5×10^{-3}	5×10^{-3}
Ge	5×10^{-3}	5×10^{-2}
As	1×10^{-3}	5×10^{-3}
Se	—	5
Br	5×10^{-3}	5×10^{-3}
Rb	5×10^{-2}	5×10^{-2}
Sr	5×10^{-3}	5×10^{-3}
Zr	1	1
Nb	5×10^{-3}	1
Mo	5×10^{-1}	1×10^{-1}
Ru	1×10^{-2}	5×10^{-2}
Rh	1×10^{-3}	5×10^{-4}
Pd	5×10^{-4}	5
Ag	5×10^{-3}	5×10^{-3}
Cd	5×10^{-2}	5×10^{-3}
In	5×10^{-5}	1×10^{-4}
Sn	5×10^{-1}	5×10^{-1}
Sb	5×10^{-3}	1×10^{-2}
Te	5×10^{-2}	5×10^{-2}
I	5×10^{-3}	1×10^{-2}
Cs	5×10^{-1}	5×10^{-1}
Ba	5×10^{-2}	1×10^{-1}
La	1×10^{-3}	5×10^{-3}
Hf	—	1
Ta	5×10^{-2}	5×10^{-1}
W	1×10^{-3}	5×10^{-3}
Re	5×10^{-4}	1×10^{-3}
Os	5×10^{-2}	—
Ir	1×10^{-4}	1×10^{-3}
Pt	5×10^{-2}	1×10^{-1}
Au	5×10^{-4}	5×10^{-4}
Hg	—	1×10^{-2}
Pb	10	—
Bi	5×10^{-1}	—

Element	β	γ
Ce	1×10^{-1}	1×10^{-1}
Pr	5×10^{-4}	5×10^{-2}
Nd	1×10^{-1}	1×10^{-1}
Sm	5×10^{-4}	5×10^{-3}
Eu	5×10^{-6}	5×10^{-4}
Gd	1×10^{-2}	5×10^{-2}
Tb	5×10^{-2}	1×10^{-1}
Dy	1×10^{-6}	5×10^{-6}
Ho	1×10^{-4}	1×10^{-4}
Er	1×10^{-3}	1×10^{-3}
Tm	1×10^{-2}	1×10^{-1}
Yb	1×10^{-3}	1×10^{-3}
Lu	5×10^{-5}	5×10^{-5}
Th	5×10^{-2}	5×10^{-2}
U	5×10^{-3}	5×10^{-3}

FIGURE 16-5 Estimated sensitivities of neutron activation methods. Upper numbers correspond to β sensitivities in micrograms; lower numbers to γ sensitivities in micrograms. In each case samples were irradiated for 1 hour or less in a thermal neutron flux of 1.8×10^{12} neutrons/cm²/sec. (Data of V. P. Guinn and H. R. Lukens, Jr., in *Trace Analysis: Physical Methods*, ed. G. H. Morrison. New York: John Wiley & Sons, Inc., 1965, p. 345. With permission.)

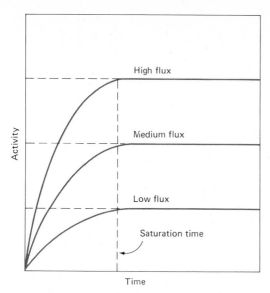

FIGURE 16-6   The effect of neutron flux upon the activity induced in a sample.

tion time. It is seen that irradiation in excess of the *saturation* time does not provide any further increase in activity; here, the rate of decay and the rate of formation of the active species are the same. Equation 16-23 indicates that the irradiation time required to reach saturation increases with increasing half-life for the product nucleus.

The efficiency of chemical recovery, if required prior to radioassay, may limit the sensitivity of an activation analysis. Other factors include the sensitivity of the detection equipment for the emitted radiation, the extent to which activity in the sample decays between irradiation and assay, the time available for counting, and the magnitude of the background count with respect to that for the sample. A high rate of decay is desirable from the standpoint of minimizing the duration of the counting period. Concomitant with high decay rates, however, is the need to establish with accuracy the time lapse between the cessation of irradiation and the commencement of counting. A further potential complication is

associated with counting rates the exceed the resolving time of the detecting system; under these circumstances, a correction must be introduced to account for the difference between elapsed (clock) and live (real) counting times.

ISOTOPIC DILUTION METHODS

Isotopic dilution methods, which predate activation procedures, have been and still are extensively applied to problems in all branches of chemistry. These methods are among the most selective available to chemists.

Both stable and radioactive isotopes are employed in the isotopic dilution technique. The latter are the more convenient, however, because of the ease with which the concentration of the isotope can be determined. We shall limit this discussion to methods employing radioactive species.

Principles of the Isotopic Dilution Procedure

Isotopic dilution methods require the preparation of a quantity of the analyte in a radioactive form. A known weight of this species is then mixed with a weighed quantity of the sample. After treatment to assure homogeneity between the active and nonactive species, a part of the analyte mixture is isolated chemically in the form of a purified compound of known composition. By counting a weighed portion of this product, the extent of dilution of the active material can be calculated and related to the amount of nonactive substance in the original sample. It is important to realize that quantitative recovery of the species is not required. Thus, in contrast to the typical analytical separation, steps can be employed to assure a highly pure product on which to base the analysis. It is this independence from the need for quantitative isolation that leads to the high selectivity of the isotopic dilution method.

Direct Isotope Dilution. Assume that W_o grams of a radioactive species having an activity of A_o are mixed with a sample containing W_x grams of the inactive substance. After separation and purification, a weight W_r of the species is found to have an activity of A_r. We may then write

$$A_r = \frac{A_o W_r}{W_o + W_x} \qquad (16\text{-}24)$$

which rearranges to

$$W_x = \frac{A_o}{A_r} W_r - W_o \qquad (16\text{-}25)$$

Thus, the weight of the species originally present is obtained from the four measured quantities on the right-hand side of Equation 16-25. Where the activity of the tracer is large, the weight W_o added can be kept small, and Equation 16-25 simplifies to

$$W_x = \frac{A_o}{A_r} W_r \qquad (16\text{-}26)$$

Substoichiometric Isotope Dilution. A substoichiometric method, analogous to that for activation analysis, can also be used in isotopic dilution experiments. Here, identical amounts W_o of the tracer are added to two solutions that are the same in every respect except that one contains the sample and the other does not. A suitable reagent is then added to isolate a quantity W_r of the species of interest from each. Care is taken to be sure that the amount of added reagent is, however, less than that required for complete removal of the species; thus, W_r is identical for the two solutions. Equation 16-24 describes the activity of the product from the solution containing the sample; for the solution having no sample, W_x is zero, and Equation 16-24 takes the form

$$A'_r = \frac{A_o W'_r}{W_o} \qquad (16\text{-}27)$$

Recall, however, that conditions have been chosen such that $W_r = W'_r$. As a consequence,

division of Equation 16-24 by Equation 16-27 and rearrangement yields

$$W_x = W_o \left(\frac{A'_r}{A_r} - 1 \right) \qquad (16\text{-}28)$$

The substoichiometric procedure is advantageous when the amount recovered, W_r, is so small that its weight is difficult to assess.

Application of the Isotopic Dilution Method

The isotopic dilution technique has been employed for the determination of about 30 elements in a variety of matrix materials.[6] Substoichiometric methods have proved useful for determining traces of several metallic elements. For example, fractions of a microgram of cadmium, copper, mercury, or zinc have been determined by a procedure in which the element is isolated for counting by extraction with a substoichiometric amount of dithizone in carbon tetrachloride.

Isotopic dilution procedures have been most widely used for the determination of compounds that are of interest in organic chemistry and biochemistry. Thus, methods have been developed for the determination of such diverse substances as vitamin D, vitamin B_{12}, sucrose, insulin, penicillin, various amino acids, corticosterone, various alcohols, and thyroxine.

Isotopic dilution analysis has had less widespread application since the advent of activation methods. Continued use of the procedure

---

[6] It is of interest that the dilution technique has also had nonchemical applications. One application has been to the estimation of the size of salmon spawning runs in Alaskan coastal streams. Here, a small fraction of the salmon are trapped, tagged, and returned to the river. A second trapping then takes place perhaps 10 miles upstream and the fraction of tagged salmon is determined. The total salmon population is readily calculated from this information and from the number originally tagged. The assumption must, of course, be made that the fish population becomes homogenized during its travel between stations.

can be expected, however, because of the relative simplicity of the equipment required. In addition, the procedure is often applicable where the activation method fails.

RADIOMETRIC METHODS[7]

One radiometric method employs a radioactive reagent of known activity to isolate the analyte from the other components of a sample. After quantitative separation, the activity of the product is readily related to the amount of the species being determined. Methods for the radiometric determination of more than 30 of the common elements have been described.

---

[7] For a review of radiometric methods, see: T. Brauer and J. Tölgyessy, *Radiometric Titrations*. New York: Pergamon Press, 1967.

Examples include the determination of chromium by formation of active silver chromate with radioactive silver ion, precipitation of magnesium or zinc by phosphate containing phosphorus-32, and the determination of fluoride ion by precipitation with radioactive calcium.

Radiometric titrations employ a radioactive compound for preparation of the standard solution. Usually, the reaction between the analyte and the standard involves precipitate formation with the activity of the supernatant liquid being monitored as the titration progresses. An example is the titration of silver ion with a bromide solution that is enriched with radioactive bromine. Until the equivalence point is reached, essentially no radioactivity is found in the supernatant liquid. After equivalence, a linear increase in count as a function of volume is observed. Clearly, precautions are needed to assure prompt coagulation and settling out of the precipitate.

PROBLEMS

1. Potassium-42 is a β-emitter with a half-life of 12.36 hr. Calculate the fraction of this isotope remaining in a sample after (a) 2 hr; (b) 5 hr; (c) 30 hr; (d) 60 hr.

2. Calculate the fraction of the following isotopes that remains after 30 hr (half-lifes are given in parentheses):
   (a) yttrium-90 (64 hr)
   (b) silicon-31 (2.6 hr)
   (c) gold-198 (2.69 days)
   (d) zirconium-95 (65 days)

3. A $BaSO_4$ sample contains 1.20 microcurie of barium-128 ($t_{1/2} = 2.4$ days). What storage period is needed to assure that its activity is less than 0.01 microcurie?

4. Estimate the standard deviation and the relative standard deviation associated with counts of (a) 50.0; (b) 500; (c) 5000; (d) 5.00×10^4.

5. Estimate the standard deviation (in absolute and relative terms) associated with a counting rate of 150 cpm that is observed for (a) 40 s; (b) 80 s; (c) 4.0 min; (d) 12.0 min.

6. Estimate the absolute and relative uncertainty associated with a measurement involving 600 counts at the
   (a) 50% confidence level.
   (b) 90% confidence level.
   (c) 99% confidence level.

7. Estimate the absolute uncertainty at the 90% confidence level for a measurement that involves a total count of (a) 64; (b) 200; (c) 422; (d) 1025.

8. Estimate the absolute and relative uncertainty at the 90% confidence level associated with the corrected counting rate obtained from a total counting rate of 250 cpm for 15 min and a background count of
   (a) 9 cpm for 2 min.
   (b) 9 cpm for 10 min.
   (c) 18 cpm for 2 min.
   (d) 40 cpm for 2 min.

9. The background activity of a laboratory when measured for 3 min was found to be approximately 11 cpm. What total count should be taken in order to keep the relative uncertainty at the 90% confidence level smaller than 5.0%, given a total counting rate of about (a) 120 cpm; (b) 250 cpm; (c) 500 cpm?

10. If a total of 25 min is available, calculate the best division of counting time between the background and sample for each of the counting rates in Problem 9. What will be the expected relative standard deviation for the analysis if the only significant uncertainty lies in the counting process?

11. A 2.00-ml solution containing 0.120 microcurie per milliliter of tritium was injected into the bloodstream of a dog. After allowing time for homogenization, a 1.00-ml sample of the blood was found to have a count corresponding to 127 disintegrations per second. Calculate the blood volume of the animal.

12. In order to determine the mercury content of a specimen of animal tissue, a 0.652-g sample of the tissue and a standard solution containing 0.102 μg of Hg as $HgCl_2$ were irradiated for 3 days in a thermal neutron flux of 10^{12} neutrons/cm^2 sec. After irradiation was complete, 20.0 mg of Hg as Hg_2Cl_2 was added to each. Both were digested in a nitric acid/sulfuric acid mixture to oxidize organic material; suitable precautions were taken to avoid loss of mercury by volatilization. Hydrochloric acid was then added, and the $HgCl_2$ formed was distilled from the reaction mixtures. The mercury in each of the distillates was deposited electrolytically on gold foil electrodes, resulting in an increase in weight of 17.6 mg for the sample and 16.5 mg for the standard. The γ activity due to ^{197}Hg was then determined. The sample was found to yield a count of 750 cpm and the standard 1009 cpm. Calculate the ppm Hg in the sample.

13. Identical electrolytic cells, fitted with silver anodes and platinum cathodes, were arranged in series. Exactly 1.00 ml of a solution containing 3.75×10^{-2} mg of KI labeled with ^{131}I (a β-emitter with a half-life of 8.0 days) and 5.00 ml of an $HOAc/OAc^-$ buffer were introduced to one cell. A 5.00-ml aliquot of an iodide-containing sample was added to the acetate buffer contained in the second cell. After passing a substoichiometric quantity of electricity, the anodes were removed

and assayed for their β activity. Calculate the weight of I^- in each milliliter of the sample solution if the activity (corrected for background) for the electrode from the cell containing the standard was 5730 cpm, while that from the cell containing the unknown was 3960 cpm.

17

MASS SPECTROMETRY

A mass spectrum is obtained by converting the compounds of a sample into rapidly moving ions (usually positive) and resolving them on the basis of their mass-to-charge ratio. The utility of mass spectrometry arises from the fact that the ionization process generally produces a family of positive particles whose mass distribution is characteristic of the parent species. As a consequence, a mass spectrum provides information that is useful for elucidating chemical structures. Mass spectral data are easier to interpret than infrared and NMR spectra in some respects, since they provide information in terms of molecular mass of the structural components of a sample; in addition, an accurate measure of the molecular weight of the analyte can usually be obtained from the data.

Mass spectra can also be employed for the quantitative analysis of complex mixtures. Here, the magnitude of ion currents at various mass settings is related to concentration.

Mass spectrometry evolved from studies, carried out at the beginning of this century, that were concerned with the behavior of positive ions in magnetic and electrostatic fields. During the following two decades, the method was refined and provided important information concerning the isotopic abundance of various elements. The first true analytical application of mass spectrometry was described in 1940 when reliable instruments became commercially available. The focus of this early analytical work was toward quantitative determination of the components in complex hydrocarbon mixtures. The mass spectrometer has since proved to be an invaluable tool for this work; it is widely used in the petroleum industry.

Beginning in about 1960, interest in mass spectrometry shifted toward its use for the identification and structural analysis of complex compounds. Within three or four years, it became recognized as a powerful and versatile tool for this purpose—on a par with or perhaps more important than infrared or nuclear magnetic resonance spectroscopy.[1]

THE MASS SPECTROMETER

The principles of mass spectral measurements are simple and easily understood; unfortunately, the simplicity does not extend to the instrumentation. Indeed, the typical high-resolution mass spectrometer is a complex electronic and mechanical device that is expensive in terms of both initial purchase as well as operation and maintenance costs.

Instrument Components

Figure 17-1 is a block diagram showing the major components of a mass spectrometer, which are similar in function to the components of the optical instruments shown in Figure 5-1 (p. 115). The mass analyzer is a dispersing device analogous to the prism or grating of an optical spectrometer; however, it disperses particles rather than electromagnetic radiation from the sample. A characteristic feature of mass spectrometry, which is not encountered in most optical methods, is the need to maintain all of the components leading up to the detector at low pressures (10^{-4} to 10^{-8} torr); thus, the elaborate vacuum systems are an important part of mass spectrometers.

Figure 17-2 shows schematically the essential parts of a typical analytical mass spectrometer. Its operation is based on the following sequence of events. (1) A micromole (or less) of sample is volatilized and allowed to leak slowly into the ionization chamber, which is maintained at a pressure of about 10^{-5} torr. (2) The molecules of the sample are ionized directly or indirectly by a stream of electrons

[1] Reference works on mass spectrometry include: R. W. Kiser, *Introduction to Mass Spectrometry and Its Applications.* Englewood Cliffs, N.J.: Prentice-Hall, 1965; J. Roboz and E. Chait, in *Physical Methods of Chemistry*, eds. A. Weissberger and B. W. Rossitor. New York: Wiley, 1977, Part VI, vol. 1, Chapter 3; J. Roboz, *Mass Spectrometry.* New York: Interscience, 1968; and *Practical Mass Spectrometry*, ed. B. S. Middleditch. New York: Plenum Press, 1979. For a brief review of the field, see: W. V. Ligon, Jr., *Science*, **205**, 151 (1979).

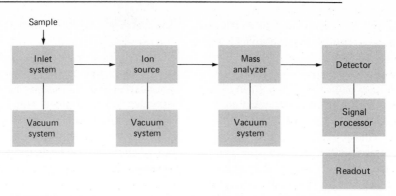

FIGURE 17-1 Components of a mass spectrometer.

flowing from the heated filament toward an anode (both positive and negative ions are formed by impact, but the former predominate; analytical methods are generally based upon positive particles). (3) The positive ions are separated from the negative by the small negative potential at slit *A* and are then accelerated by a potential of a few hundred to a few

thousand volts between *A* and *B*. A collimated beam of positive ions enters the separation area through slit *B*. (4) In the analyzer tube, which is maintained at a pressure of about 10^{-7} torr, the fast-moving particles are subjected to a strong magnetic field which causes them to describe a curved path, the radius of which depends upon their velocity and mass[2]

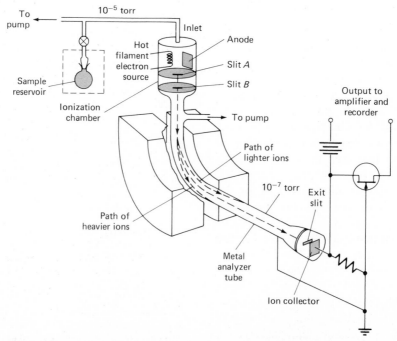

FIGURE 17-2 Schematic diagram of a mass spectrometer.

as well as upon the field strength. Particles of different mass can be focused on the exit slit by varying the accelerating potential or the field strength. (5) The ions passing through the exit slit fall upon a collector electrode; the ion current that results is amplified and recorded as a function of field strength or accelerating potential.

Sample Handling System

The purpose of the sample handling system is to introduce a representative sample of the material to be analyzed into the ion source as a gas at a low and reproducible pressure. The nature of the inlet system will differ depending upon the physical state of the sample; several systems must be available if all types of samples are to be accommodated.

The Batch Inlet. Most commonly, the sample is introduced as a gas from a 1- to 5-liter reservoir (see Figure 17-2). The pressure of the sample in the reservoir is about one to two orders of magnitude greater than that within the ionization chamber, so as to maintain a steady flow through a pinhole into the chamber; sample pressures of 0.01 torr are typical. For liquids boiling below 150°C, a suitable quantity can be evaporated into the evacuated reservoir at room temperature. For less volatile samples, the sample and reservoir may be heated, provided the compound is thermally stable; otherwise, the sample must be introduced directly into the ionization chamber, a procedure that requires special equipment. Gaseous samples are readily handled by expansion of a small volume into the reservoir.

The Direct Probe Inlet. Nonvolatile or thermally unstable materials are often introduced directly into the ion source by means of a sample probe, which is inserted through a vacuum lock. The probe consists of a holder for a small capillary tube or cup that contains a microgram (or less) of the sample. A principal advantage of the probe is the minute quantity of sample required. The probe is equipped with a heater to volatilize the sample; at the low pressure of the ion source, however, much lower temperatures are needed to produce sufficient gaseous molecules of the sample. The use of a probe permits the study of such nonvolatile materials as carbohydrates, steroids, and low-molecular-weight polymeric substances.

Gas Chromatographic Inlet Systems. Sample requirements with regard to volatility and quantity (~ 1 μmol) are similar for both gas chromatography (Chapter 26) and mass spectrometry; thus, the effluent from a chromatographic column can serve as a sample source. Enhanced sensitivity can be realized by separating the sample components from the large excess of carrier gas (usually helium) that is always present. This separation is readily achieved by passing the effluent from the chromatographic column through a narrow tube of porous glass or Teflon which is permeable to helium atoms but not to the larger sample molecules. The latter are then bled into the ion chamber through a pinhole.

The excellent separation qualities of gas chromatography, combined with the powerful identification properties of mass spectrometry, provides the chemist with a most useful tool for analyzing complex mixtures. This application of mass spectrometry is considered in Chapter 26.

Ion Sources

Numerous methods exist for converting molecules into gaseous ions; many find application in mass spectrometry.[3]

---

[2] More correctly, the dependence is on the ratio of mass to charge (m/e). Generally, however, the ions of interest bear only a single charge; thus, the mass is frequently used in lieu of the more cumbersome mass-to-charge ratio.

[3] For a review of ion sources, see: E. M. Chait, *Anal. Chem.*, **44** (3), 77A (1972).

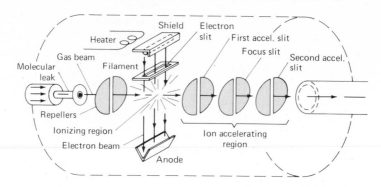

FIGURE 17-3 An ion source. (From R. M. Silverstein, G. C. Bassler, and T. C. Morrill, *Spectrometric Identification of Organic Compounds*, 3d ed. New York: John Wiley & Sons, Inc., 1974 p. 7. With permission.)

Electron Impact Source. The most common way of producing ions for mass spectrometry involves bombarding the sample with a beam of energetic electrons. Figure 17-3 is a schematic diagram of a typical electron ionization source or *ion gun*. The positive ions produced on electron impact are forced through the slit of the first accelerating plate by a small potential difference between this plate and the repeller. The high potential between the first and second accelerators gives the particles their final velocities; the third slit provides further collimation to the beam. In most spectrometers, the potential impressed between the accelerator slits provides the means whereby particles of a particular mass are focused on the collector.

The electron impact source is sufficiently energetic to cause a good deal of fragmentation of molecules, leading to a large number of positive ions of various masses. The complex mass patterns that result are useful for identification.

Chemical Ionization Source.[4] In a chemical ionization source, a reagent gas such as methane is introduced into a modified electron beam source at a pressure of perhaps 1 torr. The sample is then introduced at a concentration that is 10^{-3} to 10^{-4} times that of the reagent. Ionization of the sample is largely a consequence of collisions with reagent ions rather than electrons. Interaction of the electron beam with the reagent gas produces a host of ions, such as CH_5^+, CH_4^+, CH_2^+, H_2^+, and $C_2H_5^+$. Several of these, such as CH_5^+, are strong proton donors and thus react with the analyte molecules to form ions that are one unit larger in mass (the $M + 1$ ion). For example,

$$CH_5^+ + MH \rightarrow MH_2^+ + CH_4$$

where MH is the analyte and MH_2^+ is its $M + 1$ ion. Generally, the chemical ionization source causes less fragmentation of the analyte and thus leads to simpler and more easily interpretable mass spectra.

Chemical ionization requires modification of the spectrometer to sustain the higher pressure in the ion source region.

Spark Source. A spark source permits ion formation from nonvolatile inorganic samples such as metals, semiconductors, and minerals. Here, a radio-frequency spark source, similar to that discussed in Chapter 12, is employed in place of the electron source. Often, the sample may serve as one or both electrodes; alter-

---

[4] For a review of this type of source, see: B. Munson, *Anal. Chem.*, **49** (9), 772A (1977).

natively, it may be mixed with graphite and pressed into an electrode. A radio-frequency voltage of about 30 kV is applied across the electrodes, whereupon the sample vaporizes to form a gaseous ionic plasma. The ions in the plasma are then accelerated into the analyzer by a suitable dc potential.

Field Ionization Sources. Several other ion sources have been developed and may prove to be important. One of these is the *field ionization source*,[5] which consists of a metallic anode in the form of a sharp blade or one or more fine points (in one version, an array of 1000 points spaced 25 μm apart is used) and a cathode which also serves as a slit; the anode and cathode are located 0.5 to 2 mm apart. When a potential of 5 to 20 kV is applied, the gas phase in contact with the fine electrode is subjected to an electrical force field as large as 10^8 V/cm, which suffices to cause ionization of organic compounds.

Field ionization is a gentle technique, in the sense that fragmentation is minimized; the molecular ion and M + 1 ion are often the major products. In structural investigations, it is often advantageous to obtain spectra with both a field ionization or chemical ionization source and an electron impact source; the former provides information about the molecular weight of the substance and the latter provides fragmentation patterns that are useful for identification.

Mass Analyzer

Several arrangements exist for the resolution of ions with different mass-to-charge ratios. Ideally, the analyzer should distinguish between minute mass differences; in addition, it should produce a high level of ion currents. As with an optical monochromator, to which the separator is analogous, these two properties are incompatible; a design compromise

must always be made. Discrimination between integral mass numbers suffices for some applications. For others, however, much higher resolution is necessary; thus, for example, identification of the ions $C_2H_4^+$, CH_2N^+, N_2^+, and CO^+ (mass numbers 28.031, 28.019, 28.006, and 27.995, respectively) requires a resolving power of about 0.01 mass unit.

The main difference among the various mass spectrometers lies in their systems for separating ions.

Single-Focusing Analyzers with Magnetic Deflection. Separators of this kind employ a circular beam path of 180, 90, or 60 deg; the second is illustrated in Figure 17-2. The path described by any given particle represents a balance between the forces that are acting upon it. The magnetic centripetal force F_M is given by

$$F_M = Hev \qquad (17\text{-}1)$$

where H is the magnetic field strength, v is the particle velocity, and e is the charge on the ion. The balancing centrifugal force F_c can be expressed as

$$F_c = \frac{mv^2}{r} \qquad (17\text{-}2)$$

where m is the particle mass and r is the radius of curvature. Finally, the kinetic energy of the particle E is given by

$$E = eV = \tfrac{1}{2}mv^2 \qquad (17\text{-}3)$$

where V is the accelerating voltage applied in the ionization chamber. Note that all particles of the same charge, regardless of mass, are assumed to acquire the same kinetic energy during acceleration in the electrical field. This assumption is only approximately valid since the ions will possess a small distribution of energies before acceleration.

A particle must fulfill the condition that F_M and F_c be equal in order to traverse the circular path to the collector; thus,

$$Hev = \frac{mv^2}{r} \qquad (17\text{-}4)$$

[5] See: M. Anbar and W. Aberth, *Anal. Chem.*, **46**, 61A (1974).

Substituting Equation 17-4 into Equation 17-3 and rearranging gives

$$\frac{m}{e} = \frac{H^2 r^2}{2V} \qquad (17\text{-}5)$$

In most spectrometers, H and r are fixed; therefore, the mass-to-charge ratio of the particle reaching the slit is inversely proportional to the acceleration voltage. Most particles possess a single unit of positive charge; thus, any desired mass can be focused on the exit slit by suitable adjustment of V.

Double-Focusing Analyzers. The ability of the single-focusing instrument to discriminate between small mass differences (that is, its resolving power) is limited by the small variations in the kinetic energy of particles of a given species as they leave the ion source. These variations, which arise from the initial distribution of kinetic energies of the neutral molecules, cause a broadening of the ion beam reaching the collector and a loss of resolving power. In a double-focusing instrument, shown schematically in Figure 17-4, the beam is first passed through a radial electrostatic field. This field has the effect of focusing only particles of the same kinetic energy on slit 2, which then serves as the source for the magnetic separator. Resolution of particles differing by small fractions of a mass

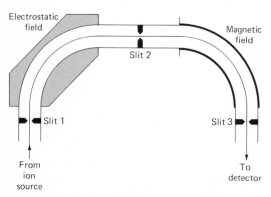

FIGURE 17-4 Design of a double-focusing separator.

unit thus becomes possible. The ion currents produced are extremely small, however, and require large amplification for detection and recording. Double-focusing mass spectrometers are available commercially; their cost is high and their maintenance difficult.

Figure 17-5 shows a second type of double-focusing instrument, which employs Mattauch-Herzog geometry. This geometry, which results in all ions of various mass-to-charge ratios being focused on a single focal plane, is particularly useful for photographic detection. The dispersion along the focal plane is linear with respect to mass-to-charge ratio; thus, identification of lines is readily accomplished if the mass of the particles responsible for two of the lines can be determined.

Time-of-Flight Analyzers. Ion separation in a time-of-flight instrument is achieved by nonmagnetic means. Here, the positive ions are produced intermittently by bombardment with brief pulses of electrons. These pulses, which are controlled by a grid, typically have a frequency of 10,000 Hz and a lifetime of $0.25\ \mu s$. The ions produced are then accelerated by an electrical field pulse that has the same frequency as, but lags behind, the ionization pulse. The accelerated particles pass into a field-free *drift tube* about a meter in length (Figure 17-6). As noted earlier, all particles entering the tube have the same kinetic energies; thus, their velocities in the drift tube must vary inversely with their masses (Equation 17-3), with the lighter particles arriving at the collector earlier than the heavier ones.

The detector in a time-of-flight mass spectrometer is an electron multiplier tube, similar in principle to a photomultiplier tube (p. 140); the output of this tube is fed across the vertical deflection plates of a cathode ray oscilloscope (p. 79), the horizontal sweep is synchronized with the accelerator pulses, and an essentially instantaneous display of the entire mass spectrum appears on the oscilloscope screen.

From the standpoint of resolution, reproducibility, and ease of mass identification, in-

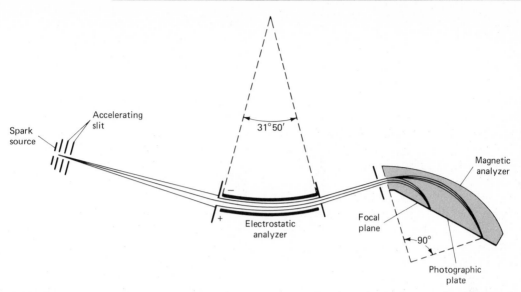

FIGURE 17-5 Double-focusing spark-source mass spectrometer, Mattauch-Herzog type.

struments employing time-of-flight mass separators are less satisfactory than those based upon magnetic focusing. On the other hand, several advantages partially offset these limitations. Included among these are the ruggedness and ease of accessibility to the ion source, which allows the ready insertion of nonvolatile or heat-sensitive samples. The instantaneous display feature is also useful for the study of short-lived species. In general, time-of-flight

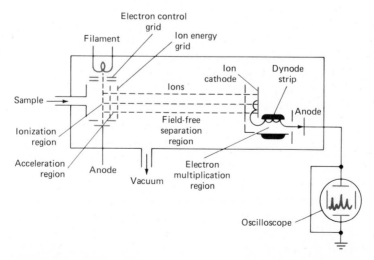

FIGURE 17-6 Schematic diagram of a time-of-flight mass spectrometer.

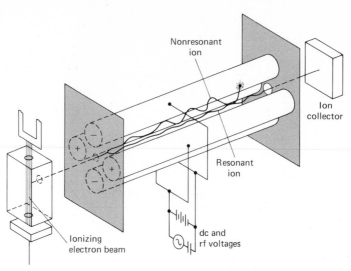

FIGURE 17-7 A quadrupole mass spectrometer. [From D. Lichtman, *Res. Dev.*, **15** (2), 52 (1964). With permission.]

instruments are smaller, more mobile, and more convenient to use than their magnetic focusing counterparts.

Quadrupole Analyzer. The *quadrupole* spectrometer employs four short, parallel metal rods arranged symmetrically around the beam (see Figure 17-7). The opposed rods are connected together, one pair being attached to the positive side of a dc source and the other pair to the negative terminal. In addition, a radiofrequency ac potential is applied to both pairs. Neither field acts to accelerate the positive particles ejected from the ion source. The combined fields, however, cause the particles to oscillate about their central axis of travel; only those with a certain mass-to-charge ratio can pass through the array without being removed by collision with one of the rods. Mass scanning is achieved by varying the frequency of the ac supply or by varying the potentials of the two sources while keeping their ratio constant. Quadrupole instruments are compact and less expensive than magnetic focusing instruments.[6]

Measurement and Display of Ion Currents

The ions from the separator pass through a slit and are collected on an electrode that is well shielded from stray ions. In many instruments, the current produced is passed through a large resistor to ground, and the resulting potential drop is impressed on a field effect transistor or the grid of an electrometer tube. The resulting current is then further amplified before being recorded. Alternatively, the ions from the separator may strike a cathode surface and cause electron emission. The electrons formed are accelerated toward a dynode and, upon impact, produce several additional electrons. This process is repeated several times, as in a photomultiplier tube, to produce a highly

---

[6] For a thorough treatment of quadrupole mass spectrometers, see: *Quadrupole Mass Spectrometry and Its Applications*, ed. P. H. Dawson. New York: Elsevier, 1977.

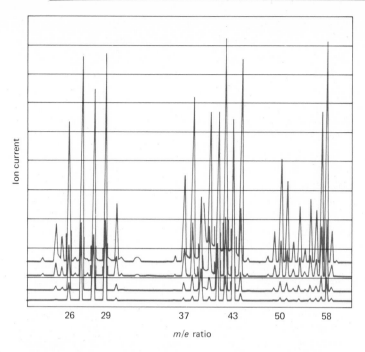

FIGURE 17-8 Mass spectra of *n*-butane recorded simultaneously by four galvanometers of different sensitivities. (From top to bottom, galvanometer sensitivities are in the ratio 30 : 10 : 3 : 1. (Courtesy of E. I. DuPont de Nemours & Company, Wilmington, Delaware. With permission.)

amplified electron current which can then be further amplified electronically and recorded.

Ion currents are ordinarily small and vary over a range of 10^{-17} to 10^{-9} A (the former current corresponds to about 60 ions reaching the detector per second). In order to record peaks of such diverse size, most mass spectrometers are equipped with several pens, with each responding to a different current range. Thus, a peak of convenient size can always be found on the chart regardless of the current magnitude. Some instruments incorporate a set of mirrored galvanometers which record the spectrum on sensitized paper; each galvanometer has a different sensitivity to permit the recording of peaks of widely different sizes. A typical tracing is shown in Figure 17-8.

Computerized Mass Spectrometers

Minicomputers and microprocessors are an integral part of most modern mass spectrometers.[7] A characteristic of a mass spectrum is the wealth of structural data that it provides. For example, a molecule with a molecular weight of 400 to 500 may be fragmented by an electron beam into as many as 200 different ions, each of which leads to a discrete spectral peak. For a structural determination, the heights and mass numbers of each peak must be determined, stored, and ultimately dis-

---

[7] For a detailed discussion of computerized mass spectrometry, see: J. R. Chapman, *Computers in Mass Spectrometry*. New York: Academic Press, 1978.

played. Because the amount of data is so large, it is essential that the data acquisition and processing be rapid; the computer is ideally suited to these tasks. Moreover, in order for mass spectral data to be useful, several instrumental variables must be closely controlled during data acquisition. Computers or microprocessors are much more efficient than a human operator in exercising such controls. These considerations have led to the widespread incorporation of mini- and microcomputers into modern commercial mass spectrometers.

The interface between a mass spectrometer and a computer usually has provisions for digitizing the amplified ion-current signal plus several other signals that are used for control of instrumental variables. Examples of the latter are source temperature, accelerating voltage, scan rate, and magnetic field strength.

The digitized ion-current signal ordinarily requires considerable processing before it is ready for display. First, the peaks must be normalized, a process whereby the height of each peak relative to some reference peak is calculated. Often the largest peak serves as the reference and is arbitrarily assigned a peak height of 100 (or sometimes 1000). Each peak must also be assigned a mass number. This assignment is frequently made on the basis of the time of its appearance and the scan rate. Periodic calibration is necessary; for this purpose, a fluorinated hydrocarbon is used. For high-resolution work, the standard may be admitted with the sample. The computer is then programmed to recognize and employ the peaks of the standard as references for mass measurement. For low-resolution instruments, the calibration must generally be obtained separately from the sample because of the likelihood of peak overlaps.

Variables other than time may be employed for mass assignments. Thus, with quadrupole spectrometers, the mass of the ion reaching the detector is proportional to the ac voltage applied to the rods. Hence, a rod voltage signal

can be digitized and superimposed on the ion-current signal. The computer is then programmed to employ the resulting signal for the assignment of mass. For magnetic instruments, a Hall probe, which measures the magnetic field strength, can serve to provide mass data.

The computer output is often displayed both in digital form and as a graph. Figure 17-9 is an example. The odd columns in the digital display list mass numbers in an increasing order. The even columns contain the corresponding ion currents normalized to the largest peak (the base peak), which is found at mass 156; the current for this particle is assigned the number 1000 and all other peak heights are relative to this one. Thus, the peak at mass 141 is 82.6% of the base peak.

As with infrared spectroscopy, computer-stored library files of mass spectra are available; some commercial instruments are programmed to search these files for spectra that match that of the analyte.

Resolution of Mass Spectrometers

The capability of a mass spectrometer to differentiate between masses is usually stated in terms of its resolution $m/\Delta m$, where m and $m + \Delta m$ are the masses of two particles that give just separable peaks of equal size. Two peaks are considered to be separated if the height of the valley between them is no more than 10% of their height.

The resolution needed in a mass spectrometer depends greatly upon its application. For example, discrimination between particles of the same nominal mass, such as N_2^+ (mass 28.006) and CO^+ (mass 27.995), requires an instrument with a resolution of several thousand. On the other hand, low-molecular-weight particles differing by a unit of mass or more (NH_3^+ and CH_4^+, for example) can be distinguished with an instrument having a resolution smaller than 50. A spectrometer with a resolution of perhaps 250 to 500 is needed for

```
PLOT MS
CCID. A    25    SCAN     32    DATE    8/ 4/71
AQRATE     10    SCTIME    2    RESPWR    600
HIMASS    600    THRESH    1

BARBITURATE

BACKGR     0    BASE      0    SUBTRT    0
IGNORE     0,    0,    0,    0
% F.S.   100    SEQUEN       191
BASE   10012  *2** 0
```

36	3	61	2	83	30	108	4	133	3	166	1
37	3	62	1	84	13	109	14	134	2	167	2
38	12	63	3	85	38	110	9	135	2	168	3
39	139	64	1	86	8	111	10	136	2	169	5
40	38	65	13	87	9	112	73	137	6	179	1
41	364	66	13	88	1	113	21	138	7	181	4
42	83	67	58	91	5	114	18	139	6	183	7
43	378	68	33	92	3	115	2	140	20	185	2
44	84	69	120	93	4	116	1	141	826	191	1
45	6	70	67	94	21	117	1	142	71	193	1
50	5	71	89	95	17	119	2	143	11	195	2
51	11	72	5	96	20	120	1	144	1	197	65
52	14	73	9	97	55	121	2	151	1	198	10
53	64	74	8	98	127	122	4	152	1	199	2
54	31	75	3	99	9	123	5	153	6	204	1
55	162	77	10	100	3	124	7	155	133	207	6
56	30	78	5	101	3	125	4	156	1000	208	2
57	17	79	12	103	1	126	12	157	273	209	1
58	5	80	22	105	2	128	14	158	29	227	6
59	1	81	18	106	3	129	12	159	3	228	1
60	5	82	13	107	2	130	2	165	1		

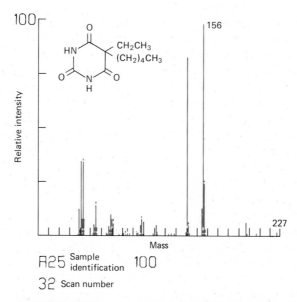

A25 Sample identification 100

32 Scan number

FIGURE 17-9   A computer display of mass-spectral data. The compound was isolated from a blood serum extract by chromatography. The spectrum showed it to be the barbituate, pentobarbital. The instrument was a DuPont Model 21-094 computerized mass spectrometer. (Courtesy of DuPont Instruments, Wilmington, Delaware.)

unit separation where the sample yields large fragments.

Table 17-1 lists typical resolutions and mass ranges for various types of commercially available mass spectrometers. The cost of these instruments is related to their resolution and range. Unit resolution to mass number 250 requires an instrument with a purchase price of roughly $30,000. A high-resolution, double-focusing instrument costs over $100,000. Low-resolution instruments with limited range are priced between $5000 and $10,000.

MASS SPECTRA

Even for relatively simple compounds, the mass spectrum generally contains an array of peaks of differing heights (Figures 17-8 and 17-9). The detailed nature of the spectrum depends upon the properties of the molecule as well as upon the ionization potential, the sample pressure, and the instrument design. Complete in-

terpretation of a spectrum is seldom, if ever, possible.

The Electron Impact Ionization Process

Generation of a mass spectrum requires some minimum electron-beam energy (7 to 15 eV for most organic compounds) to initiate the ionization process

$$M + e \rightarrow M^+ + 2e$$

where M represents the molecule and M^+ is the *molecular ion*, or *parent ion*. Small increases above the minimum in electron-beam energy produce a higher yield of molecular ions, owing to the greater probability for ion-producing collisions. Large increases in beam energy may result in a decrease in the molecular ion peak; here, the added energy causes bond rupture, with the formation of fragments that have smaller masses (and occasionally larger masses as well) than the parent molecule. Typical of these processes are the following, illustrated with the hypothetical molecule ABCD:

$$ABCD + e \rightarrow ABCD^+ + 2e \qquad \text{Molecular ion formation} \qquad (17\text{-}6)$$

$$
\begin{aligned}
ABCD^+ &\rightarrow BCD^\bullet + A^+ \\
&\rightarrow CD^\bullet + AB^+
\begin{cases}
\nearrow B^\bullet + A^+ \\
\searrow A^\bullet + B^+ \\
 D^\bullet + C^+ \nearrow
\end{cases} \\
&\rightarrow AB^\bullet + CD^+ \searrow C^\bullet + D^+
\end{aligned}
\qquad \text{Fragmentation} \qquad (17\text{-}7)
$$

$$
ABCD^+ \rightarrow ADBC^+
\begin{cases}
\nearrow BC^\bullet + AD^+ \\
\searrow AD^\bullet + BC^+
\end{cases}
\qquad \begin{array}{l} \text{Rearrangement} \\ \text{followed by} \\ \text{fragmentation} \end{array} \qquad (17\text{-}8)
$$

$$
ABCD^+ + ABCD \rightarrow (ABCD)_2^+ \rightarrow BCD^\bullet + ABCDA^+ \qquad \begin{array}{l} \text{Collision} \\ \text{followed by} \\ \text{fragmentation} \end{array} \qquad (17\text{-}9)
$$

For a molecule containing a large number of atoms, the number of different positive ions produced can be large. Their distribution depends upon the stability of the precursor ion and the energy imparted to the molecule by the electron beam. Fortunately, when the beam energy exceeds 50 to 70 eV, the pattern of products from a given molecule becomes more or less reproducible.

The neutral fragments in Equations 17-7 to 17-9 are shown as radicals, but they may also occur as molecules. Neither will reach the detector.

The Molecular Ion

The mass of the molecular ion M^+ is identical to the molecular weight of the compound from which it is generated. Thus, the mass of this ion is an important parameter in the identification of a compound. For perhaps 80 to 90% of organic substances, the molecular ion peak is readily recognizable; certain characteristics of this peak must be borne in mind, however.

Isotope Peaks. Table 17-2 lists the natural isotopic abundances of elements that frequently occur in organic compounds. Because of the presence of these elements, there beyond the molecular ion peak (the M + 1 peak), and on occasion at M + 2 as well. Compounds containing chlorine or bromine will clearly have relatively large M + 2 peaks.

Peaks for Collision Products. Ion-molecule collision can produce peaks of higher mass number than the molecular ion peak (Equation 17-9). At ordinary sample pressures, the only important reaction of this type is one in which the collision process transfers a hydrogen atom to the ion; an enhanced M + 1 peak results. The proton-transfer process is a second-order reaction, and the amount of product depends strongly upon the reactant concentration. The height of an M + 1 peak of this type increases at a more rapid rate with increased sample pressure than does the height of other peaks; thus, detection of this reaction is usually possible.

Stability of the Molecular Ion. For a given set of conditions, the intensity of a molecular ion peak depends upon the stability of the ionized particle; a minimum lifetime of about 10^{-5} s is needed for a particle to reach the collector and to be detected. The stability of the ion is strongly affected by structure; the size of molecular ion peaks will thus show great variability.

TABLE 17-1   COMPARISON OF SOME TYPICAL COMMERCIAL MASS SPECTROMETERS

Type	Approximate Mass Range	Approximate Resolution
Double focusing	2–5000	10,000–20,000
	1–240	1000–2500
Single focusing	1–1400	1500
	2–700	500
	2–150	100
Time-of-flight	1–700	150–250
	0–250	130
Quadrupole	2–100	100
	2–80	20–50

In general, the molecular ion is stabilized by the presence of π electron systems, which more easily accommodate the loss of an electron. Cyclic structures also give large parent peaks, since rupture of a bond does not necessarily produce two fragments. In general, the stability of the molecular ion decreases in the following order: aromatics, conjugated olefins, alicyclics, sulfides, unbranched hydrocarbons, mercaptans, ketones, amines, esters, ethers, carboxylic acids, branched hydrocarbons, and alcohols. These effects are illustrated in Table 17-3 (p. 492), which compares the height of the molecular ion peak for some C_{10} compounds relative to the total peak heights in the spectrum.

Figure 17-9 (p. 487) is a spectrum of a compound with an unstable molecular ion. Here, the molecular ion peak at 226 is absent and the M + 1 is barely detectable. The base peak at mass 156, on the other hand, corresponds to a particle formed by the loss of the branched aliphatic side chain.

A mass spectrometer will generally provide a detectable parent ion peak provided the peak is at least 1% of the total; from some instruments, this limit is lowered to 0.1%. Thus, with the exception of some alcohols and branched hydrocarbons, the mass of the molecular ion, and thus the molecular weight of the compound, can be determined.

The Base Peak

The largest peak in a mass spectrum is termed the *base peak*; it is common practice to report peak heights as a fraction of the base peak height. Alternatively, intensities are reported as percentages of the total peak heights, a more informative number, but one that is more laborious to calculate.

Chemical Ionizaton Spectra

Figure 17-10 contrasts the electron-impact ionization spectrum for ephedrine with its chemical ionization counterpart. The electron-impact spectrum (Figure 17-10a) consists largely of peaks corresponding to small fragments of the

TABLE 17-2   NATURAL ABUNDANCE OF ISOTOPES OF SOME COMMON ELEMENTS

Element[a]	Most Abundant Isotope	Abundance of Other Isotopes Relative to 100 Parts of the Most Abundant[b]	
Hydrogen	^1H	^2H	0.016
Carbon	^{12}C	^{13}C	1.08
Nitrogen	^{14}N	^{15}N	0.38
Oxygen	^{16}O	^{17}O	0.04
		^{18}O	0.20
Sulfur	^{32}S	^{33}S	0.78
		^{34}S	4.40
Chlorine	^{35}Cl	^{37}Cl	32.5
Bromine	^{79}Br	^{81}Br	98.0

[a] Fluorine (^{19}F), phosphorus (^{31}P), and iodine (^{127}I) have no additional naturally occurring isotopes.

[b] The numerical entries indicate the average number of isotopic atoms present for each 100 atoms of the most abundant isotope; thus, for every 100 ^{12}C atoms there will be an average of 1.08 ^{13}C atoms.

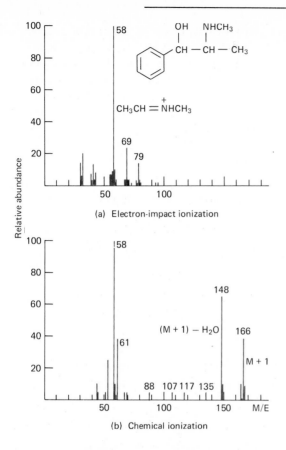

(a) Electron-impact ionization

(b) Chemical ionization

FIGURE 17-10 Mass spectra for ephedrine (molecular weight = 165) by (a) electron impact ionization and (b) chemical ionization with methane as the reagent. (From H. M. Fales, in *Mass Spectrometry*, ed. G. W. A. Milne. New York: John Wiley & Sons, Inc., 1971, p. 198. With permission.)

parent molecule. The molecular ion peak 165 is entirely absent; as a result, the molecular weight of the compound cannot be determined. The base peak at 58 is due to the fragment $(CH_3CH=NHCH_3)^+$.

Figure 17-10b is the chemical ionization spectrum for ephedrine. Here, a strong M + 1 peak is observed, which arises from a reaction such as that shown on page 480. Note also

the M − 1 peak, which is formed from a reaction such as

$$MH + C_2H_5^+ \rightarrow M^+ + C_2H_6$$

Clearly, the molecular weight of the compound can be determined from this spectrum.

QUALITATIVE APPLICATIONS OF MASS SPECTROMETRY

The mass spectrum of a pure compound provides valuable information for qualitative identification purposes. In addition, mass spectrometry has proved useful for identifying the components of simple mixtures—in particular, gaseous mixtures. Some of these applications are outlined in the paragraphs that follow.

Molecular Weight Determination

For most compounds that can be volatilized, the mass spectrometer is unsurpassed for the determination of molecular weight. As we have noted earlier, the method requires the identification of the molecular ion peak (or the M + 1 peak with chemical ionization), the mass of which gives the molecular weight to at least the nearest whole number—an accuracy that cannot be realized by other molecular weight measurements. Caution must be used, however, for occasionally the molecular ion peak may either be absent or so small that it is confused with a peak caused by an impurity (see Figure 17-10a). In addition, collision processes may produce an M + 1 peak that is more intense than the parent ion peak.

No single method is available for establishing unambiguously that the peak of highest mass number (neglecting the small isotope peaks) is indeed produced by the molecular ion. On the other hand, from a series of observations, the experienced mass spectroscopist can ordinarily make this judgment with reasonable assurance of being correct.

As mentioned earlier, it is frequently possible to identify an M + 1 peak by observing its behavior as a function of sample size. Determination of whether or not the highest mass peak is indeed the molecular ion is more troublesome. Here, a knowledge of fragmentation patterns for various types of compounds is essential. For example, a peak at M − 3 immediately casts doubt upon the peak at M being caused by the molecular ion. This pattern could only occur as the result of abstraction of three hydrogen atoms from the molecular ion; such a fragmentation is most unlikely. On the other hand, a strong peak at M − 18 (or M + 1 − 18) suggests that even a weak peak at M may be the parent ion since, for alcohols and aldehydes, the loss of water is a common occurrence; Figure 17-10b provides an example of this effect.

It is of interest to point out that the molecular weight determined by mass spectroscopy will not be identical with that calculated from atomic weights on the chemical scale if the parent compound contains certain elements. For example, methyl bromide will have a peak at mass 96 that is nearly as strong as the one at mass 94 because of the high isotopic abundance of bromine-81 (see Table 17-2).

Determination of Molecular Formulas

Partial or exact molecular formulas can be determined from the mass spectrum of a compound, provided the molecular ion peak can be identified.

Molecular Formulas from High-Resolution Instruments. A unique formula for a compound can often be derived from the exact

TABLE 17-3[a]   **VARIATION IN MOLECULAR ION PEAK WITH STRUCTURE**

Compound	Formula	Relative Peak Height (percent of total peak heights)
Naphthalene		44.3
n-Butylbenzene	$-C_4H_9$	8.3
trans-Decaline		8.2
Diamyl sulfide	$(C_5H_{11})_2S$	3.7
n-Decane	$C_{10}H_{22}$	1.41
Diamylamine	$(C_5H_{11})_2NH$	1.14
Methyl nonanoate	$C_9H_{17}COOCH_3$	1.10
Diamyl ether	$(C_5H_{11})_2O$	0.33
3,3,5-Trimethylheptane	$C_{10}H_{22}$	0.007
n-Decanol	$C_{10}H_{21}OH$	0.002

[a] Taken from K. Biermann, *Mass Spectrometry, Organic Applications*. New York: McGraw-Hill Book Company, Inc., 1962, p. 52. With permission.

mass of the parent ion peak. This application, however, requires a high-resolution instrument capable of detecting mass differences of a few thousandths of a mass unit. Consider, for example, the molecular weights of the following compounds: purine, $C_5H_4N_4$ (120.044); benzamidine, $C_7H_8N_2$ (120.069); ethyltoluene, C_9H_{12} (120.094); and acetophenone, C_8H_8O (120.157). If the measured mass of the parent ion peak is 120.069, then all but $C_7H_8N_2$ are excluded as possible formulas. Tables that list all reasonable combinations of C, H, N, and O by molecular weight to the third decimal place have been compiled.[8] In addition, a table of molecular weights (to the sixth decimal place) of all of the compounds listed in the ninth edition of the *Merck Index* is available.[9]

Formulas from Isotopic Ratios. The data from an instrument that can discriminate between whole mass numbers provides useful information about the formula of a compound, provided only that the molecular ion peak is sufficiently intense that its height and the heights of the (M + 1) and (M + 2) isotope peaks can be determined accurately. The following example illustrates this type of analysis.

EXAMPLE

Calculate the ratios of the (M + 1) to M peak heights for the following two compounds: dinitrobenzene, $C_6H_4N_2O_4$ (M = 168) and an olefin, $C_{12}H_{24}$ (M = 168).

From Table 17-2, we see that for every 100 ^{12}C atoms there are 1.08 ^{13}C atoms. Since there are six carbon atoms in nitrobenzene, however, we would expect there to be 6.48 (6 × 1.08) molecules of nitrobenzene having one ^{13}C atom for every 100 molecules having none. Thus, from this effect alone the (M + 1)

peak will be 6.48% of the M peak. The isotopes of the other elements also contribute to this peak; we may tabulate their effects as follows:

$$C_6H_4N_2O_4$$

^{13}C	6 × 1.08	=	6.48%
^2H	4 × 0.016	=	0.064%
^{15}N	2 × 0.38	=	0.76%
^{17}O	4 × 0.04	=	0.16%
	(M + 1)/M	=	7.46%

$$C_{12}H_{24}$$

^{13}C	12 × 1.08	=	12.96%
^2H	24 × 0.016	=	0.38%
	(M + 1)/M	=	13.34%

It is seen in this example that a measurement of the ratios of the (M + 1) to M peak heights would permit discrimination between two compounds that have identical whole-number mass weights.

The use of relative isotope peak heights for the determination of molecular formulas is greatly expedited with the tables developed by Beynon[10]; a portion of a modified form of his tabulations is shown in Table 17-4. Here, a listing for all reasonable combinations of C, H, O, and N is given for mass numbers 83 and 84 (the original tables extend to mass number 500); also tabulated are the heights of the corresponding (M + 1) and (M + 2) peaks reported as percentages of the M peak. If a reasonably accurate experimental determination of these percentages can be obtained from a spectrum, a likely formula can be ascertained. For example, a molecular ion peak at mass 84 and with (M + 1) and (M + 2) peaks of 5.6 and 0.3% would suggest that the formula of the compound is C_5H_8O (Table 17-4).

[8] J. H. Beynon and A. E. Williams, *Mass and Abundance Tables for Use in Mass Spectrometry.* Amsterdam: Elsevier, 1963.

[9] *Table of Molecular Weights.* Rahway, N. J.: Merck and Co., Inc., 1978.

[10] J. H. Beynon and A. E. Williams, *Mass and Abundance Tables for Use in Mass Spectrometry.* Amsterdam: Elsevier, 1963.

The isotopic ratio is particularly useful for the detection and estimation of the number of sulfur, chlorine, and bromine atoms in a compound because of the large contribution they make to the (M + 2) peak (see Table 17-2). Thus, for example, an (M + 2) peak that is about 65% of the M peak would be strong evidence for a molecule containing two chlorine atoms; an (M + 2) peak of about 4%, on the other hand, would suggest one atom of sulfur. By examination of the heights of the (M + 4) and (M + 6) peaks as well, it is sometimes feasible to identify combinations of chlorine and bromine atoms.

The Nitrogen Rule. The *nitrogen rule* also provides information concerning possible formulas of a compound whose molecular weight has been determined. This rule states that all organic compounds with an even molecular weight must contain zero or an even number of nitrogen atoms; all compounds with odd molecular weights must have an odd number of nitrogen atoms. The fragments formed by cleavage of one bond, however, have an odd mass number if they contain zero or an even number of nitrogen atoms; conversely, such fragments have an even mass number if the nitrogens are odd in number. The rule is a direct consequence of the fact that, with the exception of nitrogen, the valency and the mass number of the isotopes of elements that commonly occur in organic compounds are either both even or both odd. The nitrogen rule applies to all covalent compounds containing carbon, hydrogen, oxygen, sulfur, the halogens, phosphorus, and boron.

Identification of Compounds from Fragmentation Patterns

From Figure 17-11, it is evident that fragmentation of even simple molecules produces a large number of ions with different masses. A complex spectrum results, which often permits identification of the parent molecule or at least recognition of likely functional groups in the compound. Systematic studies of fragmentation patterns for pure substances have

TABLE 17-4   ISOTOPIC ABUNDANCE RATIOS FOR VARIOUS COMBINATIONS OF CARBON, HYDROGEN, OXYGEN, AND NITROGEN[a,b]

M = 83			M = 84		
	M + 1	M + 2		M + 1	M + 2
C_2HN_3O	3.36	0.24	$C_2H_2N_3O$	3.38	0.24
$C_2H_3N_4$	3.74	0.06	$C_2H_4N_4$	3.75	0.06
C_3HNO_2	3.72	0.45	$C_3H_2NO_2$	3.73	0.45
$C_3H_3N_2O$	4.09	0.27	$C_3H_4N_2O$	4.11	0.27
$C_3H_5N_3$	4.47	0.08	$C_3H_6N_3$	4.48	0.81
$C_4H_3O_2$	4.45	0.48	$C_4H_4O_2$	4.47	0.48
C_4H_5NO	4.82	0.29	C_4H_6NO	4.84	0.29
$C_4H_7N_2$	5.20	0.11	$C_4H_8N_2$	5.21	0.11
C_5H_7O	5.55	0.33	C_5H_8O	5.57	0.33
C_5H_9N	5.93	0.15	$C_5H_{10}N$	5.95	0.15
C_6H_{11}	6.66	0.19	C_6H_{12}	6.68	0.19

[a] Taken from R. M. Silverstein, G. C. Bassler, and T. C. Morrill, *Spectrometric Identification of Organic Compounds*, 3d ed. New York: John Wiley & Sons, Inc., 1974, p. 43. With permission.
[b] Data are given as percentages of peak height of M.

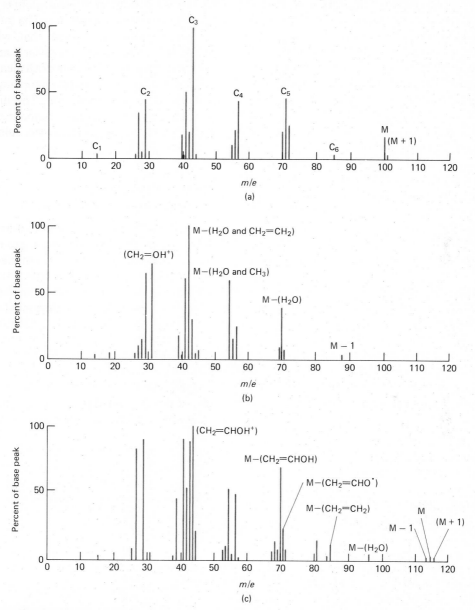

FIGURE 17-11 Mass spectra of some simple compounds. (a) *n*-Heptane; (b) 1-pentanol; (c) *n*-heptanal.

led to rational fragmentation mechanisms and a series of general rules that are helpful in interpreting spectra.[11] It is seldom possible (or desirable) to account for all of the peaks in a spectrum. Instead, characteristic patterns of fragmentation are sought. For example, the top spectrum in Figure 17-11 is characterized by clusters of peaks differing in mass by 14. Such a pattern is typical of straight-chain paraffins, in which the successive loss of a methylene group results in the observed mass decreases. This same pattern is evident in the left-hand parts of the two lower spectra as well. Quite generally, the most stable hydrocarbon fragments contain three or four carbon atoms and the corresponding peaks are thus the largest.

Alcohols usually have a very weak or nonexistent parent ion peak but lose water to give a strong peak at $(M - 18)$ (see Figure 17-11b). Cleavage of the C—C bond next to an oxygen is also common, and primary alcohols always have a strong peak at mass 31 due to the ion $CH_2 = OH^+$.

The interested reader should refer to reference literature for further generalizations concerning the identification of organic compounds from mass spectrometric data.[12]

QUANTITATIVE APPLICATIONS OF MASS SPECTROMETRY

The mass spectrometer is a powerful tool for the quantitative analysis of mixtures contain-

---

[11] See, for example: R. M. Silverstein, G. C. Bassler, and T. C. Morrill, *Spectrometric Identification of Organic Compounds*, 3d ed. New York: Wiley, 1974, p. 16.

[12] R. M. Silverstein, G. C. Bassler, and T. C. Morrill, *Spectrometric Identification of Organic Compounds*, 3d ed. New York: Wiley, 1974, Chapter 2; K. Biemann, *Mass Spectrometry, Organic Chemical Applications*. New York: McGraw-Hill, 1962; H. Budzikiewicz, C. Djerassi, and D. H. Williams, *Interpretation of Mass Spectra of Organic Compounds*. San Francisco: Holden-Day, 1964; and F. W. McLafferty, *Interpretation of Mass Spectra*, 2d ed. Menlo Park, CA: Benjamin, 1973.

ing closely related components. As mentioned earlier, the first commercial mass spectrometers were developed in about 1940 for determining the constituents in volatile hydrocarbon samples of the kind encountered in the petroleum industry. Within a decade, however, the method had been applied successfully to the analysis of a large number of other volatile compound types. More recently, mass spectroscopy has been adapted to the analysis of both inorganic and organic mixtures of low volatility.

Quantitative Analysis of Organic Mixtures

Basic Requirements. The basic requirements for a successful mass spectrometric analysis are: (1) each component must exhibit at least one peak that differs markedly from the others; (2) the contribution of each component to a peak must be linearly additive; (3) the sensitivity (ion current per unit partial pressure) must be reproducible to perhaps 1% relative; and (4) suitable standards for calibration must be available.

Calibration. Quantitative mass spectrometry is based upon empirical calibration with standards. Under appropriate conditions, mass peak heights are directly proportional to the partial pressures of the components. For complex mixtures, it is seldom possible to find a mass peak that is unique for each component; a set of simultaneous equations must, therefore, be solved to resolve the data from mixtures. That is,

$$i_{11}p_1 + i_{12}p_2 + \cdots + i_{1n}p_n = I_1$$
$$i_{21}p_1 + i_{22}p_2 + \cdots + i_{2n}p_n = I_2$$
$$\cdot \qquad \cdot \quad \cdot \quad \cdot \qquad \cdot$$
$$\cdot \qquad \cdot \quad \cdot \quad \cdot \qquad \cdot$$
$$i_{m1}p_1 + i_{m2}p_2 + \cdots + i_{mn}p_n = I_m$$

where I_m is the measured ion current at mass m in the spectrum of the mixture, and i_{mn}

refers to the ion current at mass m for component n. The partial pressure of component n in the mixture is given by p_n. The value of i_{mn} is determined for each component by calibration with a standard at a known partial pressure p_n. Substituting the measured ion currents for a mixture into the equations permits solution of the simultaneous equation to give the partial pressures of each of the components in the sample.

Precision and Accuracy. The precision of quantitative mass spectral measurements appears to range between 2 and 5 % relative. The accuracy varies considerably depending upon the complexity of the mixture being analyzed and the nature of its components. For gaseous hydrocarbon mixtures containing 5 to 10 components, absolute errors of 0.2 to 0.8 mole percent appear to be typical.

Component Analysis

The literature dealing with quantitative applications of mass spectrometry is so extensive as to make a summary difficult.[13] The listing of typical applications assembled by Melpolder and Brown demonstrates clearly the versatility of the method. For example, some of the mixtures that can be analyzed without sample heating include natural gas, C_3—C_5 hydrocarbons; C_6—C_8 saturated hydrocarbons; C_1—C_5 alcohols, aldehydes, and ketones; C_1—C_4 chlorides and iodides; fluorocarbons; thiophenes; atmospheric polutants; exhaust gases; and many others. By employing higher temperatures, successful analytical methods have been reported for C_{16}—C_{27} alcohols, aromatic acids and esters, steroids, fluorinated polyphenyls, aliphatic amides, halogenated aromatic derivatives, and aromatic nitriles.

Mass spectrometry has also been used for the characterization and analysis of high-molecular-weight polymeric materials. Here, the sample is first pyrolyzed; the volatile products are then led into the spectrometer for examination. Alternatively, heating can be performed on the probe of a direct inlet system. Some polymers yield essentially a single fragment; for example, isoprene from natural rubber, styrene from polystyrene, ethylene from polyethylene, and CF_2=CFCl from Kel-F. Other polymers yield two or more products that depend in amount and kind upon the pyrolysis temperature. Studies of temperature effects can provide information regarding the stabilities of the various bonds, as well as the approximate molecular weight distribution.

Component Type Determination

Because of the complex nature of petroleum products, quantitative data as to *types* of compounds are often more useful than analysis for individual components. Mass spectrometry can provide such information. For example, it has been found that paraffinic hydrocarbons generally give unusually strong peaks at masses 43, 57, 71, 85, and 99. Cycloparaffins and monoolefins, on the other hand, exhibit characteristically intense peaks at masses 41, 55, 69, 83, and 97. Another group of peaks is attributable to cycloolefins, diolefins, and acetylenes (67, 68, 81, 82, 95, and 96). Finally, alkylbenzenes are found to fragment to masses of 77, 78, 79, 91, 92, 105, 106, 119, 120, 133, and 134. A mathematical combination of the peak heights of a set provides an analytical parameter for assessing the concentration of each type of hydrocarbon. Type analyses have been used to characterize the properties and behavior of gasolines, fuel oils, lubricating oils, asphalts, and mixtures of paraffins, olefins, alcohols, and ketones.

---

[13] See, for example: F. W. Melpolder and R. A. Brown, in *Treatise on Analytical Chemistry*, eds. I. M. Kolthoff and P. J. Elving. New York: Interscience, 1963, Part I, vol. 4, p. 2047; and B. J. Millard, *Quantitative Mass Spectrometry*. Philadelphia: Heyden, 1977.

Inorganic Trace Analysis

The development of the spark source (p. 480) has made possible the application of mass spectrometry to the analysis of inorganic solids; in the last two decades, this technique has become important for the trace analysis of elements in metals, alloys, superconductors, and minerals.

Spark source mass spectrometry owes its growth to three inherent advantages. First is its high sensitivity, which often permits detection and semiquantitative determination of elements in the parts per billion range. Second is the simplicity of its spectrum for an element, which consists of but a single major line and a few weaker lines that occur at fractional values of the mass of the element (the latter are due to multiply charged ions). The third advantage is that line intensities for various elements are roughly equal (usually within a factor of 2 to 3).

The major disadvantage of spark source mass spectrometry can be traced to the erratic nature of the source and the consequent lack of reproducibility of the measurements. Because of the wide fluctuation in ion currents with time, measurements with an integrating detector over a period of time are required; photographic detection is usually employed for this reason.

A second disadvantage of the spark source is that it produces ions varying widely in kinetic energy. Thus, expensive double-focusing spectrometers are required.

Instruments. Several commercial spark source mass spectrometers are available. The requirement of photographic detectors has resulted in general use of the Mattauch-Herzog geometry for the design of the analyzer (Figure 17-5).

Sensitivity, Accuracy, and Precision. Under ideal conditions, spark source mass spectrometry is sensitive to concentrations in the range of a few parts per billion. In less favorable cases, on the other hand, the sensitivity limit may be one or two orders of magnitude greater than this figure.

Quantitative measurements in spark source mass spectrometry are based upon measurements of the blackening of the photographic plate with a microdensitometer; where suitable standards are available, uncertainties of 10 to 30 % relative are to be expected. Greater errors are often encountered with less favorable standard samples.

Isotope Abundance Measurement

The mass spectrometer was developed initially for the study of isotopic abundance, and the instrument continues to be the most important source for this kind of data. Information regarding the abundance of various isotopes is now employed for a variety of purposes; the determination of formulas of organic compounds cited earlier is one example. Other important applications include analysis by isotope dilution, tracer studies with isotopes, and dating of rocks and minerals by isotopic ratio measurements. The techniques are similar to those described in Chapter 16 for radioactive isotopes. Mass spectrometry, however, permits the extension of isotope detection to nonradioactive species such as ^{13}C, ^{17}O, ^{18}O, ^{15}N, ^{34}S, and others.

PROBLEMS

1. Calculate the $(M + 2)/M$ peak ratios for the following molecules.
   (a) Ethyl bromide, C_2H_5Br
   (b) Chlorobenzene, C_6H_5Cl
   (c) Dimethylsulfoxide, $C_2H_6SO_2$

2. Calculate the (M + 2)/M and (M + 4)/M peak ratios for the following compounds.
   (a) Dibromotoluene, $C_7H_6Br_2$
   (b) Methylene chloride, CH_2Cl_2
   (c) 1-Bromo-2-chloroethane, C_2H_4ClBr

3. What degree of resolution would be needed to distinguish between formaldehyde (CH_2O) and ethane (C_2H_6)? What type of mass spectrometer would be required? How else might this distinction be made?

4. Given below are the major peaks from the mass spectra of two isomers of C_2H_6O. Identify the two compounds.

(a)	m/e	% Intensity	(b)	m/e	% Intensity
	46	16		46	61
	45	50		45	100
	31	100		31	5

5. Compounds A and B both have a molecular ion peak at 58. After treatment in NaOD with D_2O, compound A has a molecular ion peak at 64, and compound B at 60. What type of compounds would show this behavior when treated with D_2O? What are A and B?

18

AN INTRODUCTION TO ELECTROANALYTICAL CHEMISTRY

Electroanalytical chemistry encompasses a group of quantitative analytical methods that are based upon the electrical properties of a solution of the analyte when it is made part of an electrochemical cell. Three types of electroanalytical methods are encountered. One includes methods that are dependent upon the direct relationship between concentration and an electrical parameter such as potential, current, resistance (or conductance), capacitance, or quantity of electricity. A second group employs one of the foregoing electrical parameters to establish the end point in a titration. A third category includes methods in which the analyte is converted to a weighable form by means of an electrical current.

Regardless of type, the intelligent application of an electroanalytical method requires an understanding of the basic theory and practical aspects of the operation of electrochemical cells. This chapter is devoted largely to these matters.[1]

ELECTROCHEMICAL CELLS

Electrochemical cells can be conveniently classified as *galvanic* if they are employed to produce electrical energy and *electrolytic* when they consume electricity from an external source. Both find use in electroanalytical chemistry. It is important to appreciate that many cells can be operated in either a galvanic or an electrolytic mode by variation of experimental conditions.

---

[1] Some reference works on electrochemistry and its applications include: J. O'M. Brockis and A. K. N. Reddy, *Modern Electrochemistry*, 2 vols. New York: Plenum Press, 1970; G. Kortum, *Treatise on Electrochemistry*, 2d ed. New York: Elsevier, 1965; J. J. Lingane, *Electroanalytical Chemistry*, 2d ed. New York: Interscience, 1958; and D. T. Sawyer and J. L. Roberts, Jr., *Experimental Electrochemistry for Chemists.* New York: Wiley, 1974.

Cell Components

An electrochemical cell consists of two metallic conductors called *electrodes*, each immersed in a suitable electrolyte solution. For electricity to flow, it is necessary: (1) that the electrodes be connected externally by means of a metal conductor and (2) that the two electrolyte solutions be in contact to permit movement of ions from one to the other. Figure 18-1 shows an example of a galvanic cell. The fritted glass disk is porous, so that Zn^{2+}, Cu^{2+}, and SO_4^{2-} ions as well as H_2O molecules can move across the junction between the two electrolyte solutions; the disk simply prevents extensive mixing of the two solutions.

Conduction in an Electrochemical Cell. Electricity is conducted by three distinct processes in various parts of the galvanic cell shown in Figure 18-1. In the copper and zinc electrodes, as well as in the external conductor, electrons serve as carriers, moving from the zinc through the conductor to the copper. Within the two solutions the flow of electricity involves migration of both cations and anions, the former away from the zinc electrode toward the copper and the latter in the reverse direction. *All* ions in the two solutions participate in this process.

A third type of conduction occurs at the two electrode surfaces. Here, an oxidation or a reduction process provides a mechanism whereby the ionic conduction of the solution is coupled with the electron conduction of the electrode to provide a complete circuit for a current. The two electrode processes are described by the equations

$$Zn(s) \rightleftharpoons Zn^{2+} + 2e$$
$$Cu^{2+} + 2e \rightleftharpoons Cu(s)$$

The net cell reaction is the sum of these two *half-cell* reactions:

$$Zn(s) + Cu^{2+} \rightleftharpoons Zn^{2+} + Cu(s)$$

Because this reaction has a strong tendency to proceed to the right, the cell is galvanic and produces a potential of about 1 V under most conditions.

Anode and Cathode. By definition, the *cathode* of an electrochemical cell is the electrode at which reduction occurs, while the *anode* is the electrode where an oxidation takes place. These definitions apply to both galvanic and electrolytic cells.

For the galvanic cell shown in Figure 18-1, the copper electrode is the cathode and the zinc electrode is the anode. Note that this cell could be caused to behave as an electrolytic cell by imposing a sufficiently large potential from an external source. Under these circumstances, the reactions occurring at the electrodes would be

$$Zn^{2+} + 2e \rightleftharpoons Zn(s)$$
$$Cu(s) \rightleftharpoons Cu^{2+} + 2e$$

Now, the roles of the electrodes are reversed; the copper electrode has become the anode and the zinc electrode the cathode.

Reactions at Cathodes. Some typical cathodic half-reactions are

$$Cu^{2+} + 2e \rightleftharpoons Cu(s)$$
$$Fe^{3+} + e \rightleftharpoons Fe^{2+}$$
$$2H^+ + 2e \rightleftharpoons H_2(g)$$
$$AgCl(s) + e \rightleftharpoons Ag(s) + Cl^-$$
$$IO_4^- + 2H^+ + 2e \rightleftharpoons IO_3^- + H_2O$$

Electrons are supplied for each of these processes from the external circuit via an electrode that does not participate directly in the chemical reaction. In the first process, copper is deposited on the electrode surface; in the second, only a change in oxidation state of a solution component occurs. The third reaction is frequently observed in aqueous solutions that contain no easily reduced species.

The fourth half-reaction is of interest because it can be considered the result of a two-step process; that is,

$$AgCl(s) \rightleftharpoons Ag^+ + Cl^-$$
$$Ag^+ + e \rightleftharpoons Ag(s)$$

Solution of the sparingly soluble precipitate occurs in the first step to provide the silver ions that are reduced in the second.

The last half-reaction has been included to demonstrate that a cathodic reaction can involve anions as well as cations.

Reactions at Anodes. Examples of typical anodic half-reactions include

$$Cu(s) \rightleftharpoons Cu^{2+} + 2e$$
$$Fe^{2+} \rightleftharpoons Fe^{3+} + e$$
$$2Cl^- \rightleftharpoons Cl_2(g) + 2e$$
$$H_2(g) \rightleftharpoons 2H^+ + 2e$$
$$2H_2O \rightleftharpoons O_2(g) + 4H^+ + 4e$$

The first half-reaction requires a copper electrode to supply Cu^{2+} ions to the solution. The remaining four half-reactions can take place at any of a variety of inert metal surfaces. To cause the fourth half-reaction to occur, it is necessary to replenish the hydrogen in the solution by bubbling the gas across the surface of the electrode (usually platinum). The reactions can then be formulated as

$$H_2(g) \rightleftharpoons H_2(sat'd)$$
$$H_2(sat'd) \rightleftharpoons 2H^+(aq) + 2e$$

The final reaction, giving oxygen as a product, is a common anodic process in aqueous solutions containing no easily oxidized species.

Liquid Junctions. Cells with a liquid junction, such as that shown at the fritted disk in Figure 18-1, are ordinarily employed to avoid direct reaction between the components of the two half-cells. If the two electrolyte solutions in Figure 18-1 were allowed to mix, a reduction in the cell efficiency would occur as a result of the direct deposition of copper on the zinc. As will be shown later, a small potential called a *junction potential* arises at the interface between two electrolyte solutions that differ in composition.

Occasionally, useful cells can be constructed in which the electrodes share a common elec-

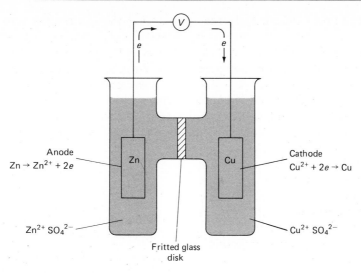

FIGURE 18-1 A galvanic cell with a liquid junction.

trolyte. An example of a cell without a liquid junction is shown in Figure 18-2. Here, the reaction at the silver cathode can be written

$$AgCl(s) + e \rightleftharpoons Ag(s) + Cl^-(aq)$$

Hydrogen is evolved at the platinum anode:

$$\tfrac{1}{2}H_2(g) \rightleftharpoons H^+(aq) + e$$

The overall cell reaction is then

$$AgCl(s) + \tfrac{1}{2}H_2(g)$$
$$\rightleftharpoons Ag(s) + H^+(aq) + Cl^-(aq)$$

The direct reaction between hydrogen and solid silver chloride is slow. As a consequence, a common electrolyte can be employed without significant loss of cell efficiency.

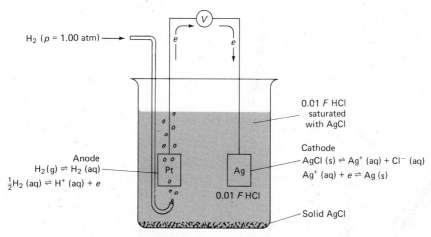

FIGURE 18-2 A galvanic cell without liquid junction.

The Salt Bridge. For reasons to be discussed later, electrochemical cells are often equipped with a *salt bridge* to separate the electrolytes in the anode and cathode compartments. This device takes a variety of forms. In Figure 18-7 (p. 514), for example, the bridge consists of a U-shaped tube filled with a saturated solution of potassium chloride. Such a cell has two liquid junctions; one is between the cathode electrolyte and one end of the bridge while the second is between the anode electrolyte and the other end of the bridge.

Schematic Representation of Cells. To simplify the description of cells, chemists often employ a shorthand notation. For example, the cells shown in Figures 18-1 and 18-2 can be described by

$$Zn\,|\,ZnSO_4(xM)\,|\,CuSO_4(yM)\,|\,Cu$$

$$Pt,\ H_2(p = 1\ atm)\,|\,H^+(0.01M),$$

$$Cl^-(0.01M),\ AgCl(sat'd)\,|\,Ag$$

By convention, *the anode and information with respect to the solution with which it is in contact is listed on the left.* Single vertical lines represent phase boundaries at which potentials may develop. Thus, in the first example, a part of the cell potential is associated with the phase boundary between the zinc electrode and the zinc sulfate solution. A small potential also develops at liquid junctions; thus, another vertical line is inserted between the zinc and copper sulfate solutions. The cathode is then represented symbolically with another vertical line separating the electrolyte solution from the copper electrode.

In the second cell, only two phase boundaries exist, the electrolyte being common to both electrodes. An equally correct representation of this cell would be

$$Pt\,|\,H_2(sat'd),\ HCl(0.01F),$$

$$Ag^+(1.8 \times 10^{-8}M)\,|\,Ag$$

Here, the molecular hydrogen concentration is that of a saturated solution (in the absence of partial pressure data, 1.00 atm is implied); the indicated molar silver ion concentration was computed from the solubility product constant for silver chloride.

The presence of a salt bridge in a cell is indicated by two vertical lines, implying that a potential difference is associated with each of the two interfaces. Thus, the cell shown in Figure 18-7 (p. 514) would be represented as

$$M\,|\,M^{2+}(yM)\,\|\,H^+(xM)\,|\,H_2(p\ atm),\ Pt$$

DC Currents in an Electrochemical Cell

As noted earlier, electricity is transported within a cell by the migration of ions. In common with metallic conductors, Ohm's law is often obeyed (departures from Ohm's law are discussed in the section devoted to polarization effects). That is,

$$I = \frac{E}{R} \qquad (18\text{-}1)$$

where I is the current in amperes, E is the potential difference in volts responsible for movement of the ions, and R is the resistance in ohms of the electrolyte to the current. The resistance depends upon the kinds and concentrations of ions in the solution.

It is found experimentally that under a fixed potential, the rate at which various ions move in a solution differs considerably. For example, the rate of movement (or *mobility*) of the proton is about seven times that for the sodium ion and five times that of the chloride ion. Thus, although all of the ions in a solution participate in conducting electricity, the fraction carried by one ion may differ markedly from that carried by another. This fraction depends upon the relative concentration of the ion as well as its inherent mobility. To illustrate, consider the cell shown in Figure 18-3a, which is divided into three imaginary compartments, each containing six hydrogen ions and six chloride ions. Six electrons are then forced into the cathode

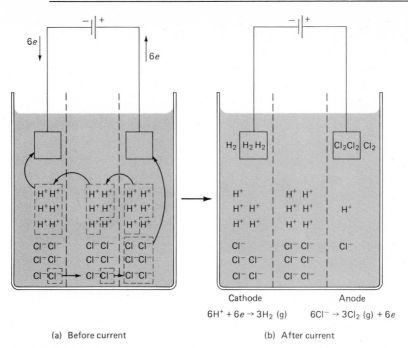

FIGURE 18-3 Changes resulting from a current made up of six electrons.

by a battery, resulting in formation of three molecules of hydrogen; three molecules of chlorine are also produced at the anode (see Figure 18-3b). The resulting charge imbalance brought about by the removal of ions from the electrode compartments is offset by migration, with positive ions moving toward the negative electrode and conversely. Because the proton is about five times more mobile than the chloride ion, however, a significant difference in concentration in the outer electrode compartments develops during electrolysis. In effect, five-sixths of the current has resulted from movement of the hydrogen ions and one-sixth from the transport of chloride ions.

It is important to appreciate that the current need not result from transport of the electrode reactants exclusively. Thus, if we were to introduce, say, 100 potassium and nitrate ions into each of the three compartments of the cell under consideration, the charge imbalance re-

sulting from electrolysis could be offset by migration of the added species as well as by the hydrogen and chloride ions. Since the added salt would represent an enormous excess, essentially all of the electricity would be carried within the cell by the potassium and nitrate ions rather than by the reactant ions; only across the electrode surfaces would the current result from the presence of hydrogen and chloride ions.

Alternating Current Conduction

When a dc potential is applied to a cell, conduction requires an oxidation reaction at the anode and a reduction at the cathode. The term *faradaic* is sometimes used to denote such currents and processes. Both faradaic and nonfaradaic conduction occurs with an ac potential.

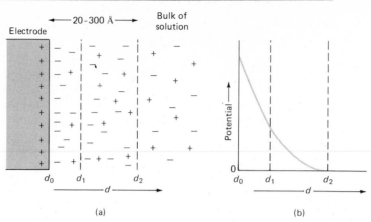

FIGURE 18-4   Electric double layer formed at electrode surface as a result of an applied potential.

Nonfaradaic Currents. Nonfaradaic currents involve the formation of an *electrical double layer* at the electrode-solution interface. When a potential is applied to a metallic electrode immersed in an electrolyte, a momentary surge of current creates an excess (or a deficiency) of negative charge at the surface of the metal. As a consequence of ionic mobility, however, the layer of solution immediately adjacent to the electrode acquires an opposing charge. This effect is shown in Figure 18-4a. The charged layer consists of two parts: (1) a compact inner layer, in which the potential decreases linearly with distance from the electrode surface; and (2) a more diffuse layer, in which the decrease is exponential; see Figure 18-4b. This assemblage of charge inhomogeneities is termed the electrical double layer.

The double layer formed by a dc potential involves the development of a momentary current which then drops to zero (that is, the electrode becomes *polarized*) unless some faradaic process occurs. With an alternating current, however, reversal of the charge relationship occurs with each half-cycle as first negative and then positive ions are attracted to the electrode surface. Electrical energy is consumed and converted to frictional heat from

this ionic movement. Thus, each electrode surface behaves as a capacitor, the capacitance of which may be remarkably large (several hundred to several thousand microfarads per cm^2). The capacitance current increases with frequency and with electrode size; by controlling these variables, it is possible to arrange conditions so that essentially all of the alternating electricity flowing through a cell is carried across the electrode interface by this nonfaradaic process.

Alternating Current in an Electrolyte Solution. For frequencies up to a few thousand cycles per second, alternating current results almost exclusively from ion movement. The direction of this motion, of course, reverses with each half-cycle. At very high frequencies, a significant fraction of the electricity is carried by a second mechanism that results from electrical polarization of the dielectric medium. Here, the voltage gradient causes *induced polarization* and *orientation polarization* of the *molecules* in the medium. In the former, distortion of the electron cloud surrounding the nucleus of a molecule causes a temporary polarized condition; in the latter, molecules with a permanent diode moment become aligned with the electrical field. Regardless of the mechanism, current results from the periodic alteration in these processes as a consequence of the alternating voltage.

The dielectric current depends upon the dielectric constant of the medium and is directly proportional to frequency; only at radio frequencies ($\sim 10^6$ Hz) does this current become important.

Reversible and Irreversible Cells

The galvanic cell shown in Figure 18-2 would develop a potential of about 0.46 V. If a battery with a potential somewhat greater than 0.46 V were inserted in the circuit, with its negative terminal connected to the platinum electrode, a reversal in direction of electron flow would occur; the reactions at the two electrodes would thus become

$$Ag(s) + Cl^- \rightleftharpoons AgCl(s) + e$$

$$H^+ + e \rightleftharpoons \tfrac{1}{2}H_2(g)$$

Now the silver electrode is the anode and the platinum electrode the cathode. A cell (or an electrode) for which a change in direction of the current causes a reversal of the electrochemical reaction is said to be *reversible*. Cells in which a current reversal results in different reactions at one or both electrodes are called *irreversible*. The cell shown in Figure 18-1 is also reversible. If, however, a small amount of dilute acid were introduced into the zinc electrode compartment, the reaction would tend to become irreversible. Here, zinc would not deposit at the cathode upon application of a potential; instead, hydrogen would form by the reaction

$$2H^+ + 2e \rightleftharpoons H_2(g)$$

Thus, the zinc electrode and the cell would be termed irreversible in the presence of acid.

CELL POTENTIALS

The galvanic cell shown in Figure 18-1 is called a *Daniell cell* or battery. Early telegraph systems in this country were powered by one form of this cell. The potential of the Daniell cell is dependent upon the concentration of the cations in the two electrolyte solutions; when these are about equal, a voltage of 1.1 V develops. As current is drawn, however, the zinc ion concentration increases while the copper ion concentration decreases by an equivalent amount. The cell potential undergoes a corresponding decrease and eventually reaches zero; at this point, equilibrium has been reached for the cell reaction

$$Zn(s) + Cu^{2+} \rightleftharpoons Zn^{2+} + Cu(s)$$

The relationship between the cell potential and the concentrations of the participants in a cell reaction is readily derived from thermodynamic considerations. A knowledge of this relationship, and its use, is vital to the understanding of all types of electroanalytical methods.

As in all thermodynamic calculations, a close agreement between calculated potential and experimental findings requires the use of activities rather than molar concentrations. Thus, it is worthwhile to review the relationship between these two concentration parameters before considering methods for the computation of cell potentials.

Activity and Activity Coefficient

The relationship between the *activity* a_M of a species and its molar concentration [M] is given by the expression

$$a_M = f_M[M] \tag{18-2}$$

where f_M is a dimensionless quantity called the *activity coefficient*. The activity coefficient, and thus the activity, of M varies with the *ionic strength* of a solution such that the employment of a_M instead of [M] in an electrode potential calculation (and in the other equilibrium calculations as well) renders the numerical value obtained independent of the ionic strength.

Here, the ionic strength μ is defined by the equation

$$\mu = \tfrac{1}{2}\left(m_1 Z_1^2 + m_2 Z_2^2 + m_3 Z_3^2 + \cdots\right) \quad (18\text{-}3)$$

where $m_1, m_2, m_3, \ldots$ represent the molar concentration of the various ions in the solution and $Z_1, Z_2, Z_3, \ldots$ are their respective charges. Note that an ionic strength calculation requires taking account of *all* ionic species in a solution, not just the reactive ones.

Properties of Activity Coefficients. Activity coefficients have the following properties:

1. The activity coefficient of a species can be thought of as a measure of the effectiveness with which that species influences an equilibrium in which it is a participant. In very dilute solutions, where the ionic strength is minimal, this effectiveness becomes constant, and the activity coefficient acquires a value of unity; the activity and molar concentration thus become numerically identical. As the ionic strength increases, an ion loses some of its effectiveness, and its activity coefficient decreases. We may summarize this behavior in terms of Equation 18-2. At moderate ionic strengths, $f_M < 1$; as the solution approaches infinite dilution, $f_M \to 1$ and thus $a_M \to [M]$.

At high ionic strengths, the activity coefficients for some species increase and may even become greater than one. The behavior of such solutions is difficult to interpret; we shall confine our discussion to regions of low to moderate ionic strengths (that is, where $\mu < 0.1$).

The variation of typical activity coefficients as a function of ionic strength is shown in Figure 18-5.

2. In dilute solutions, the activity coefficient for a given species is independent of the specific nature of the electrolyte, and depends only upon the ionic strength.

3. For a given ionic strength, the activity coefficient of an ion departs further from unity as the charge carried by the species increases. This effect is shown in Figure 18-5. The activity coefficient of an uncharged molecule is approximately one, regardless of ionic strength.

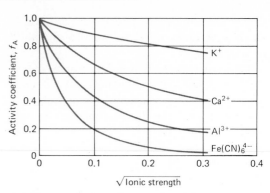

FIGURE 18-5   Effect of ionic strength on activity coefficients.

4. Activity coefficients for ions of the same charge are approximately the same at any given ionic strength. The small variations that do exist can be correlated with the effective diameter of the hydrated ions.

5. The product of the activity coefficient and molar concentration of a given ion describes its effective behavior in all equilibria in which it participates.

Evaluation of Activity Coefficients. In 1923, P. Debye and E. Hückel derived the following theoretical expression, which permits the calculation of activity coefficients of ions[2]

$$-\log f_A = \frac{0.5085 Z_A^2 \sqrt{\mu}}{1 + 0.3281 \alpha_A \sqrt{\mu}} \quad (18\text{-}4)$$

where

f_A = activity coefficient of the species A
Z_A = charge on the species A
μ = ionic strength of the solution
α_A = the effective diameter of the hydrated ion in ångström units

[2] P. Debye and E. Hückel, *Physik. Z.*, **24**, 185 (1923).

The constants 0.5085 and 0.3281 are applicable to solutions at 25°C; other values must be employed at different temperatures.

Unfortunately, considerable uncertainty exists regarding the magnitude of α_A in Equation 18-4. Its value appears to be approximately 3 Å for most single charged ions so that, for these species, the denominator of the Debye-Hückel equation reduces to approximately $(1 + \sqrt{\mu})$. For ions with higher charge, α_A may be as large as 10 Å. It should be noted that the second term of the denominator becomes small with respect to the first when the ionic strength is less than 0.01; under these circumstances, uncertainties in α_A are of little significance in calculating activity coefficients.

Kielland[3] has calculated values of α_A for numerous ions from a variety of experimental data. His "best values" for effective diameters are given in Table 18-1. Also presented are activity coefficients calculated from Equation 18-4 using these values for the size parameter.

---

[3] J. Kielland, *J. Amer. Chem. Soc.*, **59**, 1675 (1937).

TABLE 18-1 ACTIVITY COEFFICIENTS FOR IONS AT 25°C[a]

Ion	α_A Effective Diameter, Å	Activity Coefficient at Indicated Ionic Strengths				
		0.001	0.005	0.01	0.05	0.1
H_3O^+	9	0.967	0.933	0.914	0.86	0.83
Li^+, $C_6H_5COO^-$	6	0.965	0.929	0.907	0.84	0.80
Na^+, IO_3^-, HSO_3^-, HCO_3^-, $H_2PO_4^-$, $H_2AsO_4^-$, OAc^-	4–4.5	0.964	0.928	0.902	0.82	0.78
OH^-, F^-, SCN^-, HS^-, ClO_3^-, ClO_4^-, BrO_3^-, IO_4^-, MnO_4^-	3.5	0.964	0.926	0.900	0.81	0.76
K^+, Cl^-, Br^-, I^-, CN^-, NO_2^-, NO_3^-, $HCOO^-$	3	0.964	0.925	0.899	0.80	0.76
Rb^+, Cs^+, Tl^+, Ag^+, NH_4^+	2.5	0.964	0.924	0.898	0.80	0.75
Mg^{2+}, Be^{2+}	8	0.872	0.755	0.69	0.52	0.45
Ca^{2+}, Cu^{2+}, Zn^{2+}, Sn^{2+}, Mn^{2+}, Fe^{2+}, Ni^{2+}, Co^{2+}, Phthalate^{2-}	6	0.870	0.749	0.675	0.48	0.40
Sr^{2+}, Ba^{2+}, Cd^{2+}, Hg^{2+}, S^{2-}	5	0.868	0.744	0.67	0.46	0.38
Pb^{2+}, CO_3^{2-}, SO_3^{2-}, $C_2O_4^{2-}$	4.5	0.868	0.742	0.665	0.46	0.37
Hg_2^{2+}, SO_4^{2-}, $S_2O_3^{2-}$, CrO_4^{2-}, HPO_4^{2-}	4.0	0.867	0.740	0.660	0.44	0.36
Al^{3+}, Fe^{3+}, Cr^{3+}, La^{3+}, Ce^{3+}	9	0.738	0.54	0.44	0.24	0.18
PO_4^{3-}, $Fe(CN)_6^{3-}$	4	0.725	0.50	0.40	0.16	0.095
Th^{4+}, Zr^{4+}, Ce^{4+}, Sn^{4+}	11	0.588	0.35	0.255	0.10	0.065
$Fe(CN)_6^{4-}$	5	0.57	0.31	0.20	0.048	0.021

[a] From J. Kielland, *J. Amer. Chem. Soc.*, **59**, 1675 (1937).

Experimental verification of individual activity coefficients such as those shown in Table 18-1 is, unfortunately, impossible; all experimental methods give only a mean activity coefficient for the positively and negatively charged ions in a solution.[4] It should be pointed out, however, that mean activity coefficients calculated from the data in Table 18-1 agree satisfactorily with the experimental values.

The Debye-Hückel relationship, and the data in Table 18-1, give satisfactory activity coefficients for ionic strengths up to about 0.1. Beyond this value, however, the equation fails, and experimentally determined mean activity coefficients must be employed. Unfortunately, many electrochemical calculations involve solutions of high ionic strength for which no experimental activity coefficients are available. Concentrations must thus be employed instead of activities; uncertainties that vary from a few percent relative to an order of magnitude may be expected. The data shown in the example on page 524 illustrate the magnitude of uncertainties that can arise when concentrations rather than activities are employed in the calculation of cell potentials.

Molar concentrations will usually be used in lieu of activities in the discussion that follows, for the sake of convenience. Occasionally, however, it will be desirable or necessary to revert to the more exact concentration parameter.

Effect of Concentration on Cell Potentials

The effect of concentration (or activity) on cell potentials can be illustrated with the cell shown in Figure 18-2, which can be formulated in more general concentration terms as

$$\text{Pt, } H_2(x \text{ atm}) \,|\, H^+(yM), \, Cl^-(yM) \,|\, \text{Ag}$$

Here, we may describe the electrode processes by the two half-reactions, namely,

$$2AgCl(s) + 2e \rightleftharpoons 2Ag(s) + 2Cl^-$$
$$2H^+ + 2e \rightleftharpoons H_2(g)$$

The reason for writing both reactions as reductions will become apparent later. To derive a cell reaction, it is necessary to *subtract* the anodic reaction from the cathodic, thus obtaining

$$2AgCl(s) + H_2(g) \rightleftharpoons 2Ag(s) + 2H^+ + 2Cl^-$$

The equilibrium constant for this reaction is given by

$$K = \frac{[H^+]^2[Cl^-]^2}{p_{H_2}} \qquad (18\text{-}5)$$

where the bracketed terms represent molar concentrations of hydrogen and chloride ions, respectively, and p_{H_2} is the partial pressure of hydrogen in atmospheres. Note that the concentrations of silver and silver chloride are

---

[4] The mean activity of the electrolyte A_mB_n is defined as follows:

$$f_\pm = \text{mean activity coefficient} = (f_A^m \cdot f_B^n)^{1/(m+n)}$$

The mean activity coefficient can be measured in any of several ways, but it is impossible experimentally to resolve this term into the individual activity coefficients for f_A and f_B. For example, if A_mB_n is a precipitate, we can write

$$K_{sp} = [A]^m[B]^n \cdot f_A^m \cdot f_B^n = [A]^m[B]^n \cdot f_\pm^{(m+n)}$$

By measuring the solubility of A_mB_n in a solution in which the electrolyte concentration approaches zero (that is, where f_A and $f_B \to 1$), we could obtain K_{sp}. A second solubility measurement at some ionic strength, μ_1 would give values for [A] and [B]. These data would then permit the calculation of $f_A^m \cdot f_B^n = f_\pm^{(m+n)}$ for ionic strength μ_1. It is important to understand that there are insufficient experimental data to permit the calculation of the *individual* quantities f_A and f_B and that there appears to be no additional experimental information that would permit evaluation of these quantities. This situation is general; the *experimental* determination of individual activity coefficients appears to be impossible.

constant in their solid phases. Thus, these quantities are included in the constant K.[5]

It is convenient to define a second quantity Q such that

$$Q = \frac{[H^+]_a^2[Cl^-]_a^2}{(p_{H_2})_a} \qquad (18\text{-}6)$$

---

[5] An alternative way of treating the effect of pure solids as well as pure liquids and the solvent in an equilibrium system is the following: we define the *standard state* of a pure substance as its physical state at 25°C and at a pressure of one atmosphere. The activity of a substance in its standard state is then *assigned* a value of unity. Thus, the equilibrium constant for the reaction in question can be written as

$$K = \frac{[H^+]^2[Cl^-]^2[Ag]}{p_{H_2}[AgCl]} = \frac{[H^+]^2[Cl^-]^2(1.00)}{p_{H_2}(1.00)}$$

For all gases, the standard state is a pressure of one atmosphere; for a solute, it is an activity of 1.00.

For the reaction

$$Br_2(l) + H_2O \rightleftharpoons HOBr + Br^- + H^+$$

we write

$$K = \frac{[HOBr][Br^-][H^+]}{[Br_2][H_2O]}$$

Both the liquid bromine and the solvent water are in their standard states. Thus, the equilibrium constant expression reduces to

$$K = [HOBr][Br^-][H^+]$$

It is important to note that this last expression applies only if the solution is *saturated* with Br_2; if it were undersaturated, we would write

$$Br_2(aq) + H_2O \rightleftharpoons HOBr + Br^- + H^+$$

$$K_1 = \frac{[HOBr][Br^-][H^+]}{[Br_2]}$$

where the denominator is the molar concentration of Br_2 in the unsaturated solution. Note that

$$K = K_1[Br_2]_{sat'd}$$

where $[Br_2]_{sat'd}$ is the molar concentration of bromine in a saturated solution.

It should also be noted that the terms in an equilibrium constant expression are in fact ratios of the concentration of the species relative to their concentrations or pressures in the standard state (1.00 M or 1.00 atm). Thus, the equilibrium constant is effectively unitless.

Here, the subscript a is employed to indicate that the bracketed terms are instantaneous concentrations and *not equilibrium concentrations*. The quantity Q, therefore, is not a constant, but changes continuously until equilibrium is reached; at that point, Q becomes equal to K and the a subscripts are deleted.

From thermodynamics, it can be shown that the change in free energy ΔG for a cell reaction (that is, the maximum work obtainable at constant temperature and pressure) is given by

$$\Delta G = RT \ln Q - RT \ln K \qquad (18\text{-}7)$$

where R is the gas constant (8.316 J mol^{-1} deg^{-1}) and T is the temperature in degrees K; the term ln refers to the logarithm to the base e. It can also be shown that the cell potential E_{cell} is related to the free energy of the reaction by the relationship

$$\Delta G = -nFE_{cell} \qquad (18\text{-}8)$$

where F is the faraday (96,491 coulombs per chemical equivalent) and n is the number of equivalents of electricity (or moles of electrons) associated with the oxidation-reduction process (in this example, $n = 2$).

Substitution of Equations 18-6 and 18-8 into 18-7 yields

$$-nFE_{cell} = RT \ln \frac{[H^+]_a^2[Cl^-]_a^2}{(p_{H_2})_a} - RT \ln K$$
$$(18\text{-}9)$$

Let us define the standard potential E_{cell}^0 for the cell as follows:

$$E_{cell}^0 = \frac{RT}{nF} \ln K \qquad (18\text{-}10)$$

Substitution of this equation into Equation 18-9 yields, upon rearrangement,

$$E_{cell} = E_{cell}^0 - \frac{RT}{nF} \ln \frac{[H^+]_a^2[Cl^-]_a^2}{(p_{H_2})_a} \qquad (18\text{-}11)$$

Note that the standard potential is a constant, which is equal to the *cell potential when the reactants and products are at unit concentration* (*more correctly activity*) *and pressure*.

HALF-CELL OR ELECTRODE POTENTIALS

In electroanalytical work, it is often convenient to think of a cell potential as being composed of two *half-cell* or *electrode potentials*, one being associated with the cathode and the other with the anode. Thus, for the cell just considered, we may write

$$E_{cell} = E_{AgCl} - E_{H_2}$$

where E_{AgCl} is the electrode potential for the silver/silver chloride electrode (which is the cathode of the cell) and E_{H_2} is the electrode potential for the hydrogen gas electrode (here, the anode). Similarly, for the Daniell cell shown in Figure 18-1,

$$E_{cell} = E_{Cu} - E_{Zn}$$

A more general statement would be

$$E_{cell} = E_{cathode} - E_{anode} \qquad (18\text{-}12)$$

where $E_{cathode}$ and E_{anode} are *electrode potentials* for the electrodes that are acting as the cathode and anode, respectively. A rigorous definition of the term electrode potential will be forthcoming (p. 515).

Nature of Electrode Potentials

At the outset, it should be emphasized that there is *no* way of determining an absolute value for the potential of a single electrode, since all voltage-measuring devices determine only *differences* in potential. One conductor from such a device is connected to the electrode in question; in order to measure a potential difference, however, the second conductor must be brought in contact with the electrolyte solution of the half-cell in question. This latter contact inevitably involves a solid-solution interface and hence acts as a second half-cell at which a chemical reaction *must also take place* if electricity is to flow. A potential will be associated with this second reaction. Thus, an absolute value for the desired half-cell potential

is not realized; instead, what is measured is a combination of the potential of interest and the half-cell potential for the contact between the voltage-measuring device and the solution.

Our inability to measure absolute potentials for half-cell processes is not a serious handicap because relative half-cell potentials are just as useful. These relative potentials can be combined to give cell potentials; in addition, they are useful for calculating equilibrium constants of oxidation-reduction processes.

The Standard Hydrogen Electrode

The standard hydrogen electrode (SHE) is the universal reference for reporting relative half-cell potentials. It is a type of gas electrode.

The Hydrogen Electrode. The hydrogen electrode was widely used in early electrochemical studies as a reference electrode and as an indicator electrode for the determination of pH. Its composition can be formulated as

$$Pt, H_2(p \text{ atm}) \,|\, H^+(xM)$$

As suggested by the terms in parentheses, the potential developed at the platinum surface depends upon the hydrogen ion concentration of the solution and the partial pressure of the hydrogen employed to saturate it.

Figure 18-6 illustrates the components of a hydrogen electrode. The conductor is constructed from platinum foil, which has been *platinized*—that is, coated with a finely divided layer of platinum (called platinum black) by rapid chemical or electrochemical reduction of H_2PtCl_6. The platinum black provides a large surface area to assure that the reaction

$$2H^+ + 2e \rightleftharpoons H_2(g)$$

proceeds rapidly and reversibly at the electrode surface. As was pointed out earlier, the stream of hydrogen serves simply to keep the solution adjacent to the electrode saturated with respect to the gas.

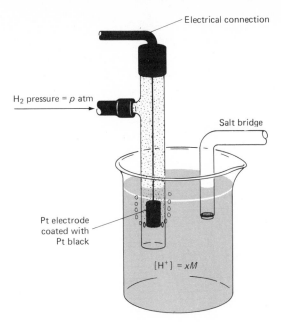

Electrical connection

H₂ pressure = p atm

Salt bridge

Pt electrode
coated with
Pt black

$[H^+] = xM$

FIGURE 18-6   The hydrogen electrode.

The hydrogen electrode may act as an anode or a cathode, depending upon the half-cell with which it is coupled. Hydrogen is oxidized to hydrogen ions at an anode; the reverse reaction takes place at a cathode. Under proper conditions, then, the hydrogen electrode is electrochemically reversible.

The potential of a hydrogen electrode depends upon the temperature, the hydrogen ion concentration (more correctly, the activity) in the solution, and the pressure of the hydrogen at the surface of the electrode. Values for these parameters must be carefully defined in order for the half-cell process to serve as a reference. Specifications for the *standard hydrogen electrode* call for a hydrogen activity of unity and a partial pressure for hydrogen of exactly one atmosphere. *By convention, the potential of this electrode is assigned the value of exactly zero volt at all temperatures.*

Electrode potentials are defined as *cell potentials* for a cell consisting of the electrode in

question and the standard hydrogen electrode. It must always be borne in mind that the electrode potentials are in fact *relative* potentials, all being referred to the common reference electrode.

Several secondary reference electrodes are more convenient for routine use and are extensively employed; some of these are described later in this chapter.

Measurement of Electrode Potentials

Although the standard hydrogen electrode is the universal standard of reference, it should be understood that the electrode, as described, can never be realized in the laboratory; that is, it is a *hypothetical* electrode to which experimentally determined potentials can be referred only by suitable computation. The reason that the electrode, as defined, cannot be prepared is that chemists lack the knowledge to produce a solution with a hydrogen ion activity of exactly unity; no adequate theory exists to permit evaluation of the activity coefficient of hydrogen ions in a solution in which the ionic strength is as great as one. Thus, the *concentration* of HCl or another acid required to give a hydrogen ion activity of unity cannot be calculated. Notwithstanding, data for more dilute solutions, where activity coefficients are known, can be used to provide potentials at unit activity. Thus, for example, the activity of hydrogen and chloride ions in the cell shown in Figure 18-2 can be calculated from the Debye-Hückel relationship; measurements can also be made at lower acid concentrations. These data can then provide, by suitable extrapolation, information about the potential of a hypothetical cell in which the hydrogen and chloride ions have unit activities. For the cell in Figure 18-2, Equation 18-11 then reduces to

$$E_{cell} = E^0_{AgCl} - \frac{RT}{nF} \ln \frac{(1.00)^2(1.00)^2}{1.00}$$

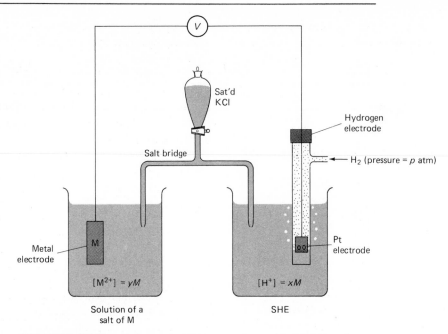

FIGURE 18-7   Schematic diagram of an arrangement for the measurement of electrode potentials against the standard hydrogen electrode.

where E^0_{AgCl} is called the *standard electrode potential* for the half-reaction

$$AgCl(s) + e \rightleftharpoons Ag(s) + Cl^- \qquad E^0 = 0.222 \text{ V}$$

$$(18\text{-}13)$$

We can also imagine a cell with a liquid junction or a salt bridge in which the hydrogen and chloride ions can be varied independently. For example,

$$Pt, H_2(1.00 \text{ atm}) \,|\, H^+(a_H = 1.00) \,\|\, Cl^-(zM),$$

$$AgCl(\text{sat'd}) \,|\, Ag$$

The left half of this cell is again not experimentally realizable, but its potential could be computed from data for a cell in which a_{H^+} was considerably smaller than one. Neglecting junction potentials, the potential for this cell can be obtained by suitable substitutions into Equation 18-11

$$E_{cell} = E_{AgCl} = E^0_{AgCl} - \frac{RT}{2F} \ln \frac{[Cl^-]^2_a (1.00)^2}{1.00}$$

or

$$E_{AgCl} = E^0_{AgCl} - \frac{RT}{F} \ln [Cl^-]_a \quad (18\text{-}14)$$

This equation shows how the electrode potential E_{AgCl} for the silver/silver chloride electrode varies as a function of chloride ion concentration.

A cell with a salt bridge such as that shown in Figure 18-7 can be used to measure electrode potentials for half-reactions involving the metal of the electrode and its ion. Here, the reference electrode is shown as a hydrogen electrode in which the acid concentration is sufficiently low to permit calculation of the hydrogen ion activity by means of the Debye-Hückel equation; alternatively, one of the secondary electrodes described at the end of this chapter, whose potential against the standard hydrogen electrode is known, can be substituted. The activity of the metal ion y is then varied and the cell potential measured. Suitable extrapolation will then provide a standard electrode potential E_M^0 for the half-reaction

$$M^{2+} + 2e = M(s) \qquad E_M^0 = zV$$

Thus, E_M^0 is the potential for the cell when the activities of M^{2+} and H^+ are exactly one. This measurement would be less accurate than the one employing a cell without a liquid junction because of the uncertainty with respect to the junction potentials that exist at the two ends of the salt bridge.

If the metal M in Figure 18-7 is cadmium and the solution is approximately 0.01 M in cadmium ions, the voltage indicated by the measuring device, V, will be about 0.5 V. Moreover, the cadmium will function as the anode; thus, electrons pass from this electrode to the hydrogen electrode via the external circuit. The half-cell reactions for this galvanic cell can be written as

$$Cd(s) \rightleftharpoons Cd^{2+} + 2e \qquad \text{Anode}$$
$$2H^+ + 2e \rightleftharpoons H_2(g) \qquad \text{Cathode}$$

The overall cell reaction is the sum of these, or

$$Cd(s) + 2H^+ \rightleftharpoons Cd^{2+} + H_2(g)$$

If the cadmium electrode is replaced by a zinc electrode immersed in a solution that is about 0.01 M in zinc ions, a potential of about 0.8 V will be observed. The metal electrode is again the anode in this cell. The larger voltage developed reflects the greater tendency of zinc to be oxidized. The difference between this potential and the one for cadmium is a quantitative measure of the relative strengths of these two metals as reducing agents.

If the half-cell in Figure 18-7 consisted of a copper electrode in a 0.01 M solution of copper(II) ions, a potential of about 0.3 V would develop. However, in distinct contrast to the previous two examples, copper would tend to deposit, and an external electron flow, if allowed, would be from the hydrogen electrode to the copper electrode. The spontaneous cell reaction, then, is the reverse of that in the two cells considered earlier:

$$Cu^{2+} + H_2(g) \rightleftharpoons Cu(s) + 2H^+$$

Thus, metallic copper is a much less effective reducing agent than either zinc, cadmium, or *hydrogen*. As before, the observed potential is a quantitative measure of this strength.

By additional measurements and suitable extrapolations, the data from the experiment just described could be made to yield computed potentials for the system when the activities of M^{2+} and H^+ are unity. These potentials for Cd^{2+}, Zn^{2+}, and Cu^{2+} have values of 0.403, 0.763, and 0.337, respectively. Note, however, that a need exists to indicate that the copper electrode behaves as a cathode while the zinc and cadmium electrodes function as anodes when coupled to the hydrogen electrode. Positive and negative signs are used to make this distinction, the potential for half-cells such as copper being provided with one sign and the other two electrodes being assigned the opposite. The choice as to which potential will be positive and which will be negative is *purely arbitrary*; however, the sign convention that is chosen must be used consistently.

Sign Conventions for Electrode Potentials

It is not surprising that the arbitrariness in specifying signs has led to much controversy and confusion in the course of the development of electrochemistry. In 1953, the International Union of Pure and Applied Chemistry (IUPAC), meeting in Stockholm, attempted to resolve this controversy. The sign convention adopted at this meeting is sometimes called the IUPAC or Stockholm convention; there is hope for its general adoption in years to come. We shall always use the IUPAC sign convention.

Any sign convention must be based upon half-cell processes written in a single way—that is, entirely as either oxidations or as reductions. According to the IUPAC convention, the term *electrode potential* (or more exactly, *relative electrode potential*) is *reserved exclusively for half-reactions written as reductions*. There is no objection to using the term *oxidation potential* to connote an electrode process written in the opposite sense, but an oxidation potential should never be called an electrode potential. The sign of an oxidation potential will always be opposite to its corresponding electrode or reduction potential.

The sign of the electrode potential is determined by the actual sign of the electrode of interest when it is coupled with a standard hydrogen electrode in a galvanic cell. Thus, a zinc or a cadmium electrode will behave as the anode from which electrons flow through the external circuit to the standard hydrogen electrode. These metal electrodes are thus the negative terminals of such galvanic cells, and their electrode potentials are *assigned* negative values. That is,

$$Zn^{2+} + 2e \rightleftharpoons Zn(s) \qquad E^0 = -0.763 \text{ V}$$
$$Cd^{2+} + 2e \rightleftharpoons Cd(s) \qquad E^0 = -0.403 \text{ V}$$

The potential for the copper electrode, on the other hand, is given a positive sign because the copper behaves as a cathode in a galvanic cell constructed from this electrode and the hydrogen electrode; electrons flow toward the copper electrode through the exterior circuit. It is thus the positive terminal of the galvanic cell and

$$Cu^{2+} + 2e \rightleftharpoons Cu(s) \qquad E^0 = +0.337 \text{ V}$$

It is important to emphasize that electrode potentials and their signs apply to half-reactions *written as reductions*. Both zinc and cadmium are oxidized by hydrogen ion; the spontaneous reactions are thus oxidations. It is evident, then, that the *sign of the electrode potential will indicate whether or not the reduction is spontaneous with respect to the standard hydrogen electrode*. That is, the positive sign for the copper electrode potential means that the reaction

$$Cu^{2+} + H_2(g) \rightleftharpoons 2H^+ + Cu(s)$$

proceeds toward the right under ordinary conditions. The negative electrode potential for zinc, on the other hand, means that the analogous reaction

$$Zn^{2+} + H_2(g) \rightleftharpoons 2H^+ + Zn(s)$$

does not ordinarily occur; indeed, the equilibrium favors the species on the left.

The IUPAC convention was adopted in 1953, but electrode potential data given in many texts and reference works are not always in accord with it. For example, in a source of oxidation potential data compiled by Latimer,[6] one finds

$$Zn(s) \rightleftharpoons Zn^{2+} + 2e \qquad E = +0.763 \text{ V}$$
$$Cu(s) \rightleftharpoons Cu^{2+} + 2e \qquad E = -0.337 \text{ V}$$

To convert these oxidation potentials to electrode potentials as defined by the IUPAC convention, one must mentally: (1) express the half-reactions as reductions; and (2) change the signs of the potentials.

The sign convention employed in a table of standard potentials may not be explicitly stated. This information is readily ascertained, however, by referring to a half-reaction with which one is familiar and noting the direction of the reaction and the sign of the potential. Whatever changes, if any, are required to convert to the IUPAC convention are then applied to the remainder of the data in the table. For example, all one needs to remember is that strong oxidizing agents such as oxygen have large positive electrode potentials under the IUPAC convention. That is, the reaction

$$O_2(g) + 4H^+ + 4e \rightleftharpoons 2H_2O$$
$$E^0 = +1.23 \text{ V}$$

tends to occur spontaneously with respect to the standard hydrogen electrode. The sign and

---

[6] W. M. Latimer, *The Oxidation States of the Elements and Their Potentials in Aqueous Solutions*, 2d ed. Englewood Cliffs, N.J.: Prentice-Hall, 1952.

direction of this reaction in a given table can then serve as a key to any changes that may be needed to convert all data to the IUPAC convention.

Effect of Concentration on Electrode Potential

Equation 18-14 shows how the electrode potential for the silver/silver chloride electrode varies as a function of chloride concentration. Turning to a more general case consider the half-reaction

$$pP + qQ + \cdots + ne \rightleftharpoons rR + sS$$

where the capital letters represent formulas of reacting species (whether charged or uncharged), e represents the electron, and the lower-case italic letters indicate the number of moles of each species (including electrons) participating in the half-cell reaction. Employing the same arguments that were used in the case of the silver/silver chloride electrode, we obtain

$$E = E^0 - \frac{RT}{nF} \ln \frac{[R]_a^r[S]_a^s \cdots}{[P]_a^p[Q]_a^q \cdots}$$

At room temperature (298°K), the collection of constants in front of the logarithm has units of joules per coulomb or volt (p. 7). That is,

$$\frac{RT}{nF} = \frac{8.316 \text{ J mol}^{-1} \text{ deg}^{-1} \times 298 \text{ deg}}{n \text{ equiv mol}^{-1} \times 96491 \text{ C equiv}^{-1}}$$

$$= \frac{2.568 \times 10^{-2} \text{ J C}^{-1}}{n}$$

Thus, upon converting from a natural to a base ten logarithm by multiplication by 2.303, the foregoing equation can be written

$$E = E^0 - \frac{0.0591}{n} \log \frac{[R]^r[S]^s \cdots}{[P]^p[Q]^q \cdots} \quad (18\text{-}15)$$

For convenience, we have also deleted the a subscripts, which were inserted earlier as a reminder that the bracketed terms represented nonequilibrium concentrations. Hereafter, we

shall not use the subscript; the student should always be alert to the fact, however, that the quotients that appear in this type of equation are *not* equilibrium constants, despite their similarity in appearance.

To summarize the meaning of the bracketed terms in Equation 18-15, when the substance R is a gas,

[R] = partial pressure in atmospheres

When R is a solute

[R] = concentration in moles per liter

or occasionally

[R] = activity of R, a_R

When R is a pure solid or liquid in excess or the solvent

[R] = 1.00

Equation 18-15 is called the *Nernst equation* in honor of a nineteenth-century electrochemist. Application of the Nernst equation is illustrated in the following examples.

1. $Zn^{2+} + 2e \rightleftharpoons Zn(s)$

$$E = E^0 - \frac{0.0591}{2} \log \frac{1}{[Zn^{2+}]}$$

2. $Fe^{3+} + e \rightleftharpoons Fe^{2+}$

$$E = E^0 - \frac{0.0591}{1} \log \frac{[Fe^{2+}]}{[Fe^{3+}]}$$

This electrode potential can be measured with an inert metal electrode immersed in a solution containing iron(II) and iron(III). The potential is dependent upon the *ratio* between the molar concentrations of the two ions.

3. $2H^+ + 2e \rightleftharpoons H_2(g)$

$$E = E^0 - \frac{0.0591}{2} \log \frac{p_{H_2}}{[H^+]^2}$$

In this example, p_{H_2} represents the partial pressure of hydrogen, expressed in atmospheres, at the surface of the electrode. Ordinarily, p_{H_2} will be very close to atmospheric pressure.

4. $Cr_2O_7^{2-} + 14H^+ + 6e \rightleftharpoons 2Cr^{3+} + 7H_2O$

$$E = E^0 - \frac{0.0591}{6} \log \frac{[Cr^{3+}]^2}{[Cr_2O_7^{2-}][H^+]^{14}}$$

Here, the potential depends not only on the concentrations of chromium(III) and dichromate ions but also on the pH of the solution.

The Standard
Electrode Potential, E^0

An examination of Equation 18-15 reveals that the constant E^0 is equal to the half-cell potential when the logarithmic term is zero. This condition occurs whenever the activity quotient is equal to unity, one such instance being when the activities of all reactants and products are unity. Thus, the standard potential is often defined as the electrode potential of a half-cell reaction (vs. SHE) when all reactants and products exist at unit activity.

The standard electrode potential is an important physical constant that gives a quantitative description of the relative driving force for a half-cell reaction. Several facts regarding this constant should be kept in mind. First, the electrode potential is temperature-dependent; if it is to have significance, the temperature at which it is determined must be specified. Second, the standard electrode potential is a relative quantity in the sense that it is really the potential of an electrochemical cell in which the anode is a carefully specified reference electrode—that is, the standard hydrogen electrode—whose potential is *assigned* a value of zero volts. Third, the sign of a standard potential is identical with that of the conductor in contact with the half-cell of interest in a galvanic cell, the other half of which is the standard hydrogen electrode. Finally, the standard potential is a measure of the intensity of the driving force for a half-reaction. As such, it is independent of the notation employed to express the half-cell process. Thus, the potential for the process

$$Ag^+ + e \rightleftharpoons Ag(s) \qquad E^0 = +0.799 \text{ V}$$

although dependent upon the concentration of silver ions, is the same regardless of whether we write the half-reaction as above or as

$$100 \, Ag^+ + 100e \rightleftharpoons 100 \, Ag(s)$$

$$E^0 = +0.799 \text{ V}$$

To be sure, the Nernst equation must be consistent with the half-reaction as it has been written. For the first of these, it will be

$$E = 0.799 - \frac{0.0591}{1} \log \frac{1}{[Ag^+]}$$

and for the second

$$E = 0.799 - \frac{0.0591}{100} \log \frac{1}{[Ag^+]^{100}}$$

Standard electrode potentials are available for numerous half-reactions. Many have been determined directly from voltage measurements of cells in which a hydrogen or other reference electrode constituted the other half of the cell. It is possible, however, to calculate E^0 values from equilibrium studies of oxidation-reduction systems and from thermochemical data relating to such reactions. Many of the values found in the literature were so obtained.[7]

For illustrative purposes, a few standard electrode potentials are given in Table 18-2; a more comprehensive table is found in Appendix 2. The species in the upper left-hand part of the equations in Table 18-2 are most easily reduced, as indicated by the large positive E^0 values; they are therefore the most effective oxidizing agents. Proceeding down the left-hand side of the table, each succeeding species is a less effective acceptor of electrons than the one above it. The half-cell reactions at the

[7] Two authoritative sources for standard potential data are: L. Meites, *Handbook of Analytical Chemistry*. New York: McGraw-Hill, 1963, pp. 5-6 to 5-14; and W. M. Latimer, *The Oxidation States of the Elements and Their Potentials in Aqueous Solutions*, 2d ed. Englewood Cliffs, N.J.: Prentice-Hall, 1952.

bottom of the table have little tendency to take place as written. On the other hand, they do tend to occur in the opposite sense, as oxidations. The most effective reducing agents, then, are those species that appear in the lower right-hand side of the equations in the table.

A compilation of standard potentials provides the chemist with qualitative information regarding the extent and direction of electron-transfer reactions between the tabulated species. On the basis of Table 18-2, for example, we see that zinc is more easily oxidized than cadmium, and we conclude that a piece of zinc immersed in a solution of cadmium ions will cause the deposition of metallic cadmium; conversely, cadmium has little tendency to reduce zinc ions. Table 18-2 also shows that iron(III) is a better oxidizing agent than triiodide ion; therefore, in a solution containing an equilibrium mixture of iron(III), iodide, iron(II), and triiodide ions, we can predict that the latter pair will predominate.

Calculation of Half-Cell Potentials from E^0 Values

Typical applications of the Nernst equation to the calculation of half-cell potentials are illustrated in the following examples.

EXAMPLE

What is the potential for a half-cell consisting of a cadmium electrode immersed in a solution that is $0.0100F$ in Cd^{2+}?

From Table 18-2, we find

$$Cd^{2+} + 2e \rightleftharpoons Cd(s) \qquad E^0 = -0.403 \text{ V}$$

Thus,

$$E = E^0 - \frac{0.0591}{2} \log \frac{1}{[Cd^{2+}]}$$

TABLE 18-2   STANDARD ELECTRODE POTENTIALS[a]

Reaction	E^0 at 25°C, V
$Cl_2(g) + 2e \rightleftharpoons 2Cl^-$	$+1.359$
$O_2(g) + 4H^+ + 4e \rightleftharpoons 2H_2O$	$+1.229$
$Br_2(aq) + 2e \rightleftharpoons 2Br^-$	$+1.087$
$Br_2(l) + 2e \rightleftharpoons 2Br^-$	$+1.065$
$Ag^+ + e \rightleftharpoons Ag(s)$	$+0.799$
$Fe^{3+} + e \rightleftharpoons Fe^{2+}$	$+0.771$
$I_3^- + 2e \rightleftharpoons 3I^-$	$+0.536$
$Hg_2Cl_2(s) + 2e \rightleftharpoons 2Hg(l) + 2Cl^-$	$+0.268$
$AgCl(s) + e \rightleftharpoons Ag(s) + Cl^-$	$+0.222$
$Ag(S_2O_3)_2^{3-} + e \rightleftharpoons Ag(s) + 2S_2O_3^{2-}$	$+0.010$
$2H^+ + 2e \rightleftharpoons H_2(g)$	0.000
$AgI(s) + e \rightleftharpoons Ag(s) + I^-$	-0.151
$PbSO_4(s) + 2e \rightleftharpoons Pb(s) + SO_4^{2-}$	-0.350
$Cd^{2+} + 2e \rightleftharpoons Cd(s)$	-0.403
$Zn^{2+} + 2e \rightleftharpoons Zn(s)$	-0.763

[a] See Appendix 2 for a more extensive list.

Substituting the Cd^{2+} concentration into this equation gives

$$E = -0.403 - \frac{0.0591}{2} \log \frac{1}{0.0100}$$

$$= -0.403 - \frac{0.0591}{2} (+2.0)$$

$$= -0.462 \text{ V}$$

The sign for the potential indicates the direction of the reaction when this half-cell is coupled with the standard hydrogen electrode. The fact that it is negative shows that the reverse reaction

$$Cd(s) + 2H^+ \rightleftharpoons H_2(g) + Cd^{2+}$$

occurs spontaneously. Note that the calculated potential is a larger negative number than the standard electrode potential itself. This follows from mass-law considerations because the half-reaction, *as written*, has less tendency to occur with the lower cadmium ion concentration.

EXAMPLE

Calculate the potential for a platinum electrode immersed in a solution prepared by saturating a $0.0100F$ solution of KBr with Br_2.

Here, the half-reaction is

$$Br_2(l) + 2e \rightleftharpoons 2Br^- \qquad E^0 = 1.065 \text{ V}$$

Note that the term (l) in the equation indicates that the aqueous solution is kept saturated by the presence of an excess of *liquid* Br_2. Thus, the overall process is the sum of the two equilibria

$$Br_2(l) \rightleftharpoons Br_2(\text{sat'd aq})$$

$$Br_2(\text{sat'd aq}) + 2e \rightleftharpoons 2Br^-$$

The Nernst equation for the overall process is

$$E = 1.065 - \frac{0.0591}{2} \log \frac{[Br^-]^2}{1.00}$$

Here, the activity of Br_2 in the pure liquid is

constant and equal to 1.00 by definition. Thus,

$$E = 1.065 - \frac{0.0591}{2} \log(0.0100)^2$$

$$= 1.065 - \frac{0.0591}{2} (-4.00)$$

$$= 1.183 \text{ V}$$

EXAMPLE

Calculate the potential for a platinum electrode immersed in a solution that is $0.0100F$ in KBr and $1.00 \times 10^{-3}F$ in Br_2.

Here, the half-reaction used in the preceding example does not apply *because the solution is no longer saturated in Br_2*. Table 18-2, however, contains the half-reaction

$$Br_2(aq) + 2e \rightleftharpoons 2Br^- \qquad E^0 = 1.087 \text{ V}$$

The term (aq) implies that all of the Br_2 present is in solution; that is, 1.087 is the electrode potential for the half-reaction when the Br^- and Br_2 *solution* activities are 1.00 mole/liter. It turns out, however, that the solubility of Br_2 in water at 25°C is only about 0.18 mole/liter. Therefore, the recorded potential of 1.087 is based on a *hypothetical system that cannot be realized experimentally*. Nevertheless, this potential is useful because it provides the means by which potentials for undersaturated systems can be calculated. Thus,

$$E = 1.087 - \frac{0.0591}{2} \log \frac{[Br^-]^2}{[Br_2]}$$

$$= 1.087 - \frac{0.0591}{2} \log \frac{(1.00 \times 10^{-2})^2}{1.00 \times 10^{-3}}$$

$$= 1.087 - \frac{0.0591}{2} \log 0.100$$

$$= 1.117 \text{ V}$$

Here, the Br_2 activity is 1.00×10^{-3} rather than 1.00, as was the situation when the solution was saturated.

Electrode Potentials in the Presence of Precipitation and Complex-Forming Reagents

As shown by the following example, reagents that react with the participants of an electrode process have a marked effect on the potential for that process.

EXAMPLE

Calculate the potential of a silver electrode in a solution that is saturated with silver iodide and has an iodide ion activity of exactly 1.00 (K_{sp} for $AgI = 8.3 \times 10^{-1}$)

$$Ag^+ + e \rightleftharpoons Ag(s) \qquad E^0 = +0.799 \text{ V}$$

$$E = +0.799 - 0.0591 \log \frac{1}{[Ag^+]}$$

We may calculate $[Ag^+]$ from the solubility product constant

$$[Ag^+] = \frac{K_{sp}}{[I^-]}$$

Substituting into the Nernst equation gives

$$E = +0.799 - \frac{0.0591}{1} \log \frac{[I^-]}{K_{sp}}$$

This equation may be rewritten as

$$E = +0.799 + 0.0591 \log K_{sp}$$
$$- 0.0591 \log [I^-] \qquad (18\text{-}16)$$

If we substitute 1.00 for $[I^-]$ and use 8.3×10^{-17} for K_{sp}, the solubility product for AgI at 25.0°C, we obtain

$$E = -0.151 \text{ V}$$

This example shows that the half-cell potential for the reduction of silver ion becomes smaller in the presence of iodide ions. Qualitatively this is the expected effect because decreases in the concentration of silver ions diminish the tendency for their reduction.

Equation 18-16 relates the potential of a silver electrode to the iodide ion concentration of a solution that is also saturated with silver iodide. *When the iodide ion activity*

is unity, the potential is the sum of two constants; it is thus the standard electrode potential for the half-reaction

$$AgI(s) + e \rightleftharpoons Ag(s) + I^- \qquad E^0 = -0.151 \text{ V}$$

where

$$E^0 = +0.799 + 0.0591 \log K_{sp} \text{ (18-17)}$$

The Nernst relationship for the silver electrode in a solution saturated with silver iodide can then be written as

$$E = E^0 - 0.0591 \log [I^-]$$
$$= -0.151 - 0.0591 \log [I^-]$$

Thus, when in contact with a solution *saturated with silver iodide*, the potential of a silver electrode can be described *either* in terms of the silver ion concentration (with the standard electrode potential for the simple silver half-reaction) *or* in terms of the iodide ion concentration (with the standard electrode potential for the silver/silver iodide half-reaction). The latter is usually more convenient.

The potential of a silver electrode in a solution containing an ion that forms a soluble complex with silver ion can be treated in a fashion analogous to the foregoing. For example, in a solution containing thiosulfate and silver ions, complex formation occurs:

$$Ag^+ + 2S_2O_3^{2-} \rightleftharpoons Ag(S_2O_3)_2^{3-}$$

$$K_f = \frac{[Ag(S_2O_3)_2^{3-}]}{[Ag^+][S_2O_3^{2-}]^2}$$

where K_f is the *formation constant* for the complex. The half-reaction for a silver electrode in such a solution is

$$Ag(S_2O_3)_2^{3-} + e \rightleftharpoons Ag(s) + 2S_2O_3^{2-}$$

The standard electrode potential for this half-reaction will be the electrode potential when both the complex and the complexing anion are at unit activity. Using the same approach as in the previous example, we find that

$$E^0 = +0.799 + 0.0591 \log \frac{1}{K_f}$$

Data for the potential of the silver electrode in the presence of selected ions are given in the tables of standard electrode potentials in Appendix 2 and Table 18-2. Similar information is also provided for other electrode systems. Such data often simplify the calculation of half-cell potentials.

Some Limitations to the Use of Standard Electrode Potentials

Standard electrode potentials are of great importance in understanding electroanalytical processes. There are, however, certain inherent limitations to the use of these data that should be clearly appreciated.

Substitution of Concentrations for Activities. As a matter of convenience, molar concentrations—rather than activities—of reactive species are generally employed in the Nernst equation. Unfortunately, the assumption that these two quantities are identical is valid only in very dilute solutions; with increasing electrolyte concentrations, potentials calculated on the basis of molar concentrations can be expected to depart from those obtained by experiment.

To illustrate, the standard electrode potential for the half-reaction

$$Fe^{3+} + e \rightleftharpoons Fe^{2+}$$

is $+0.771$ V. Neglecting activities, we would predict that a platinum electrode immersed in a solution that was *one formal* in iron(II), iron(III), and perchloric acid would exhibit a potential numerically equal to this value relative to the standard hydrogen electrode. In fact, however, a potential of $+0.732$ V is observed experimentally. The reason for the discrepancy is seen if we write the Nernst equation in the form

$$E = E^0 - 0.0591 \log \frac{[Fe^{2+}]f_{Fe^{2+}}}{[Fe^{3+}]f_{Fe^{3+}}}$$

where $f_{Fe^{2+}}$ and $f_{Fe^{3+}}$ are the respective activity coefficients. The activity coefficients of the two species are less than one in this system because of the high ionic strength imparted by the perchloric acid and the iron salts. More important, however, the activity coefficient of the iron(III) ion is smaller than that of the iron(II) ion, inasmuch as the effects of ionic strength on these coefficients increase with the charge on the ion (p. 508). As a consequence, the ratio of the activity coefficients as they appear in the Nernst equation would be larger than one and the potential of the half-cell would be smaller than the standard potential.

Activity coefficient data for ions in solutions of the types commonly encountered in oxidation-reduction titrations and electrochemical work are fairly limited; consequently, molar concentrations rather than activities must be used in many calculations. Appreciable errors may result.

Effect of Other Equilibria. The application of standard electrode potentials is further complicated by the occurrence of solvolysis, dissociation, association, and complex-formation reactions involving the species of interest. The equilibrium constants required to correct for these effects are frequently unknown. Lingane[8] cites the ferrocyanide/ferricyanide couple as an excellent example of this problem:

$$Fe(CN)_6^{3-} + e \rightleftharpoons Fe(CN)_6^{4-}$$

$$E^0 = +0.356 \text{ V}$$

Although the hydrogen ion does not appear in this half-reaction, the experimentally measured potential is markedly affected by pH. Thus, instead of the expected value of $+0.356$ V, solutions containing equiformal concentrations of the two species yield potentials of $+0.71$, $+0.56$, and $+0.48$ V with respect to the standard hydrogen electrode when the measurements are made in media that are respec-

---

[8] J. J. Lingane, *Electroanalytical Chemistry*, 2d ed. New York: Interscience, 1958, p. 59.

tively $1.0F$, $0.1F$, and $0.01F$ in hydrochloric acid. These differences are attributable to the difference in the degree of association of ferrocyanide and ferricyanide ions with hydrogen ions. The hydroferrocyanic acids are weaker than the hydroferricyanic acids; thus, the concentration of the ferrocyanide ion is lowered more than that of the ferricyanide ion as the acid concentration increases. This effect tends in turn to shift the oxidation-reduction equilibrium to the right and leads to more positive electrode potentials.

A somewhat analogous effect is encountered in the behavior of the potential of the iron(III)/iron(II) couple. As noted earlier, an equiformal mixture of these two ions in $1F$ perchloric acid has an electrode potential of $+0.73$ V. Substitution of hydrochloric acid of the same concentration alters the observed potential to $+0.70$ V; a value of $+0.6$ V is observed in $1F$ phosphoric acid. These differences arise because iron(III) forms more stable complexes with chloride and phosphate ions than does iron(II). As a result, the actual concentration of uncomplexed iron(III) in such solutions is less than that of uncomplexed iron(II), and the net effect is a shift in the observed potential.

Phenomena such as these can be taken into account only if the equilibria involved are known and constants for the processes are available. Often, however, such information is lacking; the chemist is then forced to neglect such effects and hope that serious errors do not flaw the calculated results.

Formal Potentials. In order to compensate partially for activity effects and errors resulting from side reactions, Swift[9] has proposed the use of a quantity called the *formal potential* in place of the standard electrode potential in oxidation-reduction calculations. The formal potential of a system is the potential of the half-cell with respect to the standard hydrogen electrode when the *concentrations* of reactants and products are $1F$ and the concentrations of any other constituents of the solution are carefully specified. Thus, for example, the formal potential for the reduction of iron(III) is $+0.732$ V in $1F$ perchloric acid and $+0.700$ V in $1F$ hydrochloric acid; similarly, the formal potential for the reduction of ferricyanide ion would be $+0.71$ V in $1F$ hydrochloric acid and $+0.48$ V in a $0.01F$ solution of this acid. Use of these values in place of the standard electrode potential in the Nernst equation will yield better agreement between calculated and experimental potentials, provided the electrolyte concentration of the solution approximates that for which the formal potential was measured. Application of formal potentials to systems differing greatly as to kind and concentration of electrolyte can, however, lead to errors greater than those encountered with the use of standard potentials. The table in Appendix 2 contains selected formal potentials as well as standard potentials; in subsequent chapters, we shall use whichever is the more appropriate.

Reaction Rates. It should be realized that the existence of a half-reaction in a table of electrode potentials does not necessarily imply that there is a real electrode whose potential will respond to the half-reaction. Many of the data in such tables have been obtained by calculations based upon equilibrium or thermal measurements rather than from the actual measurement of the potential for an electrode system. For some, no suitable electrode is known; thus, the standard electrode potential for the process,

$$2CO_2 + 2H^+ + 2e \rightleftharpoons H_2C_2O_4$$

$$E^0 = -0.49 \text{ V}$$

has been arrived at indirectly. The reaction is not reversible, and the rate at which carbon dioxide combines to give oxalic acid is negligibly slow. No electrode system is known whose potential varies with the ratio of activi-

[9] E. H. Swift, *A System of Chemical Analysis*. San Francisco: Freeman, 1939, p. 50.

ties of the reactants and products. Nonetheless, the potential is useful for computational purposes.

CALCULATION OF CELL POTENTIALS FROM ELECTRODE POTENTIALS

An important use of standard electrode potentials is the calculation of the potential obtainable from a galvanic cell or the potential required to operate an electrolytic cell. These calculated potentials (sometimes called *thermodynamic potentials*) are theoretical in the sense that they refer to cells in which there is essentially no current; additional factors must be taken into account where a current is involved.

Calculation of Thermodynamic Cell Potentials

As shown earlier (Equation 18-12), the electromotive force of a cell is obtained by combining half-cell potentials as follows:

$$E_{cell} = E_{cathode} - E_{anode}$$

where E_{anode} and $E_{cathode}$ are the *electrode potentials* for the two half-reactions constituting the cell.

Consider the hypothetical cell

$$Zn \,|\, ZnSO_4(a_{Zn2+} = 1.00) \|$$
$$CuSO_4(a_{Cu2+} = 1.00) \,|\, Cu$$

The overall cell process involves the oxidation of elemental zinc to zinc(II) and the reduction of copper(II) to the metallic state. Because the activities of the two ions are specified as unity, the standard potentials are also the electrode potentials. The cell diagram also specifies that the zinc electrode is the anode. Thus, using E^0 data from Table 18-2,

$$E_{cell} = +0.337 - (-0.763) = +1.100 \text{ V}$$

The positive sign for the cell potential indicates that the reaction

$$Zn(s) + Cu^{2+} \rightarrow Zn^{2+} + Cu(s)$$

occurs spontaneously and that this is a galvanic cell.

The foregoing cell, diagrammed as

$$Cu \,|\, Cu^{2+}(a_{Cu2+} = 1.00)\|$$
$$Zn^{2+}(a_{Zn2+} = 1.00) \,|\, Zn$$

implies that the copper electrode is now the anode. Thus,

$$E_{cell} = -0.763 - (+0.337) = -1.100 \text{ V}$$

The negative sign indicates the nonspontaneity of the reaction

$$Cu(s) + Zn^{2+} \rightarrow Cu^{2+} + Zn(s)$$

The application of an external potential greater than 1.100 V would be required to cause this reaction to occur.

EXAMPLE

Calculate the potentials for the following cell employing (a) concentrations and (b) activities:

$$Zn \,|\, ZnSO_4(xF), \; PbSO_4(sat'd) \,|\, Pb$$

where $x = 5.00 \times 10^{-4}$, 2.00×10^{-3}, 1.00×10^{-2}, 2.00×10^{-2}, and 5.00×10^{-2}.

(a) In a neutral solution, little HSO_4^- will be formed; thus, we may assume that

$$[SO_4^{2-}] = F_{ZnSO_4} = x = 5.00 \times 10^{-4}$$

The half-reactions and standard potentials are

$$PbSO_4(s) + 2e \rightleftharpoons Pb(s) + SO_4^{2-}$$
$$E^0 = -0.350 \text{ V}$$
$$Zn^{2+} + 2e \rightleftharpoons Zn \qquad E^0 = -0.763 \text{ V}$$

The potential of the lead electrode is given by

$$E_{Pb} = -0.350 - \frac{0.0591}{2} \log 5.00 \times 10^{-4}$$
$$= -0.252 \text{ V}$$

For the zinc half-reaction,

$$[Zn^{2+}] = 5.00 \times 10^{-4}$$

and

$$E_{Zn} = -0.763 - \frac{0.0591}{2} \log \frac{1}{5.00 \times 10^{-4}}$$

$$= -0.860 \text{ V}$$

Since the Pb electrode is specified as the cathode,

$$E_{cell} = -0.252 - (-0.860) = 0.608$$

Cell potentials at the other concentrations can be derived in the same way. Their values are tabulated at the bottom of the page.

(b) To obtain activity coefficients for Zn^{2+} and SO_4^{2-} we must first calculate the ionic strength with the aid of Equation 18-3.

$$\mu = \tfrac{1}{2}[5.00 \times 10^{-4} \times (2)^2$$
$$+ 5.00 \times 10^{-4} \times (2)^2]$$

$$= 2.00 \times 10^{-3}$$

In Table 18-1, we find for SO_4^{2-}, $\alpha_A = 4.0$ and for Zn^{2+}, $\alpha_A = 6.0$. Substituting these values into Equation 18-4 gives for sulfate ion

$$-\log f_{SO_4} = \frac{0.5085 \times 2^2 \times \sqrt{2.00 \times 10^{-3}}}{1 + 0.3281 \times 4.0\sqrt{2.00 \times 10^{-3}}}$$

$$= 8.59 \times 10^{-2}$$

$$f_{SO_4} = 0.820$$

Repeating the calculations employing $\alpha_A = 6.0$ for Zn^{2+} yields

$$f_{Zn} = 0.825$$

The Nernst equation for the Pb electrode now becomes

$$E_{Pb} = -0.350 - \frac{0.0591}{2} \times$$
$$\log 0.820 \times 5.00 \times 10^{-4}$$

$$= -0.250$$

For the zinc electrode

$$E_{Zn} = -0.763 - \frac{0.0591}{2} \times$$
$$\log \frac{1}{0.825 \times 5.00 \times 10^{-4}}$$

$$= -0.863$$

and

$$E_{cell} = -0.250 - (-0.863) = 0.613 \text{ V}$$

Values at other concentrations are listed at the bottom of the page.

It is of interest to compare the calculated cell potentials shown in the columns labeled (a) and (b) in the foregoing example with the experimental results shown in the last column. Clearly, the use of activities provides a significant improvement at the higher ionic strengths.

EXAMPLE

Calculate the potential required to initiate the deposition of copper from a solution that is $0.010F$ in $CuSO_4$ and contains sufficient sulfuric acid to give a hydrogen ion concentration of $1.0 \times 10^{-4}M$.

x	μ	(a) E(calc)	(b) E(calc)	E(exptl)[a]
5.00×10^{-4}	2.00×10^{-3}	0.608	0.613	0.611
2.00×10^{-3}	8.00×10^{-3}	0.572	0.582	0.583
1.00×10^{-2}	4.00×10^{-2}	0.531	0.549	0.553
2.00×10^{-2}	8.00×10^{-2}	0.513	0.537	0.542
5.00×10^{-2}	2.00×10^{-1}	0.490	0.521	0.529

[a] Experimental data from: I. A. Cowperthwaite and V. K. LaMer, *J. Amer. Chem. Soc.,* **53**, 4333 (1931).

The deposition of copper necessarily occurs at the cathode. Because no easily oxidizable species are present, the anode reaction will involve oxidation of H_2O to give O_2. From the table of standard potentials, we find

$$Cu^{2+} + 2e \rightleftharpoons Cu(s) \qquad E^0 = +0.337 \text{ V}$$

$$O_2(g) + 4H^+ + 4e \rightarrow 2H_2O \quad E^0 = +1.229 \text{ V}$$

Thus, for the copper electrode

$$E = +0.337 - \frac{0.0591}{2} \log \frac{1}{0.010} = +0.278 \text{ V}$$

Assuming that O_2 is evolved at 1.00 atm, the potential for the oxygen electrode is

$$E = +1.229 - \frac{0.0591}{4}$$

$$\log \frac{1}{(1.00)(1.0 \times 10^{-4})^4}$$

$$= +0.993 \text{ V}$$

The cell potential is then

$$E_{cell} = +0.278 - 0.993 = -0.715 \text{ V}$$

Thus, to initiate the reaction

$$2Cu^{2+} + 2H_2O \rightarrow O_2(g) + 4H^+ + 2Cu(s)$$

would require the application of a potential greater than 0.715 V.

Liquid Junction Potential

When two electrolyte solutions of different composition are brought in contact with one another, a potential develops at the interface. This *junction potential* arises from an unequal distribution of cations and anions across the boundary due to differences in the rates at which these species migrate.

Consider the liquid junction that exists in the system

$$HCl(1F) | HCl(0.01F)$$

Both hydrogen ions and chloride ions tend to diffuse across this boundary from the more concentrated to the more dilute solution, the

driving force for this migration being proportional to the concentration difference. The rate at which various ions move under the influence of a fixed force varies considerably (that is, their *mobilities* are different). In the present example, hydrogen ions are several times more mobile that chloride ions. As a consequence, there is a tendency for the hydrogen ions to outstrip the chloride ions as diffusion takes place; a separation of charge is the net result (see Figure 18-8). The more dilute side of the boundary becomes positively charged owing to the more rapid migration of hydrogen ions; the concentrated side, therefore, acquires a negative charge from the slower-moving chloride ions. The charge that develops tends to counteract the differences in mobilities of the two ions, and as a consequence, an equilibrium condition soon develops. The junction potential difference resulting from this charge separation may amount to 30 mV or more.

In a simple system such as that shown in Figure 18-8, the magnitude of the junction

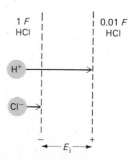

FIGURE 18-8   Schematic representation of a liquid junction showing the source of the junction potential E_j. The length of the arrows correspond to the relative mobility of the two ions.

potential can be calculated from a knowledge of the mobilities of the two ions involved. However, it is seldom that a cell of analytical importance has a sufficiently simple composition to permit such a computation.

It is an experimental fact that the magnitude of the liquid junction potential can be greatly decreased by interposition of a concentrated electrolyte solution (a *salt bridge*) between the two solutions. The effectiveness of this contrivance improves as the concentration of the salt in the bridge increases and as the mobilities of the ions of the salt approach one another in magnitude. A saturated potassium chloride solution is good from both standpoints, its concentration being somewhat greater than $4F$ at room temperature, and the mobility of its ions differing by only 4%. With such a bridge, the junction potential typically amounts to a few millivolts or less, a negligible quantity in many, but not all, analytical measurements.

EFFECT OF CURRENT ON CELL POTENTIALS

When electricity flows in an electrochemical cell, the overall potential may be influenced by three additional phenomena, namely, ohmic potential, concentration polarization, and kinetic polarization.

Ohmic Potential; *IR* Drop

To develop a current in either a galvanic or an electrolytic cell, a driving force or a potential is required to overcome the resistance of the ions to movement toward the anode or the cathode. Just as in metallic conduction, this force follows Ohm's law and is equal to the product of the current in amperes and the resistance of the cell in ohms. The force is generally referred to as the *ohmic potential*, or the *IR drop*.

The net effect of IR drop is to increase the potential required to operate an electrolytic cell and to decrease the measured potential of a galvanic cell. Therefore, the IR drop is always *subtracted* from the theoretical cell potential. That is,[10]

$$E_{\text{cell}} = E_{\text{cathode}} - E_{\text{anode}} - IR \quad (18\text{-}18)$$

EXAMPLE

1. Calculate the potential when 0.100 A is drawn from the galvanic cell

$$\text{Cd} \,|\, \text{Cd}^{2+}(0.0100M) \,\|\, \text{Cu}^{2+}(0.0100M) \,|\, \text{Cu}$$

Assume a cell resistance of 4.00 Ω.

Substitution into the Nernst equation reveals that the electrode potential for the Cu electrode is 0.278 V, while for the Cd electrode it is -0.462 V. Thus, the cell potential is

$$E = E_{\text{Cu}} - E_{\text{Cd}}$$
$$= 0.278 - (-0.462) = 0.740 \text{ V}$$
$$E_{\text{cell}} = 0.740 - IR$$
$$= 0.740 - (0.100 \times 4.00) = 0.340 \text{ V}$$

Note that the emf of this cell drops dramatically in the presence of a current.

2. Calculate the potential required to generate a current of 0.100 A in the reverse direction in the foregoing cell.

$$E = E_{\text{Cd}} - E_{\text{Cu}}$$
$$= -0.462 - 0.278 = -0.740 \text{ V}$$
$$E_{\text{cell}} = -0.740 - (0.100 \times 4.00)$$
$$= -1.140 \text{ V}$$

Here, an external potential greater than 1.140 V would be needed to cause Cd^{2+} to deposit and Cu to dissolve at a rate required for a current of 0.100 A.

---

[10] Here and in the subsequent discussion we will assume the junction potential is negligible relative to the other potentials.

Polarization Effects

The linear relationship between potential and the instantaneous current in a cell (Equation 18-18) is frequently observed experimentally when I is small; at high currents, however, marked departures from linearity occur. Under these circumstances, the cell is said to be *polarized* (see Figure 18-9). Thus, a polarized electrolytic cell requires application of potentials larger than theoretical for a given current; similarly, a polarized galvanic cell develops potentials that are smaller than predicted. The polarization of a cell may be so extreme that the current becomes essentially independent of the voltage; under these circumstances, polarization is said to be complete.

Polarization is an electrode phenomenon; either or both electrodes in a cell can be affected. Included among the factors influencing the extent of polarization are the size, shape, and composition of the electrodes; the composition of the electrolyte solution; the temperature and the rate of stirring; the magnitude of the current; and the physical states of the species involved in the cell reaction. Some of these factors are sufficiently understood to permit quantitative statements concerning their effects upon cell processes; others, however, can be accounted for on an empirical basis only.

For purposes of discussion, polarization phenomena are conveniently divided into the two categories of *concentration polarization* and *kinetic polarization* (also called *overvoltage* or *overpotential*).

Concentration Polarization. When the reaction at an electrode is rapid and reversible, the concentration of the reacting species in the layer of solution immediately adjacent to the electrode is always that which would be predicted from the Nernst equation. Thus, the cadmium ion concentration C_0 in the immediate vicinity of a cadmium electrode will always be given by

$$E = E_{Cd}^0 - \frac{0.0591}{2} \log \frac{1}{C_0}$$

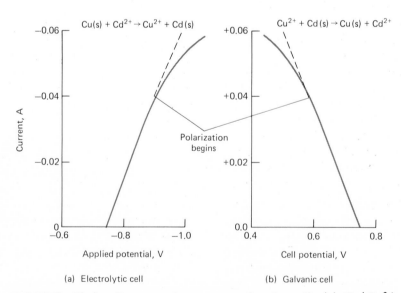

FIGURE 18-9 Current-voltage curves for the cell: (a) $Cu|Cu^{2+}$ $(1.00\ M)\|Cd^2(1.00\ M)|Cd$, (b) $Cd|Cd^{2+}(1.00\ M)\|Cu^{2+}(1.00\ M)|Cu$.

irrespective of the concentration of this cation in the bulk of the solution. The reduction of cadmium ions is rapid and reversible; consequently, *the concentration of this ion in the film of liquid surrounding the cadmium electrode is determined at any instant by the potential of that electrode* at that instant. If the potential is changed, deposition or dissolution of cadmium occurs such that the cadmium ion in the surface film rapidly attains the concentration required by the Nernst equation.

In contrast to this substantially instantaneous surface process, the rate at which equilibrium between the electrode and the bulk of the solution is attained can be very slow, depending upon the magnitude of the current as well as the volume of the solution and its solute concentration.

When a sufficient potential is applied to this electrode, cadmium ions are reduced, and an instantaneous current is generated. For current to continue at a level predicted by Equation 18-18, however, an additional supply of the cation must be transported at a suitable rate into the surface film surrounding the electrode. If the demand for reactant cannot be met by this mass transfer, concentration polarization will set in, and *lowered currents must result*. This type of polarization, then, occurs when the rate of transfer of reactive species between the bulk of the solution and the electrode surface is inadequate to maintain the current at the level required by Ohm's law. A departure from linearity such as that shown in Figure 18-9a is the consequence. Inadequate material transport will also cause concentration polarization in a galvanic cell such as that described by the curve in Figure 18-9b; here, however, the current would be limited by the transport of copper ions.

Ions or molecules can be transported through a solution by: (1) diffusion; (2) electrostatic attraction or repulsion; and (3) mechanical or convection forces. We must therefore consider the variables that influence these

forces as they relate to electrode processes.

Whenever a concentration gradient develops in a solution, molecules or ions diffuse from the more concentrated to the more dilute region. The rate at which transfer occurs is proportional to the concentration difference. In an electrolysis, a gradient is established as a result of ions being removed from the film of solution adjacent to the cathode. Diffusion then occurs, the rate being expressed by the relationship

Rate of diffusion to cathode surface

$$= k(C - C_0) \qquad (18\text{-}19)$$

where C is the reactant concentration in the bulk of the solution, C_0 is its equilibrium concentration at the cathode surface, and k is a proportionality constant. *The value of C_0 is fixed by the potential of the electrode and can be calculated from the Nernst equation.* As higher potentials are applied to the electrode, C_0 becomes smaller and smaller and the diffusion rate greater and greater.

Electrostatic forces also influence the rate at which an ionic reactant migrates to or from an electrode surface. The electrostatic attraction (or repulsion) between a particular ionic species and the electrode becomes smaller as the total electrolyte concentration of the solution is increased. It may approach zero when the reactive species is but a small fraction of the total concentration of ions with a given charge.

Clearly, reactants can be transported to an electrode by mechanical means. Thus, stirring or agitation will aid in decreasing concentration polarization. Convection currents, due to temperature or density differences, also tend to offset concentration polarization.

To summarize, then, concentration polarization occurs when the forces of diffusion, electrostatic attraction, and mechanical mixing are insufficient to transport the reactant to or from an electrode surface at a rate demanded by the theoretical current. Concentration polarization causes the potential of a galvanic cell to be smaller than the value predicted on the basis of the theoretical potential and the *IR* drop (Figure 18-9b). Similarly, in an electrolytic cell, a potential more negative than theoretical is required in order to maintain a given current (Figure 18-9a).

Concentration polarization is important in several electroanalytical methods. In some applications, steps are taken to eliminate it; in others, however, it is essential to the method, and every effort is made to promote its occurrence. The degree of concentration polarization is influenced experimentally by: (1) the reactant concentration; (2) the total electrolyte concentration; (3) mechanical agitation; and (4) the size of the electrodes; as the area toward which a reactant is transported becomes greater, polarization effects become smaller.

Kinetic Polarization. Kinetic polarization results when the rate at which the electrochemical reaction at one or both electrodes is slow; here, an additional potential (the *overvoltage* or *overpotential*) is required to overcome the energy barrier to the half-reaction. In contrast to concentration polarization, the current is controlled by the *rate of the electron-transfer process* rather than by the rate of mass transfer.

Although exceptions can be cited, some empirical generalizations can be made regarding the magnitude of overvoltage.

1. Overvoltage increases with current density (current density is defined as the amperes per square centimeter of electrode surface).
2. It usually decreases with increases in temperature.
3. Overvoltage varies with the chemical composition of the electrode, often being most pronounced with softer metals such as tin, lead, zinc, and particularly mercury.
4. Overvoltage is most marked for electrode processes that yield gaseous products such as hydrogen or oxygen; it is frequently negligible where a metal is being deposited or where an ion is undergoing a change of oxidation state.

5. The magnitude of overvoltage in any given situation cannot be predicted exactly, because it is determined by a number of uncontrollable variables.[11]

Overvoltage associated with the evolution of hydrogen and oxygen is of particular interest to the chemist. Table 18-3 presents data that depict the extent of the phenomenon under specific conditions. The difference between the overvoltage of these gases on smooth and on platinized platinum surfaces is of particular interest. This difference is primarily due to the much larger surface area associated with platinized electrodes, which results in a *real* current density that is significantly smaller than is apparent from the overall dimensions of the electrode. A platinized surface is always employed in construction of hydrogen reference electrodes, so as to lower the current density to a point where the overvoltage is negligible.

The high overvoltage associated with the formation of hydrogen permits the electrolytic deposition of several metals that require potentials at which hydrogen would otherwise be expected to interfere. For example, it is readily shown from their standard potentials that rapid formation of hydrogen should occur well before a potential sufficient for the deposition of zinc from a neutral solution can be realized. Nevertheless, quantitative deposition of zinc can be attained provided a mercury or copper electrode is used; because of the high overvoltage of hydrogen on these metals, little or no gas is evolved during the electrodeposition.

The magnitude of overvoltage can, at best, be only crudely approximated from empirical information available in the literature. Calculation of cell potentials in which overvoltage plays a part cannot, therefore, be very accurate.

---

[11] Overvoltage data for various gaseous species at different electrode surfaces are to be found in: *Handbook of Analytical Chemistry*, ed. L. Meites. New York: McGraw-Hill, p. 5-184.

As with *IR* drop, the overvoltage is subtracted from the theoretical cell potential.

REFERENCE ELECTRODES

In many electroanalytical applications, it is desirable that the half-cell potential of one electrode be known, constant, and completely insensitive to the composition of the solution under study. An electrode that fits this description is called a *reference electrode*.[12] Employed in conjunction with the reference electrode will be an *indicator electrode*, whose response depends upon the analyte concentration.

A reference electrode should be easy to as-

semble and should maintain an essentially constant and reproducible potential in the presence of small currents. Several electrode systems meet these requirements.

Calomel Electrodes

Calomel half-cells may be represented as follows:

$$\|Hg_2Cl_2(\text{sat'd}), KCl(xF)\,|\,Hg$$

where x represents the formal concentration of potassium chloride in the solution. The electrode reaction is given by the equation

$$Hg_2Cl_2(s) + 2e \rightleftharpoons 2Hg(l) + 2Cl^-$$

The potential of this cell will vary with the chloride concentration x, and this quantity must be specified in describing the electrode.

Table 18-4 lists the composition and the potentials for the three most commonly en-

---

[12] For a description of commercially available reference electrodes, see: R. D. Caton, Jr., *J. Chem. Educ.*, **50**, A571 (1973); **51**, A7 (1974).

TABLE 18-3 OVERVOLTAGE FOR HYDROGEN AND OXYGEN FORMATION AT VARIOUS ELECTRODES AT 25°C[a]

Electrode Composition	Overvoltage (V) (Current Density 0.001 A/cm²)		Overvoltage (V) (Current Density 0.01 A/cm²)		Overvoltage (V) (Current Density 1 A/cm²)	
	H_2	O_2	H_2	O_2	H_2	O_2
Smooth Pt	0.024	0.721	0.068	0.85	0.676	1.49
Platinized Pt	0.015	0.348	0.030	0.521	0.048	0.76
Au	0.241	0.673	0.391	0.963	0.798	1.63
Cu	0.479	0.422	0.584	0.580	1.269	0.793
Ni	0.563	0.353	0.747	0.519	1.241	0.853
Hg	0.9[b]		1.1[c]		1.1[d]	
Zn	0.716		0.746		1.229	
Sn	0.856		1.077		1.231	
Pb	0.52		1.090		1.262	
Bi	0.78		1.05		1.23	

[a] National Academy of Sciences, *International Critical Tables*. New York: McGraw-Hill, 1929, vol. 6, pp. 339–340. (With permission.)
[b] 0.556 V at 0.000077 A/cm²; 0.929 V at 0.00154 A/cm².
[c] 1.063 V at 0.00769 A/cm².
[d] 1.126 V at 1.153 A/cm².

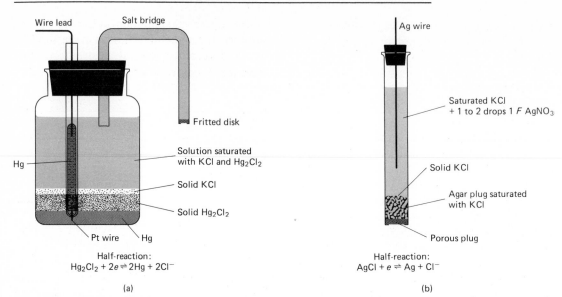

FIGURE 18-10 Two easily constructed reference electrodes. (a) A saturated calomel electrode. (b) A silver/silver chloride electrode.

countered calomel electrodes. Note that each solution is saturated with mercury(I) chloride and that the cells differ only with respect to the potassium chloride concentration. Note also that the potential of the normal calomel electrode is greater than the standard potential for the half-reaction ($E^0 = 0.268$ V) because the chloride ion *activity* in a 1F solution of potassium chloride is significantly smaller than one. The last column in Table 18-4 gives expressions that permit the calculation of elec-

trode potentials for calomel half-cells at temperatures t other than 25°C.

The saturated calomel electrode (SCE) is most commonly used by the analytical chemist because of the ease with which it can be prepared. Compared with the other two, however, its temperature coefficient is somewhat larger.

A simple, easily constructed saturated calomel electrode is shown in Figure 18-10a. The salt bridge, a tube filled with saturated potassium chloride, provides electric contact with

TABLE 18-4 SPECIFICATIONS OF CALOMEL ELECTRODES

Name	Concentration of Hg$_2$Cl$_2$	KCl	Electrode Potential (V) vs. Standard Hydrogen Electrode Hg$_2$Cl$_2$(s) + 2e $\rightleftarrows$ 2Hg(l) + 2Cl$^-$
Saturated	Saturated	Saturated	$+0.241 - 6.6 \times 10^{-4}(t - 25)$
Normal	Saturated	1.0F	$+0.280 - 2.8 \times 10^{-4}(t - 25)$
Decinormal	Saturated	0.1F	$+0.334 - 8.8 \times 10^{-5}(t - 25)$

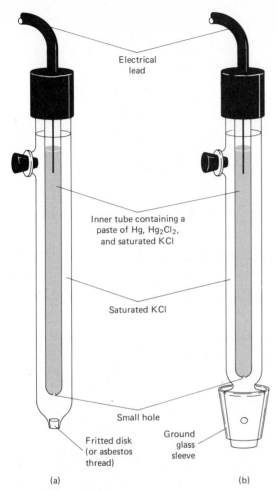

FIGURE 18-11 Typical commercial calomel reference electrodes.

is 5 to 15 cm in length and 0.5 to 1.0 cm in diameter. A mercury/mercury(I) chloride paste is contained in an inner tube that is connected to the saturated potassium chloride solution in the outer tube through a small opening. For electrode (a), contact with the second half-cell is made by means of a fritted disk or a porous fiber sealed in the end of the outer tubing. This type of junction has a relatively high resistance (2000 to 3000 Ω) and a limited current-carrying capacity; on the other hand, contamination of the analyte solution due to leakage is minimal. The electrode shown in Figure 18-11b has a much lower resistance but tends to leak small amounts of saturated potassium chloride into the sample. Before it is used, the ground glass collar of this electrode is loosened and turned so that a drop or two of the KCl solution flows from the hole and wets the entire inner ground surface. Better electrical contact to the analyte solution is thus established.

Silver/Silver Chloride Electrodes

A reference electrode system analogous to the calomel electrode consists of a silver electrode immersed in a solution of potassium chloride that has been saturated with silver chloride

$$\|AgCl(\text{sat'd}), KCl(xF)\,|\,Ag$$

The half-reaction is

$$AgCl(s) + e \rightleftharpoons Ag(s) + Cl^-$$

Normally, this electrode is prepared with a saturated potassium chloride solution, the potential at 25°C being 0.197 V with respect to the standard hydrogen electrode.

A simple and easily constructed silver/silver chloride electrode is shown in Figure 18-10b. The electrode is contained in a Pyrex tube fitted with a 10-mm fritted-glass disk. A plug of agar gel saturated with potassium chloride is formed on top of the disk to prevent loss of solution from the half-cell. The plug can be prepared by heating 4 to 6 g of pure agar in 100 ml of water until a clear solution is obtained; about 35 g of potassium chloride are then added. A

the solution surrounding the indicator electrode. A fritted disk or wad of cotton at one end of the salt bridge is often employed to prevent siphoning of the cell liquid and contamination of the solutions by foreign ions; alternatively, the tube can be filled with a 5% agar gel that has been saturated with potassium chloride.

Several convenient calomel electrodes are available commercially; typical are the two illustrated in Figure 18-11. The body of each electrode consists of an outer glass tube that

portion of this mixture, while still warm, is poured into the tube; upon cooling, it solidifies to a gel with low electrical resistance. A layer of solid potassium chloride is placed on the gel, and the tube is filled with a saturated solution of the salt. A drop or two of $1F$ silver nitrate is then added and a heavy gauge (1- to 2-mm diameter) silver wire is inserted in the solution.

Silver/silver chloride reference electrodes are also available from commercial sources. Their construction is similar to that of the electrodes shown in Figure 18-11.

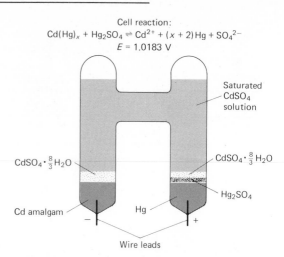

Cell reaction:
$$Cd(Hg)_x + Hg_2SO_4 \rightleftharpoons Cd^{2+} + (x+2)Hg + SO_4^{2-}$$
$$E = 1.0183 \text{ V}$$

FIGURE 18-12 A Weston standard cell (saturated).

STANDARD WESTON CELLS

The accurate measurement of potentials for electrochemical work requires the frequent use of a cell whose emf is precisely known. The *Weston cell*, which is used almost universally for this purpose, can be represented as follows:

$$Cd(Hg) \,|\, CdSO_4 \cdot 8/3H_2O(sat'd),$$
$$Hg_2SO_4(sat'd) \,|\, Hg$$

The half-reactions as they occur in the cell are

$$Cd(Hg)_x \rightarrow Cd^{2+} + xHg(l) + 2e$$
$$Hg_2SO_4(s) + 2e \rightarrow 2Hg(l) + SO_4^{2-}$$

Figure 18-12 shows a typical Weston cell. Its theoretical potential at 25°C is 1.0183 V.

The emf of a Weston cell is governed by the activities of the cadmium and mercury(I) ions in the solution. At any given temperature, these quantities are invariant, being fixed by the solubilities of the cadmium sulfate and the mercury(I) sulfate. As a result, Weston cells remain constant in voltage for remarkably long periods of time, provided large currents are not drawn from them.

The Weston cell has a temperature coefficient of about -0.04 mV/°C, due primarily to changes in solubility of the cadmium(II) and mercury(I) salts. A cell with a coefficient about one-fourth this magnitude is obtained by using a solution of cadmium sulfate that has been saturated at 4°C; no solid $CdSO_4 \cdot 8/3H_2O$ is incorporated in the cell itself. Most commercially available cells are of this type and are termed *unsaturated Weston cells*. Their potentials lie between 1.0185 and 1.0195 V.

Weston cells may be sent to the National Bureau of Standards for calibration and certification.

PROBLEMS   ✔ 1. Calculate the electrode potentials of the following half-cells.
   (a) $Ag^+(0.0300M) \,|\, Ag$
   (b) $Ni^{2+}(0.662M) \,|\, Ni$
   (c) $HCl(1.50 \times 10^{-4}F) \,|\, H_2(1.00 \text{ atm}), Pt$
   (d) $Fe^{3+}(2.35 \times 10^{-4}M), Fe^{2+}(0.100M) \,|\, Pt$
   (e) $AgBr(sat'd), Br^-(4.50M) \,|\, Ag$

2. Calculate the electrode potentials of the following half-cells.
   (a) $Fe^{2+}(0.132M)|Fe$
   (b) $BiO^+(2.00 \times 10^{-3}M)$, $H^+(3.00 \times 10^{-2}M)|Bi$
   (c) $HCl(4.50F)|H_2(1.00$ atm$)$, Pt
   (d) $HOAc(0.0100F)|H_2(1.00$ atm$)$, Pt
   (e) $IO_3^-(0.100M)$, $I_2(0.235M)$, $H^+(2.00 \times 10^{-3}M)|Pt$
   (f) $Ag_2CrO_4($sat'd$)$, $CrO_4^{2-}(0.0400M)|Ag$

3. Calculate the potential of a silver electrode in contact with the following.
   (a) $1.00 \times 10^{-4}M$ Ag^+
   (b) $0.060M$ Ag^+
   (c) a solution that is $0.0400M$ in I^- and saturated with AgI
   (d) a solution that is $0.0050M$ in CN^- and $0.0300M$ in $Ag(CN)_2^-$
   (e) the solution that results from mixing 25.0 ml of $0.0500F$ KBr with 20.0 ml of $0.100M$ Ag^+
   (f) the solution that results from mixing 25.0 ml of $0.050M$ Ag^+ with 20.0 ml of $0.100F$ KBr

4. Calculate the electrode potentials for the following systems.
   (a) $V^{3+}(3.00 \times 10^{-3}M)$, $VO^{2+}(1.50M)$, $HCl(0.100F)|Pt$
   (b) $MnO_4^-(0.200M)$, $Mn^{2+}(0.0500M)$, $H^+(0.0800M)|Pt$

5. Calculate the electrode potentials for the following systems.
   (a) $Cr_2O_7^{2-}(5.00 \times 10^{-3}M)$, $Cr^{3+}(1.00 \times 10^{-2}M)$, $H^+(0.100M)|Pt$
   (b) $UO_2^+(0.100M)$, $U^{4+}(0.200M)$, $H^+(0.300M)|Pt$

6. Calculate the theoretical potential of each of the following cells. Is the cell as written galvanic or electrolytic?
   (a) $Pt|Cr^{3+}(2.00 \times 10^{-4}M)$, $Cr^{2+}(1.00 \times 10^{-1}M)\|Pb^{2+}(4.00 \times 10^{-2}M)|Pb$
   (b) $Pt|Sn^{4+}(5.00 \times 10^{-4}M)$, $Sn^{2+}(3.00 \times 10^{-2}M)\|Ag(CN)_2^-(2.5 \times 10^{-3}M)$, $CN^-(4.00 \times 10^{-2}M)|Ag$
   (c) $Hg|Hg_2^{2+}(2.00 \times 10^{-2}M)\|H^+(3.00 \times 10^{-2}M)$, $V^{3+}(8.00 \times 10^{-2}M)$, $VO^{2+}(4.00 \times 10^{-3}M)|Pt$
   (d) $Pt|Fe^{3+}(4.00 \times 10^{-2}M)$, $Fe^{2+}(6.00 \times 10^{-5}M)\|Sn^{2+}(5.00 \times 10^{-2}M)$, $Sn^{4+}(2.00 \times 10^{-4}M)|Pt$
   (e) $Mn|Mn^{2+}(7.50 \times 10^{-3}M)\|Ag(CN)_2^-(0.0800M)$, $CN^-(0.0300M)|Ag$

7. Calculate the theoretical potential of each of the following cells. Is the cell galvanic or electrolytic as written?
   (a) $Pt|Fe(CN)_6^{4-}(4.00 \times 10^{-2}M)$, $Fe(CN)_6^{3-}(8.00 \times 10^{-4}M)\|I^-(2.00 \times 10^{-3}M)$, $I_2(3.00 \times 10^{-2}M)|Pt$
   (b) $Pt|V^{3+}(4.00 \times 10^{-3}M)$, $V^{2+}(0.800M)\|Fe^{2+}(3.00 \times 10^{-4}M)|Fe$
   (c) $Bi|BiO^+(0.0400M)$, $H^+(2.00 \times 10^{-2}M)\|I^-(0.200M)$, $AgI($sat'd$)|Ag$
   (d) $Zn|Zn^{2+}(3.00 \times 10^{-4}M)\|Fe(CN)_6^{4-}(4.00 \times 10^{-2}M)$, $Fe(CN)_6^{3-}(7.00 \times 10^{-2}M)|Pt$
   (e) Pt, $H_2(0.100$ atm$)|HCl(5.00 \times 10^{-3}F)$, $AgCl($sat'd$)|Ag$

8. Calculate the potential of each of the following cells. Is the cell galvanic or electrolytic as written?
   (a) $Cu|CuI(\text{sat'd}), I^-(0.250M)\|I^-(3.50 \times 10^{-3}M), CuI(\text{sat'd})|Cu$
   (b) $Pt, H_2(0.800 \text{ atm})|H^+(3.00 \times 10^{-6}M)\|H^+(2.00 \times 10^{-2}M)|$
   $H_2(0.200 \text{ atm}), Pt$

9. Compute E^0 for the process

$$Ni(CN)_4^- + 2e \rightleftharpoons Ni(s) + 4CN^-$$

given that the formation constant for the complex is 1.0×10^{22}.

10. The solubility product constant for TlBr is 3.4×10^{-6} at 25°C. Calculate E^0 for the process

$$TlBr(s) + e \rightleftharpoons Tl(s) + Br^-$$

11. Calculate the standard potential for the half-reaction

$$Cu(OH)_2(s) + 2e \rightleftharpoons Cu(s) + 2OH^-$$

given that K_{sp} for $Cu(OH)_2$ has a value of 5.9×10^{-15}.

12. Calculate the standard potential for the half-reaction

$$Al(C_2O_4)_2^- + 3e \rightarrow Al(s) + 2C_2O_4^{2-}$$

if the formation constant for the complex is 1.3×10^{13}.

13. Calculate the standard potential for the half-reaction

$$FeY^- + e \rightleftharpoons FeY^{2-}$$

if the formation constant for the EDTA complex of iron(III) is 1.3×10^{25} and that for the iron(II) complex is 2.1×10^{14}.

14. From the standard potentials

$$Tl^+ + e \rightleftharpoons Tl(s) \qquad E^0 = -0.336 \text{ V}$$

$$TlI(s) + e \rightleftharpoons Tl(s) + I^- \qquad E^0 = -0.761 \text{ V}$$

calculate the solubility product constant for TlI.

15. From the standard potentials

$$Ag_2SeO_4(s) + 2e \rightleftharpoons 2Ag(s) + SeO_4^{2-} \qquad E^0 = 0.355 \text{ V}$$

$$Ag^+ + e \rightleftharpoons Ag(s) \qquad E^0 = 0.799 \text{ V}$$

calculate the solubility product constant for Ag_2SeO_4.

16. From the cell potentials shown on the right below, calculate the electrode potentials for the half-cells coupled to the reference electrode.
    (a) saturated calomel electrode $\|M^{+n}|M$ $\qquad E = 0.567 \text{ V}$
    (b) normal calomel electrode $\|X^{3+}, X^{2+}|Pt$ $\qquad E = -0.642 \text{ V}$
    (c) saturated, silver/silver chloride electrode $\|MA(\text{sat'd}), A^{2-}|M$
    $E = -0.319 \text{ V}$

17. Convert each of the following electrode potentials to potentials versus the saturated calomel electrode.
    (a) $Cu^{2+} + 2e \rightleftharpoons Cu(s)$ $\qquad E^0 = 0.334 \text{ V}$

(b) $Ce^{4+} + e \rightleftharpoons Ce^{3+}$ $E = 1.44$ V (in $1F$ H_2SO_4)

(c) $Tl^+ + e \rightleftharpoons Tl(s)$ $E = -0.33$ V (in $1F$ $HClO_4$)

18. A current of 0.0510 A is to be drawn from the cell

$$Pt \,|\, V^{3+}(1.0 \times 10^{-5}M), V^{2+}(2.50 \times 10^{-1}M) \,\|\, Br^-(3.00 \times 10^{-1}M),$$
$$AgBr(sat'd) \,|\, Ag$$

As a consequence of its design, the cell has an internal resistance of 4.30 Ω. Calculate the potential to be expected initially.

19. The cell

$$Pt \,|\, V(OH)_4^+ (4.00 \times 10^{-4}M), VO^{2+}(6.73 \times 10^{-2}M),$$

$$H^+(2.75 \times 10^{-3}M) \,\|\, Cu^{2+}(3.00 \times 10^{-2}M) \,|\, Cu$$

has an internal resistance of 1.25 Ω. What will be the initial potential if 0.0300 A is drawn from this cell?

20. The resistance of the galvanic cell

$$Pt \,|\, Fe(CN)_6^{4-}(3.60 \times 10^{-2}M), Fe(CN)_6^{3-}(2.70 \times 10^{-3}M) \,\|\, Ag^+$$
$$(1.65 \times 10^{-2}M) \,|\, Ag$$

is 2.5 Ω. Calculate the initial potential when 0.0120 A is drawn from this cell.

21. Compute the chloride ion *activity* in a normal calomel electrode at 25°C.

22. Compute the chloride ion *activity* in a decinormal calomel electrode at 25°C.

19

POTENTIOMETRIC METHODS

In the preceding chapter, it was shown that the potential of an electrode is determined by the concentration (or, more correctly, the activity) of one or more species in a solution. This chapter considers how this phenomenon is applied to the quantitative determination of ions or molecules.[1]

The equipment required for a potentiometric measurement includes a *reference electrode*, an *indicator electrode*, and a *potential measuring device.* The preparation and properties of some common reference electrodes were considered in Chapter 18; the latter two components will be discussed in this chapter, in addition to applications of potentiometric measurements to quantitative analyses.

INDICATOR ELECTRODES

Indicator electrodes for potentiometric measurements are of two basic types, namely, *metallic* and *membrane*. The latter are also referred to as *specific* or *selective ion* electrodes.

Metallic Indicator Electrodes

First-Order Electrodes for Cations. A first-order electrode is employed to determine the cation derived from the electrode metal. Several metals, such as silver, copper, mercury, lead, and cadmium, exhibit reversible half-reactions with their ions and are satisfactory for construction of first-order electrodes. In contrast, other metals are not very satisfactory as indicator electrodes because they tend to develop nonreproducible potentials that are influenced by strains or crystal deformations in their structures and by oxide coatings on their surfaces. Metals in this category include iron, nickel, cobalt, tungsten, and chromium.

Second-Order Electrodes for Anions. A metal electrode is also indirectly responsive to anions that form slightly soluble precipitates or stable complexes with its cation. For the former, it is only necessary to saturate the solution under study with the sparingly soluble salt. For example, the potential of a silver electrode will accurately reflect the concentration of iodide ion in a solution that is saturated with silver iodide. Here, the electrode behavior can be described by

$$AgI(s) + e \rightleftharpoons Ag(s) + I^- \qquad E^0_{AgI} = -0.151 \text{ V}$$

Application of the Nernst equation to this half-reaction gives the relationship between the electrode potential and the anion concentration. Thus,

$$E = -0.151 - 0.0591 \log[I^-]$$
$$= -0.151 + 0.0591 \text{ pI}$$

where pI is the negative logarithm of the iodide ion concentration.[2] A silver electrode serving as an indicator for iodide is an example of an *electrode of the second-order* because it measures the concentration of an ion that is not directly involved in the electron transfer process.

---

[1] For a detailed review of potentiometric methods, see: R. P. Buck, in *Physical Methods of Chemistry*, eds. A. Weissberger and B. W. Rossiter. New York: Wiley-Interscience, 1971, Part IIA, vol. 1, Chapter 2.

[2] The results of potentiometric measurements are often expressed in terms of a parameter, the *p function*, that is directly proportional to the measured potential. The p function then provides a measure of concentration in terms of a convenient, small, and ordinarily positive, number. Thus, for a solution with a calcium ion concentration of 2.00×10^{-6}, we may write

$$pCa = -\log(2.00 \times 10^{-6}) = 5.699$$

Note that as the concentration of calcium increases, its p function decreases. Note also that because the concentration was given to three significant figures, we are entitled to keep this number of figures *to the right of the decimal point* in the computed pCa because these are the only numbers that carry information about the original 2.00. The 5 in the value for the pCa provides information about the position of the decimal point in the original number only.

An important second-order electrode for measuring the concentration of the EDTA anion Y^{4-} (p. 567) is based upon the response of a mercury electrode in the presence of a small concentration of the stable EDTA complex of Hg(II). The half-reaction for the electrode process can be written as

$$HgY^{2-} + 2e \rightleftharpoons Hg(l) + Y^{4-} \qquad E^0 = 0.21 \text{ V}$$

for which

$$E = 0.21 - \frac{0.0591}{2} \log \frac{[Y^{4-}]}{[HgY^{2-}]}$$

To employ this electrode system, it is necessary to introduce a small concentration of HgY^{2-} into the analyte solution at the outset. The complex is very stable (for HgY^{2-}, $K_f = 6.3 \times 10^{21}$); thus, its concentration remains essentially constant over a wide concentration range of Y^{4-} because dissociation to form Hg^{2+} is slight. The foregoing equation can then be written in the form

$$E = K - \frac{0.0591}{2} \log[Y^{4-}] = K + \frac{0.0591}{2} pY$$

where the constant K is equal to

$$K = 0.21 - \frac{0.0591}{2} \log \frac{1}{[HgY^{2-}]}$$

This second-order electrode is useful for establishing end points of EDTA titrations.

Indicators for Redox Systems. Electrodes fashioned from platinum or gold serve as indicator electrodes for oxidation-reduction systems. Of itself, such an electrode is inert; the potential it develops depends solely upon the potential of the oxidation-reduction systems of the solution in which it is immersed. For example, the potential of a platinum electrode in a solution containing Ce(III) and Ce(IV) ions is given by

$$E = E^0 - 0.0591 \log \frac{[Ce^{3+}]}{[Ce^{4+}]}$$

Thus, a platinum electrode can serve as the indicator electrode in a titration in which Ce(IV) serves as the standard reagent.

Membrane Indicator Electrodes[3]

For many years, the most convenient method for determining pH has involved measurement of the potential that develops across a thin glass membrane that separates two solutions with different hydrogen ion concentrations. The phenomenon, first reported by Cremer,[4] has been extensively studied by many investigators; as a result, the sensitivity and selectivity of glass membranes to pH is reasonably well understood. Furthermore, membrane electrodes have now been developed for the direct potentiometric determination of two dozen or more ions such as K^+, Na^+, Li^+, F^-, and Ca^{2+}.[5]

It is convenient to divide membrane electrodes into four categories based upon membrane composition. These include: (1) glass electrodes; (2) liquid-membrane electrodes; (3) solid-state or precipitate electrodes; and (4) gas-sensing membrane electrodes. We shall consider the properties and behavior of the glass electrode in particular detail both because of its historical and its current importance.

The Glass Electrode for pH Measurements

Figure 19-1 shows a modern cell for the measurement of pH. It consists of commercially available calomel and glass electrodes immersed in a solution whose pH is to be measured. The calomel reference electrode is simi-

[3] For further information on this topic, see: *Ion-Selective Electrodes*, ed. R. A. Durst. National Bureau of Standards Special Publication 314. Washington, D.C.: U.S. Government Printing Office, 1969; N. Lakshminarayanaiah, *Membrane Electrodes*. New York: Academic Press, 1976; J. Vesely, D. Weis, and K. Stulik, *Analysis with Ion-Selective Electrodes*. New York: Wiley, 1979; and *Ion Selective Electrodes in Analytical Chemistry*, ed. H. Freiser. New York: Plenum Press, 1978.
[4] M. Cremer, *Z. Biol.*, **47**, 562 (1906).
[5] See: G. A. Rechnitz, *Chem. Eng. News*, **45** (25), 146 (1967).

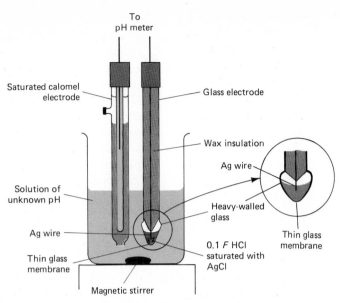

FIGURE 19-1   Typical electrode system for measuring pH.

lar to the ones described in Chapter 18 (p. 533). The glass electrode is manufactured by sealing a thin, pH-sensitive glass tip to the end of a heavy-walled glass tubing. The resulting bulb is filled with a solution of hydrochloric acid (often $0.1F$) that is saturated with silver chloride. A silver wire is immersed in the solution and is connected via an external lead to one terminal of a potential-measuring device; the calomel electrode is connected to the other terminal. Note that the cell contains *two* reference elec-

trodes, each with a potential that is constant and *independent of pH*; one reference electrode is the external calomel electrode, the other is the internal silver/silver chloride electrode (p. 533), which is a *part* of the glass electrode but is *not* the pH-sensitive component. In fact, it is the *thin membrane at the tip of the electrode* that responds to pH changes.

Figure 19-2 is a schematic representation of the cell shown in Figure 19-1. It is found experimentally that at 25°C the potential of

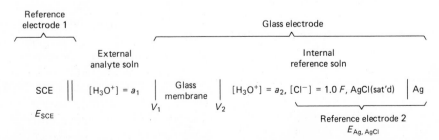

FIGURE 19-2   Schematic diagram of a glass-calomel cell for measurement of pH.

such a cell depends on the hydrogen ion activities a_1 and a_2 on either side of the membrane as follows:

$$E = Q + 0.0591 \log \frac{a_1}{a_2} \qquad (19\text{-}1)$$

The constant Q contains four components

$$Q = E_{Ag, AgCl} - E_{SCE} + E_j + E_a$$

where $E_{Ag, AgCl}$ and E_{SCE} are the potentials of the two reference electrodes, E_j is the junction potential across the salt bridge, and E_a is the *asymmetry potential*; the source and properties of E_a will be considered later.

In practice, the hydrogen ion activity of the internal solution a_2 is fixed and constant. We may thus write

$$E = L + 0.0591 \log a_1 \qquad (19\text{-}2)$$

$$= L - 0.0591 \text{ pH}$$

where L is a new constant consisting of Q and the logarithmic function of a_2.

It is important to emphasize that, in principal, the potentials E_{SCE}, $E_{Ag, AgCl}$, E_j, and E_a remain constant during a pH measurement. Thus, the source of the pH-dependent variation in E must lie *across the glass membrane*. That is, when a_1 and a_2 differ, the two surfaces of the membrane must differ in potential by some amount $V_1 - V_2$. The *only* function of the two reference electrodes is to make observation of this difference possible

Composition of Glass Membranes. Much systematic investigation has been devoted to the effects of glass composition on the sensitivity of membranes to protons and other cations, and a variety of compositions are now used commercially.[6] For many years, Corning 015 glass (consisting of approximately 22% Na_2O, 6% CaO, and 72% SiO_2) has been

widely used. This glass shows an excellent specificity toward hydrogen ions up to a pH of about 9. At higher pH values, however, the membrane becomes somewhat sensitive to sodium and other alkali ions.

Hygroscopicity of Glass Membranes. It has been shown that the surfaces of a glass membrane must be hydrated in order to have pH activity. Nonhygroscopic glasses such as Pyrex and quartz show no pH function. Even Corning 015 glass shows little pH response after dehydration by storage over a desiccant; its sensitivity is restored, however, after standing for a few hours in water. Hydration involves absorption of approximately 50 mg of water per cubic centimeter of glass.

It has also been demonstrated experimentally that hydration of a pH-sensitive glass membrane is accompanied by a chemical reaction in which certain cations of the glass are exchanged for protons of the solution. This exchange involves singly charged cations almost exclusively, inasmuch as the di- and trivalent cations in the silicate structure are much more strongly bonded. The ion-exchange reaction can be written as

$$\underset{\text{soln}}{H^+} + \underset{\text{solid}}{Na^+Gl^-} \rightleftharpoons \underset{\text{soln}}{Na^+} + \underset{\text{solid}}{H^+Gl^-} \qquad (19\text{-}3)$$

where Gl^- represents a cation bonding site in the glass. The equilibrium constant for this process favors incorporation of hydrogen ions into the silicate lattice; as a result, the surface of a well-soaked membrane will ordinarily consist of a layer of silicic acid gel which is 10^{-4} to 10^{-5} mm thick.

In all but highly alkaline media—where sodium ions may occupy an appreciable number of bonding sites—the predominant univalent cation in the outer surface of the gel is the proton. From the surface to the interior of the gel, there is a continuous decrease in the number of protons and a corresponding increase in the number of sodium ions. A schematic representation of the two surfaces of a glass membrane is shown in Figure 19-3.

---

[6] For a summary of this work, see: J. O. Isard, "The Dependence of Glass-Electrode Properties on Composition," in *Glass Electrodes for Hydrogen and Other Cations*, ed. G. Eisenman, New York: Marcel Dekker, 1967, Chapter 3.

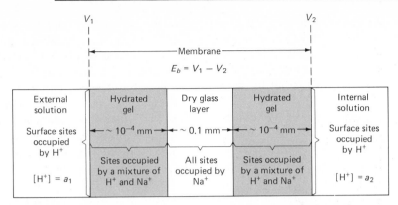

FIGURE 19-3 Schematic representation of a well-soaked glass membrane. Note that the dimensions of the three inner layers are *not* to scale.

Electrical Resistance of Glass Membranes. The membrane in a typical commercial glass electrode will be between 0.03 and 0.1 mm thick and will have an electrical resistance of 50 to 500 MΩ. Current conduction across the membrane involves migration of singly charged cations. Across each solution interface, passage of charge involves a transfer of protons; the direction of migration is from glass to solution at one interface and from solution to glass at the other. That is,

$$H^+Gl^- \rightleftharpoons Gl^- + H^+ \qquad (19\text{-}4)$$
$$\text{solid} \qquad \text{solid} \quad \text{soln}$$

and

$$H^+ + Gl^- \rightleftharpoons H^+Gl^- \qquad (19\text{-}5)$$
$$\text{soln} \quad \text{solid} \qquad \text{solid}$$

The position of these two equilibria is determined by the hydrogen ion concentrations in the two solutions. When these positions differ, the surface at which the greater dissociation has occurred will be negative with respect to the other surface. Thus, a potential develops whose magnitude depends upon the *difference* in hydrogen ion concentration on the two sides of the membrane. It is this potential that serves as the analytical parameter in a potentiometric pH measurement.

Conduction within the two silicic gel layers is due to migration of hydrogen and sodium ions. In the dry center region of the membrane, sodium ions take over this function.

Theory of the Glass Electrode Potential.[7] The potential across a glass membrane consists of a *boundary potential* and a *diffusion potential*. Ideally, only the former is affected by pH.

The boundary potential for the membrane of a glass electrode contains two components, each associated with one of the gel-solution interfaces. If V_1 is the component originating at the interface between the external solution and the gel (see Figure 19-3), and if V_2 is the corresponding potential at the inner surface, then the boundary potential E_b for the membrane is given by

$$E_b = V_1 - V_2 \qquad (19\text{-}6)$$

The potential V_1 is determined by the hydrogen ion activities both in the external solution and upon the surface of the gel; it can be considered a measure of the driving

---

[7] For a comprehensive discussion of this topic, see: G. Eisenman, *Glass Electrodes for Hydrogen and Other Cations.* New York: Marcel Dekker, 1967, Chapters 4–6.

force of the reaction shown by Equation 19-4. In the same way, the potential V_2 is related to the hydrogen ion activities in the internal solution and on the corresponding gel surface; it is a measure of the driving force of the reaction described by Equation 19-5.

It can be demonstrated from thermodynamic considerations[8] that V_1 and V_2 are related to hydrogen ion activities at each interface as follows:

$$V_1 = j_1 + \frac{RT}{F} \ln \frac{a_1}{a_1'} \qquad (19\text{-}7)$$

$$V_2 = j_2 + \frac{RT}{F} \ln \frac{a_2}{a_2'} \qquad (19\text{-}8)$$

where R, T, and F have their usual meanings; a_1 and a_2 are activities of the hydrogen ion in the *solutions* on either side of the membrane; and a_1' and a_2' are the hydrogen ion activities in each of the *gel layers* contacting the two solutions. If the two gel surfaces have the same number of sites from which protons can leave, then the two constants j_1 and j_2 will be identical; so also will be the two activities a_1' and a_2' in the gel layers, provided that all original sodium ions on the surface have been replaced by protons (that is, insofar as the equilibrium shown in Equation 19-3 lies far to the right). With these equalities, substitution of Equations 19-7 and 19-8 into 19-6 yields

$$E_b = V_1 - V_2 = \frac{RT}{F} \ln \frac{a_1}{a_2} \qquad (19\text{-}9)$$

Thus, *provided the two gel surfaces are identical,* the boundary potential E_b depends only upon the activities of the hydrogen ion in the *solutions* on either side of the membrane. If one of these activities a_2 is kept constant, then Equation 19-9 further simplifies to

$$E_b = \text{constant} + \frac{RT}{F} \ln a_1 \qquad (19\text{-}10)$$

---

[8] G. Eisenman, *Biophys. J.,* **2** Part 2, 259 (1962).

and the potential becomes a measure of the hydrogen ion activity in the external solution.

In addition to the boundary potential, a so-called *diffusion potential* also develops in each of the two gel layers. Its source lies in the difference between the mobilities of hydrogen and alkali-metal ions in the membrane. The two diffusion potentials are equal and opposite in sign insofar as the two solution-gel interfaces are the same. Under these conditions, then, the net diffusion potential is zero, and the emf across the membrane depends only on the boundary potential as given by Equation 19-10.

Asymmetry Potential. If identical solutions and identical reference electrodes are placed on either side of the membrane shown in Figure 19-3, $V_1 - V_2$ should be zero. However, it is found that a small potential, called the *asymmetry potential,* often does develop when this experiment is performed. Moreover, the asymmetry potential associated with a given glass electrode changes slowly with time.

The causes for the asymmetry potential are obscure; they undoubtedly include such factors as differences in strains established within the two surfaces during manufacture of the membrane, mechanical and chemical attack of the surfaces, and contamination of the outer face during use. The effect of the asymmetry potential on a pH measurement is eliminated by frequent calibration of the electrode against a standard buffer of known pH.

The Alkaline Error. Some glass membranes respond not only to changes in hydrogen ion concentration but also to the concentration of alkali-metal ions in solutions of pH 9 or greater. The magnitude of the resulting error for four types of glass membranes is indicated on the right-hand side of Figure 19-4. In each of these curves, the sodium ion concentration was maintained at 1 M, and the pH was varied. Note that the pH error is negative at high pH, which suggests that the electrode is responding to sodium ions as well as to protons. This observation is confirmed by data obtained for solutions with different sodium ion concentra-

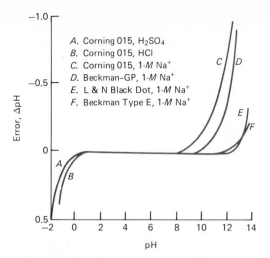

A. Corning 015, H_2SO_4
B. Corning 015, HCl
C. Corning 015, 1-M Na^+
D. Beckman–GP, 1-M Na^+
E. L & N Black Dot, 1-M Na^+
F. Beckman Type E, 1-M Na^+

FIGURE 19-4 Acid and alkaline error of selected glass electrodes at 25°C. (From R. G. Bates, *Determination of pH: Theory and Practice.* New York: John Wiley & Sons, Inc., 1964, p. 316. With permission.)

tions. Thus, at pH 12, the alkaline error for Corning 015 glass is about -0.7 pH when the sodium ion concentration is $1M$ (Figure 19-4) and only about -0.3 pH in solutions that are $0.1M$ in this ion.

All singly charged cations cause alkaline errors; the magnitude of the error varies according to the kind of metallic ion and the composition of the glass.

The alkaline error can be satisfactorily explained by assuming an exchange equilibrium between the hydrogen ions on the surface of the glass and the cations in the solution. This process, which is the reverse of the one described by Equation 19-3, can be formulated as

$$H^+Gl^- + B^+ \rightleftharpoons B^+Gl^- + H^+$$
$$\text{gel} \quad \text{soln} \quad \text{gel} \quad \text{soln}$$

where B^+ represents a singly charged cation, such as sodium ion. The equilibrium constant for this reaction is

$$K_{ex} = \frac{a_1 b_1'}{a_1' b_1} \qquad (19\text{-}11)$$

where a_1 and b_1 are activities of H^+ and B^+ in solution, and a_1' and b_1' are the activities of the same ions in the surface of the gel. The constant K_{ex} depends upon the composition of the glass membrane and is ordinarily small. Thus, the fraction of the surface sites populated by cations other than hydrogen is small except when the hydrogen ion concentration is very low and the concentration of B^+ is high.

The effect of an alkali-metal ion on the emf across a membrane can be described in quantitative terms by rewriting Equation 19-11 in the form

$$\frac{a_1}{a_1'} = \frac{(a_1 + K_{ex}b_1)}{(a_1' + b_1')} \qquad (19\text{-}12)$$

Substitution of Equation 19-12 into Equation 19-7 and subtraction of Equation 19-8 gives

$$E_b = V_1 - V_2 \qquad (19\text{-}13)$$

$$= j_1 - j_2 + \frac{RT}{F} \ln \frac{(a_1 + K_{ex}b_1)a_2'}{(a_1' + b_1')a_2}$$

If the number of sites on each side of the membrane is the same, the activity of hydrogen ions on the internal surface is approximately equal to the sum of the activities of the two cations on the external surface; that is, $a_2' \cong (a_1' + b_1')$. Further, $j_1 \cong j_2$; thus, Equation 19-13 reduces to

$$E_b = \frac{RT}{F} \ln \frac{(a_1 + K_{ex}b_1)}{a_2}$$

Finally, where a_2 is constant,

$$E_b = \text{constant} + \frac{RT}{F} \ln(a_1 + K_{ex}b_1) \quad (19\text{-}14)$$

It has been shown that not only the boundary potential but also the diffusion potential on one side of the membrane is changed when some of the surface sites are occupied by cations other than hydrogen. Under such circumstances, the diffusion potentials do not subtract out as they did previously (see p. 544), and E_b does not adequately describe the total emf that develops in the presence of another cation. It has also been shown that this effect can be accounted for by modifying Equation 19-14 as follows[9]:

$$E = \text{constant} + \frac{RT}{F} \ln \left[a_1 + K_{ex}\left(\frac{U_B}{U_H}\right) b_1 \right]$$
$$(19\text{-}15)$$

where U_B and U_H are measures of the mobilities of B^+ and H^+ in the gel.

For many glasses, the term $K_{ex}(U_B/U_H)b_1$ is small with respect to the hydrogen ion activity a_1, as long as the pH of the solution is less than about 9. Under these conditions, Equation 19-15 simplifies to Equation 19-10. With high concentrations of singly charged cations and at higher pH levels, however, this second term plays an increasingly important part in determining E. The composition of the glass membrane determines the magnitude of $K_{ex}(U_B/U_H)b_1$; this parameter is relatively small for glasses that have been designed for work in strongly basic solutions. The membrane of the Beckman Type E glass electrode, illustrated in Figure 19-4, is of this type.

The Acid Error. As shown in Figure 19-4, the typical glass electrode exhibits an error, opposite in sign to the alkaline error, in solutions of pH less than about 0.5; pH readings tend to be too high in this region. The magnitude of the error depends upon a variety of factors and is generally not very reproducible. The causes of the acid error are not well understood.

Glass Electrodes for the Determination of Other Cations

The existence of the alkaline error in early glass electrodes led to studies concerning the effect of glass composition on the magnitude of this error. One consequence of this work has been the development of glasses for which the term $K_{ex}(U_B/U_H)b_1$ in Equation 19-15 is small enough so that the alkaline error is negligible below a pH of about 12. Other studies have been directed toward finding glass compositions in which this term is greatly enhanced, in the interests of developing glass electrodes for the determination of cations other than hydrogen. This application requires that the hydrogen ion activity a_1 in Equation 19-15 be negligible with respect to the second term containing the activity b_1 of the other cation; under these circumstances, the potential of the electrode would be *independent* of pH but would vary with pB in the typical way.

A number of investigators have demonstrated that the presence of Al_2O_3 or B_2O_3 in glass causes the desired effect. Eisenman and co-workers have carried out a systematic study of glasses containing Na_2O, Al_2O_3, and SiO_2 in various proportions; they have demonstrated clearly that it is practical to prepare membranes for the selective measurement of several cations in the presence of others.[10] Glass electrodes for potassium ion and sodium ion are now available commercially.

Figure 19-5 illustrates how two types of glass electrodes respond to pH changes in solutions that are $0.1F$ in various alkali-metal ions. The one electrode was fabricated from Corning 015 glass, a composition that contains no Al_2O_3; at pH levels below 9, the term $K_{ex}(U_B/U_H)b_1$ in Equation 19-15 is small with respect to a_1. At higher pH values, the second term becomes important; note that its magnitude also depends upon the type of alkali ion present.

With the glass containing Al_2O_3, both potential-determining terms in Equation 19-15

[9] See: G. Eisenman, *Glass Electrodes for Hydrogen and Other Cations.* New York: Marcel Dekker, 1967, Chapters 4 and 5.

[10] For a summary of this work, see: G. Eisenman, in *Advances in Analytical Chemistry and Instrumentation*, ed. C. E. Reilley. New York: Wiley, 1965, vol. 4, p. 213.

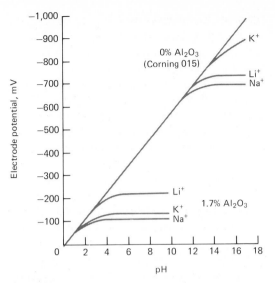

FIGURE 19-5 Response of two glass membranes in the presence of alkali-metal ions (0.1 M in each case). (Curves taken from G. Eisenman, in *Advances in Analytical Chemistry and Instrumentation*, ed. C. N. Reilley. New York: John Wiley & Sons, Inc., 1965, vol. 4, p. 219. With permission.)

are clearly important at low pH values, but at levels above about pH 5, the potential becomes *independent* of pH. Thus, $a_1 \ll K_{ex}(U_B/U_H)b_1$ in this region, with the result that the potential can be expected to vary linearly with pNa, pK, or pLi, Experimentally, this prediction is verified; such a glass can be employed to measure the concentration of these ions.

Sources for the difference in behavior between the two glasses shown in Figure 19-5 (toward Na^+, for example) include the relative affinity of anionic surface sites of the gel for hydrogen ions as compared to sodium ions, and (to a lesser extent) the mobility ratios of these two ions in the two glasses. Above pH 4, where glass containing Al_2O_3 is no longer sensitive to hydrogen ions, these anionic sites are occupied entirely by the metallic ion, and the electrode behaves as a perfect sodium electrode. With lithium ion, saturation occurs only

at lower hydrogen ion concentrations (pH > 6). Thus, the surface has less affinity for lithium than for sodium ion.

Studies such as these have led to the development of glass membranes suitable for direct potentiometric measurement of the concentrations of such singly charged species as Na^+, K^+, NH_4^+, Rb^+, Cs^+, Li^+, and Ag^+. Some of these glasses are reasonably selective, as may be seen in Table 19-1. Because of the strong bond between multiply charged cations and glass, little hope exists for the development of glass membranes for determination of these species.

Liquid Membrane Electrodes

Liquid membranes owe their response to the potential that is established across the interface between the solution to be analyzed and an immiscible liquid that selectively bonds with the ion being determined. Liquid membrane electrodes are particularly important because they permit the direct potentiometric determination of the activities of several polyvalent cations and certain anions as well.

A liquid membrane electrode differs from a glass electrode only in that the solution of known and fixed activity is separated from the analyte solution by a thin layer of an immiscible organic liquid instead of a thin glass membrane. As shown schematically in Figure 19-6 a porous, hydrophobic (that is, water-repelling), plastic disk serves to hold the organic layer between the two aqueous solutions. Wick action causes the pores of the disk or membrane to stay filled with the organic liquid from the reservoir in the outer of the two concentric tubes. The inner tube contains an aqueous standard solution of MCl_2, where M^{2+} is the cation whose activity is to be determined. This solution is also saturated with AgCl to form a Ag/AgCl reference electrode with the silver lead wire.

The organic liquid is a nonvolatile, water-immiscible, organic ion exchanger that contains acidic, basic, or chelating functional groups. Between this liquid and an aqueous solution containing a divalent cation, there is established an equilibrium that can be represented as

$$RH_{2x} + xM^{2+} \rightleftharpoons RM_x + 2xH^+$$

organic	aqueous	organic	aqueous
phase	phase	phase	phase

Repeated treatment converts the exchange liquid essentially completely to the cationic form, RM_x; it is this form that is employed in electrodes for the determination of M^{2+}.

In order to determine the pM of a solution, the electrode shown in Figure 19-6 is immersed in a solution of the analyte which also contains a reference electrode—usually a saturated calomel electrode. The potential be-

TABLE 19-1 PROPERTIES OF CERTAIN CATION-SENSITIVE GLASSES[a]

Principal Cation to Be Measured	Glass Composition	Selectivity Characteristics[b]	Remarks
Li^+	$15Li_2O$ $25Al_2O_3$ $60SiO_2$	$K_{Li^+, Na^+} \approx 3$, $K_{Li^+, K^+} > 1000$	Best for Li^+ in presence of H^+ and Na^+
Na^+	$11Na_2O$ $18Al_2O_3$ $71SiO_2$	$K_{Na^+, K^+} \approx 2800$ at pH 11 $K_{Na^+, K^+} \approx 300$ at pH 7	Nernst type of response to $\sim 10^{-5}M$ Na^+
	$10.4Li_2O$ $22.6Al_2O_3$ $67SiO_2$	$K_{Na^+, K^+} \approx 10^5$	Highly Na^+ selective, but very time-dependent
K^+	$27Na_2O$ $5Al_2O_3$ $68SiO_2$	$K_{K^+, Na^+} \approx 20$	Nernst type of response to $< 10^{-4}M$ K^+
Ag^+	$28.8Na_2O$ $19.1Al_2O_3$ $52.1SiO_2$	$K_{Ag^+, H^+} \approx 10^5$	Highly sensitive and selective to Ag^+, but poor stability
	$11Na_2O$ $18Al_2O_3$ $71SiO_2$	$K_{Ag^+, Na^+} > 1000$	Less selective for Ag^+ but more reliable

[a] Reprinted, with permission of the copyright owner, the American Chemical Society, from the June 12, 1967 issue of *Chemical and Engineering News*, p. 149.

[b] The constant $K_{i, j}$, the selectivity ratio, is a measure of the relative response of the electrode to the two species i and j. For example, for the sodium electrode, K_{Na^+, K^+} indicates that the electrode is 2800 times as sensitive to sodium ion as to potassium. Thus, 2800 times the concentration of potassium ion is required to produce the same potential as a unit concentration of sodium ion. Another method for expressing selectivity is to define a constant which is the reciprocal of the one just described; that is, $K_{Na^+, K^+} = 1/2800 = 3.6 \times 10^{-4}$. Here one can say that a $3.6 \times 10^{-4}M$ solution of sodium ion will produce the same effect as a $1M$ solution of potassium ions. Unfortunately, both types of constants are found in the literature (often without definition). Thus, the data in Table 19-2 employ a constant which is the reciprocal of the one employed in this table.

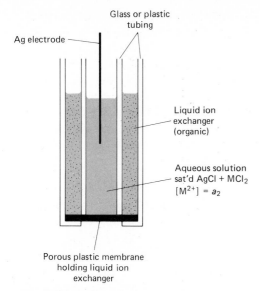

FIGURE 19-6 Liquid membrane electrode sensitive to M^{2+}.

glass electrode. The ion exchanger is an aliphatic diester of phosphoric acid dissolved in a polar solvent. The chain lengths of the aliphatic groups in the ester range from 8 to 16 carbon atoms. The diester contains a single acidic proton; thus, two molecules are required to bond a divalent cation (here, calcium). It is the selective affinity of this compound for calcium ions that imparts the selective properties to the electrode. The internal aqueous solution in contact with the exchanger contains a fixed concentration of calcium chloride; a silver/silver chloride reference electrode is immersed in this solution. The porous disk containing the ion-exchange liquid separates the analyte solution from the reference calcium chloride solution. The equilibrium established at each interface can be represented as

$$[(RO)_2POO]_2Ca \rightleftharpoons 2(RO)_2POO^- + Ca^{2+}$$

organic   organic   aqueous

Note the similarity of this equation to Equation 19-4 for the glass electrode. The potential of this electrode is given by an equation analogous to Equations 19-2 and 19-10 for the glass electrode; that is,

$$E = L + \frac{0.0591}{2} \log a_1 \qquad (19\text{-}16)$$

tween the external and the internal reference electrodes is proportional to the pM of the analyte.

Figure 19-7 shows construction details of a commercial liquid-membrane electrode that is selective for calcium ion and compares the structural features of this electrode with a

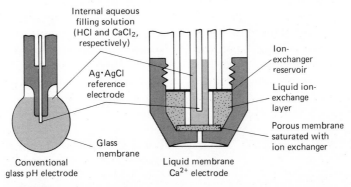

FIGURE 19-7 Comparison of a liquid membrane calcium ion electrode with a glass electrode. (Courtesy of Orion Research, Inc., Cambridge, MA.)

where a_1 here is the activity of Ca^{2+} and L is a constant whose value is determined by the measurement of E when the electrode is immersed in a standard solution.

The sensitivity of the electrode just described is reported to be 50 times greater for calcium ion than for magnesium ion, and 1000 times greater than that for sodium or potassium ions. Calcium ion activities as low as $5 \times 10^{-7}M$ can be measured. The performance of the electrode is said to be independent of pH in the range between 5.5 and 11. At lower pH levels, hydrogen ions undoubtedly replace some of the calcium ions on the exchanger to a significant extent; the electrode then becomes pH- as well as pCa-sensitive.

The calcium ion membrane electrode has proved to be a valuable tool for physiological studies because this ion plays important roles in nerve conduction, bone formation, muscle contraction, cardiac conduction and contraction, and renal tubular function. At least some of these processes are influenced more by calcium ion *activity* than calcium concentration; activity, of course, is the parameter measured by the electrode.

Another liquid specific-ion electrode of great value for physiological studies is that for potassium, because the transport of nerve signals appears to involve movement of this ion across nerve membranes. Study of the process requires an electrode that can detect small concentrations of potassium ion in the presence of much larger concentrations of sodium. A number of liquid membrane electrodes show promise of meeting these needs; one is based upon the antibiotic valinomycin, a cyclic ether that has a strong affinity for potassium ion. Of equal importance is the observation that a liquid membrane consisting of valinomycin in diphenyl ether is about 10^4 times as responsive to potassium ion as to sodium ion.[11]

Table 19-2 lists commercially available liquid membrane electrodes. The anion-sensitive electrodes employ a solution of an anion exchange resin in an organic solvent. Recently, liquid membrane electrodes for Ca^{2+}, K^+, NO_3^-, and BF_4^- have become available in which the liquid is held in the form of a polyvinylchloride gel. These electrodes have the appearance of a solid-state electrode described in the next section.

Solid-State or Precipitate Electrodes

Considerable work has been devoted to the development of solid membranes that are selective towards anions in the way that some glasses behave toward specific cations. As we have seen, the selectivity of a glass membrane is due to the presence of anionic sites on its surface that show particular affinity toward certain positively charged ions. By analogy, a membrane having similar cationic sites might be expected to respond selectively toward anions. To exploit this possibility, attempts have been made to prepare membranes of salts containing the anion of interest and a cation that selectively precipitates that anion from aqueous solutions; for example, barium sulfate has been proposed for sulfate ion and silver halides for the various halide ions. The problem encountered in this approach has been in finding methods for fabricating membranes from the desired salts that possess adequate physical strength, conductivity, and resistance to abrasion and corrosion.

Membranes prepared from cast pellets of silver halides have been successfully used in electrodes for the selective determination of chloride, bromide, and iodide ions. An electrode employing a polycrystalline Ag_2S membrane is offered by one manufacturer for the determination of sulfide ion. Silver ions are sufficiently mobile to conduct electricity through the membrane. Mixtures of PbS, CdS, and CuS with silver sulfide provide membranes

---

[11] M. S. Frant and J. W. Ross, Jr., *Science,* **167,** 987 (1970).

which are selective for Pb^{2+}, Ca^{2+}, and Cu^{2+}; in these, silver ion serves to transport electricity within the solid membrane.

A solid-state electrode selective for fluoride ion is also available from commercial sources. The membrane consists of a single crystal of lanthanum fluoride that has been doped with europium(II) to increase its electrical conductivity. The membrane, supported between a reference solution and the solution to be measured, shows the theoretical response to changes in fluoride ion activity from 10^0 to $10^{-6}M$; that is,

$$E = K - 0.0591 \log a_{F^-}$$

The electrode is reported to be selective for fluoride over other common anions by several orders of magnitude. Only hydroxide ion appears to offer serious interference.

Table 19-3 lists some of the commercially available solid-state electrodes.

Gas-Sensing Electrodes

Figure 19-8 is a schematic diagram of a so-called gas-sensing electrode, which consists of a reference electrode, a specific ion electrode, and an electrolyte solution housed in a cylindrical plastic tube. A thin, replaceable, gas-permeable membrane is attached to one end of the tube and serves to separate the internal electrolyte solution from the analyte solution. The membrane is described by its manufacturer as being a thin microporous film fabricated from a hydrophobic plastic; water and electrolytes are prevented from entering and passing through the pores of the film, owing to its water-repellent properties. Thus, the pores contain only air or other gases to which the membrane is exposed. When a solution containing a gaseous analyte such as carbon dioxide is in contact with the membrane, the CO_2 distills into the pores, as shown by

TABLE 19-2   COMMERCIAL LIQUID-MEMBRANE ELECTRODES[a]

Analyte Ion	Concentration Range, M	Preferred pH	Selectivity Constant for Interferences[b]
Ca^{2+}	10^0–5×10^{-7}	6–8	Zn^{2+}, 3.2; Fe^{2+}, 0.8; Pb^{2+}, 0.63; Mg^{2+}, 0.014
Cl^-	10^0–8×10^{-6}	2–11	I^-, 17; NO_3^-, 4.2; Br^-, 1.6; HCO_3^-, 0.19; SO_4^{2-}, 0.14; F^-, 0.10
BF_4^-	10^{-1}–3×10^{-6}	3–10	NO_3^-, 0.1; Br^-, 0.04; OAc^-, 0.004; HCO_3^-, 0.004; Cl^-, 0.001
NO_3^-	10^0–7×10^{-6}	3–10	I^-, 20; Br^-, 0.1; NO_2^-, 0.04; Cl^-, 0.004; SO_4^{2-}, 0.00003; CO_3^{2-}, 0.0002; ClO_4^-, 1000; F^-, 0.00006
ClO_4^-	10^0–2×16^{-6}	3–10	I^-, 0.012; NO_3^-, 0.0015; Br^-, 0.00056; F^-, 0.00025; Cl^-, 0.00022
K^+	10^0–10^{-6}	3–10	Cs^+, 1.0; NH_4^+, 0.03; H^+, 0.01; Ag^+, 0.001; Na^+, 0.002; Li^+, 0.001
$Ca^{2+} + Mg^{2+}$	10^{-2}–6×10^{-6}	5–8	Zn^{2+}, 3.5; Fe^{2+}, 3.5; Cu^{2+}, 3.1; Ni^{2+}, 1.35; Ba^{2+}, 0.94; Na^+, 0.015

[a] *Analytical Methods Guide*, 9th ed. Cambridge, MA.: Orion Research, Inc., December, 1978. With permission.

[b] Note that these selectivity constants are the reciprocals of the ones in Table 19-1 (see footnote *b* of Table 19-1).

the reaction

$$CO_2(aq) \rightleftharpoons CO_2(g)$$
external   membrane
solution   pores

Because the pores are numerous, a state of equilibrium is rapidly approached. The CO_2 in the pores, however, is also in contact with the internal solution and a second equilibrium reaction can readily take place; that is,

$$CO_2(g) \rightleftharpoons CO_2(aq)$$
membrane   internal
pores   solution

As a consequence of the two reactions, the external solution rapidly (in a few seconds to minutes) equilibrates with the film of internal solution adjacent to the membrane. Here, another equilibrium is established that causes the pH of the internal surface film to change; namely,

$$CO_2(aq) + 2H_2O \rightleftharpoons HCO_3^- + H_3O^+$$

A glass-reference electrode pair immersed in the film of internal solution (see Figure 19-8) then detects the pH change.

The overall reaction for the process just

TABLE 19-3   COMMERCIAL SOLID-STATE ELECTRODES[a]

Analyte Ion	Concentration Range, M	Preferred pH	Interferences
Br^-	10^0–5×10^{-6}	2–12	Max: $[S^{2-}] = 10^{-7} M$; $[I^-] = 2 \times 10^{-4}[Br^-]$
Cd^{2+}	10^{-1}–10^{-7}	3–7	Max: $[Ag^+]$, $[Hg^{2+}]$, $[Cu^{2+}] = 10^{-7} M$
Cl^-	10^0–5×10^{-5}	2–11	Max: $[S^{2-}] = 10^{-7} M$; traces of Br^-, I^-, CN^- do not interfere
Cu^{2+}	10^0–10^{-8}	3–7	Max: $[S^{2-}]$, $[Ag^+]$, $[Hg^{2+}] = 10^{-7} M$; high levels of Cl^-, Br^-, Fe^{3+}, Cd^{2+} interfere
CN^-	10^{-2}–10^{-6}	11–13	Max: $[S^{2-}] = 10^{-7} M$; $[I^-] = 0.1 [CN^-]$; $[Br^-] = 5 \times 10^3[CN^-]$; $[Cl^-] = 10^6[CN^-]$
F^-	10^0–10^{-6}	5–8	Max: $[OH^-] = 0.1[F^-]$
I^-	10^0–5×10^{-8}	3–12	Max: $[S^{2-}] = 10^{-7} M$
Pb^{2+}	10^0–10^{-6}	4–7	Max: $[Ag^+]$, $[Hg^{2+}]$, $[Cu^{2+}] = 10^{-7} M$; high levels of Cd^{2+} and Fe^{3+} interfere
Ag^+	10^0–10^{-7}	2–9	Max: $[Hg^{2+}] = 10^{-7} M$
S^{2-}	10^0–10^{-7}	13–14	Max: $[Hg^{2+}] = 10^{-7} M$
Na^+	10^0–10^{-6}	9–10	Selectivity constants: Li^+, 0.002; K^+, 0.001; Rb^+, 0.00003; Cs^+, 0.0015; NH_4^+, 0.00003; Tl^+, 0.0002; Ag^+, 350; H^+, 100
SCN^-	10^0–5×10^{-6}	2–12	Max: $[OH^-] = [SCN^-]$; $[Br^-] = 0.003[SCN^-]$; $[Cl^-] = 20[SCN^-]$; $[NH_3] = 0.13[SCN^-]$; $[S_2O_3^{2-}] = 0.01[SCN^-]$; $[CN^-] = 0.007[SCN^-]$; $[I^-]$, $[S^{2-}] = 10^{-7} M$

[a] From *Analytical Methods Guide*, 9th ed. Cambridge, MA.: Orion Research, Inc., December 1978. With permission.

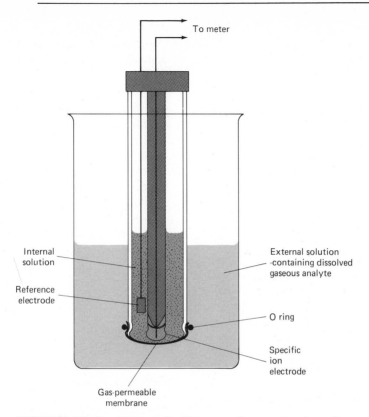

To meter

Internal solution

Reference electrode

Gas-permeable membrane

External solution -containing dissolved gaseous analyte

O ring

Specific ion electrode

FIGURE 19-8 Schematic diagram of a gas-sensing electrode. Specific ion electrode is shown as a glass electrode. The reference electrode is a Ag–AgCl electrode. Other electrode combinations are possible.

described is obtained by adding the three chemical equations to give

$$CO_2(aq) + 2H_2O \rightleftharpoons H_3O^+ + HCO_3^-$$
external   internal solution
solution

The equilibrium constant for the reaction is given by

$$\frac{[H_3O^+][HCO_3^-]}{[CO_2(aq)]_{ext}} = K$$

If the concentration of HCO_3^- in the internal solution is made relatively high so that its concentration is not altered significantly by the CO_2 which distills, then

$$\frac{[H_3O^+]}{[CO_2(aq)]_{ext}} = \frac{K}{[HCO_3^-]} = K_g \quad (19\text{-}17)$$

which may be rewritten as

$$a_1 = [H_3O^+] = K_g[CO_2(aq)]_{ext} \quad (19\text{-}18)$$

where a_1 is the internal hydrogen ion activity.

The potential of the electrode system in the internal solution is dependent upon a_1 as described by Equation 19-2. Substitution of Equation 19-18 into 19-2 yields

$$E = L + 0.0591 \log K_g[CO_2(aq)]_{ext}$$

or

$$E = L' + 0.0591 \log[CO_2(aq)]_{ext}$$

where

$$L' = L + 0.0591 \log K_g$$

Thus, the potential of the cell consisting of the internal reference and indicator electrode is determined by the CO_2 concentration of the external solution. Note that *no electrode comes directly in contact* with the analyte solution; for this reason, it would be better to call the device a gas-sensing *cell* rather than a gas-sensing electrode. Note also that the only species that will interfere with the measurement are dissolved gases that can pass through the membrane and can additionally affect the pH of the internal solution.

The possibility exists for increasing the selectivity of the gas-sensing electrode by employing an internal electrode that is sensitive to some species other than hydrogen ion; for example, a nitrate-sensing electrode could be used to provide a cell that would be sensitive to nitrogen dioxide. Here, the equilibrium would be

$$2NO_2(aq) + H_2O \rightleftharpoons NO_2^- + NO_3^- + 2H^+$$
external   internal solution
solution

This electrode should permit the determination of NO_2 in the presence of gases such as SO_2, CO_2, and NH_3 which would also alter the pH of the internal solution.

Gas-sensing electrode systems are commercially available for CO_2, NO_2, and NH_3. An oxygen sensitive cell system is also on the market; it, however, is based on a voltammetric measurement and is discussed in Chapter 21.

INSTRUMENTS FOR CELL POTENTIAL MEASUREMENT

An instrument for potentiometric measurements should draw essentially no electricity from the galvanic cell being studied. One reason for this is that a current causes changes in potential. More important is the dependency of cell potentials on current, which arises from the effects of IR drop and polarization phenomena (p. 527). The influence of IR drop is particularly significant with specific ion electrodes, which may have resistances of 100 MΩ or more. With these, currents must be limited to 10^{-12} A or smaller; this limitation requires that the potential measuring device have an internal resistance of 10^{11} Ω or more (see example, p. 31).

Two types of instruments for voltage measurement are employed in potentiometry—the potentiometer and the electronic voltmeter. Both instruments are referred to as *pH meters* when their internal resistances are sufficiently high to be used with glass and other membrane electrodes; with the advent of the many new specific ion electrodes, *pIon* or *ion meters* would perhaps be a more descriptive name.

Potentiometers

The potentiometer described in Chapter 2 (p. 33) is the classical instrument for potential measurement and has, in the past, been widely used in potentiometric studies. It is unsatisfactory for measurements with membrane electrodes. It can, however, be adapted to such measurements by replacing the relatively low-resistance galvanometer with a current detector having a high-input resistance. In the past, vacuum tube electrometers have been used for this purpose. Alternatively, the out-of-balance current of the potentiometer can be fed into a high-resistance voltage-follower amplifier and then amplified by means of a suitable operational amplifier (see Figure 3-20, p. 60). An instrument of this type with a high-precision slide wire is sensitive to better than ± 0.1 mV, which corresponds to ± 0.002 pH.

Potentiometric pH meters appear to be disappearing from the laboratory, being replaced by more convenient and rapid direct-reading electronic voltmeters or pH meters.

Direct-Reading Instruments

Numerous direct-reading pH meters are available commercially. Generally, these are solid-state devices employing a field effect transistor or a voltage follower as the first amplifier stage in order to provide the needed high internal resistance. As shown in Figure 19-9, the circuit can be relatively simple. Here, the output of the ion-selective electrode is connected to the inverting terminal of an operational amplifier, which employs a field effect transistor for the first stage of amplification. Note that the cell is incorporated in the feedback loop. The amplifier current then drives a meter that is usually calibrated in terms of both pH and millivolts. The variable resistance R_T is calibrated in terms of temperature so that correction for temperature variations is possible.

The readout of commercial pH meters is either a digital meter or a meter with a 5- to 10-in. scale that covers a range of 0 to 14 pH units. Many of the latter instruments are also equipped with scale expansion capabilities, which provide full-scale ranges of 0.5 to 2 pH units; a precision of ± 0.001 to 0.005 pH unit

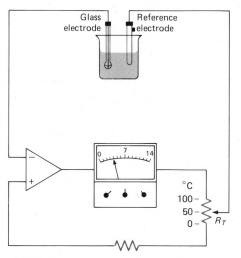

FIGURE 19-9 Simplified circuit diagram for a direct-reading pH meter.

can then be realized, depending upon amplifier noise level. It should be appreciated, however, that it is seldom, if ever, possible to measure pH with this kind of *accuracy*; indeed, uncertainties of ± 0.02 to ± 0.03 pH unit are more typical (see p. 557).

The price of pH meters ranges from less than \$100 to \$1000 or more.

DIRECT POTENTIOMETRIC MEASUREMENTS

Direct potentiometric measurements can be used to complete chemical analyses of those species for which an indicator electrode is available. The technique is simple, requiring only a comparison of the potential developed by the indicator electrode in the test solution with its potential when immersed in a standard solution of the analyte; insofar as the response of the electrode is specific for the analyte, no preliminary separation steps are required. In addition, direct potentiometric measurements are readily adapted to the continuous and automatic monitoring of analytical parameters.

Notwithstanding these attractive advantages, the user of direct potentiometric measurements must be alert to limitations that are inherent to the method. An important example is the existence in most potentiometric measurements of a *liquid junction* potential (see p. 526). For most electroanalytical methods, this junction potential is inconsequential and can be neglected. Unfortunately, however, its existence places a limitation on the accuracy that can be attained from a direct potentiometric measurement.

Equation for Direct Potentiometry

The observed potential of a cell employed for a direct potentiometric measurement can be expressed in terms of the potentials developed by the indicator electrode, the reference electrode, and a junction potential. That is

$$E_{obs} = E_{ref} - E_{ind} + E_j \quad (19\text{-}19)[12]$$

Typically, the junction potential E_j has two components, the first being located at the junction of the analyte solution and one end of a salt bridge and the second where the bridge interfaces with the reference electrode solution. The two potentials tend to cancel one another, but seldom do so completely. Thus, E_j in Equation 19-19 may be as large as 1 mV or more.

Ideally, the potential of the indicator electrode is related to the activity a_1 of M^{n+}, the ion of interest, by a Nernst-type equation

$$E_{ind} = L + \frac{0.0591}{n} \log a_1 = L - \frac{0.0591}{n} \, pM$$
$$(19\text{-}20)$$

where L is a constant. For metallic electrodes, L is often the standard potential for the indicator electrode; with membrane electrodes, however, L may also include an unknown, time-dependent asymmetry potential (p. 544).

Combination of Equation 19-20 with Equation 19-19 and rearrangement yields

$$pM = -\log a_1$$

$$= \frac{E_{obs} - (E_{ref} + E_j - L)}{0.0591/n} \quad (19\text{-}21)$$

$$= \frac{E_{obs} - K}{0.0591/n} \quad (19\text{-}22)$$

The new constant K consists of three constants, of which at least one (E_j) has a magnitude that *cannot be evaluated from theory*. Thus, K must be determined experimentally with the aid of a standard solution of M before Equation 19-22 can be used for the measurement of pM.

Several methods for performing analyses by direct potentiometry have been developed; all are based, directly or indirectly, upon Equation 19-22.[13]

Electrode Calibration Method

In the electrode calibration procedure, K in Equation 19-22 is determined by measuring E_{obs} for one or more standard solutions of known pM. The assumption is then made that K does not change during measurement of the analyte solution. Generally, the calibration operation is performed at the time that pM for the unknown is determined; recalibration may be required if measurements extend over several hours.

The electrode calibration method offers the advantages of simplicity, speed, and applicability to the continuous monitoring of pM. Two important disadvantages attend its use, however. One of these is that the results of an analysis are in terms of activities rather than concentrations; the other is that the accuracy of a measurement obtained by this procedure is limited by the inherent uncertainty caused by the junction potential; unfortunately, this uncertainty can never be totally eliminated.

Activity versus Concentration. Electrode response is related to activity rather than to analyte concentration. Ordinarily, however, the scientist is interested in concentration, and the determination of this quantity from a potentiometric measurement requires activity coefficient data. More often than not, activity coefficients will be unavailable because the ionic strength of the solution is either unknown or so high that the Debye-Hückel equation is not applicable. Unfortunately, the assumption that activity and concentration are identical may lead to serious errors, particularly when the analyte is polyvalent.

---

[12] As written, Equation 19-19 has the reference electrode acting as cathode and the indicator electrode acting as anode. If, in a particular cell, the roles are reversed, the signs of the two electrodes are likewise reversed.

[13] For additional information, see: R. A. Durst, *Ion-Selective Electrodes*. National Bureau of Standards Special Publication 314. Washington, D.C.: U.S. Government Printing Office, 1969, Chapter 11.

The difference between activity and concentration is illustrated by Figure 19-10, where the lower curve gives the change in potential of a calcium electrode as a function of calcium chloride concentration (note that the activity or concentration scale is logarithmic). The nonlinearity of the curve is due to the increase in ionic strength—and the consequent decrease in the activity coefficient of the calcium ion—as the electrolyte concentration becomes larger. When these concentrations are converted to activities, the upper curve is obtained; note that this straight line has the theoretical slope of 0.0296 (0.0591/2).

Activity coefficients for singly charged ions are less affected by changes in ionic strength than are coefficients for species with multiple charges. Thus, the effect shown in Figure 19-10 will be less pronounced for electrodes that respond to H^+, Na^+, and other univalent ions.

In potentiometric pH measurements, the pH of the standard buffer employed for calibration is generally based on the activity of hydronium ions. Thus, the resulting hydrogen ion analysis is also on an activity scale. If the unknown sample has a high ionic strength, the hydrogen ion *concentration* will differ appreciably from the activity measured.

Inherent Error in the Electrode Calibration Procedure. A serious disadvantage of the electrode calibration method is the existence of an inherent uncertainty that results from the assumption that K in Equation 19-22 remains constant after calibration. This assumption can seldom, if ever, be exactly true because the electrolyte composition of the unknown will almost inevitably differ from that of the solution employed for calibration. The junction potential contained in K will vary slightly as a consequence, even when a salt bridge is used. Often, this uncertainty will be of the order of 1 mV or more; it is readily shown (see Problems 17 and 18 at the end of this chapter) that the relative error in activity or concentration associated with a 1 mV uncertainty in K is about 4% when n in Equation 19-22 is 1, and 8% when it is 2. *It is important to appreciate that this uncertainty is characteristic of all measurements involving cells that contain a salt bridge and that this error cannot be eliminated by even the most careful measurements of cell potentials or the most sensitive measuring devices; nor does it appear possible to devise a method for completely eliminating the uncertainty in K that is the source of this error.*

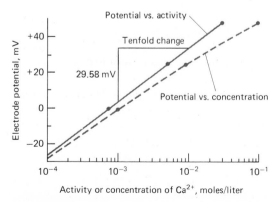

FIGURE 19-10   Response of a calcium ion electrode to variations in the calcium ion concentration and activity of solutions prepared from pure calcium chloride. (Courtesy of Orion Research, Inc., Cambridge, MA.)

Calibration Curves for Direct Potentiometry

An obvious way of correcting potentiometric measurements to give results in terms of concentration is to make use of an empirical calibration such as the lower curve in Figure 19-10. For this approach to be successful, however, it is essential that the ionic composition of the standard closely approximate that of the analyte—a condition that is difficult to realize experimentally for complex samples.

Where electrolyte concentrations are not too great, it is often helpful to swamp both the samples and the calibration standards with a measured excess of an inert electrolyte.

Under these circumstances, the added effect of the electrolyte in the sample becomes negligible, and the empirical calibration curve yields results in terms of concentration. This approach has been employed for the potentiometric determination of fluoride in public water supplies. Here, both samples and standards are diluted on a 1 : 1 basis with a solution containing sodium chloride, a citrate buffer, and an acetate buffer; the diluent is sufficiently concentrated so that the samples and standards do not differ significantly in ionic strength. The procedure permits a rapid measurement of fluoride ion in the 1-ppm range, with a precision of about 5% relative.

Standard Addition Method

In the standard addition method, the potential of the electrode system is measured before and after addition of a small volume of a standard to a known volume of the sample. The assumption is made that this addition does not alter the ionic strength and thus the activity coefficient f of the analyte. It is further assumed that the added standard does not significantly alter the junction potential.

If the potentials before and after the addition are E_1 and E_2, respectively, we may write, with the aid of Equation 19-22

$$-\log a_1 = -\log C_x f = \frac{E_1 - K}{0.0591/n}$$

where a_1 is activity of the analyte in the sample, f is its activity coefficient, and C_x is its molar concentration. Addition of V_s ml of a standard with a molar concentration of C_s to V_x ml of the sample causes the potential to acquire a value E_2; Equation 19-22 then becomes

$$-\log\left[\frac{C_x V_x + C_s V_s}{V_x + V_s} f\right] = \frac{E_2 - K}{0.0591/n}$$

Subtracting the first equation from the second gives, upon rearrangement,

$$\log\left[\frac{C_x(V_x + V_s)}{C_x V_x + C_s V_s}\right] = \frac{E_2 - E_1}{0.0591/n} = \frac{\Delta E}{0.0591/n}$$

This equation rearranges to

$$C_x = \frac{C_s V_s}{(V_x + V_s)10^{-n\,\Delta E/0.0591} - V_x} \quad (19\text{-}23)$$

Thus, C_x is readily calculated from the concentration of the standard, the two volumes, and the potential difference ΔE.

The standard addition method has been applied to the determination of chloride and fluoride in samples of commercial phosphors.[14] Here, two solid-state indicator electrodes and a reference electrode were used; the added standard contained known quantities of the two anions. The relative standard deviation for the measurement of replicate samples was found to be 0.7% for fluoride and 0.4% for chloride. When the standard addition method was not used, relative errors for the analyses appeared to range between 1 and 2%.

Potentiometric pH Measurements with a Glass Electrode[15]

The glass electrode is unquestionably the most important indicator electrode for hydrogen ion. It is convenient to use and is subject to few of the interferences that affect other pH-sensing electrodes.

Glass electrodes are available at relatively low cost and in many shapes and sizes. A common variety is illustrated in Figure 19-1 (p. 541); the reference electrode is usually a commercial saturated calomel electrode.

The glass/calomel electrode system is a remarkably versatile tool for the measurement of pH under many conditions. The electrode can be used without interference in solutions containing strong oxidants, reductants, pro-

[14] L. G. Bruton, *Anal. Chem.*, **43**, 579 (1971).

[15] For a detailed discussion of potentiometric pH measurements, see: R. G. Bates, *Determination of pH, Theory and Practice.* New York: Wiley, 1964.

teins, and gases; the pH of viscous or even semisolid fluids can be determined. Electrodes for special applications are available. Included among these are small electrodes for pH measurements in a drop (or less) of solution or in a cavity of a tooth, microelectrodes which permit the measurement of pH inside a living cell, systems for insertion in a flowing liquid stream to provide a continuous monitoring of pH, and a small glass electrode that can be swallowed to indicate the acidity of the stomach contents (the calomel electrode is kept in the mouth).

Summary of Errors Affecting pH Measurements with the Glass Electrode. The ubiquity of the pH meter and the general applicability of the glass electrode tend to lull the chemist into the attitude that any measurement obtained with such an instrument is surely correct. It is well to guard against this false sense of security since there are distinct limitations to the electrode system. These have been discussed in earlier sections and include the following:

1. **The alkaline error.** The ordinary glass electrode becomes somewhat sensitive to alkalimetal ions at pH values greater than 9.
2. **The acid error.** At a pH less than 0.5, values obtained with a glass electrode tend to be somewhat high.
3. **Dehydration.** Dehydration of the electrode may cause unstable performance and errors.
4. **Errors in unbuffered neutral solutions.** Equilibrium between the electrode surface layer and the solution is achieved only slowly in poorly buffered, approximately neutral solutions. Errors will arise unless time (often several minutes) is allowed for this equilibrium to be established. In determining the pH of poorly buffered solutions, the glass electrode should first be thoroughly rinsed with water. Then, if the unknown is plentiful, the electrodes should be placed in successive portions until a constant pH reading is obtained. Good stirring is also helpful; several minutes should be allowed for the attainment of steady readings.

5. **Variation in junction potential.** It should be reemphasized that this is a fundamental uncertainty in the measurement of pH, for which a correction cannot be applied. Absolute values more reliable than 0.01 pH unit are generally unobtainable. Even reliability to 0.03 pH unit requires considerable care. On the other hand, it is often possible to detect pH *differences* between similar solutions or pH *changes* in a single solution that are as small as 0.001 unit. For this reason, many pH meters are designed to permit readings to less than 0.01 pH unit.
6. **Error in the pH of the standard buffer.** Any inaccuracies in the preparation of the buffer used for calibration, or changes in its composition during storage, will be propagated as errors in pH measurements. A common cause of deterioration is the action of bacteria on organic components of buffers.

POTENTIOMETRIC TITRATIONS

The potential of a suitable indicator electrode is conveniently employed to establish the equivalence point for a titration (a *potentiometric titration*). A potentiometric titration provides different information from a direct potentiometric measurement. For example, the direct measurement of $0.100F$ acetic and hydrochloric acid solutions with a pH-sensitive electrode would yield widely different pH values because the former is only partially dissociated. On the other hand, potentiometric titrations of equal volumes of the two acids would require the same amount of standard base for neutralization.

The potentiometric end point is widely applicable and provides inherently more accurate data than the corresponding method employing indicators. It is particularly useful for titration of colored or turbid solutions and for detecting the presence of unsuspected species in a solution. Unfortunately, it is more time-consuming than a titration that makes use of an indicator.

Figure 19-11 shows a typical apparatus for performing a potentiometric titration. Ordinarily, the titration involves measuring and recording a cell potential (or a pH reading) after each addition of reagent. The titrant is added in large increments at the outset; as the end point is approached (as indicated by larger potential changes per addition), the increments are made smaller.

Sufficient time must be allowed for the attainment of equilibrium after each addition of reagent. Precipitation reactions may require several minutes for equilibration, particularly in the vicinity of the equivalence point. A close approach to equilibrium is indicated when the measured potential ceases to drift by more than a few millivolts. Good stirring is frequently effective in hastening the achievement of equilibrium.

The first two columns of Table 19-4 consist of typical potentiometric titration data obtained with the apparatus illustrated in Figure 19-11. The data near the end point are plotted in Figure 19-12a. Note that this experimental plot closely resembles titration curves derived from theoretical considerations.

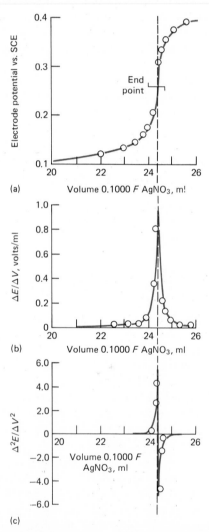

(a)

(b)

(c)

FIGURE 19-12 (a) Potentiometric titration curve for 2.433 meq of Cl^- with 0.1000 F $AgNO_3$. (b) First-derivative curve. (c) Second-derivative curve.

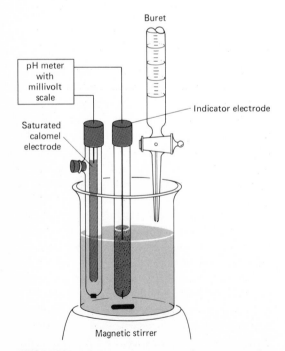

FIGURE 19-11 Apparatus for a potentiometric titration.

End-Point Determination

Several methods can be used to determine the end point for a potentiometric titration. The most straightforward involves a direct plot of potential versus reagent volume, as in Figure 19-12a. The midpoint in the steeply rising portion of the curve is then estimated visually and taken as the end point. Various mechanical methods to aid in the establishment of the midpoint have been proposed; it is doubtful, however, that these significantly improve the accuracy.

A second approach is to calculate the change in potential per unit change in volume of reagent (that is, $\Delta E/\Delta V$), as has been done in column 3 of Table 19-4. A plot of this parameter as a function of average volume leads to a sharp maximum at the end point (see Figure 19-12b). Alternatively, the ratio can be evaluated during the titration and recorded directly in lieu of the potential itself. Thus, in column 3 of Table 19-4, it is seen that the maximum is located between 24.3 and 24.4 ml; selection of 24.35 ml would be adequate for most purposes.

With both techniques, the assumption is made that the titration curve is symmetric about the true equivalence point and that the inflection in the curve therefore corresponds to that point. This assumption is valid, provided the participants in the chemical process react with one another in an equimolar ratio, and also provided the electrode process is perfectly reversible. Where these provisions are not met, an asymmetric curve results (see curve *B*, Figure 19-13). Note that the curve for the oxidation of iron(II) by cerium(IV) is symmetrical about the equivalence point. On the other hand, 5 moles of iron(II) are consumed by each

TABLE 19-4 POTENTIOMETRIC TITRATION DATA FOR 2.422 MILLIEQUIVALENTS OF CHLORIDE WITH 0.1000 *F* SILVER NITRATE[a]

Vol AgNO$_3$, ml	E vs. SCE, V	$\Delta E/\Delta V$, V/ml	$\Delta^2 E/\Delta V^2$
5.0	0.062		
		0.002	
15.0	0.085		
		0.004	
20.0	0.107		
		0.008	
22.0	0.123		
		0.015	
23.0	0.138		
		0.016	
23.50	0.146		
		0.050	
23.80	0.161		
		0.065	
24.00	0.174		
		0.09	
24.10	0.183		
		0.11	
24.20	0.194		2.8
		0.39	
24.30	0.233		4.4
		0.83	
24.40	0.316		−5.9
		0.24	
24.50	0.340		−1.3
		0.11	
24.60	0.351		−0.4
		0.07	
24.70	0.358		
		0.050	
25.00	0.373		
		0.024	
25.5	0.385		
		0.022	
26.0	0.396		
		0.015	
28.0	0.426		

[a] The electrode system consists of a silver indicator electrode and a saturated calomel electrode. The volume of the analyte solution was 50.00 ml.

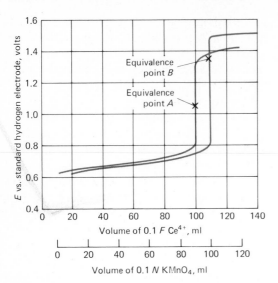

FIGURE 19-13   Curves depicting the titration of 100 ml of 0.100 N iron(II) solution with (A) cerium(IV) and with (B) permanganate. Note that the horizontal axes are displaced to permit comparison of the curves.

mole of permanganate and a highly non-symmetric titration curve results. Ordinarily, the change in potential within the equivalence-point region of these curves is sufficiently large so that a negligible titration error is introduced if the midpoint of the steeply rising portion is chosen as the end point. Only when unusual accuracy is required or where very dilute solutions are employed must account be taken of this source of uncertainty. Under such circumstances, a correction can be determined empirically by titration of a standard. Alternatively, the correct position of the equivalence point can be calculated from theoretical considerations when the error is due to a non-symmetrical reaction.[16]

---

[16] See: I. M. Kolthoff and N. H. Furman, *Potentiometric Titrations*, 2d ed. New York: Wiley, 1931, Chapters 2 and 3.

Gran's Plots. Gran[17] has suggested a method to permit evaluation of the end point with data from a region of the titration where the reaction is forced nearly to completion by the common ion effect. Such a procedure has the advantage of requiring fewer data points than a conventional plot. Furthermore, it may yield more accurate end points in instances where the rate of change in p function is small in the equivalence-point region.

As an example, we shall apply Gran's method to the data shown in Table 19-4 for the potentiometric titration of chloride ion with silver nitrate. Short of the equivalence point in this titration, the silver electrode behaves as a second-order electrode for chloride ion, and we may describe the cell potential E_{obs} by the equation

$$E_{obs} = E^0_{AgCl} - 0.0591 \log[Cl^-] - E_{SCE}$$

We have neglected the junction potential because it is of no significance in a potentiometric titration. It is convenient to rewrite this equation in the form

$$\log[Cl^-] = -16.9E_{obs} + K'$$

where 16.9 is the reciprocal of 0.0591 and K' is equal to $(E^0_{AgCl} - E_{SCE})/0.0591$. Taking the antilogarithm of both sides of this equation leads to

$$[Cl^-] = antilog(-16.9E_{obs} + K')$$

Until the end point is reached, $[Cl^-]$ is given by

$$[Cl^-] = \frac{V_{Cl}F_{Cl} - V_{Ag}F_{Ag}}{V_{Cl} + V_{Ag}}$$

where V_{Cl} and V_{Ag} are the measured volumes of the analyte and the silver nitrate solutions; F_{Cl} and F_{Ag} are the initial formal concentrations of the two solutions. Combining the two equations and rearranging gives

$$(V_{Cl}F_{Cl} - V_{Ag}F_{Ag}) = (V_{Cl} + V_{Ag})antilog$$
$$(-16.9E_{obs} + K')$$

---

[17] G. Gran, *Analyst*, **77**, 661 (1952).

For a given titration, all of the terms in this equation are constant except V_{Ag} and E_{obs}. Thus, a plot of the quantity $(V_{Cl} + V_{Ag})$anti-log$(-16.9E_{obs})$ against V_{Ag} should be linear. In addition, at equivalence, the left-hand side of the equation will equal zero. Thus, an extrapolation of the plot to zero on the antilog axis will give the equivalence-point volume. A Gran's plot of the data in Table 19-4 is shown in Figure 19-14. Note that the first four points lie on a good straight line. A departure from linearity is observed, however, for the data that lie less than about 2 ml from equivalence. For less complete reactions, the curvature occurs earlier in the titration. Clearly, a satisfactory end point can be obtained from 3 or 4 data points in contrast to the 19 that were recorded in Table 19-4.

When the concentration of the reagent is made large with respect to the analyte, $(V_{Cl} + V_{Ag})$ is equal to approximately V_{Cl} throughout the titration; the quantity antilog $(-16.9E_{obs})$ can then serve as the ordinate.

Semiantilog paper is available from Orion Research Inc.,[18] and makes it possible to obtain Gran's plots without computations. The vertical axis of this paper is skewed to eliminate the need to correct for volume change, provided the titrant volume is less than 10% of the total at the end point.

Titration to a Fixed Potential. Another procedure consists of titrating to a predetermined end point potential. The value chosen may be the theoretical equivalence-point potential calculated from formal potentials or an empirical potential obtained by the titration of standards. Such a method demands that the equivalence-point behavior of the system be entirely reproducible.

Precipitation Titrations

Electrode Systems. For a precipitation titration, the indicator electrode is often the

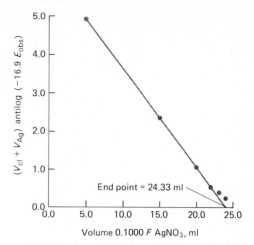

FIGURE 19-14 Gran's plot of the data in Table 19-4.

metal from which the reacting cation is derived. Membrane electrodes that are sensitive to one of the ions involved in the titration process may also be employed. Occasionally, an inert platinum electrode can serve as the indicator. Thus, for example, the titration of zinc ion with ferrocyanide may be followed with a platinum electrode, provided a quantity of ferricyanide ion is also present. The potential developed by this electrode is determined by the half-reaction

$$Fe(CN)_6^{3-} + e \rightleftharpoons Fe(CN)_6^{4-}$$

With the first excess of reagent, the ferrocyanide concentration increases rapidly and causes a corresponding change in the potential of the electrode.

The most widely used volumetric precipitation reagent is silver nitrate, which is employed for the determination of the halogens, the halogenoids, mercaptans, sulfides, arsenates, phosphates, and oxalates. For all of these, a silver indicator electrode is used. When dilute solutions are being titrated or where highest precision is required, reference electrodes such as those shown in Figure 18-11 cannot be used directly because of leakage of

[18] Orion Research Inc., 380 Putman Ave., Cambridge, MA. 02139.

chloride ion from the salt bridge. This problem can be avoided by immersing the calomel electrode in a concentrated potassium nitrate solution; this solution is then connected to the analyte solution by means of an agar bridge that contains about 3% potassium nitrate. For reagent and analyte concentrations of $0.1F$ or greater, a calomel electrode such as that shown in Figure 18-11a can be used directly in the analyte solution without incurring significant errors.

Titration Curves. Figure 19-15 shows theoretically derived titration curves for chloride, bromide, and iodide ions with a silver/saturated calomel electrode system. The ordinate on the left is calibrated in terms of the cell potential in volts while that on the right is in terms of pAg. As shown by the dotted lines, the experimental curve for iodide departs somewhat from the theoretical as a consequence of adsorption. Just before the equivalence point, iodide ions are held on the surface of the solid, thus decreasing the iodide concentration in the solution and increasing the potential slightly. Beyond equivalence, adsorption acts to decrease the concentration of silver ions in the solution; lower potentials result. This effect is less pronounced for bromide and chloride ion.

Figure 19-15 demonstrates that the magnitude of change in electrode potential in the equivalence-point region increases as the reaction product becomes less soluble—that is, as the reaction becomes more complete. This

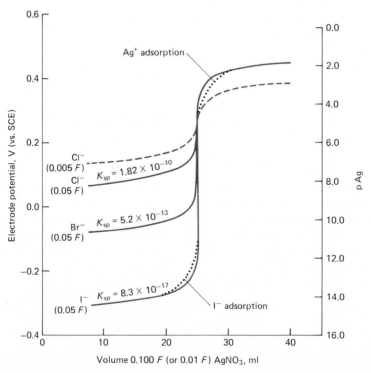

FIGURE 19-15 Theoretical titration curves for 50.0 ml of $0.0500\ F$ (or $0.005\ F$) sodium halide. Dotted lines show experimental deviations from theoretical curve for iodide due to adsorption of I^- and Ag^+ on the AgI precipitate.

latter effect is general and is encountered in complex formation, neutralization, and oxidation-reduction titrations.

It is evident from the two curves for chloride ion shown in Figure 19-15 that less sharp end points are observed as the analyte and reagent solutions become more dilute. For chloride, titrations with solutions as dilute as $0.001F$ can be performed with a relative uncertainty of about 1%. Smaller concentrations of bromide and iodide can be titrated successfully because their reactions with silver ion are more complete. The example that follows illustrates how these curves are derived.

EXAMPLE

An electrode system consisting of a silver indicator electrode and a saturated calomel electrode was employed for the potentiometric titration of 100 ml of $0.0200F$ sodium iodide with $0.100F$ silver nitrate. Calculate the cell potential at: (a) 1.00 ml before the equivalence point; (b) the equivalence point; and (c) 1.00 ml beyond the equivalence point. The required electrode potential data are

$$AgI(s) + e \rightleftharpoons Ag(s) + I^- \qquad E^0 = -0.151 \text{ V}$$

$$Hg_2Cl_2(s) + 2e \rightleftharpoons 2Hg(l) + 2Cl^- \text{ (sat'd KCl)}$$

$$E = +0.241 \text{ V}$$

(a) The titration will require 20.0 ml of $AgNO_3$. After 19.0 ml of the reagent have been added, the formal concentration of I^- will be

$$F_{NaI} = \frac{100 \times 0.0200 - 19.0 \times 0.100}{119}$$

$$= 8.40 \times 10^{-4}$$

An additional source of iodide results from the finite solubility of the precipitate. The concentration of iodide ions from this source will be just equal to the silver ion concentration. Thus,

$$[I^-] = 8.40 \times 10^{-4} + [Ag^+]$$

It is reasonable to assume, however, that

$$[Ag^+] \ll 8.40 \times 10^{-4}$$

Therefore, the potential of the indicator electrode will be given by

$$E_{Ag} = -0.151 - 0.0591 \log 8.40 \times 10^{-4}$$

$$= 0.031 \text{ V}$$

If the calomel electrode is treated as the anode,

$$E_{cell} = 0.031 - 0.241 = -0.210 \text{ V}$$

(b) At the equivalence point

$$[Ag^+] = [I^-]$$

The solubility product for silver iodide is

$$K_{sp} = [Ag^+][I^-] = 8.3 \times 10^{-17}$$

At equivalence, $[Ag^+] = [I^-]$. Thus,

$$[I^-] = \sqrt{8.3 \times 10^{-17}} = 9.1 \times 10^{-9}$$

Therefore,

$$E_{Ag} = -0.151 - 0.0591 \log 9.1 \times 10^{-9}$$

$$= 0.324$$

Again treating the calomel cell as the anode,

$$E_{cell} = 0.324 - 0.241 = 0.083 \text{ V}$$

It is interesting to note the polarity of the electrodes has changed between the initial addition and the equivalence point. This behavior is not often encountered.

(c) When the equivalence point has been exceeded by 1.00 ml, the concentration of excess Ag^+ is given by

$$F_{AgNO_3} = \frac{1.00 \times 0.100}{121} = 8.26 \times 10^{-4}$$

and

$$[Ag^+] = 8.26 \times 10^{-4} + [I^-] \cong 8.26 \times 10^{-4}$$

Here, it is easier to describe the behavior of the silver electrode in terms of

$$Ag^+ + e \rightleftharpoons Ag \qquad E^0 = 0.799 \text{ V}$$

and

$$E_{Ag} = 0.799 - 0.0591 \log \frac{1}{[Ag^+]}$$

$$= 0.799 - 0.0591 \log \frac{1}{8.26 \times 10^{-4}}$$

$$= 0.617 \text{ V}$$

$$E_{cell} = 0.617 - 0.241 = 0.376 \text{ V}$$

Titration of Mixtures. An important advantage of the potentiometric method is that it often permits discrimination among components of a mixture that react with a common titrant. Figure 19-16 illustrates this type of application for the determination of iodide, bromide, and chloride ions in an approximately equimolar mixture. During the early stages of the titration, only silver iodide, the least soluble of the three silver halides, precipitates. Indeed, it is readily shown that precipitation of silver bromide will not in theory begin until all but about 0.02% of the

iodide originally present has precipitated.[19] Thus, to this point, the titration curve will be identical to that for iodide by itself (see Figure 19-15). Precipitation of silver bromide then occurs; here again, it can be shown that in theory no silver chloride forms until all but about 0.3% of the bromide ion has been removed from solution. Therefore, the titration curve in this region is similar to that for bromide ion alone. Finally, chloride starts to precipitate, and the remainder of the curve is similar to the chloride titration curve in Figure 19-15.

Figure 19-16 suggests that the individual components of a halide mixture can be determined by a single potentiometric titration; in fact, the feasibility of such an analysis has been demonstrated experimentally.[20] For the titration of any pair of these ions, it is generally found that the volume of silver nitrate required to reach the first end point is somewhat greater than that predicted by theory; the total volume, however, approaches the correct amount. These observations can be explained by assuming that coprecipitation of the more soluble silver halide occurs during formation of the less-soluble compound; an overconsumption of reagent in the first part of the titration results.

Despite the coprecipitation error, the potentiometric method is useful for the analysis of halide mixtures. When approximately equal quantities of halides are present, relative errors can be kept to about 1 to 2%.

Complex Formation Titrations

Both metal and membrane electrodes can be used to detect end points for reactions that involve formation of soluble complexes. By

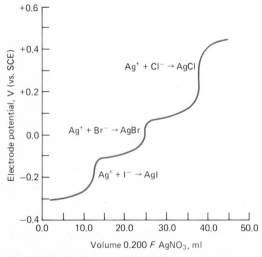

FIGURE 19-16 Potentiometric titration of 2.50 mfw of I^-, Br^-, and Cl^- with 0.200 F $AgNO_3$.

[19] See: D. A. Skoog and D. M. West, *Fundamentals of Analytical Chemistry*, 3d ed. New York: Holt, Rinehart and Winston, 1976, p. 176.

[20] For further details, see: I. M. Kolthoff and N. H. Furman, *Potentiometric Titrations*, 2d ed. New York: Wiley, 1931, pp. 154–158.

far, the most important reagents for complexometric titrations are a group of amino carboxylic acids, of which ethylene-diaminetetraacetic acid is an example. This reagent, which is commonly abbreviated as EDTA and given the formula H_4Y, has the structure

$$\begin{array}{l} HOOCCH_2 \\ \qquad\qquad N-CH_2-CH_2-N \\ HOOCCH_2 \end{array} \begin{array}{l} CH_2COOH \\ \\ CH_2COOH \end{array}$$

The four carboxylate functional groups, as well as the two amine nitrogens, participate in bond formation with metal ions. Regardless of the charge on the cation, the complex formation reaction can be formulated as

$$M^{n+} + H_4Y \rightleftharpoons MY^{(n-4)} + 4H^+ \qquad (19\text{-}24)$$

An important example of the application of this reagent is the determination of calcium and magnesium ions in hard water. The reaction for calcium is

$$Ca^{2+} + H_4Y \rightarrow CaY^{2-} + 4H^+$$

Indicator Electrodes. The use of mercury as a second-order electrode for EDTA anions was considered earlier (p. 540). Figure 19-17 is a schematic diagram of a typical commercially available mercury electrode.

Application of Potentiometric EDTA Titrations. Reilley and co-workers have made a systematic theoretical and experimental study of the application of the mercury electrode to the potentiometric determination of 29 di-, tri-, and quadrivalent cations with EDTA as the reagent.[21] In these studies, 5 to 500 mg quantities of the cation were titrated with 0.05 or $0.005F$ reagent solutions. One drop of a $10^{-3}F$ solution of HgY^{2-} was employed for each titration.

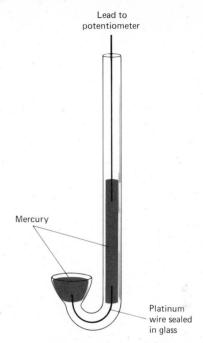

Lead to potentiometer

Mercury

Platinum wire sealed in glass

FIGURE 19-17 A typical mercury electrode. (Electrodes of this type can be obtained from Kontes Manufacturing Corp., Vineland, New Jersey.)

Figure 19-18 illustrates an application of their procedure to the determination of bismuth, cadmium, and calcium in a mixture. Bismuth(III) is first titrated at a pH of 1.2. At this acidity, neither cadmium nor calcium ions react to any significant extent. That is, for these two ions the equilibrium shown by Equation 19-24 lies far to the left. Bismuth, on the other hand, forms a complex that is sufficiently stable to provide a satisfactory end point. After bismuth has been titrated, the solution is brought to a pH of 4 by addition of an acetate/acetic acid buffer, and the titration is continued to an end point for cadmium. Calcium ions do not react appreciably at this pH but can be subsequently titrated in a basic solution obtained by addition of an ammonia/ammonium chloride buffer.

[21] See: C. N. Reilley and R. W. Schmid, *Anal. Chem.*, **30**, 947 (1958); and C. N. Reilley, R. W. Schmid, and D. W. Lamson, *Ibid.*, 953 (1958).

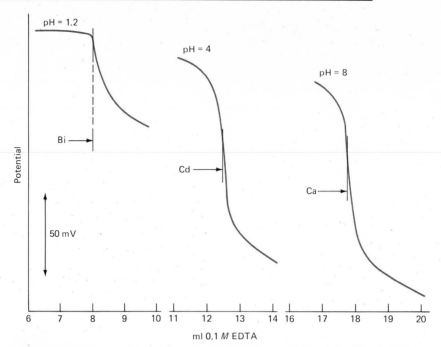

FIGURE 19-18   Potentiometric EDTA titration of a mixture of three cations. [Reprinted with permission from C. N. Reilley, R. W. Schmid, and D. W. Lamson, *Anal. Chem.*, **30**, 953 (1958). Copyright by the American Chemical Society.]

Neutralization Titrations

Potentiometric acid-base titration curves are readily obtained with a glass/calomel electrode system. End points from such curves are particularly useful when colored or turbid solutions must be titrated.

   Titration Curves for Weak Acids (or Weak Bases). Figure 19-19a shows theoretical curves for acids having different dissociation constants when titrated with standard sodium hydroxide. Figure 19-19b shows the effect of analyte and reagent concentrations on the titration curve for one of these acids. Experimental curves, obtained with a glass/calomel electrode system, approximate these theoretical curves closely. Clearly, as the strength of the acid becomes less (and thus the reaction with base less complete) and the concentra-

tion becomes lower, the ease of end point detection becomes poorer. Furman[22] has indicated that in order to obtain an accuracy of 1% relative in the potentiometric titration of a weak acid with a strong base, it is necessary that

$$F_{\text{NaA}} K_a \geq 10^{-5}$$

where F_{NaA} is the end point concentration of the sodium salt of the acid being titrated and K_a is the dissociation constant for the acid.

   The foregoing considerations apply equally to the titration of solutions of weak bases with strong acids.

[22] N. H. Furman, in *Treatise on Analytical Chemistry*, eds. I. M. Kolthoff and P. J. Elving. New York: Wiley-Interscience, 1967, Part I, vol. 4, p. 2287.

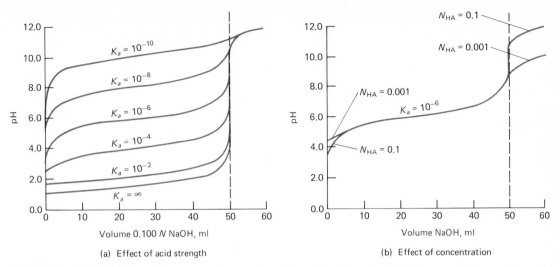

(a) Effect of acid strength

(b) Effect of concentration

FIGURE 19-19 Titration curves for weak acids. (a) 50.0 ml of 0.100 N acid in each case. (b) 50.0 ml of 0.100 N and 0.001 N HA with 0.100 N and 0.00100 N NaOH.

Titration of Mixtures of Acids (or Bases). Potentiometric titrations are advantageously employed for the titration of mixtures of acids (or bases). If the dissociation constants differ sufficiently, it is possible to determine the concentrations of the individual components in such mixtures. This application is demonstrated by the titration curves shown in Figure 19-20. Note that the individual concentrations of trichloroacetic and acetic acid in a mixture could be derived from a potentiometric titration curve. On the other hand, only the total acid concentration of the iodoacetic/acetic acid mixture could be ascertained because the strengths of these two acids are not sufficiently different. Generally, the concentrations of the two components of a mixture of acids (or bases) can be determined if the ratio of their dissociation constants is 10^4 or greater.

Titration Curves of Polyprotic Acids or Bases. Figure 19-21 shows titration curves for three common acids that have more than one titratable proton. Sulfuric acid exhibits only a single end point (curve C) because the degree of dissociation of the two protons is not

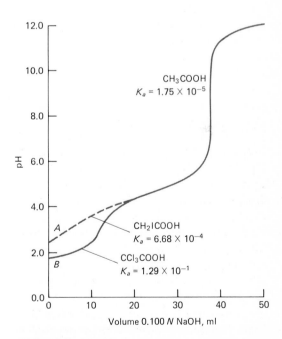

FIGURE 19-20 Titration curves for 50.0 ml of two mixtures. Both are 0.0500 F in acetic acid. Mixture A is also 0.0250 F in iodoacetic acid; similarly B; is 0.0250 F in trichloroacetic acid.

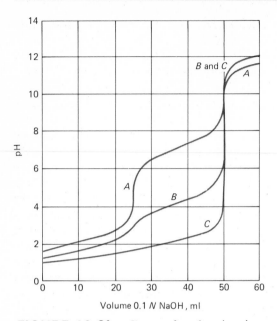

FIGURE 19-21 Curves for the titration of polybasic acids. A 0.100 N NaOH solution is used to titrate 25.0 ml of: 0.100 F H_3PO_4 (Curve A), 0.100 F oxalic acid (Curve B), 0.100 F H_2SO_4 (Curve C).

sufficiently different (one proton is completely dissociated while K_a for the second is 1.2×10^{-2}). The ratio between K_1 and K_2 for oxalic acid is about 1000; the curve for the titration (curve B) shows an inflection corresponding to the first equivalence point. However, the magnitude of the pH change is too small to permit accurate determination of this end point; the second is clearly well suited for a quantitative determination of the acid.

For phosphoric acid, the ratio of K_1 to K_2 is 10^5; two well-defined end points are obtained. Thus, analysis of phosphoric acid, sodium dihydrogen phosphate, and mixtures of these two compounds would be possible by means of a single potentiometric titration. The third hydrogen of phosphoric acid is so slightly dissociated $(K_3 = 4.2 \times 10^{-13})$ that it does not yield an end point of practical value. Its presence, however, is reflected in the some-

what lower pH values for curve A compared with curves B and C.

Determination of Dissociation Constants for Acids (or Bases). An approximate numerical value for the dissociation constant of a weak acid or base can be estimated from potentiometric titration curves. In theory, this quantity can be obtained from any point along the curve; as a practical matter, it is most easily found from the pH at the point of half-neutralization. For example, in the titration of the weak acid HA, we may ordinarily assume that at the midpoint

$$[HA] \cong [A^-]$$

and, therefore,

$$K_a = \frac{[H^+][A^-]}{[HA]} = [H^+]$$

or

$$pK_a = pH$$

It is important to note that a dissociation constant determined in this way may differ from that shown in a table of dissociation constants by a factor of 2 or more because the latter is based upon activities while the former is not. Thus, if we write the dissociation constant expression in its more exact form, we obtain

$$K_a = \frac{a_{H_3O^+} \cdot a_{A^-}}{a_{HA}} = \frac{a_{H_3O^+} \cdot [A^-] \cdot f_{A^-}}{[HA] \cdot f_{HA}}$$

Again, we assume that $[A^-]$ and $[HA]$ are approximately equal; furthermore, the potential of a glass electrode gives a good approximation for $a_{H_3O^+}$. Therefore,

$$K_a = \frac{a_{H_3O^+} \cdot [A^-] \cdot f_{A^-}}{[HA] \cdot f_{HA}} = \frac{a_{H_3O^+} \cdot f_{A^-}}{f_{HA}}$$

Taking the negative logarithm of the two sides of the equation yields

$$-\log K_a = -\log a_{H_3O^+} - \log \frac{f_{A^-}}{f_{HA}}$$

Upon converting to p functions and rearranging, we obtain

$$(pK_a)_{expt} = pH = pK_a + \log \frac{f_{A^-}}{f_{HA}}$$

Thus, the true pK_a will differ from the experimental pK_a by the logarithm of the ratio of activities. Ordinarily, the ionic strength during a titration will be 0.1 or greater. Thus, the ratio of f_{A^-} to f_{HA} would be at least 0.75 (see Table 18-1) if HA is uncharged. For solutes such as $H_2PO_4^-$ and HPO_4^{2-}, this ratio will be even larger.

A value for the equivalent weight, the number of titratable protons, and the approximate dissociation constant of a pure sample of an unknown acid can be obtained from a single potentiometric titration; this information is frequently sufficient to identify the acid.

Titrations in Nonaqueous Solvents. Acid-base titrations in aqueous solvents are limited to substances with acidic or basic dissociation constants greater than about 10^{-9}. For weaker acids or bases, the reactions with the reagent are so incomplete that the change in pH at the equivalence point is not sufficient to permit accurate establishment of the end point. Often, however, analysis becomes feasible if a nonaqueous solvent is substituted.[23] For example, aqueous solutions of aniline ($K_b \sim 10^{-10}$) are so weakly basic that titration with standard perchloric acid is unsatisfactory. On the other hand, if glacial acetic acid solutions of aniline and perchloric acid are employed, the reaction between the two is nearly complete at the equivalence point, and a large change in pH occurs as a consequence. Similarly, the reaction of phenol ($K_a \sim 10^{-10}$) with a base in a nonaqueous solvent such as methyl isobutyl ketone is sufficiently favorable to make titration feasible; in water, the end point for this titration is not satisfactory.

The potentiometric method has proved particularly useful for signaling end points of titrations in nonaqueous solvents. The ordinary glass/calomel electrode system can be used; the electrodes must be stored in water between titrations to prevent dehydration of the glass and precipitation of potassium chloride in the salt bridge. Ordinarily, the millivolt scale rather than the pH scale of the potentiometer should be employed because the potentials in nonaqueous solvents may exceed the pH scale. Furthermore, the pH scale based upon aqueous buffers has no significance in a nonaqueous environment. The titration curves are thus empirical; however, they provide a useful and satisfactory means of end point detection.

Oxidation-Reduction Titrations

A platinum electrode responds rapidly to many important oxidation-reduction couples and develops a potential that depends upon the concentration (strictly, activity) ratio of the reactants and the products of such half-reactions. For example, a platinum electrode system can be used to determine the end point in the titration of iron(II) with a standard solution of potassium permanganate (see Figure 19-13). The platinum electrode is responsive to both oxidation-reduction systems that exist in the solution throughout the titration. That is,

$$MnO_4^- + 8H^+ + 5e \rightleftharpoons Mn^{2+} + 4H_2O$$

$$E^0 = 1.51 \text{ V}$$

$$Fe^{3+} + e \rightleftharpoons Fe^{2+} \qquad E^0 = 0.771 \text{ V}$$

After each addition of reagent, interaction among the species occurs until reactant and product concentrations are such that the *electrode potentials for the two half-reactions are*

[23] For a more complete discussion of acid-base titrations in nonaqueous solvents, see: D. A. Skoog and D. M. West, *Fundamentals of Analytical Chemistry*, 3d ed. New York: Holt, Rinehart and Winston, 1976, Chapter 12.

identical. That is, at equilibrium,

$$E_{Pt} = 1.51 - \frac{0.0591}{5} \log \frac{[Mn^{2+}]}{[MnO_4^-][H^+]^8}$$

$$= 0.771 - \frac{0.0591}{1} \log \frac{[Fe^{2+}]}{[Fe^{3+}]}$$

Thus, the measured potential can be thought of as arising from *either* of the two half-cell systems.

The change in the ordinate function in the equivalence-point region of an oxidation-reduction titration becomes larger as the reaction becomes more complete; this effect is identical with that encountered for other reaction types. Thus, in Figure 19-22, data are plotted for titrations involving a hypothetical analyte that has a standard potential of 0.2 V with several reagents that have standard potentials ranging from 0.4 to 1.2 V; the corresponding equilibrium constants for the reaction range from about 2×10^3 to 9×10^{16}. Clearly, an increase in completeness is accompanied by an in-

creased change in the electrode potential in the end point region.

The curves in Figure 19-22 were derived for reactions in which the oxidant and reductant each exhibit a one-electron change; where both reactants undergo a two-electron transfer, the change in potential in the region of 24.9 to 25.1 ml is larger by about 0.14 V.

Solutions containing two oxidizing agents or two reducing agents will yield titration curves that contain two inflection points, provided the standard potentials for the two species are sufficiently different. If this difference is greater than about 0.2 V, the end points are usually distinct enough to permit analysis for each component. This situation is quite comparable to the titration, with the same reagent, of two acids with different dissociation constants or of two ions that form precipitates of different solubilities.

Differential Titrations

We have seen that a derivative curve generated from the data of a conventional potentiometric titration curve (Figure 19-12b) reaches a distinct maximum in the vicinity of the equivalence point. It is also possible to acquire titration data directly in derivative form by means of suitable apparatus.

A differential titration requires the use of two *identical* indicator electrodes, one of which is well shielded from the bulk of the solution. Figure 19-23 illustrates a typical arrangement. Here, one of the electrodes is contained in a small sidearm test tube. Contact with the bulk of the solution is made through a small (~ 1 mm) hole in the bottom of the tube. Because of this restricted access, the composition of the solution surrounding the shielded electrode will not be immediately affected by an addition of titrant to the bulk of the solution. The resulting difference in solution composition gives rise to a difference in potential ΔE between the electrodes. After each potential measurement, the solution is

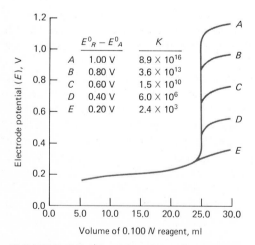

FIGURE 19-22 Titration of 50.0 ml of 0.0500 N A. E_A^0 is assumed to be 0.200 V. From the top, E_R^0 for the reagent is 1.20, 1.00, 0.80, 0.60, and 0.40 V, respectively. Both the reagent and analyte are assumed to undergo a one-electron change.

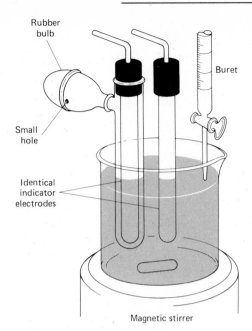

FIGURE 19-23 Apparatus for differential potentiometric titrations.

homogenized by squeezing the rubber bulb several times, whereupon ΔE again becomes zero. If the volume of solution in the tube that shields the electrode is kept small (say, 1 to 5 ml), the error arising from failure of the final addition of reagent to react with this portion of the solution can be shown to be negligibly small.

For oxidation-reduction titrations, two platinum wires can be employed, with one enclosed in an ordinary medicine dropper.

The main advantage of a differential method is the elimination of the need for the reference electrode and salt bridge.

Automatic Titrations

In recent years, several automatic titrators based on the potentiometric principle have become available from commercial sources. These instruments are useful where many

routine analyses are required.[24] Automatic titrators cannot yield results that are more accurate than those obtained by manual potentiometric techniques; however, they do decrease the operator time needed to perform the titrations and thus may offer significant economic advantages.

Flow Control and Measurement of Reagent Volume. Several methods exist for automatic control of the addition of a reagent and the measurement of its volume. The simplest employs an ordinary buret in which the stopcock is replaced with an electromagnetic pincer device. Here, an elastic plastic tube is inserted between the buret body and the tip. Flow is prevented by pinching the tube between a spring-loaded soft iron piece and a metal wedge. Titrant is introduced by passage of electricity through a solenoid that surrounds the pinching device. In another type of valve, a small piece of iron is sealed into a glass or plastic tube that fits inside the outflow tube of a buret. The two surfaces are ground to form a stopper. Current in a solenoid unseats the stopper and allows a flow of reagent.

The most widely used apparatus for automatic reagent addition consists of a calibrated syringe that is activated by a motor-driven micrometer screw. Such a device is shown in Figure 19-25.

Preset End-Point Titrators. Figure 19-24 is a schematic diagram of the simplest and least expensive type of automatic titrator. Here, a preset equivalence point potential is applied across the electrodes by means of a calibrated potentiometer. If a difference exists between this potential and that of the electrodes, an "error" signal results. This signal is amplified and closes an electronic switch that permits

---

[24] For a comprehensive discussion of automatic titrators, see: G. Svehla, *Automatic Potentiometric Titrations.* New York: Pergamon Press, 1978; and J. K. Foreman and P. B. Stockwell, *Automatic Chemical Analysis.* New York: Wiley, 1975, pp. 44–62.

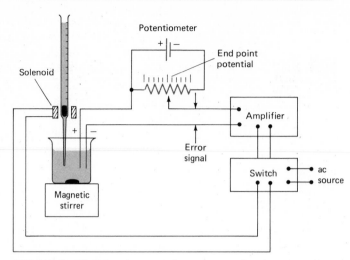

FIGURE 19-24   Automatic preset end-point titrator.

a flow of electricity through the solenoid-operated valve of the buret. As the error signal approaches zero, current to the solenoid is switched off, and flow of titrant ceases.

Titrators such as the one shown in Figure 19-24 do not respond instantaneously and tend to overshoot the end point. To overcome this problem, some instruments are equipped to superimpose a square-wave signal upon the error signal. The solenoid switch is then set to close only when the net signal exceeds that of the square wave. Titrant is then added in increments that can be controlled by the frequency of the square-wave signal.

Second-Derivative Titrators. Second-derivative titrators can also be relatively simple devices; they have the advantage that no pre-knowledge of the equivalence-point potential is required. The signal processor of these devices contains two electronic derivative circuits (p. 66) in series to convert the amplified signal from the electrode to a voltage proportional to the second derivative of the electrode potential of the indicator electrode. The output is then similar in form to that shown in Figure 19-12c, where the sign of the signal changes at

the equivalence point. This change in sign then causes a switching device to turn off the flow of titrant.

Recording Titrators. Recording titrators carry a titration beyond the equivalence point while recording a curve for the analyte. In some instruments, the rate of reagent addition is held constant and is synchronized with the chart drive of a millivolt recorder. The pen then records the amplified output potential of the cell as a function of time, which is proportional to the reagent volume.

Often, recording titrators are equipped to perform a titration the way a skilled chemist does—that is, to add titrant rapidly before and after the end point but in small increments as the end point is approached and passed. Figure 19-25 is a simplified block diagram of such an instrument. The switching device in this instrument operates from an amplified error signal that develops when the potential of the electrodes differs by some minimum amount from the potential applied by the potentiometer. The contact of this potentiometer is driven by the motor, which also positions the pen of the recorder. When

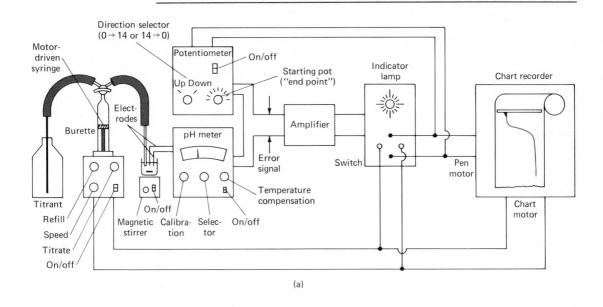

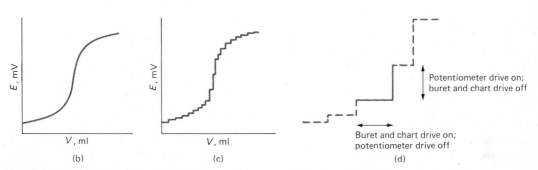

FIGURE 19-25 (a) An automated, curve-recording titrator. (b) Theoretical titration curve. (c) Recorded curve with end-point anticipation. (d) Enlarged portion of (c) with explanation. (From G. Svehla, *Automatic Potentiometric Titrations.* New York: Pergamon Press, 1978, p. 176. With permission.)

the error signal is small, the electronic switch activates the motor of the syringe-type buret and the chart drive of the recorder; simultaneously, the potentiometer and pen drive motor is switched off. When sufficient reagent has been added to cause the error signal to exceed some predetermined level, the switching is reversed. Now, the buret and chart drives are off and the potentiometer and pen

drives are turned on. Thus, the reagent is added in a series of steps, as shown by the recorded titration curve in Figure 19-25c.

Fully Automatic Titrators. A fully automatic titrator is equipped with a turntable that holds a series of samples for titration. After a titration has been completed, the solution is discarded, the vessel and electrodes are rinsed, the buret is refilled, the sample table is rotated, a measured volume of a new sample is introduced into the system, and the titration process is resumed. Such instruments are controlled by microprocessors and usually have computing facilities for calculating and printing out the analytical results. Instruments of this kind are expensive, but their costs can be easily justified for laboratories that must perform large numbers of routine titrations regularly.

PROBLEMS

1. (a) Calculate the standard potential for the reaction

$$CuBr(s) + e \rightleftharpoons Cu(s) + Br^-$$

For CuBr, $K_{sp} = 5.2 \times 10^{-9}$.

(b) Give a schematic representation of a cell with a copper indicator electrode as an anode and a saturated calomel electrode as a cathode that could be used for the determination of Br^-.

(c) Derive an equation that relates the measured potential of the cell in (b) to pBr (assume that the junction potential is zero).

(d) Calculate the pBr of a bromide-containing solution that is saturated with CuBr and contained in the cell described in (b) if the resulting potential is 0.0176 V.

2. (a) Calculate the standard potential for the reaction

$$AgSCN(s) + e \rightleftharpoons Ag(s) + SCN^-$$

For AgSCN, $K_{sp} = 1.1 \times 10^{-12}$.

(b) Give a schematic representation of a cell with a silver indicator electrode as the cathode and a saturated calomel electrode as an anode that could be used for determining SCN^-.

(c) Derive an equation that relates the measured potential of the cell in (b) to pSCN (assume that the junction potential is zero).

(d) Calculate the pSCN of a solution that is saturated with AgSCN and contained in the cell described in (b) if the resulting potential is 0.308 V.

3. Give a schematic representation of each of the following cells and an equation for the relationship between the cell potential and the desired quantity. Assume that the junction potential is negligible, and specify any necessary concentrations as $1.00 \times 10^{-4}M$. Treat the indicator electrode as the cathode in each case.

(a) A cell with a Hg indicator electrode for the determination of pBr. (Hg_2Br_2 has a low solubility.)

(b) A cell with a Ag indicator electrode for the determination of pC_2O_4. ($Ag_2C_2O_4$ has a low solubility.)

(c) A cell with a platinum electrode for the determination of pSn(II).

4. Give a schematic representation of each of the following cells and an equation for the relationship between the cell potential and the desired quantity. Assume that the junction potential is negligible, and specify any necessary concentration as $1.00 \times 10^{-4}M$. Treat the indicator electrode as the anode in each case.

(a) A cell with a Pb indicator electrode for the determination of pSO_4.

(b) A cell with a Ag indicator electrode for the determination of pCO_3. (Ag_2CO_3 has a slow solubility.)

(c) A cell with a platinum electrode for the determination of $pTl(I)$.

5. The following cell was employed for the determination of $pCrO_4$:

$$SCE \| Ag_2CrO_4(sat'd), CrO_4^{2-}(xM) | Ag$$

Calculate $pCrO_4$ if the cell potential is 0.356 V.

6. Calculate the potential of the cell (neglecting the junction potential)

$$Indicator\ electrode \| SCE$$

where the indicator electrode is mercury immersed in a solution that is

(a) $4.00 \times 10^{-6}M$ Hg^{2+}.

(b) $4.00 \times 10^{-6}M$ Hg_2^{2+}.

(c) saturated with Hg_2Br_2 ($K_{sp} = 5.8 \times 10^{-23}$) and is $3.00 \times 10^{-3}M$ in Br^-.

(d) $2.50 \times 10^{-5}F$ in $Hg(NO_3)_2$ and $0.0400F$ in KCl.

$$Hg^{2+} + 2Cl^- \rightleftharpoons HgCl_2(aq) \qquad K_f = 6.1 \times 10^{12}$$

(e) $2.50 \times 10^{-5}F$ in $Hg(NO_3)_2$ and $0.00400F$ in KCl.

7. The formation constant for the mercury(II) cyanide complex is

$$Hg^{2+} + 2CN^- \rightleftharpoons Hg(CN)_2(aq) \qquad K_f = 5.0 \times 10^{34}$$

Calculate the standard potential for the half-reaction

$$Hg(CN)_2 + 2e \rightleftharpoons Hg(l) + 2CN^-$$

8. The standard electrode potential for the reduction of the mercury complex of EDTA is given by

$$HgY^{2-} + 2e \rightleftharpoons Hg(l) + Y^{4-} \qquad E^0 = 0.21\ V$$

Calculate the formation constant for the reaction

$$Hg^{2+} + Y^{4-} \rightleftharpoons HgY^{2-}$$

9. Calculate the potential of the cell (neglecting the junction potential)

$$Hg | HgY^{2-}(4.50 \times 10^{-5}M), Y^{4-}(xM) \| SCE$$

where Y^{4-} is the EDTA anion, and the concentration of Y^{4-} is
(a) $3.33 \times 10^{-1}M$, (b) $3.33 \times 10^{-3}M$, and (c) $3.33 \times 10^{-5}M$.

$$HgY^{2-} + 2e \rightleftharpoons Hg(l) + Y^{4-} \qquad E^0 = 0.21\ V$$

10. The following cell was found to have a potential of 0.209 V when the solution in the left compartment was a buffer of pH 5.21:

$$Glass\ electrode | H^+(a = x) \| SCE$$

The following potentials were obtained when the buffered solution was replaced with unknowns. Calculate the pH and the hydrogen ion activity of each unknown.

(a) 0.064 V   (c) 0.510 V
(b) 0.329 V   (d) 0.677 V

11. The following cell was found to have a potential of 0.411 V:

$$\text{Membrane electrode for } Mg^{2+} \,|\, Mg^{2+}(a = 1.77 \times 10^{-3}M)\|SCE$$

When the solution of known magnesium activity was replaced with an unknown solution, the potential was found to be 0.439 V. What was the pMg of this unknown solution? Neglect the junction potential.

12. The following cell was found to have a potential of 0.893 V:

$$Cd\,|\,CdX_2(\text{sat'd}), \; X^-(0.0200M)\|SCE$$

Calculate the solubility product of CdX_2, neglecting the junction potential.

13. The following cell was found to have a potential of 0.693 V:

$$Pt, \, H_2(1.00 \text{ atm})\,|\,HA(0.200F), \; NaA(0.300F)\|SCE$$

Calculate the dissociation constant of HA, neglecting the junction potential.

14. A 40.00-ml aliquot of $0.1000N$ U^{4+} is diluted to 75.0 ml and titrated with $0.0800N$ Ce^{4+}. The pH of the solution is maintained at 1.00 throughout the titration. (Use 1.44 V for the formal potential of the cerium system.)

(a) Calculate the potential of the indicator cathode with respect to a saturated calomel reference electrode after the addition of 5.00, 10.00, 15.00, 25.00, 40.00, 49.00, 50.00, 51.00, 55.00, and 60.00 ml of cerium(IV).

(b) Draw a titration curve for these data.

15. Calculate the potential of a mercury cathode (vs. SCE) after the addition of 5.00, 15.00, 25.00, 30.00, 35.00, 39.00, 40.00, 41.00, 45.00, and 50.00 ml of $0.0500F$ $Hg_2(NO_3)_2$ to 50.00 ml of $0.0800F$ NaCl. Construct a titration curve from these data K_{sp} Hg_2Cl_2, 1.3×10^{-18}.

16. Calculate the potential (vs. SCE) of a lead anode after the addition of 0.00, 10.00, 20.00, 24.00, 24.90, 25.00, 25.10, 26.00, and 30.00 ml of $0.2000F$ $NaIO_3$ to 50.00 ml of $0.0500F$ $Pb(NO_3)_2$. For $Pb(IO_3)_2$, $K_{sp} = 3.2 \times 10^{-13}$.

17. A glass-calomel electrode system was found to develop a potential of 0.0620 V when used with a buffer of pH 7.00; with an unknown solution the potential was observed to be 0.2794 V.

(a) Calculate the pH and $[H^+]$ of the unknown.

(b) Assume that K is uncertain by ± 0.001 V as a consequence of a difference in the junction potential between standardization and measurement. What is the range of $[H^+]$ associated with this uncertainty?

    (c) What is the relative error in $[H^+]$ associated with the uncertainty in E_j?

18. The following cell was found to have a potential of 0.3674 V:

    Membrane electrode for Mg^{2+} | $Mg^{2+}(a = 6.87 \times 10^{-3}M)$‖SCE

    (a) When the solution of known magnesium activity was replaced with an unknown solution, the potential was found to be 0.4464 V. What was the pMg of this unknown solution?

    (b) Assuming an uncertainty of ± 0.002 V in the junction potential, what is the range of Mg^{2+} activities within which the true value might be expected?

    (c) What is the relative error in $[Mg^{2+}]$ associated with the uncertainty in E_j?

19. The sodium ion concentration of a solution was determined by measurements with a glass-membrane electrode. The electrode system developed a potential of 0.2331 V when immersed in 10.0 ml of the unknown. After addition of 1.00 ml of a standard solution that was $2.00 \times 10^{-2}M$ in Na^+, the potential decreased to 0.1846 V. Calculate the sodium ion concentration and the pNa of the original solution.

20. The calcium ion concentration of a solution was determined by measurements with a liquid-membrane electrode. The electrode system developed a potential of 0.4965 V when immersed in 25.0 ml of the sample. After addition of 2.00 ml of $5.45 \times 10^{-2}F$ $CaCl_2$, the potential changed to 0.4117 V. Calculate the calcium concentration and the pCa of the sample.

20

ELECTROGRAVIMETRIC AND COULOMETRIC METHODS

Three related electroanalytical methods, namely, *electrogravimetric analysis, constant-potential coulometry*, and *coulometric titrations*, are discussed in this chapter. Each involves an electrolysis that is carried on for a sufficient length of time to assure quantitative oxidation or reduction of the analyte. In electrogravimetric methods, the product of the electrolysis is weighed as a deposit on one of the electrodes. In the two coulometric procedures, on the other hand, the quantity of electricity needed to complete the electrolysis serves as a measure of the amount of analyte present.

The three methods generally have moderate selectivity, sensitivity, and speed; in many instances, they are among the most accurate and precise methods available to the chemist, with uncertainties of a few tenths percent relative being not uncommon. Finally, in contrast to all of the other methods discussed in this text, these three require no calibration against standards; that is, the functional relationship between the quantity measured and the weight of analyte can be derived from theory.

CURRENT-VOLTAGE RELATIONSHIP DURING AN ELECTROLYSIS

It is useful to consider the changes in current, voltage, and time when an electrolytic cell is operated in three different modes: (1) the applied cell potential is held constant; (2) the cell current is kept constant; and (3) the potential of one of the electrodes (the working electrode, which involves the analyte) is held constant. For all three, the behavior of the cell is governed by the relationship

$$E_{appl} = E_c - E_a + \Pi_1 + \Pi_2 - IR \quad (20\text{-}1)$$

where E_{appl} is the applied potential from the external source and E_c and E_a are the reversible or thermodynamic potentials associated with the cathode and anode, respectively; their

values can be calculated from standard potentials by means of the Nernst equation. The terms Π_1 and Π_2 are potentials associated with concentration and kinetic polarization (p. 528), respectively. Both are always negative, thus implying that an additional applied potential is required to overcome their effects. Furthermore, Π_1 and Π_2 are each made up of two terms, one associated with each electrode. Thus,

$$\Pi_1 = \Pi_{1c} + \Pi_{1a} \quad (20\text{-}2)$$

and

$$\Pi_2 = \Pi_{2c} + \Pi_{2a} \quad (20\text{-}3)$$

Here, the subscripts a and c again refer to anode and cathode.

Of the seven potentials shown in the three equations, only E_c and E_a can be derived from theory; the others must be evaluated empirically.

Operation of a Cell at a Fixed Applied Potential

One method of operating an electrolytic cell is to hold the applied potential at some fixed predetermined level through the entire electrolysis. This method, while simple, has some distinct limitations. In order to understand these limitations, it is necessary to consider how the current and the potential of the working electrode (the electrode at which the analytical reaction occurs) vary as a function of time.

To illustrate current-voltage relationships during an electrolysis at fixed potential, consider a cell consisting of two platinum electrodes, each with a surface area of 150 cm², immersed in 200 ml of a solution that is $0.0220M$ with respect to copper(II) ion and $1.00M$ with respect to hydrogen ion. The cell has a resistance of 0.50 Ω. When electricity is forced through the cell, the analyte copper is deposited upon the cathode, and oxygen is evolved at a partial pressure of 1.00 atm at

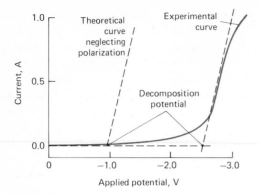

FIGURE 20-1   Current-voltage curve for the electrolysis of a copper(II) solution.

the anode. The overall cell reaction can be expressed as

$$Cu^{2+} + H_2O \rightarrow Cu(s) + \tfrac{1}{2}O_2(g) + 2H^+$$

Decomposition Potential. The reversible decomposition potential can be computed from standard potential data for the half-reactions

$$Cu^{2+} + 2e \rightleftharpoons Cu(s) \qquad E^0 = 0.34 \text{ V}$$

$$\tfrac{1}{2}O_2 + 2H^+ + 2e \rightleftharpoons H_2O \qquad E^0 = 1.23 \text{ V}$$

and has a value of -0.94. No current would be expected at less negative potentials; at greater applied potentials, the current would theoretically be determined by the magnitude of the cell resistance.

The dotted lines in Figure 20-1 illustrate the theoretical current-voltage behavior of this cell; the decomposition potential is seen to be the intersection of two straight lines. The linear rise in current immediately beyond the decomposition potential is a reflection of Ohm's law. The experimental current-voltage relationship for this system will, however, more closely resemble the solid curve in Figure 20-1. Here, the oxygen overvoltage at the anode has the effect of displacing the curve to more negative potentials. Moreover, a small current is observed with the first application of a potential. A part of this current is due to the reduc-

tion of trace impurities such as oxygen and iron(III), which are inevitably present in the solution. In addition, however, small amounts of copper are deposited at potentials lower than the decomposition potential. This apparent departure from theoretical behavior arises because it was assumed that the activity of copper was unity in calculating the decomposition potential. Experiments have shown, however, that the activity of a metal deposit that only partially covers a platinum surface is less than one[1]; under these circumstances, then, the behavior of the cathode is more correctly described by

$$E = E^0_{Cu^{2+}} - \frac{0.0591}{2} \log \frac{[Cu^{2+}]}{[Cu]}$$

where [Cu] is infinitely small at the outset and approaches one only after sufficient current has passed to coat the platinum surface completely.

Finally, it should be noted that the experimental curve departs from linearity at high currents owing to the onset of concentration polarization (see p. 528).

Calculation of Required Potential. For the cell under consideration, kinetic polarization occurs only at the anode where oxygen is evolved; the cathodic reaction is rapid and reversible. Thus, Equation 20-3 simplifies to $\Pi_2 = \Pi_{2a}$. Furthermore, concentration polarization at the anode is negligible because the anodic reactant and products are in large excess compared with concentration changes brought about by the electrolysis; therefore, it is unlikely that the surface layer will become depleted in these species, and Equation 20-2 simplifies to $\Pi_1 = \Pi_{1c}$. Finally, it is readily shown by suitable substitutions into the Nernst equation that the increase in hydrogen ion concentration due to the anodic reaction results in a negligible change in electrode potential (<0.01 V); that is, the anode potential

---

[1] See: L. B. Rogers et al., *J. Electrochem. Soc.*, **95**, 25, 33, 129 (1949); **98**, 447, 452, 457 (1951).

will be 1.23 V throughout the electrolysis. Thus, the required applied potential at any time during the electrolysis is given by (Equation 20-1)

$$E_{\text{appl}} = E_c - 1.23 + \Pi_{1c} + \Pi_{2a} - 0.50I \tag{20-4}$$

provided that the cell resistance remains constant throughout the electrolysis.

Let us now assume that we wish to operate the cell initially at a current of 1.5 A, which corresponds to a current density of 0.010 A/cm^2. From Table 18-3, it is seen that Π_{2a}, the initial oxygen overvoltage, will be about -0.85 V. Initially, the concentration polarization will be zero because the concentration of copper ions is high, and Equation 20-4 becomes

$$E_{\text{appl}} = 0.29 - 1.23 - 0.00 - 0.85$$
$$- 1.50 \times 0.50$$
$$= -2.54 \text{ V}$$

Thus, a reasonable estimate of the potential required to produce an initial current of 1.5 A is -2.5 V.

Current Changes During an Electrolysis at Constant Applied Potential. It is useful to consider the changes in current in the foregoing cell when the potential is held constant at -2.5 V throughout the electrolysis. Under these circumstances, the current would be found to decrease with time owing to depletion of copper ions in the solution and an increase in the degree of concentration polarization. In fact, with the onset of concentration polarization, the current decrease becomes exponential in time. That is,

$$I_t = I_0 e^{-kt}$$

where I_t is the current t min after the onset of polarization and I_0 is the current immediately before polarization. Lingane[2] has shown that values for the constant k can be computed from the relationship

$$k = \frac{25.8DA}{V\delta}$$

where D is the diffusion coefficient in cm^2/s or the rate at which the reactant diffuses under a unit concentration gradient. The quantity A is the electrode surface area in cm, V is the volume of the solution in cm^3, and δ is the thickness of the surface layer in which the concentration gradient exists. Typical values for D and δ are 10^{-5} cm^2/s and 2×10^{-3} cm. (The constant 25.8 includes the factor of 60 for converting D to cm^2/min, thus making k compatible with the units of t in the equation for I_t.)

When the initial applied potential is -2.5 V, it is found that concentration polarization, and thus an exponential decrease in current, occurs essentially immediately after application of the potential. Figure 20-2a depicts this behavior; the curve shown was derived for the cell under consideration with the aid of the foregoing two equations. After 30 min, the current has decreased from the initial 1.5 A to 0.08 A; by this time, approximately 96% of the copper has been deposited.

Potential Changes During an Electrolysis at Constant Applied Potential. It is instructive to consider the changes in some of the potentials in Equation 20-4 as an electrolysis proceeds. Figure 20-2b depicts these changes during the copper deposition under consideration.

As mentioned earlier, the thermodynamic anode potential remains substantially unchanged throughout the electrolysis. The reversible cathode potential E_c, on the other hand, becomes smaller (more negative) as the copper concentration decreases. The curve for E_c in Figure 20-2b was derived by substituting the calculated copper concentration after various electrolysis periods into the Nernst equation. Note that the potential decrease is approximately linear with time over a considerable period.

The IR drop shown in Figure 20-2b parallels the current changes shown in Figure 20-2a.

---

[2] See: J. J. Lingane, *Electroanaytical Chemistry*, 2d ed. New York: Interscience, 1958, pp. 223–229.

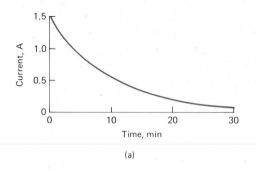

(a)

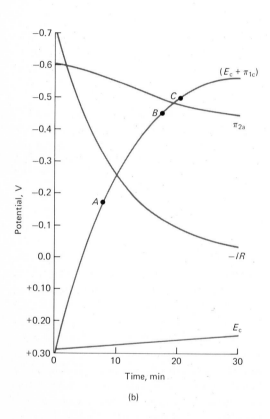

(b)

FIGURE 20-2 Changes in (a) current and (b) potentials during the electrolytic deposition of Cu^{2+}. Points A and B are potentials at which Pb and Cd would begin to codeposit if present. Point C is the potential at which H_2 might begin to form at the cathode. Any of these processes would distort the curve A.

The negative sign is employed for consistency with Equation 20-4.

As shown by the topmost curve of Figure 20-2b, the oxygen overvoltage also becomes less negative as the current, and thus the current density, falls. The data for this curve were obtained from more extensive compilations that are similar to Table 18-3.

The most significant feature of Figure 20-2b is the curve representing the change in *total cathode potential* $(E_c + \Pi_{1c})$ as a function of time. It is evident that as IR and Π_{2a} become less negative, one or more of the other potentials in Equation 20-4 must become more negative. Because of the large excess of reactant and product at the anode, its potential remains substantially constant. Thus, the *only* potentials that can change are those associated with the cathode; as seen from a comparison of the curves labeled E_c and $(E_c + \Pi_{1c})$, it is evident that most of this negative drift is a consequence of a rapid increase in concentration polarization Π_{1c}. That is, even with vigorous stirring, copper ions are not brought to the electrode surface at a sufficient rate to prevent polarization. The result is a rapid decrease in IR and a corresponding negative drift of the cathode potential.

The rapid shift in cathode potential that accompanies concentration polarization often leads to codeposition of other species and loss of selectivity. For example, points A and B on the cathode potential curve indicate where lead and cadmium ions would begin to deposit if present in concentrations about equal to that of the copper ion. Another event that probably would occur would be the evolution of hydrogen at about point C (this process was not taken into account in deriving the curve in Figure 20-2a).

The interferences just described could be avoided by a decrease in the applied potential by several tenths of a volt so that the negative drift of the cathode potential could never reach a level at which the interfering ions precipitate. The consequence, however, is a diminution in current and, ordinarily, an enormous increase in the time required for an analysis.

At best, an electrolysis at constant cell potential can only be employed to separate an easily reduced cation from those that are more difficult to reduce than hydrogen ion. Evolution of hydrogen can also be expected near the end of the electrolysis unless certain precautions are taken to prevent it.

Constant-Current Electrolysis

The analytical electrodeposition under consideration, as well as others, can be carried out by maintaining the current, rather than the applied potential, at a more or less constant level. Here, periodic increases in the applied potential are required as the electrolysis proceeds.

In the preceding section, it was shown that concentration polarization at the cathode causes a decrease in current. Initially, this effect can be partially offset by increasing the applied potential; the enhanced electrostatic attraction will cause copper ions to migrate more rapidly, thus maintaining a constant current. Shortly, however, the solution becomes sufficiently depleted in copper ions so that the forces of diffusion and electrostatic attraction cannot keep the electrode surface supplied with sufficient copper ion to maintain the desired current. When this occurs, further increases in E_{appl} cause rapid changes in Π_1 and the cathode potential; codeposition of hydrogen (or other reducible species) then takes place. The cathode potential ultimately becomes stabilized at a level fixed by the standard potential and the overvoltage for the new electrode reaction; further large increases in the cell potential are no longer necessary to maintain a constant current. Copper continues to deposit as copper(II) ions reach the electrode surface; the contribution of this process to the total current, however, becomes smaller and smaller as the deposition becomes more and more nearly complete. The alternative process, such as reduction of hydrogen or nitrate ions, soon

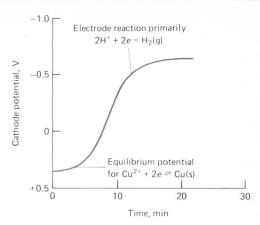

FIGURE 20-3 Changes in cathode potential during the deposition of copper with a constant current of 1.5 A. Here, the cathode potential is equal to $(E_c + \pi_{1c})$.

predominates. The changes in cathode potential under conditions of constant current are shown in Figure 20-3.

Constant Cathode Potential Electrolysis

From the Nernst equation, it is seen that a tenfold decrease in the concentration of an ion being deposited requires a negative shift in potential of only $0.0591/n$ V. Electrogravimetric methods, therefore, are potentially highly selective. In the present example, the copper concentration is decreased from $0.02M$ to $10^{-6}M$ as the thermodynamic cathode potential E_c changes from an initial value of $+0.29$ V to $+0.16$ V. In theory, then, it should be feasible to separate copper from any element that does not deposit within this 0.13-V potential range. Species that deposit quantitatively at potentials more positive than $+0.29$ V could be eliminated with a prereduction; ions that require potentials smaller than $+0.16$ V would not interfere with the copper deposition. Thus, if we are willing to accept a reduction in analyte concentration to $10^{-6}M$ as a quantitative separation, it follows that divalent ions differing in standard potentials by about 0.15 V or greater can, theoretically, be separated quantitatively by electrodeposition, provided their initial concentrations are about the same. Correspondingly, about 0.3- and 0.1-V differences are required for univalent and trivalent ions, respectively.

An approach to these theoretical separation values, with a reasonable electrolysis period, requires a more sophisticated technique than the ones thus far discussed because concentration polarization at the cathode, if unchecked, will prevent all but the crudest of separations. The change in cathode potential is governed by the decrease in IR drop (Figure 20-2b). Thus, where relatively large currents are employed at the outset, the change in cathode potential can ultimately be expected to be large. On the other hand, if the cell is operated at low current levels so that the variation in cathode potential is lessened, the time required for completion of the deposition may become prohibitively long. An obvious answer to this dilemma is to initiate the electrolysis with an applied cell potential that is sufficiently high to ensure a reasonable current; the applied potential is then continuously decreased to keep the cathode potential at the level necessary to accomplish the desired separation. Unfortunately, it is not feasible to predict the required changes in applied potential on a theoretical basis because of uncertainties in variables affecting the deposition, such as overvoltage effects and perhaps conductivity changes. Nor, indeed, does it help to measure the potential across the working electrodes, since such a measurement gives only the overall cell potential E_{appl}. The alternative is to measure the cathode potential against a third electrode whose potential in the solution is known and constant—that is, a reference electrode. The potential impressed across the *working* electrodes can then be adjusted to the level that will impart the desired potential to the cathode with respect to the reference elec-

trode. This technique is called *controlled cathode potential electrolysis*.

Experimental details for performing a controlled cathode potential electrolysis are presented in a later section. For the present, it is sufficient to note that the potential difference between the reference electrode and the cathode is measured with a potentiometer or electronic voltmeter. The potential applied between the working electrodes is controlled with a voltage divider so that the cathode potential is maintained at a level suitable for the separation. Figure 20-4 is a schematic diagram of an apparatus that would permit deposition at a constant cathode potential.

An apparatus of the type shown in Figure 20-4 can be operated at relatively high initial applied potentials to give high currents. As the electrolysis progresses, however, a lowering of the applied potential across AC is required. This decrease, in turn, diminishes the current. Completion of the electrolysis will be indicated by the approach of the current to zero. The changes that occur in a typical constant cathode potential electrolysis are

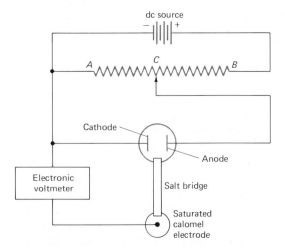

FIGURE 20-4 Apparatus for electrolysis at a controlled cathode potential. Contact C is continuously adjusted to maintain the cathode potential at the desired level.

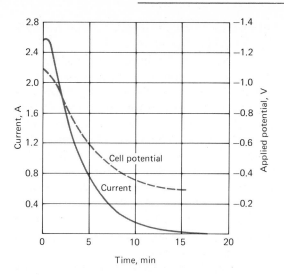

FIGURE 20-5 Changes in applied potential and current during a controlled cathode potential electrolysis. Deposition of copper upon a cathode maintained at −0.36 V vs. SCE. [Experimental data from J. J. Lingane, *Anal. Chim. Acta*, **2**, 589 (1949). With permission.]

depicted in Figure 20-5. In contrast to the electrolytic methods described earlier, this technique demands constant attention during operation. Usually, some provision is made for automatic control; otherwise, the operator time required represents a major disadvantage to the controlled cathode potential method.

ELECTROGRAVIMETRIC METHODS OF ANALYSIS

Electrolytic precipitation has been used for over a century for the gravimetric determination of metals. In most applications, the metal is deposited on a weighed platinum cathode, and the increase in weight is determined. Important exceptions to this procedure include the anodic depositions of lead as lead dioxide on platinum and chloride as silver chloride on silver.

Physical Properties of Electrolytic Precipitates

Ideally, an electrolytic deposit should be strongly adherent, dense, and smooth so that the processes of washing, drying, and weighing can be performed without mechanical loss or reaction with the atmosphere. Good metallic deposits are fine-grained and have a metallic luster; spongy, powdery, or flaky precipitates are likely to be less pure and less adherent.

The principal factors that influence the physical characteristics of deposits include current density, temperature, and the presence of complexing agents. Ordinarily, the best deposits are formed at current densities that are less than 0.1 A/cm^2. Stirring generally improves the quality of a deposit. The effects of temperature are unpredictable and must be determined empirically.

It is also found that many metals form smoother and more adherent films when deposited from solutions in which their ions exist primarily as complexes. Cyanide and ammonia complexes often provide the best deposits. The reasons for this effect are not obvious.

Codeposition of hydrogen during electrolysis is likely to cause the formation of nonadherent deposits, which are unsatisfactory for analytical purposes. The evolution of hydrogen can be avoided by introduction of a *cathode depolarizer*—a substance that is reduced in preference to hydrogen ion. Nitrate functions in this manner, being reduced to ammonium ion

$$NO_3^- + 10H^+ + 8e \rightleftharpoons NH_4^+ + 3H_2O$$

Instrumentation

The apparatus for an analytical electrodeposition consists of a suitable cell and a direct-current power supply.

Cells. Figure 20-6 shows a typical cell for the deposition of a metal on a solid electrode. Tall-form beakers are ordinarily employed,

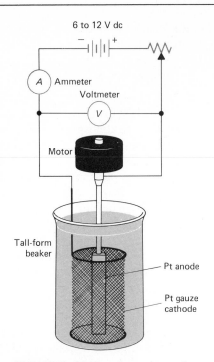

FIGURE 20-6 Apparatus for electrodeposition of metals.

analysis. For example, copper, nickel, cobalt, silver, and cadmium are readily separated from ions such as aluminum, titanium, the alkali metals, and phosphates. The precipitated elements dissolve in the mercury; little hydrogen evolution occurs even at high applied potentials because of large overvoltage effects. Ordinarily, no attempt is made to determine the elements deposited in the mercury, the goal being simply their removal from solution. A cell such as that shown in Figure 20-7 is used.

Power Supplies. The apparatus shown in Figure 20-6 is typical of that employed for most electrolytic analyses. The dc power supply may consist of a storage battery, a generator, or an alternating-current rectifier (p. 53). A rheostat is used to control the applied potential; an ammeter and a voltmeter are provided to indicate the approximate current and applied voltage.

Figure 20-8 shows an apparatus suitable for carrying out a deposition at a controlled

and mechanical stirring is provided to minimize concentration polarization; frequently, the anode is rotated with an electric motor.

Electrodes. Electrodes are usually constructed of platinum, although copper, brass, and other metals find occasional use. Platinum electrodes have the advantage of being relatively nonreactive; moreover, they can be ignited to remove any grease, organic matter, or gases that could have a deleterious effect on the physical properties of the deposit. Certain metals (notably bismuth, zinc, and gallium) cannot be deposited directly onto platinum without causing permanent damage; a protective coating of copper should always be deposited on a platinum electrode before the electrolysis of these metals is undertaken.

The Mercury Cathode. A mercury cathode is particularly useful in removing easily reduced elements as a preliminary step in an

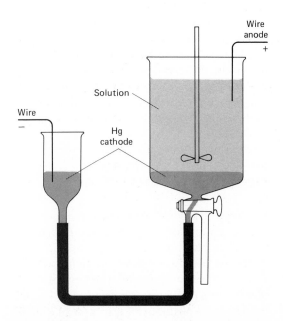

FIGURE 20-7 Mercury cathode for the electrolytic removal of metal ions from solution.

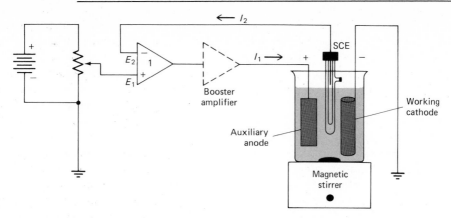

FIGURE 20-8   A potentiostat for constant cathode potential electrolysis.

cathode (or anode) potential (see also p. 63). A voltage E_1 is impressed on the noninverting terminal of the operational amplifier, the output of which is fed into a booster amplifier, which provides the large current needed for the electrolysis. The saturated calomel electrode is in the feedback circuit to the operational amplifier and controls the current output I_1 to the anode. The cathode is then connected to the common ground.

In order to understand the cathode potential control exercised by this system, consider first the operation of the circuit in the absence of the booster amplifier. Note that the input E_1 is to the noninverting terminal of amplifier 1. On page 62, it was shown that in a noninverting circuit of this type the potentials at the two inputs are always equal; that is, $E_1 = E_2$. Thus, the potential applied to the cell consisting of the calomel electrode and the working cathode is E_1. That is,

$$E_2 = E_1 = E_c - E_{SCE} - I_2 R_2 \cong E_c - E_{SCE}$$

where E_c and E_{SCE} are the reversible potentials for the two electrodes. Because the internal resistance of the amplifier R_2 is high, current I_2 (and thus $I_2 R_2$) will be negligible, as indicated.

The potential for the cell consisting of the auxiliary electrode and the cathode is given by

$$E_1 = E_c - E_{aux} - I_1 R_1$$

where R_1 is the internal resistance of the cell. If the potential of the cathode E_c begins to rise, as a consequence of concentration polarization or an increase in the cell resistance, the amplifier responds by decreasing its output current I_1 until E_1 again equals E_2.

The booster amplifier shown in Figure 20-8 is a noninverting type that provides a much larger current than can be obtained from the single operational amplifier. Its presence has no effect on the cathode control circuit.

An electronic instrument, such as that shown in Figure 20-8 is called a *potentiostat*. Several instrument suppliers offer potentiostats for sale.

Applications of Constant-Current Electrolysis

Without control of the cathode potential, electrolytic methods suffer from a lack of specificity. Despite this limitation, several applications of practical importance make use of this relatively unrefined technique. In general, the analyte must be the only component in the solution that is more readily reduced than the hydrogen ion. Any interfering species should be eliminated by a chemical precipitation or prevented from depositing by complexation with a ligand that does not appreciably influence the electrochemical behavior of the analyte.

Constant-current deposition with a mercury cathode is also useful in removing easily reduced ions from solution prior to completion of an analysis by some other method. The deposition of interfering heavy metals prior to the quantitative determination of the alkali metals is an example of this application.

Table 20-1 lists the common elements that can be determined by electrogravimetric procedures for which control over the cathode potential is not required.

Application of Controlled Electrode Potential Electrolysis[3]

The controlled cathode potential method is a potent tool for the direct analysis of solutions containing a mixture of metallic elements. Such control permits quantitative separation of elements with standard potentials that differ by only a few tenths of a volt. For example, Lingane and Jones[4] developed a method for the successive determination of copper, bismuth, lead, and tin. The first three can be deposited from a nearly neutral tar-

trate solution. Copper is first reduced quantitatively by maintaining the cathode potential at -0.2 V with respect to a saturated calomel electrode. After weighing, the copper-plated cathode is returned to the solution, and bismuth is removed at a potential of -0.4 V. Lead is then deposited quantitatively by increasing the cathode potential to -0.6 V. Throughout these depositions, the tin is retained in solution as a very stable tartrate complex. Acidification of the solution after deposition of the lead is sufficient to decompose the complex by converting tartrate ion to the undissociated acid; tin can then be readily deposited at a potential of -0.65 V. This method can be extended to include zinc and cadmium as well. Here, the solution is made ammoniacal after removal of the copper, bismuth, and lead. Cadmium and zinc are then successively deposited and weighed. Finally, the tin is determined after acidification, as before.

A procedure such as this is particularly attractive for use with a potentiostat because of the small operator time required for the complete analysis.

Table 20-2 lists some other separations that have been performed by the controlled cathode method.

COULOMETRIC METHODS OF ANALYSIS

Coulometry encompasses a group of analytical methods which involve measuring the quantity of electricity (in coulombs) needed to convert the analyte quantitatively to a different oxidation state. In common with the gravimetric method, coulometry offers the advantage that the proportionality constant between coulombs and the weight of analyte can be derived from known physical constants; thus, a calibration or standardization step is not ordinarily required. Coulometric methods are often as accurate as gravimetric or volumetric procedures; they are usually faster and

---

[3] This method was first suggested by H. J. S. Sand, *Trans. Chem. Soc.*, **91**, 373 (1907). For many of its applications, see: H. J. S. Sand, *Electrochemistry and Electrochemical Analysis.* Glasgow: Blackie & Son, Ltd., 1940, vol. 2. An excellent discussion of applications of automatic control to the method can be found in J. J. Lingane, *Electroanalytical Chemistry*, 2d ed. New York: Interscience, 1958, Chapters 13–16. See also: G. A. Rechnitz, *Controlled-Potential Analysis.* New York: Macmillan, 1963.

[4] J. J. Lingane and S. L. Jones, *Anal. Chem.*, **23**, 1798 (1951).

more convenient than the former. Finally, coulometric procedures are readily adapted to automation.[5]

Units for Quantity of Electricity

The quantity of electricity is measured in units of the *coulomb* (C) and the *faraday* (F). The

coulomb is that amount of electricity that flows during the passage of a constant current of one ampere for one second. Thus, for a constant current of I amperes flowing for t seconds, the number of coulombs Q is given by the expression

$$Q = It \qquad (20\text{-}5)$$

For a variable current, the number of coulombs is given by the integral

$$Q = \int_0^t I\,dt \qquad (20\text{-}6)$$

The faraday is the quantity of electricity that will produce one equivalent of chemical change at an electrode. Since the equivalent in an

---

[5] For summaries of coulometric methods, see: H. L. Kies, *J. Electroanal. Chem.*, **4**, 257 (1962); J. J. Lingane, *Electroanalytical Chemistry*, 2d ed. New York: Interscience, 1958, Chapters 19–21; G. W. C. Milner and G. Phillips, *Coulometry in Analytical Chemistry*. New York: Pergamon Press, 1967; and J. T. Stock, *Anal. Chem.*, **50**, 1R (1978); **48**, 1R (1976).

TABLE 20-1 COMMON ELEMENTS THAT CAN BE DETERMINED BY ELECTROGRAVIMETRIC METHODS

Ion	Weighed As	Conditions
Cd^{2+}	Cd	Alkaline cyanide solution
Co^{2+}	Co	Ammoniacal sulfate solution
Cu^{2+}	Cu	HNO_3/H_2SO_4 solution
Fe^{3+}	Fe	$(NH_4)_2C_2O_4$ solution
Pb^{2+}	PbO_2	HNO_3 solution
Ni^{2+}	Ni	Ammoniacal sulfate solution
Ag^+	Ag	Cyanide solution
Sn^{2+}	Sn	$(NH_4)_2C_2O_4/H_2C_2O_4$ solution
Zn^{2+}	Zn	Ammoniacal or strong NaOH solution

TABLE 20-2 SOME APPLICATIONS OF CONTROLLED CATHODE POTENTIAL ELECTROLYSIS

Element Determined	Other Elements That May Be Present
Ag	Cu and heavy metals
Cu	Bi, Sb, Pb, Sn, Ni, Cd, Zn
Bi	Cu, Pb, Zn, Sb, Cd, Sn
Sb	Pb, Sn
Sn	Cd, Zn, Mn, Fe
Pb	Cd, Sn, Ni, Zn, Mn, Al, Fe
Cd	Zn
Ni	Zn, Al, Fe

oxidation-reduction reaction corresponds to the change brought about by one mole of electrons, the faraday is equal to 6.02×10^{23} electrons. One faraday is also equal to 96,491 C.

EXAMPLE

A constant current of 0.800 A was passed through a solution for 15.2 min. Calculate the grams of copper deposited at the cathode and the grams of O_2 evolved at the anode, assuming that these are the only products formed.

The equivalent weights are determined from consideration of the two half-reactions

$$Cu^{2+} + 2e \rightarrow Cu(s)$$

$$2H_2O \rightarrow 4e + O_2(g) + 4H^+$$

From Equation 20-5, we find

Quantity of electricity = 0.800 A

$$\times 15.2 \text{ min} \times 60 \text{ s/min}$$

$$= 729.6 \text{ A s}$$

$$= 729.6 \text{ C}$$

or

$$\frac{729.6 \text{ C}}{96,491 \text{ C/F}} = 7.56 \times 10^{-3} \text{ F}$$

From the definition of the faraday, 7.56×10^{-3} equivalent of copper is deposited on the cathode; a similar quantity of oxygen is evolved at the anode. Therefore,

$$\text{wt Cu} = 7.56 \times 10^{-3} \text{ eq Cu} \times \frac{63.5 \text{ g Cu/mol}}{2 \text{ eq Cu/mol}}$$

$$= 0.240 \text{ g}$$

and

$$\text{wt O}_2 = 7.56 \times 10^{-3} \text{ eq O}_2 \times \frac{32.0 \text{ g O}_2\text{/mol}}{4 \text{ eq O}_2\text{/mol}}$$

$$= 0.0605 \text{ g}$$

Types of Coulometric Methods

Two general techniques are used for coulometric analyses. The first involves maintaining the potential of the working electrode at a constant level such that quantitative oxidation or reduction of the analyte occurs without involvement of less reactive species in the sample or solvent. Here, the current is initially high but decreases rapidly and approaches zero as the analyte is removed from the solution (see Figure 20-5). The quantity of electricity required is measured with a chemical coulometer or by integration of the current-time curve. A second coulometric technique makes use of a constant current that is continued until an indicator signals completion of the reaction. The quantity of electricity required to attain the end point is then calculated from the magnitude of the current and the time of its passage. The latter method has enjoyed wider application than the former; it is frequently called a *coulometric titration*.

A fundamental requirement of all coulometric methods is that the species determined interact with 100% current efficiency. That is, each faraday of electricity must bring about a chemical change corresponding to one equivalent of the analyte. This requirement does not, however, imply that the analyte must necessarily participate directly in the electron-transfer process at the electrode. Indeed, more often than not, the substance being determined is involved wholly or in part in a reaction that is secondary to the electrode reaction. For example, at the outset of the oxidation of iron(II) at a platinum anode, all of the current results from the reaction

$$Fe^{2+} \rightleftharpoons Fe^{3+} + e$$

As the concentration of iron(II) decreases, however, concentration polarization may cause the anode potential to rise until decomposition of water occurs as a competing process. That is,

$$2H_2O \rightleftharpoons O_2(g) + 4H^+ + 4e$$

The current required to complete the oxidation of iron(II) would then exceed that demanded by theory. To avoid the consequent error, an unmeasured excess of cerium(III) can be introduced at the start of the electrolysis. This ion is oxidized at a lower anode potential than is water:

$$Ce^{3+} \rightleftharpoons Ce^{4+} + e$$

The cerium(IV) produced diffuses rapidly from the electrode surface, where it then oxidizes an equivalent amount of iron(II):

$$Ce^{4+} + Fe^{2+} \rightarrow Ce^{3+} + Fe^{3+}$$

The net effect is an electrochemical oxidation of iron(II) with 100% current efficiency even though only a fraction of the iron(II) ions are directly oxidized at the electrode surface.

The coulometric determination of chloride provides another example of an indirect process. Here, a silver electrode serves as the anode and silver ions are produced by the current. These cations diffuse into the solution and precipitate the chloride. A current efficiency of 100% with respect to the chloride ion is achieved even though this ion is neither oxidized nor reduced in the cell.

Coulometric Methods at Constant Electrode Potential

In variable current coulometry, the potential of the working electrode is maintained at a constant level that will cause the analyte to react quantitatively without involvement of other components in the sample. A constant electrode potential or variable current analysis of this kind possesses all the advantages of an electrogravimetric method and is not subject to the limitation imposed by the need for a weighable product. The technique can therefore be applied to systems that yield deposits with poor physical properties as well as to reactions that yield no solid product at all. For example, arsenic may be determined coulometrically by the electrolytic oxidation of arsenous acid (H_3AsO_3) to arsenic acid (H_3AsO_4) at a platinum anode. Similarly, the analytical conversion of iron(II) to iron(III) can be accomplished with suitable control of the anode potential.

Apparatus. A constant electrode potential coulometric analysis requires a potentiostat such as that shown in Figure 20-8, an instrument for measuring current as a function of time, and an integrating device for deriving the quantity of electricity by means of Equation 20-6.

In early work, *chemical coulometers*, placed in series with the working cell, were employed as current-time integrators. Here, the quantity of electricity was determined by a volumetric or gravimetric measurement of the extent of chemical change occurring in the coulometer during the electrolysis. An example is the hydrogen/oxygen coulometer, which consists of a pair of platinum electrodes immersed in a solution of potassium sulfate; the oxygen and hydrogen evolved at the two electrodes are collected in a gas buret and their combined volume is measured.

Modern integrators are generally electronic and employ a circuit similar to that shown in Figure 3-26 (p. 66). Often such integrators are a part of a recorder, which provides a plot of current as a function of time. Several controlled potential coulometric instruments are available from commercial sources.

Application.[6] Controlled potential coulometric methods have been widely used for the determination of various metal ions. Mercury appears to be favored as the cathode and methods for the deposition of two dozen metals at this electrode have been described. The method has found widespread use in the nuclear energy field for the relatively interference-free analysis of uranium and plutonium.

---

[6] For a summary of the applications, see: J. E. Harrar, *Electroanalytical Chemistry*, ed. A. J. Bard. New York: Marcel Dekker, 1975, vol. 8.

The controlled potential coulometric procedure also offers possibilities for the electrolytic determination (and synthesis) of organic compounds. For example, Meites and Meites[7] have demonstrated that trichloroacetic acid and picric acid are quantitatively reduced at a mercury cathode whose potential is suitably controlled:

$$Cl_3CCOO^- + H^+ + 2e =$$

$$Cl_2HCCOO^- + Cl^-$$

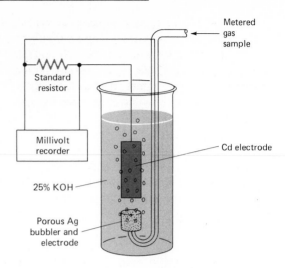

FIGURE 20-9 An instrument for continuously recording the O_2 content of a gas stream.

Coulometric measurements permit the analysis of these compounds with an accuracy of a few tenths of a percent.

Variable current coulometric methods are frequently used to monitor continuously and automatically the concentration of constituents in gas or liquid streams. An important example is the determination of small concentrations of oxygen.[8] A schematic diagram of the apparatus is shown in Figure 20-9. The porous silver cathode serves to break up the incoming gas into small bubbles; the reduction of oxygen takes place quantitatively within the pores. That is,

$$O_2(g) + 2H_2O + 4e \rightleftharpoons 4OH^-$$

The anode is a heavy cadmium sheet; here the half-cell reaction is

$$Cd(s) + 2OH^- \rightleftharpoons Cd(OH)_2(s) + 2e$$

Note that a galvanic cell is formed so that no external power supply is required. The electricity produced is passed through a standard resistor and the potential drop is recorded on a millivolt recorder. The oxygen concentration is proportional to the recorded potential, and the chart paper can be made to display the instantaneous oxygen concentration directly. The instrument is reported to provide oxygen concentration data in the range from 1 ppm to 1%.

Coulometric Titrations

A coulometric titration employs an electrolytically generated titrant for reaction with the analyte. In some analyses, the active electrode process involves only generation of the reagent; an example is the titration of a halide by electrolytically generated silver ions. In other titrations, the analyte may also be directly involved at the generator electrode; an

[7] T. Meites and L. Meites, *Anal. Chem.*, **27**, 1531 (1955); **28**, 103 (1956).

[8] For further details, see: F. A. Keidel, *Ind. Eng. Chem.*, **52**, 491 (1960).

example of the latter is the coulometric oxidation of iron(II)—in part by electrolytically generated cerium(IV) and in part by direct electrode reaction (p. 592). Under any circumstance, the net process must approach 100% current efficiency with respect to a single chemical change in the analyte.

The current in a coulometric titration is carefully maintained at a constant and accurately known level; the product of this current in amperes and the time in seconds required to reach an end point yields the number of coulombs and thus the number of equivalents involved in the electrolysis. The constant current aspect of this operation precludes the quantitative oxidation or reduction of the unknown species entirely at the generator electrode; as the solution is depleted of analyte, concentration polarization is inevitable. The electrode potential must then rise if a constant current is to be maintained (p. 585). Unless this potential rise produces a reagent that can react with the analyte, the current efficiency will be less than 100%. In a coulometric titration, then, at least part (and frequently all) of the analytical reaction occurs away from the surface of the working electrode.

A coulometric titration, like a more conventional volumetric titration, requires some means of detecting the point of chemical equivalence. Most of the end points applicable to volumetric analysis are equally satisfactory here; color changes of indicators, potentiometric, amperometric (p. 635), and conductance measurements have all been successfully applied.

The analogy between a volumetric and a coulometric titration extends well beyond the common requirement of an observable end point. In both, the amount of analyte is determined through evaluation of its combining capacity—in the one case, for a standard solution and, in the other, for a quantity of electricity. Similar demands are made of the reactions; that is, they must be rapid, essentially complete, and free of side reactions.

Electrical Apparatus. Coulometric titrators are available from several laboratory supply houses. In addition, they can be readily assembled from components available in most laboratories.

Figure 20-10 depicts the principal components of a typical coulometric titrator. Included are a source of constant current and a switch that simultaneously initiates the current and starts an electric stopclock. Also required is the means for accurately measuring the current; in Figure 20-10, the potential drop across the standard resistor, R_{std} is used for this measurement.

Many electronic or electromechanical constant current sources are described in the literature. The ready availability of inexpensive operational amplifiers makes their construction a relatively simple matter (for example, see Figure 3-24, p. 64).

An ordinary electric stopclock is inadequate for determining the total electrolysis time because the motor tends to coast when stopped and lag when started; the resulting accumulated error can be appreciable. Stopclocks with solenoid operated brakes or electronic timers eliminate this problem.

It is useful to point out the close analogy between the various components of the apparatus shown in Figure 20-10 and the apparatus and solutions employed in a conventional volumetric analysis. The constant current source of known magnitude serves the same function as the standard solution in a volumetric method. The electric clock and switch correspond closely to the buret, the switch performing the same function as a stopcock. During the early phases of a coulometric titration, the switch is kept closed for extended periods; as the end point is approached, however, small additions of "reagent" are achieved by closing the switch for shorter and shorter intervals. The similarity to the operation of a buret is obvious.

Cells for Coulometric Titrations. A typical coulometric titration cell is shown in Figure 20-11. It consists of a generator electrode at

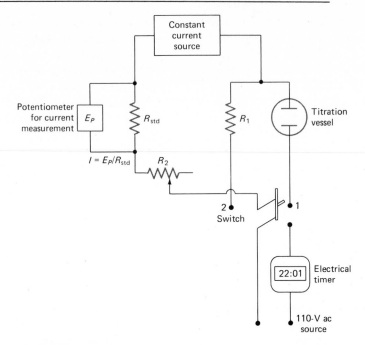

FIGURE 20-10 Schematic diagram of a coulometric titration apparatus.

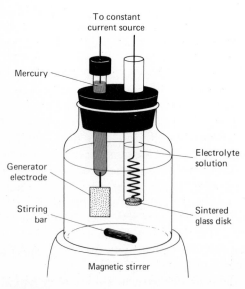

FIGURE 20-11 A typical coulometric titration cell.

which the reagent is formed and an auxiliary electrode to complete the circuit. The generator electrode, which should have a relatively large surface area, is often a rectangular strip or a wire coil of platinum; a gauze electrode such as that shown on page 588 can also be employed.

The products formed at the second electrode frequently represent potential sources of interference. For example, the anodic generation of oxidizing agents is often accompanied by the evolution of hydrogen from the cathode; unless this gas is allowed to escape from the solution, reaction with the oxidizing agent becomes a likelihood. To eliminate this type of difficulty, the second electrode is isolated by a sintered disk or some other porous medium.

As an alternative to isolation of the auxiliary electrode, a device such as that shown in Figure 20-12 may be employed to generate

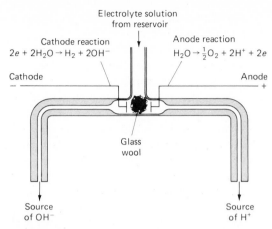

Electrolyte solution
from reservoir

Cathode reaction
$2e + 2H_2O \rightarrow H_2 + 2OH^-$

Anode reaction
$H_2O \rightarrow \frac{1}{2}O_2 + 2H^+ + 2e$

Cathode
−

Anode
+

Glass
wool

Source
of OH⁻

Source
of H⁺

FIGURE 20-12 A cell for external generation of acid and base.

the reagent externally. The apparatus is so arranged that flow of the electrolyte continues briefly after the current is discontinued, thus flushing the residual reagent into the titration vessel. Note that the apparatus shown in Figure 20-12 provides either hydrogen or hydroxide ions depending upon which arm is used. The apparatus has also been used for generation of other reagents such as iodine produced by oxidation of iodide at the anode.

Applications of Coulometric Titrations[9]

Coulometric titrations have been developed for all types of volumetric reactions. Selected applications are described in the following paragraphs.

Neutralization Titrations. Both weak and strong acids can be titrated with a high degree of accuracy using hydroxide ions generated by the reaction

$$2H_2O + 2e \rightarrow 2OH^- + H_2(g)$$

[9] Applications of the coulometric procedure are summarized in J. J. Lingane, *Electroanalytical Chemistry*, 2d ed. New York: Interscience, 1958, pp. 536–613; H. L. Kies, *J. Electroanal. Chem.*, **4**, 257 (1962); and J. T. Stock, *Anal. Chem.*, **50**, 1R (1978); **48**, 1R (1976).

The cells shown in Figures 20-11 and 20-12 can be employed. A convenient alternative involves substitution of a silver wire as the anode and the addition of chloride or bromide ions to the solution of the analyte. The anode reaction then becomes

$$Ag + Br(s)^- \rightleftharpoons AgBr(s) + e$$

Clearly, the silver bromide will not interfere with the neutralization reaction as would the hydrogen ions that are formed at most anodes.

Both potentiometric and indicator end points can be employed for these titrations. The problems associated with the estimation of the equivalence point are identical with those encountered in a conventional volumetric analysis. A real advantage to the coulometric method, however, is that interference by carbonate ion is far less troublesome; it is only necessary to eliminate carbon dioxide from the solution containing the analyte by aeration with a carbon-dioxide-free gas before beginning the analysis.

The coulometric titration of strong and weak bases can be performed with hydrogen ions generated at a platinum anode.

$$H_2O \rightleftharpoons \tfrac{1}{2}O_2(g) + 2H^+ + 2e$$

Here, the cathode must be isolated from the solution or external generation must be employed to prevent interference from the hydroxide ions produced at that electrode.

Precipitation and Complex-Formation Titrations. A variety of coulometric titrations involving anodically generated silver ions have been developed (see Table 20-3). A cell, such as that shown in Figure 20-11, can be employed with a generator electrode constructed from a length of heavy silver wire. End points are detected potentiometrically or with chemical indicators. Similar analyses, based upon the generation of mercury(I) ion at a mercury anode, have been described.

An interesting coulometric titration makes use of a solution of the ammine mercury(II) complex of ethylenediaminetetraacetic acid

(H_4Y).[10] The complexing agent is released to the solution as a result of the following reaction at a mercury cathode:

$$HgNH_3Y^{2-} + NH_4^+ + 2e$$
$$\rightleftharpoons Hg + 2NH_3 + HY^{3-} \qquad (20\text{-}7)$$

Because the mercury chelate is more stable than the corresponding complexes with calcium, zinc, lead, or copper, complexation of these ions will not occur until the electrode process frees the ligand.

Oxidation-Reduction Titrations. Table 20-4 indicates the variety of reagents that can be generated coulometrically and the analyses to which they have been applied. Electro-generated bromine has proved to be particularly useful among the oxidizing agents and forms the basis for a host of methods. Of interest also are some of the unusual reagents not ordinarily encountered in volumetric analysis because of the instability of their solutions; these include dipositive silver ion, tri-

positive manganese, and the chloride complex of unipositive copper.

Comparison of Coulometric and Volumetric Titrations. Some real advantages can be claimed for a coulometric titration in comparison with the classical volumetric process. Principal among these is the elimination of problems associated with the preparation, standardization, and storage of standard solutions. This advantage is particularly important with labile reagents such as chlorine, bromine, or titanium(III) ion; owing to their instability, these species are inconvenient as volumetric reagents. Their utilization in coulometric analysis is straightforward, however, because they undergo reaction with the analyte immediately after being generated.

Where small quantities of reagent are required, a coulometric titration offers a considerable advantage. By proper choice of current, micro quantities of a substance can be introduced with ease and accuracy; the equivalent volumetric process requires small volumes of very dilute solutions, a recourse that is always difficult.

A single constant-current source can be employed to generate precipitation, oxidation-

---

[10] C. N. Reilley and W. W. Porterfield, *Anal. Chem.*, **28**, 443 (1956).

TABLE 20-3   SUMMARY OF APPLICATIONS OF COULOMETRIC TITRATIONS INVOLVING NEUTRALIZATION, PRECIPITATION, AND COMPLEX-FORMATION REACTIONS

Species Determined	Generator-Electrode Reaction	Secondary Analytical Reaction
Acids	$2H_2O + 2e \rightleftharpoons 2OH^- + H_2$	$OH^- + H^+ \rightleftharpoons H_2O$
Bases	$H_2O \rightleftharpoons 2H^+ + \frac{1}{2}O_2 + 2e$	$H^+ + OH^- \rightleftharpoons H_2O$
Cl^-, Br^-, I^-	$Ag \rightleftharpoons Ag^+ + e$	$Ag^+ + Cl^- \rightleftharpoons \underline{AgCl(s)}$, etc.
Mercaptans	$Ag \rightleftharpoons Ag^+ + e$	$Ag^+ + RSH \rightleftharpoons \underline{AgSR(s)} + H^+$
Cl^-, Br^-, I^-	$2Hg \rightleftharpoons Hg_2^{2+} + 2e$	$Hg_2^{2+} + 2Cl^- \rightleftharpoons \underline{Hg_2Cl_2(s)}$, etc.
Zn^{2+}	$Fe(CN)_6^{3-} + e \rightleftharpoons Fe(CN)_6^{4-}$	$3Zn^{2+} + 2K^+ + 2Fe(CN)_6^{4-} \rightleftharpoons$ $\underline{K_2Zn_3[Fe(CN)_6]_2(s)}$
Ca^{2+}, Cu^{2+}, Zn^{2+} and Pb^{2+}	See Equation 20-7	$HY^{3-} + Ca^{2+} \rightleftharpoons CaY^{2-} + H^+$, etc.

reduction, or neutralization reagents. Furthermore, the coulometric method is readily adapted to automatic titrations, because current control is easily accomplished.

Coulometric titrations are subject to five potential sources of error: (1) variation in the current during electrolysis; (2) departure of the process from 100% current efficiency; (3) error in the measurement of current; (4) error in the measurement of time; and (5) titration error due to the difference between the equivalence point and the end point. The last of these difficulties is common to volumetric methods as well; where the indicator error is the limiting factor, the two methods are likely to have comparable reliability.

With simple instrumentation, currents constant to 0.2% relative are easily achieved; with somewhat more sophisticated apparatus, control to 0.01% is obtainable. In general, then, errors due to current fluctuations are seldom of importance.

Although generalizations concerning the magnitude of uncertainty associated with the electrode process are difficult, current efficiencies of 99.5 to better than 99.9% are often reported in the literature. Currents are readily measured to $\pm 0.1\%$ relative or better. Similarly, a good quality timer permits measurement of time to within $\pm 0.1\%$ relative.

To summarize, then, the current-time measurements required for a coulometric titration are inherently as accurate or more accurate than the comparable volume-normality measurements of a classical volumetric analysis, particularly where small quantities of reagent are involved. Often, however, the accuracy of a titration is not limited by these measurements but by the sensitivity of the end point; in this respect, the two procedures are equivalent.

Automatic Coulometric Titrators

A number of instrument manufacturers offer automatic coulometric titrators. Most of these employ the potentiometric end point and are

TABLE 20-4   SUMMARY OF APPLICATIONS OF COULOMETRIC TITRATIONS INVOLVING OXIDATION-REDUCTION REACTIONS

Reagent	Generator Electrode Reaction	Substance Determined
Br_2	$2Br^- \rightleftharpoons Br_2 + 2e$	As(III), Sb(III), U(IV), Tl(I), I^-, SCN^-, NH_3, N_2H_4, NH_2OH, phenol, aniline, mustard gas; 8-hydroxyquinoline
Cl_2	$2Cl^- \rightleftharpoons Cl_2 + 2e$	As(III), I^-
I_2	$2I^- \rightleftharpoons I_2 + 2e$	As(III), Sb(III), $S_2O_3^{2-}$, H_2S
Ce^{4+}	$Ce^{3+} \rightleftharpoons Ce^{4+} + e$	Fe(II), Ti(III), U(IV), As(III), I^-, $Fe(CN)_6^{4-}$
Mn^{3+}	$Mn^{2+} \rightleftharpoons Mn^{3+} + e$	$H_2C_2O_4$, Fe(II), As(III)
Ag^{2+}	$Ag^+ \rightleftharpoons Ag^{2+} + e$	Ce(III), V(IV), $H_2C_2O_4$, As(III)
Fe^{2+}	$Fe^{3+} + e \rightleftharpoons Fe^{2+}$	Cr(VI), Mn(VII), V(V), Ce(IV)
Ti^{3+}	$TiO^{2+} + 2H^+ + e \rightleftharpoons Ti^{3+} + H_2O$	Fe(III), V(V), Ce(IV), U(VI)
$CuCl_3^{2-}$	$Cu^{2+} + 3Cl^- + e \rightleftharpoons CuCl_3^{2-}$	V(V), Cr(VI), IO_3^-
U^{4+}	$UO_2^{2+} + 4H^+ + 2e \rightleftharpoons U^{4+} + 2H_2O$	Cr(VI), Ce(IV)

similar in construction to the automatic titrators discussed in the previous chapter (p. 573). Here, however, the error signal controls a flow of electricity rather than a flow of liquid reagent. Some of the commercial instruments are multipurpose and can be used for the determination of a variety of species. Others are designed for a single analysis. Examples of the latter include chloride titrators in which silver ion is generated coulometrically, sulfur dioxide monitors, where anodically generated bromine oxidizes the analyte to sulfate ions, and water titrators in which Karl Fischer reagent is generated electrolytically.

PROBLEMS

1. Bismuth is to be deposited at a cathode from a solution that is $0.150M$ in BiO^+ and $0.600F$ in $HClO_4$. Oxygen is evolved at a pressure of 0.800 atm at a 20-cm^2 platinum anode. The cell has a resistance of 1.30 Ω.
   (a) Calculate the thermodynamic potential of the cell.
   (b) Calculate the IR drop if a current of 0.200 A is to be used.
   (c) Estimate the O_2 overvoltage.
   (d) Estimate the total applied potential required to begin electrodeposition at the specified conditions.
   (e) What potential will be required when the BiO^+ concentration is $0.0800M$?

2. Nickel is to be deposited at a cathode from a solution that is $0.200M$ in Ni^{2+} and $0.400F$ in $HClO_4$. Oxygen is evolved at a pressure of 0.800 atm at a 15-cm^2 platinum anode. The cell has a resistance of 2.10 Ω.
   (a) Calculate the thermodynamic potential of the cell.
   (b) Calculate the IR drop if a current of 0.150 A is to be used.
   (c) Estimate the O_2 overvoltage.
   (d) Estimate the total applied potential required to begin electrodeposition at the specified conditions.
   (e) What potential will be required when the Ni^{2+} concentration is $0.100M$?

3. It is desired to separate and determine bismuth and lead in a solution that is $0.0800M$ in BiO^+, $0.0500M$ in Pb^{2+}, and $1.00F$ in $HClO_4$.
   (a) Using $1.00 \times 10^{-6}M$ as the criterion for quantitative removal, determine whether or not a separation is feasible by controlled cathode potential electrolysis.
   (b) If a separation is feasible, evaluate the range (vs. SCE) within which the cathode potential should be controlled.
   (c) Calculate the potential needed to deposit the second ion quantitatively after removal of the first.

4. It is desired to separate and determine bismuth, copper, and silver in a solution that is $0.0800M$ in BiO^+, $0.242M$ in Cu^{2+}, $0.106M$ in Ag^+, and $1.00F$ in $HClO_4$.

(a) Using $1.00 \times 10^{-6}M$ as the criterion for quantitative removal, determine whether or not separation of the three species is feasible by controlled cathode potential electrolysis.

(b) If separations are feasible, evaluate the range (vs. SCE) within which the cathode potential should be controlled for the deposition of each.

(c) Calculate the potential needed to deposit the third ion quantitatively after removal of the first two.

5. Halide ions can be deposited at a silver anode, the reaction being

$$Ag(s) + X^- \rightarrow AgX(s) + e$$

(a) Determine whether or not it is theoretically feasible to separate I^- and Br^- ions from a solution that is $0.0400M$ in each ion by controlling the potential of the silver anode. Take $1.00 \times 10^{-6}M$ as the criterion for quantitative removal of one ion.

(b) Is a separation of Cl^- and I^- theoretically feasible?

(c) If a separation is feasible in either (a) or (b), what range of anode potentials (vs. SCE) should be employed?

6. What cathode potential (vs. SCE) would be required to lower the total nickel concentration of the following solutions to $1 \times 10^{-5}F$:

(a) a perchloric acid solution of Ni^{2+}?

(b) a solution having an equilibrium CN^- concentration of $0.0100M$?

$$Ni(CN)_4^{2-} + 2e \rightleftharpoons Ni(s) + 4CN^- \qquad E^0 = -0.82$$

7. What cathode potential (vs. SCE) would be required to lower the Hg^{2+} concentration of the following solutions to $1.00 \times 10^{-6}F$:

(a) an aqueous solution of Hg^{2+}?

(b) a solution with an equilibrium SCN^- concentration of $0.100M$?

$$Hg^{2+} + 2SCN^- \rightleftharpoons Hg(SCN)_2(aq) \qquad K_f = 1.8 \times 10^7$$

(c) a solution with an equilibrium Br^- concentration of $0.250M$?

$$HgBr_4^- + 2e \rightleftharpoons Hg(l) + 4Br^- \qquad E^0 = 0.223 \text{ V}$$

8. The cadmium and zinc in a 1.06-g sample was dissolved and subsequently deposited from an ammoniacal solution with a mercury cathode. When the cathode potential was maintained at -0.95 V (vs. SCE), only the cadmium deposited. When the current ceased at this potential, a hydrogen/oxygen coulometer in series with the cell was found to have evolved 44.6 ml of hydrogen plus oxygen (corrected for water vapor) at a temperature of $21.0°C$ and a barometric pressure of 773 torr. The potential was raised to about -1.3 V, whereupon zinc ion was reduced. Upon completion of this electrolysis, an additional 31.3 ml of gas had been produced under the same conditions. Calculate the percent Cd and Zn in the ore.

9. A 1.74-g sample of a solid containing $BaBr_2$, KI, and an inert species was dissolved, made ammoniacal, and placed in a cell equipped with a

silver anode. By maintaining the potential at -0.06 V (vs. SCE), I^- was quantitatively precipitated as AgI without interference from Br^-. The volume of H_2 and O_2 formed in a gas coulometer in series with the cell was 39.7 ml (corrected for water vapor) at 21.7°C and at a pressure of 748 torr.

After precipitation of iodide was complete, the solution was acidified, and the Br^- was removed from solution as AgBr at a potential of 0.016 V. The volume of gas formed under the same conditions was 23.4 ml. Calculate the percent $BaBr_2$ and KI in the sample.

10. The nitrobenzene in a 210-mg sample of an organic mixture was reduced to phenylhydroxylamine at a constant potential of -0.96 V (vs. SCE) applied to a Hg cathode:

$$C_6H_5NO_2 + 4H^+ + 4e \rightarrow C_6H_5NHOH + H_2O$$

The sample was dissolved in 100 ml of methanol; after electrolysis for 30 min, the reaction was judged complete. An electronic coulometer in series with the cell indicated that the reduction required 26.74 C. Calculate the percent nitrobenzene in the sample.

11. At a potential of -1.0 V (vs. SCE), carbon tetrachloride in methanol is reduced to chloroform at a Hg cathode:

$$2CCl_4 + 2H^+ + 2e + 2Hg(l) \rightarrow 2CHCl_3 + Hg_2Cl_2(s)$$

At -1.80 V, the chloroform further reacts to give methane:

$$2CHCl_3 + 6H^+ + 6e + 6Hg(l) \rightarrow 2CH_4 + 3Hg_2Cl_2(s)$$

A 0.750-g sample containing CCl_4, $CHCl_3$, and inert organic species was dissolved in methanol and electrolyzed at -1.0 V until the current approached zero. A coulometer indicated that 11.63 C had been used. The reduction was then continued at -1.80 V; an additional 44.24 C were required to complete the reaction. Calculate the percent CCl_4 and $CHCl_3$ in the mixture.

12. A 0.1309-g sample containing only $CHCl_3$ and CH_2Cl_2 was dissolved in methanol and electrolyzed in a cell equipped with a Hg cathode; the potential of the cathode was held constant at -1.80 V (vs. SCE). Both compounds were reduced to CH_4 (see Problem 11 for the reaction type). Calculate the percent $CHCl_3$ and CH_2Cl_2 if 306.7 C were required to complete the reduction.

13. The iron in a 0.854-g sample of ore was converted to the $+2$ state by suitable treatment and then oxidized quantitatively at a Pt anode maintained at -1.0 V (vs. SCE). The quantity of electricity required to complete the oxidation was determined with a chemical coulometer equipped with a platinum anode immersed in an excess of iodide ion. The iodine liberated by the passage of current required 26.3 ml of 0.0197-N sodium thiosulfate to reach a starch end point. What was the percentage of Fe_3O_4 in the sample?

14. An apparatus similar to that shown in Figure 20-9 was employed to determine the oxygen content of a light hydrocarbon gas stream having a density of 0.00140 g/ml. A 20.0-liter sample of the gas consumed 3.13 C of electricity. Calculate the parts per million O_2 in the sample on a weight basis.

15. The odorant concentration of household gas can be monitored by passage of a fraction of the gas stream through a solution containing an excess of bromide ion. Electrogenerated bromine reacts rapidly with the mercaptan odorant:

$$2RSH + Br_2 \rightarrow RSSR + 2H^+ + 2Br^-$$

Continuous analysis is made possible by means of a potentiometric electrode system that signals the need for additional Br_2 to oxidize the mercaptan. Ordinarily, the current required to react with the odorant is automatically plotted as a function of time. Calculate the average odorant concentration (as ppm C_2H_5SH) from the following information:

Average density of the gas	0.00185 g/ml
Rate of gas flow	9.4 liters/min
Average current during a 10.0-min sampling period	1.35 mA

16. The phenol content of water downstream from a coking furnace was determined by coulometric analysis. A 100-ml sample was rendered slightly acidic and an excess of KBr was introduced. To produce Br_2 for the reaction

$$C_6H_5OH + 3Br_2 \rightarrow Br_3C_6H_2OH(s) + 3HBr$$

a steady current of 0.0313 A for 7 min and 33 s was required. Express the results of this analysis in terms of parts phenol per million parts of water (assume that the density of water = 1.00 g/ml).

17. The calcium content of a water sample was determined by adding an excess of an $HgNH_3Y^{2-}$ solution to a 25.0-ml sample. The anion of EDTA was then generated at a mercury cathode (see Equation 20-7, (p. 598). A constant current of 20.1 mA was needed to reach an end point after 2 min and 56 s. Calculate the milligrams of $CaCO_3$ per liter of sample.

18. The cyanide concentration in a 10.0-ml sample of a plating solution was determined by titration with electrogenerated hydrogen ions to a methyl orange end point. A color change occurred after 3 min and 22 s with a current of 43.4 mA. Calculate the number of grams of NaCN per liter of solution.

19. A 6.39-g sample of an ant-control preparation was decomposed by wet-ashing with H_2SO_4 and HNO_3. The arsenic in the residue was reduced to the trivalent state with hydrazine. After the excess reducing

agent had been removed, the arsenic(III) was oxidized with electrolytically generated I_2 in a faintly alkaline medium:

$$HAsO_3^{2-} + I_2 + 2HCO_3^- \rightarrow HAsO_4^{2-} + 2I^- + 2CO_2 + H_2O$$

The titration was complete after a constant current of 98.3 mA had been passed for 13 min and 12 s. Express the results of this analysis in terms of the percentage As_2O_3 in the original sample.

20. The H_2S content of a water sample was assayed with electrolytically generated iodine. After 3.00 g of potassium iodide had been introduced to a 25.0-ml portion of the water, the titration required a constant current of 66.4 mA for a total of 7.25 min. Reaction:

$$H_2S + I_2 \rightarrow S(s) + 2H^+ + 2I^-$$

Express the concentration of H_2S in terms of mg/liter of sample.

21. A 0.0145-g sample of a purified organic acid was dissolved in an alcohol/water mixture and titrated with coulometrically generated hydroxide ions. With a current of 0.0324 A, 251 s were required to reach a phenolphthalein end point. Calculate the equivalent weight of the acid.

22. The chromium deposited on one surface of a 10.0-cm^2 test plate was dissolved by treatment with acid and oxidized to the $+6$ state with ammonium peroxodisulfate:

$$3S_2O_8^{2-} + 2Cr^{3+} + 7H_2O = Cr_2O_7^{2-} + 14H^+ + 6SO_4^{2-}$$

The solution was boiled to remove the excess peroxodisulfate, cooled, and then subjected to coulometric titration with Cu(I) generated from 50 ml of $0.10F$ Cu^{2+}. Calculate the weight of chromium that was deposited on each square centimeter of the test plate if the titration required a steady current of 32.5 mA for a period of 7 min and 33 s.

21

VOLTAMMETRY AND POLAROGRAPHY

Voltammetry comprises a group of electro-analytical methods in which information about the analyte is derived from current-voltage curves—that is, plots of current as a function of applied potential—obtained under conditions that encourage polarization of the indicator or working electrode. Generally, the working electrodes in voltammetry are characterized by their small surface area (usually a few square millimeters), which enhances polarization. Such electrodes are generally referred to as *microelectrodes*.

Historically, the field of voltammetry developed from the discovery of polarography by the Czechoslovakian chemist Jaroslav Heyrovsky[1] in the early 1920s. Polarography, which is still the most widely used of all voltammetric methods, differs from the others in the respect that a *dropping mercury electrode* (DME) serves as the microelectrode. The unique properties of this electrode are discussed in a later section.

By 1950, voltammetry appeared to be a mature and fully developed technique. The decade from 1955 to 1965, however, was marked by the appearance of several major modifications of the original method, which served to overcome many of its limitations. Following this, the advent of low-cost operational amplifiers in the early 1960s made possible the development of relatively inexpensive, commercially available instruments that incorporated many of these important modifications. The result has been a recent resurgence of interest in the applications of voltammetric methods to the qualitative and quantitative determination of a host of organic and inorganic species.[2]

In the sections that follow, the theory, practice, and limitations of the conventional polarographic method are first considered; subsequently, the modifications to the original method, which have led to the enhancement of its usefulness, are presented.

POLAROGRAPHY[3]

Virtually every element, in one form or another, is amenable to polarographic analysis. In addition, the method can be extended to the determination of several organic functional groups. Because the polarographic behavior of any species is unique for a given set of experimental conditions, the technique offers attractive possibilities for selective analysis.

Most polarographic analyses are performed in aqueous solution, but other solvent systems may be substituted if necessary. For quantitative analyses, the optimum concentration range lies between 10^{-2} and $10^{-4}M$; through suitable modifications, however, concentration determinations in the parts-per-billion range become possible. An analysis can be easily performed on 1 to 2 ml of solution; with a little effort, a volume as small as one drop is sufficient. The polarographic method is thus particularly useful for the determination of quantities in the milligram to microgram range.

Relative errors ranging between 2 and 3% are to be expected in routine polarographic work.

---

[1] J. Heyrovsky, *Chem. Listy*, **16**, 256 (1922). Heyrovsky was awarded the 1959 Nobel Prize in chemistry for his discovery and development of polarography.

[2] For a brief summary of the new voltammetric techniques, see: J. B. Flato, *Anal. Chem.*, **44**, 75A (1972).

[3] The principles and applications of polarography are considered in detail in a number of monographs; see, for example: I. M. Kolthoff and J. J. Lingane, *Polarography*, 2d ed. New York: Interscience, 1952; L. Meites, *Polarographic Techniques*, 2d ed. New York: Interscience, 1965; J. Heyrovsky and J. Kůta, *Principles of Polarography*. New York: Academic Press, 1966; P. Zuman, *Topics in Organic Polarography*. New York: Pergamon Press, 1970; *Electroanalytical Chemistry*, ed. H. W. Nurnberg. New York: Wiley, 1974, Chapters 1–5; and O. H. Müller, in *Physical Methods of Chemistry*, eds. A. Weissberger and B. W. Rossitor. New York: Wiley-Interscience, 1971, Part IIA, vol. 1, Chapter 5.

A Brief Description of Polarographic Measurements

Polarographic data are obtained by measuring current as a function of the potential applied to a special type of electrolytic cell. A plot of the data gives current-voltage curves, called *polarograms*, which provide both qualitative and quantitative information about the composition of the solution in which the electrodes are immersed.

Polarographic Cells. A polarographic cell consists of a small easily polarized microelectrode, a large nonpolarizable reference electrode, and the solution to be analyzed. The microelectrode, at which the analytical reaction occurs, is a mercury surface with an area of a few square millimeters; the metal is forced by gravity through a very fine capillary to provide a continuous flow of identical droplets with a maximum diameter between 0.5 and 1 mm. The lifetime of a drop is typically 2 to 6 s. We shall see that the dropping mercury electrode has properties that make it particularly well suited for voltammetric studies.

The reference electrode in a polarographic cell is massive relative to the microelectrode so that its behavior remains essentially constant with the passage of small currents; that is, it remains unpolarized during the analysis. A saturated calomel electrode and salt bridge, arranged in the manner shown in Figure 21-1, is frequently employed; other common reference electrodes consist of a large pool of mercury or a silver/silver chloride electrode.

Polarograms. A polarogram is a plot of current as a function of the potential applied to a polarographic cell. The microelectrode is ordinarily connected to the negative terminal of the power supply; *by convention* the applied potential is given a negative sign under these circumstances. By convention also, the currents are designated as positive when the flow of electrons is from the power supply into the microelectrode—that is, when that electrode behaves as a cathode.

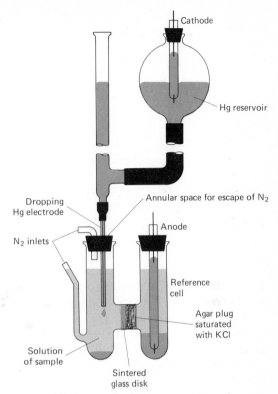

FIGURE 21-1 A dropping mercury electrode and cell. Reprinted with permission from J. J. Lingane and H. A. Laitinen, *Ind. Eng. Chem., Anal. Ed.,* **11**, 504 (1939). (Copyright by the American Chemical Society.)

Figure 21-2 shows two polarograms. The lower one is for a solution that is $0.1F$ in potassium chloride; the upper one is for a solution that is additionally $1 \times 10^{-3} F$ in cadmium chloride. A step-shaped current-voltage curve, called a *polarographic wave*, is produced as a result of the reaction

$$Cd^{2+} + 2e + Hg \rightleftharpoons Cd(Hg)$$

where $Cd(Hg)$ represents elementary cadmium dissolved in mercury. The sharp increase in current at about -2 V in both plots is associated with reduction of potassium ions to give a potassium amalgam.

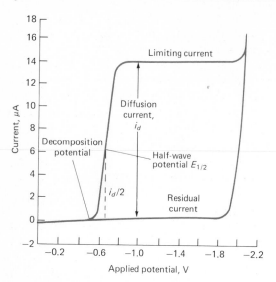

FIGURE 21-2 Polarogram for cadmium ion. The upper curve is for a solution that is 1×10^{-3} F with respect to Cd^{2+} and 0.1 F with respect to KCl. The lower curve is for a solution that is 0.1 F in KCl only.

For reasons to be considered presently, a polarographic wave suitable for analysis is obtained only in the presence of a large excess of a *supporting electrolyte*; potassium chloride serves this function in the present example. Examination of the polarogram for the supporting electrolyte alone reveals that a small current, called the *residual current*, passes through the cell even in the absence of cadmium ions.

A characteristic feature of a polarographic wave is the region in which the current, after increasing sharply, becomes essentially independent of the applied voltage; this constant current is called a *limiting current*. The limiting current is the result of a restriction in the rate at which the participant in the electrode process can be brought to the surface of the microelectrode; with proper control over experimental conditions, this rate is determined exclusively for all points on the wave by the velocity at which the reactant diffuses. A

diffusion-controlled limiting current is given a special name, the *diffusion current*, and is assigned the symbol i_d. Ordinarily, the diffusion current is directly proportional to the concentration of the reactive constituent and is thus of prime importance from the standpoint of analysis. As shown in Figure 21-2, the diffusion current is the difference between the limiting and the residual currents.

One other important quantity, the *half-wave potential*, is the potential at which the current is equal to one-half the diffusion current. The half-wave potential is usually given the symbol $E_{1/2}$; it may permit qualitative identification of the reactant.

Interpretation of Polarographic Waves

To account for the source of the typical polarographic wave, consider the following generalized half-reaction at a dropping electrode:

$$Ox(aq) + ne \rightleftharpoons Red(aq) \qquad (21-1)$$

where Ox and Red are the formulas for the oxidized and reduced form of an electroreactive substance. In contrast to the previous example, note that Red is not an amalgam; the treatment that follows is readily extended to such a reduction as well.

It will be assumed that the half-reaction shown by Equation 21-1 is reversible. In the context of polarography, reversibility implies that the electron transfer process is sufficiently rapid that the concentrations (more exactly, the activities) of reactants and products in the liquid film at the interface between the solution and the mercury electrode are determined by the electrode potential alone; thus, for the reversible reduction of Ox, it may be assumed that the concentrations of reactant and product in the film are given at any instant by

$$E_{DME} = E^0 - \frac{0.0591}{n} \log \frac{[Red]_0}{[Ox]_0} \qquad (21-2)$$

Here, the bracketed terms are concentrations of the reactant and product, E^0 is the standard potential for the half-reaction shown by Equation 21-1, and E_{DME} is the potential of the dropping mercury electrode (vs. SHE). The subscript zeros are employed to denote that the concentration terms apply to the *surface film only*; the concentrations in the bulk of the solution will ordinarily be substantially different from these surface values.

If the anode is a saturated calomel electrode, the applied potential is given by the equation

$$E_{appl} = E_{DME} - E_{SCE} - IR$$

In general, polarographic currents are so small that IR is negligible with respect to the other two potential terms; so also is the junction potential. Under these conditions, substitution of Equation 21-2 yields

$$E_{appl} = E^0 - \frac{0.0591}{n} \log \frac{[\text{Red}]_0}{[\text{Ox}]_0} - E_{SCE}$$
$$(21\text{-}3)$$

Consider now what occurs when E_{appl} is made more negative than the decomposition potential (Figure 21-2). Because the half-reaction is reversible, an essentially instantaneous decrease in $[\text{Ox}]_0$ and a corresponding increase in $[\text{Red}]_0$ takes place until the relationship shown by Equation 21-3 is again satisfied; a momentary surge of current results. This current would rapidly decay to zero were it not for the fact that Ox particles are mobile in the aqueous medium and migrate to the surface of the mercury electrode as a consequence. Because the reduction reaction is instantaneous, *the magnitude of the current depends upon the rate at which the Ox particles move from the bulk of the solution to the surface where reaction occurs*. That is,

$$i = k' \times r_{Ox}$$

where i is the current at an applied potential E_{appl}, r_{Ox} is the rate of migration of Ox particles, and k' is a proportionality constant.

We have noted (p. 529) that the ions or molecules in a cell migrate under the influence of diffusion, convection (thermal or mechanical), and electrostatic attraction. In polarography, every effort is made to eliminate the latter two effects. Thus, vibration or stirring of the solution is avoided, and an excess of a nonreactive electrolyte is employed; if the concentration of this supporting electrolyte is 50 times (or more) greater than that of the reactant, the attractive (or repulsive) force between the electrode and the charged reactant becomes negligible.

With the effects of mechanical mixing and electrostatic attraction eliminated, the only force responsible for the transport of Ox to the electrode surface is diffusion. Since the rate of diffusion is directly proportional to the concentration (strictly activity) difference between two parts of a solution, we may write

$$r_{Ox} = k''([\text{Ox}] - [\text{Ox}]_0)$$

where $[\text{Ox}]$ is the concentration *in the bulk of the solution* from which ions are diffusing and $[\text{Ox}]_0$ is the concentration in the aqueous film surrounding the electrode. If diffusion is the only process bringing Ox particles to the surface, it follows that

$$i = k'r_{Ox} = k'k''([\text{Ox}] - [\text{Ox}]_0)$$
$$= k([\text{Ox}] - [\text{Ox}]_0) \qquad (21\text{-}4)$$

Note that $[\text{Ox}]_0$ becomes smaller as E_{appl} is made more negative. Thus, the rate of diffusion as well as the current increases with increases in applied potential. Ultimately, the applied potential becomes so negative that the concentration of Ox particles in the surface film approaches zero with respect to the concentration in the bulk of the solution; under this circumstance, the rate of diffusion and thus the current becomes constant. That is, when

$$[\text{Ox}]_0 \ll [\text{Ox}]$$

the expression for current becomes

$$i_d = k[\text{Ox}] \qquad (21\text{-}5)$$

where i_d is the potential-independent diffusion current. Note that *the magnitude of the diffusion current is directly proportional to the concentration of the reactant in the bulk of the solution.* Quantitative polarography is based upon this fact.

When the current in a cell is limited by the rate at which a reactant can be brought to the surface of an electrode, a state of *complete concentration polarization* is said to exist. With a microelectrode, the current required to achieve this condition is small—typically 3 to 10 μA (microampere) for a $10^{-3}M$ solution. Such current levels do not significantly alter the reactant concentration, as shown by the following example.

EXAMPLE

The diffusion current for a $1.00 \times 10^{-3}M$ solution of Zn^{2+} was found to be 8.4 μA. Calculate the percent decrease in the Zn^{2+} concentration after passage of this current for 8.0 min through 10.0 ml of the solution.

$$q = 8.0 \text{ min} \times 60 \ \frac{s}{\text{min}} \times 8.4 \times 10^{-6} \text{ A}$$

$$= 4.03 \times 10^{-3} \text{ C}$$

No. of meq Zn^{2+} consumed $= \dfrac{4.03 \times 10^{-3} \text{ C}}{96,491 \text{ C/F}}$

$$\times \frac{10^3 \text{ meq}}{F}$$

$$= 4.18 \times 10^{-5}$$

No. of millimoles Zn^{2+} consumed

$$= 2.09 \times 10^{-5}$$

% decrease in Zn^{2+} concentration

$$= \frac{2.09 \times 10^{-5}}{1.00 \times 10^{-3} \times 10} \times 100 = 0.21$$

Equation for the Polarographic Wave

To obtain an equation relating current and potential, let us subtract Equation 21-4 from

21-5, which yields, upon rearrangement,

$$[\text{Ox}]_0 = \frac{i_d - i}{k}$$

The concentration of Red in the surface film will be proportional to the current. That is,

$$i = k_R[\text{Red}]_0$$

or

$$[\text{Red}]_0 = \frac{i}{k_R}$$

Substitution of these quantities into Equation 21-3 gives, upon rearrangement,

$$E_{\text{appl}} = E^0 - E_{\text{SCE}} - \frac{0.0591}{n} \log \frac{k}{k_R}$$

$$- \frac{0.0591}{n} \log \frac{i}{i_d - i} \qquad (21\text{-}6)$$

By definition, $E_{\text{appl}} = E_{1/2}$ when

$$i = \frac{i_d}{2}$$

Substituting this relationship in Equation 21-6 reveals that

$$E_{1/2} = E^0 - E_{\text{SCE}} - \frac{0.0591}{n} \log \frac{k}{k_R} \qquad (21\text{-}7)$$

Thus, the half-wave potential is a constant that is related to the standard potential for the half-reaction, k, and k_R. The latter two terms are constants that depend upon the diffusion coefficients for the two species and certain characteristics of the dropping electrode, which will be discussed later. Substitution of Equation 21-7 into 21-6 yields

$$E_{\text{appl}} = E_{1/2} - \frac{0.0591}{n} \log \frac{i}{i_d - i} \qquad (21\text{-}8)$$

As will be shown subsequently, i_d can be calculated from the characteristics of the particular electrode and the diffusion coefficient for Ox. Thus the current i can be calculated at any desired value of E_{appl}.

Examination of Equation 21-7 reveals that the half-wave potential is a reference point on a polarographic wave; it is independent of the reactant concentration but directly related to the standard potential for the half-reaction. In practice, the half-wave potential can be a useful quantity for identification of the species responsible for a given polarographic wave.

It is important to note that the half-wave potential may vary considerably with concentration for electrode reactions that are slow and thus irreversible; in fact, Equation 21-8 no longer describes such waves.

Effect of Complex Formation on Polarographic Waves. We have already seen (Chapter 18) that the potential for the oxidation or reduction of a metallic ion is greatly affected by the presence of species that form complexes with that ion. It is not surprising, therefore, that similar effects are observed with polarographic half-wave potentials. The date in Table 21-1 indicate that the half-wave potential for the reduction of a metal complex is generally more negative than that for reduction of the corresponding simple metal ion.

Lingane[4] has shown that the shift in half-wave potential as a function of the concentration of complexing agent can be employed to

----

[4] J. J. Lingane, *Chem. Rev.*, **29**, 1 (1941).

determine the formula and the formation constant for a complex, provided that the cation involved reacts reversibly at the dropping electrode. Thus, for the reactions

$$M^{n+} + Hg + ne \rightleftharpoons M(Hg)$$

and

$$M^{n+} + xA^- \rightleftharpoons MA^{(n-x)+}$$

he derived the relationship

$$(E_{1/2})c - E_{1/2} = -\frac{0.0591}{n} \log K_f$$

$$- \frac{0.0591x}{n} \log F_A \qquad (21\text{-}9)$$

where $(E_{1/2})c$ is the half-wave potential when the formal concentration of A is F_A, while $E_{1/2}$ is the half-wave potential in the absence of complexing agent; K_f is the formation constant of the complex.

Equation 21-9 makes it possible to evaluate the formula for the complex. Thus, a plot of the half-wave potential against $\log F_A$ for several formal ligand concentrations gives a straight line, the slope of which is $0.0591x/n$. If n is known, the combining ratio of ligand to metal ion is thus obtained. Equation 21-9 can then be employed to calculate K_f.

TABLE 21-1   EFFECT OF COMPLEXING AGENTS ON POLAROGRAPHIC HALF-WAVE POTENTIALS AT THE DROPPING MERCURY ELECTRODE

Ion	Noncomplexing Media	1F KCN	1F KCl	1F NH₃, 1F NH₄Cl
Cd^{2+}	-0.59	-1.18	-0.64	-0.81
Zn^{2+}	-1.00	NR^a	-1.00	-1.35
Pb^{2+}	-0.40	-0.72	-0.44	-0.67
Ni^{2+}	—	-1.36	-1.20	-1.10
Co^{2+}	—	-1.45	-1.20	-1.29
Cu^{2+}	$+0.02$	NR^a	$+0.04$	-0.24
			-0.22	-0.51

[a] No reduction occurs before involvement of the supporting electrolyte.

Because of the effect of complexing reagents on half-wave potentials, the electrolyte content of the solution should be carefully controlled whenever polarographic data are used for the qualitative identification of the constituents of a solution.

Polarograms for Irreversible Reactions. Many polarographic electrode processes, particularly those associated with organic systems, are irreversible; drawn-out and less well-defined waves result. The quantitative description of such waves requires an additional term in Equation 21-8 (involving the activation energy of the reaction) to account for the kinetics of the electrode process. Although half-wave potentials for irreversible reactions ordinarily show a dependence upon concentration, diffusion currents remain linearly related to this variable, and such processes are readily adapted to quantitative analysis.

The Dropping Mercury Electrode

The dropping mercury electrode possesses unique properties that make it particularly useful for voltammetric studies.

Current Variations During the Lifetime of a Drop. The current in a cell containing a dropping electrode undergoes periodic fluctuations corresponding in frequency to the drop rate. As a drop breaks, the current falls to zero (see Figure 21-3a); it then increases rapidly as the electrode area grows because of the greater surface to which diffusion can occur. In order to determine the average current, it is necessary to reduce the large fluctuations in the current by a damping device or by sampling the current near the end of each drop, where the change in current with time is relatively small. Damping is ordinarily achieved by employing a well-damped galvanometer or a low-pass filter (p. 23). As shown in Figure 21-3b, damping limits the oscillations to a reasonable magnitude; the average current (or, alternatively, the maximum current) is then readily

determined, provided the drop rate t is reproducible. Note the effect of irregular drops in the center of the limiting current region, probably caused by vibration of the apparatus.

Advantages and Limitations of the Dropping Mercury Electrode. The dropping mercury electrode offers several advantages over other types of microelectrodes. The first is the large overvoltage for the formation of hydrogen from hydrogen ions; as a consequence, the reduction of many substances from acidic solutions can be studied without interference. Second, because a new metal surface is continuously generated, the behavior of the electrode is independent of its past history; thus, reproducible current-voltage curves are obtained regardless of how the electrode has been used previously. A third unique feature of the dropping electrode is that reproducible average currents are immediately achieved at any given applied potential.

The most serious limitation to the dropping electrode is the ease with which mercury is oxidized; this property severely restricts the use of the electrode as an anode. At applied potentials much above $+0.4$ V formation of mercury(I) occurs; the resulting current masks the polarographic waves of other oxidizable species in the solution. Thus, the dropping mercury electrode can be employed only for the analysis of reducible or very easily oxidizable substances. Other disadvantages of the dropping mercury electrode are that it is cumbersome to use and tends to malfunction as a result of clogging.

Polarographic Diffusion Currents

The Ilkovic Equation. In 1934, D. Ilkovic[5] derived a fundamental equation relating the

---

[5] D. Ilkovic, *Collect. Czechoslov. Chem. Commun.*, **6**, 498 (1934).

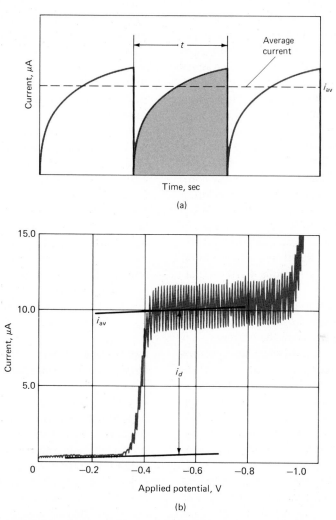

FIGURE 21-3 Effect of drop growth on polarographic currents. (a) Current-time relationship during the lifetime t of drops. The shaded area represents the microcoulombs of electricity associated with each drop. Note that the current increases by only about 20% during the second half of the drop lifetime. (b) A recorded polarogram in which the current fluctuations have been damped, thus permitting the measurement of the average current i_{av}.

various parameters that determine the magnitude of the diffusion currents obtained with a dropping mercury electrode. He showed that at 25°C,

$$i_d = 607 \, nD^{1/2}m^{2/3}t^{1/6}C \qquad (21\text{-}10)$$

Here, i_d is the time-average diffusion current in microamperes (see Figure 21-3) during the lifetime of a drop; n is the number of faradays per mole of reactant, D is the diffusion coefficient for the reactive species (expressed in units of square centimeters per second), m is the rate of mercury flow in milligrams per second, t is the drop time in seconds, and C is the concentration of the reactant in *millimoles* per liter. The quantity 607 represents the combination of several constants (for the maximum current, this constant takes a value of 709).

The Ilkovic equation is based upon certain assumptions that cause discrepancies of a few percent between calculated and experimental diffusion currents. Corrections to the equation have been derived that yield better correspondence[6]; for most purposes, however, the simple equation gives a satisfactory accounting of the factors that influence the current.

Capillary Characteristics. The product $m^{2/3}t^{1/6}$ in the Ilkovic equation, called the *capillary constant*, describes the influence of dropping electrode characteristics upon the diffusion current; since both m and t are readily evaluated experimentally, comparison of diffusion currents from different capillaries is thus possible.

Two factors, other than the geometry of the capillary itself, play a part in determining the magnitude of the capillary constant. The head that forces the mercury through the capillary influences both m and t such that the diffusion current is directly proportional to the square root of the column height. The drop time t of a given electrode is also affected by the applied potential since the interfacial tension between the mercury and the solution varies with the charge on the drop. Generally, t passes through a maximum at about -0.4 V (vs. SCE) and then falls off fairly rapidly; at -2.0 V, t may be only one-half of its maximum value. Fortunately, the diffusion current varies only as the one-sixth power of the drop time so that, over small potential ranges, the decrease in current due to this variation is negligibly small.

Temperature. Temperature affects several of the variables that govern the diffusion current for a given species, and its overall influence is thus complex. The most temperature-sensitive factor in the Ilkovic equation is the diffusion coefficient, which ordinarily can be expected to change by about 2.5% per degree. As a consequence, temperature control to a few tenths of a degree is needed for accurate polarographic analysis.

Polarograms for Mixtures of Reactants

Ordinarily, the reactants of a mixture will behave independently of one another at a microelectrode; a polarogram for a mixture is thus simply the summation of the waves for the individual components. Figure 21-4 shows the polarogram of a five-component cation mixture. Clearly, a single polarogram may permit the quantitative determination of several elements. Success depends upon the existence of a sufficient difference between succeeding half-wave potentials to permit evaluation of individual diffusion currents. Approximately 0.2 V is required if the more reducible species undergoes a two-electron reduction; a minimum of about 0.3 V is needed if the first reduction is a one-electron process. The analysis of mixtures is considered in a later section.

[6] See: J. J. Lingane and B. A. Loveridge, *J. Amer. Chem. Soc.*, **72**, 438 (1950).

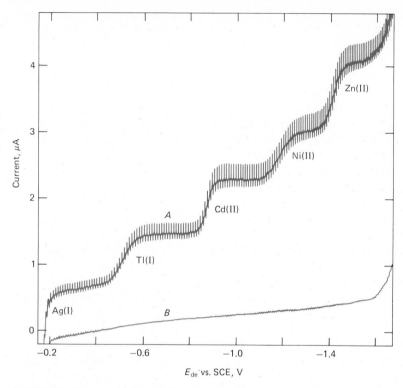

FIGURE 21-4 Polarograms of A: approximately 0.1 mM each of silver(I), thallium(I), cadmium(II), nickel(II), and zinc(II), listed in the order in which their waves appear, in 1F amonia/1F ammonium chloride containing 0.002% Triton X-100; B: the supporting electrolyte alone. (From L. Meites, *Polarographic Techniques*, 2d ed. New York: John Wiley & Sons, Inc., 1967, p. 164. With permission.)

Anodic Waves and Mixed Anodic-Cathodic Waves

Anodic waves as well as cathodic waves are encountered in polarography. The former are less common because of the relatively small range of anodic potentials that can be covered with the dropping mercury electrode before oxidation of the electrode itself commences. An example of an anodic wave is illustrated in curve A of Figure 21-5, where the electrode reaction involves the oxidation of iron(II) to iron(III) in the presence of citrate ion. A diffusion current is obtained at about $+0.1$ V,

which is due to the half-reaction

$$Fe^{2+} \rightleftharpoons Fe^{3+} + e$$

As the potential is made more negative, a decrease in the anodic current occurs; at about -0.02 V, the current becomes zero because the oxidation of iron(II) ion has ceased.

Curve C represents the polarogram for a solution of iron(III) in the same medium. Here, a cathodic wave results from reduction of the iron(III) to the divalent state. The half-wave potential is identical with that for the anodic wave, indicating that the oxidation

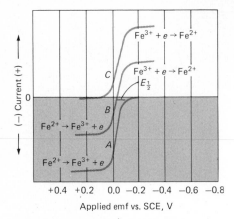

FIGURE 21-5 Polarographic behavior of iron(II) and iron(III) in a citrate medium. Curve A: anodic wave for a solution in which $[Fe^{2+}] = 1 \times 10^{-3}$. Curve B: anodic-cathodic wave for a solution in which $[Fe^{2+}] = [Fe^{3+}] = 0.5 \times 10^{-3}$. Curve C: cathodic wave for a solution in which $[Fe^{3+}] = 1 \times 10^{-3}$.

and reduction of the two iron species are perfectly reversible at the dropping electrode.

Curve B is the polarogram of an equiformal mixture of iron(II) and iron(III). The portion of the curve below the zero-current line corresponds to the oxidation of the iron(II); this reaction ceases at an applied potential equal to the half-wave potential. The upper portion of the curve is due to the reduction of iron(III).

Current Maxima

The shapes of polarograms are frequently distorted by so-called *current maxima* (see Figure 21-6), which are troublesome because they interfere with the accurate evaluation of diffusion currents and half-wave potentials. Although the cause or causes of maxima are not fully understood, there is considerable empirical knowledge of methods for eliminating them. Generally, the addition of traces of such high molecular weight substances as gelatin, Triton X-100 (a commercial surface-active agent), methyl red, or other dyes will cause a maximum to disappear. Care must be taken to avoid large amounts of these reagents, however, because the excess may reduce the magnitude of the diffusion current. The proper amount of suppressor must be determined by trial and error; the amount required varies widely from species to species.

Residual Current

The typical residual current curve, such as curve B in Figure 21-4, has two sources. The first of these is the reduction of trace impurities that are almost inevitably present in the blank solution; contributors here include small amounts of dissolved oxygen, heavy metal ions from the distilled water, and impurities present in the salt used as the supporting electrolyte.

A second component of the residual current is the so-called *charging* or *condenser current* resulting from a flow of electrons that charges the mercury droplets with respect to the solution; this current may be either negative or positive. At potentials more negative than about -0.4 V, an excess of electrons provides the surface of each droplet with a negative charge. These excess electrons are carried down with the drop as it breaks; since

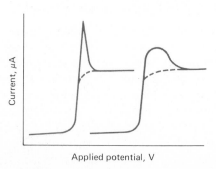

FIGURE 21-6 Typical current maxima.

each new drop is charged as it forms, a small but steady current results. At applied potentials smaller than about -0.4 V, the mercury tends to be positive with respect to the solution; thus, as each drop is formed, electrons are repelled from the surface toward the bulk of mercury, and a negative current is the result. At about -0.4 V, the mercury surface is uncharged, and the condenser current is zero (see Figure 21-4, curve *B*). The charging current is classed as a nonfaradaic current in the sense that electricity is carried across an electrode-solution interface without an accompanying oxidation-reduction process.

Ultimately, the accuracy and sensitivity of the polarographic method depend upon the magnitude of the residual current and the accuracy with which a correction for its effect can be determined.

Supporting Electrolyte

An electrode process will be diffusion-controlled only if the solution contains a sufficient concentration of a supporting electrolyte to diminish the attractive or repulsive force between the electrode and the analyte to a value that approaches zero. Generally, this condition is realized when the supporting electrolyte concentration exceeds that of the reactive species by a factor of 50 to 100. Under these circumstances, the fraction of charge carried through the solution by the species of interest is negligibly small (p. 504) because of the large excess of other ions. The limiting current then becomes a diffusion current, which is independent of electrolyte concentration.

Oxygen Waves

Dissolved oxygen is readily reduced at the dropping mercury electrode; an aqueous solution saturated with air exhibits two distinct waves attributable to this element (see Figure 21-7). The first results from the reduction of oxygen to peroxide:

$$O_2(g) + 2H^+ + 2e \rightleftharpoons H_2O_2$$

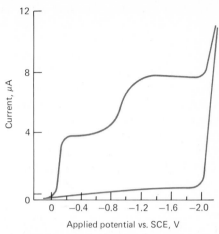

FIGURE 21-7 Polarogram for the reduction of oxygen in an air-saturated 0.1 *F* KCl solution. The lower curve is for oxygen-free 0.1 *F* KCl.

The second corresponds to the further reduction of the hydrogen peroxide:

$$H_2O_2 + 2H^+ + 2e \rightleftharpoons 2H_2O$$

As would be expected from stoichiometric considerations, the two waves are of equal height.

These polarographic waves are convenient for determining the concentration of dissolved oxygen; however, the presence of this element often interferes with the accurate determination of other species. Thus, oxygen removal is ordinarily the first step in a polarographic analysis. Aeration of the solution for several minutes with an inert gas accomplishes this end; a stream of the same gas, usually nitrogen, is passed over the surface during the analysis to prevent reabsorption.

APPLICATIONS OF POLAROGRAPHY

Apparatus

Cells. A typical general-purpose cell for polarographic analysis is shown in Figure 21-1

(p. 607). The solution to be analyzed is separated from the calomel electrode by a sintered disk backed by an agar plug that has been rendered conducting by the addition of potassium chloride. Such a bridge is readily prepared and lasts for extended periods, provided a potassium chloride solution is kept in the reaction compartment when the cell is not in use. A capillary sidearm permits the passage of nitrogen or other gas through the solution. Provision is also made for blanketing the solution with the gas during the analysis.

In a simpler arrangement, a mercury pool in the bottom of the sample container can be used as the nonpolarizable electrode. Here, the observed half-wave potentials differ somewhat from published values, which are based upon a saturated calomel reference electrode.

Dropping Electrodes. A dropping electrode, such as that shown in Figure 21-1, can be purchased from commercial sources. A 10-cm capillary ordinarily has a drop time of 3 to 6 s under a mercury head of about 50 cm. The tip of the capillary should be as nearly perpendicular to the length of the tube as possible, and care should be taken to assure a vertical mounting of the electrode; otherwise, erratic and nonreproducible drop times and sizes will be observed.

With reasonable care, a capillary can be used for several months or even years. Such performance, however, requires the use of scrupulously clean mercury and the maintenance of a mercury head, no matter how slight, at all times. If solution comes in contact with the inner surface of the tip, malfunction of the electrode is to be expected. For this reason, the head of mercury should always be increased to provide a good flow before the tip is immersed in a solution.

Storage of an electrode always presents a problem. One method is to rinse the capillary thoroughly with water, dry it, and then carefully decrease the head until the flow of mercury in air just ceases. Care must be taken to avoid lowering the mercury too far. Before

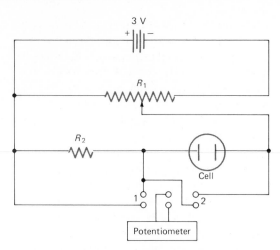

FIGURE 21-8 A simple circuit for polarographic measurements. [Reprinted with permission from J. J. Lingane, *Anal. Chem.*, **21**, 45 (1949). Copyright by the American Chemical Society.]

use, the head is increased, the tip is immersed in 1 : 1 nitric acid for a minute or so, and then rinsed with distilled water.

Electrical Apparatus. The electrical apparatus for polarographic measurements can be relatively simple. A voltage divider that permits continuous variation of an applied voltage from 0 to about ± 2.5 V is needed; the applied voltage must be known to about ± 0.01 V. In addition, it must be possible to measure currents over the range between 0.01 and perhaps 100 μA with a precision of ± 0.01 μA. A manual instrument that meets these requirements is easily constructed from equipment available in most laboratories. More elaborate devices for recording polarograms automatically are commercially available.

Figure 21-8 shows a circuit diagram of a simple instrument for polarographic work. Two 1.5-V batteries provide a voltage across the 100-Ω voltage divider R_1, by means of which the potential applied to the cell can be varied. The magnitude of this voltage can be determined with the potentiometer when the double-pole, double-throw switch is in position 2. The current is measured by determining the potential drop across the precision 10,000-Ω resistance R_2 with the same potentiometer and the switch in position 1. The current oscillations associated with the dropping electrode require that the null-detecting galvanometer be damped with a suitable resistance.

The apparatus shown in Figure 21-8 provides satisfactory data for routine quantitative work; in such applications, current measurements at only two voltages are needed to define i_d (one preceding the decomposition potential and one in the limiting-current region). On the other hand, where the entire polarogram is required, a point-by-point determination of the curve with a manual instrument is tedious and time-consuming; for this type of work, automatic recording is of great value. Several simple recording instruments are available in the range of $2000 to $3000. In such instruments, the contact of the potential divider R_1 in Figure 21-8 is driven by a motor synchronized with the paper drive motor of a millivolt recorder; the voltage drop across resistance R_2 is then recorded.

Treatment of Data

It is common practice to use either the average or the maximum value of the galvanometer or recorder oscillations.

Determination of Diffusion Currents. For quantitative work, limiting currents must always be corrected for the residual current. One method involves the use of a blank to derive an experimental residual-current curve. The diffusion current is then evaluated from the difference between the two at some potential in the limiting-current region.

Because the residual current usually increases nearly linearly with applied voltage, it is often possible to make the correction by extrapolation; this technique is illustrated in Figure 21-3 (p. 613).

Analysis of Mixtures. The quantitative determination of each species in a multicomponent mixture from a single polarogram is theoretically feasible, provided the half-wave potentials of the various species are sufficiently different (see Figure 21-4). Where a major constituent is more easily reduced than a minor one, however, the accuracy with which the latter can be determined may be poor because its diffusion current can occupy only a small fraction of the current scale of the instrument. Under such circumstances, a small error in current measurement leads to a large relative error in the analysis.

Several methods exist for treating mixtures containing species in unfavorable concentration ratios. One method is to alter the supporting electrolyte so that the wave for the minor constituent appears first; with the variety of complexing agents available, this technique is often feasible. Alternatively, a preliminary chemical separation can be employed. Finally, the current due to the major constituent can be decreased to zero (or a very small value) by application of a counter emf in the current-measuring circuit. The current sensitivity can then be increased to give a satisfactory signal for the reduction of the minor component. Most modern polarographs are equipped with such compensating devices.

Concentration Determination. The best and most straightforward method for quantitative polarographic analysis involves preliminary calibration with a series of standard solutions; as nearly as possible, these standards should be identical with the samples being analyzed, and should cover a concentration range within which the unknown samples will likely fall. The linearity of the current-concentration relationship can be assessed from such data; if the relationship is nonlinear, the analysis can be based upon the calibration curve.

Another useful technique is the standard-addition method. Here, the diffusion current for an accurately known volume of sample solution is measured. Then, a known amount of the species of interest is introduced (ordinarily as a known volume of a standard solution), and the diffusion current is again evaluated. Provided the relationship between current and concentration is linear, the increase in wave height will permit calculation of the concentration in the original solution. The standard-addition method is particularly effective when the diffusion current is sensitive to other components of the solution that are introduced with the sample.

Inorganic Polarographic Analysis

The polarographic method is generally applicable to the analysis of inorganic substances. Most metallic cations, for example, are reduced at the dropping electrode to form a metal amalgam or an ion of lower oxidation state. Even the alkali metals and alkaline-earth metals are reducible, provided the supporting electrolyte does not react at the high potentials required; here, the tetraalkyl ammonium halides are useful.

The successful polarographic analysis of cations frequently depends upon the supporting electrolyte that is used. To aid in this selection, tabular compilations of half-wave potential data should be consulted.[7] The judicious choice of anion often enhances the selectivity of the method. For example, with potassium chloride as a supporting electrolyte, the waves for iron(III) and copper(II) interfere with one another; in a fluoride medium, however, the half-wave potential of the former is shifted by about -0.5 V, while that for the latter is altered by only a few hundredths of a volt. The presence of fluoride thus results in the appearance of separate waves for the two ions.

The polarographic method is also applicable to the analysis of such inorganic anions as bromate, iodate, dichromate, vanadate, selenite, and nitrite. In general, polarograms for these substances are affected by the pH of the solution because the hydrogen ion is a participant in the reduction process. As a consequence, strong buffering to some fixed pH is necessary to obtain reproducible data for these species.

For further applications of polarography to inorganic analysis, the reader is referred to the monographs by Kolthoff and Lingane and by Meites.[8]

Organic Polarographic Analysis

Almost from its inception, the polarographic method has been used for the study and analysis of organic compounds, and many papers have been devoted to this subject. Several common functional groups are oxidized or reduced at the dropping electrode; compounds containing these groups are thus subject to polarographic analysis.[9]

In general, the reactions of organic compounds at a microelectrode are slower and more complex than those for inorganic cations. Thus, theoretical interpretation of the data is more difficult and perhaps impossible; moreover, a much stricter adherence to detail is required for quantitative work. Despite these handicaps, organic polarography has proved

---

[7] See, for example, the references cited on page 606; another extensive source is *Handbook of Analytical Chemistry*, ed. L. Meites. New York: McGraw-Hill, 1963.

[8] I. M. Kolthoff and J. J. Lingane, *Polarography*, 2d ed. New York: Interscience, 1952, vol. 2; and L. Meites, *Polarographic Techniques*, 2d ed. New York: Interscience, 1965.

[9] For a detailed discussion of organic polarographic analysis, see: P. Zuman, *Organic Polarographic Analysis*. Oxford: Pergamon Press, 1964; *Polarography of Molecules of Biological Significance*, ed. W. F. Smyth. New York: Academic Press, 1979; and *Topics in Organic Polarography*, ed. P. Zuman. New York: Plenum Press, 1970.

fruitful for the determination of structure, the qualitative identification of compounds, and the quantitative analysis of mixtures.

Effect of pH on Polarograms. Organic electrode processes ordinarily involve hydrogen ions, the most common reaction being represented as

$$R + nH^+ + ne = RH_n$$

where R and RH_n are the oxidized and reduced forms of the organic molecule. Half-wave potentials for organic compounds are therefore markedly pH-dependent. Furthermore, alteration of the pH often results in a change in the reaction products. For example, when benzaldehyde is reduced in a basic solution, a wave is obtained at about -1.4 V, attributable to the formation of benzyl alcohol:

$$C_6H_5CHO + 2H^+ + 2e \rightleftharpoons C_6H_5CH_2OH$$

If the pH is less than 2, however, a wave occurs at about -1.0 V that is just half the size of the foregoing one; here, the reaction involves the production of hydrobenzoin:

$$2C_6H_5CHO + 2H^+ + 2e$$

$$\rightleftharpoons C_6H_5CHOHCHOHC_6H_5$$

At intermediate pH values, two waves are observed, indicating the occurrence of both reactions.

It should be emphasized that an electrode process that consumes or produces hydrogen ions tends to alter the pH of the solution *at the electrode surface*; unless the solution is well-buffered, marked changes in pH can occur in the surface film *during* the electrolysis. These changes affect the reduction potential of the reaction and cause drawnout, poorly defined waves. Moreover, where the electrode process is altered by the pH, nonlinearity in the diffusion current-concentration relationship must also be expected. Thus, in organic polarography good buffering is vital for the generation of reproducible half-wave potentials and diffusion currents.

Solvents for Organic Polarography. Solubility considerations frequently dictate the use of some solvent other than pure water for organic polarography; aqueous mixtures containing varying amounts of such miscible solvents as glycols, dioxane, alcohols, Cellosolve, or acetic acid have been employed. Anhydrous media such as acetic acid, formamide, diethylamine, and ethylene glycol have also been investigated. Supporting electrolytes are often lithium salts or tetraalkyl ammonium salts.

Reactive Functional Groups. Organic compounds containing any of the following functional groups can be expected to produce one or more polarographic waves.

1. *The carbonyl group*, including aldehydes, ketones, and quinones, produce polarographic waves. In general, aldehydes are reduced at lower potentials than ketones; conjugation of the carbonyl double bond also results in lower half-wave potentials.

2. *Certain carboxylic acids* are reduced polarographically, although simple aliphatic and aromatic monocarboxylic acids are not. Dicarboxylic acids such as fumaric, maleic, or phthalic acid, in which the carboxyl groups are conjugated with one another, give characteristic polarograms; the same is true of certain keto and aldehydo acids.

3. *Most peroxides and epoxides* yield polarograms.

4. *Nitro, nitroso, amine oxide, and azo groups* are generally reduced at the dropping electrode.

5. *Most organic halogen groups* produce a polarographic wave which results from replacement of the halogen group with an atom of hydrogen.

6. *The carbon/carbon double bond* is reduced when it is conjugated with another double bond, an aromatic ring, or an unsaturated group.

7. *Hydroquinones and mercaptans* produce anodic waves.

In addition, a number of other organic groups cause catalytic hydrogen waves that can be used for analysis. These include amines, mercaptans, acids, and heterocyclic nitrogen compounds. Numerous applications to biological systems have been reported.[10]

VOLTAMMETRY AT SOLID ELECTRODES

Two limitations to the dropping mercury electrode have led chemists to search for substitutes. One undesirable property is the ease with which mercury is oxidized; as a consequence, oxidation processes that occur at potentials greater than about +0.4 V vs. SCE cannot be studied (in the presence of halide and certain other ions, the electrode is restricted to potentials more negative than 0.0 V); nor can the reduction of strong oxidizing agents be investigated because these substances oxidize mercury.

The second limitation of a dropping electrode is that of sensitivity, which prevents the analysis of solutions more dilute than about $10^{-5}M$. At lower concentrations, the diffusion current is of the same magnitude or smaller than the residual current; corrections for the background current then lead to large errors that ultimately limit the sensitivity. Most of the residual currents are nonfaradaic (p. 506) and cannot be decreased to less than a few hundredths of a microampere. Their main sources are the charging current (p. 616) and noise currents, which originate in the cell circuit of the instrument itself and from other electrical sources as well (see p. 71); the effects of the latter increase as the resistance of the cell becomes larger.

Most of the substitutes for the dropping mercury electrode are solid electrodes fabricated from platinum, gold, or graphite.[11] Two operational modes are employed. In one, the electrode is stationary, and every effort is made to avoid vibrational and convectional stirring. In the second, the electrode is rotated or vibrated at a constant rate; alternatively, the solution is stirred in a reproducible manner. In either case, the electrode and solution are in motion with respect to one another.

Voltammetry at Stationary Electrodes

Stationary microelectrodes have been employed for the analysis of a variety of organic compounds, usually by anodic oxidation at the indicator electrode. In addition, such electrodes have made possible the determination of strong inorganic oxidizing agents; the most important application of this type is to the determination of dissolved oxygen in water.

Electrode Construction. A platinum electrode is formed by sealing a wire into the end of a soft glass tube. The platinum may be filed and polished flush with the glass to provide a planar rather than a cylindrical surface; such electrodes, as well as graphite and gold electrodes, are available from commercial sources.

No solid electrode yields as reproducible current-voltage behavior as does the dropping mercury electrode, which continuously provides a new and fresh electrode surface. In contrast, the behavior of a solid electrode is often affected by adsorption, deposition, or

[10] M. Brezine and P. Zuman, *Polarography in Medicine, Biochemistry and Pharmacy.* New York: Interscience, 1958; and *Polarography of Molecules of Biological Significance*, ed. W. F. Smyth. New York: Academic Press, 1979.

[11] For discussions of voltammetry at solid electrodes, see: R. N. Adams, in *Treatise on Analytical Chemistry*, eds. I. M. Kolthoff and P. J. Elving. New York: Interscience, 1963, Part I, vol. 4, Chapter 47; and S. Piekarski and R. N. Adams, in *Physical Methods of Chemistry*, eds. A. Weissberger and B. W. Rossitor. New York: Wiley-Interscience, 1971, Part IIA, vol. 1, Chapter 7.

oxide film formation. These processes are time-dependent; thus, the electrode behavior often depends upon the history of its use.

Currents with Stationary Electrodes. When a potential sufficient to cause an electrode reaction is applied to a solid microelectrode, a large initial current is observed, which then decays to a constant and reproducible level after several minutes. The initial current is associated with the formation of a concentration gradient that ultimately extends several tenths of a millimeter from the electrode surface. Within this gradient, the reactant concentration increases continuously from the equilibrium value at the electrode surface to that of the bulk of the solution. Constant currents, which are applicable for analysis, are obtained only when the slope of the relationship between concentration and distance from the electrode surface becomes essentially constant; ordinarily, 2 to 5 min are required to reach this condition. These same current changes are also observed with a dropping electrode.[12]

Generally, no attempt is made to reach steady currents for each new voltage setting that is applied to a stationary electrode. Instead, a *linear voltage sweep* is employed in which the applied potential is varied at a constant rate, and the instantaneous current is recorded. Current-voltage curves such as those shown in Figure 21-9 are obtained. Here, the voltage sweep was begun at about $+0.4$ V (vs. SCE) and extended to $+0.7$ to 1.0 V; the sweep rate was 74 mV/min. The voltammograms result from the oxidation of N,N-tetramethylbenzidine at a micro platinum wire electrode in an aqueous medium. In contrast to a polarogram, a current maximum instead of a constant limiting current is observed. The height of the maximum is proportional to concentration. The position of the peak along the potential axis is characteristic of the electroactive species.

The cause of the initial current increases observed in Figure 21-9 is the same as that already described for a conventional polarographic wave. At the peak, the concentration of the ions or molecules of the reactive species at the electrode surface approaches zero, and the current becomes limited by their diffusion rate. As the sweep is continued, the time of electrolysis is longer, and the distance over which the reactive particles must move to reach the surface is greater. Thus, the rate of their arrival at the surface is lower; so also is the current.

The peak height with a stationary electrode is proportional to the square root of the voltage sweep rate and depends upon the direction of the sweep.

The Oxygen Electrode. An important application of stationary solid electrodes is to the determination of oxygen in water, sewage effluents, blood, and other fluids.[13] The electrode consists of a platinum or gold surface on the end of a support rod that is covered with an oxygen-permeable membrane of Teflon or polyurethane. The arrangement is often similar to the gas-sensing electrode shown in Figure 19-8 (p. 553) with the exception that the glass electrode is replaced with a micro platinum electrode. The reference electrode is ordinarily a silver/silver chloride or other silver based electrode. A potential of about -0.8 V (vs. SCE) is applied across the pair of electrodes, whereupon reduction of oxygen occurs in the film of liquid adjacent to the platinum surface. The electrode process is:

$$O_2 + 2H_2O + 2e \rightleftharpoons H_2O_2 + 2OH^-$$

---

[12] With a dropping electrode, however, the solution becomes homogenized by local stirring as a drop breaks, and for each new drop, conditions are identical to those encountered by the previous drop. Thus, the current fluctuations are perfectly reproducible, and the average current is independent of the time when it is measured.

[13] For a thorough treatment of the oxygen electrode, see: I. Fatt, *Polarographic Oxygen Sensors*. Cleveland: CRC Press, 1976.

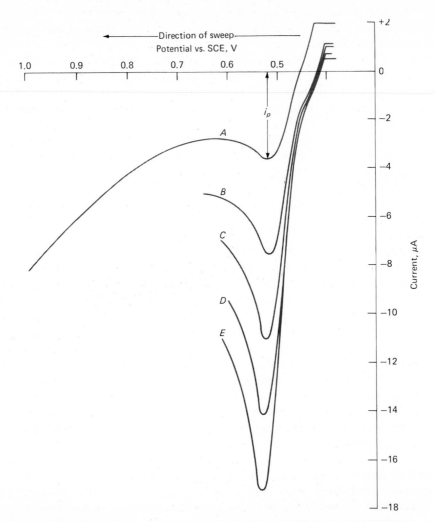

FIGURE 21-9 Linear sweep voltammograms for N,N'-tetramethyl-benzidine with a platinum wire microelectrode. Curve A: $0.4 \times 10^{-4} F$; Curve B: $0.8 \times 10^{-4} F$; Curve C: $1.2 \times 10^{-4} F$; Curve D: $1.6 \times 10^{-4} F$; Curve E: $2.0 \times 10^{-4} F$. (Adapted from R. N. Adams, in *Treatise on Analytical Chemistry*, eds. I. M. Kolthoff and P. J. Elving. New York: John Wiley & Sons, Inc., 1963, Part I, vol. 4, p. 2407. With permission.)

The current is proportional to the concentration of oxygen in the internal solution, which in turn is proportional to the oxygen concentration in the external solution adjacent to the electrode. Transport of oxygen occurs by the mechanism described earlier for CO_2 (p. 551).

Oxygen electrodes are available commercially. They could more aptly be termed oxygen sensitive voltammetric cells.

Voltammetry at Rotating Platinum Electrodes

Most voltammetric applications of solid electrodes are based upon the *rotating platinum electrode*, an example of which is shown in Figure 21-21 (p. 637). The microelectrode consists of a short length of platinum wire sealed into the side of a glass tube; mercury within the tube provides electrical contact between the wire and the lead to the polarograph. The tube is held in the hollow chuck of a synchronous motor and rotated at a constant speed in excess of 600 rpm. Commercial models of the rotating electrode are available.

Voltammetric waves, which are similar to those observed with the dropping electrode, can be obtained with the rotating platinum electrode. Here, however, the reactive species is brought to the electrode surface not only by diffusion but also by mechanical mixing. As a consequence, the limiting currents are as much as 20 times larger than those obtained with a microelectrode that is supplied by diffusion only; thus, the rotating electrode has the potential for enhancing sensitivity. Rotating electrodes, like their dropping electrode counterparts, provide steady currents instantaneously. This behavior is in distinct contrast to that of a solid microelectrode in the absence of stirring.

Several factors restrict the widespread application of the rotating platinum electrode for voltammetry. The low hydrogen overvoltage prevents its use as a cathode in acidic solutions.

In addition, the high currents obtained cause the electrode to be particularly sensitive to traces of oxygen in the solution. These two factors have largely confined its employment to anodic reactions. Limiting currents are often influenced by the previous history of the rotating platinum electrode and are seldom as reproducible as the diffusion currents obtained with a dropping electrode.

Rotating electrodes have been used extensively in studying the oxidation of organic compounds and as indicator electrodes for amperometric titrations (p. 636).

POLAROGRAPHY WITH POTENTIOSTATIC CONTROL

Figure 21-10 illustrates the effect of cell resistance on the polarographic wave for a reversible system. At 100 Ω, the IR drop is so small that it does not influence the shape of the wave significantly. With high resistances, on the other hand, the waves become drawn out and ultimately are so poorly defined as to be of little use. The cause of this effect is easily understood by reference to the equation

$$E_{appl} = E_{ind} - E_{SCE} - IR$$

Here, E_{appl} is varied in a regular way while E_{SCE} is constant. If IR is small, the indicator

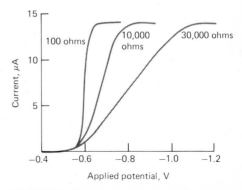

FIGURE 21-10   Effect of cell resistance on a reversible polarographic wave.

electrode then faithfully reflects the variation in E_{appl}. On the other hand, if R is large, a greater and greater fraction of E_{appl} is required to overcome the IR drop with increasing I. Thus, E_{ind} is no longer directly proportional to E_{appl}. A drawn-out wave results.

The use of potentiostatic control has permitted the extension of polarography to solvents with high electrical resistances. Here, three electrodes are employed: a dropping electrode (or other microelectrode), a counter or auxiliary electrode, and a reference electrode. The counter electrode and the microelectrode serve the same purpose as the two electrodes in ordinary polarography. The reference electrode, on the other hand, is employed to measure and control the potential of the microelectrode. The arrangement here is analogous to that described for controlled potential electrolysis (see Figure 20-4 or 20-8). The reference electrode is positioned as close as possible to the dropping electrode, and the potential between the two is then employed to control the applied potential E_{appl} such that the variation of E_{ind} is linear with time. Thus, the abscissa of the polarographic wave becomes E_{ind} rather than E_{appl}. The effect is to produce a wave form similar to that for the lowest resistance curve in Figure 21-10.

Figure 21-11 is a circuit diagram of an all-electronic instrument for three-electrode polarography. It consists of four components: a *ramp generator*, a potentiostat, a cell containing three electrodes, and a current amplifier and recorder. The purpose of the ramp generator is to provide a continuously varying potential or linear sweep across the indicator-saturated calomel electrode system. Note that the generator circuit is similar to the integration circuit shown in Figure 3-26 (p. 66). The relationship between the constant input potential E_i and the resulting output potential E_o can be obtained from Equation 3-20. That is,

$$E_o = -\frac{E_i}{R_i}\int_0^t dt = -\frac{E_i}{R_i}t$$

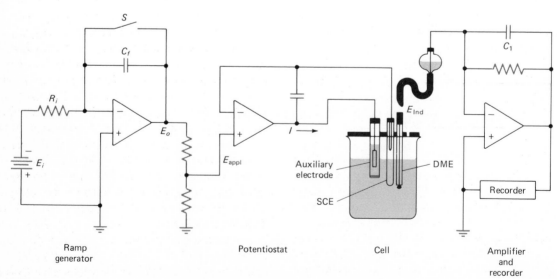

FIGURE 21-11 Schematic diagram showing the components of a modern, three-electrode polarograph and cell.

Thus, the output potential is determined by E_i and R_i and at any instant is directly proportional to the time that has elapsed after closure of the switch S. A fraction of E_o is applied to the noninverting terminal of the operational amplifier associated with the potentiostat.

The potentiostat shown in Figure 21-11 is similar in design to the one shown in Figure 20-8 and described on page 589.

The cell consists of a dropping mercury electrode, a saturated calomel electrode (which is part of the control circuit), and an auxiliary or working electrode. Essentially no current develops in the circuit that includes the reference and indicator electrodes because of the high internal resistance of the operational amplifier. The measured current is thus the output current I from the operational amplifier, which is determined by the potential of the dropping electrode *relative to the saturated calomel electrode*.

The auxiliary electrode in polarographic studies with three electrodes is frequently a pool of mercury in the bottom of the cell. In Figure 21-11, it is depicted as a platinum electrode separated from the analyte solution by a fritted disk.

The capacitor C_1 in the current amplifier circuit is employed to damp the current fluctuations associated with the dropping electrode.

MODIFIED VOLTAMMETRIC METHODS

As mentioned earlier (p. 622), classical polarography with a dropping mercury electrode is limited to solutions more concentrated than $10^{-5}M$; in addition, potentials more positive than 0.4 V versus the SCE are not accessible because of the ready oxidation of mercury. Several modifications of the classical polarographic procedure overcome these limitations to various degrees. Some of these modern adaptations of the classical method are summarized in Figure 21-12 and are described briefly in the paragraphs that follow.[14]

Differential Pulsed Polarography

In differential pulsed polarography, a dc potential, which is increased linearly with time, is applied to the polarographic cell. As in classical polarography, the rate of increase is perhaps 5 mV/s. In contrast, however, a dc pulse of an additional 20 to 100 mV is applied at regular intervals of about 1 to 3 s; the length of the pulse is about 60 ms and terminates with detachment of the mercury drop from the electrode. To synchronize the pulse with the drop, the latter is detached by an appropriately timed mechanical tap or movement of the electrode. The voltage program is shown in Figure 21-13.

Two current measurements are made alternatively—one just prior to the dc pulse and one near the end of the pulse (see Figure 21-13). The *difference in current per pulse* (Δi) is recorded as a function of the linearly increasing voltage. A differential curve results consisting of a current peak (see Figure 21-14); the height of this peak is directly proportional to concentration.

One advantage of the derivative-type polarogram is that individual peak maxima can be observed for substances with half-wave potentials differing by as little as 0.04 to 0.05 V; in contrast, classical polarography requires a potential difference of at least 0.2 V for resolution of waves. More important, however, differential pulse polarography increases the sensitivity of the polarographic method by about three orders of magnitude. This enhancement is illustrated in Figure 21-14. Note that a classical polarogram for a solution containing 180 ppm of the antibiotic tetracycline gives barely discernible waves; pulsed polarography,

---

[14] For additional details, see: J. B. Flato, *Anal. Chem.*, **44**, (11), 75A (1972).

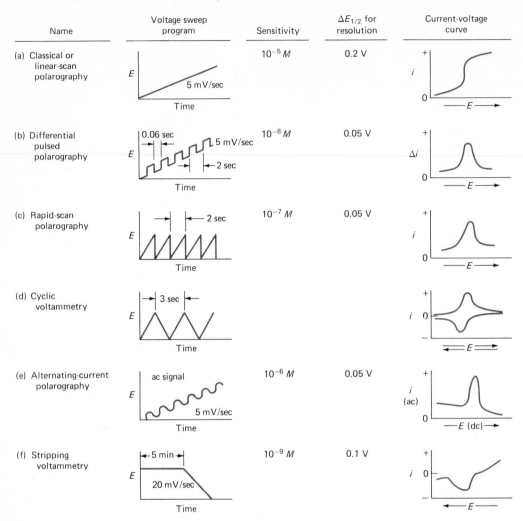

Name	Voltage sweep program	Sensitivity	$\Delta E_{1/2}$ for resolution	Current-voltage curve
(a) Classical or linear-scan polarography	5 mV/sec	$10^{-5} M$	0.2 V	
(b) Differential pulsed polarography	0.06 sec 5 mV/sec 2 sec	$10^{-8} M$	0.05 V	
(c) Rapid-scan polarography	2 sec	$10^{-7} M$	0.05 V	
(d) Cyclic voltammetry	3 sec			
(e) Alternating-current polarography	ac signal 5 mV/sec	$10^{-6} M$	0.05 V	
(f) Stripping voltammetry	5 min 20 mV/sec	$10^{-9} M$	0.1 V	

FIGURE 21-12 Modified polarographic procedures.

in contrast, provides two well-defined peaks at a concentration level that is 2×10^{-3} that for the classic wave, or 0.36 ppm. Note also that the current scale for Δi is in nA (nanoamperes) or $10^{-3}\ \mu A$.

The greater sensitivity of pulse polarography can be attributed to two sources. The first is an enhancement of the faradaic current, and the second is a decrease in the nonfaradaic charging current. To account for the former, let us consider the events that

must occur in the surface layer around an electrode as the potential is suddenly increased by 20 to 100 mV. If a reactive species is present in this layer, there will be a surge of current that lowers the reactant concentration to that demanded by the new potential. As the equilibrium concentration for that potential is approached, however, the current decays to a level just sufficient to counteract diffusion; that is, to the diffusion-controlled current. In classical polarography, the initial

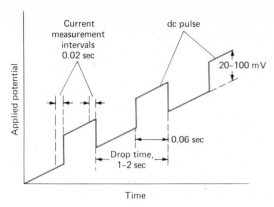

FIGURE 21-13 Voltage program for differential-pulsed polarography.

surge of current is not observed because the time scale of the measurement is long relative to the lifetime of the momentary current. On the other hand, in pulse polarography, the current measurement is made before the surge has completely decayed. Thus, the current measured contains both a diffusion-controlled component and a component that has to do with reducing the surface layer to the concentration demanded by the Nernst expression; the total current is typically several times larger than the diffusion current.

When the potential pulse is first applied to the electrode, a nonfaradaic current surge also occurs, which increases the charge on the

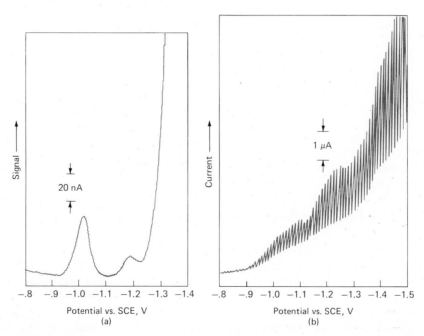

FIGURE 21-14 (a) Differential-pulsed polarogram. 0.36 ppm tetracycline · HCl in 0.1 M acetate buffer, pH 4, PAR Model 174 polarographic analyzer, dropping mercury electrode, 50-mV pulse amplitude, 1-sec drop. (b) DC Polarogram. 180 ppm tetracycline · HCl in 0.1 M acetate buffer, pH 4, similar conditions. [Reprinted with permission from J. B. Flato, *Anal. Chem.*, **44** (11), 75A (1972). Copyright by the American Chemical Society.]

drop (p. 616). This current, however, decays exponentially with time and approaches zero near the end of the life of a drop when its surface area is changing only slightly (see Figure 21-3a). Thus, by measuring currents at this time only, the residual current is virtually eliminated and the signal-to-noise ratio is larger. Enhanced sensitivity results.

Reliable instruments for pulse polarography are now available commercially at reasonable cost. The method has thus become an electro-analytical tool of considerable importance.

Rapid-Scan (Oscillographic) Polarography

In rapid-scan polarography, the applied potential is swept over a range of perhaps 0.5 V during the lifetime (or part of the lifetime) of a single mercury drop. An oscilloscope is required to display the current-voltage relationship; for a permanent record, the image on the oscilloscope screen may be photographed. Alternatively, the data can be stored in a computer for subsequent plotting.

The most successful rapid-scan procedure appears to be one in which the voltage scan occurs over the final 2 to 3 s of a drop that has a total lifetime of 6 to 7 s. As mentioned in the previous section, the residual current under these circumstances is small when compared with those encountered in ordinary polarography.

As will be seen from Figure 21-15, a rapid-scan polarogram also takes the form of a peak or a summit. At the summit, the current is made up of two components. One is the initial current surge required to adjust the concentration of the surface film to an equilibrium state demanded by the Nernst equation. The second is the normal diffusion-controlled current. The first current then decays as equilibrium is approached and ultimately only the diffusion-controlled current remains.

It can be shown that at 25°C the *summit potential* E_s for a reversible process is related

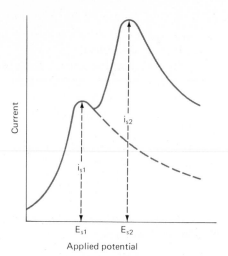

FIGURE 21-15 Rapid-span polarogram of a two-component mixture. (Taken from H. Schmidt and M. von Stackelberg, *Modern Polarographic Methods*, New York: Academic Press, 1963, p. 27. With permission.)

to the half-wave potential as follows:

$$E_s = E_{1/2} - 1.1 \frac{RT}{nF} = E_{1/2} - \frac{0.0282}{n}$$

$$(21\text{-}16)$$

The current i_s at the summit is given by

$$i_s = 2.72 \times 10^5 n^{3/2} A D^{1/2} C v^{1/2} \quad (21\text{-}17)$$

where A is the area of the electrode in cm^2, v is the rate of scan in volts per second, and the remaining terms are the same as in the Ilkovic equation (p. 612). Generally, a summit current exceeds a diffusion current by a factor of ten or more.

Perhaps the most important advantage that can be cited for rapid-scan polarography is its enhanced sensitivity. Not only are summit currents larger than diffusion currents, but the residual currents are also smaller; therefore, the signal-to-background ratio is considerably more favorable than in normal polarography. Quantitative analysis thus becomes possible for solutions as dilute as 10^{-6} to $10^{-7}M$. In addition, the shapes of rapid-scan polarograms are such that accurate current measurements are possible even when the half-wave potentials of two species differ by as little as 0.05 V.

Not surprisingly, rapid-scan equipment is more complex and more expensive than that required for ordinary polarography.

Cyclic Voltammetry

The apparatus and techniques of cyclic polarography or voltammetry are similar to those for rapid-scan polarography, which were considered in the previous section. The difference between the two lies in the voltage sweep program (compare Figures 21-12c and 21-12d). In cyclic voltammetry, the sweep takes a triangular form; that is, the potential is first increased linearly to a peak and then decreased to its starting point at the same rate. Generally, the cycle is completed in a fraction of a second to a few seconds. With a dropping electrode, the sweep is timed to occur near the end of the lifetime of the drop to minimize the nonfaradaic current.

Figure 21-16 shows three cyclic voltammetric curves. The solid curve is for a reversible reaction. The cause of cathodic peak A is the same as that for the peaks encountered in rapid-scan polarography. When the potential is first reversed (point B), the current remains positive and is largely due to the diffusion-controlled reduction of the analyte. Ultimately, however, potential C is reached at which the analyte is no longer reduced; the current here is zero. With further positive changes in the potential, oxidation of the previously reduced species begins and proceeds until its concentration reaches zero. The anodic peak D results. For a reversible reaction, the theoretical potential difference between peaks A and D is $(2 \times 0.0282/n)$ V (Equation 21-16).

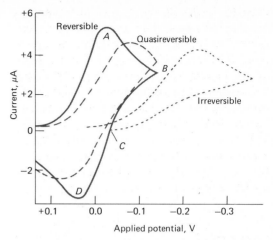

FIGURE 21-16   Cyclic voltammograms.

As the electrode reaction becomes slower or less reversible, this difference becomes greater. The effect of a slow electrode reaction rate is shown in the other two curves in Figure 21-16.

Cyclic voltammetry has become an important tool for the study of mechanisms and rates of oxidation-reduction processes, particularly in organic and metal-organic systems. Usually platinum is used for fabrication of the microelectrode.

Alternating-Current Polarography

In this modification, a constant ac potential (sine wave) of a few millivolts is superimposed upon the dc potential normally employed in polarography. As in the conventional experiment, the dc potential is varied over its usual range; here, however, the *alternating current is measured*.

Consider the application of this technique to a solution containing a single, easily reduced species Ox, which, at a suitable potential, reacts rapidly and reversibly as follows:

$$\text{Ox} + ne \rightleftharpoons \text{Red}$$

Assume also that an excess of supporting electrolyte is present.

We must first focus our attention on the *faradaic* alternating current in this cell; that is, the current that results from the oxidation-reduction process at the interface. In order for such a current to exist, it is clearly necessary to have both an oxidizable and a reducible species at the microelectrode surface. Significant concentrations of Red occur, however, only when the *applied dc potential* corresponds to the steeply rising portion of the dc polarogram. Thus, no faradaic alternating current will be observed before the decomposition potential has been reached. Beyond the decomposition potential, but short of the half-wave potential, the following relationship exists between the *surface* concentration of Ox and Red

$$0 < [\text{Red}]_0 < [\text{Ox}]_0$$

and the magnitude of the alternating current will be determined by $[\text{Red}]_0$. At applied dc potentials slightly greater than the half-wave potential,

$$[\text{Red}]_0 > [\text{Ox}]_0 > 0$$

and here, the magnitude of the alternating current is controlled by $[\text{Ox}]_0$. At the half-wave potential,

$$[\text{Red}]_0 = [\text{Ox}]_0$$

and the faradaic current will have reached its maximum value. When a potential corresponding to the diffusion current region is reached, $[\text{Ox}]_0 \to 0$, and the alternating current will, for lack of a reducible species, again approach zero. Thus, a typical ac polarogram consists of one or more peaks, the maxima occurring at the half-wave potentials for each of the reactive species present. The relationship between ac and classical dc polarograms is shown in Figure 21-17.

As noted in Chapter 18 (p. 506), alternating currents also result from a nonfaradaic process that involves charging of the double

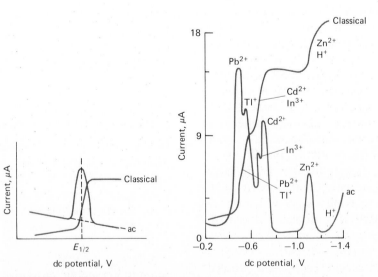

FIGURE 21-17 Comparison of ac and classical polarograms. [The right-hand plot is taken from B. Breyer, F. Gutmann, and S. Hacobian, *Australian J. Sci. Research*, **A3**, 567 (1950). With permission.]

layer surrounding an electrode. Thus, ac polarographic peaks are superimposed upon a base line of nonfaradaic current. The double layer structure is, however, affected by the dc component; as a consequence, the base line of an ac polarogram is not horizontal. Fortunately, over a small potential range, the nonfaradaic current is nearly linear, and a satisfactory correction can be made by linear extrapolation between the base lines on either side of a peak (see Figure 21-17).

The height of an ac polarographic peak is proportional to bulk concentration of the analyte and, for a reversible reaction, to the square root of frequency. For reversible electrode reactions, ac polarography is more sensitive than the classical procedure and permits analysis of solutions as dilute as $10^{-6}M$.

No ac peak is observed for a totally nonreversible half-reaction (that is, where the rate of the half-reaction is very slow in one direction), since the product generated by the dc current does not provide sufficient reactant for transport of the alternating current during the oxidation half-cycle. For electrode processes where the rate of one or the other half-reaction is less than instantaneous, but still appreciable, the magnitude of the alternating current becomes dependent upon the frequency of the ac source. As the frequency is increased, a condition is ultimately reached in which the time lapse of a half-cycle is insufficient for completion of the slow half-reaction; the alternating current will then be limited by the reaction rate, and the current maximum will be smaller than for an entirely reversible process. Thus, ac polarography is useful for kinetic studies of electrode reactions.

Alternating-current polarography with current having a square wave rather than a sinusoidal form has also been investigated (*square-wave polarography*). The advantage of the square wave form arises from the exponential decay of the nonfaradaic current with time and its approach to zero near the end of each half-cycle. By intermittent measurement

of the current at the end of each half-cycle only, the base line current becomes vanishingly small.

Alternating-current polarography has the advantage of enhanced sensitivity, and in contrast to classical polarography, permits the ready determination of a trace of a reducible species in the presence of a much larger concentration of a substance that gives a wave at lower potentials. The technique is also useful for the study of electrode kinetics. The disadvantage of the procedure is, of course, the complex nature of the equipment required.

Stripping Methods

Stripping analysis encompasses a variety of electrochemical methods having a common, characteristic initial step.[15] In all of these procedures, the analyte is first collected by electrodeposition at a mercury or a solid electrode; it is then redissolved (*stripped*) from the electrode to produce a more concentrated solution than originally existed. The analysis may then be based either upon electrical measurements made during the actual stripping process or upon some electroanalytical measurement of the more concentrated solution.

Stripping methods are of prime importance in trace work because the concentration aspects of the electrolysis permit the determination of minute amounts of an analyte with reasonable accuracy. Thus, the analysis of solutions in the 10^{-6} to $10^{-9}M$ range becomes feasible by methods that are both simple and rapid.

The stripping method that has had the

---

[15] For detailed discussions of stripping methods, see: F. Vydra, K. Stulik, and E. Julakova, *Electrochemical Stripping Analysis.* New York: Halsted, 1977; I. Shain, *Stripping Analysis*, in *Treatise on Analytical Chemistry*, eds. I. M. Kolthoff and P. J. Elving. New York: Interscience, 1963, Part 1, vol. 4, Chapter 50; W. D. Ellis, *J. Chem. Educ.*, **50**, A131 (1973); and T. R. Copeland and R. K. Skogerboe, *Anal. Chem.*, **46** (14), 1257A (1974).

most widespread application makes use of a micro mercury electrode for the deposition process, followed by anodic voltammetric measurement of the analyte; discussion in the paragraphs that follow will be largely confined to this particular application.

Electrodeposition Step. Ordinarily, only a fraction of the analyte is deposited during the electrodeposition step; hence, quantitative results depend not only upon control of electrode potential but also upon such factors as electrode size, length of deposition, and stirring rate for both the sample and a standard solution employed for calibration.

Mercury electrodes of various forms have been most widely used for the electrolysis, although platinum or other inert metals can also be employed. Generally, it is desirable to minimize the volume of the mercury to enhance the concentration of the deposited species. Several

methods have been devised to produce a microelectrode of reproducible dimensions, which is essential for quantitative work. Among these is the *hanging drop* electrode, which is illustrated in Figure 21-18. Here, an ordinary dropping mercury capillary provides a means of transferring a reproducible quantity of mercury (usually 1 to 3 drops) to a Teflon scoop. Note that the dropping mercury capillary does *not* serve as an electrode in this application. The hanging drop electrode is then formed by rotating the scoop and bringing the mercury into contact with a platinum wire sealed in a glass tube. The drop adheres strongly enough so that the solution can be stirred; it can, however, be dislodged by tapping the electrode at the completion of the electrolysis.

To use this apparatus, the drop is formed, stirring is begun, and a potential is applied that is a few tenths of a volt more negative

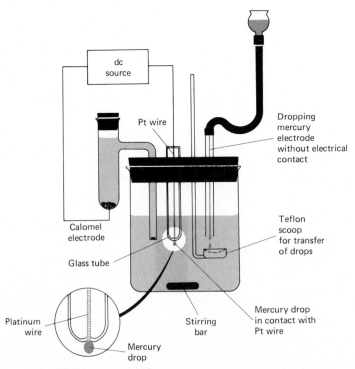

FIGURE 21-18 Apparatus for stripping analysis.

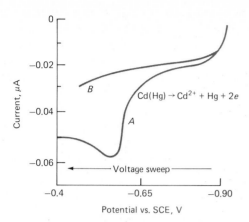

FIGURE 21-19 Curve *A*: Current-voltage curve for anodic stripping of cadmium. Curve *B*: Residual current curve for blank. [Reprinted with permission and adapted from R. D. DeMars and I. Shain, *Anal. Chem.*, **29**, 1825 (1957). Copyright by the American Chemical Society.]

than the half-wave potential for the ion of interest. Deposition is allowed to occur for a carefully measured period; 5 min usually suffice for solutions that are $10^{-7}M$ or greater, 15 min for $10^{-8}M$ solutions, and 60 min for those that are $10^{-9}M$. It should be emphasized that these times seldom result in complete removal of the ion. The electrolysis period is determined by the sensitivity of the method ultimately employed for completion of the analysis.

Voltammetric Completion of the Analysis. The analyte collected in the hanging drop electrode can be determined by a voltammetric measurement. Here, stirring is discontinued for perhaps 30 s after termination of the deposition. The voltage is then decreased *at a fixed rate* from its original cathodic value toward the anodic, and the resulting anodic current is recorded as a function of the applied voltage. This linear scan produces a curve of the type shown in Figure 21-19. In this experiment, cadmium was first deposited

from a $1 \times 10^{-8}M$ solution by application of a potential of about -0.9 V (vs. SCE), which is about 0.3 V more negative than the half-wave potential for this ion. After 15 min of electrolysis, stirring was discontinued; 30 s later, the potential was decreased at a rate of 21 mV/s. A rapid increase in anodic current occurred about -0.65 V as a result of the reaction

$$Cd(Hg) \rightarrow Cd^{2+} + Hg + 2e$$

The current decayed after reaching a maximum, owing to depletion of elemental cadmium in the hanging drop. The peak current, after correction for the residual current (curve *B* in Figure 21-19), was directly proportional to the concentration of cadmium ions over a range of 10^{-6} to $10^{-9}M$ and inversely proportional to deposition time. This analysis was based on calibration with standard solutions of cadmium ion. With reasonable care, an analytical precision of about 2% relative can be obtained. Mixtures can be resolved and analyzed by controlling the potential of deposition.

Many other variations of stripping analysis have been proposed. For example, a number of metals have been determined by electrodeposition on a platinum cathode. The quantity of electricity required to remove the deposit is then measured coulometrically. Here again, the method is particularly advantageous for trace analysis.

AMPEROMETRIC TITRATIONS

Voltammetric methods can be employed to estimate the equivalence point of titrations, provided at least one of the participants or products of the reaction involved is oxidized or reduced at a microelectrode. Here, the current at some fixed potential is measured as a function of the reagent volume (or of time if the reagent is generated by a constant-current coulometric process). Plots of the data on

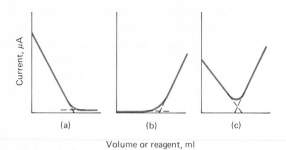

FIGURE 21-20 Typical amperometric titration curves. (a) Analyte is reduced, reagent is not. (b) Reagent is reduced, analyte is not. (c) Both reagent and analyte are reduced.

either side of the equivalence point are straight lines with differing slopes; the end point is established by extrapolation to their intersection.

An amperometric titration is inherently more accurate than voltammetric methods and less dependent upon the characteristics of the microelectrode and the supporting electrolyte. Furthermore, the temperature need not be fixed accurately, although it must be kept constant during the titration. Finally, the substance being determined need not be reactive at the electrode; a reactive reagent or product is equally satisfactory.[16]

Titration Curves

Amperometric titration curves typically take one of the forms shown in Figure 21-20. Figure 21-20a represents a titration in which the analyte reacts at the electrode while the reagent does not. The titration of lead with sulfate or oxalate ions may be cited as an example. The potential that is applied is sufficient to give a diffusion current for lead; a linear decrease in current is observed as

lead ions are removed from the solution by precipitation. The curvature near the equivalence point reflects the incompleteness of the precipitation reaction in this region. The end point is obtained by extrapolation of the linear portions, as shown.

Figure 21-20b is typical of a titration in which the reagent reacts at the microelectrode and the analyte does not; an example is the titration of magnesium with 8-hydroxyquinoline. A diffusion current for the latter is obtained at -1.6 V, whereas magnesium ion is inert at this potential.

Figure 21-20c corresponds to the titration of lead ion with a chromate solution at an applied potential greater than -1.0 V. Both lead and chromate ions give diffusion currents, and a minimum in the curve signals the end point. This system would yield a curve resembling that in Figure 21-20b at zero applied potential, since only chromate ions are reduced under these conditions.

Amperometric titrations in which the microelectrode serves as anode, or perhaps as anode for one reactant and cathode for another, have also been reported in the literature. The curves for such titrations are interpreted in the same way as the examples cited in the previous paragraphs. It should be recalled that an anodic current is provided with a minus sign by convention.

Apparatus and Techniques

Accurate amperometric titrations can be achieved with relatively simple apparatus.

Cells. Figure 21-21 shows a typical cell for an amperometric titration. A calomel half-cell is usually employed as the nonpolarizable electrode; the indicator electrode may be a dropping mercury electrode or a rotating microwire electrode, as shown. The cell should have a capacity of 75 to 100 ml.

Volume Measurements. For linear plots, it is necessary to correct for volume changes due to the added titrant. The measured currents can be multiplied by the quantity

---

[16] For a detailed discussion of amperometric titration, see: J. T. Stock, *Amperometric Titrations*. New York: Interscience, 1965.

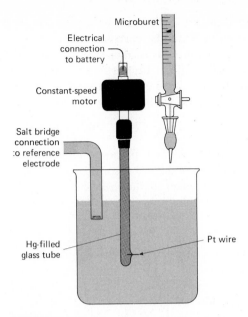

FIGURE 21-21 Typical cell arrangement for amperometric titrations with a rotating platinum electrode.

$(V + v)/V$, where V is the original volume and v is the titrant volume. An alternative, which is often satisfactory, is to use a reagent that is 20 (or more) times as concentrated as the solution being titrated. Under these circumstances, v is so small with respect to V that the correction is negligible. This approach requires the use of a microburet so that a total reagent volume of 1 or 2 ml can be measured with a suitable accuracy. The microburet should be so arranged that its tip can be touched to the surface of the solution after each addition of reagent to permit removal of the fraction of a drop that tends to remain attached.

Electrical Measurements. A simple manual polarograph is entirely adequate for amperometric titrations.

Electrodes. Many amperometric titrations can be carried out conveniently with a dropping mercury electrode. For reactions involving oxidizing agents that attack mercury [bromine, silver ion, cerium(IV), among others], a rotating platinum electrode is preferable.

Application of Amperometric Titrations

As shown in Table 21-2, the amperometric end point has been largely confined to titrations in which a precipitate or a stable complex is the product. A notable exception is the application of the rotating platinum electrode to titrations with bromate ion in the presence of bromide and hydrogen ions. For example, styrene in an acidic 75% methanol solution can be titrated with a standard solution of $KBrO_3$ that contains an excess of potassium bromide. The reactions are

$$BrO_3^- + 5Br^- + 6H^+ \rightarrow 3Br_2 + 3H_2O$$

$$C_6H_5CH{=}CH_2 + Br_2 \rightarrow C_6H_5CHBrCH_2Br$$

A titration curve similar to that in Figure 21-20b is observed.

AMPEROMETRIC TITRATIONS WITH TWO POLARIZED MICROELECTRODES

A convenient modification of the amperometric method involves the use of two identical stationary microelectrodes immersed in a well-stirred solution of the sample. A small potential (say, 0.1 to 0.2 V) is applied between these electrodes, and the resulting current is followed as a function of the volume of added reagent. The end point is marked by a sudden current rise from zero, a decrease in the current to zero, or a minimum (at zero) in a V-shaped curve.

Although the use of two polarized electrodes for end point detection was first proposed before 1900, almost 30 years passed before chemists came to appreciate the potentialities of the method.[17] The name *dead-stop end point* was used to describe the technique, and this term is still occasionally encountered. It was not until about 1950 that a clear interpretation of dead-stop titration curves was made.[18]

Oxidation-Reduction Titrations

Twin polarized platinum microelectrodes are conveniently used for end point detection for oxidation-reduction titrations. Figure 21-22 illustrates three types of titration curves that are commonly encountered. Curve (a) is observed when both reactant systems are reversible with respect to the electrodes. Curves (b)

---

[17] C. W. Foulk and A. T. Bawden, *J. Amer. Chem. Soc.*, **48**, 2045 (1926).

[18] For an excellent analysis of this type of end point, see: J. J. Lingane, *Electroanalytical Chemistry*, 2d ed. New York: Interscience, 1958, pp. 280–294.

and (c) are obtained when only one of the reactants exhibits reversible behavior.

Titration Curves When Both Systems Are Reversible. The titration of iron(II) with cerium(IV) is an example of a fully reversible system. That is, the equilibrium half-reaction

$$Ce^{4+} + e \rightleftharpoons Ce^{3+} \qquad E^0 = 1.44 \text{ V}$$

is established at a platinum electrode essentially instantaneously; so also is the equilibrium described by the equation

$$Fe^{3+} + e \rightleftharpoons Fe^{2+} \qquad E^0 = 0.771 \text{ V}$$

When a potential of 0.1 V is applied to a pair of platinum electrodes immersed in a solution containing both iron(II) and iron(III), a current is observed, owing to the following electrode reactions:

$$\text{Cathode} \quad Fe^{3+} + e \rightleftharpoons Fe^{2+}$$

$$\text{Anode} \quad Fe^{2+} \rightleftharpoons Fe^{3+} + e$$

These processes take place essentially instan-

TABLE 21-2 APPLICATIONS OF AMPEROMETRIC TITRATIONS

Reagent	Reaction Product	Type Electrode[a]	Substance Determined
K_2CrO_4	Precipitate	DME	Pb^{2+}, Ba^{2+}
$Pb(NO_3)_2$	Precipitate	DME	SO_4^{2-}, MoO_4^{2-}, F^-, Cl^-
8-Hydroxyquinoline	Precipitate	DME	Mg^{2+}, Zn^{2+}, Cu^{2+}, Cd^{2+}, Al^{3+}, Bi^{3+}, Fe^{3+}
Cupferron	Precipitate	DME	Cu^{2+}, Fe^{3+}
Dimethylglyoxime	Precipitate	DME	Ni^{2+}
α-Nitroso-β-naphthol	Precipitate	DME	Co^{2+}, Cu^{2+}, Pd^{2+}
$K_4Fe(CN)_6$	Precipitate	DME	Zn^{2+}
$AgNO_3$	Precipitate	RP	Cl^-, Br^-, I^-, CN^-, RSH
EDTA	Complex	DME	Bi^{3+}, Cd^{2+}, Cu^{2+}, Ca^{2+}, etc.
$KBrO_3$, KBr	Substitution, addition, or oxidation	RP	Certain phenols, aromatic amines, olefins; N_2H_4, As(III), Sb(III)

[a] DME: dropping mercury electrode; RP: rotating platinum electrode.

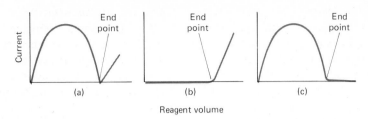

FIGURE 21-22 End points for amperometric oxidation-reduction titrations with twin-polarized electrodes. (a) Both reactants behave reversibly at the electrode. (b) Only reagent behaves reversibly. (c) Only analyte behaves reversibly.

taneously in the surface film surrounding the electrodes; that is, overvoltage effects are negligible. Since the electrodes are also small, concentration polarization will surely occur at the potential that has been applied; the magnitude of the current will thus be determined by the rate at which the reactant in *lesser* concentration is brought to the electrode surface. For example, if the concentration of iron(II) is smaller than that for iron(III), concentration polarization will occur at the anode, and the magnitude of the current will be determined by the concentration of the former. If iron(II) is in excess, cathodic polarization will occur, and the current will depend upon the iron(III) concentration. If the iron(III) concentration is zero (as it would be at the initial point in the titration), no current will be observed unless some alternative cathode process can occur at the small applied potential. A possible reaction would be

$$H^+ + e \rightleftharpoons \tfrac{1}{2}H_2$$

The overvoltage of hydrogen on platinum is sufficiently great, however, to rule out this prospect. Thus, no current is observed because of complete polarization of the cathode.

The behavior of the twin electrode system in a solution containing both cerium(III) and cerium(IV) is analogous to that in the iron system. Here, the electrode processes are

$$\text{Cathode} \quad Ce^{4+} + e \rightarrow Ce^{3+}$$

$$\text{Anode} \quad Ce^{3+} \rightarrow Ce^{4+} + e$$

The rates of these reactions are also large, and the magnitude of the observed currents depends upon the concentration of the cerium species that is present in lesser amount.

Let us now consider the curve for the titration of iron(II) with cerium(IV) as shown in Figure 21-22a. At the outset, no current is observed because no suitable cathode reactant is available. With addition of cerium(IV), a mixture of iron(III) and iron(II) is produced, which permits the passage of current. Initially, the magnitude of the current is determined by the iron(III) concentration. Beyond the midpoint in the titration, the concentration of this species exceeds that of iron(II); the current is then regulated by the decreasing iron(II) concentration. At the equivalence point, the concentrations of both iron(II) and cerium(IV) are vanishingly small. Here, possible electrode reactions are

$$\text{Cathode} \quad Fe^{3+} + e \rightleftharpoons Fe^{2+}$$

$$H^+ + e \rightleftharpoons \tfrac{1}{2}H_2$$

$$\text{Anode} \quad Ce^{3+} \rightleftharpoons Ce^{4+} + e$$

$$H_2O \rightleftharpoons \tfrac{1}{2}O_2 + 2H^+ + 2e$$

A potential considerably greater than 0.1 V is required to cause any possible combination of these reactions to occur. Thus, the system is completely polarized and the current is zero.

Beyond the equivalence point, depolarization occurs because of the excess cerium(IV) ions. Here, the electrode processes are

$$\text{Cathode} \quad Ce^{4+} + e \rightleftharpoons Ce^{3+}$$

$$\text{Anode} \quad Ce^{3+} \rightleftharpoons Ce^{4+} + e$$

and the current increases as more and more cerium(IV) is added to the solution.

Titration Curves When Only One System is Reversible. Let us now consider the behavior of a twin electrode system in the titration of an analyte that reacts only slowly at the electrodes (that is, a nonreversible system). An example is encountered in the titration of arsenic(III) with iodine. In the early stages of this titration, the solution contains appreciable concentrations of H_3AsO_3, H_3AsO_4, and I^-. Possible electrode processes that might occur upon application of a potential are

Cathode

$$H_3AsO_4 + 2H^+ + 2e \rightleftharpoons H_3AsO_3 + H_2O$$

Anode

$$H_3AsO_3 + H_2O \rightleftharpoons H_3AsO_4 + 2H^+ + 2e$$

$$2I^- \rightleftharpoons I_2 + 2e$$

The reactions involving the two arsenic species are slow at a platinum surface, however, and can be made to occur only by applying an overpotential of several tenths of a volt. Thus, until the equivalence point is passed, no significant current is observed (see Figure 21-22b).

Beyond the equivalence point, depolarization of the cell can occur at 0.1 V by virtue of the reactions

$$\text{Cathode} \quad I_2 + 2e \rightarrow 2I^-$$

$$\text{Anode} \quad 2I^- \rightarrow I_2 + 2e$$

Here, the current depends upon the concentration of excess iodine.

Figure 21-22c shows the titration of a dilute solution of iodine with thiosulfate ion. In the initial stages, iodine and iodide are present, and both react reversibly at the electrodes; the current depends upon the concentration of the species present in lesser amount. Thiosulfate does not behave reversibly at the electrodes; therefore, the current remains at zero beyond the equivalence point.

Precipitation Titrations

Twin silver microelectrodes have been employed for end point detection in titrations involving silver ion. For example, in the titration of silver ion with a standard solution of chloride ion, currents proportional to the metal ion concentration would result from the reactions

$$\text{Cathode} \quad Ag^+ + e \rightleftharpoons Ag(s)$$

$$\text{Anode} \quad Ag(s) \rightleftharpoons Ag^+ + e$$

With the effective removal of silver ion by the analytical reaction, cathodic polarization would occur, and the current would approach zero at the end point.

Applications

The amperometric method with two microelectrodes has been widely applied to titrations involving iodine; it is also useful with reagents such as bromine, titanium(III), and cerium(IV). An important use is in the titration of water with the Karl Fischer reagent. The technique has also been applied to end point detection for coulometric titrations.

The principal advantage of the twin microelectrode procedure is its simplicity. No reference electrode is required, and the only instrumentation needed is a simple voltage divider, powered by a dry cell, and a galvanometer or microammeter for current detection.

PROBLEMS

1. Calculate the Ni concentration (mg/liter) on the basis of the following polarographic data:

Solution	Current at -1.1 V, μA
25.0 ml of 0.2F NaCl diluted to 50.0 ml	8.4
25.0 ml of 0.2F NaCl plus 10.0 ml of sample diluted to 50.0 ml	46.3
25.0 ml of 0.2F NaCl, 10.0 ml of sample, 5.00 ml of $2.30 \times 10^{-2}M$ Ni^{2+} diluted to 50.0 ml	68.4

2. Calculate the concentration of Al (mg/liter) on the basis of the following polarographic data:

Solution	Current at -1.7 V, μA
20.0 ml of 0.20F HCl plus 20.0 ml of H_2O	10.2
20.0 ml of 0.20F HCl plus 10.0 ml of sample plus 10.0 ml of H_2O	33.3
20.0 ml of 0.20F HCl plus 10.0 ml of sample plus 10.0 ml of $6.32 \times 10^{-3}M$ Al^{3+}	52.0

3. The mercury from a dropping electrode was collected for 100 s and found to weigh 0.196 g. The time required for 10 drops of mercury to form was 43.2 s. When this electrode was employed with a standard $1.00 \times 10^{-3}M$ solution of Pb^{2+}, a current of 8.76 μA was observed. Subsequently, an unknown lead solution produced a current of 16.31 μA with a new electrode which had a drop time of 6.13 s and a flow rate of 3.85 mg/s. Calculate the molar concentration of Pb^{2+} in the unknown.

4. The following data were collected for three dropping electrodes. Complete the data for electrodes A and C.

	A	B	C
Flow rate, mg/s	1.89	4.24	3.11
Drop time, s	2.12	5.84	3.87
i_d/C, μA liter $mmol^{-1}$		4.86	

5. Electrode C in Problem 4 was employed to study the reduction of an organic compound known to have a diffusion coefficient of 7.3×10^{-6} cm^2/s. A $5.00 \times 10^{-4}M$ solution of the compound yielded a diffusion current of 8.6 μA. Calculate n for the reaction.

6. An organic compound underwent a two-electron reduction at dropping electrode A in Problem 4. A diffusion current of 9.6 μA was produced by a $1.17 \times 10^{-3}M$ solution of the compound. Calculate the diffusion coefficient for the compound.

7. The half-wave potential for Ni(II) in a 0.10F NaClO$_4$ solution was found to be -1.02 V (vs. SCE). In a solution that was 0.10F in NaClO$_4$ and 0.100F in ethylenediamine (en), the half-wave potential was -1.60 V; calculate the formation constant of the complex between the two species assuming its formula is Ni(en)$_3^{2+}$.

8. The following polarographic data were obtained for the reduction of Pb^{2+} to its amalgam from solutions that were $2.00 \times 10^{-3}M$ in Pb^{2+}, 0.100F in KNO$_3$, and that also had the following concentrations of the anion A$^-$. From the half-wave potentials, derive the formula of the complex as well as its formation constant.

Concn A$^-$, M	$E_{1/2}$ vs. SCE, V
0.0000	-0.405
0.0200	-0.473
0.0600	-0.507
0.1007	-0.516
0.300	-0.547
0.500	-0.558

9. A $1.00 \times 10^{-3}M$ solution of europium(III) in 0.100F KNO$_3$ has a polarographic wave for the reversible reduction of Eu^{3+} to the amalgam. The following data show the effect of increasing concentration of the anion X^{2-} on the half-wave potential.

Concn X^{2-}, M	$E_{1/2}$ vs. SCE, V
0.000	-0.692
0.0200	-1.083
0.0600	-1.113
0.100	-1.128
0.300	-1.152
0.500	-1.170

Derive the formula of the complex as well as its formation constant.

22

CONDUCTOMETRIC METHODS

Conduction of electricity through an electrolyte solution involves migration of positively charged species toward the cathode and negatively charged ones toward the anode. The *conductance*, which is a measure of the current that results from the application of a unit electrical force, depends directly upon the number of charged particles in the solution. All ions contribute to the conduction process, but the fraction of current carried by any given species is determined by its relative concentration and its inherent mobility in the medium.

The application of direct conductance measurements to analysis is limited because of the nonselective nature of this property. The principal uses of direct measurements have been confined to the analysis of binary water/electrolyte mixtures and to the determination of total electrolyte concentration. The latter measurement is particularly useful as a criterion of purity for distilled water.

On the other hand, *conductometric titrations*, in which conductance measurements are used for end-point detection, can be applied to the determination of numerous substances.[1]

The principal advantage to the conductometric end point is its applicability to the titration of very dilute solutions and to systems in which the reaction between the titrant and analyte is relatively incomplete. Thus, for example, the conductometric titration of an aqueous phenol $(K_a \approx 10^{-10})$ solution is feasible, even though the change in pH at the equivalence point is insufficient for either a potentiometric or an indicator end point.

Conductometric titrations become less accurate and less satisfactory with increasing total electrolyte concentration. Indeed, the change in conductance due to the addition of titrant can become largely masked in solutions with high salt concentrations; under these circumstances, the method cannot be used.

ELECTROLYTIC CONDUCTANCE

Upon application of a potential, ions in a solution are essentially instantly accelerated toward the electrode of opposite charge. The rate at which they migrate, however, is limited by the frictional forces generated by their motion. As in a metallic conductor, the velocity of the particles is linearly related to the applied field; Ohm's law is thus obeyed by electrolyte solutions.

Some Important Relationships

Conductance G. The conductance of a solution is the reciprocal of the electrical resistance and has the units of ohm^{-1}. That is,

$$G = \frac{1}{R} \qquad (22\text{-}1)$$

where R is the resistance in ohms.

Specific Conductance k. Conductance is directly proportional to the cross-sectional area A and inversely proportional to the length l of a uniform conductor; thus,

$$G = k\frac{A}{l} \qquad (22\text{-}2)$$

where k is a proportionality constant called the *specific conductance*. Clearly, it is the conductance when A and l are numerically equal. If these parameters are based upon the centimeter, k is the conductance of a cube of liquid one centimeter on a side. The dimensions of specific conductance are then $ohm^{-1}\ cm^{-1}$.

Equivalent Conductance. The equivalent conductance Λ is defined as the conductance of one gram equivalent of solute contained be-

---

[1] For further discussion of conductometric methods, see: J. W. Loveland, in *Treatise on Analytical Chemistry*, eds. I. M. Kolthoff and P. J. Elving. New York: Interscience, 1963, Part I, vol. 4, Chapter 51; and T. Shedlovsky and L. Shedlovsky, in *Physical Methods of Chemistry*, eds. A. Weissberger and B. W. Rossitor. New York: Wiley-Interscience, 1971, vol. 1, Part IIA, Chapter 3.

tween electrodes spaced one centimeter apart.[2] Neither the volume of the solution nor the area of the electrodes is specified; these vary to satisfy the conditions of the definition. For example, a $1.0N$ solution (1.0 gram equivalent per liter) would require electrodes with individual surface areas of 1000 cm^2; a $0.1N$ solution would need 10,000-cm^2 electrodes. The direct measurement of equivalent conductance is thus seldom, if ever, undertaken because of the experimental inconvenience associated with such relatively large electrodes. Instead, this quantity is determined indirectly from specific conductance data. By definition, Λ will be equal to G when one gram equivalent of solute is contained between electrodes spaced one centimeter apart. The volume V of the solution (cm^3) that will contain one gram equivalent of solute is given by

$$V = \frac{1000}{C}$$

where C is the concentration in equivalents per liter. This volume can also be expressed in terms of the dimensions of the cell

$$V = lA$$

With l fixed by definition at one centimeter,

$$V = A = \frac{1000}{C}$$

Substitution into Equation 22-2 thus gives

$$\Lambda = \frac{1000k}{C} \qquad (22\text{-}3)$$

Equation 22-3 permits calculation of the equivalent conductance from the experimental value of k for a solution of known concentration.

---

[2] In the context of electrolytic conductance, the equivalent is defined in terms of the number of charges carried by an ion and not upon its fate in a specific reaction. Thus, the weight of one gram equivalent of Ba^{2+} is equal to the atomic weight of barium divided by two.

Equivalent Conductance at Infinite Dilution. The mobility of an ion in solution is governed by four forces. An *electrical force*, equal to the product of the potential of the electrode and the charge of the ion, tends to move the particle toward one of the electrodes. This effect is partially balanced by a *frictional force* that is a characteristic property for each ion. These are the only two effects that play a significant role in determining the conductivity of a dilute solution; under these circumstances, the equivalent conductance of a salt is independent of its concentration.

At finite concentrations, two other factors, the *electrophoretic* effect and the *relaxation* effect, become important and cause the equivalent conductance of a substance to decrease as its concentration increases. The behavior of a sodium chloride solution is typical and is shown in Table 22-1.

The electrophoretic effect stems from the motion of the oppositely charged ions surrounding the ion of interest. These ions carry with them molecules of solvent; the motion of the primary particle is thus retarded by the flow of solvent in the opposite direction. The relaxation effect also owes its genesis to movement of the ionic atmosphere surrounding a given particle. Here, however, the ion is slowed by the charge of opposite sign that builds up behind the moving particle.

For strong electrolytes, a linear relationship exists between equivalent conductance

---

TABLE 22-1   EFFECT OF CONCENTRATION ON EQUIVALENT CONDUCTANCE

Concentration of NaCl, eq/liter	Λ
0.1	106.7
0.01	118.5
0.001	123.7
Infinite dilution	126.4(Λ_0)

and square root of the concentration. Extrapolation of this straight-line relationship to zero concentration yields a value for the equivalent conductance at infinite dilution Λ_0. A similar plot for a weak electrolyte is nonlinear, and direct evaluation of Λ_0 is difficult.

At infinite dilution, interionic attractions become nil; the overall conductance of the solution then consists of the sum of the individual equivalent ionic conductances

$$\Lambda_0 = \lambda_+^0 + \lambda_-^0$$

where λ_+^0 and λ_-^0 are the equivalent ionic conductances of the cation and the anion of the salt at infinite dilution. Individual ionic conductances can be determined from other electrolytic measurements; values for a number of common ions are given in Table 22-2. Note that symbols such as $\frac{1}{2}Mg^{2+}$, $\frac{1}{3}Fe^{3+}$, and $\frac{1}{2}SO_4^{2-}$ are used to emphasize that the concentration units are in *equivalents* per liter.

The differences that exist among the equivalent ionic conductances of various species (Table 22-2) arise primarily from differences in their size and the degree of their hydration.

The equivalent ionic conductance is a measure of the mobility of an ion under the influence of an electric force field and is thus a gauge of its capacity for transporting electricity. For example, the ionic conductance of a potassium ion is nearly the same as that of a chloride ion; therefore, electricity passing through a potassium chloride solution is carried nearly equally by the two species. The situation is quite different with hydrochloric acid; because of the greater mobility of the hydrogen ion, a larger fraction of the electricity $[350/(350 + 76) = 0.82]$ is carried by that species in an electrolysis.

Ionic conductance data permit comparison of the relative conductivity of various solutes. Thus, we are justified in saying that $0.01F$ hydrochloric acid will have a greater conductivity than $0.01F$ sodium chloride because of the very large ionic conductance of the hydrogen ion. Such conclusions are important in predicting the course of a conductometric titration.

Alternating Currents in a Cell

As was pointed out earlier (p. 505), conduction of dc electricity across a solution/electrode interface is a faradaic process in which oxidation and reduction must occur at the two electrodes. An alternating current, on the other hand, requires no electrochemical reaction at the electrodes (p. 506); here, electricity flows as a consequence of nonfaradaic processes. Because the changes associated with faradaic conduction can materially alter the electrical characteristics of a cell, conductometric measurements are advantageously based upon nonfaradaic processes.

At low frequencies, conduction of ac electricity through an electrolyte involves periodic motion of the ions toward and away from the electrodes, as noted earlier (p. 506); however, at radio frequencies, a significant fraction of ac electricity is transported in the form of a dielectric current which results from induced and orientation polarization. Most

TABLE 22-2 EQUIVALENT IONIC CONDUCTANCES AT 25°C

Cation	λ_+^0	Anion	λ_-^0
H_3O^+	349.8	OH^-	199.0
Li^+	38.7	Cl^-	76.3
Na^+	50.1	Br^-	78.1
K^+	73.5	I^-	76.8
NH_4^+	73.4	NO_3^-	71.4
Ag^+	61.9	ClO_4^-	67.3
$\frac{1}{2}Mg^{2+}$	53.1	$C_2H_3O_2^-$	40.9
$\frac{1}{2}Ca^{2+}$	59.5	$\frac{1}{2}SO_4^{2-}$	80.0
$\frac{1}{2}Ba^{2+}$	63.6	$\frac{1}{2}CO_3^{2-}$	69.3
$\frac{1}{2}Pb^{2+}$	69.5	$\frac{1}{2}C_2O_4^{2-}$	74.2
$\frac{1}{3}Fe^{3+}$	68.0	$\frac{1}{4}Fe(CN)_6^{4-}$	110.5
$\frac{1}{3}La^{3+}$	69.6	—	—

conductance measurements are made at sufficiently low frequencies so that the latter mechanisms are not important. *Oscillometry*, on the other hand, makes use of radio frequencies; here, the dielectric current becomes important. Oscillometric measurements are discussed at the end of this chapter.

THE MEASUREMENT OF CONDUCTANCE

A conductance measurement requires a source of electrical power, a cell to contain the solution, and a suitable bridge to measure the resistance of the solution.

Power Sources

Use of an alternating current source eliminates the effect of faradaic currents. There are, however, both upper and lower limits to the frequencies that can be employed; audio oscillators that produce signals of about 1000 Hz are the most satisfactory.

For less refined work, an ordinary 60-cycle current, stepped down from 110 to perhaps 10 V, is also used. With such a source, however, faradaic processes often limit the accuracy of the conductance measurements. On the other hand, the convenience and ready availability of 60-cycle power justify its use for many purposes.

Power sources with frequencies much greater than 1000 Hz create problems in conductance measurements with a bridge. Here, the cell capacitance and stray capacitances in other parts of the circuit cause phase changes in the current for which compensation is difficult.

Resistance Bridges

The Wheatstone bridge arrangement, shown in Figure 2-24 (p. 35), is typical of the apparatus used for conductance measurements. The power source S provides an alternating current in the frequency range of 60 to 1000 Hz at a potential of 6 to 10 V. The resistances R_{AC} and R_{BC} can be calculated from the position of C. The cell, of unknown resistance R_x, is placed in the upper left arm of the bridge and a precision variable resistance R_s is placed in the right-hand side. A null detector ND is used to indicate the absence of current between D and C. The detector may consist of a pair of ordinary headphones, since the ear is responsive to frequencies in the 1000 Hz range; alternatively, it may be a "magic eye" tube, a cathode ray tube, or a microammeter.

Capacitance effects within R_x cause an alternating-current device such as this to suffer a loss in sensitivity when very high resistances are measured; in practice, a variable capacitor across R_s compensates for this effect.

Conductance can also be determined by means of a simple electronic circuit such as that shown in Figure 3-21a (p. 62). Here, the conductance is read directly from a dc meter or a recorder.

Cells

Figure 22-1 depicts three common types of cells for the measurement of conductivity. Each contains a pair of electrodes firmly fixed in a constant geometry with respect to one another. The electrodes are ordinarily platinized to increase their effective surface and thus their capacitances; faradaic currents are minimized as a result.

Determination of the Cell Constant. According to Equation 22-2, the specific conductance k is related to the measured conductance G by the ratio of the distance separating the electrodes to their surface area. This ratio has a fixed and constant value in any given cell and is known as the *cell constant*. Its value is seldom determined directly. Instead, the conductance of a solution whose specific conductance is reliably known is measured; the cell constant can then be calculated. Solutions of potassium chloride are commonly chosen for

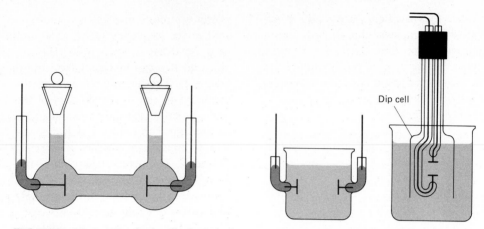

FIGURE 22-1 Typical cells for conductometric measurements.

cell calibration.[3] Typical data are shown in Table 22-3.

Once the value of the cell constant has been determined, conductivity data can be easily converted to terms of specific conductance with the aid of Equation 22-2.

Temperature Control. The temperature coefficient for conductance measurements is about 2%/°C; as a consequence, some temperature control is ordinarily required during a conductometric measurement. Although a constant temperature is necessary, control at

some specific level is not required for a successful conductometric titration. For many purposes, it is sufficient to immerse the cell in a reasonably large bath of water or oil, which is at room temperature.

CONDUCTOMETRIC TITRATIONS

Conductometric measurements provide a convenient means for locating end points in titrations. Sufficient measurements (three to four before and after the equivalence point) are needed to define the titration curve. After being corrected for volume change, the conductance data are plotted as a function of titrant volume.

---

[3] G. Jones and B. C. Bradshaw, *J. Amer. Chem. Soc.*, **55**, 1780 (1933).

TABLE 22-3   CONDUCTANCE OF SOLUTIONS FOR CELL CALIBRATION

Grams KCl per 1000 g of Solution in Vacuum	Specific Conductance at 25°C, ohm^{-1} cm^{-1}
71.1352	0.111342
7.41913	0.0128560
0.745263	0.00140877

The two linear portions are then extrapolated, the point of intersection being taken as the equivalence point.

Because reactions fail to proceed to absolute completion, conductometric titration curves invariably show departures from strict linearity in the region of the equivalence point. Curved regions become more pronounced as the reaction in question becomes less favorable and as the solution becomes more dilute. The linear portions of the curve are best defined by measurements sufficiently removed from the equivalence point so that the common ion effect forces the reaction more nearly to completion. It is in this respect that the conductometric technique appears to best advantage; in contrast to potentiometric or indicator

methods, which require observations under conditions where reaction is least complete, a conductometric analysis can be employed successfully for titrations based upon relatively unfavorable equilibria.

The conductometric end point is completely nonspecific. Although it is potentially adaptable to all types of volumetric reactions, the number of its useful applications to oxidation-reduction systems is limited; the substantial excess of hydrogen ions typically needed for such reactions tends to mask conductivity changes associated with the volumetric reaction.

Acid-Base Titrations

Neutralization titrations are particularly well adapted to the conductometric end point because of the large ionic conductances of hydrogen and hydroxide ions compared with the conductances of the species that replace them in solution.

Titration of Strong Acids or Bases. The solid line in Figure 22-2 represents a curve (corrected for volume change) obtained when hydrochloric acid is titrated with sodium hydroxide. Also plotted are the calculated contributions of the individual ions to the conductance of the solution. During neutralization, hydrogen ions are replaced by an equivalent number of less mobile sodium ions; the conductance changes to lower values as a result of this substitution. At the equivalence point, the concentrations of hydrogen and hydroxide ions are at a minimum and the solution exhibits its lowest conductance. A reversal of slope occurs past the end point as the sodium ion and hydroxide ion concentrations increase. With the exception of the immediate equivalence-point region, an excellent linearity exists between conductance and the volume of base added; as a result, only three or four observations on each side of the equivalence point are needed for an analysis.

The percentage change in conductivity dur-

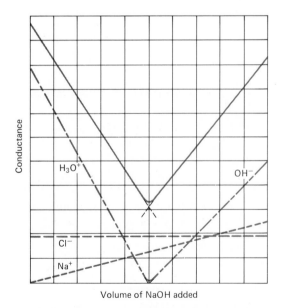

FIGURE 22-2 Conductometric titration of a strong acid with a strong base. The solid line represents the titration curve, corrected for volume change. Broken lines indicate the contribution of the individual species, also corrected for volume change, to the conductance of the solution.

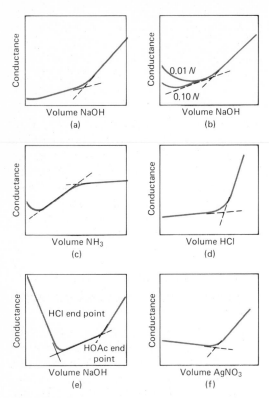

FIGURE 22-3 Typical conductometric titration curves. Titration of (a) a very weak acid $(K_a \approx 10^{-10})$ with sodium hydroxide; (b) a weak acid $(K_a \approx 10^{-5})$ with sodium hydroxide (Note that for 0.01 N solutions, conductance × 10 is plotted); (c) a weak acid $(K_a \approx 10^{-5})$ with aqueous ammonia; (d) the salt of a weak acid; (e) a mixture of hydrochloric and acetic acids with sodium hydroxide; and (f) chloride ion with silver nitrate.

ing the course of the titration of a strong acid or base is the same regardless of the concentration of the solution. Thus, very dilute solutions can be analyzed with an accuracy comparable to more concentrated ones.

Titration of Weak Acids or Bases. Figure 22-3a illustrates application of the conductometric end point to the titration of boric acid $(K_a = 6 \times 10^{-10})$ with strong base. This reaction is so incomplete that a potentiometric or visual indicator end point is unsatisfactory. In the early stages of the titration, a buffer is rapidly established that imparts to the solution a relatively small and nearly constant hydrogen ion concentration. The added hydroxide ions are consumed by this buffer and thus do not directly contribute to the conductivity. A gradual increase in conductance does result, however, owing to the increase in concentration of sodium and of borate ions. With attainment of the equivalence point, no further borate is produced; additions of base then cause a more rapid increase in conductance due to the highly mobile hydroxide ions.

Figure 22-3b illustrates the titration of a moderately weak acid, such as acetic acid $(K_a \cong 10^{-5})$, with sodium hydroxide. Nonlinearity in the early portions of the titration curve creates problems in establishing the end point; with concentrated solutions, however, the titration is feasible. As before, we can interpret this curve in light of the changes in composition that occur. Here, the solution initially has a moderate concentration of hydrogen ions $(\sim 10^{-3} M)$. Addition of base results in the establishment of a buffer system and a consequent diminution in the hydrogen ion concentration. Concurrent with this decrease is an increase in the concentration of sodium ions as well as the conjugate base of the acid. These two factors act in opposition to one another. At first, the decrease in hydrogen ion concentration predominates and a decrease in conductance is observed. As the titration progresses, however, the pH becomes stabilized (in the buffer region); the increase in salt content then becomes the more important factor, and a linear increase in conductance finally results. Beyond the equivalence point, the curve steepens because of the greater ionic conductance of hydroxide ion.

In principle, all titration curves for weak acids or bases contain the general features of Figure 22-3b. The ionization of very weak species is so slight, however, that little or no curvature occurs with the establishment of the buffer region (see Figure 22-3a, for example). As the strength of the acid (or base) becomes greater, so also does the extent of the curvature in the early portions of the titration curve. For weak acids or bases with dissociation constants greater than about 10^{-5}, the curvature becomes so pronounced that an end point cannot be distinguished.

Figure 22-3c illustrates the titration of the same weak acid as in Figure 22-3b, but with aqueous ammonia instead of sodium hydroxide. Here, because the titrant is a weak electrolyte, the curve is essentially horizontal past the equivalence point. Use of ammonia as titrant actually provides a curve that can be extrapolated with less uncertainty than the corresponding curve based upon titration with sodium hydroxide.

Figure 22-3d represents the titration curve for a weak base, such as acetate ion, with a standard solution of hydrochloric acid. The addition of strong acid results in formation of sodium chloride and undissociated acetic acid. The net effect is a slight rise in conductance due to the greater mobility of the chloride ion over that of the acetate ion it replaces. After the end point has been passed, a sharp rise in conductance attends the addition of excess hydrogen ions. The conductometric method is convenient for the titration of salts whose acidic or basic character is too weak to give satisfactory end points with indicators.

Figure 22-3e is typical of the titration of a mixture of two acids that differ in degree of dissociation. The conductometric titration of such mixtures frequently leads to more accurate results than those obtained with a potentiometric method.

Precipitation and Complex-Formation Titrations. Figure 22-3f illustrates the conductance changes that occur during the titration of sodium chloride with silver nitrate. The initial additions of reagent in effect cause a substitution of chloride ions by the some-what less mobile nitrate ions of the reagent; a slight decrease in conductance results. After the reaction is complete a rapid increase occurs, owing to the addition of excess silver nitrate.

Conductometric methods based upon precipitation or complex-formation reactions are not so useful as those involving neutralization processes. Conductance changes during these titrations are seldom as large as those observed with acid-base reactions because no other ion approaches the conductance of either the hydrogen or the hydroxide ion. Such factors as slowness of reaction and coprecipitation represent further sources of difficulty with precipitation reactions.

APPLICATIONS OF DIRECT CONDUCTANCE MEASUREMENTS

Direct conductometric measurements suffer from a lack of selectivity, since any charged species contributes to the total conductance of a solution. On the other hand, the high sensitivity of the procedure makes it an important analytical tool for certain applications. As we have noted, an important use of the method has been for estimating the purity of distilled or deionized water. The specific conductance of pure water is only about 5×10^{-8} ohm^{-1} cm^{-1}; traces of an ionic impurity will increase the conductance by an order of magnitude or more.

Conductance measurements are also employed for determining the concentration of solutions containing a single strong electrolyte, such as solutions of the common alkalis or acids. A nearly linear increase in conductance with concentration is observed for solutions containing as much as 20% by weight of solute. Analyses are based upon calibration curves.

Conductance measurements are also widely used to measure the salinity of sea water in oceanographic work.

Finally, conductance measurements yield valuable information about association and dissociation equilibria in aqueous solutions—provided, of course, that one or more of the reacting species is ionic.

OSCILLOMETRY

As mentioned earlier (p. 506), the conduction of ac radio-frequency (10^5 to 10^7 Hz) electricity through a solution is a complex process that involves not only ionic motion but also induced and orientation polarization of the molecules in the medium. As a result of these various conduction mechanisms, the impedance of a solution to the flow of current varies in an intricate way with the dielectric constant, the electrolyte concentration, and the frequency. Interpretation of the relationships among these variables is difficult, and the use of high-frequency currents could hardly be justified were it not for the fact that the experimental measurement does not require direct contact between the electrodes and the solution.

Instruments

As shown in Figure 22-4, a typical cell for high-frequency measurements consists of a glass container designed to fit snugly between a pair of cylindrical metal electrodes. The electrode and cell are then made part of the resonant circuit of a sine-wave oscillator (p. 24); the behavior of this resonant circuit can be interpreted by assuming that the solution in the cell acts like a parallel capacitor and resistance. The capacitive reactance depends upon frequency and the dielectric constant of the solution; the resistive impedance is also frequency-dependent and is related to the number and kinds of ions in the solution as well. At high frequencies, the resistive impedance is generally lower than that measured with currents in the 1000-Hz range.

Instruments are available commercially that are based upon either the change in frequency or the change in oscillator circuit current (at constant frequency) caused by the presence of

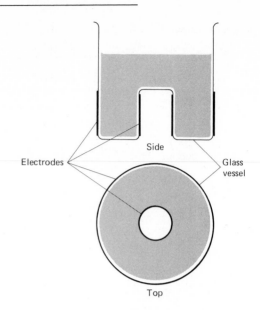

FIGURE 22-4   Sample cell and electrodes for high-frequency measurements.

the sample and cell. In the former, one or more variable capacitors are wired in parallel with the cell and the circuit is tuned to its resonance frequency in the absence of sample. The sample is then introduced and the capacitance that must be subtracted (by means of a calibrated capacitor) to restore resonance is determined. For instruments based on the resistive impedance of the sample, the change in oscillator current resulting from the introduction of sample is measured.

Because of the limited scope of oscillometry, we shall not dwell further on the somewhat complicated instrumentation and complex interrelations associated with this technique; the reader is referred to the treatment of the subject by Reilley[4] and Pungor.[5]

---

[4] C. N. Reilley, in *New Instrumental Methods in Electrochemistry*, ed. P. Delahay. New York: Interscience, 1954, Chapter 15.

[5] E. Pungor, *Oscillometry and Conductometry*. New York: Pergamon Press, 1965.

Applications

Analysis of Binary Mixtures. Oscillometric measurements have been employed for the determination of binary mixtures of nonionic species. Such analyses are based upon the dielectric behavior of the mixtures and require that the dielectric constants for the two components differ significantly. For this application, an instrument based on frequency changes is employed. Empirical calibration curves permit conversion of instrument readings to concentration ratios. Examples of this type of application include the analysis of mixtures of ethanol and nitrobenzene, benzene and chlorobenzene, alcohol and water, and *o*- and *p*-xylene.

Titrations. Oscillometric measurements have been employed to locate the end point in various types of titrations. The methods are similar to those for conventional low-frequency conductometric titrations. The appearance of the titration curves is often quite different, however; both V-shaped curves and inverted V-type end points are encountered. Curvature is also frequently observed; data must, therefore, be collected in the immediate end-point region. For a given reaction, the titration curve depends upon the concentration of the substance titrated as well as upon the parameter measured (that is, the change in frequency or the oscillator current).

The high-frequency method does not appear to yield titration data of higher accuracy than the classical method, and suffers the drawbacks of greater empiricism and of requiring measurements near the end point. Thus, except where the presence of electrodes interferes with a titration, oscillometry offers little advantage.

PROBLEMS   For each of the following problems, derive an approximate titration curve of specific conductance versus volume of reagent, employing the assumption that

  (a) 100.0 ml of $1.00 \times 10^{-3}N$ solution is being titrated with $1.00 \times 10^{-2}N$ reagent;

  (b) the volume change occurring as the titration proceeds can be neglected;

  (c) the equivalent ionic conductances of the various ions are not significantly different from their equivalent ionic conductances at infinite dilution (Table 22-2).

Calculate theoretical specific conductances of the mixtures after the following additions of reagent: 0.00, 2.00, 4.00, 6.00, 8.00, 10.0, 12.0, 14.0, 16.0 ml.

1. $Ba(OH)_2$ with HCl
2. Phenol with KOH (use $K_a = 1.0 \times 10^{-10}$ and $\lambda^0_- = 30$ for $C_6H_5O^-$)
3. Sodium acetate with $HClO_4$ (assume K_a for acetic acid $= 1.8 \times 10^{-5}$)
4. Propanoic acid ($\lambda^0_- = 35.8$ for $CH_3CH_2COO^-$ and $K_a = 1.3 \times 10^{-5}$) with NaOH
5. Propanoic acid with NH_3
6. $Pb(NO_3)_2$ with NaCl
7. $Pb(NO_3)_2$ with LiCl

23

THERMAL METHODS

Thermal methods are based upon the measurement of the dynamic relationship between temperature and some property of a system such as mass, heat of reaction, or volume. Wendlandt lists twelve thermal methods; four of the most important of these will be considered in this chapter,[1] namely, *thermogravimetric methods*, *differential thermal methods*, *differential scanning calorimetry*, and *thermometric titrations*.

THERMOGRAVIMETRIC METHODS

In a thermogravimetric analysis, the mass of sample is recorded continuously as its temperature is increased linearly from ambient to as high as 1200°C. A plot of mass as a function of temperature (a *thermogram*) provides both qualitative and quantitative information.

Apparatus

The apparatus required for a thermogravimetric analysis includes: (1) a sensitive analytical balance; (2) a furnace; (3) a furnace temperature controller and programmer; and (4) a recorder that provides a plot of sample mass as a function of temperature. Often, auxiliary equipment is needed to provide an inert atmosphere for the sample.

The Balance. The Cahn electromagnetic balance is used in the thermogravimetric systems sold by several manufacturers. The sample holder of this balance is suspended from a lever arm attached to a D'Arsonval galvanometer coil (p. 28). Deflection of the beam from its rest position by a change in mass is detected photoelectrically. The resulting photocurrent is then amplified and fed into the galvanometer coil in such a direction as to restore the beam to its original position. The amplified current also determines the position of a recorder pen. Two

balances of this type are offered, one with a maximum load of 2.5 g and the other 100 g; the former is the more common. Typically, mass changes of 200 mg, or some fraction thereof, can be followed with a precision of about $\pm 0.1\%$ relative.

Figure 23-1 is a schematic diagram of the Mettler thermobalance. The sample holder is housed in a furnace that is thermally isolated from the remainder of the balance. A change in the sample mass causes a deflection of the beam, which interposes a light shutter between a lamp and one of two photodiodes. The resulting imbalance in the photodiode current is amplified and fed into coil E, which is situated between the poles of the permanent magnet F. The magnetic field generated by the current in the coil restores the beam to its original position. The amplified photodiode current also determines the position of the pen of a recorder. The Mettler instrument has several weight ranges (1, 10, 100, and 1000 mg) and has a reproducibility of ± 10 μg.

Furnaces. The furnace of a thermogravimetric apparatus is generally programmed to increase the temperature linearly at predeter-

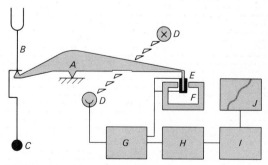

FIGURE 23-1 Components of a thermal balance. *A*, Beam; *B*, sample cup and holder; *C*, counterweight; *D*, lamp and photodiodes; *E*, coil; *F*, magnet; *G*, control amplifier; *H*, tare calculator; *I*, amplifier; and *J*, recorder. (Courtesy Mettler Instrument Corp., Hightstown, New Jersey.)

---

[1] For a thorough treatment of thermal methods, see: W. W. Wendlandt, *Thermal Methods of Analysis*, 2d ed. New York: Wiley, 1974.

mined rates (typically, from 0.5 to 25°C/min). The temperature range for most instruments is from ambient to 1200°C. Temperatures are determined by a thermocouple located as close as possible to the sample. Insulation and cooling of the exterior of the furnace is required to avoid heat transfer to the balance.

Applications

Figure 23-2 is a recorded thermogram obtained by increasing the temperature of pure $CaC_2O_4 \cdot H_2O$ at a rate of 5°C/min. The clearly defined horizontal regions correspond to temperature ranges in which the indicated calcium compounds are stable. This figure illustrates one of the important applications of thermogravimetry, namely, that of defining thermal conditions necessary to produce a pure weighing form for the gravimetric determination of a species.

Figure 23-3b illustrates an application of thermogravimetry to the quantitative analysis of a mixture of calcium, strontium, and barium. The three ions are first precipitated as the monohydrated oxalates. The mass in the temperature range between 250 and 260°C is that of the three anhydrous compounds, CaC_2O_4,

SrC_2O_4, and BaC_2O_4, while the mass between about 560 and 620°C corresponds to the weight of the three carbonates. The weight change in the next two steps results from the loss of carbon dioxide, as first CaO and then SrO are formed. Clearly, sufficient data are available in the thermogram to calculate the weight of each of the three elements present in the sample.

Figure 23-3a is the derivative of the thermogram shown in (b). Many modern instruments are equipped with electronic circuitry to provide such a curve as well as the thermogram itself. The derivative curve may reveal information that is not detectable in the ordinary thermogram. For example, the three peaks at 140, 180, and 205°C suggest that the three hydrates lose moisture at different temperatures. However, all appear to lose carbon monoxide simultaneously and thus yield a single sharp peak at 450°C.

Perhaps the most important applications of thermogravimetric methods are found in the study of polymers. Thermograms provide information about decomposition mechanisms for various polymeric preparations. In addition, the decomposition patterns are characteristic for each kind of polymer and can be used for identification purposes.

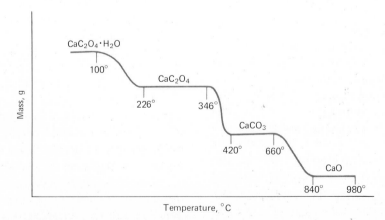

FIGURE 23-2 A thermogram for decomposition of $CaC_2O_4 \cdot H_2O$.

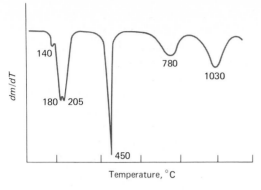

(a) Differential thermogram

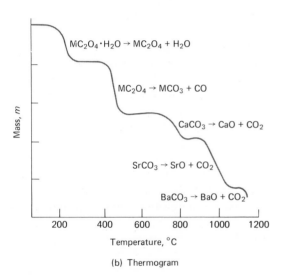

(b) Thermogram

FIGURE 23-3   Decomposition of $CaC_2O_4 \cdot H_2O$, $SrC_2O_4 \cdot H_2O$, and $BaC_2O_4 \cdot H_2O$.

DIFFERENTIAL THERMAL ANALYSIS AND DIFFERENTIAL SCANNING CALORIMETRY

In *differential thermal analysis*, the heat absorbed or emitted by a chemical system is observed by measuring the temperature difference between that system and an inert reference compound (often alumina, silicon carbide, or glass beads) as the temperatures of both are increased at a constant rate.[2] In *differential scanning calorimetry*, the sample and a reference substance are also subjected to a continuously increasing temperature; here, however, heat is added to the sample or to the reference as necessary to maintain the two at identical temperatures. The added heat, which is recorded, compensates for that lost or gained as a consequence of endothermic or exothermic reactions occurring in the sample.

General Principles

Figure 23-4 is a differential thermogram obtained by heating calcium oxalate monohydrate in a flowing stream of air. The two minima indicate that the sample becomes cooler than the reference material as a consequence of the heat absorbed by two endothermic processes; equations for these decomposition reactions are shown below the minima. The single maximum indicates that the reaction to give calcium carbonate and carbon dioxide is exothermic. It is noteworthy that when the differential thermogram is obtained in an inert atmosphere, all three reactions are endothermic, and the maximum is replaced by a minimum; here, the reaction product from the decomposition of calcium oxalate is carbon monoxide rather than carbon dioxide (see Figure 23-2).

Sources of Differential Thermogram Peaks. The maxima and minima, such as those appearing in Figure 23-4, are termed peaks. Those appearing above zero on the ordinate scale are the consequence of exothermic processes; those below zero are for endothermic ones. These heat changes may be the result of physical or chemical phenomena. Physical processes that are endothermic include fusion, vaporization, sublimation, absorption, and desorption. Adsorption is generally an exothermic physical

---

[2] For a concise review of this method and its applications, see: M. I. Pope and M. D. Judd, *Differential Thermal Analysis*. Philadelphia: Heyden, 1977.

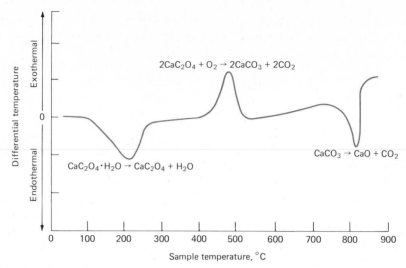

FIGURE 23-4   Differential thermogram of $CaC_2O_4 \cdot H_2O$ in the presence of O_2; the rate of temperature increase was 8°C/min. (From *Handbook of Analytical Chemistry*, ed. L. Meites. New York: McGraw-Hill, 1963, p. **8**–14. With permission.)

change, while crystalline transitions may be either exothermic or endothermic.

Chemical reactions also produce differential peaks; both exothermic and endothermic processes are, of course, possible (for example, see Figure 23-4).

Peak Areas. The peak areas for differential thermograms depend upon the mass of the sample m, the heat or enthalpy ΔH of the chemical or physical process, and certain geometric and heat conductivity factors. These variables are related by the equation[3]

$$A = -\frac{Gm\,\Delta H}{k} = -k'm\,\Delta H \quad (23\text{-}1)$$

where A is the peak area ($\Delta T \times$ time), G is a calibration factor that depends upon the sample geometry, and k is a constant related to the thermal conductivity of the sample. The enthalpy ΔH is given a negative sign for

an exothermic reaction and a positive one for an endothermic process.

For a given species, k' remains constant provided that a number of variables such as heating rate, particle size of the sample, and placement of the thermocouples are closely controlled. Under these circumstances, Equation 23-1 can be employed to calculate the mass of the analyte from peak areas; here, $k'\,\Delta H$ can be determined by calibration. Alternatively, Equation 23-1 permits the determination of ΔH for species when k' and m have been measured.

The thermograms obtained by differential scanning calorimetry are similar in appearance to differential thermograms. In this instance, however, the constant k' in Equation 23-1 becomes independent of the temperature at which the reaction occurs.

Apparatus

The furnace, heating programs, and recording devices used for differential methods are simi-

---

[3] See: W. W. Wendlandt, *Thermal Methods of Analysis*, 2d ed. New York: Wiley, 1974, p. 172.

lar to those employed in thermogravimetry. Indeed, several commercial instruments are designed to permit all three types of thermal analysis.

Figure 23-5 shows a commercial differential thermal-analyzer system. Weighed quantities of the sample and reference material are held in the small pans labeled S and R. The thermocouple on the right (labeled control TC) controls the rate at which the furnace must be heated in order to provide a linear temperature increase. The sample and reference thermocouples are connected in series. Any current due to a temperature difference between the two is amplified and used to determine the position of a recorder pen. With the switch in position T_S, the sample thermocouple is connected not only to the reference thermocouple but also to a reference junction,

which may be at room or ice-bath temperature. The output of this circuit provides a measure of the sample temperature at any instant.

Generally, the sample and reference chamber in a differential thermal apparatus is designed to permit the circulation of inert or reactive gases. Some systems also have the capability for operation at high or low pressures.

For differential scanning calorimetry, individual heaters, located as close as possible to the sample and reference vessels, are provided. When the thermocouples indicate a temperature difference, heat is added to the cooler of the two until temperature equality is restored. The rate of heating required to keep the temperatures equal is recorded as a function of sample temperature. The ordinate of the differential thermogram can then be expressed in units of calories or millicalories per second.

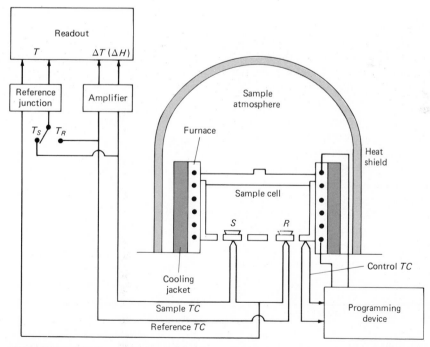

FIGURE 23-5   Schematic diagram of the Fisher Series 300 differential thermal analyzer system. (Courtesy of Fisher Scientific Co., Pittsburgh, Pennsylvania.)

Differential scanning calorimetry thermograms are similar in appearance to Figure 23-4.

Applications

Differential thermal methods find widespread use in determining the composition of naturally occurring and manufactured products. The number of applications is impressive and can be appreciated by an examination of a two-volume monograph[4] devoted largely to uses of the two methods. A few illustrative applications follow.

Inorganic Substances. Differential thermal measurements have been widely used for studies involving the thermal behavior of such inorganic compounds as silicates, ferrites, clays, oxides, ceramics, and glasses. Information is provided about such processes as fusion, desolvation, dehydration, oxidation, reduction, adsorption, degradation, and solid-state reactions. One of the most important applications is to the generation of phase diagrams and the study of phase transitions. An example is shown in Figure 23-6, which is a differential thermogram for pure sulfur. Here, the peak at 113°C corresponds to a solid-phase change from the rhombic to the monoclinic form, while the peak at 124°C corresponds to the melting point of the element. Liquid sulfur is known to exist in at least three forms, and the peak at 179°C apparently involves a transition among these. The peak at 446°C corresponds to the boiling point of sulfur.

Organic Compounds. The differential thermal method provides a simple and accurate way of determining the melting, boiling, and decomposition points for organic compounds. Generally, the data appear to be more consistent and reproducible than those obtained with a hot stage, oil bath, or capillary tube. Figure 23-7*A* shows thermograms for benzoic

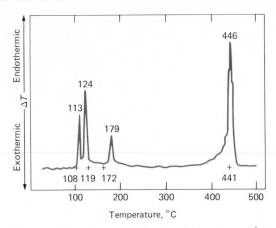

FIGURE 23-6 Differential thermogram for sulfur. [Reprinted with permission from J. Chiu, *Anal. Chem.*, **35**, 933 (1963). Copyright by The American Chemical Society.]

acid at atmospheric pressure and at 200 psi. The first peak corresponds to the melting point and the second to the boiling point of the acid.

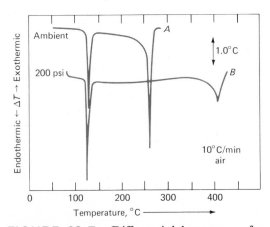

FIGURE 23-7 Differential thermogram for benzoic acid. Curve *A*: at atmospheric pressure; Curve *B*: at 200 lbs/in². [From P. F. Levy, G. Nieuweboer, and L. C. Semanski, *Thermochim. Acta*, **1**, 429 (1970). With permission.]

---

[4] *Differential Thermal Analysis*, ed. R. C. Mackenzie. New York: Academic Press, 1970, vols. 1 and 2.

Card indexes, useful for identification of organic compounds by differential thermal methods, are available from several commercial sources. The Sadtler Research Laboratories collection[5] includes thermograms for 1000 pure organic compounds, 450 commercial compounds, 150 pharmaceutical and steroidal substances, and 360 pure inorganic compounds.

Polymers. Differential thermal methods have been widely applied to the study and characterization of polymeric materials. Figure 23-8 is an idealized thermogram that illustrates the various types of transitions that may be encountered during heating of a polymer. Figure 23-9 is a thermogram of a physical mixture of seven commercial polymers. Each peak corresponds to the characteristic melting point of one of the components. Polytetrafluoroethylene (PTFE) has an additional low-temperature peak which arises from a crystal-

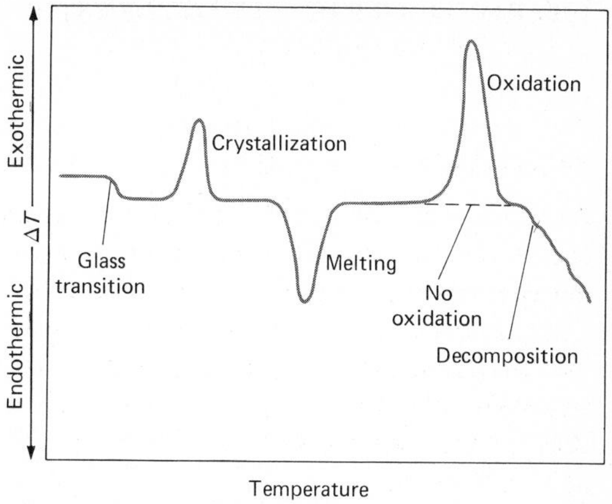

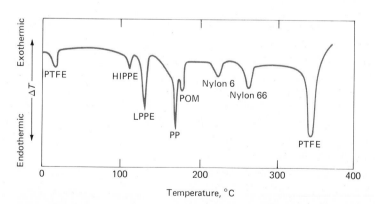

FIGURE 23-8 Schematic differential thermogram showing types of changes encountered with polymeric materials. [From R. M. Schulken, Jr., R. E. Boy, Jr., and R. H. Cox, *J. Polymer Sci.*, Part C, **6**, 1717 (1964). Reprinted by permission of John Wiley & Sons, Inc.]

[5] Sadtler Research Laboratories, Philadelphia, PA 19104.

line transition. Clearly, differential thermal methods can be useful for qualitative analysis of polymer mixtures.

FIGURE 23-9 Differential thermogram for a mixture of seven polymers. PTFE = polytetrafluoroethylene; HIPPE = high-pressure polyethylene; LPPE = low-pressure polyethylene; PP = polypropylene; POM = polyoxymethylene. [From J. Chiu, *DuPont Thermogram*, **2**, (3), 9 (1965). With permission.]

THERMOMETRIC TITRATIONS

The end point in a thermometric titration is obtained from a plot of the temperature of the solution being titrated as a function of the volume of standard reagent added.[6]

---

[6] For more details on thermometric titrations, see: J. Jordan, in *Treatise on Analytical Chemistry*, eds. I. M. Kolthoff and P. J. Elving. New York: Interscience, 1968, Part I, vol. 8, Chapter 91; and H. V. Tyrell and A. E. Beezer, *Thermometric Titrimetry*. London: Chapman and Hall, 1968.

Principles of Thermometric Titration

The temperature changes observed during a thermometric titration are the consequence of the heat evolved or absorbed by the reaction between the analyte and the reagent. The heat or enthalpy of a reaction ΔH is given by the familiar thermodynamic expression

$$\Delta H = \Delta G + T\,\Delta S$$

where T is the temperature, ΔG is the free energy change, and ΔS is the entropy change for the reaction.

Most titrimetric end points, such as the potentiometric ones discussed in Chapter 19, re-

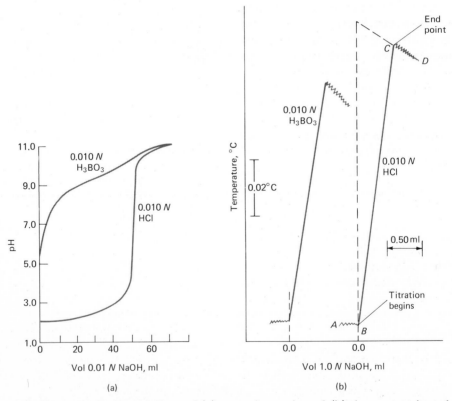

FIGURE 23-10 Comparison of (a) potentiometric and (b) thermometric end points. In each case, 50.0 ml of acid were titrated. [Thermometric titration curves from J. Jordan and W. H. Dumbaugh, Jr., *Anal. Chem.*, **31**, 212 (1959). Reprinted with permission. Copyright by the American Chemical Society.]

quire that ΔG in the foregoing equation be a large negative number; only then will the equilibrium between the analyte and reagent lie sufficiently far to the right. This condition is necessary to make the potential changes in the equivalence-point region large enough for accurate location of chemical equivalence.

In contrast, the success of a thermometric titration depends not only upon the magnitude of ΔG but of $T\,\Delta S$ as well. Thus, if $T\,\Delta S$ is a large negative number, successful end points may still be realized even though ΔG is zero or even positive. A classic example of this situation is illustrated by a comparison of the potentiometric and thermometric end points for boric and hydrochloric acids shown in Figure 23-10. For boric acid, ΔG is so small that neutralization is relatively incomplete at chemical equivalence, and no detectable potentiometric end point is observed. On the other hand, ΔH for the neutralization of boric acid is about -10.2 kcal/mol compared with -13.5 for hydrochloric. As a consequence, sharp thermometric end points are realized for both acids (Figure 23-10b).

The change in temperature ΔT during a thermometric titration is given by

$$\Delta T = -\frac{n\,\Delta H}{k}$$

where n is the number of moles of reactant and k is the heat capacity of the system. Thus, the total change in temperature is directly proportional to the number of moles of analyte.

Apparatus

Reproducible thermometric titration curves require the use of an automatic instrument. Figure 23-11 shows a typical apparatus.

Reagent Delivery System. Reagent is added by means of a screw-driven syringe powered by a motor that is synchronized with the paper drive of the recorder. Generally, the reagent is

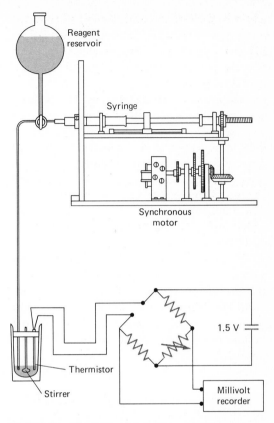

FIGURE 23-11 Schematic diagram of an apparatus for thermometric titrations. (Adapted from J. Jordan, in *Treatise on Analytical Chemistry*, eds. I. M. Kolthoff and P. J. Elving. New York: Wiley-Interscience, 1968, Part I, Volume 8, p. 5201. With permission.)

50 to 100 times more concentrated than the sample, so that titrant volumes are, at the most, a milliliter or two. Under these circumstances, corrections for dilution or temperature difference between the analyte solution and the reagent are unnecessary.

Titration Vessel. Titrations must be carried out under conditions that approach adiabatic; that is, without gain or loss of heat from the surroundings. Thus, the titration vessel is generally a Dewar flask or a beaker surrounded by a thick layer of styrofoam insulation. Efficient stirring must be provided. In addition, the titration must be completed in a brief period (less than 5 min).

Temperature Measurements. Thermistors are generally employed as temperature sensors for thermometric titrations because of their large temperature coefficients (about 10 times greater than a thermocouple), their small size, and their rapid response to temperature changes. A thermistor is a sintered metal oxide semiconductor that has, in contrast to other temperature sensors, a negative temperature coefficient of resistance.

As shown in Figure 23-11, changes in resistance appear as a voltage difference across the bridge circuit; the difference is then recorded on a millivolt recorder. For a thermistor having a resistance of about 2.0 kΩ, the output voltage would correspond to about 0.16 mV per 0.01°C.

Applications

Figure 23-10b reveals that a typical thermometric titration curve is made up of three parts. The region from A to B gives the temperature of the system before addition of reagent. Theoretically, the slope here should be but seldom is zero because of small gains or losses of heat due to stirring and imperfections in the insulation. At point B, reagent addition is begun at a controlled rate of several microliters per minute. Point C corresponds to the end point in the titration. Ordinarily, conditions are arranged so that point D is reached in less than 5 min in order to avoid excess heat leakage to the surroundings. Beyond C, a straight line is observed. Its slope may be negative or positive depending upon whether the dilution process is exothermic or endothermic.

Table 23-1 shows some typical applications of thermometric titrations.

TABLE 23-1 SOME TYPICAL APPLICATIONS OF THERMOMETRIC TITRATIONS[a]

Analyte	Titrant	Minimum Titratable Concentration, M	Precision, Relative %	$\Delta H°$ kcal/mol
Acids, $K_a \geq 10^{-10}$	Strong base	0.005	1	-10 to -15
Bases, $K_b \geq 10^{-10}$	Strong acid	0.005	1	—
Divalent cations	EDTA	0.001 to 0.01	0.1 to 1	-13 to $+5$
Ag^+	HCl	0.1	0.3	—
Ca^{2+}	$(NH_4)_2C_2O_4$	0.01	1	-6.1
Fe^{2+}	Ce^{4+}	0.001	1	-24
	$Cr_2O_7^{2-}$	0.006	0.5	-27
	MnO_4^-	0.003	1	-28
$Fe(CN)_6^{4-}$	Ce^{4+}	0.001	1	-10
Ti^{3+}	Ce^{4+}	0.002	2	-30

[a] Data taken from J. Jordan and G. J. Ewing, in *Handbook of Analytical Chemistry*, ed. L. Meites. New York: McGraw-Hill, 1963, pp. **8**-5 to **8**-7; and J. Jordan, in *Treatise on Analytical Chemistry*, eds. I. M. Kolthoff and P. J. Elving. New York: Wiley-Interscience, 1968, Part I, vol. 8, p. 5198.

PROBLEMS

1. Describe what quantity is measured and how the measurement is performed for each of the following techniques: (a) thermogravimetric analysis, (b) differential thermal analysis, (c) differential scanning calorimetry, and (d) thermometric titration.

2. A 562.46 mg sample of a mixture containing the monohydrates of calcium, strontium, and barium oxalate only was subjected to a thermogravimetric analysis. At the plateau near 260°C (Figure 23-3) the weight was 509.80 mg. The weight was reduced to 427.93 mg at the plateau near 600°C. Find the weight percent of strontium in the original mixture.

3. Why are the two low temperature endotherms in Figure 23-7 coincident whereas the high temperature peaks are displaced from each other?

4. The reaction of Ca^{2+} with ethylenediaminetetraacetic acid (EDTA) is exothermic by 5.7 kcal/mol. The reaction of Mg^{2+} with EDTA is endothermic by 5.5 kcal/mol. The stability constants for the EDTA complexes of Ca^{2+} and Mg^{2+} are $10^{11.0}$ and $10^{9.1}$, respectively. Draw a schematic diagram showing the expected thermometric titration curve for the reaction of an equimolar mixture of Ca^{2+} and Mg^{2+} titrated with EDTA. Show how you would locate the two end points.

5. The iron-transport protein, transferrin, found in human blood serum, has a molecular weight of 81,000 and is capable of binding 0, 1, or 2 iron(III) ions. When subjected to differential scanning calorimetry, transferrin (and many other proteins) exhibits endotherms corresponding to thermal denaturation (unfolding) of part or all of the protein. In the absence of iron, a solution of transferrin exhibited two endotherms at 62° and 72°C. When one mole of iron per mole of transferrin was added in the form of ferric nitrilotriacetate at pH 7.5, the thermogram of this solution exhibited endotherms at 76° and 88°C, with only a minor endotherm at 62°C. When a second mole of iron was added, the sample produced just a single endotherm at 89°C. Suggest some possible interpretations of these results in terms of structural equivalence of the sites and the sequence of iron binding to the sites.

24

AN INTRODUCTION TO CHROMATOGRAPHY

The physical and chemical properties upon which analytical methods are based are seldom, if ever, entirely specific. Instead, these properties are far more likely to be shared by numerous species; as a consequence, the elimination of interferences in quantitative analyses is more often the rule than the exception.

The most general method for dealing with an interference involves its physical separation from the analyte. Well-known methods for accomplishing separations include distillation, crystallization, solvent extraction, and chemical or electrolytic precipitation. Without question, however, the most widely used means of eliminating interferences is by *chromatography*, a separation procedure that finds application to all branches of science.

Chromatography was invented and named by the Russian botanist Mikhail Tswett shortly after the turn of the century. He employed the technique to separate various plant pigments such as chlorophylls and xanthophylls by passing a solution of these compounds through a glass column packed with finely divided calcium carbonate. The separated species appeared as colored bands on the column, which accounts for the name he chose for the method.

The applications of chromatography have grown explosively in the last four decades, owing not only to the development of several new types of chromatographic techniques, but also to the growing need by scientists for better methods for separating complex mixtures. The tremendous impact of these methods on science is attested by the 1952 Nobel prize that was awarded to A. J. P. Martin and R. L. M. Synge for their discoveries in the field.

A GENERAL DESCRIPTION OF CHROMATOGRAPHY

Chromatography encompasses a diverse and important group of separation methods that permit the scientist to separate, isolate, and identify closely related components of complex mixtures; many of these separations are impossible by other means.[1]

The term chromatography is difficult to define rigorously, owing to the variety of systems and techniques to which it has been applied. All of these methods, however, employ a *stationary phase* and a *mobile phase*. Components of a mixture are carried through the stationary phase by the flow of the mobile one; separations are based on differences in migration rates among the sample components.

Types of Stationary Phases

In chromatography, the components to be separated must be soluble in the mobile phase. They must also be capable of interacting with the stationary phase either by dissolving in it, being adsorbed by it, or reacting with it chemically. As a consequence, during the separations, the components become distributed between the two phases.

In some applications, the stationary phase is a finely divided solid held in a narrow glass or metal tube. The mobile phase, which may be a liquid or a gas, is then forced through the solid under pressure or allowed to percolate through it by gravity. This type of method is termed *column chromatography*. In *planar chromatography*, the stationary phase may be porous paper or a finely ground solid that has been spread on a glass plate; here, the mobile phase moves through the solid either by capil-

---

[1] General references on chromatography include: E. Heftmann, *Chromatography*, 3d ed. New York: Van Nostrand Reinhold, 1975; B. L. Karger, L. R. Snyder, and C. Horvath, *An Introduction to Separation Science.* New York: Wiley, 1973; J. M. Miller, *Separation Methods in Chemical Analysis.* New York: Wiley, 1975; R. Stock and C. B. F. Rice, *Chromatographic Methods*, 3d ed. London: Chapman & Hall, 1974; and *Chromatographic and Allied Methods*, ed. O. Mikeš. New York: Wiley, 1979.

lary action or under the influence of gravity. The stationary phase can also be an immobilized liquid which is immiscible with the mobile phase. Several procedures are employed to fix the stationary liquid in place. For example, a finely divided solid, coated with a thin layer of liquid, may be held in a glass or metal tube through which the mobile phase flows or percolates. Ordinarily the solid plays no direct part in the separation, functioning only to hold the stationary liquid phase in place by adsorption. Alternatively, the inner walls of a capillary tube can be coated with a thin layer of liquid; a gaseous mobile phase is then caused to flow through the tube. Finally, the stationary liquid phase can be held in place on the fibers of paper or on the surface of finely ground particles held on a glass plate.

Table 24-1 classifies common chromatographic methods according to the nature of the stationary and mobile phases. Most of the theoretical discussions in this chapter will be concerned with partition and gas-liquid chromatography. With suitable modification, these concepts can be adapted to the other type as well.

Linear Chromatography

All chromatographic separations are based upon differences in the extent to which solutes are partitioned between the mobile and the stationary phase. The equilibria involved can be described quantitatively by means of a temperature-dependent constant, the *partition coefficient K*:

$$K = \frac{C_S}{C_M} \qquad (24\text{-}1)$$

where C_S is the analytical concentration of a solute in the stationary phase and C_M is its concentration in the mobile phase. In the ideal case, the partition ratio is constant over a wide range of solute concentrations; that is, C_S is directly proportional to C_M. More often than not, however, nonlinear relationships occur. Some typical distribution curves are shown in Figure 24-1.

The ideal relationship, shown by curve C in this figure, is often approximated by distribution equilibria between two immiscible liquids, provided association or dissociation reactions do not occur in one of the solvents. Where such equilibria do exist, a relationship

TABLE 24-1 CLASSIFICATION OF CHROMATOGRAPHIC SEPARATIONS

Name	Type Mobile Phase	Type Stationary Phase	Method of Fixing the Stationary Phase
Gas-liquid	Gas	Liquid	Adsorbed on a porous solid held in a tube or adsorbed on the inner surface of a capillary tube
Gas-solid	Gas	Solid	Held in a tubular column
Partition	Liquid	Liquid	Adsorbed on a porous solid held in a tubular column
Adsorption	Liquid	Solid	Held in a tubular column
Paper	Liquid	Liquid	Held in the pores of a thick paper
Thin layer	Liquid	Liquid or solid	Finely divided solid held on a glass plate; liquid may be adsorbed on particles
Gel	Liquid	Liquid	Held in the interstices of a polymeric solid
Ion exchange	Liquid	Solid	Finely divided ion-exchange resin held in a tubular column

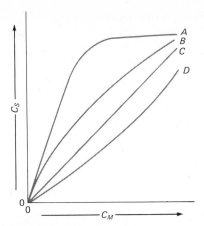

FIGURE 24-1 Typical distribution curves. C_S is the concentration of solute in the stationary phase and C_M its concentration in the mobile phase.

ent of the solute concentration. Curves of this type can be described by the relationship

$$C_S = kC_M^n$$

where k and n are constants.

The assumption of a constant K will frequently be made in the derivation of working equations for fractionation processes. The plots in Figure 24-1 suggest that this approximation will not cause serious errors in limited regions of the lower concentration ranges, where all curves are roughly linear. For wide concentration ranges, however, the derived expressions must be modified to take into account the variation in K with solute concentration. Fortunately, the concentrations in a chromatographic column are ordinarily low. Chromatography carried out under conditions such that K is constant is called *linear chromatography*. We shall treat only this type.

Linear Elution Chromatography

In elution chromatography, a single portion of the sample, dissolved in the mobile phase, is introduced at the head of a column (see Figure 24-2), whereupon the components of the sample distribute themselves between the two phases. Introduction of additional mobile phase (the *eluent*) forces the solvent containing a part of the sample down the column, where further partition between the mobile phase and fresh portions of the stationary phase occurs. Simultaneously, partitioning between the fresh solvent and the stationary phase takes place at the site of the original sample. Continued additions of solvent carry solute molecules down the column in a continuous series of transitions between the mobile and the stationary phases. Because solute movement can only occur in the mobile phase, however, the average *rate* at which a solute migrates *depends upon the fraction of time it spends in that phase*. This fraction is small for solutes with partition ratios that favor retention in the stationary phase and large where

similar to B or D is more likely. For example, if the stationary phase is water and the mobile phase is benzene, a curve of type B is observed for a solute consisting of a weak organic acid. Only the undissociated acid is soluble in benzene. In the aqueous solution, however, appreciable amounts of both the undissociated acid and its conjugate base are present; moreover, the ratio between these species is concentration-dependent. Thus, at low concentrations, a smaller fraction of the total acid is available for distribution between the two solvents, and the partition ratio is greater. If, on the other hand, water represented the mobile phase, a curve such as D would be expected. Curves of type B are also commonly encountered in vapor-liquid equilibria, where the vapor is the mobile phase.

Curve A is a typical *adsorption isotherm* which relates the amount of solute adsorbed on the surface of a solid to the solute concentration in the solution that contacts the solid. At high solute concentrations, all adsorption sites on the solid surface are occupied; the extent of adsorption then becomes independ-

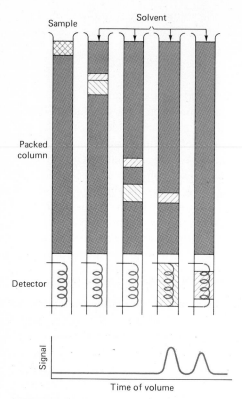

FIGURE 24-2 Schematic diagram of an elution chromatographic separation of a two-component mixture.

retention in the mobile phase is more likely. Ideally, the resulting differences in rates cause the components in a mixture to separate into bands located along the length of the column (see Figures 24-2 and 24-3). Isolation can then be accomplished by passing a sufficient quantity of mobile phase through the column to cause these various bands to pass out the end, where they can be collected. Alternatively, the column packing can be removed and divided into portions containing the various components of the mixture.

The process whereby the solute is washed through the column by addition of fresh solvent is called *elution*. If a detector that re-

sponds to solute concentration is placed at the end of the column and its signal is plotted as a function of time (or of volume of the added mobile phase), a series of symmetric peaks is obtained, as shown in the lower part of Figure 24-2. Such a plot, called a *chromatogram*, is useful for both qualitative and quantitative analysis. The positions of peaks may serve to identify the components of the sample; the areas under the peaks can be related to concentration.

Theories of Elution Chromatography

Figure 24-3 shows concentration profiles for solutes A and B on a chromatographic column at an early and a late state of elution. The partition coefficient for A is the larger of the two; thus, A lags behind during the migration process. It is apparent that movement down the column increases the distance between the two peaks. At the same time, however, broadening of both bands takes place, which lowers the efficiency of the column as a separating device. Zone broadening is unavoidable; fortunately, however, it occurs more slowly than zone separation. Thus, as shown in Figure 24-3, a clean separation of species is possible provided the column is sufficiently long.

It is important to appreciate that a useful theory of chromatography must be able to account not only for the rate at which solutes migrate but also the rate at which the zones broaden during migration, since the cleanness of separation is equally dependent upon the two phenomena. The original theory of chromatography, called the *plate theory*, was able to describe migration rates in quantitative terms. Unfortunately, however, its utility was limited because it failed to describe the effects of the numerous variables responsible for zone broadening. As a consequence, the plate theory has now been supplanted by the *kinetic* or *rate theory*, which is capable of accounting for these variables.

It is useful to consider the plate theory briefly in order to indicate the genesis of two

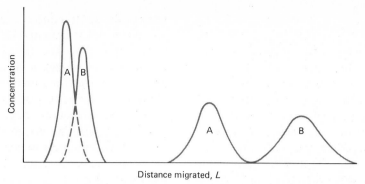

FIGURE 24-3 Concentration profiles for solutes A and B at different points in their migration down a column.

terms employed in the kinetic theory. The plate theory of chromatography, which was originally developed by Martin and Synge,[2] envisages a chromatographic column as being composed of a series of discrete but contiguous, narrow, horizontal layers called *theoretical plates*. At each plate, equilibration of the solute between the mobile and the stationary phase is assumed to take place. Movement of the solute and solvent is then viewed as a series of stepwise transfers from one plate to the next.

The efficiency of a chromatographic column as a separation device improves as the number of equilibrations increases—that is, as the number of theoretical plates increases. Thus, the *number of theoretical plates N* is used as a measure of column efficiency. A second term, the *height equivalent of a theoretical plate H*, also serves this purpose. The relationship between these two parameters is

$$N = \frac{L}{H} \qquad (24\text{-}2)$$

where L is the length of the column packing.

Note that H decreases as the efficiency of a column becomes greater. That is, as H becomes smaller, the number of equilibrations that occur in a given length of column becomes larger. It is important to note that H and N are retained as efficiency parameters in the rate theory and that Equation 24-2 continues to apply. It should be appreciated, however, that *a plate as a physical entity does not exist* in a column. Thus, the plate and the plate height should be viewed as criteria for column efficiency.

THE RATE THEORY OF CHROMATOGRAPHY

The rate theory of chromatography successfully describes the effects of variables which affect the width of an elution band as well as its time of appearance at the end of a column.[3]

[2] A. J. P. Martin and R. L. M. Synge, *Biochem. J.*, **35**, 1358 (1941).

[3] For a detailed presentation of the rate theory, see: J. C. Giddings, *J. Chem. Educ.*, **44**, 704 (1967); J. C. Giddings, *Dynamics of Chromatography*. New York: Marcel Dekker, 1965, Part I; and J. C. Giddings, in *Chromatography*, 3d ed., ed. E. Heftmann. New York: Van Nostrand Reinhold, 1975, Chapter 3.

Zone Shapes

Examination of a typical chromatogram (Figures 24-2 and 24-3) reveals a similarity to normal error or Gaussian curves (see Figure A-1, Appendix 1) that are obtained when replicate values of a measurement are plotted as a function of the frequency of their occurrence. Such plots can be rationalized by assuming that the uncertainty associated with any single measurement is the summation of a much larger number of small, individually undetectable, and random uncertainties, each of which may be positive or negative in sign. The most common occurrence is for these uncertainties to cancel one another, thus leading to the mean value. With less likelihood, the summation may lead to results that are greater or smaller than the mean. The consequence is a symmetric distribution of data around the mean value. In a similar way, the typical Gaussian shape of a chromatogram can be attributed to the additive combination of the random motions of the myriad solute particles in the chromatographic band or zone.

Let us first consider the behavior of an individual solute particle, which, during migration, undergoes many thousands of transfers between the stationary and the mobile phase. The time it spends in either phase after a transfer is highly irregular and depends upon its accidentally gaining sufficient thermal energy from its environment to accomplish a reverse transfer. Thus, in some instances, the residence time in a given phase may be transitory; in others, the period may be relatively long. Recall that the particle can move *only during residence in the mobile phase*; as a result, its migration down the column is also highly irregular. Because of variability in the residence time, the average rate at which individual particles move relative to the mobile phase varies considerably. Certain individuals travel rapidly by virtue of their accidental inclusion in the mobile phase for a majority of the time. Others, in contrast, may lag because they happen to

have been incorporated in the stationary phase for a greater-than-average time. The consequence of these random individual processes is a symmetric spread of velocities around the mean value, which represents the behavior of the average and most common particle.

The breadth increases as the zone moves down the column because more time is allowed for migration to occur. Thus, the zone breadth is directly related to residence time in the column and inversely related to the velocity at which the mobile phase flows.

Standard Deviation as a Measure of Zone Breadth. The breadth of a Gaussian curve is conveniently related to a single parameter called the standard deviation σ (Figure A-1, Appendix 1); approximately 96% of the area under such a curve lies within plus or minus two standard deviations ($\pm 2\sigma$) of its maximum. Thus, a standard deviation derived from a chromatogram serves as a convenient quantitative measure of zone broadening.

It is important to understand that the standard deviation for a chromatographic peak can be expressed in units of time or, alternatively, in distance along the length of the column. Thus, a standard deviation derived from the zone profiles shown in Figure 24-3 would be in terms of length (usually in cm). More commonly, the abscissa of a chromatogram is in units of seconds or minutes (see Figure 24-4), and a standard deviation derived from such a curve would carry time units. It is useful to employ a symbol to indicate which units are being employed; thus, we shall use σ for a standard deviation in units of length and τ when the units are time.

Figure 24-4 illustrates a simple means for approximating τ or σ from an experimentally derived chromatogram. Tangents to the two sides of the Gaussian curve are extended to form a triangle with the abscissa. The intercepts then occur at approximately $\pm 2\tau$ from the maximum; that is, $W = 4\tau$.

Calculation of H and N from Zone Breadth. The broadening of a chromatographic zone is

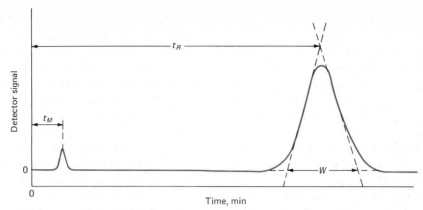

FIGURE 24-4   Determination of the standard deviation τ from a chromatographic peak. Here, $W = 4\tau$; t_R is the retention time for a solute that is retained by the column packing, and t_M is the time for one that is not. Thus, t_M is equal approximately to the time required for a molecule of the mobile phase to pass through the column.

conveniently expressed in terms of the square of the standard deviation (the *variance*) per unit length of column. By definition,

$$H = \frac{\sigma^2}{L} \qquad (24\text{-}3)$$

where H, the plate height in centimeters, is the measure of efficiency and L is the length of column, also in centimeters; note that σ must carry units of length (cm) to be consistent with H and L. Substitution of Equation 24-2 and rearrangement provides an alternative way of describing the separation characteristics of a column. That is,

$$N = \frac{L^2}{\sigma^2} \qquad (24\text{-}4)$$

where N is the number of plates contained in a column of length L. Clearly, column efficiency increases as the number of plates increases and the plate height decreases.

Experimental Evaluation of N and H. Most experimental chromatograms, such as that in Figure 24-4, are obtained with time as the abscissa. Thus, the standard deviation derived from such a curve will be in units of time rather than length. From Figure 24-4, it is seen that t_R, the *retention time*, is the time required after sample injection for the solute peak to appear at the end of the chromatographic column. The average rate at which the solute particles travel is then L/t_R. Thus, the relationship between the standard deviation in units of length and units of time is given by the quotient of the standard deviation in centimeters and the rate of migration in centimeters per second, or

$$\tau = \frac{\sigma}{L/t_R} \qquad (24\text{-}5)$$

Here τ is the standard deviation in units of time and t_R is the time required for 1σ of the zone to emerge from the column. As noted earlier, however, for a Gaussian curve such as that in Figure 24-4, $W = 4\tau$. Substitution of this relationship into (24-5) yields, upon rearrangement,

$$\sigma = \frac{LW}{4t_R}$$

Substitution into Equation 24-3 gives

$$H = \frac{LW^2}{16t_R^2} \qquad (24\text{-}6)$$

To obtain N, we substitute into Equation 24-2 and rearrange, giving

$$N = 16\left(\frac{t_R}{W}\right)^2 \qquad (24\text{-}7)$$

Thus, N can be calculated from the two time measurements t_R and W; to obtain H, the length of the column must also be known.

Sources of Zone Broadening

Chromatographic peaks are generally broadened by three kinetically controlled processes, *eddy diffusion*, *longitudinal diffusion*, and *nonequilibrium mass transfer*. The magnitudes of these effects are determined by such controllable variables as flow rate, particle size of packing, diffusion rates, and thickness of the stationary phase. A number of equations have been developed that relate the efficiency of chromatographic columns to the extent to which these three processes occur.[4] The earliest and simplest of these, which is known as the *van Deemter* equation, was derived for gas-liquid chromatography; it provides an approximate relationship between the flow rate u and the plate height.

$$H = A + \frac{B}{u} + Cu \qquad (24\text{-}8)$$

Here, the quantity A is associated with eddy diffusion, B with longitudinal diffusion, and C with nonequilibrium mass transfer.

Eddy Diffusion. Zone broadening from eddy diffusion arises from the multitude of pathways by which a molecule can find its way through a packed column. As shown in Figure 24-5,

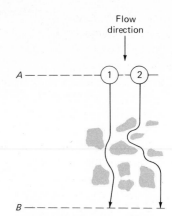

FIGURE 24-5 Typical pathways of two solute molecules during elution. Note that distance traveled by molecule 2 is greater than that traveled by molecule 1. Thus, molecule 2 would arrive at B later than molecule 1.

the lengths of these pathways differ; thus, the residence times in the column for molecules of the same species are also variable. Solute molecules then reach the end of the column over a time interval that tends to broaden the elution band.

The quantity A in Equation 24-8 describes the effect of eddy diffusion and can be related to particle size, geometry, and tightness of packing of the stationary phase. As a first approximation, A is independent of flow rate, and its value is given by

$$A = 2\lambda d_R$$

where d_R is the average particle diameter and λ is the packing factor. Values of λ are determined by the range of particle sizes in the stationary phase and how they are packed. A typical value for 20- to 40-mesh particles is about 1; for 200- to 400-mesh solids, λ has a value of about 8. Band broadening due to eddy diffusion is minimized by careful packing of a column with small spherical particles

---

[4] For a summary of these various equations, see: J. M. Miller, *Separation Methods in Chemical Analysis*. New York: Wiley, 1975, pp. 120–126.

possessing a limited range of sizes. Particular care is needed to avoid open channels. With carefully packed columns, broadening due to eddy diffusion is believed to be minimal.

Longitudinal Diffusion. Longitudinal diffusion results from the tendency of molecules to migrate from the concentrated center part of a band toward more dilute regions on either side. This type of diffusion, which can occur in both the mobile and the stationary phase, causes further band broadening. Longitudinal diffusion is most important where the mobile phase is a gas, because diffusion rates in the gas phase are several orders of magnitude greater than those in liquids. The amount of diffusion increases with time; thus, the extent of broadening increases as the flow rate decreases.

The second term (the longitudinal diffusion term) in Equation 24-8 is seen to be inversely proportional to flow rate. The constant B is related to the diffusion coefficient D_M of the solute in the mobile phase by the equation

$$B = 2\psi D_M$$

where ψ is the obstruction factor, a measure of the hindrance to free molecular diffusion caused by the presence of the particles of the stationary phase; typical values for ψ range from 0.5 to 0.9, depending upon the type of packing. Broadening due to longitudinal diffusion can be reduced by decreasing the temperature (thus reducing D_M) and increasing the velocity of flow.

Nonequilibrium Mass Transfer. Chromatographic bands are also broadened because the flow of the mobile phase is ordinarily so rapid that true equilibrium between phases cannot be realized. For example, at the front of the zone, where the mobile phase encounters fresh stationary phase, equilibrium is not instantly achieved, and solute is therefore carried somewhat farther down the column than would be expected under true equilibrium conditions. Similarly, at the end of the zone, solutes in the stationary phase encounter fresh mobile phases. Again, the rate of transfer of solute molecules is not instantaneous; thus, the tail of the zone is more drawn out than it would be if time existed for equilibration. The net effect is a broadening at both ends of the solute band.

The effects of nonequilibrium mass transfer become smaller as the flow rate is decreased (Equation 24-8) because more time is available for equilibrium to be approached. Furthermore, a closer approach to true equilibrium is to be expected if the channels through which the mobile phase flows are narrow so that solute molecules do not have far to diffuse in order to reach the stationary phase. For the same reason, the layers of immobilized liquid on a stationary phase should be as thin as possible.

The last term in Equation 24-8, which is of considerable importance at the high flow rates of a gaseous mobile phase, describes the effect of nonequilibrium on band broadening. The value for C is made up of two terms, the first having to do with the rate of mass transfer in the stationary phase and the second with the rate in the mobile phase. Thus,

$$C = \frac{qR(1 - R)d_f^2}{D_S} + \frac{\omega d_R^2}{D_M}$$

where d_f is the thickness of the film or stationary particle having a diameter of d_R. The quantity R is the retention ratio and is equal to t_M/t_R (see Figure 24-4). The terms D_S and D_M are diffusion coefficients for the solutes in the stationary and mobile phases, respectively. The quantities q and ω are constants whose values are determined by the nature and configuration of the packing. The latter has a value of about 1.3. For a uniform film of liquid on a solid support, a typical value of q is about 0.7; for a liquid immobilized by paper, q is about 0.5. The most important of the variables affecting C is d_f. A marked improvement in column efficiency is obtained with very thin layers of

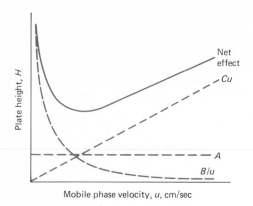

FIGURE 24-6 Effect of variables in Equation 24-8 on plate height.

liquid phase. Equilibrium is also more closely approached at high temperatures and with low solvent viscosities.

Net Effect of the Zone Broadening Processes. Figure 24-6 shows the contribution of each term in the van Deemter equation as a function of mobile-phase velocity (broken lines) as well as their net effect (solid line) on H. Clearly, the optimum efficiency is realized at a flow rate corresponding to the minimum in the solid curve. From experimental curves of this type, A, B, and C are readily evaluated for any column. Such data provide hints for improving the performance of a given type of packing.

The van Deemter equation provides only a first approximation of plate height, and several modifications have been developed that give a more precise description of the variables affecting column efficiency (see footnote 3, p. 671).

SEPARATIONS ON COLUMNS

The discussion thus far has focused upon the efficiency of a chromatographic column—that is, how many plates it contains. We must now give consideration to the relationship between the number of plates in a column and the time required for any given separation.

Rates of Solute Migration

For the chromatogram shown in Figure 24-4, zero on the time axis corresponds to the moment the sample was injected onto the column and elution was started. The peak arising at t_M is for a species (often air) that is *not* retained by the column packing; its rate of motion will be the same as the average rate for the molecules of the mobile phase. The *retention time* t_R for the solute responsible for the second peak is the time required for that peak to reach the detector at the end of a column.

The average rate of migration $\bar{v}$ of the solute is given by

$$\bar{v} = \frac{L}{t_R} \qquad (24\text{-}9)$$

Similarly, the average rate of movement u of molecules of the mobile phase will be

$$u = \frac{L}{t_M} \qquad (24\text{-}10)$$

The rate of migration of the solute can also be expressed as a fraction of the velocity of the mobile phase. That is,

$$\bar{v} = u\left(\begin{array}{c}\text{fraction of time the solute} \\ \text{spends in the mobile phase}\end{array}\right)$$

This fraction, however, equals the average number of moles of the solute in the mobile phase at any instant compared with the total number of moles in the column. That is,

$$\bar{v} = u \times \frac{\text{no. moles solute in mobile phase}}{\text{total number moles solute}}$$

or

$$\bar{v} = u \times$$

$$\frac{C_M V_M}{C_M V_M + C_S V_S} = u\left(\frac{1}{1 + C_S V_S / C_M V_M}\right)$$

where C_M and C_S are the concentrations of the solute in the mobile and stationary phases respectively; similarly, V_M and V_S are the total volumes of the two phases contained in the column.

Substitution of Equation 24-1 into this expression gives

$$\bar{v} = u\left(\frac{1}{1 + KV_S/V_M}\right)$$

or

$$\bar{v} = u\left(\frac{1}{1 + k'}\right) \qquad (24\text{-}11)$$

where k', the *capacity factor*, is related to the partition coefficient for the solute. That is,

$$k' = \frac{KV_S}{V_M} \qquad (24\text{-}12)$$

Substitution of Equations 24-9 and 24-10 into 24-11 yields, upon rearrangement,

$$t_R = t_M(1 + k') = t_M\left(1 + \frac{KV_S}{V_M}\right) \quad (24\text{-}13)$$

Note that the retention time for a solute increases with the partition coefficient of the solute and the volume of the stationary phase with respect to the mobile.

Rearrangement of Equation 24-13 provides an expression that permits calculation of the capacity factor for a solute from the experimental parameters, t_R and t_M. Thus,

$$k' = \frac{t_R - t_M}{t_M} \qquad (24\text{-}14)$$

Column Resolution

The ability of a column to resolve two solutes is of prime interest in chromatography. A quantitative term for expressing the resolving power of a column is the column *resolution* R_s, which is defined as

$$R_s = \frac{2\,\Delta Z}{W_X + W_Y} = \frac{2[(t_R)_Y - (t_R)_X]}{W_X + W_Y} \quad (24\text{-}15)$$

where W_X and W_Y are the widths of the peaks (in units of time) at their bases and ΔZ is the difference in time of their arrival at the detector (see Figure 24-7). If W is taken as the average width of the two peaks, Equation 24-15 reduces to

$$R_s = \frac{\Delta Z}{W}$$

For peaks that are close together (as they will be when there is interest in R_s), W_X and W_Y will often be sufficiently alike that only one needs to be measured.

Figure 24-7 indicates that a resolution (R_s) of 1.5 gives an essentially complete separation of X and Y, whereas a resolution of 0.75 does not. At a resolution of 1.0, zone X contains about 4% of Y, and conversely; at a resolution of 1.5, the overlap is about 0.3%. For the same kind of packing, the resolution can be improved by lengthening the column and thus increasing the number of theoretical plates.

Relationship Between R_s and Column Properties. It is useful to be able to relate R_s to certain properties of a chromatographic column. The following derivation shows how this is possible.

EXAMPLE

Derive a relationship between R_s and $(t_R)_X$, and $(t_R)_Y$ where the latter two terms are the measured retention times shown in Figure 24-7. As we have mentioned, usually

$$W = W_Y \cong W_X$$

Thus, we may write Equation 24-15 in the form

$$R_s = \frac{(t_R)_Y - (t_R)_X}{W_Y}$$

Equation 24-7 permits expression of W_Y in terms of $(t_R)_Y$ and N. Thus,

$$R_s = \frac{(t_R)_Y - (t_R)_X}{(t_R)_Y} \times \frac{\sqrt{N}}{4}$$

Substitution of Equation 24-13 permits ex-

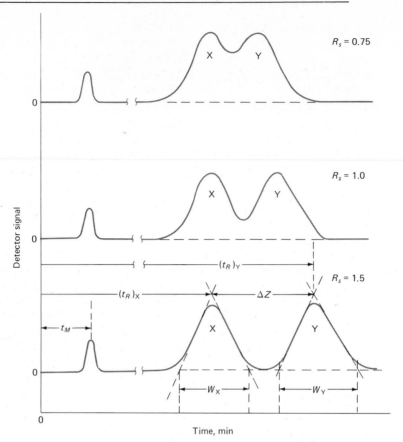

FIGURE 24-7 Separations at three resolutions. Here, $R_s = 2\Delta Z/(W_X + W_Y)$.

pression of R_s in terms of capacity factors for X and Y. Thus,

$$R_s = \frac{(k'_Y - k'_X)}{(1 + k'_Y)} \times \frac{\sqrt{N}}{4}$$

Let us now define k'_Y/k'_X as the *selectivity factor* α. That is,

$$\alpha = \frac{k'_Y}{k'_X} \qquad (24\text{-}16)$$

Employing this relationship to eliminate k'_X from the preceding equation gives, with rearrangement,

$$R_s = \frac{\sqrt{N}}{4}\left(\frac{\alpha - 1}{\alpha}\right)\left(\frac{k'_Y}{1 + k'_Y}\right) \qquad (24\text{-}17)$$

Often, it is desirable to calculate the number of plates required to achieve a desired resolution. Such an expression is obtained by rearrangement of Equation 24-17, which gives

$$N = 16R_s^2\left(\frac{\alpha}{\alpha - 1}\right)^2\left(\frac{1 + k'_Y}{k'_Y}\right)^2 \qquad (24\text{-}18)$$

Properties of the Selectivity Factor α. Before examining the application of Equations 24-17 and 24-18, it is worthwhile noting the properties of the selectivity factor α. First, it is readily derived from a chromatogram such as that shown in Figure 24-7. This property can be seen by substitution of Equation 24-14 into 24-16, which yields

$$\alpha = \frac{(t_R)_Y - t_M}{(t_R)_X - t_M} \qquad (24\text{-}19)$$

Furthermore, substitution of Equation 24-12 into Equation 24-16 reveals that

$$\alpha = \frac{K_Y}{K_X} \qquad (24\text{-}20)$$

Resolution and Separation Time. One important performance characteristic is absent from Equations 24-17 and 24-18, namely, the time required to complete the separation. In order to include this important variable, Equation 24-9 is employed to obtain the velocity of the slower-moving solute Y. That is,

$$\bar{v}_Y = \frac{L}{(t_R)_Y}$$

Combination of this relationship with Equations 24-11 and 24-2 yields

$$(t_R)_Y = \frac{NH(1 + k_Y')}{u}$$

where $(t_R)_Y$ is the time required to elute Y when the velocity of the mobile phase is u. Substitution of Equation 24-18 into the foregoing expression gives

$$(t_R)_Y = \frac{16R_s^2 H}{u} \left(\frac{\alpha}{\alpha - 1} \right)^2 \frac{(1 + k_Y')^3}{(k_Y')^2} \qquad (24\text{-}21)$$

Optimization of Column Performance

Equations 24-17 and 24-21 are of considerable importance in column chromatography because they serve as guides to the choice of conditions that will lead to a desired degree of resolution with a minimum expenditure of time. An examination of these equations reveals that each is made up of three parts. The first describes the efficiency of the column in terms of $\sqrt{N}$ or H. The second, which is the quotient containing α, is a selectivity term and depends solely on the properties of the two solutes. The third component is the capacity term, which is the quotient containing k_Y'; this term depends on the properties of both the solute and the column.

Column Efficiency. It is evident from Equation 24-17 that resolution is proportional to the square root of the number of plates in the column. Increasing the number of plates to achieve a separation can be expensive in terms of time, however, unless the increase is accomplished by a reduction in H (Equation 24-21) and not by lengthening the column. The earlier section on rate theory described the variables that can be controlled to maximize efficiency or minimize H.

Variation in the Capacity Factor. The capacity factor k_Y' for a solute can frequently be varied over a considerable range by altering the composition of the mobile or the stationary phase. The former is usually easier and is the one that is often changed to optimize a given separation. From Equation 24-12, it is evident that k_Y' can also be increased or decreased by a change in the ratio of the volume of the stationary phase to the mobile. An increase in k_Y' accompanies a reduction in particle size of the stationary support because the consequent increase in surface area results in an increase in V_S.

Increases in k_Y' enhance resolution but at the expense of elution time. In order to understand the two effects, it is convenient to write Equations 24-17 and 24-21 in the form

$$R_s = Q \frac{k_Y'}{1 + k_Y'}$$

and

$$(t_R)_Y = Q' \frac{(1 + k_Y')^3}{(k_Y')^2}$$

where Q and Q' contain the rest of the terms in the equations. Figure 24-8 is a plot of R_s/Q and $(t_R)_Y/Q'$ as a function of k_Y', assuming that Q and Q' remain approximately constant. It is clear that values of k_Y' greater than about 10 are to be avoided because they provide little increase in resolution but markedly increase the time required for separations. The

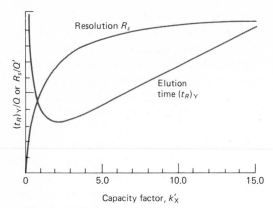

FIGURE 24-8 Effect of capacity factor k'_X on resolution R_s and elution time $(t_R)_Y$. It is assumed that Q and Q' remain constant with variation in k'_Y.

minimum in the elution time curve occurs at values of k'_Y between 2 and 3. Often, then, the optimal value of k'_Y lies in the range of 2 to 5.

Variation in the Selectivity Factor. It is evident from Equation 24-17 that resolution is improved as the selectivity factor becomes larger. When this parameter is greater than 2, a clean separation is generally possible in a minimal time. On the other hand, as α approaches 1, resolution is lost unless a very long column is employed. As shown in Equation 24-21, however, α values that approach 1 increase enormously the time required for separation. For example, an 84-fold increase in time is required for the same resolution when α is equal to 1.01 as compared with 1.1.

EXAMPLE

Substances A and B were found to have retention times of 16.40 and 17.63 min, respectively, on a 30.0-cm column. An unretained species passed through the column in 1.30 min. The peak widths were 1.11 and 1.21 mm. Calculate

(a) the column resolution.

(b) the average number of plates in the column.

(c) the plate height.

(d) the length of column required to achieve a resolution of 1.5.

(e) the time required to elute substance B on the longer column.

Employing Equation 24-15, we find

(a) $R_s = 2(17.63 - 16.40)/(1.11 + 1.21)$
$\quad = 1.06$

(b) Equation 24-7 permits computation of N. Thus,

$$N = 16\left(\frac{16.40}{1.11}\right)^2 \quad \text{and} \quad N = 16\left(\frac{17.63}{1.21}\right)^2$$

$$= 3493 \qquad\qquad\qquad = 3397$$

$$N_{av} = (3493 + 3397)/2 = 3445 = 3.44 \times 10^3$$

(c) $H = L/N = 30.0/3445 = 8.708 \times 10^{-3}$

$$cm = 8.71 \times 10^{-3} \text{ cm}$$

(d) k' and α do not change with increasing N and L. Thus, substituting N_1 and N_2 into Equation 24-18 and dividing one of the resulting equations by the other yields

$$\frac{N_1}{N_2} = \frac{(R_s)_1^2}{(R_s)_2^2}$$

where subscripts 1 and 2 refer to the original and the longer columns, respectively. Substituting the appropriate values for N_1, $(R_s)_1$, and $(R_s)_2$ gives

$$\frac{3445}{N_2} = \frac{(1.06)^2}{(1.5)^2}$$

$$N_2 = 3445 \times 2.25/1.124$$

$$= 6.90 \times 10^3$$

But $L = 6.90 \times 10^3 \times 8.71 \times 10^{-3}$

$$= 60.1 \text{ cm}$$

(e) From Equation 24-21, we may write

$$\frac{(t_R)_1}{(t_R)_2} = \frac{(R_s)_1^2}{(R_s)_2^2} = \frac{17.63}{(t_R)_2} = \frac{(1.06)^2}{(1.5)^2}$$

and

$$(t_R)_2 = 35.3 \text{ min.}$$

SUMMARY OF IMPORTANT RELATIONSHIPS FOR CHROMATOGRAPHY

The number of quantities, terms, and relationships employed in chromatography is large and often confusing. Table 24-2 serves to summarize the most important definitions and equations used in this text.

QUALITATIVE AND QUANTITATIVE ANALYSIS BY CHROMATOGRAPHY

Chromatography has grown to be the premiere method for separating closely related chemical species. In addition, it can be employed for qualitative identification and quantitative determination of separated species. This section considers some of the general characteristics of chromatography as a tool for completion of an analysis.

Qualitative Analysis

A chromatogram provides only a single piece of qualitative information about each species in a sample, namely, its retention time or its position on the stationary phase after a certain elution period. Additional data can, of course, be derived from chromatograms involving different mobile and stationary phases and various elution temperatures. Still, the number of data points for a species obtainable by chromatography is small compared with the number provided by a single IR, NMR, or mass spectrum. Furthermore, spectral abscissa data (λ, δ, or m/e) can be determined with a much higher accuracy than can their chromatographic counterpart (t_R).

The foregoing should not be interpreted to mean that chromatography lacks important qualitative applications. Indeed, it is a widely used tool for identifying the components of mixtures containing a limited number of possible species whose identities are known. For example, the presence or absence of 30 or more amino acids in a protein hydrolysate can be ascertained with a relatively high degree of certainty from their positions after development on a thin layer chromatographic plate. Even here, however, confirmation of identity would require spectral or chemical investigation of the isolated components.

It is important to add that positive spectroscopic identification would ordinarily be impossible on as complex a sample as the foregoing without a preliminary chromatographic separation. Thus, chromatography is often a vital precursor to a qualitative spectroscopic analyses.

Quantitative Analysis

Chromatography owes its precipitous growth during the past two decades in part to its speed, simplicity, relatively low cost, and wide applicability as a separating tool. It is doubtful, however, if its use would have become as widespread had it not been for the fact that it can also provide useful quantitative information about the separated species. It is important, therefore, to discuss some of the general aspects of quantitative chromatography.

Quantitative chromatography is based upon a comparison of either the height or the area of the chromatographic peak of the analyte with that of one or more standards. If conditions are properly controlled, both of these parameters vary linearly with concentration.

Analyses Based on Peak Height. The height of a chromatographic peak is obtained by connecting the base lines on either side of the peak by a straight line and measuring the perpendicular distance from this line to the peak. This measurement can ordinarily be made with reasonably high precision and yields accurate results, provided variations in column conditions do not alter the peak widths during the

TABLE 24-2 SUMMARY OF THE MOST IMPORTANT CHROMATOGRAPHIC QUANTITIES AND RELATIONSHIPS

EXPERIMENTAL QUANTITIES

Name	Symbol	Determined from
Retention time, mobile phase	t_M	Chromatogram (see Figure 24-7)
Retention times, species X and Y	$(t_R)_X, (t_R)_Y$	Chromatogram (see Figure 24-7)
Peak widths, species X and Y	W_X, W_Y	Chromatogram (see Figure 24-7)
Length of column packing	L	Direct measurement
Flow rate	F	Direct measurement
Volume of stationary phase	V_S	Packing preparation data
Concentration of solute in mobile and stationary phases	C_M, C_S	Analysis and preparation data

DERIVED QUANTITIES

Name	Calculation of Derived Quantities	Relationship to Other Quantities
Mobile phase velocity	$u = L/t_M$	
Volume of mobile phase	$V_M = t_M F$	
Capacity factor	$k' = (t_R - t_M)/t_M$	$k' = \dfrac{K V_S}{V_M}$
Partition coefficient	$K = \dfrac{k' V_M}{V_S}$	$K = \dfrac{C_S}{C_M}$
Selectivity factor	$\alpha = \dfrac{(t_R)_Y - t_M}{(t_R)_X - t_M}$	$\alpha = \dfrac{k'_Y}{k'_X} = \dfrac{K_Y}{K_X}$
Resolution	$R_s = \dfrac{2[(t_R)_Y - (t_R)_X]}{W_X + W_Y}$	$R_s = \dfrac{\sqrt{N}}{4}\left(\dfrac{\alpha - 1}{\alpha}\right)\left(\dfrac{k'_Y}{1 + k'_Y}\right)$
Number of plates	$N = 16\left(\dfrac{t_R}{W}\right)^2$	$N = \dfrac{L}{H} = 16 R_s^2 \left(\dfrac{\alpha}{\alpha - 1}\right)^2 \left(\dfrac{1 + k'_Y}{k'_Y}\right)^2$
Retention time	$(t_R)_Y = \dfrac{16 R_s^2 H}{u}\left(\dfrac{\alpha}{\alpha - 1}\right)^2 \dfrac{(1 + k'_Y)^3}{(k'_Y)^2}$	

period required to obtain chromatograms for sample and standards. The variables that must be controlled closely are column temperature, the eluent flow rate, and the rate of sample injection. In addition, care must be taken to avoid overloading the column. The effect of sample injection rate is particularly critical for the early peaks of a chromatogram. Here, relative errors of 5 to 10% due to this cause are not unusual with syringe injection.

Analyses Based on Peak Areas. Peak areas are independent of broadening effects due to the variables mentioned in the previous paragraph. From this standpoint, therefore, areas are a more satisfactory analytical parameter than peak heights. On the other hand, peak heights are more easily measured and for narrow peaks, more accurately determined.

Many modern chromatographic instruments are equipped with ball and disc or electronic integrators (p. 66), which permit precise estimation of peak areas. When such equipment is not available, a manual estimate must be made. A simple method, which works well for symmetric peaks of reasonable widths, is to multiply the height of the peak by its width at one-half the peak height. Other methods involve the use of a planimeter or cutting out the peak and determining its weight relative to the weight of a known area of recorder paper. McNair and Bonelli measured the precision of these various techniques on chromatograms for ten replicate samples.[5] They reported the following standard deviations: electronic integration, 0.44%; ball and disc integration, 1.3%; weight of paper, 1.7%; height times width at one-half height 2.6%; and planimeter, 4.1%.

Calibration with Standards. The most straightforward method for quantitative chromatographic analyses involves the preparation of a series of standard solutions that approximate the composition of the unknown. Chromatograms for the standards are then obtained and peak heights or areas are plotted as a function of concentration. A plot of the data should yield a straight line passing through the origin; analyses are based upon this plot. Frequent restandardization is necessary for highest accuracy.

The most important source of error in analyses by the method just described is usually the uncertainty in the volume of sample; occasionally the rate of injection is also a factor. Ordinarily, samples are small (~ 1 μliter), and the uncertainties associated with injection of a reproducible volume of this size with a microsyringe may amount to several percent relative. The situation is exacerbated in gas-liquid chromatography, where the sample must be injected into a heated sample port; here, evaporation from the needle tip may lead to large variations in the volume injected.

Errors in sample volume can be reduced to perhaps 1 to 2% relative by means of a rotary gas sample valve such as that shown in Figure 24-9. Here, the sample loop ACB in (a) is filled with sample; rotation of the valve by 45 deg then introduces a reproducible volume of sample (the volume originally contained in ACB) into the mobile-phase stream.

The Internal Standard Method. The highest precision for quantitative chromatography is obtained by use of an internal standard because the uncertainties introduced by sample injection are avoided. In this procedure, a carefully measured quantity of an internal standard substance is introduced into each standard and sample, and the ratio of analyte to internal standard peak areas (or heights) serves as the analytical parameter. For this method to be successful, it is necessary that the internal standard peak be well separated from the peaks of all other components of the sample ($R_s > 1.25$); the standard peak should, on the other hand, appear close to the analyte peak. With a suitable internal standard, precisions of 0.5 to 1% relative are reported.

[5] H. M. McNair and E. J. Bonelli, *Basic Gas Chromatography*. Walnut Creek, CA: Varian Aerograph, 1968, p. 158.

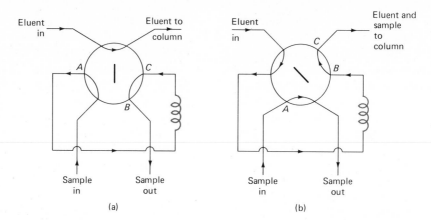

Eluent in

Eluent to column

Sample in

Sample out

(a)

Eluent in

Eluent and sample to column

Sample in

Sample out

(b)

FIGURE 24-9   A rotary sample valve. (a) Valve position for filling sample loop *ACB* and (b) for introduction of sample into column.

The Area Normalization Method. Another approach that avoids the uncertainties associated with sample injection is the area normalization method. Complete elution of all components of the sample is required, a restriction that has limited the procedure to gas-liquid chromatography. In the normalization method, the areas of all eluted peaks are computed; after correcting these areas for differences in the detector response to different compound types, the concentration of the analyte is found from the ratio of its area to the total area of all peaks. The following example illustrates the procedure.

EXAMPLE

The following area data were obtained from a chromatogram of a mixture of butyl alcohols (the detector sensitivity corrections were obtained in separate experiments with known amounts of pure alcohols).

Alcohol	Peak Area, cm^2	Detector Response Factor	Corrected Areas, cm^2
n-butyl	2.74	0.603	1.652
i-butyl	7.61	0.530	4.033
s-butyl	3.19	0.667	2.128
t-butyl	1.66	0.681	1.130
			8.943

Each entry in column 4 is the product of the data in columns 2 and 3. To normalize,

% *n*-butyl alcohol = 1.652 × 100/8.943 = 18.5
% *i*-butyl alcohol = 4.033 × 100/8.943 = 45.1
% *s*-butyl alcohol = 2.128 × 100/8.943 = 23.8
% *t*-butyl alcohol = 1.130 × 100/8.943 = 12.6
                                          ────
                                          100.0

Computerized Chromatography

Most modern instruments for chromatography are equipped with microprocessors for controlling such operating parameters as column temperature, eluent flow rate, sample injection time, and sample temperature. As a consequence, chromatograms can be obtained with little or no human control. The most sophisticated instruments are completely automatic. Here, several dozen samples are contained in a sample turntable. After each analysis, the column is returned automatically to its initial condition, the sample turntable is rotated, a new sample is withdrawn and injected into the column, and the chromatogram is obtained and stored in a computer memory. The data are presented as chromatogram and in tabular form with retention times and relative peak areas printed out. Some instruments also print out the names of possible compounds giving each peak.

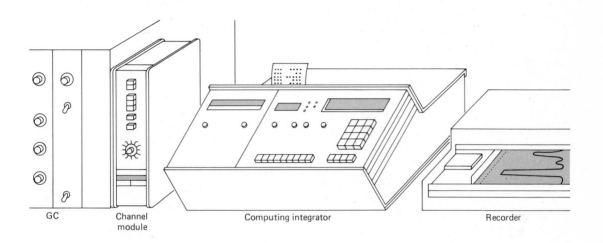

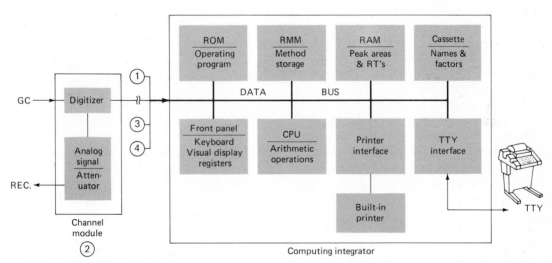

FIGURE 24-10   A computing integrator for chromatography. (From J. M. Gill, *Chromatography.* Greens Farms, CT: International Scientific Communications, Inc., 1974, Series 1, vol. 1, p. 188. With permission.)

Several instrument manufacturers offer small, modular computers that are specifically designed for processing the output signals from chromatographic instruments. Figure 24-10 is a schematic diagram showing the components of one of these units, which is capable of handling data from four chromatographic columns. The analog signal, after digitilization, passes into an ROM (read only memory) operating program, which constantly monitors and operates on the data. Here, peaks are detected and their areas and retention times determined. The arithmetic operations are performed by the CPU. These processed data are then transferred to the RAM (random access memory) module for storage until the end of the analysis. The RMM (read mostly memory) module contains programs for optimizing the data treatment. Instructions as to the choice of programs are entered by means of the front panel keyboard. Each of the three memories consists of a single chip. The contents of the ROM are fixed and cannot be changed. The contents of the RMM are also protected but can be changed under certain circumstances. Data can be readily entered or removed from the random access memory.

After an analysis is complete, the stored data are printed out as shown in Figure 24-11. The first set of data gives the channel or column number and the sample identification. These entries are followed by a coded description of the operating conditions, which were entered at the outset by means of the keyboard. Retention times in seconds and relative areas for peaks then follow.

VIDAR	AUTOLAB	
	SYSTEM - 4	
	4	CH
	2185	ID
	21	XD
RUN	PARAMTRS	
---	------	
	23	T 1
	2567	T 2
	3000	T 3
	10	P W
	150	S S
	1000	M A
TIME	AREA	
---	------	
26	.37	%/
142	64.72	%/
563	1.60	%/
1269	21.12	%⊥
2122	7.39	%/
2968	4.80	%/

FIGURE 24-11 Printer output from the computing integrator shown in Figure 24-10. (From J. M. Gill, *Chromatography*. Green Farms, CT: International Scientific Communications, Inc., 1974, Series 1, vol. 1, p. 188. With permission.)

PROBLEMS

1. The following data apply to a column for partition chromatography:

Length of packing	22.6 cm
Flow rate	0.287 ml/min
V_M	1.26 ml
V_S	0.148 ml

A chromatogram of a mixture of species A, B, C, and D provided the following data:

	Retention Time, min	Width of Peak Base (W), min
Nonretained	4.2	—
A	6.4	0.45
B	14.4	1.07
C	15.4	1.16
D	20.7	1.45

Calculate
(a) the number of plates from each peak.
(b) the mean and the standard deviation for N.
(c) the plate height for the column.

2. From the data in Problem 1, calculate for A, B, C, and D
(a) the capacity factor.
(b) the partition coefficient.

3. From the data in Problem 1 for species B and C, calculate
(a) the resolution.
(b) the selectivity factor α.
(c) the length of column necessary to give a resolution of 1.5.
(d) the time required to separate B and C with a resolution of 1.5.

4. From the data in Problem 1 for species C and D, calculate
(a) the resolution.
(b) the length of column required to give a resolution of 1.5.
(c) the time required to separate C and D with a resolution of 1.5.

5. The following data were obtained by gas-liquid chromatography on a 40-cm packed column:

Compound	t_R, min	W, min
Air	2.5	—
Methylcyclohexane	10.7	1.3
Methylcyclohexene	11.6	1.4
Toluene	14.0	1.8

Calculate
(a) an average number of plates from the data.
(b) the standard deviation for the average in (a).
(c) an average plate height for the column.

6. Referring to Problem 5, calculate the resolution for
(a) methylcyclohexene and methylcyclohexane.
(b) methylcyclohexene and toluene.
(c) methylcyclohexane and toluene.

7. If a resolution of 1.5 was required to resolve methylcyclohexane and methylcyclohexene in Problem 5,
   (a) how many plates would be required?
   (b) how long would the column have to be if the same packing were employed?
   (c) what would be the retention times for the three compounds on the column in Problem 7b?
   (d) what would be the resolution for the last two compounds on the column in Problem 7b?

8. If V_S and V_M for the column in Problem 5 were 20.6 and 64.2 ml, respectively, and a nonretained air peak appeared after 2.5 min, calculate the
   (a) capacity factor for each of the three compounds.
   (b) partition coefficient for each of the three compounds.
   (c) selectivity factor for methylcyclohexane and methylcyclohexene.
   (d) selectivity factor for methylcyclohexene and toluene.

9. A column was found to have the following constants for Equation 24-8 $A = 0.01$ cm, $B = 0.30$ cm^2/s, and $C = 0.015$ s. What is the optimum flow rate and the corresponding minimum plate height?

10. What would be the effect on a chromatographic peak of introducing the sample at too slow a rate?

11. From distribution studies species M and N are known to have partition coefficients between water and hexane of 6.50 and 6.31 ($K = [M]_{H_2O}/[M]_{hex}$). The two species are to be separated by elution with hexane in a column packed with silica gel with water on its surface. The ratio V_S/V_M for the packing is known to be 0.422.
   (a) Calculate the capacity factor ratio for each of the solutes.
   (b) Calculate the selectivity factor.
   (c) How many plates will be needed to provide a resolution of 1.5?
   (d) How long a column is needed if the plate height of the packing is 1.02×10^{-2} cm?
   (e) If a flow rate of 7.10 cm/s is employed, what time will be required to elute each of the species?

12. Repeat the calculations in Problem 11 assuming the two constants have values of 6.50 and 6.11.

25

LIQUID
CHROMATOGRAPHY

Five of the methods listed in Table 24-1 fall in the category of liquid chromatography because a liquid mobile phase is employed in each. Included in this group are partition (or liquid-liquid), adsorption (or liquid-solid), ion-exchange, paper, and thin-layer chromatography. The first three employ a glass or metal tube to hold the stationary phase and are thus further classified as column procedures. The latter two are carried out on a flat surface and are sometimes referred to as plane or planar chromatographic procedures as a consequence.

This chapter treats the five liquid chromatographic methods[1] and, in addition, presents a brief discussion of *electrochromatography*, a procedure in which chromatographic separations are enhanced by application of an electrical field.

COLUMN CHROMATOGRAPHY

The classical liquid chromatographic method employed a glass tube with a diameter of perhaps 10 to 50 mm to hold a 50- to 500-cm column of solid particles of the stationary phase. To assure reasonable flow rates, the particles of the solid were kept larger than 150 to 200 μm in diameter. Ordinarily, the head of liquid above the packing sufficed to force the mobile phase down the column. Flow rates were, at best, a few tenths of a milliliter per minute; thus, separations tended to be time consuming.

Attempts to speed up the classical procedure

by application of vacuum or by pumping were not effective because within the range of conditions that were feasible, plate heights were well beyond the minimum in the curve shown in Figure 24-6 (p. 676). Thus, increased flow rates were accompanied by marked increases in plate heights and decreased efficiencies.

Early in the history of liquid chromatography, it was realized that large increases in column efficiency could be expected to accompany decreases in the particle size of packings. It was not until the late 1960s, however, that the technology of producing and using packings with particle diameters as small as 10 μm was developed. This technology required sophisticated instruments that contrasted markedly with the simple devices employed in classical liquid chromatography. The name *high-performance liquid chromatography* (HPLC) is often employed to distinguish these newer procedures from the classical methods, which still find considerable use for preparative purposes.

High-Performance Liquid Chromatography

Figure 25-1 shows plots of experimental liquid chromatographic data relating plate height to flow rate and particle diameter of packing materials. In none of the plots is the minimum shown by Figure 24-6 reached; generally, in liquid chromatography, this minimum is observed only at prohibitively low flow rates. Furthermore, Equation 24-8 does not adequately describe the relationship between efficiency and velocity of the mobile phase; here, a considerably more complex equation, developed by Giddings, must be employed.[2]

It is apparent from Figure 25-1 that separation efficiency is greatly enhanced at low particle diameters. Reasonable flow rates with

---

[1] A large number of books on liquid chromatography are available. Among these are: B. L. Karger, L. R. Snyder, and C. Horvath, *An Introduction to Separation Science.* New York: Wiley, 1973; *Practical High-Performance Liquid Chromatography*, ed. C. F. Simpson. London: Heyden, 1977; and R. P. W. Scott, *Contemporary Liquid Chromatography*, in *Techniques of Chemistry*, ed. A. Weissberger. New York: Wiley-Interscience, 1976, vol. XI.

[2] J. C. Giddings, *Anal. Chem.*, **35**, 1338 (1963).

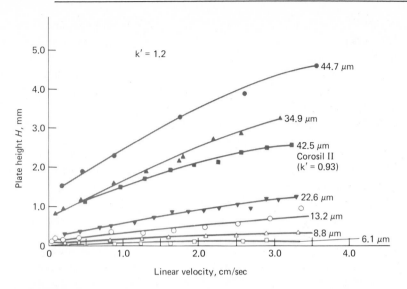

FIGURE 25-1 Effect of particle size of packing and flow rate upon plate height H. Column dimensions: 30 cm × 2.4 mm. Solute: N,N-diethyl-p-aminoazobenzene. Mobile phase: mixture of hexane, methylene chloride, and isopropyl alcohol. The packing labeled Corosil®II is a pellicular silica gel (p. 694). [Reproduced from R. E. Majors, *J. Chromatogr. Sci.*, **11**, 88 (1973). With permission.]

packings consisting of such particles can only be realized, however, by pumping at high pressures. As a consequence, the equipment required for HPLC is considerably more elaborate than the simple gravity-fed column used in classical liquid chromatography. Figure 25-2 is a block diagram showing the components of such equipment. Each of these components is discussed in the paragraphs that follow.[3]

Solvent Reservoir and Degassing System. A modern HPLC apparatus is equipped with

one or more glass or stainless steel reservoirs, each of which contains 1 to 2 liters of a solvent. Ordinarily, the reservoirs are equipped with a means of removing dissolved gases—usually oxygen and nitrogen—that interfere by forming bubbles in the column and the detector systems. These bubbles cause band spreading; in addition, they often interfere with the performance of the detector. Degassers may consist of a vaccum pumping system, a distillation system, or a device for heating and stirring the solvent.

A separation that employs but a single solvent is termed an *isocratic* elution. Frequently, separation efficiency is greatly enhanced by *gradient elution*. Here two (and sometimes more) solvents, which differ significantly in polarity, are employed. After elution is begun, the ratio of the two solvents is varied in a programmed way, sometimes continuously

---

[3] For a detailed discussion of HPLC systems, see: *Instrumentation for High-Performance Liquid Chromatography*, ed. J. F. K. Huber. New York: Elsevier, 1978; H. Engelhardt, *High Performance Liquid Chromatography. Chemical Laboratory Practice*. New York: Springer-Verlag, 1979; H. Veening, *J. Chem. Educ.*, **50**, A429, A481, A529 (1973); and H. Kern and K. Imhof, *Amer. Lab.*, **10** (2) 131 (1978).

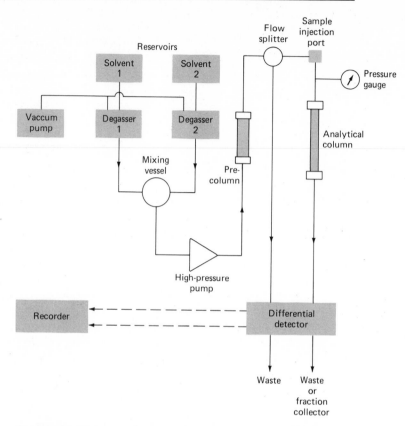

FIGURE 25-2   Schematic diagram showing liquid flow in a typical liquid chromatographic instrument. Many instruments do not employ a flow splitter and produce only a single output stream.

and sometimes in a series of steps. Modern HPLC equipment is often equipped with devices that introduce solvents from two or more reservoirs into a mixing chamber at rates that vary continuously; the volume ratio of the solvents may then be altered linearly or exponentially with time.

Figure 25-3 illustrates the improvement in separation brought about by gradient elution. For Figure 25-3b, elution was carried out with a 50 : 50 (v/v) methanol/water solution. In the case of Figure 25-3a, elution was initiated with a 40:60 ratio of the two solvents; the methanol concentration was then increased at the rate of 8%/min. Note that gra-

dient elution shortened the time of separation significantly without sacrifice in resolution of the early peaks.

Pumps. Most HPLC pumps have outputs of at least 1000 psi (lbs/in^2), and preferably 4000 to 6000 psi, with a flow delivery rate of at least 3 ml/min. The flow rate should be constant to $\pm 2\%$. Two types of mechanical pumps are employed; one is a screw-driven syringe type and the other a reciprocating pump. The former produces a pulse-free delivery that is readily controlled; it suffers from lack of capacity and inconvenience in changing solvents. Reciprocating pumps, which are much more widely used, produce a pulsed flow which must

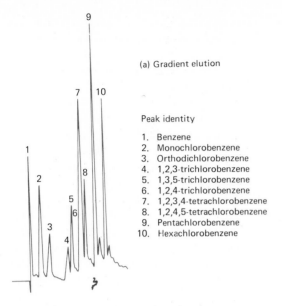

(a) Gradient elution

Peak identity

1. Benzene
2. Monochlorobenzene
3. Orthodichlorobenzene
4. 1,2,3-trichlorobenzene
5. 1,3,5-trichlorobenzene
6. 1,2,4-trichlorobenzene
7. 1,2,3,4-tetrachlorobenzene
8. 1,2,4,5-tetrachlorobenzene
9. Pentachlorobenzene
10. Hexachlorobenzene

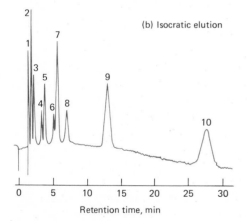

(b) Isocratic elution

Retention time, min

FIGURE 25-3 Improvement in separation efficiency by gradient elution. Column: 1 m × 2.1 mm id, precision-bore stainless; packing: 1% Permaphase® ODS. Sample: 5 μl of chlorinated benzenes in isopropanol. Detector: UV photometer (254 nm). Conditions: temperature, 60°C, pressure, 1200 psi. (From J. J. Kirkland, *Modern Practice of Liquid Chromatography*. New York: Wiley-Interscience, 1971, p. 88. With permission.)

be subsequently damped. The advantage of the reciprocating pump is that it can be employed with unlimited volumes of solvent; furthermore, its low internal volume makes it ideal for gradient elution.

Pneumatic pumps are also employed. In the simplest of these, the mobile phase is contained in a collapsible container housed in a vessel that can be pressurized by a compressed gas. Pumps of this type are simple, inexpensive, and pulse-free; they suffer from limited capacity and pressure output as well as a dependence of flow rate on solvent viscosity. In addition, they are not amenable to gradient elution.

It should be noted that the high pressures generated by the pumping devices do not constitute a hazard because liquids are not very compressible. Thus, rupture of a component of the system results only in solvent leakage.

Precolumns. Some HPLC instruments are equipped with a so-called precolumn, which contains a packing chemically identical to that in the analytical column. The particle size is much larger, however, so that the pressure drop across the precolumn is negligible with respect to the rest of the system. The purpose of the precolumn is to remove impurities from the solvent, and thus prevent contamination of the analytical column. In addition, the precolumn saturates the mobile phase with the liquid making up the stationary phase; in this way, stripping of the stationary phase from the packing of the analytical column is eliminated.

Sample Injection Systems. Rotary sampling valves, similar in construction to that shown in Figure 24-9, are often employed in HPLC. Slider valves, such as the one shown in Figure 25-4, are also common. The slider is manufactured from Kel-F and moves in a plane between two pieces of polytetrafluoroethylene. The sample is drawn into the valve by means of a syringe. The slider is then moved from left to right so that the sample is placed in the solvent stream. Valves holding various volumes from 2 to 100 μl are available.

Sample injection can also be accomplished by means of a syringe and a self-sealing

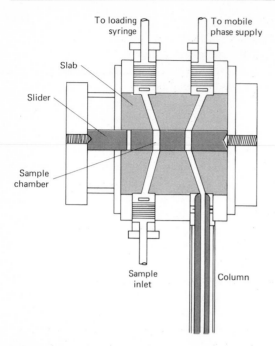

FIGURE 25-4 A sample injection valve. (From R. P. W. Scott, *Contemporary Liquid Chromatography*. New York: Wiley, 1976, p. 131. With permission.)

septum of silicone, neoprene, or Teflon. Alternatively, flow through the column can be stopped momentarily, a cap on the column removed, and the sample injected directly onto the head of the column with a syringe.

Columns. Columns for HPLC are manufactured from heavy-walled glass tubing or precision-bore stainless steel; the former are limited to pressures below about 600 psi. The typical column ranges from 15 to 150 cm in length with an inside diameter of about 2 to 3 mm. Longer columns (up to a meter or more) are usually prepared by connecting sections of shorter columns. Coiled columns are also employed, although some loss in efficiency results from this configuration.

The most common type of packing for HPLC columns is finitely divided silica gel. Oc-

casionally, alumina and Celite (a diatomaceous earth) are employed. Uniform particle sizes in the 5- to 10-μm range are desirable.

A second type of packing material, called *pellicular* particles, consists of small beads (often of glass and having diameters of about 40 μm), coated with a 1- to 3-μm layer of a porous material such as silica gel, alumina, or an ion exchange resin. The rate at which equilibrium between phases is approached is high in these thin layers (p. 675); thus, improved efficiency results. Pellicular packings have the disadvantage of limited sample capacity (roughly one-tenth that of porous packings).

Temperature Control. Most liquid chromatography is performed at room temperature and without thermostating. Water jacketed columns are available commercially when precise temperature control is desired.

Detectors. No highly sensitive, universal detector system, such as those found in gas chromatography (p. 720), is available for HPLC. Thus, various systems must be employed, depending upon the nature of the sample. Table 25-1 lists the common detectors and some of their properties.

Perhaps the most common detectors for HPLC are based upon absorption of ultraviolet or visible radiation. Both photometers and spectrophotometers are available commercially. The former usually employ the 254- and 280-nm lines from a mercury source; at these wavelengths many organic functional groups absorb (see pp. 172 to 175). Spectrophotometric detectors are considerably more versatile than photometers because the radiation can be tailored to the anticipated absorption peaks of the sample components. Obviously, photometric detectors require that the sample components absorb and the solvent transmit radiation within the wavelength range of the instrument.

Figure 25-5 is a schematic diagram of an ultraviolet photometric detector.

Infrared detectors are also available; these are similar in construction to the infrared instrument shown in Figure 8-8 (p. 226).

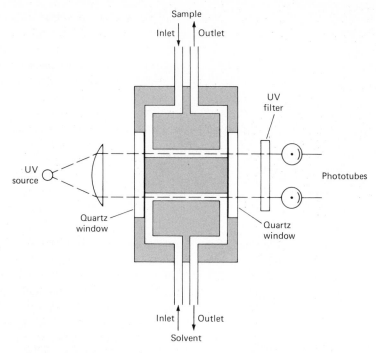

FIGURE 25-5   Ultraviolet detector for HPLC.

TABLE 25-1   CHARACTERISTICS OF LIQUID CHROMATOGRAPHY DETECTORS[a]

Detector Basis	Type[b]	Maximum Sensitivity[c]	Flow Rate Sensitive?	Temperature Sensitivity	Useful with Gradient?	Available Commercially?
UV absorption	S	5×10^{-10}	No	Low	Yes	Yes
IR absorption	S	10^{-6}	No	Low	Yes	Yes
Fluorometry	S	10^{-10}	No	Low	Yes	Yes
Refractive index	G	5×10^{-7}	No	$\pm 10^{-4}$ °C	No	Yes
Conductometric	S	10^{-8}	Yes	± 1°C	No	Yes
Moving wire	G	10^{-8}	Yes	None	Yes	Yes
Mass spectrometry	S	10^{-10}	No	None	—	Yes
Polarography	S	10^{-10}	Yes	± 1°C	—	No
Radioactivity	S	—	No	None	Yes	No

[a] Most of these data were taken from: L. R. Snyder and J. J. Kirkland, *Introduction to Modern Liquid Chromatography*. New York: Wiley-Interscience, 1964, p. 165. With permission.

[b] G = general; S = selective.

[c] Sensitivity for a favorable sample in grams per milliliter.

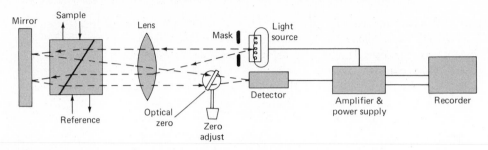

FIGURE 25-6   Schematic diagram of a differential refractive-index detector. (Courtesy of Waters Associates, Inc., Milford, Massachusetts 01757.)

Figure 25-6 is a schematic diagram of a differential refractive index detector. Here, the solvent and analyte solutions are separated by a glass plate mounted at an angle such that bending of the incident beam occurs if the two solutions differ in refractive index. As a result, the position of the beam changes with respect to the photoelectric surface; the output signal varies as a consequence. The electrical signal is then amplified and recorded to provide the chromatogram. In contrast to absorbance detectors, the refractive index indicator is general rather than selective and responds to the presence of all solutes.

Another general type detector employs a continuous moving wire loop to transport a portion of the eluate to a flame ionization detector (p. 721). The wire first passes through the eluate and carries it into an oven where the solvent is evaporated. The sample is then passed into a furnace and is pyrolyzed. The volatile products are detected by the ionization detector.

Mass spectrometry has also been employed as a sensitive specific detector.[4] One commercial instrument employs a stainless-steel or polyimide belt to transport the eluate to the

ion source. Before reaching the source, the belt passes under an infrared evaporator to remove most of the solvent, through two differential vacuum chambers, and into a quadrupole type mass spectrometer. Here, the sample is flash vaporized into the ion chamber. Detection limits as low as 0.2 to 1 ng are reported.

Adsorption Chromatography

All of the early work in chromatography was of the adsorption type, in which the stationary phase was the surface of a finely divided solid. In such a packing, the solute competes with the eluting solvent for sites on the surface of the solid; retention is the result of adsorption forces. Currently, liquid-solid chromatography is the most widely used of the HPLC techniques for the separation of neutral organic compounds.

Ideally, the distribution of a solute between a solvent and an adsorbing solid is described by an adsorption isotherm such as that shown as curve *A* in Figure 24-1 (p. 669). Note that the curve is nearly a straight line at low concentrations; under these circumstances, elution is linear and peak distortion is minimal.

Stationary and Mobile Phases. Silica gel is by far the most widely employed adsorbent for liquid-solid chromatography. Alumina also finds considerable use. Packing materials of specified particle diameter ranges are avail-

---

[4] W. H. McFadden, D. C. Bradford, D. E. Games, and J. L. Gower, *Amer. Lab.,* **9** (10), 55 (1977); and P. J. Arpino and G. Guiochon, *Anal. Chem.,* **51**, 682A (1979).

able from several commercial sources; typical for HPLC is a 10 μm silica gel packing in which 80% of the particles lie in the range of 8 to 12 μm. One company offers columns packed with specially prepared spherical silica gel particles with 95% of these particles having diameters within ± 1 μm of the nominal particle size of 6 μm.

The choice of mobile phase is all-important for successful liquid-solid chromatography; by varying the solvent, the capacity factor (k') for solutes can be varied until they fall in the ideal range of 1 to 10 (p. 680). The effect of a given solvent on k' depends upon its *eluent strength* ε^0, which can be measured relative to some reference solvent. Thus, for a solute X, it has been shown that[5]

$$\log \frac{(K_X)_1}{(K_X)_2} = A_X(\varepsilon_2^0 - \varepsilon_1^0)$$

where $(K_X)_1$ and $(K_X)_2$ are the partition coefficients for X on the solid adsorbent when solvents 1 and 2 are employed. The quantity A_X is the molecular size of the solute and ε_1^0 and ε_2^0 are the eluent strengths of solvent 1 and 2, respectively. Table 25-2 shows the eluent

strengths for a number of solvents (relative to *n*-pentane) when alumina is the adsorbent. The ε^0 values for silica gel are equal to about 0.77 that for alumina. A solvent with a high eluent strength will elute adsorbed species more rapidly than one with a low value.

Often, gradient elution is useful in adsorption chromatography. Here, the easily eluted compounds are removed with a solvent having a low ε^0. Then, an increasing amount of a solvent with a high ε^0 is added to remove the more tightly bound solutes from the column.

Applications. Large differences exist in the tendencies of compounds to be adsorbed, and these differences serve as the basis of adsorption chromatography. For example, a positive correlation can be discerned between adsorption properties and the number of hydroxyl groups in an organic molecule. A similar correlation exists with double bonds. Compounds containing certain functional groups are more strongly held than others. The tendency to be adsorbed decreases in the following order: acid > alcohol > carbonyl > ester > hydrocarbon. The nature of the adsorbent is also influential in determining the order of adsorption. Much of the available knowledge in this field is empirical; the choice of adsorbent and solvent for a given separation frequency must be made on a trial-and-error basis.

---

[5] L. R. Snyder, *Principles of Adsorption Chromatography.* New York: Dekker, 1968.

TABLE 25-2   ELUENT STRENGTH OF SOME COMMON SOLVENTS WITH ALUMINA ADSORBENTS

Solvent	Eluent Strength, ε^0	Solvent	Eluent Strength, ε^0
Fluoroalkanes	−0.25	Chloroform	0.40
n-Pentane	0.00	Methylethyl ketone	0.51
Petroleum ether	0.01	Acetone	0.56
Cyclohexane	0.04	Diethylamine	0.63
Carbon tetrachloride	0.18	Pyridine	0.71
n-Propyl chloride	0.30	*n*-Propanol	0.82
Benzene	0.32	Methanol	0.95

A

B

C

| 0 | 6 | 12 |

Minutes

FIGURE 25-7 A liquid-solid chromatogram for adrenal steroid-cortisones. Peaks: (*A*) cortisol; (*B*) cortisone; (*C*) corticosterone. Column: 25 cm × 4.6 mm packed with 10 μm silica gel. Pressure: 225 psi. Flow rate: 59.8 ml/hr. Mobile phase: 75 % heptane/25 % ethanol. (Courtesy of Whatman, Inc., Clifton, N.J.)

Figure 25-7 shows a typical liquid-solid separation by HPLC.

Partition Chromatography

Partition or liquid-liquid chromatography was originated in 1941 in the Nobel prize-winning work of Martin and Synge.[6] In their studies, they were able to demonstrate that a properly prepared column can have a plate height as small as 0.002 cm. Thus, a 10-cm column of this type may contain as many as 5000 theoretical plates. High separation efficiencies are to be expected even with relatively short columns.

[6] A. J. P. Martin and R. L. M. Synge, *Biochem. J.*, **35**, 1358 (1941).

Solid Supports. The most widely used solid support for partition chromatography has been silicic acid or silica gel. This material adsorbs water strongly; the stationary phase is often aqueous as a consequence. For some separations, the inclusion of a buffer or a strong acid (or base) in the water film has proved helpful. Polar solvents such as aliphatic alcohols, glycols, or nitromethane, alone or mixed with water, have also served as the stationary phase on silica gel. Other support media include alumina, diatomaceous earth, starch, cellulose, and powdered glass; water and a variety of organic liquids have been used to coat these solids.

The mobile phase may be a pure solvent or a mixture of solvents; its polarity must be markedly different from the stationary liquid so that the two are immiscible. Ordinarily, the more polar of the two solvents is incorporated on the solid and serves as the stationary phase; *reverse-phase* chromatography employs a nonpolar stationary phase and polar mobile one. The choice of liquid pairs is largely empirical. As mentioned earlier, gradient elution is frequently employed to enhance separation efficiency.

Bonded Phase Packings. A type of packing that is becoming increasingly popular for HPLC consists of pure silica gel particles onto which an organic group has been attached chemically. As an example, a hydrocarbon surface can be formed by the reaction of chlorooctadecyl silane with the OH groups on the surface of silica gel. That is,

$$\text{SiOH} + \text{Cl}-\underset{\underset{\text{R}}{|}}{\overset{\overset{\text{R}}{|}}{\text{Si}}}-\text{R} \longrightarrow \text{Si}-\text{O}-\underset{\underset{\text{R}}{|}}{\overset{\overset{\text{R}}{|}}{\text{Si}}}-\text{R}$$

where R is the octodecyl group and the Si in the circle represents one of many SiOH groups on the surface of the gel particle. Other groups that have been bonded to silica gel include aliphatic amines, ethers, and nitrates as well as aromatic hydrocarbons.

The behavior of the chemically bonded silica surface appears to be intermediate between a solid surface at which adsorption occurs and an immobilized liquid at which a liquid-liquid equilibrium exists. Chemically bonded surfaces offer a considerable advantage over ordinary solid-supported liquids in that the stationary phase cannot be stripped of its liquid by the mobile phase. On the other hand, chemically bonded surfaces suffer from limited loading capacities.

Applications. Partition chromatography has become a powerful tool for the separation of closely related substances. Typical examples include the resolution of the numerous amino acids formed in the hydrolysis of a protein, the separation and analysis of closely related aliphatic alcohols, and the separation of sugar derivatives. Figure 25-8 illustrates an application of the procedure to the separation of amino acids. Here, the chemically bonded silica gel described in the previous section was employed as the support. Note that the separation required less than 2 min.

Ion-Exchange Chromatography

Ion-exchange resins are among the most widely used packings in column chromatography. In contrast to the other column methods, the solvent is generally water and the species to be separated are ions. Thus, ion-exchange chromatography finds wide use in inorganic chemistry. It also has been used extensively for the separation of amino acids and other organic acids and bases.[7]

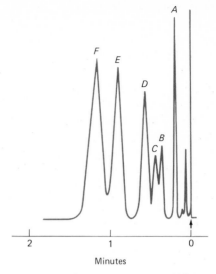

Minutes

PTH amino acid derivatives

FIGURE 25-8   Separation of amino acids by partition chromatography. Peaks: *A*, proline; *B*, leucine; *C*, isoleucine; *D*, valine; *E*, phenylalanine; *F*, methionine. Column: 10 cm × 4.6 mm; packed with 20 μm chemically bonded silica gel. Mobile phase: 50-50 heptane/chloroform. Pressure: 80 psi. Flow rate: 1.9cm^3/ min. Sample size: 1 μl. (Courtesy of Gow-Mac Instrument Co., Bridgewater, New Jersey 08807. With permission.)

---

[7] For a detailed treatment of ion-exchange chromatography, see: W. Rieman, III and H. F. Walton, *Ion Exchange in Analytical Chemistry.* New York: Pergamon Press, 1970; P. R. Brown and A. Krstulovic, *Separation and Purification,* in *Techniques of Chemistry,* eds. E. S. Perry and A. Weissberger. New York: Wiley-Interscience, 1978, vol. 12, Chapter 4.

Ion exchange is a process involving an interchange of ions of like sign between a solution and an essentially insoluble solid in contact with the solution. Many substances, both natural and synthetic, act as ion exchangers. The ion-exchange properties of clays and zeolites have been recognized and studied for more than a century. Synthetic ion-exchange resins were first produced in 1935 and have since found widespread laboratory and industrial application for water softening, water deionization, solution purification, and ion separation.

Synthetic ion-exchange resins are high molecular weight polymeric materials containing large numbers of an ionic functional group per molecule. For cation exchange, there is a choice between strong acid resins containing sulfonic acid groups $(RSO_3^-H^+)$ and weak acid resins containing carboxylic acid groups (RCOOH); the former have wider application. Anion-exchange resins contain basic functional groups, generally amines, attached to the polymer molecule. Strong base exchangers are quaternary amines $(RN(CH_3)_3^+OH^-)$; weak base types contain secondary or tertiary amines.

A cation-exchange process is illustrated by the equilibrium

$$xRSO_3^-H^+ + M^{x+} \rightleftharpoons$$
$$\text{solid} \qquad \text{solution}$$

$$(RSO_3^-)_xM^{x+} + xH^+$$
$$\text{solid} \qquad \text{solution}$$

where M^{x+} represents a cation and R represents *a part* of a resin molecule. The analogous process involving a typical anion-exchange resin can be written

$$xRN(CH_3)_3^+OH^- + A^{x-} \rightleftharpoons$$
$$\text{solid} \qquad \text{solution}$$

$$[RN(CH_3)_3^+]_xA^{x-} + xOH^-$$
$$\text{solid} \qquad \text{solution}$$

where A^{x-} is an anion.

Ion-exchange resins have been successfully employed as the stationary phase in elution chromatography. For example, Beukenkamp and Reiman separated sodium and potassium ions on a column packed with a sulfonic acid resin in its acidic form.[8] When the sample was introduced at the top of the column, the following exchange equilibrium was established for each of the alkali ions:

$$RH + B^+ \rightleftharpoons RB + H^+ \qquad (25\text{-}1)$$

where B^+ represents either Na^+ or K^+.

Equilibrium constants for ion-exchange reactions take the form

$$K = \frac{[RB][H^+]}{[RH][B^+]} \qquad (25\text{-}2)$$

where [RB] and [RH] are the concentrations (strictly the activities) of the alkali and hydrogen ions in the solid resin phase. Equation 25-2 can be rewritten in the form

$$\frac{[RB]}{[B^+]} = \frac{K[RH]}{[H^+]} = K_D \qquad (25\text{-}3)$$

where K_D is defined as the distribution coefficient. Under the conditions extant during elution with hydrochloric acid, K_D is approximately constant because the hydrogen ion concentration of the eluate is large relative to the concentration of potassium and sodium ions. Also, the resin has an enormous number of exchange sites relative to the number of alkali-metal ions in the sample. Thus, the overall concentrations $[H^+]$ and [RH] are not affected significantly by shifts in the equilibrium (Equation 25-1). Therefore, under conditions where [RH] and $[H^+]$ are large with respect to [RB] and $[B^+]$, Equation 25-3 can be employed in the same way as Equation 24-1 (p. 668), and the general theory of chromatography, which we have already described, can be applied to a stationary phase consisting of an ion-exchange resin.

Note that K_D in Equation 25-3 represents the affinity of the resin for the ion B^+ *relative* to another ion (here, H^+). Where K_D is large, a strong tendency exists for the solid phase to retain ion B; where K_D is small, the reverse obtains. By selecting a common reference ion such as H^+, distribution ratios for different ions on a given type of resin can be compared. Such experiments reveal that polyvalent ions are much more strongly held than singly

---

[8] J. Beukenkamp and W. Reiman III, *Anal. Chem.*, **22**, 582 (1950).

charged species. Within a given charge group, however, differences appear that are related to the size of the hydrated ion as well as other properties. Thus for a typical sulfonated cation-exchange resin, values for K_D decrease in the order $Cs^+ > Rb^+ > K^+ > NH_4^+ > Na^+ > H^+ > Li^+$. For divalent cations, the order is $Ba^{2+} > Pb^{2+} > Sr^{2+} > Ca^{2+} > Cd^{2+} > Cu^{2+} > Zn^{2+} > Mg^{2+}$.

The techniques for fractionation of ions with K_D values that are relatively close to one another are analogous to those described for adsorption and partition chromatography. For example, Figure 25-9a shows an HPLC separation of the alkali-metal ions on a pellicular sulfonic acid cation-exchange resin; the

original solution was 0.01M in each cation. Figure 25-9b illustrates the use of an anion-exchange column for separating the anions in a boiler blow-down water; here, the ion concentrations were in the low parts-per-million range. In both separations, quantitative data were obtained by comparison of peak heights to heights for appropriate standards. It is of interest to note that a conductivity detector was employed in both determinations. In order to prevent the conductivity of the eluent ions from swamping that of the analyte, the analytical column was followed by a stripping column. For the cation separation, the stripper was packed with the hydroxide form of an anion-exchange resin that absorbed the HCl

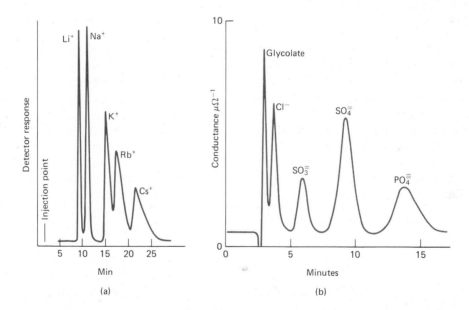

FIGURE 25-9 Separation of inorganic ions by ion-exchange chromatography. (a) Column: 250×9 mm with a sulfonic acid. Pellicular packing. Eluent: 0.01 N HCl. Sample size: 0.1 ml. [Reprinted with permission from H. Small, T. S. Stevens, and W. C. Bauman, *Anal. Chem.*, **47**, 1803 (1975). Copyright by the American Chemical Society.] (b) Column: 500×2.8 mm. Flow rate: 118 ml/hr. Packing: pellicular anion-exchange resin. Eluent: 0.0015 M $NaHCO_3$ for glycolate + Cl^- and 0.003 M $NaHCO_3$ + 0.0024 M Na_2CO_3 for other anions. Sample size: 100 μliters. [Reprinted with permission from T. M. Stevens, V. T. Turkelson, and W. A. Albe, *Anal. Chem.*, **49**, 1176 (1977). Copyright by the American Chemical Society.]

eluent without affecting the cations. The reaction in the stripper is

$$H^+ + Cl^- + ROH \rightarrow RCl + H_2O$$

For the anion separations, the stripper reaction can be written as

$$Na^+ + HCO_3^- + RH \rightarrow RNa + H_2CO_3$$

Here, the H_2CO_3 has no significant effect on the conductivity of the solvent.

Ion-exchange chromatography finds widespread use in amino acid analyzers. For example, Figure 25-10 shows a chromatogram of a mixture containing 1.0×10^{-8} mol each of 17 amino acids. Separation time was only 42 min. Gradient elution with buffers of different pH was employed.

Gel Chromatography

Gel chromatography is a technique in which fractionation is based, at least in part, upon the molecular size and shape of the species in the sample.[9] Several names have been given to this procedure, including gel-permeation chromatography, exclusion chromatography, and molecular-sieve chromatography.

Ordinarily, gel chromatography is performed on a column by the elution method. Here, the degree of retardation depends upon the extent to which the solute molecules or ions can penetrate that part of the solution phase which is held within the pores of the highly porous gel-like packing material. Molecules or ions larger than the pores of the gel are completely excluded from the interior, while smaller species are more or less free to enter these regions. Thus, larger molecules pass quickly through the column, while smaller ones are retarded to a greater or lesser

---

[9] For further information, see: S. D. Abbott, *Amer. Lab.*, **9** (9), 8 (1977); and H. Determann, *Gel Chromatography.* New York: Springer-Verlag, 1965.

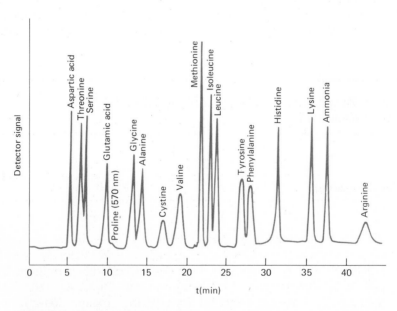

FIGURE 25-10   Separation of amino acids on an ion-exchange column. Packing: cation exchange with particle size of 8 μm. Pressure: 2700 psi. [Reprinted J. R. Benson, from *American Laboratory*, **4** (10) 53, 1972. Copyright 1972 by International Scientific Communications, Inc., Fairfield, CT.]

extent depending upon their size, their shape, and sometimes, upon their tendency to be adsorbed by the gel.

Column Packing. The stationary phase in gel chromatography consists of beads of a porous polymeric material that readily absorbs water (and in some instances, other solvents) and swells as a consequence. The resulting solid contains a large volume of solvent held in the interstices of the polymeric network. The average size of the resulting pores or interstices is directly related to the quantity of solvent adsorbed; this in turn is determined by the amount of cross-linking in the polymer molecules.

One of the most widely used polymers is prepared by cross-linking the polysaccharide dextran with epichlorhydrin. By varying the amount of the latter, a series of resins with differing pore sizes is obtained. These gels sell under the trade name of Sephadex®. Table 25-3 describes the properties of some of these Sephadex gels. Another group of commercial resins consists of polyacrylamide cross-linked with methylene bisacrylamide.

Theory of Gel Chromatography. The total volume of a column packed with a gel that has been swelled by water (or other solvent) is given by

$$V_t = V_g + V_i + V_o$$

where V_g is the volume occupied by the solid matrix of the gel, V_i is the volume of solvent held in its interstices, and V_o is the free volume outside the gel particles. Assuming no mixing or diffusion, V_o also represents the volume of solvent required to carry through the column those components too large to enter the pores of the gel. In fact, however, some mixing and diffusion will occur, and as a consequence the components will appear in a Gaussian-shaped band with a concentration maximum at V_o. For components small enough to enter freely into the interstices of the gel, band maxima will appear at the end of the column at a volume corresponding to $(V_i + V_o)$. Generally, V_i, V_o, and V_g are of the same order of magnitude; thus, a gel column permits separation of the large molecules of a sample from the small with a minimal volume of wash liquid.

TABLE 25-3   PROPERTIES OF COMMERCIAL Sephadex[a] GELS

Gel Designation	Volume Relationships,[b] ml/g original gel			Approximate Exclusion Limit, mol wt
	V_g	V_i	V_o	
G-10	0.6	1.0	0.9	700
G-15	0.6	1.5	0.9	1,500
G-25	0.5	2.5	2	5,000
G-50	1	5	4	10,000
G-75	1	7	5	50,000
G-100	1	10	6	100,000
G-150	1	15	8	150,000
G-200	1	20	9	200,000

[a] ®Pharmacia Fine Chemicals, Inc., Piscataway, NJ.
[b] V_g = volume occupied by solid gel matrix.
V_i = volume of solvent held interstitially.
V_o = liquid volume between the gel beads.

Molecules of intermediate size are able to transfer into some fraction K_d of the interstitially held solvent; for these, the elution volume V_e is

$$V_e = V_o + K_d V_i \qquad (25\text{-}4)$$

Equation 25-4 describes the behavior of a gel column with respect to all solutes. For molecules too large to enter the gel pores, $K_d = 0$ and $V_e = V_o$; for molecules that can enter the pores unhindered, $K_d = 1$ and $V_e = (V_o + V_i)$. In arriving at Equation 25-4, the assumption was made that no interaction, such as adsorption, occurs between the solute molecules and the gel surfaces. If such interaction does occur, the amount of interstitially held solute will be increased; for interacting solutes that can freely enter the pores, K_d will be greater than unity.

Table 25-4 gives experimentally determined values of K_d for some dextrant-type gels.

Applications of Gel Chromatography. Tightly cross-linked gels with small pores, such as Sephadex G-25 and G-50, have found wide application to the *desalting* or removal of low molecular weight molecules from high molecular weight natural-product molecules. For example, it is evident from Table 25-4 that the G-25 gel will effectively separate salts and amino acids from most proteins. Here the proteins, all with K_d values of zero, are completely excluded from the gel interior, while the low molecular weight species have K_d values approaching unity. As a result, clean separation can be achieved even with large sample volumes (as much as 20 to 30% of the total volume of the bed). This type of separation performs the same function as a dialysis through a cellophane membrane, but is more rapid and convenient to perform.

Sephadex G-25 and G-50 gels have also proved useful for the fractionation of peptides having a size range intermediate between the proteins shown in Table 25-4 and the low molecular weight species. These compounds have K_d values greater than zero but less than one. Elution techniques for small samples are similar to those used in other types of chromatography.

The more porous gels shown in Table 25-4 have found wide application to the fractionation and the purification of macromolecules such as proteins, nucleic acids, and polysac-

TABLE 25-4   K$_d$ VALUES WITH SEPHADEX GELS[a]

| Substance | Approximate mol wt | K_d | | | |
		Sephadex G-25	Sephadex G-75	Sephadex G-100	Sephadex G-200
Ammonium sulfate	132	0.9	—	—	—
Potassium chloride	74	1.0	—	—	—
Tryptophan	204	2.2	1.2	—	—
Glycine	76	0.9	1.0	—	—
Ribonuclease	13,000	0	0.4	—	—
Trypsin	24,000	0	0.3	0.5	0.7
Serum albumin	75,000	0	0	0.2	0.4
Fibrinogen	330,000	0	0	0.0	0.0

[a] Pharmacia Fine Chemicals, Inc., Piscataway, N.J. From B. Gelotte, "Fractionation of Proteins, Peptides, and Amino Acids by Gel Filtration," in *New Biochemical Separations*, eds. A. T. James and L. J. Morris. Princeton: Van Nostrand, 1964. See Table 25-3 for the properties of these gels.

charides. For example, Sephadex G-75 permits the clean separation of ribonuclease, trypsin, pepsin, and cytochrome C from higher molecular weight proteins such as serum albumin, hemoglobin, and fibrinogen (see Table 25-4).

Finally, it should be noted that gel-permeation chromatography has been used by polymer chemists and biochemists for estimation of the molecular weights of large molecules. Here, the elution volume of the unknown is compared with elution volumes for a series of standard compounds that possess the same chemical characteristics.

Comparison of High-Performance Liquid Chromatography with Gas-Liquid Chromatography

Table 25-5 provides a comparison between high-performance liquid chromatography (HPLC) and gas-liquid chromatography, which is discussed in the next chapter. Where either is applicable, gas-liquid chromatography offers the advantage of speed and simplicity of equipment. On the other hand, HPLC is applicable to nonvolatile substances (including inorganic ions) and thermally un-

stable materials whereas gas-liquid chromatography is not. Often, the two methods are complementary.

PLANAR CHROMATOGRAPHY

Planar chromatography takes two forms. In thin-layer chromatography, separation takes place on a layer of a finely divided solid that is fixed on a flat surface.[10] A sheet or strip of heavy filter paper serves as the separatory medium for paper chromatography.[11] Historically, paper chromatography was first used for separations in the middle of the nineteenth century. It was not until the late 1940s, however, that the usefulness of the technique became fully appreciated by scientists. Thin-layer

---

[10] References: J. G. Kirchner, in *Thin Layer Chromatography*, ed. A. Weissberger. New York: Wiley-Interscience, 1978; and J. T. Touchstone and M. R. Dobbins, *Practice of Thin Layer Chromatography*. New York: Wiley, 1978.

[11] References: R. Stock and C. B. F. Rice, *Chromatographic Methods*, 3d ed. London: Chapman and Hall, 1974, Chapter 3; and J. Sherma and G. Zweig, *Paper Chromatography*. New York: Academic Press, 1971.

TABLE 25-5 COMPARISON OF HIGH-PERFORMANCE LIQUID AND GAS–LIQUID CHROMATOGRAPHY

Characteristics possessed by both methods:
    Efficient, highly selective, and widely applicable
    Only small sample required
    Ordinarily nondestructive of sample
    Readily adapted to quantitative analysis

Particular advantages of high-performance liquid chromatography:
    Can accommodate nonvolatile and thermally unstable samples
    Generally applicable to inorganic ions

Particular advantages of gas-liquid chromatography:
    Simple and inexpensive equipment
    Rapid

chromatography was developed in the late 1950s and by now has become the more important of the two planar methods.

Both paper and thin-layer chromatography provide remarkably simple and inexpensive means for separating and identifying the components of small samples of complex inorganic, organic, and biochemical substances. Furthermore, the methods, particularly thin-layer chromatography, permit reasonably accurate quantitative determination of the concentrations of the components of such mixtures.

General Considerations

In planar chromatography, a drop of a solution containing the sample is introduced at some point on the planar surface of the stationary phase. After evaporation of the solvent, the chromatogram is *developed* by the flow of a mobile phase (the developer) across the surface. Movement of the developer is caused by capillary forces. In *ascending development*, the motion of the mobile phase is upward; in *radial development*, it is outward from a central spot. Gravity also contributes to solvent flow in *descending development*, where movement is in a downward direction.

The equilibrium upon which planar chromatography is based is most often of the liquid-liquid type, although liquid-solid and ion-exchange equilibria are also encountered. Usually, the stationary phase is water or some other polar liquid; reverse-phase separations also find application, however.

The general principles developed earlier for column chromatography appear to apply equally well to paper and thin-layer media. Within these media, repeated transfer of solute between the stationary and mobile phases occurs. The rate of movement of the solute depends upon its partition coefficient

$$K = C_S/C_M$$

It is customary to describe the travel of a particular solute in terms of its retardation factor R_F, which is defined as

$$R_F = \frac{\text{distance of solute motion}}{\text{distance of solvent motion}}$$

To compensate for uncontrolled variables, the distance traveled by a solute is frequently compared with that for a standard substance under identical conditions. The ratio of these distances is designated as R_{std}.

Thin-Layer Chromatography

Stationary Phase. The solid absorbents (or sometimes adsorbents) employed in thin-layer chromatography are similar in chemical composition and particle size to those already described in the discussion of the various types of column chromatography. The most widely used substance is silica gel, which often functions as a support for water or other polar solvents for liquid-liquid separations. If, however, the silica layer is oven-dried after preparation, much of the moisture is lost and the surface becomes a predominately solid one that serves as an adsorbent for liquid-solid separations. In this latter mode, care must be taken to avoid exposing the surface to the atmosphere, for it has been demonstrated that water adsorption from air occurs in a few minutes; the surface then reverts to a support for a liquid.

Ion-exchange resins and Sephadex® gels are also employed as the stationary phase in thin-layer chromatography.

Preparation of Thin-Layer Plates. A thin-layer plate can be prepared by spreading an aqueous slurry of the finely ground solid onto the clean surface of a glass or mylar plate or microscope slide. A binder often is incorporated into the slurry to enhance adhesion of the solid particles to the glass and to one another. The plate is then allowed to stand until the layer has set up and adheres tightly to the surface; for some purposes, it may be heated in an oven for several hours.

The cleanness of separations on the thin layer depends upon having a narrow range of particle sizes and a uniform layer thickness. Several spreading devices have been developed for forming a layer of constant thickness; some of these are available commercially. Prepared plates can also be purchased from a number of sources.

Plate Development. Typical arrangements for the development of thin-layer plates are shown in Figure 25-11. A drop of the sample is placed near one end of the plate (most plates have dimensions of 5 × 20 or 20 × 20 cm), and its position is marked with a pencil. After the sample solvent has evaporated, the plate is placed in a closed container saturated with vapors of the developer. One end of the plate is then wetted with the developing solvent by means of one of the arrangements shown in Figure 25-11. (Note that the sample is not immersed in the developer.) After the developer has traversed the length of the plate, the latter is removed, dried, and the position of the components determined in any of several ways.

Figure 25-12 illustrates the separation of

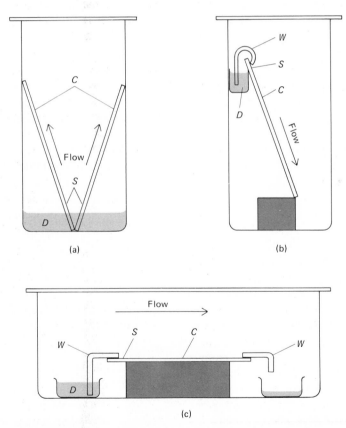

FIGURE 25-11 Three types of apparatus for thin-layer chromatography. (a) Ascending flow; (b) descending flow; and (c) horizontal flow. *S*: Initial position of sample; *D*: developer; *C*: chromatographic surface; *W*: cotton wick.

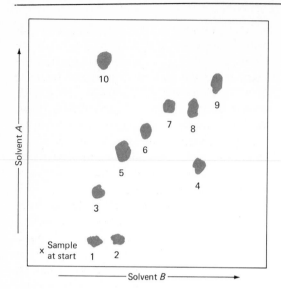

FIGURE 25-12 Two-dimensional thin-layer chromatogram (silica gel) of some amino acids. Solvent *A*: toluene-2-chloroethanol-pyridine. Solvent *B*: chloroform-benzyl alcohol-acetic acid. Amino acids: (1) aspartic acid; (2) glutamic acid; (3) serine; (4) β-alanine; (5) glycine; (6) alanine; (7) methionine; (8) valine; (9) isoleucine; and (10) cysteine.

amino acids in a mixture by development in two directions (*two-dimensional planar chromatography*). The sample was placed in one corner of a square plate and development was performed in the ascending direction with solvent *A*. This solvent was then removed by evaporation, the plate was rotated 90 deg, following which ascending development with solvent *B* was carried out. After solvent removal, the positions of the amino acids were determined by spraying with ninhydrin, a reagent that forms a pink to purple product with amino acids. The spots were identified by comparison of their positions with those of standards.

Identification of Species. Several methods are employed to locate the various sample components after separation. Two common methods, which can be applied to most organic mixtures, involve spraying with solutions of iodine or sulfuric acid; both react with organic compounds to yield dark reaction products. Several specific reagents (such as ninhydrin) are also useful for locating separated species. Fluorescent substances can be detected with an ultraviolet lamp. Alternatively, a fluorescent material is incorporated into the stationary phase. After development, the plate is examined with ultraviolet light. Here, all of the plate fluoresces except where the nonfluorescing sample components are located.

Quantitative Analysis. A semiquantitative estimate of the amount of a component present can be obtained by comparing the area of a spot with that of a standard. Better data can be obtained by scraping the spot from the plate, extracting the analyte from the resulting solid, and measuring the analyte by a suitable physical or chemical method. Finally, a scanning densitometer can be employed to measure radiation emitted from the spot by fluorescence or reflection.[12]

Paper Chromatography

Separations by paper chromatography are performed in the same way as those on thin-layer plates. Usually, special papers, which are highly purified and reproducible as to porosity and thickness, are employed. Such papers contain sufficient adsorbed water that paper chromatography can be classified as a liquid-liquid type. Other liquids can be made to displace the water, however, thus providing a different type of stationary phase. For example, paper treated with silicone or paraffin oil permits reverse-phase paper chromatography, in which the mobile phase is a polar solvent. Also available commercially are special papers that contain

---

[12] For a description of commercially available densitometers as well as other thin-layer equipment, see: P. F. Lott and R. J. Hurtubise, *J. Chem. Educ.*, **48**, A437 (1971); and D. Rogers, *Amer. Lab.*, **11** (5), 77 (1979).

an adsorbent or an ion-exchange resin, thus permitting adsorption and ion-exchange paper chromatography.

ELECTROPHORESIS AND ELECTROCHROMATOGRAPHY

Electrophoresis is defined as the migration of particles through a solution under the influence of an electrical field. Historically, the particles referred to were colloidal and owed their charge to adsorbed ions. This definition has now become too restrictive, and the term electrophoresis is presently applied both to the migration of individual ions and to colloidal aggregates. Electrophoretic methods provide a powerful means of fractionating the components of a mixture, be they aggregated or monodispersed.

Electrophoretic methods can be subdivided into two categories, depending upon whether the separation is carried out in the absence or presence of a supporting or stabilizing medium. In the *free-solution method*, the sample solution is introduced as a band at the bottom of a U-tube filled with a buffered liquid. A field is applied by means of electrodes located near the tube ends; the differential movement of the charged particles toward one or the other electrode is observed. Separations occur as a result of differences in migration rates; these rates in turn are related to the charge-to-mass ratios and the inherent mobilities of the species in the medium. The free-solution method was perfected and applied to the separation of proteins by Tiselius; for this work, he received the 1948 Nobel prize. This procedure has been of signal importance in the development of biochemistry; nevertheless, its widespread application has been hindered by such experimental problems as the tendency of the separated components to mix by convection, as a consequence of thermal and density gradients, as well as mechanical vibrations. In addition, elaborate optical systems are often needed to detect the bands of the separated species.

Many of the experimental difficulties associated with free-solution electrophoresis are avoided if the separations are carried out in a stabilizing medium such as a paper, a layer of finely divided solid, or a column packed with a suitable solid. Here, the experimental techniques closely resemble the various chromatographic methods discussed earlier, with the additional parameter of the superimposed electrical field. Depending upon the properties of the medium, the separations may result primarily from the electrophoretic effect or from a combination of electrophoresis and adsorption, ion exchange, or other distribution equilibria. Methods based upon electrophoresis in a stabilizing medium bear a variety of names, including *electrochromatography*, *zone electrophoresis*, *electromigration*, and *ionophoresis*. Our discussion is limited to electrochromatography.[13]

Experimental Methods of Electrochromatography

The solid media employed in electrochromatography are as numerous and varied as those encountered in the other chromatographic methods. Examples include paper, cellulose acetate membranes, cellulose powders, starch gels, ion-exchange resins, glass powders, and agar gels. Depending upon the physical nature of the solid, separations are performed on strips of paper or membranes, in columns, in trays, or in thin layers supported by glass or plastic. Despite certain disadvantages, filter paper and cellulose acetate are the most widely used stabilizing materials; we will focus our attention on paper electrochromatography.

Figure 25-13 is a schematic diagram illustrating three of the many ways for performing a paper electrophoresis. In Figure 25-13a, a

---

[13] For a detailed discussion, see: E. Heftmann, *Chromatography*, 3d ed. New York: Reinhold, 1974, Chapters 10 and 11; and J. R. Sargent and S. G. George, *Methods in Zone Electrophoresis*, 3d ed. Poole, England: BDH Chemicals, Ltd., 1975.

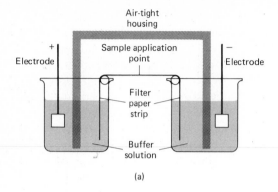

(a)

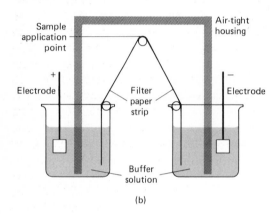

(b)

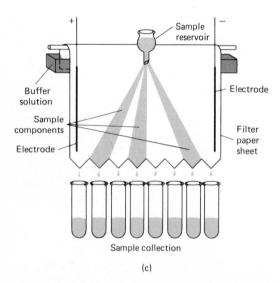

(c)

FIGURE 25-13 Some types of apparatus for paper electrochromatography.

strip of paper is held horizontally between two containers filled with a buffer mixture; the paper is well soaked with buffer and evaporation is prevented by housing the apparatus in an air-tight container. The sample is introduced as a band at the center of the strip, and a dc potential of 100 to 1000 V is applied across the two electrodes. The latter are sufficiently isolated from the paper to prevent the electrode reactions from altering the composition of the buffer on the paper. Currents in the milliampere range are ordinarily observed. After a suitable electrolysis period, the paper is removed, dried, and the component bands are detected by applying suitable colorimetric reagents.

The inverted V arrangement shown in Figure 25-13b is also widely used. Here, the sample is introduced at the apex of the V; cationic species migrate down one arm of the V and anionic constituents down the other.

Many other modifications of the apparatus shown in Figures 25-13a and 25-13b have been described, and several are available commercially. In some, the paper is supported (either horizontally or at some suitable angle) between two glass or plastic plates; the plates may be cooled to dissipate the heat generated by the current. In all of these devices, the rate at which components migrate is controlled primarily by their charge-to-mass ratios and their mobilities, with adsorption or other equilibria having only a secondary influence.

Figure 25-13c illustrates an apparatus for a type of two-dimensional electrochromatography. Here, the paper is held vertically and the sample components are carried down the sheet by the flow of a buffered solvent. Separation then occurs in the vertical direction as a consequence of differences in distribution ratios between the mobile and the fixed phases. In addition, however, a field is applied at right angles to the solution flow, and differential electromigration along the horizontal axis occurs as a consequence. Thus, the various species in the sample describe a radial path from the point at which the sample is introduced.

The technique illustrated in Figure 25-13c has been employed for both analytical and preparatory purposes. In the former, an amount of buffer just equal to that retained by the paper is allowed to pass before the experiment is discontinued; the paper is then removed, dried, and the detection reagents are applied. In preparatory applications, the buffer flow is continued and the various fractions are collected as shown in Figure 25-13c.

Applications of Electrochromatography

Electrochromatographic methods are indispensible to the clinical chemist and the biochemist, who use them for fractionating an amazing variety of biological materials. The widest application has perhaps been in clinical diagnosis; here, electrochromatography makes possible the separation of proteins and other large molecules contained in serum, urine, spinal fluid, gastric juices, and other body fluids.

Electrochromatography has also been applied widely by biochemists to fractionate smaller molecules such as alkaloids, antibiotics, nucleic acids, vitamins, natural pigments, steroids, amino acids, carbohydrates, and organic acids.

It has also been shown that electrochromatography provides a convenient means for separating inorganic ions. Figure 25-14a illustrates the separation of six metallic ions from a complexing medium. Note that both anodic and cathodic migration have occurred depending upon the charge of the complex ion. Note also that the movement of the nickel(II)

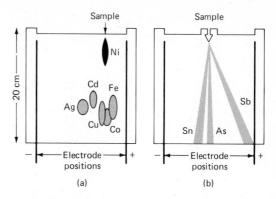

FIGURE 25-14 Electrochromatograms for inorganic ions. (a) Cations (0.05 M) in 0.1 M tartaric acid; wash liquid was 0.01 M in ammonium tartrate, 0.005 M in dimethylglyoxime, and 4 M in ammonia. Spots were detected with hydrogen sulfide: 160 V, 100 mamp, 20 min. (b) Continuous separation. Solution 0.005 M each in As(III), Sb(III), and Sn(II); solution was also 0.02 M in tartaric acid, 0.04 M in lactic acid, and 0.04 M in dl-alanine. Paths were detected with hydrogen sulfide. 300 V, 95 mamp. [Reprinted with permission from H. H. Strain and J. C. Sullivan, *Anal. Chem.*, **23**, 816 (1951). Copyright by the American Chemical Society.]

ion has been retarded by the precipitant dimethylglyoxime.

Figure 25-14b illustrates the path traveled by the components of a sample when subjected to continuous electrophoresis and elution. By collecting the eluate at appropriate places on the bottom edge of the paper, solutions of the separated species can be obtained.

PROBLEMS

1. Define the following terms: (a) isocratic elution, (b) gradient elution, (c) pellicular beads, (d) eluent strength, (e) desalting, and (f) retardation factor (R_F).

2. What is the purpose of the precolumn in Figure 25-2?

3. What are the relative advantages and disadvantages of pellicular beads as compared with porous packing material?

4. In preparing a benzene/acetone gradient for an alumina HPLC column, is it desirable to increase or decrease the proportion of benzene as the column is eluted?

5. Would a strong acid resin or weak acid resin be more appropriate for cation exchange if the solvent is $0.01M$ HCl?

6. Referring to Table 25-4, what can be deduced about the interaction of tryptophan with Sephadex G-25?

7. Explain the trend in K_d values for trypsin across Table 25-4.

8. Gel permeation chromatography is widely used as an approximate means of finding the molecular weight of globular proteins. It is frequently found that a graph of K_d vs. log(molecular weight) is reasonably linear. To find K_d, it is necessary to know V_o, V_e, and V_i. The quantity V_o is measured with a very large molecule (such as the dye blue dextran) which is completely excluded from the gel. To find V_i, V_t is calculated from the column dimensions and V_g is usually assumed to be negligible. A column of Sephadex G-75 with a diameter of 2.5 cm and a length of 36 cm gave the following results:

Substance	Molecular weight	V_e, ml
Blue dextran	2,000,000	53.1
Aldolase	158,000	55.0
Ovalbumin	45,000	71.8
Chymotrypsinogen A	25,000	90.3
Ribonuclease A	13,700	109.5

From these data, estimate the molecular weight of a protein with $V_e = 76.5$ ml.

9. The most common way to estimate the molecular weight of a protein is by electrophoresis in a polyacrylamide gel containing sodium dodecyl sulfate (SDS). The latter reagent coats polypeptide chains and denatures proteins such that they adopt a random coil conformation. The negatively charged SDS also imparts a nearly constant charge-to-mass ratio to the coated protein. As a result, the electrophoretic mobility in this medium depends mainly on the molecular weight of the protein. Proteins with subunits which are held together by noncovalent bonds break apart into their subunits in SDS. Proteins with subunits held together by disulfide bonds may be cleaved into subunits by treatment with the reducing agent, 2-mercaptoethanol. The relative electrophoretic mobility of a protein is measured by comparing its distance of migration to that of a low molecular weight ionic

dye run in the same gel. It is found that a graph of log(molecular weight) vs. relative mobility is a reasonably linear for most proteins.
(a) From the data below,[14] estimate the molecular weight of the unknown.

Protein	Molecular weight	Relative mobility
Lysozyme	14,400	0.90
Trypsin	23,300	0.73
Lactate dehydrogenase	36,000	0.57
Ovalbumin	43,000	0.54
Catalase	62,000	0.38
Bovine serum albumin	68,000	0.40
Unknown	?	0.68

(b) Estimate the relative mobility expected for the dimer of the unknown protein. The dimer can be prepared by chemically crosslinking two molecules.

10. A cation exchange resin tends to exclude anions and an anion exchange resin tends to exclude cations. As a consequence, the concentration of electrolyte inside of the resin is lower than the concentration outside of the resin. Consider a cation exchange resin in the K^+ form immersed in a solution of KCl. Let $[K^+]_i$ and $[Cl^-]_i$ be the concentrations of species inside the resin and let $[K^+]_o$ and $[Cl^-]_o$ be the concentrations outside of the resin. From thermodynamic considerations it can be shown that the condition which must be satisfied at equilibrium is

$$[K^+]_o[Cl^-]_o = [K^+]_i[Cl^-]_i$$

To preserve electroneutrality we can also write

$$[K^+]_o = [Cl^-]_o$$

and

$$[K^+]_i = [Cl^-]_i + [R^-]$$

where $[R^-]$ is the concentration of anionic sites in the resin.
(a) Derive an expression relating $[Cl^-]_o$ to $[Cl^-]_i$.
(b) If the concentration of anionic sites in the resin is $10M$ and $[Cl^-]_i = 0.001M$, calculate $[Cl^-]_o$. Find $[Cl^-]_o$ if $[Cl^-]_i = 0.1M$.
(c) Does the resin exclude electrolyte more effectively at high or low electrolyte concentration?

11. Use the principle of Problem 10 to explain the mechanism of *ion exclusion chromatography*. In this technique an electrolyte (such as

[14] O. A. Moe, Jr. and G. Smith, *J. Chem. Educ.*, **54**, 392 (1977).

NaCl) can be separated from a nonelectrolyte (such as glycerol) by passage through an ion exchange column. *Hint:* Consider the analogy to gel permeation chromatography.

12. Suggest a procedure for measuring the exchange capacity (ion exchange sites per unit mass or volume of the resin) of a cation exchange resin using standard solutions of HCl and NaOH and any other reagents you wish.

26

GAS-LIQUID CHROMATOGRAPHY

In gas chromatography, the components of a vaporized sample are fractionated as a consequence of partition between a mobile *gaseous* phase and a stationary phase held in a column. *Gas-solid chromatography* employs a solid stationary phase; the partition process, then, involves gaseous adsorption equilibria. In *gas-liquid chromatography* (GLC), the stationary phase is a liquid, supported on an inert solid matrix; here, gas-liquid equilibria are important.

Gas-solid chromatography has had somewhat limited application owing to tailing of elution peaks (a direct result of the nonlinear character of adsorption isotherms) and semipermanent retention of active gases on the stationary phase. For a decade or more, gas-liquid chromatography has been the most important and widely used of all of the column chromatographic methods. Recently, however, HPLC has been growing rapidly in importance and now rivals gas-liquid chromatography in numbers of applications.

The concept of gas-liquid chromatography was first described in 1941 by Martin and Synge, who were also responsible for liquid-partition chromatography. More than a decade was to elapse, however, before the value of this method was demonstrated experimentally.[1] Since that time, the growth in applications of the procedure has been phenomenal.[2]

[1] A. J. James and A. J. P. Martin, *Analyst*, **77**, 915 (1952); *Biochem. J.*, **50**, 679 (1952).

[2] For detailed discussions of GLC, see: *Modern Practice of Gas Chromatography*, ed. R. L. Grob. New York: Wiley-Interscience, 1977; A. B. Littlewood, *Gas Chromatography*, 2d ed. New York: Academic Press, 1970; J. Q. Walker, M. T. Jackson, Jr., and J. B. Maynard, *Chromatographic Systems*, 2d ed. New York: Academic Press, 1977, Part II; and H. M. McNair and E. J. Bonelli, *Basic Gas Chromatography*. Walnut Creek, CA: Varian Aerograph, 1971.

PRINCIPLES OF GAS–LIQUID CHROMATOGRAPHY

In principle, gas-liquid and liquid-liquid partition chromatography differ only in that the mobile phase of the former is a carrier gas rather than a liquid. The sample is introduced as a vapor at the head of the column, whereupon those components having a finite solubility in the stationary liquid phase distribute themselves between this phase and the gas according to the equilibrium law. Elution is accomplished by forcing an inert carrier gas, such as nitrogen or helium, through the column. The rates at which the various components move along the column depends upon their tendency to dissolve in the stationary liquid phase. A distribution coefficient favoring the immobilized liquid results in a low rate; on the other hand, those components whose solubility in the liquid phase is negligible move rapidly through the column. Ideally, bell-shaped (Gaussian) elution curves are obtained.

The various relationships that were developed in Chapter 24 and are summarized in Table 24-2 are all applicable to gas chromatography and can be used to calculate column efficiency, resolution, and elution time.

Specific Retention Volume

The *specific retention volume* V_g for a species is a parameter that, unlike retention time, is independent of many of the column variables but is dependent upon the nature of the solute and the stationary phase. In addition, V_g is readily derived from easily obtained experimental data. Thus, V_g is useful for qualitative analysis.

To obtain the specific retention volume for a compound, the average flow rate F within the column must be known. The flow rate at the end of the column F_m is readily obtained by means of a soap-bubble meter (p. 719); F

is then given by the expression

$$F = F_m \times \frac{T_C}{T} \times \frac{P - P_{H_2O}}{P} \qquad (26\text{-}1)$$

where T_C is the column temperature in degrees K, T is the temperature of the gas at the bubble meter, P is the pressure at the end of the column, and P_{H_2O} is the vapor pressure of water. The quantities P and T are normally the pressure and temperature of the room. The retention volume V_R for a species is then given by

$$V_R = t_R F$$

where t_R is the retention time for the species (p. 673). For a component that is not retained on the column (usually air in gas chromatography), we may also write

$$V_M = t_M F$$

where t_M is the time required for it to pass through the column.

Both V_R and V_M depend upon the average pressure *within the column*—a quantity that lies between the inlet pressure P_i and the outlet pressure P (atmospheric pressure). *Corrected retention volumes* V_R^0 and V_M^0, which correspond to volumes at the average column pressure are obtained from the relationships

$$V_R^0 = jt_R F \quad \text{and} \quad V_M^0 = jt_M F \quad (26\text{-}2)$$

The quantity j is readily derived from the relationship

$$j = \frac{3[(P_i/P)^2 - 1]}{2[(P_i/P)^3 - 1]} \qquad (26\text{-}3)$$

The specific retention volume is then defined as

$$V_g = \frac{V_R^0 - V_M^0}{W} \times \frac{273}{T_C} = \frac{jF(t_R - t_M)}{W} \times \frac{273}{T_C}$$

$$(26\text{-}4)$$

where W is the weight of the mobile phase, a quantity determined at the time of column preparation.

Relationship Between V_g and K

It is of interest to relate V_g to the partition coefficient K. To do so, we substitute the expression relating t_R and t_M to k' (Equation 24-14 p. 677) into 26-4, which gives

$$V_g = \frac{jFt_M k'}{W} \times \frac{273}{T_C} = \frac{V_M^0 k'}{W} \times \frac{273}{T_C}$$

Substituting Equation 24-12 yields (here, V_M^0 and V_M are identical)

$$V_g = \frac{KV_S}{W} \times \frac{273}{T_C} \qquad (26\text{-}5)$$

The density ρ_S of the liquid on the stationary phase is given by

$$\rho_S = \frac{W}{V_S}$$

Thus,

$$V_g = \frac{K}{\rho_S} \times \frac{273}{T_C} \qquad (26\text{-}6)$$

Note that V_g at 273°K depends only upon the partition coefficient of the solute and the density of the liquid making up the stationary phase. As such, it should be a useful parameter for identifying species.

The literature contains large numbers of specific retention volumes; unfortunately these data are widely scattered and often unreliable.

APPARATUS

Numerous instruments, varying in sophistication and ranging in price from several hundreds to several thousands of dollars, are available for gas chromatography. The basic components of these instruments are shown in Figure 26-1. A description of each component follows.

Carrier Gas Supply

Carrier gases, which must be chemically inert, include helium, argon, nitrogen, and hydrogen;

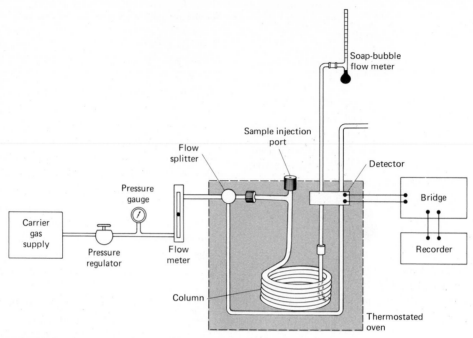

FIGURE 26-1 Schematic diagram of a gas chromatograph.

helium is the most widely used. All of these gases are available commercially in pressurized tanks. Associated with the gas supply are pressure regulators, gauges, and flowmeters. In addition, the carrier gas system often contains a molecular sieve to remove water or other impurities.

Flow rates are controlled by a pressure regulator. Inlet pressures usually range from 10 to 50 psi (above room pressure), which lead to flow rates of 25 to 50 ml/min. Generally, it is assumed that the flow rate will be constant if the inlet pressure remains constant. Flow rates can be established by a rotameter at the column head; this device, however, is not so accurate as a simple soap-bubble meter, which is located at the end of the column.

Figure 26-2 is a schematic diagram of a soap-bubble flow meter. A soap film is formed in the path of the gas when a rubber bulb containing an aqueous solution of soap or detergent is squeezed; the time required for this film to move between two graduations on the buret is measured and converted to a flow rate.

Sample Injection System

Column efficiency requires that the sample be of suitable size and be introduced as a "plug" of vapor; slow injection or oversized samples cause band spreading and poor resolution. A microsyringe is used to inject liquid samples through a rubber or silicone diaphragm into a heated sample port located at the head of the column (the sample port is ordinarily about 50°C above the boiling point of the analyte). For ordinary analytical columns, sample sizes vary from a few tenths to 10 μliter. Capillary columns require much smaller samples ($\sim 10^{-3}$ μliter); here, a sample splitter system is employed to deliver only a small fraction of the injected sample to the column head, the remainder going to waste.

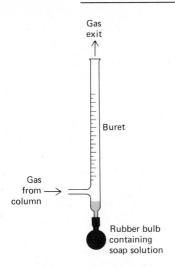

Gas exit

Buret

Gas from → column

Rubber bulb containing soap solution

FIGURE 26-2 Soap-bubble meter.

Gas samples are best introduced by means of a sampling valve such as that shown in Figure 24-9 (p. 684). Solid samples are introduced as solutions, or alternatively, are sealed into thin-walled vials that can be inserted at the head of the column and crushed from the outside.

Columns

Two types of columns are employed in gas-liquid chromatography. One type is fabricated from capillary tubing (0.3 to 0.5 mm i.d.), the bore of which is coated with a very thin film ($\sim 1\ \mu m$) of the liquid phase. Capillary columns have a negligible pressure drop and can thus be of great length (10 to 100 m or more); the ratio of V_S/V_M (Equation 24-12) for these columns ranges from 100 to 300, which in part accounts for their high efficiencies. Columns of several hundred thousand theoretical plates have been described. These columns, however, have very low sample capacities ($< 0.01\ \mu$liter). The capacity of capillary columns can be increased by coating the inside of the tube with a porous material such as graphite, a metal oxide, or a silicate. The added surface area increases the amount of liquid retained by the tube, and thus the capacity of the column.

Packed columns are fabricated from glass or metal tubes 1 to 8 mm in inside diameter; they are designed to hold solid packings that range from 2 to 20 m in length. The tubes ordinarily are folded or coiled so that they can be conveniently fitted into a thermostat. Typically, columns have V_S/V_M ratios of 15 to 20 and contain 100 to 1000 theoretical plates per foot. The best packed columns have a total of 20,000 or more theoretical plates.

Solid Support for Packed Columns. The ideal solid support would consist of small, uniform, spherical particles with good mechanical strength and a specific surface area of at least 1 m²/g. In addition, the material should be inert at elevated temperatures and be readily wetted by the liquid phase to give a uniform coating. No substance that meets all of these criteria perfectly is yet available.

The most common supports are derived from diatomaceous earths. Two types are available. *Firebrick*, which is sold under trade names such as Chromosorb P, C 22, and Sterchamol, has the better strength and the larger specific surface area (~ 4 m²/g); its disadvantage lies in the fact that it is more active and, therefore, cannot be employed with polar compounds. Kieselguhr is more fragile and has a smaller specific surface area (~ 1 m²/g) but is less reactive; it is sold under such trade names as Chromosorb W, Celite, Embacel, and Celatom.

Liquid Phase. Desirable properties for the immobilized, liquid phase in a gas-liquid chromatographic column include: (1) *low volatility*; ideally, the boiling point of the liquid should be at least 200° higher than the maximum operating temperature for the column; (2) *thermal stability*; (3) *chemical inertness*; and (4) *solvent characteristics* such that α and k' (p. 682) values for the solutes to be resolved fall within a suitable range.

No single liquid meets all of these requirements, the last in particular. As a consequence, it is common practice to have available several interchangeable columns, each with a different stationary phase. The choice among these is often a matter of trial and error, although some qualitative guidelines exist; factors affecting this choice are considered in the section devoted to applications.

Column Preparation. The support material is first screened to limit the particle size range. It is then made into a slurry with a volatile solvent that contains an amount of the stationary liquid calculated to produce a thin coating (5 to 10 μm) on all of the particles. After the solvent has been evaporated, the particles appear dry and are free-flowing.

Columns are fabricated from glass, stainless steel, copper, or aluminum. They are filled by slowly pouring the coated support into the straight tube with gentle tapping or shaking to provide a uniform packing. Care must be taken to avoid channeling. After the column has been packed, it is bent or coiled into an appropriate shape to fit the oven (see Figure 26-1).

A properly prepared column may be employed for several hundred analyses. Numerous types of packed columns are available commercially.

Column Thermostating. Column temperature is an important variable that must be controlled to a few tenths of a degree for precise work. Thus, the column is ordinarily housed in a thermostated oven. The optimum column temperature depends upon the boiling point of the sample and the degree of separation required. Roughly, a temperature equal to or slightly above the average boiling point of a sample results in an elution period of reasonable time (2 to 30 min). For samples with a broad boiling range, it is often desirable to employ temperature programming, whereby the column temperature is increased either continuously or in steps as the separation proceeds.

In general, optimum resolution is associated with minimal temperature; the cost of lowered temperature, however, is an increase

in elution time and therefore the time required to complete an analysis. Figure 26-3 illustrates this principle.

Detection Systems

Detection devices for a gas-liquid chromatograph must respond rapidly and reproducibly to the low concentrations of solutes emitted from the column. The solute concentration in a carrier gas at any instant is only a few parts per thousand at most; frequently, the detector is called upon to respond to concentrations that are smaller than this by one or two orders of magnitude (or more). In addition, the interval during which a peak passes the detector is usually one second or less; therefore, the detector must be capable of exhibiting its full response during this brief period.

Other desirable properties of the detector include linear response, good stability over extended periods, and uniform response for a wide variety of compounds. No single detector meets all of these requirements, although more than a dozen different types have been proposed. We shall describe the most widely used of these.

Thermal Conductivity Detectors. A relatively simple and broadly applicable detection system is based upon changes in the thermal conductivity of the gas stream; an instrument employed for this purpose is sometimes called a *katharometer*. The sensing element of this device is an electrically heated source whose temperature at constant electrical power depends upon the thermal conductivity of the surrounding gas. The heated element may consist of a fine platinum or tungsten wire or, alternatively, a semiconducting thermistor. The resistance of the wire or thermistor gives a measure of the thermal conductivity of the gas; in contrast to the wire detector, the thermistor has a negative temperature coefficient.

In chromatographic applications, a double detector is always employed, one element being placed in the gas stream *ahead* of the sample

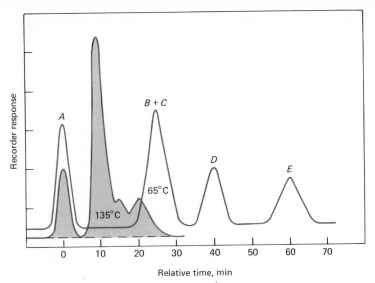

FIGURE 26-3 Effect of temperature on the separation of hexane isomers: *A*, 2,2-dimethylbutane; *B*, 2,3-dimethylbutane; *C*, 2-methylpentane; *D*, 3-methylpentane; *E*, *n*-hexane. (From C. E. Bennett, S. Dal Nogare, and L. W. Safranski, in *Treatise on Analytical Chemistry*, eds. I. M. Kolthoff and P. J. Elving, Part I, vol. 3. New York: Wiley-Interscience, 1961, p. 1690. With permission.)

injection chamber and the other immediately beyond the column. In this way, the thermal conductivity of the carrier gas is canceled, and the effects of variation in flow rate, pressure, and electrical power are minimized. The resistances of the twin detectors are usually compared by incorporating them into two arms of a simple Wheatstone bridge circuit such as that shown in Figure 26-4.

The thermal conductivities of hydrogen and helium are roughly six to ten times greater than those of most organic compounds. Thus, the presence of even small amounts of organic materials causes a relatively large decrease in the thermal conductivity of the column effluent; the detector undergoes a marked rise in temperature as a result. The conductivities of nitrogen and carbon dioxide more closely resemble those of organic constituents; thus, detection by thermal conductivity is less sensitive when these substances are used as carrier gases.

Thermal conductivity detectors are simple, rugged, inexpensive, nonselective, accurate, and nondestructive of the sample. They are not so sensitive, however, as some of the other devices to be described.

Flame Ionization Detectors. Most organic compounds, when pyrolyzed at the temperature of a hydrogen/air flame, produce ionic intermediates that provide a mechanism by which electricity can be carried through the flame. By employing an apparatus such as that shown in Figure 26-5, these ions can be collected and the resulting ion current measured. The electrical resistance of a flame is very high (perhaps 10^{12} Ω), and the resulting currents are therefore minuscule; an electrometer must be employed for their measurement.

The ionization of carbon compounds in a

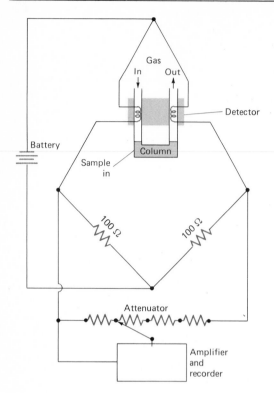

Gas
In   Out

Detector

Battery

Column

Sample
in

100 Ω   100 Ω

Attenuator

Amplifier
and
recorder

FIGURE 26-4 Schematic diagram of a thermal conductivity detector-recorder system.

X-radiation (p. 440). Here, the effluent from the column is passed over a β-emitter such as nickel-63 or tritium (adsorbed on platinum or titanium foil). An electron from the emitter causes ionization of the carrier gas (often nitrogen) and the production of a burst of electrons. In the absence of organic species, a constant standing current between a pair of electrodes results from this ionization process. The current decreases, however, in the presence of organic molecules that tend to capture electrons. The response is nonlinear unless the potential across the detector is pulsed.

The electron-capture detector is selective in its response and is highly sensitive toward electronegative functional groups such as halogens, peroxides, quinones, and nitro groups. It is insensitive to compounds such as amines, alcohols, and hydrocarbons. An important application of the electron-capture detector has been for the detection and determination of chlorinated insecticides.

Electron-capture detectors are highly sensitive and possess the advantage of not altering the sample significantly (in contrast to the flame detector).

APPLICATIONS OF GAS–LIQUID CHROMATOGRAPHY

In evaluating the importance of gas-liquid chromatography, it is necessary to distinguish between the two roles the method plays in its application to chemical problems. The first is as a tool for performing separations; in this capacity, it is unsurpassed when applied to complex organic, metal-organic, and biochemical systems. The second, and distinctly different function, is that of providing the means for completion of an analysis. Here, retention times or volumes are employed for qualitative identification, while peak heights or areas provide quantitative information. In its analytical role, gas-liquid chromatography is considerably more limited than some of the other methods considered in earlier chapters.

flame is a poorly understood process, although it is known that the number of ions produced is roughly proportional to the number of reduced carbon atoms in the flame. Functional groups such as carbonyl, alcohol, and amine produce fewer ions or none at all.

The hydrogen flame detector currently is one of the most popular and most sensitive detectors. It is more complicated and more expensive than the thermal conductivity detector, but has the advantage of higher sensitivity. In addition, it has a wide range of linear reponse. It is, of course, destructive of the sample.

Electron-Capture Detectors. Electron-capture detectors operate in much the same way as a proportional counter for measurement of

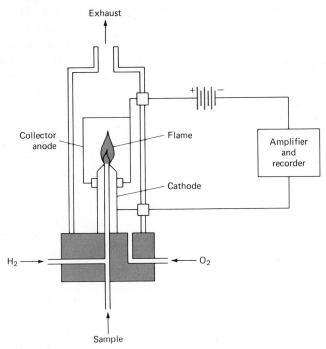

FIGURE 26-5 Hydrogen flame-ionization detector.

As a consequence, an important trend in the field appears to be in the direction of combining the remarkable fractionation qualities of gas-liquid chromatography with the superior analytical properties of such instruments as mass, ultraviolet, infrared, and NMR spectrometers.

Choice of the Stationary Liquid Phase

Several hundred liquids have been suggested for use as the stationary phase in gas-liquid chromatography. The successful separation of closely related compounds is often critically dependent upon the proper choice among these.

The retention time for a solute depends directly upon its partition coefficient (Equation 24-13), which, in turn, is related to the nature of the stationary phase. Clearly, to be useful in gas-liquid chromatography the immobilized liquid must generate different partition coefficients among solutes; additionally, however, these coefficients must be neither extremely small nor extremely large. Solutes with small coefficients pass through the column so rapidly that no significant separation occurs. On the other hand, when the coefficients are large, the time required to remove solutes from the columns becomes inordinate.

To have a reasonable residence time in the column, a solute must show at least some degree of compatability (solubility) with the stationary phase. Thus, the polarities of the two substances should be at least somewhat alike. For example, a stationary liquid such as squalane (a high molecular weight saturated hydrocarbon) might be chosen for separation of members of a nonpolar homologous series such as hydrocarbons or ethers. On the other hand, a more polar liquid such as polyethylene glycol would probably be more effective for separating alcohols or amines. For aromatic hydrocarbons, benzyldiphenyl might prove appropriate.

Among solutes of similar polarity, the elution order usually follows the order of boiling points; where these differ sufficiently, clean separations are feasible. Solutes with nearly identical boiling points but different polarities frequently require a liquid phase that will selectively retain one (or more) of the components by dipole interaction or by adduct formation. The data in Table 26-1 illustrate some of these effects. Here, the retention times (relative to ethane) for a series of paraffinic and olefinic hydrocarbons are compared for three liquid phases. The first, triisobutylene, is nonpolar; the other two are polar. The third solvent, in addition, contains silver nitrate, which selectively forms loose adducts with specific olefins. Note that the retention times on triisobutylene correlate closely with the boiling points of the liquids and that little or no separation is realized among the compounds with similar volatilities. Acetonyl acetone, on the other

hand, is moderately polar and capable of inducing polarization in the olefins; selective retention results. Note also that the retention times for the nonpolarizable paraffins are significantly less in the polar solvent than in the nonpolar one. The selectivity of adduct formation between olefins and silver nitrate permits some useful separation as well; for example, isobutene and butene-1 are not separated on the first two columns but are resolved in the presence of silver nitrate; improved separation of *cis*- and *trans*-butene-2 also occurs.

Another important interaction that often enhances selectivity is hydrogen bond formation. For this effect to operate, the solute must have a polar hydrogen atom and the solvent must have an electronegative group (oxygen, fluorine, or nitrogen)—or the converse must be true.

Several systems have been developed to optimize the choice of a stationary phase for a separation within any given group of compounds. One characterizes stationary phases by comparing retention indices (p. 726) for benzene, ethanol, methylethyl ketone, nitromethane, and pyridine relative to those on

TABLE 26-1   SEPARATION OF HYDROCARBONS WITH VARIOUS STATIONARY LIQUIDS[a]

Compound	Boiling Point, °C	Relative Retention Time On		
		Triiso-butylene	Acetonyl Acetone	$AgNO_3$ in Glycol
Ethane	−104	1.0	1.0	1.0
Isobutane	−11.7	5.6	2.2	0.75
Isobutene	−6.9	6.7	4.75	3.25
Butene-1	−6.3	6.7	4.75	6.25
Butadiene	−4.4	6.7	10.0	10.0
n-Butane	−0.5	7.5	3.0	0.75
trans-Butene-2	0.88	8.0	5.85	1.75
cis-Butene-2	3.72	8.9	6.8	5.5

[a] Data from B. W. Bradford, D. Harvey and D. E. Chalkley, *J. Inst. Petroleum*, **41**, 80 (1955). With permission.

squalene.[3] The resulting data provide a quantitative means for classifying stationary media and predicting the performance of individual packings for various types of separations. Unfortunately, the effort required to obtain the required data is usually greater than that necessary to determine empirically which of several columns will bring about the required separation. Thus, most chemists have several columns available, and, by trial and error, determine which is the best for solving a particular problem.

Table 26-2 lists a few of the most widely used stationary liquid phases.

Qualitative Analysis

Gas chromatography, like liquid chromatography, provides the scientist with a powerful method for separating complex mixtures. It is much less useful, however, as a tool for identifying separated components.

Retention Time or Retention Volume. A single retention-time measurement is not a very satisfactory parameter upon which to base a qualitative analysis because of its dependence on such variables as column temperature, flow rate, pressure, and composition of the stationary phase. To be sure, if the retention times for an unknown and a standard are found to vary similarly as these operating conditions are varied, the probability of identity of the two is enhanced. Thus, if similar retention times for the standard and analyte are observed on two or three columns at two or three temperatures, it is ordinarily reasonable to assume that the analyte and standard are the same. Usually, however, the separated component is more easily identified by spectroscopic measurements than through several chromatograms.

[3] W. R. Supina and L. P. Rose, *J. Chromatog. Sci.*, **8**, 214 (1970).

TABLE 26-2   SOME COMMON STATIONARY PHASES

Name	Chemical Composition	Maximum Temperature, °C	Polarity[a]	Type of Separation
Squalene	$C_{30}H_{62}$	150	NP	Hydrocarbons
OV-1	Polymethyl siloxane	350	NP	General purpose nonpolar
DC 710	Polymethylphenyl siloxane	300	NP	Aromatics
QF-1	Polytrifluoropropyl methyl siloxane	250	P	Amino acids, steroids, nitrogen compounds
XE-30	Polycyanomethyl siloxane	275	P	Alkaloids, halogenated compounds
Carbowax 20M	Polyethylene glycol	250	P	Alcohols, esters, essential oils
DEG adipate	Diethylene glycol adipate	200	SP	Fatty acids, esters
	Dinonyl phthalate	150	SP	Ketones, ethers, sulfur compounds

[a] NP = nonpolar; SP = semipolar; P = polar.

As was noted earlier, the specific retention volume is a better parameter upon which to base a qualitative analysis than is the retention time, because it depends only upon the partition coefficient of the analyte, the temperature of the column, and the density of the stationary phase (Equation 26-6).

Relative Retention time. The selectivity factor α is a measure of relative retention time. As shown in Chapter 24, α is defined as

$$\alpha = \frac{(t_R)_Y - t_M}{(t_R)_X - t_M}$$

Thus, by introducing an internal standard Y into a sample containing the analyte X, the readily measured analytical parameter α is obtained. For a given column, this parameter should be independent of the column variables, provided that X and Y are chemically similar. Clearly, the peaks for X and Y must not overlap and the two species must be chemically compatible with each other and the mobile phase. The selectivity factor has the advantage of requiring only three simple measurements $(t_R)_X$, $(t_R)_Y$, and t_M.

The Retention Index. The *retention index* (*I*) was proposed by Kovats[4] as a qualitative parameter for general use in reporting chromatographic data. The retention index is based upon a comparison between the position of an analyte peak and the peaks for two or more normal paraffins. By definition, the retention index for a normal paraffin is equal to 100 times the number of carbon atoms in the compound, *regardless of the columns used or the chromatographic conditions.* Thus, *I* for *n*-pentane is *always* 500; *n*-octane it is 800.

It has been observed that, within a homologous series, a plot relating the logarithm of retention time (or volume) to the number of carbon atoms is linear, provided the lowest member of the series is excluded. As a consequence, a plot of $(t_R - t_M)$ versus the retention

index for a series of *n*-paraffins will be a straight line.

The retention index for species X is then defined as

$$I_X = 100\left[\frac{\log(t'_R)_X - \log(t'_R)_n}{\log(t'_R)_{n+1} - \log(t'_R)_n}\right] + 100n$$

$$(26\text{-}7)$$

where (t'_R) is the corrected retention time for each species; that is

$$t'_R = (t_R - t_M)$$

In Equation 26-7, the subscript X refers to the analyte and subscripts n and $(n + 1)$ refer to normal paraffins containing n and $(n + 1)$ carbon atoms, respectively. Generally, the hydrocarbons are chosen so that their retention times bracket that of the analyte.

EXAMPLE

From the following data, calculate the retention index for benzene on a squalene column operated at 100°C.

Substance	$(t_R - t_M)$, min	I
n-hexane	3.43	600
benzene	4.72	?
n-heptane	6.96	700

Employing Equation 26-7, we find

$$I = 100\left[\frac{\log 4.72 - \log 3.43}{\log 6.96 - \log 3.43}\right] + 100 \times 6$$

$$= 45.1 + 600 = 645$$

The retention index system has the advantage of having readily available reference materials that cover a wide boiling range. In addition, the temperature dependence of retention indices is relatively small. Furthermore, the change in retention index ΔI between a polar and a nonpolar stationary phase provides a

[4] E. Kovats, *Helv. Chim. Acta,* **41**, 1915 (1958).

measure of the relative polarity of different stationary phases.[5]

Retention Time Plots. The linear relationship between log t_R and the number of carbon atoms for members of a homologous series (p. 726) can be useful for qualitative work, since it is often possible to separate one homologous series from another by chromatography (if the two have different polarities) or by other physical or chemical means

An even more useful relationship can sometimes be obtained by injecting portions of the sample into two columns with stationary phases that differ in polarity. For a homologous series a plot of the retention time on one column versus that on the other (employing a log scale) yields a straight line (see Figure 26-6); more important, the intercepts of the plots differ from one series to another. Thus, resolution of species with similar boiling points but differing polar characteristics frequently can be achieved by obtaining chromatograms for the sample on two columns. In addition, the peaks may be identified with considerably less ambiguity. This technique is quite analogous to two-dimensional paper chromatography.

Summary. A single gas chromatogram does not provide sufficient data to permit the unambiguous identification of a compound. On the other hand, identical retention index data on three columns of differing polarity would provide a good basis for assuming a standard and an analyte were the same species.

Gas chromatograms are widely used as criteria of purity for volatile compounds. Contaminants, if present, are revealed by the appearance of additional peaks; the areas under these peaks provide rough estimates of the extent of contamination. The technique is also useful for evaluating the effectiveness of purification processes.

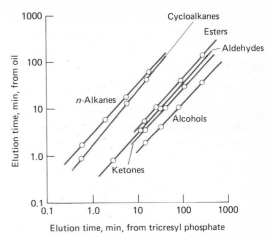

FIGURE 26-6 Behavior of several homologous series on two columns. Points from left to right correspond to the following:
n-alkanes: butane, pentane, hexane, heptane
cycloalkanes: cyclopropane, cyclopentane, cyclohexane
esters: methyl-, *n*-propyl-, *n*-butyl- acetates
aldehydes: ethanal, *n*-propanal, *n*-butanol
ketones: acetone, 2-butanone, 2-pentanone
alcohols: methanol, ethanol, *n*-propanol, *n*-butanol
[Reprinted with permission from J. S. Lewis, H. W. Patton and W. L. Kaye, *Anal. Chem.*, **28**, 1370 (1956). Copyright by the American Chemical Society.]

Quantitative Analysis

The detector signal from a gas-liquid chromatographic column has had wide use for quantitative and semiquantitative analyses. Under carefully controlled conditions, an accuracy of 1 to 3% relative is attainable. As with most analytical tools, reliability is directly related to the amount of effort spent in calibration and control of variables; the nature of the sample also plays a part in determining the potential accuracy.

The general discussion of quantitative chromatographic analysis given in Chapter 24 applies to gas chromatography as well as to other

[5] For details, see: W. R. Supina and L. P. Rose, *J. Chromatog. Sci.*, **8**, 214 (1970); and W. O. McReynolds, *J. Chromatog. Sci.*, **8**, 685 (1970).

types; therefore, no further consideration of this topic is needed here.

Programmed Temperature Gas Chromatography

In programmed temperature gas chromatography, the column temperature is increased (usually linearly) as elution proceeds. Such a procedure has many of the advantages of gradient elution in liquid chromatography. It is used to simplify, improve, and accelerate separations.

Figure 26-7 illustrates the effectiveness of the programmed temperature approach to the separation and determination of a mixture of normal paraffins. With the isothermal method, the low molecular weight peaks are bunched,

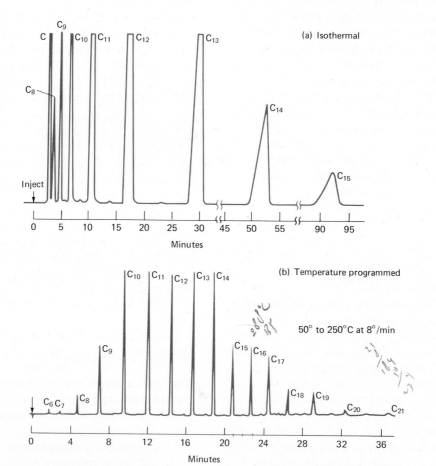

FIGURE 26-7 Comparison of (a) isothermal and (b) temperature-programmed chromatograms for some normal paraffins. Columns: 20 ft by 1/16 in with 3% Apiezon on 100-120 mesh VarAport 30. Flow rate: 10 ml/min with He. Temperature: (a) 150°C and (b) 50 to 250°C at 8°/min. (From H. M. McNair and E. J. Bonelli, *Basic Gas Chromatography*. Walnut Creek, CA: Varian Aerograph, 1969, p. 190. With permission.)

while peaks for the highest molecular weight components have not appeared even after 95 min of elution. Furthermore, the shapes of the higher molecular weight peaks that do appear are not very satisfactory for quantitative analysis.

In contrast, the programmed temperature chromatogram provides well-spaced peaks with heights that are readily measured.

Programmed temperature work requires a device for varying the temperature linearly and reproducibly at a rate varying from about 0.25 to 20°C per minute. Furthermore, the procedure requires separate heaters for the injection port and the detector in order to keep these components at constant temperature while the column temperature is being varied. In order to achieve a constant flow rate, a flow controller is needed because a temperature rise causes an increase in gas viscosity and column resistance; both of these variables tend to reduce the flow. The result of changes in flow rate is an unstable base line and changes in detector response.

Gas—Liquid Chromatography with Selective Detectors

Most of the detection systems that we have considered thus far are nonselective in the sense that they respond in more or less the same way to all of the solutes emerging from a column (ideally, the nonselective detector would respond identically to every compound). The employment of selective or specific detectors, however, makes it possible to combine the tremendous separatory power of gas-liquid chromatography with the superior analytical qualities of the instruments we have considered in earlier chapters.

Two general approaches are possible. In the first, the solute vapors are collected as separate fractions in a cold trap, a nondestructive and nonselective detector being employed to indicate their appearance. The fractions are then identified by NMR, IR, mass, or other spectroscopic techniques. The main limitation to this approach lies in the very small (usually micromolar) quantities of solute contained in a fraction; nonetheless, the general procedure has proved useful in the qualitative analyses of components of many complex mixtures.

A second general method involves the use of a selective detector to monitor the column effluent continuously. Some examples will illustrate this approach.

Automatic Titration Detectors. In their first publication on gas chromatography, James and Martin employed an automatic titration cell as a detector for volatile fatty acids; the same device was later applied to mixtures of aromatic and aliphatic amines. In this method, the effluent from the column was led directly into a cell containing an acid-base indicator solution. The absorbance of the solution was monitored photometrically, the output from the cell being employed to control continuously the addition of titrant; the volume of the latter was recorded on a chart to produce an integral curve of volume versus time. Both qualitative and quantitative information could be retrieved from the recorded curve.

A coulometric cell has been employed as the detector in the analysis of the various chlorinated compounds contained in pesticide residues.

Coupling of Gas Chromatography to Mass Spectroscopy.[6] Chromatographic columns have been directly interfaced to rapid-scan mass spectrometers, thus permitting the instantaneous display of the spectrum of each species as it leaves the column. Generally, these instruments are also interfaced with a computer so that each spectrum is digitalized

---

[6] For additional information, see: F. W. Karasek, *Anal. Chem.*, **44** (4), 32A (1972); and C. Fenselau, *Anal. Chem.*, **49**, 563A (1977). For information on liquid chromatography/mass spectroscopy, see: W. H. McFadden, D. C. Bradford, D. E. Ganes, and J. L. Gower, *Amer. Lab.*, **9** (10), 55 (1977).

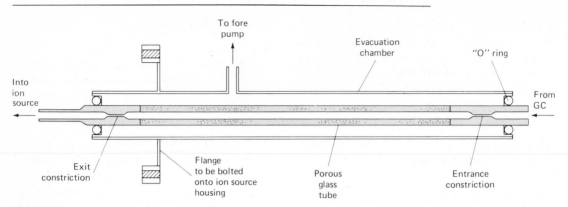

FIGURE 26-8 Pressure reduction system, the link between gas chromatograph and mass spectrometer. [Reprinted with permission from J. T. Watson and K. Biemann, *Anal. Chem.*, **37**, 844 (1965). Copyright by the American Chemical Society.]

and stored for later reproduction on paper. Instruments of this type make possible the identification of the hundreds of components that may be present in natural and biological systems. For example, the interfacing of chromatography with spectroscopy has permitted characterization of the odor and flavor components of foods, identification of pollutants, medical diagnosis based on breath components, and studies of drug metabolites.

All major manufacturers of mass spectrometers offer units for interfacing their equipment with gas chromatography. These units range from $20,000 to $125,000, with computerization adding 50 to 100% to these costs.

A major problem in interfacing of a gas chromatograph with a mass spectrometer arises from the presence of the carrier gas, which dilutes the eluted components enormously and also tends to swamp the pumping system of the spectrometer. Several methods have been developed for overcoming this problem; one is shown in Figure 26-8. Here, the exit gases flow through a fritted glass tube situated in an evacuated chamber. The smaller atoms or molecules of the carrier gas (He or H_2) diffuse readily through the walls of the tube and are pumped away, leaving the molecules of the

eluted sample; these are then led directly into the ion source of the mass spectrometer. Both quadrupole and small double-focusing spectrometers have been employed for analysis of the resulting ionic beam.

Infrared Spectroscopy/Chromatography.[7] Instruments have also been developed that will provide infrared spectra of chromatographic peaks. Both rapid-scan and Fourier transform instruments have been employed. Generally, the effluent from the column is led into a long narrow tube equipped with infrared-transparent windows. Scanning is triggered by the output from a nondestructive chromatographic peak detector and begins after a brief delay to allow the component to travel from the detector region to the infrared cell. The spectral data are digitalized and stored in a computer from which printed spectra are ultimately derived.

Difficulty is sometimes encountered when attempts are made to compare spectra for the

---

[7] For additional information, see: K. L. Kizer, *Amer. Lab.*, **5** (6), 40 (1973); F. D. Mercaldo, *Amer. Lab.*, **6**, 63 (1974); R. H. Shaps, W. Simons, and A. Varano, *Amer. Lab.*, **9** (3), 95 (1977); and V. Rossiter, *Amer. Lab.*, **11** (5), 59 (1979).

gaseous effluents from a column with library spectra that have been obtained with liquid or solid samples. Gaseous spectra contain rotational fine structure while liquid or solid spectra do not; significant differences in appearance are the result.

GAS–SOLID CHROMATOGRAPHY

Gas-solid chromatography is based upon adsorption of gaseous substances on solid surfaces. Distribution coefficients are generally much larger than those for gas-liquid chromatography. Thus, gas-solid chromatography is useful for the separation of species that are not retained in a gas-liquid column, such as the components of air, hydrogen sulfide, carbon disulfide, nitrogen oxides, carbon monoxide, carbon dioxide, and the rare gases.

The most common adsorbents are molecular sieves and certain porous polymers.

Molecular Sieves

Molecular sieves are zeolites of carefully controlled pore size. These pores or holes are of molecular dimensions; molecules that are smaller than the holes penetrate into the interior of the sieve particles where they can be adsorbed. For such molecules, the surface area is enormous when compared with the area available to molecules that cannot reach the interior surfaces. Thus, sieves can be used to separate small molecules from larger ones.

Partition coefficients among small molecules on molecular sieves differ considerably, thus making possible the chromatographic separation of such species as oxygen, nitrogen, carbon monoxide, hydrogen, and methanol, some of which are not retained to any significant extent on liquid columns. One problem with molecular sieves is that polar compounds, such as carbon dioxide, may become more or less permanently adsorbed, thus blocking the surface to other species.

Porous Polymers

Porous polymer beads of uniform size are manufactured from styrene cross-linked with divinylbenzene. These materials are sold under the trade name of Porapak.[8] The pore size of these materials is uniform and is controlled by the method of manufacture. The application of Porapaks to gas-solid chromatography is analogous to the application of Sephadex beads in gel-permeation chromatography (p. 703).

The Porapaks have found considerable use in the separation of gaseous polar species such as hydrogen sulfide, oxides of nitrogen, water, carbon dioxide, methanol, and vinyl chloride.

EXAMPLES OF APPLICATIONS OF GAS CHROMATOGRAPHY

Figure 26-9 illustrates some typical applications of gas chromatography to analytical problems. All are based upon gas-liquid equilibria except the one shown in Figure 26-9a. Here, a gas-liquid column is followed by a gas-solid molecular sieve. The former retains only the carbon dioxide and passes the remaining gases at rates corresponding to the carrier rate. When the carbon dioxide is eluted from the first column, a switch directs the flow around the second column so that the gas is not permanently adsorbed in the molecular sieve. After the carbon dioxide signal has returned to zero, the flow is switched back through the second column, thereby permitting elution of the remainder of the sample components.

Data on the conditions employed for obtaining the various chromatograms in Figure 26-9 are found in Table 26-3.

---

[8] Waters Associates, Inc., Milford, Massachusetts.

(a) Exhaust mixture: A, 35% H_2; B, 25% CO_2; C, 1% O_2; D, 1% N_2; E, 1% C_2; F, 30% CH_4; G, 3% CO; H, 1% C_3; I, 1% C_4; J, 1% i-C_5; K, 1% n-C_5.

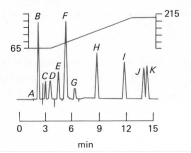

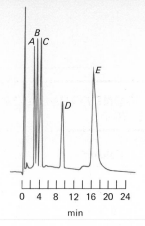

(b) Pesticides: A, Lindane; B, Heptachlor; C, Aldrin; D, Dieldrin; E, DDT. A–D, 0.3 ng; E, 3.0 ng.

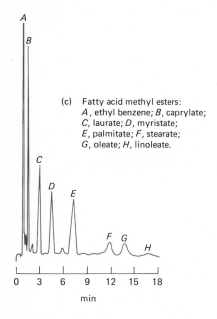

(c) Fatty acid methyl esters: A, ethyl benzene; B, caprylate; C, laurate; D, myristate; E, palmitate; F, stearate; G, oleate; H, linoleate.

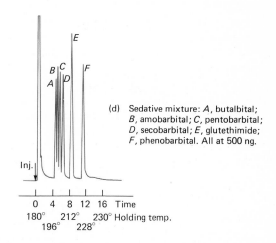

(d) Sedative mixture: A, butalbital; B, amobarbital; C, pentobarbital; D, secobarbital; E, glutethimide; F, phenobarbital. All at 500 ng.

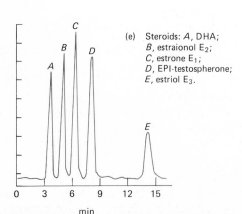

(e) Steroids: A, DHA; B, estraionol E_2; C, estrone E_1; D, EPI-testospherone; E, estriol E_3.

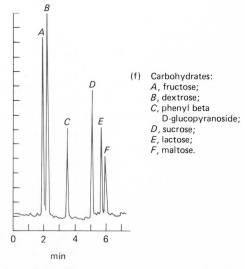

(f) Carbohydrates: A, fructose; B, dextrose; C, phenyl beta D-glucopyranoside; D, sucrose; E, lactose; F, maltose.

FIGURE 26-9 Some examples of gas-chromatographic separations.

TABLE 26-3   CONDITIONS FOR CHROMATOGRAMS SHOWN IN FIGURE 26-8

Chromatogram	Column[a]	Packing	Detector[b]	Temperature, °C	Carrier	Flow, ml/min
(a)	10′ × $\frac{1}{8}$″ S 5′ × $\frac{1}{8}$″ S	Chromosorb 102 Molecular sieve 5A	TCD	65–200	He	30
(b)	6′ × $\frac{1}{4}$″ G	1.5% OV-17 on Chromosorb G	ECD	22		
(c)	6′ × $\frac{1}{8}$″ S	Chromosorb W	TCD	190	He	30
(d)	6′ × $\frac{1}{4}$″ G	1.5% OV-17 on HP Chromosorb G	FID	180–230		
(e)	6′ × 3.4 mm G	5% OV-210 + 2.5% OV-17 on Supelcoport	ECD	260	A	
(f)	6′ × 3.4 mm	2% OV-17 on Chromosorb W	FID	105-325	N$_2$	40

[a] S = stainless steel; G = glass.
[b] TCD = thermal conductivity; ECD = electron capture; FID = flame ionization.

PROBLEMS

1. A GLC column was operated under the following conditions:

   Column: 1.20 m × 2.0 mm packed with Chromosorb P; weight of stationary liquid added, 1.35 g; density of liquid, 1.12 g/ml

   Pressures: inlet, 25.1 psi above room; room, 748 torr

   Measured outlet flow rate: 27.3 ml/min

   Temperature: room, 21.2°C; column, 102.0°C

   Retention times: air, 24.0 s; methyl acetate, 2.12 min; methyl propionate, 4.32 min; methyl *n*-butyrate, 8.63 min

   Peak widths at base: 0.21, 0.42, and 0.85 min, respectively

   Calculate

   (a) the average flow rate in the column.

   (b) the corrected retention volumes for air and the three esters.

   (c) the specific retention volumes for the three components.

   (d) partition coefficients for each of the esters.

   (e) a retention volume and a retention time for methyl *n*-hexanoate.

2. From the data in Problem 1, calculate

   (a) k' for each compound.

   (b) α values for each adjacent pair of compounds.

   (c) the average number of theoretical plates and plate height for the column.

   (d) the resolution for each adjacent pair of compounds.

3. For the chromatogram described in Problem 1, the areas of the three peaks were found to be 18.1, 43.6, and 29.9 in the order of increasing retention time. Calculate the percent of each compound if the relative

detector response for the three species was 0.60, 0.78, and 0.88, respectively.

4. The stationary liquid in the column described in Problem 1 was didecylphthalate, a solvent of intermediate polarity. If a nonpolar solvent such as a silicone oil had been used instead, would the retention times for the three compounds be larger or smaller? Why?

5. Corrected retention times for ethyl, n-propyl, and n-butyl alcohols on a column employing a packing coated with silicone oil are 0.69, 1.51, and 3.57. Predict retention times for the next two members of the homologous series.

6. A chromatographic column was operated under the following conditions:

    Column: 1.00 m × 2.0 mm packed with Chromosorb P; weight of paraffin stationary liquid, 2.10 g; density, 0.796 g/ml
    Pressures: inlet, 18.3 psi above room; room, 768 torr
    Outlet flow rate: 33.6 ml/min
    Temperatures: room, 19.6°C; column, 90.3°C
    Retention time air: 0.42 min
    Retention times and peak widths: i-propylamine, 4.59 and 0.36 min; n-propylamine, 4.91 and 0.39 min

    Calculate
    (a) the average flow rate in the column.
    (b) the corrected retention volumes for air and the two amines.
    (c) the specific retention volumes for the two amines.
    (d) partition coefficients for the two amines.

7. From the data in Problem 6, calculate
    (a) k' for each amine.
    (b) an α value.
    (c) an average number of theoretical plates in the column.
    (d) the plate height for the column.
    (e) the resolution for the two amines.
    (f) the length of column required to achieve a resolution of 1.5.
    (g) the time required to achieve a resolution of 1.5, assuming the original linear flow rate.

8. For a gas-chromatographic column, the values for A, B, and C in the van Deemter equation were 0.15 cm, 0.36 cm^2 s^{-1}, and 4.3 × 10^{-2} s. Calculate the minimum plate height and the best flow rate.

9. What would be the effect of the following on the plate height of a column? Explain.
    (a) increasing the weight of the stationary phase relative to the packing weight.
    (b) decreasing the rate of sample injection.
    (c) increasing the injection port temperature.
    (d) increasing the flow rate.
    (e) reducing the particle size of the packing.
    (f) decreasing the column temperature.

10. What would be the effect of the variations listed in Problem 9 on the retention time for a chromatographic peak? Explain.

11. For a gas chromatographic column, the values of A, B, and C in the van Deemter equation were 0.060 cm, 0.57 cm^2 s^{-1}, and 0.13 s. Calculate the minimum plate height and the optimum flow rate.

12. Calculate the retention index for each of the following compounds:

Compound	$(t_R - t_M)$
(a) Propane	1.29
(b) *n*-Butane	2.21
(c) *n*-Pentane	4.10
(d) *n*-Hexane	7.61
(e) *n*-Heptane	14.08
(f) *n*-Octane	25.11
(g) Toluene	16.32
(h) Butane-2	2.67
(i) *n*-Propanol	7.60
(j) Methylethyl ketone	8.40
(k) Cyclohexane	6.94
(l) *n*-Butanol	9.83

13. Predict the corrected retention times for ethanol and *n*-hexanol on the column in Problem 12.

APPENDIX 1: PROPAGATION OF UNCERTAINTIES IN PHYSICAL MEASUREMENTS

Whenever measurements are made with an instrument having a sensitivity that is not limited by its readout device, small variations among replicate measurements are observed. These variations represent the accumulation of a large number of random extraneous signals (noise) that develop in various instrument components. In some instances, the direction of the noise from most of the components will, by chance, be positive, and a higher than average reading will result; at other times, negative noise signals may predominate, leading to a result that is less than average. The most probable event, however, is for the number and size of the negative and positive noise signals to be about the same, thus giving a reading that approaches an average or mean value.

For a large number of replicate measurements, the accumulated uncertainties due to random noise cause the results to be distributed in the symmetrical way shown in Figure A-1. This distribution, which is termed *Gaussian*, can be described in terms of three parameters; that is,

$$y = \frac{e^{-(x_i - \mu)^2/2\sigma^2}}{\sigma\sqrt{2\pi}} = \frac{e^{-z^2/2}}{\sigma\sqrt{2\pi}} \qquad (A-1)$$

In this equation, x_i represents values of individual measurements, and μ is the arithmetic mean for an infinite number of such measurements. The quantity $(x_i - \mu)$ is thus the deviation from the mean; y is the frequency of occurrence for each value of $(x_i - \mu)$. The symbol π has its usual meaning, and e is the base for Napierian logarithms. The parameter σ is called the *standard deviation* and is a constant that has a unique value for any set of data comprising a large number of measurements. The breadth of the normal error curve, which increases as the precision of a measurement decreases, is directly related to σ. Thus, σ is widely employed to define the precision of instruments.

The exponential term in Equation A-1 can be simplified by introducing the variable

$$z = \frac{x_i - \mu}{\sigma} \qquad (A-2)$$

which then gives the deviation from the mean in units of standard deviations.

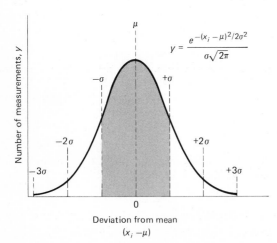

FIGURE A-1 A normal or Gaussian distribution curve.

THE STANDARD DEVIATION

Equation A-1 indicates that a unique distribution curve exists for each value of the standard deviation. Regardless of the size of σ, however, it can be shown that 68.3% of the area beneath the curve lies within one standard deviation ($\pm 1\sigma$) of the mean, μ (shaded region, Figure A-1). Thus, 68.3% of the results from replicated measurements would be expected to lie within these boundaries. Approximately 95.5% of all values should lie within $\pm 2\sigma$; 99.7% should appear within $\pm 3\sigma$. Values of $(x_i - \mu)$ corresponding to $\pm 1\sigma$, $\pm 2\sigma$, and $\pm 3\sigma$ are indicated by broken vertical lines in Figure A-1.

These properties of a Gaussian curve are

useful because they permit statements to be made about the probable magnitude of the uncertainty for any given measurement, *provided the standard deviation for the output of an instrument is known.* Thus, if σ for an instrument is available, one can say that the chances are 68.3 out of 100 that the uncertainty associated with any *single* measurement is smaller than $\pm 1\sigma$, that the chances are 95.5 out of 100 that the error is less than $\pm 2\sigma$, and so forth. Clearly, the standard deviation for a method of measurement is a useful quantity for estimating and reporting the probable size of instrumental uncertainties arising from random noise.

For a very large set of data the standard deviation is given by

$$\sigma = \sqrt{\frac{\sum_{i=1}^{N} (x_i - \mu)^2}{N}} \qquad \text{(A-3)}$$

Here, the sum of the squares of the individual deviations from the mean $(x_i - \mu)$ is divided by the total number of measurements in the set, N. Extraction of the square root of this quotient gives σ.

Equations A-1 and A-3 apply exactly only as the number of measurements approaches infinity. When N is small, a better estimate for the standard deviation is given by

$$s = \sqrt{\frac{\sum_{i=1}^{N} (x_i - \bar{x})^2}{N - 1}} \qquad \text{(A-4)}$$

Note that the symbol s is used to distinguish the experimentally realizable estimate of the standard deviation from the theoretical σ. Furthermore, the experimental mean of this set $\bar{x}$ can only approximate the true mean for an infinite number of measurements μ. When N is greater than 20 to 30, it is usually safe to assume that $s \to \sigma$.

EXAMPLE

The following data were obtained for the determination of mercury in the flesh of a trout by a fluorescence method: 1.67, 1.63,

and 1.70 ppm. Calculate the standard deviation for the measurement.

x_i	$x_i - \bar{x}$	$(x_i - \bar{x})^2$
1.67	0.003	0.00001
1.63	-0.037	0.00137
1.70	0.033	0.00109
3 $\overline{|\,5.00}$		$\sum_{i=1}^{3} (x_i - \bar{x})^2 = 0.00247$

$x = 1.667$

$$s = \sqrt{\frac{0.00247}{3 - 1}} = 0.351 = 0.04$$

Generally, in a calculation of this type, it is wise to avoid rounding until the final result has been obtained.

Another precision term widely employed by statisticians is the *variance*, which is equal to σ^2. Most experimental scientists prefer to employ σ rather than σ^2 because the units of the standard deviation are the same as those of the quantity measured.

CONFIDENCE INTERVALS

The true mean value (μ) of a measurement is a constant that must always remain unknown. With the aid of statistical theory, however, limits may be set about the experimentally determined mean $(\bar{x})$ within which we may expect to find the true mean with a given degree of probability; the limits obtained in this manner are called *confidence limits.* The interval defined by these limits is known as the *confidence interval.* The confidence interval can be determined from the value of z in Equation A-2. For example, if z is given a value of $+1$, then Equation A-2 takes the form

$$x_i - \mu = \sigma$$

and if z is equal to -1,

$$x_i - \mu = -\sigma$$

As mentioned earlier, approximately 68% of

the area under a normal distribution curve lies within the limits of $\pm 1\sigma$. Thus, there exists a 68% probability that the deviation from the mean $(x_i - \mu)$ for any *single* measurement will be within this interval. Similarly, when z is equal to ± 2,

$$x_i - \mu = \pm 2\sigma$$

Here, 96% of the area under the curve lies within these limits; thus there exists a 96% probability that the deviation from the mean of an individual measurement will lie within this range. Areas in terms of percentages are termed *confidence levels*. Confidence levels for a few values of z are given below.

Confidence Level	50%	68%	90%
z	± 0.67	± 1.00	± 1.64
Confidence Level	95%	96%	99%
z	± 1.96	± 2.00	± 2.58

The confidence limit for a single measurement is obtained by rearranging Equation A-2 to

$$\text{confidence limit for } \mu = x_i - z\sigma$$

When x_i is replaced by $\bar{x}$, the mean for several measurements, the confidence limit is decreased by $\sqrt{N}$. Thus, a more general expression is

$$\text{confidence limit for } \mu = \bar{x} - \frac{z\sigma}{\sqrt{N}} \quad \text{(A-5)}$$

The example that follows illustrates the calculation of confidence limits.

EXAMPLE

From 25 replicate measurements, the standard deviation for the fluorescence method used in the earlier example was found to be 0.07 ppm Hg. Calculate the 50 and 95% confidence limits for the mean of the three analyses.

For the 50% confidence level, $z = \pm 0.67$, and

$$50\% \text{ confidence limit} = 1.67 \pm \frac{0.67 \times 0.07}{\sqrt{3}}$$

$$= 1.67 \pm 0.03$$

Similarly,

$$95\% \text{ confidence limit} = 1.67 \pm \frac{1.96 \times 0.07}{\sqrt{3}}$$

$$= 1.67 \pm 0.08$$

Thus, the chances are 50 in 100 that the true mean μ will lie in the interval between 1.64 and 1.70 ppm Hg; there is a 95% chance that the true mean lies between 1.59 and 1.75 ppm.

It should be noted that the standard deviation from the three measurements themselves was *not* employed in calculating the confidence limit in the foregoing example. With such a small number of measurements, there is no assurance that $s \rightarrow \sigma$; indeed, s for the three data in question is considerably smaller (0.04) than the value of 0.07 that was obtained from 25 replicate measurements. Without this sure knowledge of the standard deviation, the confidence limit would have had to be widened.[1]

PROPAGATION OF UNCERTAINTIES IN A CALCULATED RESULT

The result of an analysis is typically calculated from two or more experimental data, each of which has associated with it an uncertainty due to random noise. It is worthwhile exploring the ways in which these various uncertainties accumulate in the final result. For

---

[1] For methods of treating data when σ is not known, see D. A. Skoog and D. M. West, *Fundamentals of Analytical Chemistry*, 3d ed. New York: Holt, Rinehart and Winston, 1976, p. 64.

the purpose of such an analysis, let us assume that the measured quantity x is dependent upon the variables $p, q, r, \ldots$, which fluctuate in a random and independent manner. That is, x is a function f of $p, q, r, \ldots$, so that we may write

$$x = f(p, q, r, \ldots) \qquad \text{(A-6)}$$

The uncertainty dx_i (that is, the deviation from the mean) in the ith measurement of x will depend upon the size and signs of the corresponding uncertainties $dp_i, dq_i, dr_i, \ldots$. That is,

$$dx_i = f(dp_i, dq_i, dr_i, \ldots) = (x_i - \mu)$$

The variation in dx as a function of the uncertainties in $p, q, r, \ldots$, can be derived by taking the total differential of Equation A-6. That is,

$$dx = \left(\frac{\delta x}{\delta p}\right)_{q, r, \ldots} dp + \left(\frac{\delta x}{\delta q}\right)_{p, r, \ldots} dq$$
$$+ \left(\frac{\delta x}{\delta r}\right)_{p, q, \ldots} dr + \cdots \qquad \text{(A-7)}$$

In order to relate the various terms in Equation A-7 to the standard deviation of $x, p, q,$ and r as given by Equation A-3, it is necessary to square the foregoing equation. Thus,

$$(dx)^2 = \left[\left(\frac{\delta x}{\delta p}\right)_{q, r, \ldots} dp + \left(\frac{\delta x}{\delta q}\right)_{p, r, \ldots} dq\right.$$
$$\left. + \left(\frac{\delta x}{\delta r}\right)_{p, q, \ldots} dr + \cdots\right]^2 \qquad \text{(A-8)}$$

Then, the resulting equation must be summed between the limits of $i = 1$ to $i = N$, where N again is the total number of replicate measurements (Equation A-3).

In squaring Equation A-7, two types of terms from the right-hand side of the equation become evident. The first are type 1 terms, such as

$$\left[\left(\frac{\delta x}{\delta p}\right) dp\right]^2, \left[\left(\frac{\delta x}{\delta q}\right) dq\right]^2, \left[\left(\frac{\delta x}{\delta r}\right) dr\right]^2$$

Because they are squares, type 1 terms *will always be positive. Therefore, these terms can*

never cancel. In contrast, type 2 terms (called *cross terms*) may be either positive or negative in sign. Examples of these terms are

$$\left(\frac{\delta x}{\delta p}\right)\left(\frac{\delta x}{\delta q}\right) dp\,dq, \left(\frac{\delta x}{\delta p}\right)\left(\frac{\delta x}{\delta r}\right) dp\,dr$$

If $dp, dq,$ and dr are independent and random, some of the cross terms will be negative and others positive. Thus, the *summation of all such terms should approach zero.*

As a consequence of the canceling tendency of type 2 terms, the square of Equation A-8 can be readily obtained and summed from $i = 1$ to N. Thus,

$$\sum_{i=1}^{N} (dx_i)^2 = \left(\frac{\delta x}{\delta p}\right)^2 \sum_{i=1}^{N} (dp_i)^2 + \left(\frac{\delta x}{\delta q}\right)^2$$
$$\sum_{i=1}^{N} (dq_i)^2 + \left(\frac{\delta x}{\delta r}\right)^2 \sum_{i=1}^{N} (dr_i)^2 + \cdots$$

Dividing through by N gives

$$\frac{\sum (dx_i)^2}{N} = \left(\frac{\delta x}{\delta p}\right)^2 \frac{\sum (dp_i)^2}{N} + \left(\frac{\delta x}{\delta q}\right)^2$$
$$\frac{\sum (dq_i)^2}{N} + \left(\frac{\delta x}{\delta r}\right)^2 \frac{\sum (dr_i)^2}{N} + \cdots$$
$$\text{(A-9)}$$

From Equation A-3, however, we see that

$$\frac{(dx_i)^2}{N} = \frac{(x_i - \mu)^2}{N} = \sigma_x^2$$

where σ_x^2 is the variance of x. Similarly,

$$\frac{(dp_i)^2}{N} = \sigma_p^2$$

and so forth. Thus, Equation A-9 can be written in terms of the variances of the quantities; that is,

$$\sigma_x^2 = \left(\frac{\delta x}{\delta p}\right)^2 \sigma_p^2 + \left(\frac{\delta x}{\delta q}\right)^2 \sigma_q^2 + \left(\frac{\delta x}{\delta r}\right)^2 \sigma_r^2 + \cdots$$
$$\text{(A-10)}$$

The example that follows demonstrates an application of Equation A-10.

EXAMPLE

The chloride in a 0.1200-g ($\pm$0.0002) sample was determined by coulometric titration with silver ions (p. 597). An end point was reached after 167.4 ($\pm$0.3) s with a current of 20.00 mA ($\pm$0.04). Titration of a blank required 13.2 ($\pm$0.3) s with the same current. The numbers in parentheses are absolute standard deviations associated with each measurement. What is the relative and absolute standard deviation for the percent chloride?

The percent chloride is given by

$$x = \% \; Cl = \frac{k(T - T_0)I}{W} \qquad \text{(A-11)}$$

where T and T_0 are the times (in seconds) for titration of the sample and blank, respectively, I is the current in mA, W is the sample weight in g, and k is a constant whose value (3.6742×10^{-5}) is known with a high degree of precision.

To employ Equation A-10, we first take the partial derivative of percent chloride with respect to T, holding the other variables constant. Thus,

$$\left(\frac{\delta x}{\delta T}\right)_{T_0, I, W} = \frac{kI}{W}$$

Similarly,

$$\left(\frac{\delta x}{\delta T_0}\right)_{T, I, W} = -\frac{kI}{W}$$

and

$$\left(\frac{\delta x}{\delta I}\right)_{T, T_0, W} = \frac{k(T - T_0)}{W}$$

Finally,

$$\left(\frac{\delta x}{\delta W}\right)_{T, T_0, I} = -\frac{k(T - T_0)I}{W^2}$$

Applying Equation A-10 gives

$$\sigma_x^2 = \left(\frac{kI}{W}\right)^2 \sigma_T^2 + \left(-\frac{kI}{W}\right)^2 \sigma_{T_0}^2$$

$$+ \left[\frac{k(T - T_0)}{W}\right]^2 \sigma_I^2 + \left[-\frac{k(T - T_0)I}{W^2}\right]^2 \sigma_W^2$$

Dividing this equation by the square of Equation A-11 gives

$$\left(\frac{\sigma_x}{x}\right)^2 = \left(\frac{\sigma_T}{T - T_0}\right)^2 + \left(-\frac{\sigma_{T_0}}{T - T_0}\right)^2$$

$$+ \left(\frac{\sigma_I}{I}\right)^2 + \left(-\frac{\sigma_W}{W}\right)^2$$

or

$$\left(\frac{\sigma_x}{x}\right)^2 = \frac{(\sigma_T)^2 + (-\sigma_{T_0})^2}{(T - T_0)^2}$$

$$+ \left(\frac{\sigma_I}{I}\right)^2 + \left(-\frac{\sigma_W}{W}\right)^2 \qquad \text{(A-12)}$$

Substitution of the numerical data into Equation A-11 gives

$$x = \% \; Cl$$

$$= \frac{3.6742 \times 10^{-5}(167.4 - 13.2) \times 20.00}{0.1200}$$

$$= 0.94427$$

Substitution of numerical values into Equation A-12 yields

$$\left(\frac{\sigma_x}{x}\right)^2 = \frac{(0.3)^2 + (-0.3)^2}{(167.4 - 13.2)^2} + \left(\frac{0.04}{20.00}\right)^2$$

$$+ \left(-\frac{0.0002}{0.1200}\right)^2$$

$$= 7.57 \times 10^{-6} + 4.00 \times 10^{-6}$$

$$+ 2.78 \times 10^{-6}$$

$$\frac{\sigma_x}{x} = \sqrt{1.435 \times 10^{-5}} = 3.8 \times 10^{-3}$$

Thus, the relative standard deviation for the analysis would be expected to be about 4 ppt. To obtain the absolute standard deviation, we write

$$\sigma_x = 0.9943 \times 3.8 \times 10^{-3} = 0.0038\% \; Cl$$

and the result could be reported as

$$\% \; Cl = 0.994(\pm 0.004)$$

The foregoing example illustrates two important, general statistical relationships.

(1) The *absolute* variance of a sum or difference is the sum of the individual *absolute* variances.

(2) The *relative* variance of a product or quotient is the sum of the individual *relative* variances. Thus, for the relation $y = a + b$ or $y = a - b$,

$$\sigma_y^2 = \sigma_a^2 + \sigma_b^2$$

and the standard deviation of y is

$$\sigma_y = \sqrt{\sigma_a^2 + \sigma_b^2}$$

In contrast, when $y = a \cdot b$ or $y = a/b$,

$$\left(\frac{\sigma_y}{y}\right)^2 = \left(\frac{\sigma_a}{a}\right)^2 + \left(\frac{\sigma_b}{b}\right)^2$$

or

$$\frac{\sigma_y}{y} = \sqrt{\left(\frac{\sigma_a}{a}\right)^2 + \left(\frac{\sigma_b}{b}\right)^2}$$

Note in the example that it was first necessary to add the absolute variances of T and T_0 to give the absolute variance of the difference between the two numbers. After this the *relative* variance was calculated. It was then combined with the *relative* variances of the other two numbers (I and W) making up the quotient to give the *relative* variance of the quotient.

APPENDIX 2:
SOME STANDARD AND FORMAL ELECTRODE POTENTIALS

Half-reaction[a]	E^0, V	Formal potential, V
$Ag^+ + e \rightleftharpoons Ag(s)$	+0.799	0.228, 1 F HCl; 0.792, 1 F HClO$_4$; 0.77, 1 F H$_2$SO$_4$
$AgBr(s) + e \rightleftharpoons Ag(s) + Br^-$	+0.073	
$AgCl(s) + e \rightleftharpoons Ag(s) + Cl^-$	+0.222	0.228, 1 F KCl
$Ag(CN)_2^- + e \rightleftharpoons Ag(s) + 2CN^-$	−0.31	
$Ag_2CrO_4(s) + 2e \rightleftharpoons 2Ag(s) + CrO_4^{2-}$	+0.446	
$AgI(s) + e \rightleftharpoons Ag(s) + I^-$	−0.151	
$Ag(S_2O_3)_2^{3-} + e \rightleftharpoons Ag(s) + 2S_2O_3^{2-}$	+0.01	
$Al^{3+} + 3e \rightleftharpoons Al(s)$	−1.66	
$H_3AsO_4 + 2H^+ + 2e \rightleftharpoons H_3AsO_3 + H_2O$	+0.559	0.577, 1 F HCl, HClO$_4$
$Ba^{2+} + 2e \rightleftharpoons Ba(s)$	−2.90	
$BiO^+ + 2H^+ + 3e \rightleftharpoons Bi(s) + H_2O$	+0.32	
$BiCl_4^- + 3e \rightleftharpoons Bi(s) + 4Cl^-$	+0.16	
$Br_2(l) + 2e \rightleftharpoons 2Br^-$	+1.065	1.05, 4 F HCl
$Br_2(aq) + 2e \rightleftharpoons 2Br^-$	+1.087[b]	
$BrO_3^- + 6H^+ + 5e \rightleftharpoons \frac{1}{2}Br_2(l) + 3H_2O$	+1.52	
$Ca^{2+} + 2e \rightleftharpoons Ca(s)$	−2.87	
$C_6H_4O_2 \text{ (quinone)} + 2H^+ + 2e \rightleftharpoons C_6H_4(OH)_2$	+0.699	0.696, 1 F HCl, HClO$_4$, H$_2$SO$_4$
$2CO_2(g) + 2H^+ + 2e \rightleftharpoons H_2C_2O_4$	−0.49	
$Cd^{2+} + 2e \rightleftharpoons Cd(s)$	−0.403	
$Ce^{4+} + e \rightleftharpoons Ce^{3+}$		1.70, 1 F HClO$_4$; 1.61, 1 F HNO$_3$; 1.44, 1 F H$_2$SO$_4$; 1.28, 1 F HCl
$Cl_2(g) + 2e \rightleftharpoons 2Cl^-$	+1.359	
$HClO + H^+ + e \rightleftharpoons \frac{1}{2}Cl_2(g) + H_2O$	+1.63	
$ClO_3^- + 6H^+ + 5e \rightleftharpoons \frac{1}{2}Cl_2(g) + 3H_2O$	+1.47	
$Co^{2+} + 2e \rightleftharpoons Co(s)$	−0.277	
$Co^{3+} + e \rightleftharpoons Co^{2+}$	+1.842	
$Cr^{3+} + e \rightleftharpoons Cr^{2+}$	−0.41	
$Cr^{3+} + 3e \rightleftharpoons Cr(s)$	−0.74	
$Cr_2O_7^{2-} + 14H^+ + 6e \rightleftharpoons 2Cr^{3+} + 7H_2O$	+1.33	
$Cu^{2+} + 2e \rightleftharpoons Cu(s)$	+0.337	
$Cu^{2+} + e \rightleftharpoons Cu^+$	+0.153	
$Cu^+ + e \rightleftharpoons Cu(s)$	+0.521	
$Cu^{2+} + I^- + e \rightleftharpoons CuI(s)$	+0.86	
$CuI(s) + e \rightleftharpoons Cu(s) + I^-$	−0.185	

[a] Sources for E^0 values: A. J. deBethune and N. A. S. Loud, *Standard Aqueous Electrode Potentials and Temperature Coefficients at 25°C*. Skokie, Ill.: Clifford A. Hampel, 1964. Source of formal potentials: E. H. Swift and E. A. Butler, *Quantitative Measurements and Chemical Equilibria*. San Francisco: W. H. Freeman and Company. Copyright © 1972.

[b] These potentials are hypothetical because they correspond to solutions that are 1.00 M in Br$_2$ or I$_2$. The solubilities of these two compounds at 25°C are 0.21 M and 0.0133 M, respectively. In saturated solutions containing an excess of Br$_2$(l) or I$_2$(s), the standard potentials for the half-reactions Br$_2$(l) + 2$e \rightleftharpoons$ 2Br$^-$ or I$_2$(s) + 2$e \rightleftharpoons$ 2I$^-$ should be used. On the other hand, at Br$_2$ and I$_2$ concentrations less than saturation, these hypothetical electrode potentials should be employed.

Half-reaction[a]	E^0, V	Formal potential, V
$F_2(g) + 2H^+ + 2e \rightleftharpoons 2HF(aq)$	$+3.06$	
$Fe^{2+} + 2e \rightleftharpoons Fe(s)$	-0.440	
$Fe^{3+} + e \rightleftharpoons Fe^{2+}$	$+0.771$	0.700, 1 F HCl; 0.732, 1 F HClO$_4$; 0.68, 1 F H$_2$SO$_4$
$Fe(CN)_6^{3-} + e \rightleftharpoons Fe(CN)_6^{4-}$	$+0.36$	0.71, 1 F HCl; 0.72, 1 F HClO$_4$, H$_2$SO$_4$
$2H^+ + 2e \rightleftharpoons H_2(g)$	0.000	-0.005, 1 F HCl, HClO$_4$
$Hg_2^{2+} + 2e \rightleftharpoons 2Hg(l)$	$+0.789$	0.274, 1 F HCl; 0.776, 1 F HClO$_4$; 0.674, 1 F H$_2$SO$_4$
$2Hg^{2+} + 2e \rightleftharpoons Hg_2^{2+}$	$+0.920$	0.907, 1 F HClO$_4$
$Hg^{2+} + 2e \rightleftharpoons Hg(l)$	$+0.854$	
$Hg_2Cl_2(s) + 2e \rightleftharpoons 2Hg(l) + 2Cl^-$	$+0.268$	0.242, sat'd KCl; 0.282, 1 F KCl; 0.334, 0.1 F KCl
$Hg_2SO_4(s) + 2e \rightleftharpoons 2Hg(l) + SO_4^{2-}$	$+0.615$	
$HO_2^- + H_2O + 2e \rightleftharpoons 3OH^-$	$+0.88$	
$I_2(s) + 2e \rightleftharpoons 2I^-$	$+0.5355$	
$I_2(aq) + 2e \rightleftharpoons 2I^-$	$+0.620^b$	
$I_3^- + 2e \rightleftharpoons 3I^-$	$+0.536$	
$ICl_2^- + e \rightleftharpoons \frac{1}{2}I_2(s) + 2Cl^-$	$+1.06$	
$IO_3^- + 6H^+ + 5e \rightleftharpoons \frac{1}{2}I_2(s) + 3H_2O$	$+1.195$	
$IO_3^- + 6H^+ + 5e \rightleftharpoons \frac{1}{2}I_2(aq) + 3H_2O$	$+1.178^b$	
$IO_3^- + 2Cl^- + 6H^+ + 4e \rightleftharpoons ICl_2^- + 3H_2O$	$+1.24$	
$H_5IO_6 + H^+ + 2e \rightleftharpoons IO_3^- + 3H_2O$	$+1.60$	
$K^+ + e \rightleftharpoons K(s)$	-2.925	
$Li^+ + e \rightleftharpoons Li(s)$	-3.045	
$Mg^{2+} + 2e \rightleftharpoons Mg(s)$	-2.37	
$Mn^{2+} + 2e \rightleftharpoons Mn(s)$	-1.18	
$Mn^{3+} + e \rightleftharpoons Mn^{2+}$		1.51, 7.5 F H$_2$SO$_4$
$MnO_2(s) + 4H^+ + 2e \rightleftharpoons Mn^{2+} + 2H_2O$	$+1.23$	1.24, 1 F HClO$_4$
$MnO_4^- + 8H^+ + 5e \rightleftharpoons Mn^{2+} + 4H_2O$	$+1.51$	
$MnO_4^- + 4H^+ + 3e \rightleftharpoons MnO_2(s) + 2H_2O$	$+1.695$	
$MnO_4^- + e \rightleftharpoons MnO_4^{2-}$	$+0.564$	
$N_2(g) + 5H^+ + 4e \rightleftharpoons N_2H_5^+$	-0.23	
$HNO_2 + H^+ + e \rightleftharpoons NO(g) + H_2O$	$+1.00$	
$NO_3^- + 3H^+ + 2e \rightleftharpoons HNO_2 + H_2O$	$+0.94$	0.92, 1 F HNO$_3$
$Na^+ + e \rightleftharpoons Na(s)$	-2.714	
$Ni^{2+} + 2e \rightleftharpoons Ni(s)$	-0.250	
$H_2O_2 + 2H^+ + 2e \rightleftharpoons 2H_2O$	$+1.776$	
$O_2(g) + 4H^+ + 4e \rightleftharpoons 2H_2O$	$+1.229$	
$O_2(g) + 2H^+ + 2e \rightleftharpoons H_2O_2$	$+0.682$	
$O_3(g) + 2H^+ + 2e \rightleftharpoons O_2(g) + H_2O$	$+2.07$	
$Pb^{2+} + 2e \rightleftharpoons Pb(s)$	-0.126	-0.14, 1 F HClO$_4$; -0.29, 1 F H$_2$SO$_4$
$PbO_2(s) + 4H^+ + 2e \rightleftharpoons Pb^{2+} + 2H_2O$	$+1.455$	
$PbSO_4(s) + 2e \rightleftharpoons Pb(s) + SO_4^{2-}$	-0.350	

Half-reaction[a]	E^0, V	Formal potential, V
$PtCl_4^{2-} + 2e \rightleftharpoons Pt(s) + 4Cl^-$	$+0.73$	
$PtCl_6^{2-} + 2e \rightleftharpoons PtCl_4^{2-} + 2Cl^-$	$+0.68$	
$Pd^{2+} + 2e \rightleftharpoons Pd(s)$	$+0.987$	
$S(s) + 2H^+ + 2e \rightleftharpoons H_2S(g)$	$+0.141$	
$H_2SO_3 + 4H^+ + 4e \rightleftharpoons S(s) + 3H_2O$	$+0.45$	
$S_4O_6^{2-} + 2e \rightleftharpoons 2S_2O_3^{2-}$	$+0.08$	
$SO_4^{2-} + 4H^+ + 2e \rightleftharpoons H_2SO_3 + H_2O$	$+0.17$	
$S_2O_8^{2-} + 2e \rightleftharpoons 2SO_4^{2-}$	$+2.01$	
$Sb_2O_5(s) + 6H^+ + 4e \rightleftharpoons 2SbO^+ + 3H_2O$	$+0.581$	
$H_2SeO_3 + 4H^+ + 4e \rightleftharpoons Se(s) + 3H_2O$	$+0.740$	
$SeO_4^{2-} + 4H^+ + 2e \rightleftharpoons H_2SeO_3 + H_2O$	$+1.15$	
$Sn^{2+} + 2e \rightleftharpoons Sn(s)$	-0.136	-0.16, 1 F HClO$_4$
$Sn^{4+} + 2e \rightleftharpoons Sn^{2+}$	$+0.154$	0.14, 1 F HCl
$Ti^{3+} + e \rightleftharpoons Ti^{2+}$	-0.37	
$TiO^{2+} + 2H^+ + e \rightleftharpoons Ti^{3+} + H_2O$	$+0.1$	0.04, 1 F H$_2$SO$_4$
$Tl^+ + e \rightleftharpoons Tl(s)$	-0.336	-0.551, 1 F HCl; -0.33, 1 F HClO$_4$, H$_2$SO$_4$
$Tl^{3+} + 2e \rightleftharpoons Tl^+$	$+1.25$	0.77, 1 F HCl
$UO_2^{2+} + 4H^+ + 2e \rightleftharpoons U^{4+} + 2H_2O$	$+0.334$	
$V^{3+} + e \rightleftharpoons V^{2+}$	-0.255	-0.21, 1 F HClO$_4$
$VO^{2+} + 2H^+ + e \rightleftharpoons V^{3+} + H_2O$	$+0.361$	
$V(OH)_4^+ + 2H^+ + e \rightleftharpoons VO^{2+} + 3H_2O$	$+1.00$	1.02, 1 F HCl, HClO$_4$
$Zn^{2+} + 2e \rightleftharpoons Zn(s)$	-0.763	

INDEX

Conversion Factors for Electromagnetic Radiation

(To convert data in units of x shown in the first column to the units indicated in the remaining columns, multiply or divide as shown.)

Units of x	Frequency, Hz	Wave-number, cm^{-1}	Energy			Wave-length, cm
			kcal/mol	erg	eV	
Hz	$1.00\,x$	$3.34 \times 10^{-11}\,x$	$9.54 \times 10^{-14}\,x$	$6.63 \times 10^{-27}\,x$	$4.14 \times 10^{-15}\,x$	$\dfrac{3.00 \times 10^{10}}{x}$
cm^{-1}	$3.00 \times 10^{10}\,x$	$1.00\,x$	$2.86 \times 10^{-3}\,x$	$1.99 \times 10^{-16}\,x$	$1.24 \times 10^{-4}\,x$	$\dfrac{1.00}{x}$
kcal/mol	$1.05 \times 10^{13}\,x$	$3.50 \times 10^{2}\,x$	$1.00\,x$	$6.95 \times 10^{-14}\,x$	$4.34 \times 10^{-2}\,x$	$\dfrac{2.86 \times 10^{-3}}{x}$
erg	$1.51 \times 10^{26}\,x$	$5.04 \times 10^{15}\,x$	$1.44 \times 10^{13}\,x$	$1.00\,x$	$6.24 \times 10^{11}\,x$	$\dfrac{1.99 \times 10^{-16}}{x}$
eV	$2.42 \times 10^{14}\,x$	$8.07 \times 10^{3}\,x$	$2.31 \times 10^{1}\,x$	$1.60 \times 10^{-12}\,x$	$1.00\,x$	$\dfrac{1.24 \times 10^{-4}}{x}$
cm	$\dfrac{3.00 \times 10^{10}}{x}$	$\dfrac{1.00}{x}$	$\dfrac{2.86 \times 10^{-3}}{x}$	$\dfrac{1.99 \times 10^{-16}}{x}$	$\dfrac{1.24 \times 10^{-4}}{x}$	$1.00\,x$
nm	$\dfrac{3.00 \times 10^{17}}{x}$	$\dfrac{1.00 \times 10^{7}}{x}$	$\dfrac{2.86 \times 10^{4}}{x}$	$\dfrac{1.99 \times 10^{-9}}{x}$	$\dfrac{1.24 \times 10^{3}}{x}$	$1.00 \times 10^{-7}\,x$